Handbook of Research on Smart Computing for Renewable Energy and Agro-Engineering

Valeriy Kharchenko
Federal Scientific Agroengineering Center VIM, Russia

Pandian Vasant
Universiti Teknologi Petronas, Malaysia

A volume in the Advances in Environmental Engineering and Green Technologies (AEEGT) Book Series

IGI Global
DISSEMINATOR OF KNOWLEDGE

Published in the United States of America by
IGI Global
Engineering Science Reference (an imprint of IGI Global)
701 E. Chocolate Avenue
Hershey PA, USA 17033
Tel: 717-533-8845
Fax: 717-533-8661
E-mail: cust@igi-global.com
Web site: http://www.igi-global.com

Copyright © 2020 by IGI Global. All rights reserved. No part of this publication may be reproduced, stored or distributed in any form or by any means, electronic or mechanical, including photocopying, without written permission from the publisher. Product or company names used in this set are for identification purposes only. Inclusion of the names of the products or companies does not indicate a claim of ownership by IGI Global of the trademark or registered trademark.

Library of Congress Cataloging-in-Publication Data

Names: Kharchenko, Valeriy, 1938- editor. | Vasant, Pandian, editor.
Title: Handbook of research on smart computing for renewable energy and agro-engineering / Valeriy Kharchenko, Pandian Vasant, editors.
Description: Hershey, PA : Engineering Science Reference, 2019. | Includes bibliographical references and index. | Summary: "This book examines the benefits and barriers of implementing computational models to agricultural production and energy sources as well as how these models can produce more cost-effective and sustainable solutions"-- Provided by publisher.
Identifiers: LCCN 2019026560 (print) | LCCN 2019026561 (ebook) | ISBN 9781799812166 (hardcover) | ISBN 9781799812173 (ebook)
Subjects: LCSH: Electricity in agriculture--Data processing. | Renewable energy sources--Data processing. | Remote area power supply systems. | Artificial intelligence.
Classification: LCC TK4018 .H335 2019 (print) | LCC TK4018 (ebook) | DDC 630.2/086--dc23
LC record available at https://lccn.loc.gov/2019026560
LC ebook record available at https://lccn.loc.gov/2019026561

This book is published in the IGI Global book series Advances in Environmental Engineering and Green Technologies (AEEGT) (ISSN: 2326-9162; eISSN: 2326-9170)

British Cataloguing in Publication Data
A Cataloguing in Publication record for this book is available from the British Library.

All work contributed to this book is new, previously-unpublished material. The views expressed in this book are those of the authors, but not necessarily of the publisher.

For electronic access to this publication, please contact: eresources@igi-global.com.

Advances in Environmental Engineering and Green Technologies (AEEGT) Book Series

Sang-Bing Tsai
University of Electronic Science and Technology of China Zhongshan Institute, China
Ming-Lang Tseng
Lunghwa University of Science and Technology, Taiwan
Yuchi Wang
University of Electronic Science and Technology of China Zhongshan Institute, China

ISSN:2326-9162
EISSN:2326-9170

Mission

Growing awareness and an increased focus on environmental issues such as climate change, energy use, and loss of non-renewable resources have brought about a greater need for research that provides potential solutions to these problems. Research in environmental science and engineering continues to play a vital role in uncovering new opportunities for a "green" future.

The **Advances in Environmental Engineering and Green Technologies (AEEGT)** book series is a mouthpiece for research in all aspects of environmental science, earth science, and green initiatives. This series supports the ongoing research in this field through publishing books that discuss topics within environmental engineering or that deal with the interdisciplinary field of green technologies.

Coverage

- Contaminated Site Remediation
- Waste Management
- Cleantech
- Sustainable Communities
- Pollution Management
- Water Supply and Treatment
- Renewable Energy
- Biofilters and Biofiltration
- Policies Involving Green Technologies and Environmental Engineering
- Alternative Power Sources

IGI Global is currently accepting manuscripts for publication within this series. To submit a proposal for a volume in this series, please contact our Acquisition Editors at Acquisitions@igi-global.com or visit: http://www.igi-global.com/publish/.

The Advances in Environmental Engineering and Green Technologies (AEEGT) Book Series (ISSN 2326-9162) is published by IGI Global, 701 E. Chocolate Avenue, Hershey, PA 17033-1240, USA, www.igi-global.com. This series is composed of titles available for purchase individually; each title is edited to be contextually exclusive from any other title within the series. For pricing and ordering information please visit http://www.igi-global.com/book-series/advances-environmental-engineering-green-technologies/73679. Postmaster: Send all address changes to above address. ©© 2020 IGI Global. All rights, including translation in other languages reserved by the publisher. No part of this series may be reproduced or used in any form or by any means – graphics, electronic, or mechanical, including photocopying, recording, taping, or information and retrieval systems – without written permission from the publisher, except for non commercial, educational use, including classroom teaching purposes. The views expressed in this series are those of the authors, but not necessarily of IGI Global.

Titles in this Series

For a list of additional titles in this series, please visit: www.igi-global.com/book-series

Novel Advancements in Electrical Power Planning and Performance
Smita Shandilya (Sagar Institute of Research, Technology and Science, India) Shishir Kumar Shandilya (Vellore Institute of Technology, India) Tripta Thakur (Maulana Azad National Institute of Technology, India) and Atulya K. Nagar (Liverpool Hope University, UK)
Engineering Science Reference • ©2020 • 388pp • H/C (ISBN: 9781522585510) • US $225.00

Global Food Politics and Approaches to Sustainable Consumption Emerging Research and Opportunities
Luke Amadi (University of Port Harcourt, Nigeria) and Fidelis Allen (University of Port Harcourt, Nigeria)
Engineering Science Reference • ©2020 • 242pp • H/C (ISBN: 9781799801252) • US $185.00

Fuzzy Expert Systems and Applications in Agricultural Diagnosis
A.V. Senthil Kumar (Hindusthan College of Arts and Science, India) and M. Kalpana (Tamil Nadu Agricultural University, India)
Engineering Science Reference • ©2020 • 335pp • H/C (ISBN: 9781522591757) • US $245.00

Handbook of Research on Environmental and Human Health Impacts of Plastic Pollution
Khursheed Ahmad Wani (Government Degree College Bijbehara, India) Lutfah Ariana (Indonesian Institute of Sciences, Indonesia) and S.M. Zuber (Government Degree College Bijbehara, India)
Engineering Science Reference • ©2020 • 600pp • H/C (ISBN: 9781522594529) • US $285.00

Decision Support Methods for Assessing Flood Risk and Vulnerability
Ahmed Karmaoui (Southern Center for Culture and Sciences (SCCS), Morocco)
Engineering Science Reference • ©2020 • 332pp • H/C (ISBN: 9781522597711) • US $215.00

Handbook of Research on Energy-Saving Technologies for Environmentally-Friendly Agricultural Development
Valeriy Kharchenko (Federal Scientific Agroengineering Center VIM, Russia) and Pandian Vasant (Universiti Teknologi Petronas, Malaysia)
Engineering Science Reference • ©2020 • 554pp • H/C (ISBN: 9781522594208) • US $295.00

Hydrology and Water Resources Management in Arid, Semi-Arid, and Tropical Regions
Christopher Misati Ondieki (Kenyatta University, Kenya) and Johnson Utu Kitheka (South Eastern Kenya University, Kenya)
Engineering Science Reference • ©2019 • 356pp • H/C (ISBN: 9781799801634) • US $215.00

IGI Global
DISSEMINATOR OF KNOWLEDGE

701 East Chocolate Avenue, Hershey, PA 17033, USA
Tel: 717-533-8845 x100 • Fax: 717-533-8661
E-Mail: cust@igi-global.com • www.igi-global.com

Editorial Advisory Board

Ugo Fiore, *Parthenope University of Naples, Italy*
Igor Judaev, *Don State Agricultural University, Russia*
Aganijas Jumayev, *State Energy Institute of Turkmenistan, Turkmenistan*
Pinar Karagoz, *Computer Engineering Department, Middle East Technical University, Turkey*
Utku Kose, *Suleyman Demirel University, Turkey*
Vladimir V. Kozyrsky, *National University of Bioresources and Nature Management of Ukraine, Ukraine*
Igor Litvinchev, *Nuevo Leon State University, Mexico*
Jose Antonio Marmolejo, *Panamerican University, Mexico*
Saim Memon, *Division of Electrical and Electronic Engineering, School of Engineering, London South Bank University, UK*
Weerakorn Ongsakul, *Asian Institute of Technology, Thailand*
Gilberto Peres Lechuga, *Autonomous University of the Hidalgo State, Mexico*
Vitali Postolati, *Academy of Sciences of Moldova, Moldova*
Aleksandr Soloviev, *Lomonosov Moscow State University, Russia*
Gulom Uzakov, *Engineering and Economic Institute, Karshi, Uzbekistan*
Alexey N. Vasilyev, *Federal Scientific Agroengineer Center VIM, Russia*
Gerhard-Wilhelm Weber, *Poznan University of Technology, Poland*

List of Contributors

Table of Contents

Detailed Table of Contents

The chapter discusses various phase-transition heat-accumulating materials. Their application in heat pumps and their use for heat supply are presented. Phase-transition heat-accumulating materials in heat pipes are also considered, various types of heat pipes are presented. The installation of heat storage with phase-transition materials is presented. Along with phase-transition heat-accumulating materials, the chapter considers thermochemical heat-accumulating material of photochemical reactions of energy storage. As an example of a thermochemical heat-accumulating material, a solar power plant for thermochemical energy storage is presented. A developed computer program for the description of thermochemical reactions allows examining chemical and thermochemical reactions in multicomponent reciprocal systems, which can be carried out in the course of the reciprocal multicomponent systems. The computer program allows identifying thermochemical reactions occurring in mutual multicomponent systems, regardless of component and dependent on temperature.

The ozonation of fruits and vegetables may increase their storage life nearly twice as long. The chapter presents the results of research completed on the ozone treatment of vegetable storehouses with the use of ozone generators, with barrier and corona types of discharge, and reveals factors affecting disinfection process. The results show that ozone concentration depends on the dynamic balance of the two basic processes – the ozone-air mixture supply and ozone decomposition. Formulas for calculating ozone generator capacity in the recirculation mode have been derived. The automated system for controlling the disinfection process in a vegetable storehouse is described. The experimental test has shown that the ozone effect is long-term and the efficiency of the sterilization of surface microflora contaminants is 100%.

Intelligent data analysis techniques such as data mining or statistical/machine learning algorithms are applied to diverse domains, including energy informatics. These techniques have been successfully employed in order to solve different problems within the energy domain, particularly forecasting problems such as renewable energy and energy consumption forecasts. This chapter elaborates the use of intelligent data analysis techniques for the facilitation of renewable energy monitoring and forecast. First, a review of the literature is presented on systems and forecasting approaches applied to the renewable energy domain. Next, a generic and large-scale renewable energy monitoring and forecast system based on intelligent data analysis is described. Finally, a genuine implementation of this system for wind energy is presented as a case study, together with its performance analysis results. This chapter stands as a significant reference for renewable energy informatics, considering the provided conceptual and applied system descriptions, heavily based on smart computing techniques.

The reliability of electrical consumers depends on the presence of sectioning devices in electrical network distribution. Determining the number and installation sites of sectioning machines is an important optimization problem, based on a comparison of the cost of installing devices and reducing the damage to consumers from an undersupply of electricity. A mathematical model is proposed to determine the cost-effectiveness of installing partitioning switching devices for networks with a means of increasing reliability. This allows for the choice when installing switching devices not only meets an economic criterion (in the form of reducing damage to the power supply system as a whole), but also takes into account the change in the duration of downtime of local consumers.

The chapter discusses the simulation of thermal operating conditions and the optimization of the design of solar photovoltaic thermal modules. As a realization of the developed method, two photovoltaic thermal modules with one-sided solar cells with one-sided heat removal and two-sided solar cells with two-sided heat removal are presented. The components of the developed models of solar modules must be optimized on the basis of the required indicators of the thermal mode of operation of the modules. For this task, a method has been developed for visualizing thermal processes using the Ansys system of finite element analysis, which has been used to research thermal modes of operation and to optimize the design of the modules created. With the help of the developed method, the temperature fields of the module components, coolant velocity and its flow lines in the developed models of a planar photovoltaic thermal roofing panel and a concentrator photovoltaic thermal two-sided module are visualized.

The study analyzes the various methods of gainful utilization of excess heat from power transformers. The ways to reduce heat loss inside the tank power transformer are found. The potential amount of heat emitted by power transformers of different capacities is calculated. New ways of combining the functions of electric transformation and heating in a power transformer are described. A system has been developed to use the excess heat of power transformers in the agricultural power systems. There are the structural and schematic diagrams of the system and a method for calculating its main elements. An improved design of the power transformer cooling system has been developed to combine the functions of electric transformation and heating. Experiments to verify the effectiveness of decisions are described. A feasibility study of the implementation of the developed system was carried out.

In the presented materials, a system of principles and approaches for studying processes of microwave-convective grain treatment is discussed. The essence of this research is described starting from the initial phase of designing structural diagrams of technological processes. Based on the explicitly-defined process target, a functional-physical analysis has been performed that made it possible to specify the most essential factors of process. The morphological analysis of the processes was carried out in several stages. At each stage, the specification of the goal and excluded variants of factors that at this stage, the development is impractical to implement. It is advisable to carry out a check of the effectiveness of the options selected from the morphological table using modeling. To this end, developed a mathematical and computer models of the process. The results of modeling the drying of grain in the microwave - convective zone confirmed the feasibility and efficiency of using the obtained model.

Sergey Nikolaevich Devyanin, Russian State Agrarian University – Moscow Timiryazev Agricultural Academy, Russia
Vladimir Anatolyevich Markov, Moscow State Technical University Named After N.E. Bauman, Russia
Andrei Alexandrovich Savastenko, Moscow State Technical University, Russia

In recent years, an increased interest has been shown in fuels derived from renewable energy resources of animal and vegetable origin, raw materials of which are practically unlimited. The use of biofuels makes it possible not only to replace petroleum motor fuels with alternative ones, but also to improve exhaust gas toxicity indicators (as a rule, there is a marked decrease in emissions of toxic exhaust gas components). This work is devoted to the study of the possibilities of solving these problems in order to adapt diesel to biogas, a theoretical study of ways to create an experimental model of a power plant for generating electricity from biogas, a theoretical study on the processes occurring in the main components of the plant being developed and the creation of this plant.

Jorge Vinna Sabrejos, The Antenor Orrego Private University, Peru
Alexander Anatolievich Belov, Federal Scientific Agroengineering Center VIM, Russia
Alexey N. Vasilyev, Federal Scientific Agroengineering Center VIM, Russia
Victor Nikolayevich Toporkov, Federal Scientific Agroengineering Center VIM, Russia
Andrey Anatolievich Musenko, Federal Scientific Agroengineering Center VIM, Russia

The authors reveal the relevance of modeling electrohydraulic technology in terms of determining the parameters that cannot be measured by existing methods and technical means. The purpose of the chapter is modeling of electrohydraulic technology in agriculture. By modeling, the authors are developing an installation for obtaining nutrient solutions under irrigation conditions in order to increase the yield in greenhouse vegetable combines while ensuring the environmental friendliness of the process. The authors used the methods of research, analysis and calculation of technical means and control systems applied to high-voltage electro-technological installations of agricultural use. Mathematical patterns are established and a logically based relationship is established between the changes in electrical conductivity of water as a result of electrohydraulic processing.

Any harvest is primarily the realization of the potential productivity of plants in certain conditions, depending on the parameters and operating modes of the sowing machine. The result of the operation of mechanized sowing will be the parameters of the nutrition area of each plant is the shape and numerical value of these area. The resulting schemes of plant placement on the field correspond to certain characteristics and parameters of the areas of plant nutrition, determining the availability of all elements of mineral nutrition, photosynthetically active radiation, and the relationship of units of agrophytocenosis. This is one of the decisive conditions for the harvest. The development of theoretical ideas about the uniformity of plant placement and the use of appropriate criteria and characteristics will create the necessary understanding of the role of the sowing machines in crop production during cultivated plants in the framework of mechanized technology.

Variants of seed response to external disturbances were studied to determine the principles that should be followed during pre-sowing treatment. The approach that involves analysis of seeds' adaptive response has been applied. The basic organizational rules have been specified for external disturbance application to seeds. The use of informational method for studying the biological response of seeds on external actions made it possible to find the rules that have to be followed when pre-sawing treatment is performed with the use of electrophysical excitation. The essence of these rules is that the following sequence of operations shall be carried out in the process of pre-sowing treatment: to define the lowest energy level of external disturbance corresponding to the seed's sensitivity threshold, to organize cyclic action of external factors on seeds, to vary the amplitude of external factor within cycles, to vary the mode of external disturbance. The field experiments were carried out for several years, during which the pre-sowing treatment of seeds was determined.

Engineering systems are currently plagued by various complexities and uncertainties. Metaheuristics have emerged as an essential tool for effective engineering design and operations. Nevertheless, conventional metaheuristics still struggle to reach optimality in the face of highly complex engineering problems. Aiming to further boost the performance of conventional metaheuristics, strategies such as hybridization and various enhancements have been added into the existing solution methods. In this work, swarm intelligence techniques were employed to solve the real-world, large-scale biofuel supply chain problem. Additionally, the supply chain problem considered in this chapter is multiobjective (MO) in nature. Comparative analysis was then performed on the swarm techniques. To further enhance the search capability of the best solution method (GSA), the Lévy flight component from the Cuckoo Search (CS) algorithm was incorporated into the Gravitational Search Algorithm (GSA) technique; developing the novel Lévy-GSA technique. Measurement metrics were then utilized to analyze the results.

The rural distribution network has deteriorated. This is due to the high failure rate of electrical equipment, high maintenance costs, reduced power quality, and the increased duration of power outages in agricultural production. This leads to a short supply of electricity, downtime of processing equipment, loss of production or production of low-grade products, as well as excessive energy losses during transmission. The important issue is the development of advanced methods for assessing the feasibility and effectiveness for component replacement of power transmission equipment with newer and more modern, reducing electrical energy loss in the distribution network. To solve these problems, various strategies have been developed and studied to improve the reliability of 10 kV overhead power lines by using modern insulators, wires, and supports.

The chapter reveals new opportunities for the use of digital video surveillance in the agro-industrial sector. It presents a methodological approach and practical guidelines for modeling and managing changes in agrotechnology, in the conditions of digital transformation of large amounts of data. The proposed technical solutions are aimed at creating better products and gaining competitive advantages, thanks to the improved relationship of specialists and the transition to predictive data analysis. Video surveillance as the most accessible way to obtain information involves the creation of large amounts of data, their long-term storage, fast retrieval, and diverse analysis. The possibilities of video surveillance systems widely used to ensure the safety of industrial buildings and the protection of areas equipped with automatic video analytics modules and integrated computer programs are considered. The organization of managing geographically-distributed agricultural production using mobile and remote video surveillance systems is shown. By using the proposed project of integrating digital video surveillance into a big data system, the organization can be transformed to improve strategy, make system decisions, improve marketing and sales results, improve efficiency and productivity of business processes, improve the performance of production personnel, minimize unprofitable risks, and maximize management efficiency, which will lead to the emergence of new opportunities.

The problem of hydrothermal coordination is a classic example of nonlinear and large-scale dynamic mathematical optimization that involves the problem of the economic dispatch plus other constraints associated with the coordination of a pair of systems integrated into a network (i.e., the production of energy from thermoelectric and hydroelectric plants). A large number of methods and models that represent this problem, from classical methods of optimization to modern methods based on metaheuristics, exist. In this chapter, the classic structure of the problem is analyzed and new constraints are proposed, including those based on the level of customer service. Some recent modern formulations of the model are shown, and the existing literature is reviewed showing the modern methods of optimization based on heuristics. The authors illustrate it by optimizing an instance consisting of three thermo plants and three hydro plants through random search and genetic algorithm.

In this chapter, the authors present a mathematical model to calculate the exact quantities of animals that must compose a herd given the water, food, and land conditions available for their breeding. In the same model, the optimal calculation of vegetables that can be cultivated in the available land spaces is incorporated considering the same restrictions of water, nutrients, and area available for planting. Both models focus primarily from a deterministic perspective. Subsequently, the randomness of the same is uncovered through the uncertainty in the availability of water. Therefore, the first part of the proposal is made through a simple model of linear mathematical programming. The stochastic model is constructed from a two-phase mathematical programming model. The novelty of the proposal and its contribution consists of illustrating, step by step, the construction and solution of the scenarios of the stochastic model for a problem related to agriculture and animal husbandry.

Foreword

SMART COMPUTING AND INNOVATIVE OPTIMIZATION FOR RENEWABLE ENERGY AND AGRO-ENGINEERING

The number of earth's inhabitants increases steadily, putting on the agenda many old and new challenges and problems. Among these challenges there is the need to offer to a continuously growing population all the needed commodities, like food, clothes, medicine and many other goods. *Decision-making*, supported by the key technologies *Optimization* and Optimal Control, chooses and assigns new lands and territories, particularly in the rural countryside, which should be fitting for economic activities as required by decision makers such as organizations and governments. In these modern times, we encounter situations wherein one territory there is an over-population, while in other territories there is emptiness and land is not yet used for the production of goods necessary for human existence. Territories like these are given not only in countries which possess vast areas, but also in countries that possess small territories only, where overpopulated zones are neighboring to deserted, sparsely inhabited areas.

Empty and deserted territories are not accessed by a majority of the population and their needs, since it is often expelled from centralized sources of finance and power supply, and independent energy supply is complicated because of the high cost of infrastructure, facilities and items employed for these goals. At the same time, in recent years the ideal of a distributed, socially-oriented economy has spread significantly. This has given reason to think that a distributed placing of capacities, habitation and resources (human included) in the upcoming future shall strongly change our minds and approaches towards the development of territories, towards existing financial, educational, technological and communication devices and techniques, towards their ethical and "spiritual" reference points, landmarks and connotations, and towards decision-making methodologies, optimization tools, and their frameworks and systems.

Mathematical modeling and solution of this global challenge asks for the creation of preconditions both technological and organizational ones. Our present days' levels in science and technology allow to offer new tools and equipment for the power supply needed in such territories, and perspectives of electro-technologies of transporting electrical power and its effective employment. This use should enable for manufacturing of agricultural products and for maintenance of people's comfortable living.

As a matter of fact, quite often, underlying information needed is not accessible by the leaders of local authorities, by businessmen and the population; hence, it is has stayed unemployed. The present work, as part of a successful series of books, strives to hand over relevant information, a representation of widely and publicly advanced results and successes in the areas of electro-technologies, of efficient involving electrical resources and renewable energy sources. This ought to be useful to a very broad range of readers and, hopefully, exercise a precious impulse to ensuring sustainable development in new territories. This book has been inviting to cover the vast problem complexities in the decision making related with the formation of a distributed economy and the cultivation of new territories as a central element of this challenge.

In this compendium, emerging innovative results and techniques which provide an effective production, transfer and utilization of electrical energy, permitting a successful development of new territories and realization of the concept of a distributed economy, have been elaborated. A special interest has been devoted to the fact that a remarkable part of the material included in the book was not published so far in Anglo-Saxon literature and, therefore, was hardly available to a wide spectrum of interested readers and users in many countries of our world.

The addressed readership of this work has been composed of practitioners and investigators who are employed in the vast growing field of Energy Efficiency and Saving Technologies, RES-based Energy Supplied Systems and numerous other kinds of Renewable Technologies. What is more, this book intents to ensure deep insights and support executives who are working on the development of agricultural and industrial companies and endeavors on new territories and for sustainable development of these territories.

The topics of this book's chapters were open, but not limited to:

- Solar thermal energy installations,
- PV solar,
- Wind-based energy generation installations and systems,
- Small hydropower plants,
- Specific RES potential estimation methodologies in condition of meteo data absence,
- Cogeneration PV-thermal modules fabrication and application,
- RES application best practices,
- RES potential in different countries and remote territories,
- Diagnostic and monitoring methods for energy equipment status,
- Providing parameters of solar power plant with given generation graph,
- Innovative 10–110 kV compact controlled overhead lines,
- Resonant electric power systems for renewables-based electric grids,
- Resonant system of electricity transfer of high power,
- Energy saving system for illumination of greenhouse plants,
- Energy efficient electric equipment for heat supply in agriculture,
- Electric pulse cultivation,
- Pre-sowing seed treatment in a magnetic field,
- Use of microwave energy at thermal treatment of grain crops,
- Scientific and methodological support of electric system operation,
- Biological objects reactions management,
- Solutions for diesel, wind, and solar power plants joint use for autonomous power supply.

We emphasize the great importance and chance of this book for all its forthcoming readers. Indeed, we esteem that in future *Optimization* and *Optimal Control* will become real centers of support to the *Decision-Making* help and support that this book intends to facilitate and strengthen. In that framework, we especially draw attention to *Stochastic Optimal Control* and *Stochastic Games* under various forms of *Memory*, such as *Delay*, to *Regime Switch* and *Paradigm Shift*, and *Information Incompleteness, Partiality* and *Asymmetry*. We and our research partners are scientifically working on it, for a further reflection of real-word cases and scenarios, with any forms of *Uncertainty* and *"Human Factors"*, in the needed *Optimal and Fair Decision-Making*.

To all the authors of this valuable book, we convey our appreciation and gratitude for having shared their excellence and devotion with the academic community and with humankind. Furthermore, we extend our gratitude to the two editors of this compendium, Prof. Dr. Valeriy Kharchenko and Prof. Dr. Pandian Vasant, for their hard-work and vision, of having gathered a remarkable variety of rich contributions. Eventually, we are very grateful to the publishing house of IGI Global, for ensuring and making become reality a premium book of real-life significance and of high potential impact on tomorrow's world, for this and following generations.

Now, it is up to us to wish you all a lot of pleasure and enrichment when browsing and reading this exciting work, and we hope that a high gain will be return from it both academically and societally.

Gerhard-Wilhelm Weber
Faculty of Engineering Management, Poznan University of Technology, Poznan, Poland

Nadi Serhan Aydın
Department of Industrial and Systems Engineering, College of Engineering, Istinye University, Istanbul, Turkey

Ioannis Baltas
Financial and Management Engineering, University of the Aegean, Chios Island, Greece

Emel Savku
École Polytechnique, Paris, France
July 2019

Gerhard-Wilhelm Weber *is a Professor at Poznan University of Technology, Poznan, Poland, at Faculty of Engineering Management, in the Chair of Marketing and Economic Engineering. His research is on mathematics, operational research, finance, economics, optimization and control, neuro-, bio- and earth-sciences, medicine, data mining, analytics, artificial intelligence, education and development; he is involved in the organization of scientific life internationally. He received Diploma and Doctorate in Mathematics, and Economics / Business Administration, at RWTH Aachen, and Habilitation at TU Darmstadt (Germany). He replaced Professorships at University of Cologne, and TU Chemnitz, Germany. At Institute of Applied Mathematics, Middle East Technical University, Ankara, Turkey, he was a Professor in both Financial Mathematics and Scientific Computing, and Assistant to the Director, and has been a member of five further graduate schools, institutes and departments of METU. G.-W. Weber has different departmental and institutional affiliations at Universities of Siegen (Germany), Federation University (Ballarat, Australia), University of Aveiro (Portugal), University of North Sumatra (Medan, Indonesia), Malaysia University of Technology, Chinese University of Hong Kong, KTO Karatay University (Konya, Turkey),Mazandaran University of Science and Technology (Babol, Iran), Istinye University (Istanbul, Turkey), at Georgian International Academy of Sciences, at EURO (Association of European OR Societies; "Advisor to EURO Conferences") and IFORS (International Federation of OR Societies), where he is member in many national OR societies or working groups and chair of IFORS OR Online Resources, at Pacific Optimization Research Activity Group, etc. G.-W. Weber has (co-)supervised many MSc. and PhD. students, (co-) authored numerous books, special issues and articles, and given many presentations in a diversity of subject areas, on theory, methods and practice. He has been a member of many international editorial and award boards; he participated at and was* *in charge of numerous research projects; G.-W. Weber received various recognitions and awards by students, universities, conferences and scientific organizations.*

Nadi Serhan Aydın *is an Assistant Professor at the Department of Industrial and Systems Engineering at Istinye University in Istanbul, Turkey. He obtained a BSc. in Management Engineering from Istanbul Technical University (2006), followed by an MSc. (2011) and a PhD. (2016) in Financial Mathematics from the Institute of Applied Mathematics at Middle East Technical University in Ankara, Turkey. Dr. Aydin carried out his doctoral thesis studies at the Financial Signal Processing (FSP) Lab at Imperial College London. During his graduate studies, he was also granted research fellowships at Heidelberg and Ulm Universities. Between 2011 and 2016, he worked as a senior research economist at an inter-governmental organization and managed multi-stakeholder projects focusing on financial sector development. Prior to Istinye University, Dr. Aydinw as part-time faculty at the Department of Industrial Engineering at TED University in Ankara, Turkey. He is a designated Financial Risk Manager (FRM)and continues to offer professional training courses in partnership with the industry. He is also serving on the editorial board of a number of international journals and review services. Dr. Aydin's research interest lies in computational finance, stochastic risk modeling, financial optimization, predictive data analytics, pattern recognition, statistical/machine learning, and financial signal processing.*

Ioannis Baltas *is an Assistant Professor of Financial Engineering at the department of Financial and Management Engineering at the University of the Aegean, Chios island, Greece. His research focuses on the applications of Stochastic Analysis (especially, stochastic optimal control, robust control theory and stochastic differential games) to Financial Mathematics, Actuarial Mathematics, Modern Risk Management and is published in well-respected journals. Ioannis graduated from the department of Statistics, Actuarial and Financial Mathematics at the University of the Aegean, Samos island, Greece. He also holds a Master's in Science diploma in Financial and Actuarial Mathematics (with distinction) from the same department. In June 2014, he was awarded a Doctorate in Stochastic Analysis by the department of Statistics at the Athens University of Economics and Business. Prior to his appointment as an Assistant Professor, Ioannis spent two years as a Post-Doctoral researcher (Athens University of Economics and Business) working on stochastic differential games and one year, as an Adjunct Lecturer of Financial Engineering (University of the Aegean). During his academic career, he has presented his work at numerous conferences/seminars/summer schools, both in Greece and abroad. He serves the international scientific community by acting as a reviewer for a wide variety of prestigious scientific journals. Ioannis is a member of Bernoulli Society and an active collaborator of the Stochastic Modeling and Applications laboratory (Athens University of Economics and Business).*

Emel Savku *is a Post-Doctorate at Ecole Polytechnique, Centre de Mathematiques Appliquées (CMAP), Palaiseau, France by March 2019. Her research is in mathematics, Stochastic Optimal Control, Markov Regime-Switches, Stochastic Differential Equations with Delay, Anticipated Backward Stochastic Differential Equations, Stochastic Game Theory, Market Microstructure, Financial Mathematics. She got her doctorate degree in Financial Mathematics at Middle East Technical University, Institute of Applied Mathematics in Turkey. She got Master and Doctorate scholarship awards from the Scientific and Research Council of Turkey. (TUBITAK). She visited CMAP by French Embassy Young Visiting Researcher Fellowship in 2018 for two months. She worked at Bilkent University and Turk Egitim Dernegi Universty as a lecturer in 2017-2018. She presented results of her works at several prestigious conferences. Her web-site is: https://quantreg.com/people/emel-savku/.*

Preface

The modern world community has a number of distinctive tendencies, one of which is population growth and the resulting growth in the consumption of food and goods that support the livelihoods of people. And this growth is due not only to the increase in population, but also to the increase in the level of consumption by each person. To meet these requirements along with a large number of measures of a different kinds constant involving in economic circulation of new lands and territories, suitable for the economic activities organization is required

As a rule, these territories and pieces of lands disposed far of main population's activity and in the same time from centralized centers of energy supply. The most suitable activity for these territories is agriculture.

Start a business in such territories is easiest if you are engaged in the production of agricultural products on small-scale enterprises, with a small number of employed people. Small scale production makes it possible to quickly deploy work, easier and faster to ensure the sale of products.

Such activities require special technical support measures both for the livelihood of people living there and for organizing a highly efficient and cost-effective production process. Naturally, in such territories it is easiest and quickest to deploy activities related to the production of agricultural products. Such enterprises are easier to provide with energy resources, in particular, using renewable energy sources.

Existing technologies of energy supply, cropping and processing of crops cannot always be easily adapted to use in small and medium-sized enterprises.

There was such a situation that information of this kind is needed not only directly by the worker in the village. The manufacturers and developers of such technical facilities and technologies and scientific organizations that are engaged in the development of technologies and equipment for the power industry of the village, production processes of various types, as well as the problems of housing and communal solutions, are showing great interest in new efficient means of conducting agricultural business, especially on a small scale, problems of the rural population.

At the same time, the implementation in practice of new technologies, both renewable energy sources and agro-production technologies at all production stages, has a number of barriers and obstacles, for the overcoming of which it is necessary to carry out a complex of scientific research or experimental design work. Their implementation requires a lot of money and time.

Significantly accelerate these processes may be by using modern approaches such as Intelligent Computing, including such approaches as Artificial intelligence, Quantum computing, Artificial Neural Networks, Self-Organizing, Fuzzy logic, Membrane computing, Evolutionary computation, Learning theory, Probabilistic methods, Image processing Computer vision etc., and Optimization, comprising such approaches as Ant colony optimization, Artificial bee colony, Artificial immune systems, Artificial neural networks, Automatic computing, Bacterial foraging, Combinatorial Optimization, Computational intelligence, continuous optimization, Cultural algorithms, Direct Search, Fuzzy optimization, Hybrid algorithms, Local and global search, Memetic algorithms and many, many others.

It seems advisable to apply these approaches for the promotion of new technologies mentioned above, that's why it would be advisable to present and consider new technologies abilities through above mentioned computerized approaches gathered together in one book.

The book includes a description of a series of case study for investigation of new technological methods and technical means that ensure the organization of production of various products with application of above-mentioned computerized approaches, optimization and modeling processes what should permit to accelerate and make cheaper the way from new idea to its practical implementation. The main attention is paid to the specified technologies and technical means adapted to application in production of small scale, at the enterprises of small and medium business. Ways for accelerating of implementation of new technical solutions designed to provide vital activity in new territories with a high level of comfort and high standards of living conditions are also in some degree considered. Modern technologies of energy supply of life and production activity at remote, first-time territories are also considered through the prism of optimization and computing modeling. Definite attention is paid to works aimed at organizing an efficient production process and social comfort in the face of scarcity of energy resources and the need to rationally use them.

The book includes 18 chapters.

Chapter 1 discusses various phase-transition heat-accumulating materials. Their application in heat pumps and their use for heat supply are presented. Phase-transition heat-accumulating materials in heat pipes are also considered, various types of heat pipes are presented. The installation of heat storage with phase-transition materials is presented. Along with phase-transition heat-accumulating materials, the chapter considers thermochemical heat-accumulating materials on the example of photochemical reactions of energy storage. As an example of a thermochemical heat-accumulating material, a solar power plant for thermochemical energy storage is presented. Developed computer program for the description of thermochemical reactions allows describing chemical and thermochemical reactions in multicomponent reciprocal systems, which can be carried out in the course of the reciprocal multicomponent systems. Computer program allows identifying thermochemical reactions occurring in mutual multicomponent systems, regardless of component and depending on temperature.

Chapter 2 is dedicated to the consideration the results of research of ozone treatment of vegetable storehouses with the use of ozone generators with barrier and corona types of discharge, and reveals factors affecting disinfection process. As it is known, ozonation of fruits and vegetables may increase their storage life nearly twice. It has been found out that ozone concentration depends on dynamic balance of the two basic processes – ozone-air mixture supply and ozone decomposition. Formulas for calculation of the ozone generator capacity in the recirculation mode have been derived. The automated system for control of disinfection process in vegetable storehouse is described. Experimental test has shown that ozone effect is long-term and efficiency of sterilization of surface contaminated with microflora amounts to 100%.

In Chapter 3, intelligent data analysis techniques such as data mining or statistical/machine learning algorithms are applied to diverse domains, including energy informatics. These techniques have been successfully employed in order to solve different problems within the energy domain, particularly forecasting problems such as renewable energy and energy consumption forecasts. This chapter elaborates the use of intelligent data analysis techniques for the facilitation of renewable energy monitoring and forecast. First, a review of the literature is presented on systems and forecasting approaches applied to the renewable energy domain. Next, a generic and large-scale renewable energy monitoring and forecast system based on intelligent data analysis is described. Finally, a genuine implementation of this system for wind energy is presented as a case study, together with its performance analysis results. This chapter stands as a significant reference for renewable energy informatics, considering the provided conceptual and applied system descriptions, heavily based on smart computing techniques.

Chapter 4 considers optimization of sectionalization parameters of distributive electric networks. Reliability of power supply of electricity consumers depends on the presence of sectioning devices in distribution electrical networks. Determining the number and installation sites of sectioning machines is an important optimization problem, based on a comparison of the cost of installing devices and reducing the damage to consumers from undersupply of electricity. A mathematical model is proposed to determine the cost-effectiveness of installing partitioning switching devices for networks with means of increasing reliability. The possibility of using when choosing a place to install switching devices not only an economic criterion (in the form of reducing damage to the power supply system as a whole), but also a criterion that takes into account the change in the duration of downtime of certain consumers is indicated.

Chapter 5 discusses the simulation of thermal operating conditions and the optimization of the design of solar photovoltaic thermal modules. As a realization of the developed method, two photovoltaic thermal modules with one-sided solar cells with one-sided heat removal and two-sided solar cells with two-sided heat removal are presented. The components of the developed models of solar modules must be optimized on the basis of the required indicators of the thermal mode of operation of the modules. For this task, a method has been developed for visualizing thermal processes using the Ansys system of finite element analysis, which has been used to research thermal modes of operation and to optimize the design of the modules created. With the help of the developed method, the temperature fields of the module components, coolant velocity and its flow lines in the developed models of a planar photovoltaic thermal roofing panel and a concentrator photovoltaic thermal two-sided module are visualized.

Chapter 6 analyzed various methods of gainful utilization of excess heat from power transformers. The ways to reduce heat loss inside the tank power transformer are found. The potential amount of heat emitted by power transformers of different capacities is calculated. New ways of combining the functions of electric transformation and heating in a power transformer are described. A system has been developed to use the excess heat of power transformers in the agricultural power systems. There are the structural and schematic diagrams of the system and a method for calculating its main elements. An improved design of the power transformer cooling system has been developed to combine the functions of electric transformation and heating. Experiments to verify the effectiveness of decisions are described. A feasibility study of the implementation of the developed system was carried out.

Chapter 7 considers a system of principles and approaches for studying processes of microwave-convective grain treatment. The essence of this research is described starting from the initial phase of designing structural diagrams of technological processes. Based on the explicitly defined process target, functional-physical analysis has been performed that made it possible to specify the most essential factors of process. The morphological analysis of the processes was carried out in several stages. At each stage, the specification of the goal and excluded variants of factors that at this stage, the development is impractical to implement. It is advisable to carry out a check of the effectiveness of the options selected from the morphological table using modeling. To this end, developed mathematical and computer models of the process. The results of modeling the drying of grain in the microwave - convective zone confirmed the feasibility and efficiency of using the obtained model.

Chapter 8 considers obtaining and using of biogas in agricultural production. In recent years, an increased interest has been shown in fuels derived from renewable energy resources of animal and vegetable origin, the raw materials of which are practically unlimited. The use of biofuels makes it possible not only to replace petroleum motor fuels with alternative ones, but also to improve exhaust gas toxicity indicators (as a rule, there is a marked decrease in emissions of toxic exhaust gas components). This work is devoted to the study of the possibilities of solving these problems in order to adapt diesel to biogas, a theoretical study of ways to create an experimental model of a power plant for generating electricity from biogas, a theoretical study on the processes occurring in the main components of the plant being developed and the creation of this plant.

In Chapter 9, the authors reveal the relevance of modeling electrohydraulic technology in terms of determining the parameters that cannot be measured by existing methods and technical means. The purpose of the chapter is modeling of electrohydraulic technology in agriculture. By modeling, the authors are developing an installation for obtaining nutrient solutions under irrigation conditions in order to increase the yield in greenhouse vegetable combines while ensuring the environmental friendliness of the process. We used the methods of research, analysis and calculation of technical means and control systems applied to high voltage electro technological installations of agricultural use. Mathematical patterns are established, and a logically based relationship is established between changes in electrical conductivity of water as a result of electrohydraulic processing.

Chapter10 is dedicated to consideration of rational placement plants on the field area by seeding machines (methodical aspects assessment rationality). Any harvest is primarily the realization of the potential productivity of plants in certain conditions, depending on the parameters and operating modes of the sowing machine. The result of the operation of mechanized sowing will be the parameters of the nutrition area of each plant is the shape and numerical value of these area. The resulting schemes of plant placement on the field correspond to certain characteristics and parameters of the areas of plant nutrition, determining the availability of all elements of mineral nutrition, photosynthetically active radiation and the relationship of units of agrophytocenosis. This is one of the decisive conditions for the harvest. The development of theoretical ideas about the uniformity of plant placement and the use of appropriate criteria and characteristics will create the necessary understanding of the role of the sowing machines in crop production during cultivated plants in the framework of mechanized technology.

Chapter 11 considers very interesting question: how to manage the effectiveness of presowing treatment of seeds. Variants of seed response to external disturbances were studied to determine the principles that should be followed during pre-sowing treatment. The approach that involves analysis of seeds' adaptive response has been applied. The basic organizational rules have been specified for external disturbances application to seeds. The use of informational method for studying biological response of seeds on external actions made it possible to find the rules that have to be followed when pre-sawing treatment is performed with the use of electrophysical excitation. The essence of these rules is that the following sequence of operations shall be carried out in the process of pre-sawing treatment: to define the lowest energy level of external disturbance corresponding to seed's sensitivity threshold, to organize cyclic action of external fac-tor on seeds, to vary the amplitude of external factor within cycles, to vary the mode of external disturbance. Field experiments were carried out for several years, during which the presowing treatment of seeds was determined.

Chapter 12 is dedicated to consideration Lévy-Enhanced Swarm Intelligence for optimizing a multiobjective biofuel supply chain. Engineering systems are currently plagued by various complexities and uncertainties. Metaheuristics have emerged as an essential tool for effective engineering design and operations. Nevertheless, conventional metaheuristics still struggle to reach optimality in the face of highly complex engineering problems. Aiming to further boost the performance of conventional metaheuristics, strategies such as hybridization and various enhancements have been added into the existing solution methods. In this work, swarm intelligence techniques were employed to solve the real-world, large-scale biofuel supply chain problem. Additionally, the supply chain problem considered in this chapter is multiobjective (MO) in nature. Comparative analysis was then performed on the swarm techniques. To further enhance the search capability of the best solution method (GSA), the Lévy flight component from the CS algorithm was incorporated into the GSA technique; developing the novel Lévy-GSA technique. Measurement metrics were then utilized to analyze the results.

In Chapter 13, multi-criteria evaluation of reconstruction strategies for distribution power networks designed for rural power supply is considered. The deterioration of rural distribution networks now exceeds 70%. This is due to the high failure rate of electrical equipment, high maintenance costs, reduced power quality and increased duration of power outages in agricultural production. This leads to a short supply of electricity, downtime of process equipment, loss of production or production of low-grade products, as well as excessive energy losses during transmission. An important issue is the development of advanced methods for assessing the feasibility and effectiveness of component replacement of power transmission equipment with new and modern, providing a reduction in electrical energy losses in distribution networks. To solve these problems, various strategies have been developed and studied to improve the reliability of 10 kV overhead power lines by reconstruction using modern insulators, wires and supports, which are the main components of rural power grids.

Chapter 14 considers video capsule management of agro-technological processes (Video capsule Management. Nowadays agricultural production has had (unused) unclaimed information and management resources for the development and improvement of agro-technological processes and service of agricultural enterprises so far. New opportunities to improve the comfort and intelligence of labor can be provided in the design of management systems geographically distributed agricultural production using mobile and remote video surveillance, unmanned video service automation, video bots, remote web clients-included in a single integrated system of self – organization of agricultural enterprises.

Chapter 15 is dedicated to consideration of modern methods of optimization in models of hydrothermal coordination and emission of contaminating particles in power generating plants. In the chapter authors describe succinctly the structure of a mathematical model that defines the popular problem of planning the operation of a hydrothermal system in the short term. In particular, we are interested in addressing some of the different ways in which it is presented in real contexts, the authors and the modern methods used for its optimization. One area of special emphasis is the constraints associated with it as well as its meaning. Finally, we present an application case or instance where we compare two totally opposite methods for the solution the same: 1) a simple finite search algorithm and 2) the well-known search method using Genetic Algorithms. We also make an approximation to the particle emission model consistent with the system operation plan. We hope to contribute to the subject.

Chapter 16 considers a model for planning the sowing of agricultural crops and raising animals through two-stage mathematical programming. In this chapter a proposal is developed which finds the optimal scenarios associated with a problem of planting vegetables and raising animals under an environment of uncertainty. The model is developed under different contexts explaining the models derived from each decision under the ``Here and now'' scheme. The results of the model are reported, and the models illustrate the way they are obtained. The uncertain parameters of yield amount and price level are calculated using real data. We present an instance that illustrates the use of the two-phase program of allocation applied to the distribution of land and the allocation of livestock raising areas. The variants presented here correspond, firstly, to deterministic linear structures for the case of the management of animal diets and, secondly, to linear stochastic programming models in two phases for the problem of land distribution and allocation of areas for the raising of cattle. The uncertainty of the model is given in terms of rainfall and the availability of water and food for the crop and cattle respectively.

Chapter 17 describes theoretical studies of the influence of the elastic-damping mechanism in the tractor transmission (on the functioning of the machine-tractor unit). It presents a mathematical model of the machine-tractor unit with an elastic damping mechanism, which is installed between the engine and the tractor transmission, when performing agricultural operations (sowing). The description of work of EDM (elastic damping mechanism) shows its advantages. The theoretical influence of the transparency degree description on the machine-tractor performance unit on sowing operation is also given. The dependence of transparency degree graph on various agricultural operations depending on frequency of fluctuations of traction loading of P and at various works for a tractor is given. The purpose of this work is to conduct theoretical studies of the influence of the elastic damping mechanism in the tractor transmission on the performance of the sowing unit. The optimal parameters of the elastic-damping mechanism installed in the transmission of the tractor for sowing operations are determined. The substantiation of expediency of installation of the elastic-damping mechanism in the tractor transmission when performing the sowing of grain crops is given.

Chapter 18 presents the results of the development of two systems. The first is a photovoltaic system parallel to the power supply network of led lamps. The algorithms of the system operation for both working and emergency lighting are shown. The basic operating modes of the system are considered taking into account the criterion of the minimum cost of electricity generated. These modes provide the most complete use of solar energy in the working day with minimal additional costs, allowing the consumer to save on electricity and increase the reliability of the emergency lighting system. The second system is a solar photovoltaic module built into a standard double-glazed window (730x700 mm), which is designed to charge a block of lithium-ion batteries with a capacity of 6.8 Ah with an output voltage of 5.25 V, the energy of which can be used to power any device having a USB 2.0 connector. The results of calculation of the required peak photovoltaic power of the module are presented; the technology of its sealing is described.

The book is intended to a wide range of persons, such as students of power and agroengineering specialties, experts and heads of municipal unions, managers of the ministries and other organizations responsible for development of agrarian industry, the businessmen dealing with such a business and many others as for organizations like big consortium, Inter-regional public organization for assistance to advancement of new technologies in social sphere “Sustainable development”

The editors strongly believe that the information got out of the book will undoubtedly be interesting and useful for a wide circle of readers from starting students to famous scientists. This edition should provide dissemination of rather new information among interested scholars, engineers, professors and students involved into the agro-engineering industry as well as in smart computing and optimization activities in different fields of science and engineering.

Valeriy Kharchenko
Federal Scientific Agroengineering Center VIM, Russia

Pandian Vasant
Universiti Teknologi Petronas, Malaysia

Acknowledgment

The editors of the published book would like first of all to express sincere gratitude to all the authors of the chapters presented in the book for their hard work in preparing the material for publication. Their work is commendable.

Special thanks deserve those of them who most actively participated in the review of submission since this book was prepared for the publication largely by the authors of this book. Especially it should be noted as active, professional reviewers of such authors as D. Budnikov, S. Senkevich, V. Bolshev, P. Karagoz, U. Fiore, A. Vasilyev etc.

A lot of thanks to the editorial advisory board members which have made a lot to provide the high quality of the book.

It is necessary to note the important role that being a specialist of the highest level in the subject matter played a respected professor, Dr. Gerhard-Wilhelm Weber, so that the book could be developed and meet high international standards. Editors express deep appreciation to this respectable scientist.

It would be wrong has not mention with deep gratitude the role that IGI Global publishing house has played in order that interesting and important information presented by authors from different countries was published and communicated to a wide circle of readers.

Finally, the editors would like to express many thanks to staff of IGI Global publishing for permanent attention.

Valeriy Kharchenko
Federal Scientific Agroengineering Center VIM, Russia

Pandian Vasant
Universiti Teknologi Petronas, Malaysia

Chapter 1
Development and Research of Phase-Transition and Thermochemical Materials for Heat Accumulation

Baba Dzhabrailovich Babaev
FSBEI HE, Dagestan State University, Russia

Valeriy Vladimirovich Kharchenko
https://orcid.org/0000-0003-3725-2976
Federal Scientific Agroengineering Center VIM, Russia

Vladimir Panchenko
https://orcid.org/0000-0002-4689-843X
Russian University of Transport, Russia

ABSTRACT

The chapter discusses various phase-transition heat-accumulating materials. Their application in heat pumps and their use for heat supply are presented. Phase-transition heat-accumulating materials in heat pipes are also considered, various types of heat pipes are presented. The installation of heat storage with phase-transition materials is presented. Along with phase-transition heat-accumulating materials, the chapter considers thermochemical heat-accumulating material of photochemical reactions of energy storage. As an example of a thermochemical heat-accumulating material, a solar power plant for thermochemical energy storage is presented. A developed computer program for the description of thermochemical reactions allows examining chemical and thermochemical reactions in multicomponent reciprocal systems, which can be carried out in the course of the reciprocal multicomponent systems. The computer program allows identifying thermochemical reactions occurring in mutual multicomponent systems, regardless of component and dependent on temperature.

DOI: 10.4018/978-1-7998-1216-6.ch001

Copyright © 2020, IGI Global. Copying or distributing in print or electronic forms without written permission of IGI Global is prohibited.

INTRODUCTION

In recent years, work has been intensively carried out in many countries around the world to develop energy storage technology based on phase-transitions of materials, which is regarded as a promising direction for creating efficient heat storage systems (HSS) (Babaev Kharchenko, Panchenko & Vasant, 2019; Adadurov, 2005; Babaev, 2016a; Babaev, 2014; Duffy & Beckman 2013; Magomedov, 2000; Shishkin, 2005). Thermal accumulation using heat of phase-transitions of the first kind of inorganic compounds and eutectic compositions is currently one of the most promising ways to save energy.

Recently, more and more attention has been paid to thermochemical batteries, which ensure the storage of energy as a result of the passage of some reversible chemical reaction. Accumulation of heat in this case occurs when a chemical reaction proceeds to the side in which it goes with heat absorption, and the heat return to the consumer is due to the reverse reaction (Babaev, 2016b, Babaev & Danilin, 2002; Duffy & Beckmann, 2013; Popel & Fortov, 2015; Twidell & Weir, 1990).

BACKGROUND

The success of the creation and introduction of thermochemical heat accumulators, which can be considered as seasonal, depend on the solution of thermophysical problems related to the selection of the necessary working substances and the economic problems associated with their cost parameters.

Chemical reagents and reactions for a chemical heat accumulator must meet the following requirements:

- Cheapness and accessibility;
- High volumetric density of stored energy;
- If possible, reaction without catalyst;
- Be chemically inert with respect to the structural materials of the thermal battery;
- Ease in implementing control over the course of a chemical reaction by means of temperature or pressure;
- A large number of thermochemical cycles;
- Have a large and developed external surface to intensify the course of the chemical reaction.

Of great importance is the search for new energy-intensive thermochemical heat-accumulating materials. Literary analysis (Magomedov, 2000; Engelsht & Muratalieva, 2013) shows the wide possibilities of their application. In addition, the chemical reactions used to store energy can convert low-potential energy into a high-potential energy. A similar feature is not possessed either by batteries based on phase transitions, nor, especially, by heat capacity.

Chemical energy can be stored either directly in the form of the internal energy of the molecules (for example, obtaining under the action of light enriched energy conformers, in particular valence isomers), or in reaction products that can interact with each other, releasing energy in a useful form.

To accumulate solar energy, in particular, it is possible to use any endoenergetic reactions that under the action of solar radiation (SR), accompanied by a rearrangement of chemical bonds or molecular structure. At the same time, two groups of processes that convert solar energy into chemical energy-photochemical and thermochemical can be distinguished from the mechanism of action of SR.

HEAT ACCUMULATORS WITH PHASE-TRANSITION HEAT-ACCUMULATING MATERIALS

Heat accumulators (HA) with phase-transition heat-accumulating materials (PTHAM) are based on the fact that when charging the battery heat is expended on any phase transition (most often to melting) of the working substance. When the heat is removed from the battery, the working substance solidifies again. Such batteries have two advantages. First, they are characterized by higher specific energy values than the heat-capacity accumulators of variable temperature. Secondly, the difference in the temperatures of the heat transfer medium that transfers energy to the consumer and the source of heat that charges the battery is determined here only by the desired intensity of heat transfer, whereas in capacitive it in most cases involves a large change in the temperature of the working substance of the battery. The disadvantage of batteries with phase transitions is that for each given temperature, a new working substance must be used in them, for the melting point of any substance is a quite definite value.

Phase transitions can be associated not only with melting or evaporation, sublimation, but also with structural transformations (for example, transition from one crystalline modification of matter to another), or by changing the amount of crystal-bound water. An example of a solid-solid phase transition (s-s) is solid sulfur, which can exist in two different modifications: low-temperature orthorhombic and high-temperature monoclinic (Sorensen, 2011). From the values given in the work (Sorensen, 2011) (table 1), the highest transition temperature and the largest latent heat of transition, the solid phase-solid phase is Li_2SO_4. The structure of pure Li_2SO_4 at 578 °C turns from monoclinic to face-centered cubic; The latent heat of this transition is 214 kJ/kg (which is much higher than the heat of fusion: 67 kJ/kg at 860 °C).

From the principle of operation of a battery of a first-order phase transition, it is clear that its specific energy is simply equal to the specific heat of the phase transition (fusion).

The melting enthalpy of most compounds was experimentally studied and described by the equation:

$$\Delta H_m = T_m \Delta S_m \quad (1)$$

In the absence of reliable experimental data on the enthalpy, it can be estimated from the additivity of the melting entropies of the compound (Magomedov, 2000):

$$\Delta H_{AmBn} = \Delta S_{AmBn} T_{AmBn} \quad (2)$$

Table 1. Enthalpies ΔH_{ss}, of some transitions of the s-s type

Substance	Transition Temperature, T_t (°C)	Latent Heat of Transition ΔH_{ss} (kJ/kg)
V_2O_2	72	50
FeS	138	50
KHF_2	196	135
Na_2SO_4	210,247	80
Li_2SO_4	578	214

(Sorensen, 2011)

$$\Delta S_{AmBn} = m\Delta S_{A} + n\Delta S_{B} \qquad (3)$$

where S_{A} и S_{B} – are the melting entropies of the elements *A* and *B*, respectively, and *m* and *n* are the stoichiometric coefficients in the chemical formula for the compound $A_{m}B_{n}$..

In a number of cases, in calculating the entropy of melting, disordering of the structure during melting is taken into account (Magomedov, 2000):

$$\Delta S_{AmBn} = m\Delta S_{A} + n\Delta S_{B} + \psi \qquad (4)$$

where ψ is a correction that takes into account an increase in the disorder of the structure of the compound upon melting:

$$\psi = 4,573\text{K}\left(m+n\right)(m\,/\,(m+n)\lg(m\,/\,(m+n)) + n\,/\,(m+n)\lg(n\,/\,(m+n))) \qquad (5)$$

The correction factor *K* takes into account the degree of structure disorder during melting and varies from zero – for compounds that completely retain their short-range order during melting, up to unity – for compounds that radically change their structure.

For a rough estimate of the heat of fusion, known empirical rules can be used, according to which the ratio of the molar heat of fusion ΔH_{m}^{mole} to the melting point is a constant value for each of the following classes of substances:

- for atomic substances $\Delta H_{m}^{mole} / T_{m}$ = 2 ÷ 3 kcal/(kmol·K), or 8,4 – 12,6 kJ/(kmol·K)
- for inorganic compounds $\Delta H_{m}^{mole} / T_{m}$ = 5 ÷ 7 kcal/(kmol·K), or 20,9 – 29,3 kJ/(кмоль·K)
- for inorganic compounds $\Delta H_{m}^{mole} / T_{m}$ = 10 ÷ 16 kcal/(kmol·K), or 41,9 – 67,0 kJ/(kmol·K)

From the above rule two conclusions can be drawn: the higher the melting point of a substance, the higher its heat of fusion; the lower the relative molecular mass of the substance, the higher is its specific heat of fusion.

For nonpolar and moderately polar compounds, in which the vapor pressure at the triple point does not exceed 10^4 Pa, a calculation using the simple formula (in J/mol) $\Delta H = 6,147T_{m}^{1,333}$.

For highly polar compounds, a linear correction is observed between the heat and the melting point, described by the formula (in J/mol) $\Delta H_{m} = 44,4T_{m} - 4400$.

The accuracy of this formula was within ± 15%. There are precise thermodynamic methods for calculating the melting enthalpy. For example, according to the Clapeyron-Clausius equation:

$$dp\,/\,dT = \Delta H_{m}\,/\,T\,(V_{l} - V_{s}).$$

Hence:

$$\Delta H_{m} = dp\,/\,dT(T(V_{l} - V_{s})) = (V_{l} - V_{s})(dp\,/\,d\ln T).$$

However, published experimental data on the dependence of the melting point on pressure are known for an extremely small number of compounds, which makes this method inapplicable for most salts.

If the melting temperature of the compound is known, then, starting from the Schroeder equation of ideal solubility:

$$d(\ln x) / dT = \Delta H_m / RT^2$$

and integrating it from T_1 to T_m, obtain the expression

$$\Delta H_m = \ln(1 / x)R(T_m T_1) / (T_m - T_1),$$

where x is the solubility of a given compound in mole fractions at a temperature of T_1 in some solvent. To increase the accuracy of x should be fairly close to unity.

The amount of heat accumulated by the accumulator during the heating of the PTHAM from t_0 to t_k, melting temperature t_{mt} can be found by formula:

$$W_{ac} = m\left(c_{ac}^{s}\left(t_{pt} - t_0\right) + \Delta H + c_{ac}^{l}\left(t_k - t_{pt}\right)\right), \text{ kJ} \quad (6)$$

where m is the mass of HSM, kg; ΔH - enthalpy of the phase transition, kJ/kg; c_{ac}^{s} and c_{ac}^{l} - the heat capacity of HSM in solid and liquid states, kJ/(kg °C).

There are a large number of eutectic compositions of two-, three-, four- and five-component systems (Dibirov, 2012; Zamaldinova, Garkushin, Garkushin & Parfenova, 2012; Magomedov, 2000), which can be used as PTHAM. A large volume of work on the search for energy-intensive PTHAM was carried out in (Magomedov, 2000; Son, 2013; Stankus, Khairulin, Martynets, Bezverkhiy, 2013, Shishkin, 2005).

Some of the two-component eutectic compositions that can be used as PTHAM and their specific heat of fusion are presented in (Babaev, 2014; Babaev, 2016b).

The greatest prospects for low-temperature accumulation can have organic acids and paraffins (tables 2 and 3). The most affordable and commercially available PTHAM are paraffins. As is known, the main component of paraffins is the aliphatic hydrocarbon C_nH_{2n+2}. In solid form, paraffins are in two allotropic states: softened at a temperature slightly below the phase transition temperature t_{pt} (the crystals are stretched into a chain) and solid at $t<t_{pt}$ (the crystals have a disk-like shape). The physical properties of technical paraffins according to data (Shishkin, 2005) are given in table 3.

When the number of carbon atoms n_c varies from 18 to 50, the temperature of the phase transition (melting point) t_m increases from 28 to 68 °C. Paraffins have a relatively high heat of phase transition (heat of fusion) ΔH = 144 - 189 MJ/m^3. The price of technical paraffins is low and is currently 15 - 35 rubles/kg.

Table 2. Main characteristics of organic heat-accumulating materials

HSM	T_m, °C	The Density r, g/sm³		Thermal Conductivity l, W/(m-K)		Heat Cap-acity, kJ/(kg-K)		Phase Transition Enthalpy, ΔH	
		r_s	r_l	L_s	l_l	C_s	C_l	kJ/kg	MJ/m³
Lauric acid	44,0	-	0,91	0,4	0,2	-	-	175,3	159,6
Myristic acid	54,1	-	0,87	-	-	1,60	2,26	187,8	162,8
Palmitic acid	65,0	-	0,88	-	-	1,80	2,73	184,5	162,9
Paraffin	42,0	0,91	0,77	-	-	2,08	-	187,8	144,0

Table 3. Thermophysical properties of paraffins

n_c	t_{pt}, °C	ΔH		ρ, kg/m³		c, kJ/kg K	λ, W/mK
		kJ/kg	MJ/m³	t=20°C	t=70°C		
18	28	244	189	814	774	2,1	0,15
13 – 24*	22 – 24	189	144	900	760	2,1	0,21
16 – 28	42 – 44	189	145	910	765	2,1	0,21
20 – 33	48 – 50	189	145	912	769	2,1	0,21
22 – 45	58 – 60	189	150	920	795	2,1	0,21
24 – 50	66 – 68	189	157	930	830	2,1	0,21

(Shishkin, 2005)
Note: * - with an admixture of 20% oil.

PHOTOCHEMICAL REACTIONS OF ENERGY ACCUMULATION

In photochemical processes, photons of sunlight are directly used for the realization of endoenergetic chemical reactions. A decay or transformation, accompanied by the release of stored energy, are carried out if necessary, introducing the appropriate catalyst.

The flow of many chemical reactions can be initiated by the absorption of a photon of visible or ultraviolet light. In this case, the corresponding molecule passes into a new, excited state, which may radically differ from the ground state. As a rule, the energy received by the molecule upon irradiation exceeds the typical activation barriers of thermal reactions, and in many cases it is even higher than the dissociation energy of chemical bonds (250-500 kJ/mol). Nevertheless, photochemical processes often proceed much more selectively than one would expect, given the high energies of the excited states. The reason for this circumstance is that the absorption of radiation energy is a quantum process, in contrast to the absorption of heat in chemical reactions, and the photoexcited state of matter has a completely definite structure.

To evaluate the efficiency of light radiation on the chemical process, the concept of the quantum yield is used η: the number of decomposed (or formed) product molecules, $\eta = n_{mol}/n_{ph}$, where n_{mol} is the number of decomposed (or formed) product molecules; n_{ph} is the number of absorbed quanta.

In some reactions, the reagent molecule does not absorb the photon, since its electron spectrum is outside the spectral range of the incident light. In this situation, a mediator is needed a sensitizer that can absorb light, and then transfer its energy to a potentially reactive molecule. This phenomenon is called photosensitization. In organic systems, aromatic ketones such as acetophenone ($C_6H_5COCH_3$) and benzophenone ($C_6H_5CO\ C_6H_5$) are very useful sensitizers.

High-energy heat carriers in closed systems can be obtained with the help of photochemical reactions, storing the light energy in the form of energy-saturated valence isomers or conformers of organic molecules. A typical example of the first type of transducers are systems based on an organic compound of norbornadiene, dissolved in non-polar solvents and capable of being converted into a kinetically stable energy-saturated quadricyclene (figure 1) with ultraviolet light in the presence of special sensitizers (for example, copper (I) compounds) reserve about 20 kcal/mol of thermal energy (about 250 kcal per 1 kg of the main substance). The release of thermal energy occurs in the reverse reaction of the formation of norbornodiene from a quadricyclene. This reaction can be carried out in the presence of special catalysts.

However, these processes are carried out in dilute solutions, as a result of which their specific indices are low. The main task, on the solution of which the prospects for the further development of this direction depend, is the search for new photochromic systems that would have high specific characteristics and at the same time would be quite cheap.

It also discusses the possibility of creating converters and accumulating energy in closed systems in the form of dissociation products of substances excited by light (table 4).

Figure 1. Scheme of photochemical reaction

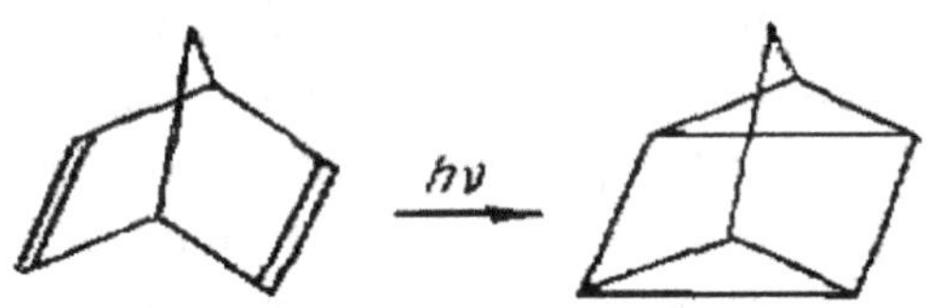

Table 4. Thermodynamic characteristics of energy storage due to photochemical reactions of dissociation

Reaction	ΔG^0_{298}	ΔH^0_{298}	λ_0, Nm
	kcal/mol		
1. $NO_{2(g)} \leftrightarrow NO + (1/2)0_2$	*8,3*	*13,5*	*435*
2. $NOCl_{(g)} \leftrightarrow NO + (1/2)Cl_2$	*4,8*	*9,0*	*637*
3. $AgCl_{(s)} \leftrightarrow Ag_{(s)} + (1/2)Cl_2$	*26,2*	*30,4*	*405*
4. $AgBr_{(s)} \leftrightarrow Ag_{(s)} + (1/2)Br_2$	*22,9*	*23,8*	*460*
5. $AgI_{(s)} \leftrightarrow Ag_{(s)} + (1/2)I_2$	*15,9*	*14,9*	*440*
6. $Ag_2O_{(s)} \leftrightarrow 2Ag_{(s)} + (1/2)O_2$	*2,6*	*7,3*	-
7. $BiOCl_{(s)} \leftrightarrow BiO_{(s)} + (1/2)Cl_2$	*33,5*	*37,4*	*>350*
8. $(3/2)O_{2(g)} \leftrightarrow O_{3(g)}$	*39,1*	*34,0*	*254*
9. $CO_{2(g)} \leftrightarrow CO + (1/2)O_2$	*61,5*	*67,7*	*169*

Note: λ_0 is the wavelength of the "red" boundary for the photochemical reaction.

Figure 2. Heat pipe: 1 - housing; 2 - wick

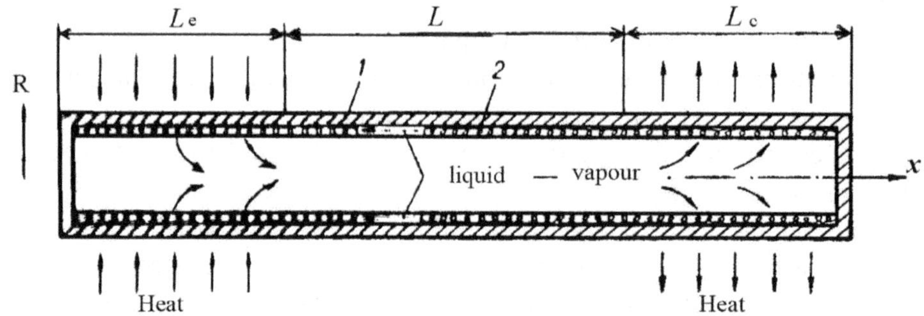

PHASE-TRANSITION HEAT-ACCUMULATING MATERIALS IN THERMAL PIPES, TYPES OF HEAT PIPES

Phase-transition evaporating heat-accumulating materials have found wide application in heat pipes (HP). HP is a closed evaporative-condensing heat transfer device. In the simplest case, it is a sealed vacuum cylinder body with a wick on the inner wall impregnated with a heat carrier (figure 2).

When the heat pipe section (the evaporation zone L_e) is heated, the heat carrier evaporates. Steam as a result of the increase in pressure moves to the cooled zone, where it condenses. Return of condensate to the evaporation zone is carried out by capillary-porous material (CPM) due to capillary forces. CPM (wick) is made from a fine mesh net, fibrous material, sintered fine particles, or in the form of longitudinal grooves made in the wall of the hull, as well as various combinations of these structures (figure 3).

The most common heat pipes consist of only three elements: the body, the working fluid, the capillary-porous material. Each of them is subject to stringent requirements.

The housing, which ensures the isolation of the working fluid from the external environment, must be sealed, withstand the pressure difference between the internal and external environments, provide heat supply to the working fluid and remove heat from it. The material of the case is usually stainless steel, aluminum alloys, copper, glass, bronze; plastics (flexible HP), ceramics (high-temperature HP). The section of the body can be round or rectangular. The minimum diameter of the HP should be such that the internal diameter of the transport zone of the steam excludes the effect of capillary forces, i.e. that the steam channel does not turn into a capillary channel; maximum – no fundamental limitations.

Phase transition material – the working fluid must: have a liquid-vapor phase transition point in the required operating temperature range; have a sufficiently large latent heat of vaporization; good to wet the wick and body material; have a low viscosity value of the liquid and vapor phases; have high thermal conductivity and high surface tension and should not decompose at the required operating temperature range.

As a phase-transition working fluid, the most diverse substances, which are brought to the liquid phase – from liquefied gases to metals: helium (-271 ... -269 °C), ammonia (-60 ... +100 °C), freon-11 (-40 ... +120 °C), acetone (0 ... +120 °C), water (30 ... 200 °C), mercury (250 ... 650 °C), sodium (600 ... 1200 °C), silver (1800 ... 2300 °C), and so on.

Figure 3. Possible designs of capillary structures of heat pipes: a – mesh, b - sintered powders, metal felt, etc ., c - longitudinal grooves, d - annular gap, e - crescent gap, f – artery, g - grooves covered by a grid, h, i - a variety of arteries; a - f - homogeneous structures, g - i - composite structures

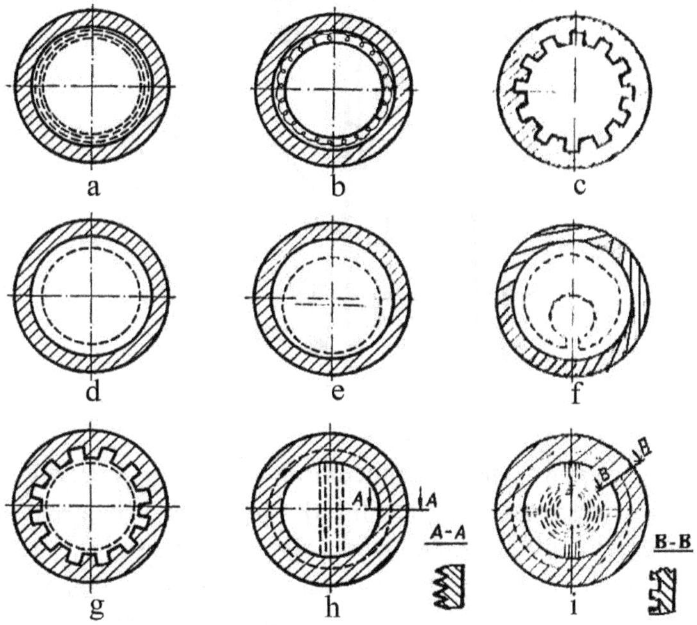

Capillary-porous material (wick) provides the creation of a capillary effect for moving liquid from the condenser to the evaporator and uniform distribution of it throughout the evaporation zone. The most commonly used metal mesh, metal fabric twill weave, sintered metal powders, metal felt. Metals: copper, nickel, titanium, phosphorous bronze, stainless steel.

Along with the capillary forces, other forces are used to return the condensate to the evaporation zone of the HP. Gravitational return of condensate takes place under conditions where the evaporation zone is below the condensation zone. Such pipes are called diphasic thermosyphons (figure 4).

In addition to gravitational (thermosyphon) and capillary (classical HP) forces, modern types of HP use centrifugal, electrostatic, magnetic, osmotic and other types of fields to return condensate (figure 5).

For normal operation of the HP, it is necessary that the operating modes of all its elements do not reach critical values.

The heat input can be carried out by any known method (open flame, heat exchange with a heated body, electric current, IR rays, electron bombardment, etc.). In this case, due to the thermal resistance of the wall, there is a normal temperature difference between the heat source and the working fluid. The temperature limit depends only on the thermal (cold) resistance of the housing material.

Evaporation zone. The density of the heat flux should not exceed a certain value at which the heat transfer crisis may occur (drying the wick, when the capillary-porous material does not have time to feed new portions of liquid into the evaporation zone).

Figure 4. Thermosiphon: 1 - zone of evaporation; 2 - condensation zone; 3 - liquid; 4 - housing; 5 - steam; 6 - return of liquid (condensate)

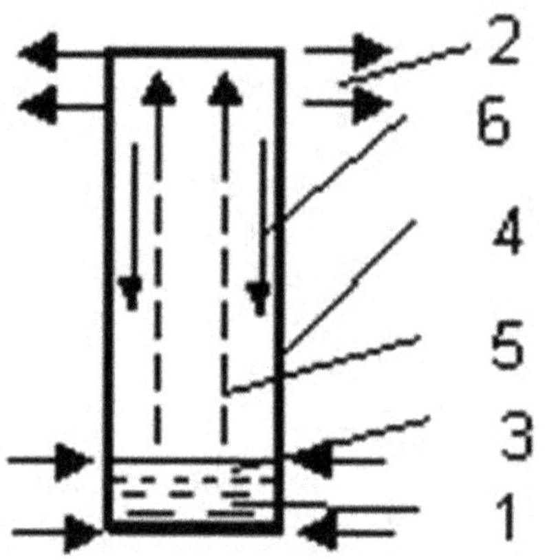

Condensation zone. There are two types of condensation – a drop or, more often, a film one. Typically, the vapor condenses on the surface of the liquid film. Between this surface and the outer surface of the wall there is a small temperature difference (i.e., the wall temperature is not much, but differs from the vapor temperature). The surface of the HP in the condensation zone has an interesting property - isothermicity, because this surface works practically at a constant temperature (close to the temperature of the vapor-liquid phase transition). If a local heat flow occurs on a site, the amount of condensed steam in that place increases and the temperature is maintained at the same level.

Figure 5. HP with the use of different forces for returning condensate: a) centrifugal (1 - evaporation zone, 2 - condensation zone, 3 - vapor flow, 4 - return of liquid); b) osmotic (1 - evaporation zone, 2 - semipermeable zone, 3 - condensation zone, 4 - steam channel, 5 - fluid return channel); c) electro-osmotic (1 - evaporation zone, 2 - condensation zone, 3 - sintered glass beads, 4 - porous electrodes, 5 - current source); d) electrohydrodynamic (1 - body HP, 2 - steam channel, 3 - ribbon electrode, 4 - working fluid); e) ultrasonic (1 - case, 2 - evaporation zone, 3 - condensation zone, 4 - concentrator, 5 - magnetostrictive type emitter, 6 - terminals to the ultrasound generator)

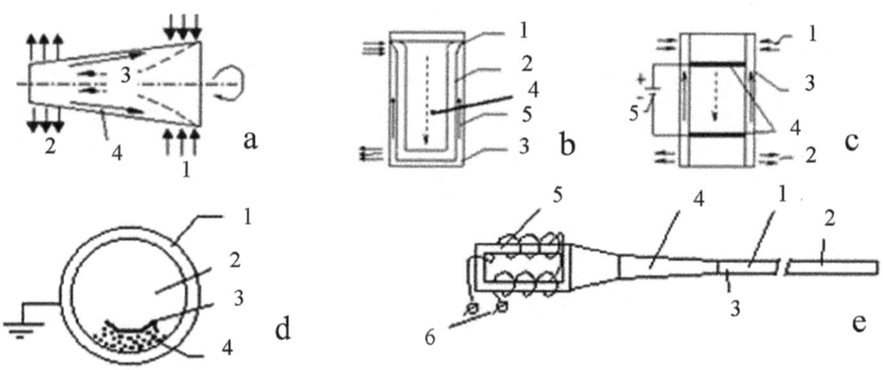

Transport zone of steam. In the evaporation zone, an excessive vapor saturation pressure is created, which is expended to accelerate the steam to an axial velocity, to overcome the frictional force of the steam flow on the surface of the wick and to overcome the vapor viscosity forces. As a result, the pressure along the length of the HP, from the evaporation zone drops. In the middle section of the pipe, the pressure stabilizes (adiabatic region). In the condensation zone, the pressure is restored almost to the value of the pressure in the evaporation zone. In practice, pressure losses in simple HP are related only to the overcoming of viscous forces.

An important element in creating a HP operating at a given temperature is the choice of an effective working fluid, which is an evaporating heat accumulator. A criterion *M* is proposed for the selection of the working substance of the HP, which, taking into account the connection between kinematic and dynamic viscosity, can be represented as:

$$M = \sigma r d \,/\, \eta \tag{7}$$

where σ - surface tension; η - dynamic viscosity; d - the density; r - the specific heat of evaporation.

PHASE-TRANSITION HEAT-ACCUMULATING MATERIALS IN HEAT PUMPS AND THEIR USE FOR HEAT-WATER SUPPLY

The heat pump (HP) is a device that allows the accumulation of heat from low-potential heat sources, which uses the effect of a phase transition of liquids to steam at low temperatures (freons boiling in the temperature range -9 to -30 °C).

The HP consists of an evaporator, a condenser, a throttling device, a compressor and a compressor drive. As a rule, the compressor drive is an electric motor (figure 6).

Figure 6. Hydropneumatic scheme of the stand ''Heat pump''

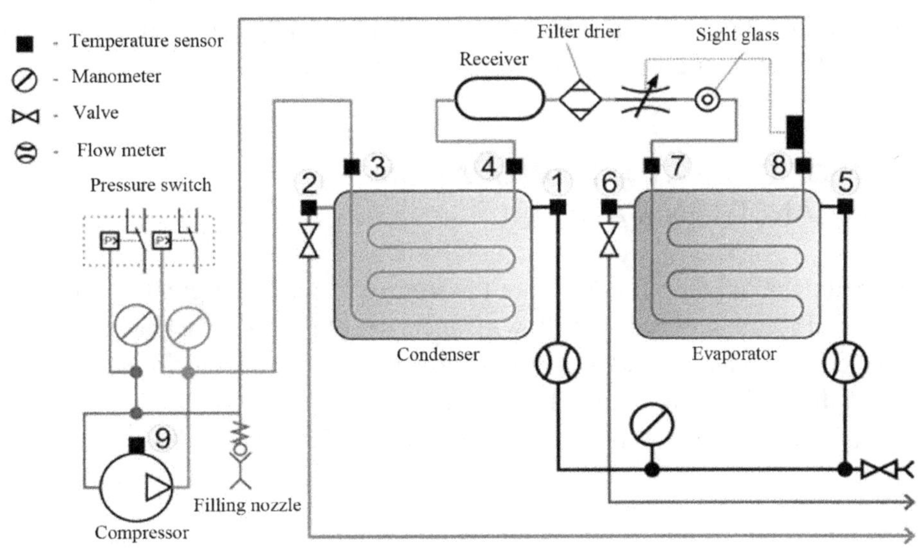

In the world, HP is widely used to heat buildings and structures based on low-potential heat sources (ground, artesian, river, lake, sea water, recycling water systems, etc.).

The study of the soil as a source of heat, carried out in Europe, showed that the heat flux to the evaporator from the ground is 20-25 W/m, the minimum value for Europe is 10 W/m, the maximum 50-60 W/m (Popel & Fortov, 2015). If the power of the ground heat exchanger is not sufficient to cover heat loss in the high-temperature circuit, a peak closer is included. If the water content in the soil is high, the indicators will increase due to increased thermal conductivity and good contact with the tubes. A greater concentration in the soil of the gravel will cause deterioration in characteristics. It is also interesting that the minimum temperature of the ground is always higher than that of air, and is reached two months later, when the required heating power is reduced.

It should also be noted that vertical tubes can be used for heat exchange in the ground. At the same time vertical tubes take up less space and allow in a sense to use the heat accumulated in the summer months, which gives them economic advantages. It is of interest to use schemes – solar collectors with HPs and heat accumulators. But there can be a complete combination of a collector with an evaporator (Babaev, 2015; Babaev, 2016b). In addition to the use of ground heat, the most attractive for use in household applications is a "free" source of heat (exhaust gases from kitchen stoves or even kitchens, waste water, etc.) to create comfortable conditions inside the house.

Low-mineralized geothermal waters, solar energy, stored with the help of solar heaters and heat accumulators, can also serve as a source of low-potential heat. However, the main sources of heat for large HP should be considered artificial sources – thermal waste. For example, thermal power plants, including nuclear power plants, are discharged with cooling water 50-55% of fuel energy.

Industrial enterprises consume huge amounts of water for cooling machines and working bodies in various technological processes. These "heat rivers" have a temperature of 20 - 40 °C all the year round, which practically does not allow the use of heat directly, and are cooled in cooling towers or other evaporative coolers, giving part of the water together with heat.

So, the negative factors of traditional heat supply can be solved by the useful use of diffuse low-temperature (5 - 30 °C) natural heat or waste industrial heat for heat supply with the help of HPs.

The HP in domestic heat supply and air conditioning can be most widely used, especially in the southern regions of Russia, where year-round air conditioning is required: cooling in the summer months and heating in winter. The HP-conditioner can be installed individually in each room with its compressor and condenser, an external or internal heat source for the evaporator.

INSTALLATION OF ENERGY ACCUMULATION WITH PHASE-TRANSITION MATERIALS

With the use of accumulation, the very design of the storage battery is of great importance.

The known thermal accumulators do not allow efficient use of PTHAM for heat accumulation because of the low thermal conductivity of solid phase-transition materials in a large volume of the tank and low heat exchange with the coolant throughout the volume.

An improved heat accumulator (Babaev, 2015) is shown in figures 7 and 8.

Figure 7. The main view of the thermal battery (a) and the attachment of the electric heater (b)

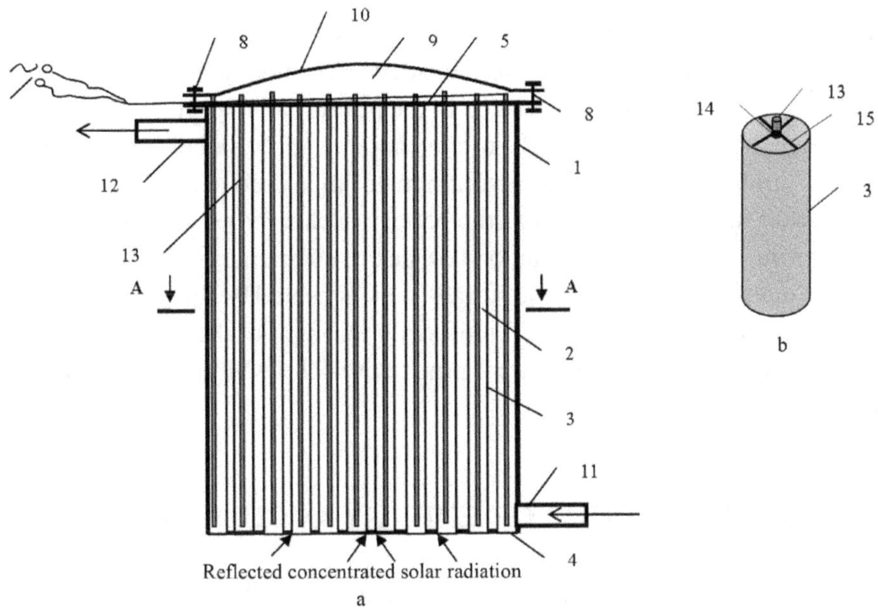

Figure 8. Cross section A-A (left) and the plan of the intermediate cover of the storage tank (right)

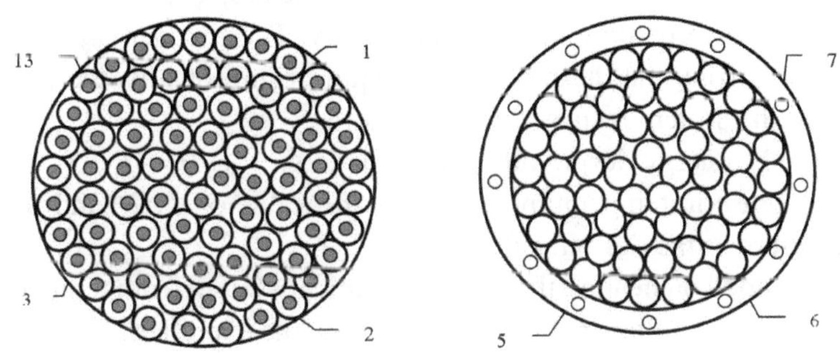

The heat accumulator with a phase-transition material contains a body 1, a working substance (PTHAM) 2, which fills vertical tubes 3, which are fixed to the bottom 4, which can serve as a receiver for concentrated SR, an intermediate cover 5 with holes 6 for passing tubes and 7 for fixing bolts 8, from below the tubes are closed, and from above are opened for volumetric expansion of PTHAM 2 into the free chamber 9 under the sealed cover 10, which is fastened to the main body of the storage tank by nuts and bolts 8 through the seals. To the tank of the storage tank 1, the branch pipes 11 and the coolant outlet 12 are welded. Electric heating elements 13 are inserted into the vertical tubes 3 with PTHAM, which are connected by electric wires to a source of electrical energy, for example, a wind power plant. The electric heating elements 13 are screwed into a threaded cage 14 which is attached to the vertical tubes 3 with PTHAM bridges 15.

To start the battery, the heat accumulator lid 10 is removed and the tubes 3 filled with PTHAM 2 and the lid is closed. Charging the battery with the phase-transition material by heat occurs due to the solar energy directed by the mirrors to the bottom of the tank-accumulator-receiver SR, while due to the high thermal conductivity of the steel tubes 3 and the casing of the electroheating elements 13, intense melting of the working substance 2 occurs in them, the expansion of the working substance takes place in the free chamber-chamber 9 under the sealed cover 10. Charging can occur and at the expense of electrical energy when connecting the source to the electric heating element 13 in vertical tubes 3 with PTHAM 2. At the same time, the coolant is heated in the intertubular space of the storage tank, which is fed from below through the connection pipe 11 of the coolant supply and discharged through the branch pipe 12. If the heat accumulator with phase-transition material is used in traditional heat supply systems on organic fuel, the working substance melts, (the battery is charged) due to the coolant (hot water) in the annular space. During the increase of heat load or lack of solar energy, the battery is discharged, and the heat carrier in the intertubular space of the storage tank is heated by the phase transition (crystallization) of the working substance 2 in the tubes 3 uniformly throughout the accumulator tank. Enclosures of electric heating elements installed in vertical pipes with PTHAM and in the absence of electrical energy act as heat exchangers.

THERMOCHEMICAL REACTIONS OF ENERGY ACCUMULATION

In thermochemical processes, as a result of phase transitions and endothermic chemical reactions, energy-intensive salt formulations are obtained. In this case, sunlight is used as a source of heat. And in the course of thermochemical reactions, solar energy is converted into chemical energy not directly (Kharchenko, Panchenko, Tikhonov & Vasant, 2018; Panchenko, Kharchenko & Vasant, 2019), but through an intermediate transformation into thermal energy. Thermochemical transformation of solar energy is carried out by shifting the thermodynamic equilibrium towards the product of the reaction when the temperature of the system is increased by light.

The change in the Gibbs energy during the reaction is $\Delta G = \Delta H - T\Delta S$. As can be seen, in the case of $\Delta S > 0$, an increase in temperature decreases the value of ΔG, and this leads to a shift of thermodynamic equilibrium to the desired side.

The flux of SR is expended on heating the thermochemical system from temperature T_0 to $T(Q)$ and to a chemical reaction at $T(Q(T))$ (Babaev, 2016a). Therefore, the overall efficiency of the thermochemical transformation of solar energy (η) is determined by the efficiency of conversion of thermal energy into chemical energy at a temperature $T(\eta_1)$ and the efficiency of conversion of solar energy into thermal energy with temperature $T(\eta_2)$.

$$\eta = \eta_1 \cdot \eta_2 \tag{8}$$

For η_1 one can take the ratio of the stored free energy $\Delta G(T_0)$ to the spent quantity $Q(T)$ of thermal energy by the chemical reaction at temperature T,

$$\eta_1 = \frac{\Delta G(T_0)}{Q(T)} \tag{9}$$

The heat energy $Q(T_0T)$, absorbed by the chemical system when heated from T_0 to T and at T in the course of the reaction, is useful from the SR flowing to the surface of the receiver with a system characteristic of a given locality $Q_{(C)}$ with the speed V_T, $Q(T_0T) = Q + Q(T)$

$$\eta_2 = \frac{Q(T_0T) \cdot V_T}{Q_{(C)}} \tag{10}$$

Substituting in the formula (10) the values of η_1 and η_2 from (8) and (9), obtain:

$$\eta = \eta_1 \cdot \eta_2 = \frac{\Delta G(T_0)}{Q(T)} = \frac{Q(T_0T) \cdot V_T}{Q_{(C)}} \tag{11}$$

The cost of heat energy for heating the initial reactants from T_0 to T can be ignored when calculating the efficiency of thermochemical converters (these costs can be largely recovered when cooling the reaction products), i.e., $Q(T) \approx Q(TT_0)$. Then formula (11) takes the form (Babaev, 2016a):

$$\eta = \frac{\Delta G(T_0)V_T}{Q_{(C)}} = \frac{[\Delta H(T_0) - T_0\Delta S(T_0)]V_T}{Q_{(C)}} \tag{12}$$

Both the photochemical and thermochemical method can provide a sufficiently high efficiency of conversion of solar energy into chemical energy. At the same time, for simple thermochemical transformations, the efficiency can be 50-60%, and for simple photochemical converters 20-30%.

The simplest thermochemical conversion methods can produce low-potential heat, suitable for heating buildings. More high-potential heat can be obtained in closed thermochemical cycles using catalytic dissociation and conversion processes (table 5).

The considered processes are widely discussed in connection with the problems of conversion of thermal energy from nuclear power plants.

Any chemical reactions that result in the formation of combustible gaseous products - hydrogen and carbon monoxide - can be classified as high-temperature accumulation. Therefore, all the thermochemical cycles of hydrogen production from water, its photocatalytic decomposition, as well as the processes of processing natural fuels and rapidly growing plants with concentrated SR should be included here.

It is possible that processing (upgrading) of natural fuel can become one of the main directions in solar energy. This is due to the facts that at present huge amounts of natural gas and coal are processed to produce hydrogen, and heating in these processes is carried out by burning the same fuel. The use of solar energy in these processes will not only ensure fuel economy, but will also allow the accumulation of solar energy in processing products in the form of high-potential heat. In general, can state that when processing fuel with the help of solar energy, a significant increase in the heat of combustion of the initial fuel occurs. Table 6 shows the data of calculations that confirm the expediency of such processing.

Table 5. Thermodynamic characteristics of thermochemical processes proposed for conversion of solar energy into chemical energy in closed systems

Reaction	ΔG^o_{298}	ΔH^o_{298}	T, K
	kcal/mol		
1. $SO_3 \leftrightarrow SO_2 + 1/2O_2$	16,7	23,5	1030
2. $NH_3 \leftrightarrow 1/2N_2 + 3/2H_2$	4,0	11,0	470
3. $CH_4 + H_2O \leftrightarrow CO + 3H_2$	36,0	59,8	960
4. $CH_4 + CO_2 \leftrightarrow 2CO + 2H_2$	40,8	59,1	960
5. $C_6H_{12} \leftrightarrow C_6H_6 + 3H_2$	23,4	49,3	570
6. $H_2SO_4(98\%) \leftrightarrow H_2SO_4(60\%) + H_2O$	12	20	>400
7. $N_2O_4 \leftrightarrow 2NO_2$	1,29	13,9	330

Note: T is the temperature at which the Gibbs thermodynamic potentials for the initial reactants and reaction products are equal.

Table 6. Accumulation of solar energy in the processing of organic fuels

The Processing Process (Indicating ΔH^0_{298}, kcal/mol)	Heat of Combustion, kcal/mol		Accumulated Amount of Thermal Energy,10^{-6} kJ/t carbon	Enrichment of Fuel With Energy, %
	Source Fuel	Products of Fuel Processing		
Steam coal conversion: $C + H_2O = CO + H_2$ - 31,4	94,0	125,4	12	33
Carbon dioxide conversion of coal: $C + CO_2 = 2CO$ - 41,2	94,0	135,3	14	44
Steam conversion of natural gas: $CH_4 + H_2O = 3H_2 + CO$ - 49,3	191,8	241,0	13	26
Carbon dioxide conversion of natural gas: CH4 + CO2 = 2CO + 2H2 - 59,1	191,8	250,9	15.5	31
Iron-smelting coal processing: Fe + H2O = FeO + H2 + 5,2 3FeO + H2O = Fe3O4 + H2 + 20,3 2C + 2H2O = 2CO + 2H2 - 62,8 Fe3O4 + CO = 3FeO + CO2 - 10,3 FeO + CO = Fe + CO2 + 4,6 2C + 2CO2 = 4CO - 82,5 4C + 4H2O = 4CO + 4H2 - 125,6	94,0	125,4	12	33

SOLAR POWER PLANT WITH THERMOCHEMICAL ENERGY ACCUMULATION

Figure 9 schematically shows a solar installation for chemical reactions (Babaev, 2014). The porous material *5* can differ both in the material used (as a porous body, graphite, platinum or other heat-resistant, inert material can be used) and the dimensions of the cells in it and, depending on the initial reaction reagents, can be changed in the apparatus as a packing.

The SR, focused by the concentrator *1*, passing through the transparent window *3* into the chamber *2*, heats the porous body *5*. In the chamber *2* to the porous body *5*, solid powder or liquid initial reagents enter through the branch pipe *6*. The porous body *5* has a high absorption coefficient of SR. In

Figure 9. Scheme of the solar power plant for chemical reactions

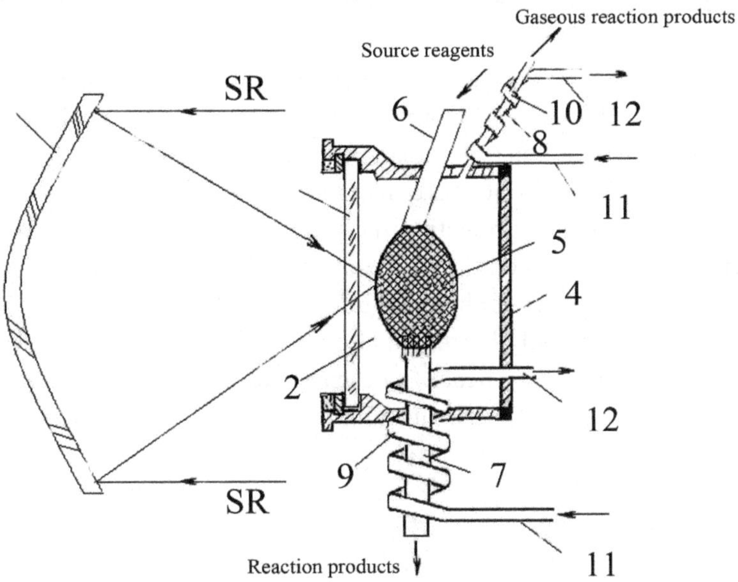

the porous body, solid initial substances melt and flow high-temperature reactions. To stop reversible reactions and increase the yield of the desired product, the reaction products immediately after exiting the porous body *5* enter the outlet pipes *7* and *8*, where they are cooled by the heat exchangers *9* and *10*. Cooling in the heat exchangers occurs due to the coolant (water) of the supply and discharge pipes *11* and *12*, respectively.

In the plant, reactions can be conducted between gaseous, liquid and solid reagents.

With gaseous starting materials, they are fed through the branch pipe *7*, and the reaction products are discharged through the pipe *8* during the operation of the heat exchanger *10* (the heat exchanger *9* is switched off).

With solid phase starting materials they are fed along the branch pipe 6 in the form of a powder to the porous body *5*, passing through which the starting materials are heated by SR and react. On the outlet branch pipe *7* where they are cooled, the reaction products are discharged during the operation of the heat exchanger *9* (the heat exchanger *10* is switched off).

If liquid, solid and gaseous products are formed during the reactions between liquid or solid reactants, the liquid and solid are discharged through the branch pipe *7*, and the gaseous gases are discharged through the additional branch pipe *8*, and both heat exchangers *9* and *10* operate.

Advantage of the proposed solar power plant in its simplification: due to the shape of the working chamber; use of a porous body, instead of layers; and efficiency: by expanding the area (use) of the reaction between single phase states of the reagents (solid - solid, solid - liquid, liquid - gaseous, etc.); an increase in the output due to the use of a heat exchanger in the housing; the focus of the concentrator makes it possible to raise the temperature to 900 - 1100 °C.

Thus, the solar power plant can be used as a thermochemical solar energy accumulator, since in the absence of SR, the components cool down, undergo a phase transition liquid - solid and in the solid phase, reactions take place with the release of thermal energy. In it, the endo-, exothermic reactions given above can be carried out.

EXPERIMENTAL METHODS FOR THE DEVELOPMENT OF PHASE-TRANSITION AND THERMOCHEMICAL HEAT-ACCUMULATING MATERIALS

Multicomponent systems (MCS) are the chemical-technological base for creating new energy-intensive heat-accumulating materials (HAM). The search for PTHAM is based on the theory and practice of physicochemical analysis of the MCS in order to identify eutectic compositions, since the eutectic compositions of the MCS are more energy-intensive than individual substances.

Studies of MCS by conversion and projection-thermographic methods to identify the eutectic composition require obtaining information on the temperatures of all phase transitions and confirmation of the phase composition of the products of chemical interaction. Therefore, differential thermal analysis (DTA) in combination with visual-polythermal analysis (VPA) are the main methods of physical and chemical analysis. The magnitude of the heat of phase transition for a nonvariant composition is determined by the method of quantitative DTA. The installation is calibrated according to the phase transition temperatures of individual salts and eutectic mixtures (table 7).

Differential Thermal Analysis

To record the cooling (heating) curves, an installation based on an electronic automatic potentiometer is used, the circuit of which is shown in figure 10.

Research is carried out in crucibles using thermocouples. As a reference substance, a refractory substance is used, for example, alumina.

Table 7. Phase transition temperatures of individual salts and eutectic mixtures

Composition	KNO_3	45% KCI + 55% Na_2SO_4	30,55% NaCI + 69,5% Na_2SO_4	KCI	NaCI	Na_2SO_4	NaF	K_2SO_4
T_m, °C	336	517	627	774	800	884	990	1069

Figure 10. Block diagram of the installation of integrated DTA: I - cold junctions of thermocouples; II - block gain; III - control unit; IV - registration unit: 1 - simple thermocouple, 2 - differential thermocouple, 3 - chromel-alumel electrodes, 4 - voltage comparator; 5 - resistance box, 6 - regulated voltage source, 7 - transformer, 8 and 9 - electronic potentiometers

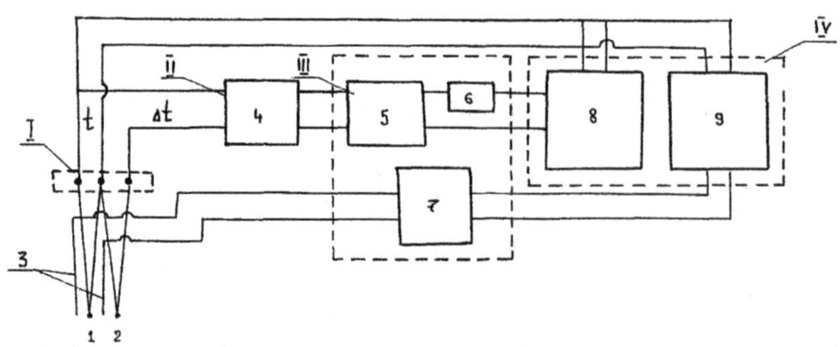

Visual and Polythermal Analysis

VPA is carried out in shaft furnaces. VPA is mainly used to clarify the primary crystallization temperatures of poorly crystallizing melts, as well as to confirm the DTA data on non-invariant compositions.

The Method of Quantitative DTA

This method estimates the value of the phase transition heat of the eutectic composition detected. To calculate the amount of released heat, eutectic mixtures of a certain mass, placed in porcelain crucibles, are immersed in a shaft furnace preheated 30 - 40 °C above the crystallization temperature of the eutectic mixture. After immersion in the test substance at a strictly defined thermocouple depth, the furnace is cooled evenly. As a reference, it is necessary to take salt mixtures with known values of the heats of phase transitions and with similar melting points, which are also immersed in a preheated furnace and cooled at the same rate. Thermocouples are immersed in substances at the same depth.

The described experimental methods investigated two-, three-, four- and five-component systems that are part of the system Li,Na,Ca,Ba//F,MoO4 (Babaev, 1996), with the aim of identifying new phase-transition materials for high-temperature accumulation.

During crystallization, organic crystalline substances increase in volume to 20%. When placed in a container, it is necessary to allow the phase-transition substance to expand. Abnormal fluids (water, gallium, bismuth) expand during crystallization, and the destruction of the container is possible in the absence of free volume. To prevent this, it is necessary to take special measures, for example, to place dampers that compensate for the change in volume of the composition during crystallization. One-component (simple substances) and eutectic compositions of MCSs can be used as PTHAMs.

In a one-component system, only three phases can coexist simultaneously. The triple point corresponds to the nonvariant (non-invariant) system. An infinitely slow transition from the solid to the liquid phase and back takes place in a single-component system at a constant temperature. This predetermines the possibility of using crystalline substances for heat storage at a given temperature. The liquid phase can be in a metastable (unstable) state in a solid region.

During phase transitions of certain substances, a supercooling effect occurs, the magnitude of which depends on the probability of the formation of a nucleus of the crystalline phase. The presence of large supercooling makes phase-transition materials unsuitable for heat storage. Reduction of supercooling is achieved by the introduction of soluble and insoluble impurities by trial and error. However, the use of semi-empirical relations arising from thermodynamic ideas about the influence of various factors on the reduction of supercooling allows reducing the number of experiments.

Nonvariant transitions from the liquid to the crystalline phase are possible not only for simple substances, but mixtures and complex substances (chemical compounds). In this regard, it is advisable to consider the melting patterns of condensed MCS with nonvariant transformations. These substances are of interest as heat-accumulating materials, since they do not create excessive pressures and can be used in sealed containers that are not designed for elevated pressure.

Nonvariant transformations are eutectic, monotectic, as well as congruent and incongruent melting of chemical compounds. Mixtures of eutectic composition also melt isothermally, as well as simple matter at a constant temperature, since the eutectic corresponds to the simultaneous crystallization of two solid phases in a two-component system.

In the case of a eutectic mixture, the enthalpy of melting is close to additive. In the case of the end of the monotectic transformation, the mixture contains, in addition to the solid component, a liquid solution, i.e. monotectic has a melting enthalpy less than additive values. Congruently melting compounds melt and crystallize as pure substance with nonvariant transformation. The incongruently melting compound melts with decomposition. During crystallization, the formation of a metastable composition of the mixtures is possible, which causes irreproducibility of the processes of melting and crystallization, and this is unacceptable when using the substance as a HAM. A disadvantage of an incongruent compound is also a lower melting enthalpy, since at the beginning of crystallization the mixture already contains the solid phase of one of the components.

In compounds with a peritectic composition (the intersection point of the liquidus of one of the components and the compound), the incongruently melting compound melts as congruent, i.e. only the compound melts and crystallizes. The difference is that in the equilibrium state there is no compound in the liquid, but only the individual components dissolved in each other.

COMPUTER PROGRAM FOR THE DESCRIPTION OF THERMOCHEMICAL REACTIONS IN MULTICOMPONENT MUTUAL SYSTEMS "THERMAL EFFECT DEPENDING ON TEMPERATURE"

The work (Babaev, 2016b) describes the method, algorithm, block diagram and program for describing chemical and thermochemical reactions in multicomponent reciprocal systems, which allows to quickly identify possible chemical and thermochemical reactions that occur in reciprocal MCSs with any number of compounds and depending on temperature. The developed program is universal, as it determines the volumetric changes of systems during chemical transformations. And by entering the values of the salts included in the MCSs, it is possible to calculate the "costs" of the reactions, i.e. determine the best way to synthesize compounds. The program allows identifying the interaction of communities with abiotic habitat, including those created and modified as a result of construction and economic activity, and to establish patterns of substance and energy transformations in the processes of biotic circulation, as well as the influence of environmental conditions on humans (Babaev & Volshanik, 2012). The program allows obtaining the equations of chemical reactions occurring in the limiting systems included in the MCSs, or throughout the MCS (figure 11).

As an example, the program identified thermochemical reactions occurring in real systems Li,Na,Ca,Ba//F,MoO_4; Li,Na,K,Mg//F,Cl,Br,SO_4. The most energy-intensive thermochemical reactions occurring in mutual systems at T = 298,15 K included in Li,Na,K,Mg//F,Cl,Br,SO_4 and Na,K,Ca,Ba//F,Cl in ranked form are given given in table 8.

The values of thermal effects of some thermochemical reactions in mutual systems entering into Li,Na,K,Mg//F,Cl,Br,SO_4, change their sign with increasing temperature from T = 298K to T = 500K (table 9). They can be used for thermochemical energy storage.

Thus, the developed program for describing thermochemical interactions (Babaev & Halilullaev, 2005) greatly facilitates the search for energy-intensive chemical reactions for thermochemical accumulation. It allows identifying thermochemical reactions occurring in mutual MCSs, regardless of component and depending on temperature. The program determines the temperature at which the reaction has the maximum thermal effect, and also allows determining volumetric changes during the melting of the MCS.

Figure 11. Screen of the program "Thermal effect depending on temperature" after revealing the thermochemical equations of the reactions occurring in the system A,B,C,D//X,Y at T = 500 K.

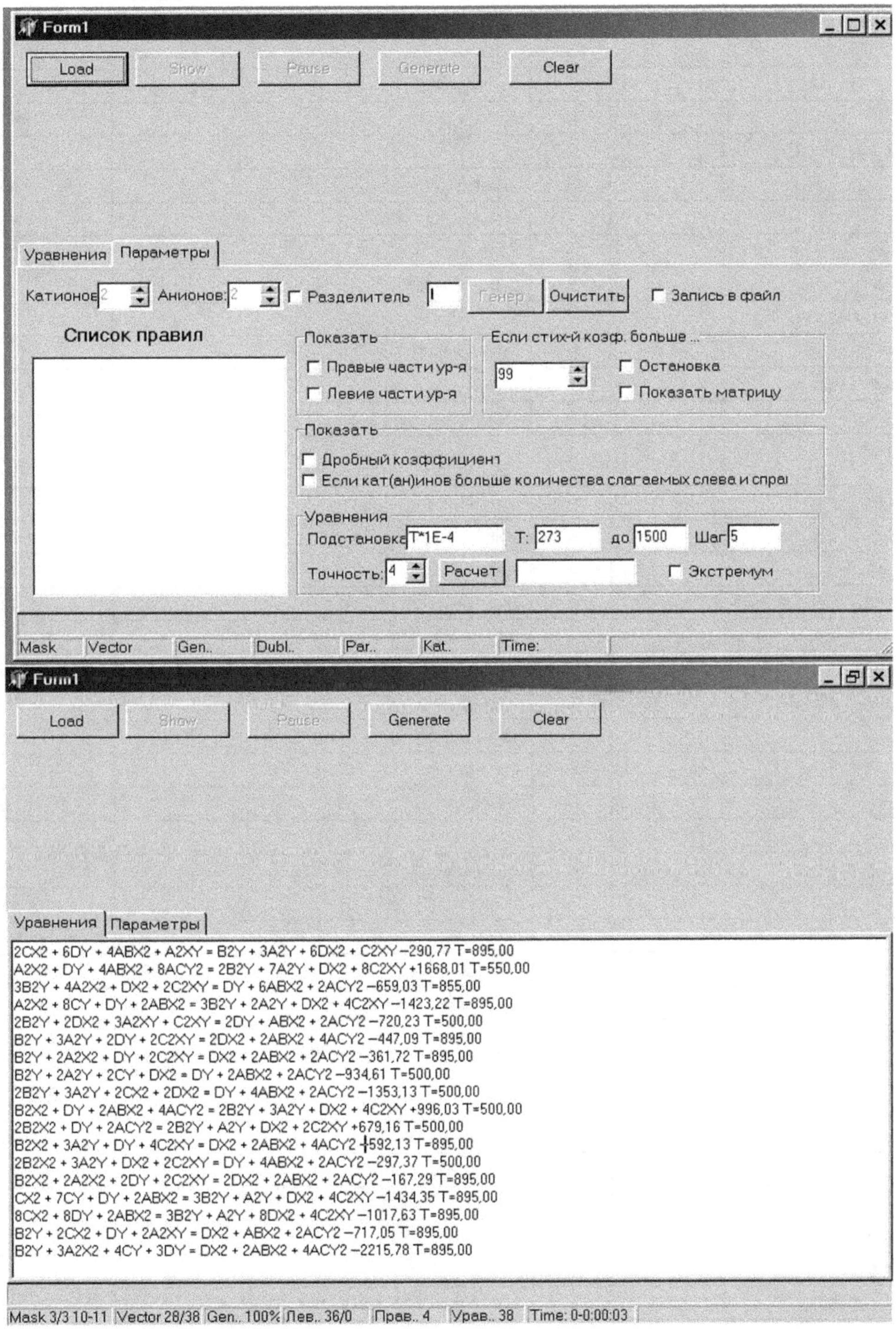

Table 8. The most energy-intensive thermochemical reactions occurring in reciprocal systems Li,Na,K,Mg// F,Cl,Br,SO_4 and Na,K,Ca,Ba//F,Cl

System Li,Na,K,Mg//F,Cl,Br,SO_4 K//A 2//2, mask 0-1			$-\Delta H_{reac}$, kJ/mol
1. $Li_2SO_4 + 3NaF$	=	$Na_3FSO_4 + 2LiF$	886,45
2. $Na_2SO_4 + 3KF$	=	$K_3FSO_4 + 2NaF$	813,46
K//A 3//2, masks 0-1, 00-11, 10-11			
3. $NaCl + KMgF_3$	=	$NaMgF_3 + KCl + MgF_2$	893,45
4. $NaCl + KMgF_3$	=	$NaMgF_3 + KCl + NaF$	342,08
5. $2LiCl + KF + NaF$	=	$NaCl + KCl + 2LiF$	121,66
6. $LiCl + 2KF + NaCl$	=	$2KCl + NaF + LiF$	108,65
K//A 4//2, masks 0-1, 00-11, 10-11			
7. $LiCl + 3KF + 2NaCl+MgF_2$	=	$NaMgF_3 + 3KCl + NaF + LiF$	157,1
8. $LiCl + 2KF + NaCl + MgF_2$	=	$NaMgF_3 + 2KCl + LiF$	125,22
9. $2LiCl + KF + NaMgF_3$	=	$NaCl + KCl + MgF_2 + 2LiF$	105,09
System Na,K,Ca,Ba//F,Cl K//A 2//2, mask 0-1			
10. $2KCaF_3 + 3Ca_2F_2Cl_2$	=	$6CaF_2 + 2KCaCl_3$	602,6
11. $2KF + 3Ca_2F_2Cl_2$	=	$4CaF_2 + 2KCaCl_3$	573,4
12. $3CaCl_2 + 2KCaF_3$	=	$3CaF_2 + 2KCaCl_3$	495,2
13. $2KF + 4CaCl_2$	=	$2KCaCl_3 + Ca_2F_2Cl_2$	430,2
14. $BaF_2 + CaBaCl_4$	=	$CaF_2 + 2BaCl_2$	360,4
15. $2KF + 2K_2BaCl_4$	=	$6KCl + Ba_2F_2Cl_2$	313,56
16. $2BaF_2 + Ca_2F_2Cl_2$	=	$2CaF_2 + Ba_2F_2Cl_2$	304
17. $2KF + 2BaCl_2$	=	$2KCl + Ba_2F_2Cl_2$	276,92
18. $CaCl_2 + 2BaF_2$	=	$CaF_2 + Ba_2F_2Cl_2$	268,2
19. $2KF + 3BaCl_2$	=	$K_2BaCl_4 + Ba_2F_2Cl_2$	258,6
20. $CaF_2 + CaBaCl_4$	=	$BaCl_2 + Ca_2F_2Cl_2$	242,4
21. $2KCaF_3 + Ca_2F_2Cl_2$	=	$2KCl + 4CaF_2$	238,12
22. $2NaF + 2BaCl_2$	=	$2NaCl + Ba_2F_2Cl_2$	213,16
23. $2KF + Ca_2F_2Cl_2$	=	$2KCl + 2CaF_2$	208,92
24. $BaF_2 + K_2BaCl_4$	=	$2KCl + Ba_2F_2Cl_2$	204,32
25. $CaCl_2 + 2KCaF_3$	=	$2KCl + 3CaF_2$	202,32
26. $KCl + Ca_2F_2Cl_2$	=	$CaF_2 + KCaCl_3$	182,24
K//A 3//2, masks 0-1, 00-11, 10-11			
27. $2KF + CaCl_2 + 3BaF_2 + CaBaCl_4$	=	$2KCl + 2CaF_2 + 2Ba_2F_2Cl_2$	905,52
28. $2KF + KCl + 3CaCl_2 + CaBaCl_4$	=	$CaF_2 + BaCl_2 + 3KCaCl_3$	890,64
29. $2KF + 6CaCl_2 + BaF_2 + K_2BaCl_4$	=	$2CaF_2 + 2BaCl_2 + 4KCaCl_3$	859,4
30. $8KF + CaCl_2 + CaBaCl_4$	=	$6KCl + BaF_2 + 2KCaF_3$	686,16
31. $9KF + CaF_2 + CaCl_2 + CaBaCl_4$	=	$6KCl + BaF_2 + 3KCaF_3$	671,56
K//A 4//2, masks 0-1, 00-11, 10-11			
32. $8NaF + KCaCl_3 + 2CaBaCl_4$	=	$8NaCl + KCl + 3CaF_2 + Ba_2F_2Cl_2$	951,2
33. $NaCl + CaCl_2 + 7KCaF_3 + CaBaCl_4$	=	$NaF + 7KCl + 9CaF_2 + BaF_2$	849,44
34. $NaCl + BaCl_2 + 7KCaF_3 + CaBaCl_4$	=	$NaF + 7KCl + 8CaF_2 + 2BaF_2$	767,24
35. $NaCl + 7KF + CaCl_2 + CaBaCl_4$	=	$NaF + 7KCl + 2CaF_2 + BaF_2$	747,24

Table 9. The values of thermal effects of some thermochemical reactions in mutual systems entering into Li,Na,K,Mg//F,Cl,Br,SO_4, at temperatures T = 298K and T = 500K

Reactions	$-\Delta H_{reac}$, kJ/mol, at Temperature T	
	298 K	500 K
2//3		
Na2Cl2 + Na2Br2 + 2K2SO4 = 2Na2SO4 + K2Cl2 + K2Br2	+7,13 (+10,9)	-3,0 (-4,5)
3//3		
3Li2Cl2 + Na2Br2 + K2Br2 + K2SO4 = 2Li2Br2 + Li2SO4 + Na2Cl2 + 2K2Cl2	+2,54 (+2,9)	-5,31 (-6,1)
2Li2Cl2 + Na2Cl2 + 2K2Br2 + K2SO4 = Li2Br2 + Li2SO4 + Na2Br2 + 3K2Cl2	+3,39 (+3,62)	-4,16 (-4,44)

Note: in parentheses the values of thermal effects of the reactions are given per 1 kg of the reaction mass.

Based on literary analysis (Grigorev, 2003; Magomedov, 2000; Trunin, 1997) and the developed program (Babaev & Halilullaev, 2005), promising combinations of anions with cations were obtained when selecting materials for heat accumulation due to phase transitions and thermochemical reactions (table 10).

CONCLUSION

As a result of the research done, heat accumulators with PTHAMs are proposed. Thermodynamic characteristics of energy storage due to photochemical reactions of dissociation are described. Considered various PTHAMs in thermal pipes, as well as types of heat pipes. Also presented PTHAMs in heat pumps and their heat-water supply. An installation of energy accumulation with phase-transition

Table 10. Combinations of anions and cations of salts, promising for use as phase-transition and thermochemical HAM

Anions	Cations												
	Li^+	Na^+	K^+	Cu^+	Be^+	Mg^+	Ca^+	Sr^+	Ba^+	Zn^+	Al^+	Mn^+	Fe^+
H^-	+	-	-	-	-	-	-	-	-	-	-	-	-
OH^-	+	+	+	-	-	-	+	+	+	-	-	-	-
F^-	+	+	+	+	-	+	+	+	+	+	+	+	+
Cl^-	+	+	+	-	-	-	+	+	+	-	-	-	-
CO_3^{2-}	+	+	+	-	-	+	+	+	+	+	+	-	-
SO_4^{2-}	+	+	+	-	-	+	+	+	+	+	+	-	-
NO_2^-	+	+	+	-	-	-	-	-	-	-	-	-	-
NO_3^-	+	+	+	-	-	-	+	+	+	-	-	-	-
MoO_4^{2-}	+	+	+	-	-	+	+	+	+	-	-	-	-
WO_4^{2-}	+	+	+	-	-	+	+	+	+	-	-	-	-

Note: + – promising for use as a HAM; - – unpromising for use as a HAM.

materials design is proposed. Considered different thermochemical reactions of energy accumulation. It is described accumulation of solar energy in the processing of organic fuels and proposed solar power plant with thermochemical energy accumulation. Experimental methods for the development of phase-transition and thermochemical heat-accumulating materials are described. Developed computer program for the description of thermochemical reactions allows describing chemical and thermochemical reactions in multicomponent reciprocal systems, which can be carried out in the course of the reciprocal MCSs. Developed program for describing thermochemical interactions greatly facilitates the search for energy-intensive chemical reactions for thermochemical accumulation. It allows identifying thermochemical reactions occurring in mutual MCSs, regardless of component and depending on temperature. The program determines the temperature at which the reaction has the maximum thermal effect, and also allows determining volumetric changes during the melting of the MCS. Promising combinations of anions with cations were obtained when selecting materials for heat accumulation due to phase transitions and thermochemical reactions.

REFERENCES

Adadurov, E. A. (2005). Povyshenie effektivnosti ispol'zovaniya akkumulyatorov teploty s vozobnovlyaemymi istochnikami energii [Increasing the efficiency of using heat accumulators with renewable energy sources] [Dissertation]. (in Russian)

Babaev, B. D. (1996). Himicheskie prevrashcheniya i fazovye ravnovesiya sistemy [Chemical transformations and phase equilibria of the system Li,Na,Ca,Ba//F,MoO4] [Dissertation]. (in Russian)

Babaev, B.D. (2014). Principy teplovogo akkumulirovaniya i ispol'zuemye teploakkumuliruyushchie materialy [Principles of thermal accumulation and used heat storage materials]. *Teplofizika vysokih temperatur [Thermophysics of High Temperatures]*, 52(5), 760-776. (in Russian).

Babaev, B.D. (2015). *Patent No. 2547680. Russian Federation.* (in Russian)

Babaev, B. D. (2016a). Akkumulyatory tepla pri ispol'zovanii vozobnovlyaemyh istochnikov energii [Heat accumulators using renewable energy sources]. *Energetik [Power engineer]*, *3*, 19–22.

Babaev, B. D. (2016b). Razrabotka i issledovanie energosistem na osnove vozobnovlyaemyh istochnikov s fazoperekhodnym akkumulirovaniem tepla [Development and investigation of power systems based on renewable sources with phase-transition heat accumulation] [Dissertation]. (in Russian)

Babaev, B. D., & Danilin, V. N. (2002). *Energoakkumuliruyushchie ustanovki [Energy storage facilities].* Makhachkala. (in Russian)

Babaev, B.D., & Halilullaev, G.M. (2005). *Certificate of state registration of computer program No. 2005610201 "Description of thermochemical reactions in multicomponent reciprocal systems "Thermal effect depending on temperature."*

Babaev, B.D., Kharchenko, V., Panchenko, V., & Vasant, P. (2019). Materials and Methods of Thermal Energy Storage in Power Supply Systems. In *Renewable Energy and Power Supply Challenges for Rural Regions* (pp. 115-135). Academic Press. doi:10.4018/978-1-5225-9179-5.ch005

Babaev, B.D. & Volshanik, V.V. (2012). Prognozirovanie ekologicheskoj obstanovki razrabotannoj programmoj EVM vyyavleniya himicheskih vzaimodejstvij [Predicting the environmental situation developed by the computer program to detect chemical interactions]. *Vestnik RUDN, seriya: Inzhenernye issledovaniya [Bulletin of RUDN, a series: Engineering researches]*, 4, 135-140.

Dibirov, Y.A. (2012). Phase equilibria and thermophysical compositions in the system Ca//F,Cl,SO4,MoO4. Actual problems of development of renewable energy resources. *Proceedings of the V-th School of Young Scientists*. Academic Press.

Duffy, J., & Beckmann, U. A. (2013). Fundamentals of solar thermal power. Educational - reference manual. 888 p.

Engelsht, V.S. & Muratalieva, V.Z. (2013). Exothermic effect in the interaction of calcium and silicon oxides. *Thermal physics of high temperatures*, *51*(5), 717.

Grigorev, V. A. (2003). Razrabotka akkumulyatorov teploty s zernistym teplonositelem i metoda ih rascheta na osnove matematicheskogo modelirovaniya (Development of heat accumulators with a granular coolant and methods for calculating them on the basis of mathematical modeling) [Dissertation]. (in Russian)

Kharchenko, V., Panchenko, V., Tikhonov, P. V., & Vasant, P. (2018). Cogenerative PV Thermal Modules of Different Design for Autonomous Heat and Electricity Supply. In Handbook of Research on Renewable Energy and Electric Resources for Sustainable Rural Development (pp. 86-119). Academic Press. doi:10.4018/978-1-5225-3867-7.ch004

Magomedov, M. M. (2000). Kompleksnaya metodologiya razrabotki fazoperekhodnyh teploakkumuliruyushchih materialov na osnove mnogokomponentnyh system [Complex methodology of development of phase-transition heat-accumulating materials on the basis of multicomponent systems] [Dissertation]. (in Russian)

Panchenko, V., Kharchenko, V., & Vasant, P. (2019). *Modeling of Solar Photovoltaic Thermal Modules. In Intelligent Computing & Optimization* (pp. 108–116). Springer Nature Switzerland. doi:10.1007/978-3-030-00979-3_11

Popel, O. S., & Fortov, V. E. (2015). Renewable energy in the modern world. Publishing house of Moscow Power Engineering Institute.

Shishkin, N. D. (2005). Kombinirovanie i effektivnoe ispol'zovanie istochnikov teplovoj energii v avtonomnyh teploenergeticheskih kompleksah: vklyuchaya vozobnovlyaemye istochniki [Combination and effective use of thermal energy sources in autonomous heat and power complexes, including renewable sources] [Dissertation]. (in Russian)

Son, E.E. (2013). Modern studies of the thermophysical properties of substances. *Thermophysics of High Temperatures*, *51*(3), 392.

Sorensen, B. (2011). Transformation, transmission and accumulation of energy. Academic Press.

Stankus, S.V., Khairulin, R.A., Martynets, V.G., & Bezverkhiy, P.P. (2013). Studies of the thermophysical properties of substances and materials at the Novosibirsk Scientific Center of the SB RAS in 2002-2012. *Thermophysics of High Temperatures*, *51*(5), 769.

Trunin, A. S. (1997). *Complex research methodology for multicomponent systems.*

Twidell, D., & Weir, A. (1990). *Renewable energy sources.*

Zamaldinova, G.I., Garkushin, I.K., Garkushin, A.I., Parfenova, S.N. (2012). Analytical description of the temperatures and compositions of low-melting eutectics in the series LiF–MF (M=Na,K,Rb,Cs) and prediction of the eutectic characteristics in the system LiF–FrF. *Journal of Inorganic Chemistry, 57*(6), 961.

KEY TERMS AND DEFINITIONS

Heat Accumulator: Device for heat accumulation for the purpose of its further use.

Heat Storage Material: Material used for the accumulation of thermal energy and may be classified according to the material class, the way of accumulation and heat release as well as from the cyclic operation.

Heat Storage System: The system of devices intended for temporary storage of thermal energy by means of various heat-storage materials and including a number of auxiliary equipment including heat exchangers, pumps, pipelines, fittings, test equipment and automatic devices, as well as other devices serving and providing the work of the heat storage system.

Phase-Transfer Heat-Accumulating Material: The type of heat-accumulating material using latent heat of phase transitions of the heat-accumulating material.

Thermochemical Heat Accumulator: The type of heat accumulator based on the use of binding energy of reversible chemical reactions of chemical accumulation.

Underground Heat Accumulator: The thermal energy accumulator located under the ground surface with different types of thermal insulation and heat storage materials.

Chapter 2
Optimization of Processing Modes of Disinfection of Vegetable Storehouses With the Use of Ozone

Alexander Smirnov
https://orcid.org/0000-0002-9236-2281
Federal Scientific Agroengineering Center VIM, Russia

Victoria Ukhanova
Federal Scientific Agroengineering Center VIM, Russia

Irina Ershova
https://orcid.org/0000-0003-1126-3837
Federal Scientific Agroengineering Center VIM, Russia

Bibigul Koshoeva
Kyrgyz State Technical University, Kyrgyzstan

ABSTRACT

The ozonation of fruits and vegetables may increase their storage life nearly twice as long. The chapter presents the results of research completed on the ozone treatment of vegetable storehouses with the use of ozone generators, with barrier and corona types of discharge, and reveals factors affecting disinfection process. The results show that ozone concentration depends on the dynamic balance of the two basic processes – the ozone-air mixture supply and ozone decomposition. Formulas for calculating ozone generator capacity in the recirculation mode have been derived. The automated system for controlling the disinfection process in a vegetable storehouse is described. The experimental test has shown that the ozone effect is long-term and the efficiency of the sterilization of surface microflora contaminants is 100%.

DOI: 10.4018/978-1-7998-1216-6.ch002

Copyright © 2020, IGI Global. Copying or distributing in print or electronic forms without written permission of IGI Global is prohibited.

INTRODUCTION

At storage of vegetables and fruits, the attention is focused on maintenance of storehouse microclimate parameters and electrically driven storage processes development. Capacity of a heating system is found based on storehouse thermal balance, at calculation of which the following aspects are taken into account: heat income from equipment; heat losses through walling; heat loss or heat input through grounds; heat losses with removed ventilating air; heat dissipation of potatoes at a 50% filled-in storehouse. In a storehouse for supporting of microclimate parameters, a system of active ventilation – including one with use of artificial freeze – must use the natural cold as much as possible. A refrigeration system is intended for removal of heat input from a storage object and other sources including that coming from products (of physiological and accumulated nature); containers; fences and walling; work of electric engines; illumination; ventilation with outside air; doors opening; workers; and so on. Choosing of a refrigeration system is determined by terms of products placement and distribution, technological modes of refrigeration, assumed temperatures of outside air, products heat and moisture production in a storage premise, etc. Depending on all those factors, a storehouse owner can adopt either a system with artificial refrigeration or a combined system with use of the natural cold. At present time for artificial refrigeration in storehouses, both centralized and decentralized systems of cold supply are used. Regardless of a specific engineering solution, it is evident that an artificial maintenance of microclimate parameters requires energy expenses.

For example, in the Fig. 1, the diagram is given of electric energy consumption of the storehouse of the agrarian firm Slava Kartofelyu, Ltd. intended for storage of 5000 tons of potatoes.

Analysis showed that for maintenance of microclimate parameters, for example, in winter period of time, the electric energy consumption reaches 5000 kWt·h/month including illumination. In this connection with purpose of electric energy saving in spring, potatoes producers should not use their existing electric equipment, which results in dramatic worsening of potatoes quality. That is why for maintenance of the microclimate parameters in a storehouse of potatoes, there are needed alternative energy & resources saving installations or systems of disinfection allowing conserving yield with proper quality as long as possible.

Figure 1. Electricity consumption graph

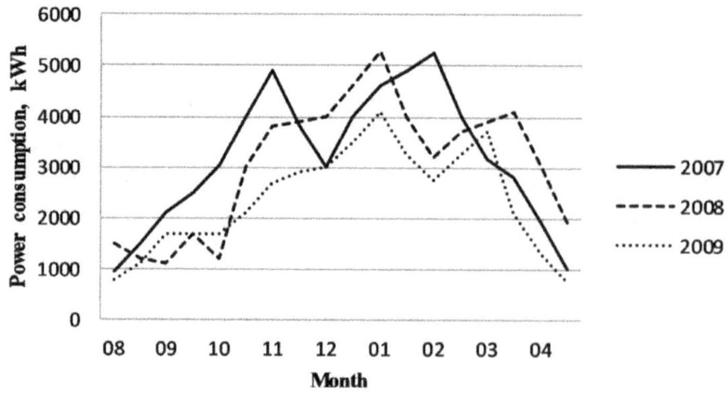

Vegetables and fruits have highest consumer appeal if they are sold fresh. However, on their surface various microorganisms may reside – Escherichia coli, saprophytes, cocci, moulds and other microflora bringing about deterioration and producing toxins. Moreover, preliminary treatment of premises for packaging and storage is also required. In storage of fruits and vegetables average losses are 30-40%. Without automation of microclimate support, losses can amount to 50%.

Today, the considerable part of sanitary-precautionary measures in storehouses for vegetables is fulfilled by way of disinfection chemicals use. Most known are substances of general purpose such as lime chloride, carbolic acid, sodium hydroxide, etc. For grain detoxification, propionic acid and phosphonic acid found application, while for disinfection pretreatment such preparations as kfugugo, dividend star, vitaros, etc., are used.

Also, other disinfection-aimed preparations exist of more limited intention, use of which is determined by specific properties of both the preparations themselves and objects of treatment. Besides, periodically, the chemicals require replacement because of possible habituation of biological objects to a specific preparation and immunity production at them.

Obtainment of disinfection-aimed chemicals represents an energy-intensive production based on electrolysis. However, with this, expenses are not finished yet. To final consumer, products are delivered after passing through a many-link system of intermediate storehouses and bases. Along with this, their transportation requires certain measures of safety and special vehicles, while storage in its turn requires territories, presence of walling structures, security officers and maintenance of a certain thermal mode. As a rule, disinfection chemicals are produced, transported and stored in a form of concentrated aqueous solutions; and from point of view of their property's conservancy, their freezing is unacceptable. As a result, for a final consumer, the cost of the disinfection chemicals turns out to be excessively high.

Recently, a growing interest has been seen in safe but effective methods of disinfection and disinsection including the use of gaseous or aqueous ozone (Rice, 1999). Ozonation of fruits and vegetables increases their storage life 1.5 – 2.5 times (Belov, Metkin, & Semenova, 2018). Even at insignificant ozone concentration (0.01 – 0.04 mg/m^3), the air in vegetable storehouse premises becomes fresh and pleasant, and the odors disappear.

The ozone use for disinfection ensures the treatment effectiveness growth and the environmental friendliness of the process. Also, it provides farms with tangible economical effect in comparison with other methods. The universal nature and the high effectiveness of ozone action on microorganisms lead to wider ozone application for sanitary purposes and health friendliness improvement of the aerial environment. In the Table 1, the data is given showing the ozone action on some species of microflora.

Ozone is the allotropic modification of oxygen. Its molecule is formed by three atoms of oxygen. The molecule itself in projection represents virtually correct equilateral triangle, i.e. in fact it is distinguished with the "golden ratio"; see the Figure 2. Its chemical formulation is O_3. An electric discharge in aerial environment is accompanied with the ozone formation: the specific smell appears. The "ozone" means "having smell".

The ozone possesses unique properties. The reaction of the ozone formation is reversible and according to the Le Chatelier principle, the greater pressure is, the more intensive the ozone formation must be:

$$3O_2 + Q = 2O_3$$

$$O_3 = O_2 + O + 68\ \text{kcal}$$

Table 1. Effects of ozone on microflora

Microflora	Threshold of Sensitivity, mg / l	Concentration of Ozone, mg / l		Time Min
		Air	Water	
Viruses	0,0002	0,002	-	60
Mushrooms	0,03	-	1,5	120
Mold	0,03	1,5÷2,0	-	20
Staphylococcus	-	1,0	-	60
Lactic acid	0,002	-	15÷20	10÷15
E. coli	-	2,0	-	10
Dysenteric	-	-	-	-
N°1402 Flexner	-	-	0,3	20
N°687 Newcastle	-	-	0,3	20

Figure 2. Ozone molecule structure

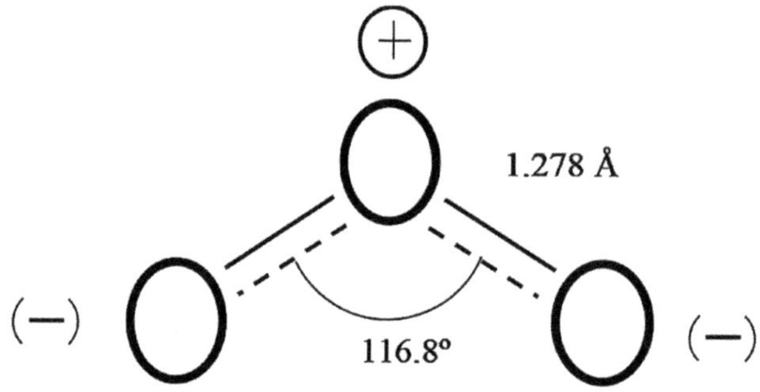

In the aerial environment, an electric discharge results in ions formation of oxygen and nitrogen oxides:

$$O_2 + e = O + O^- + e$$

$$N_2 + O_2 = 2\,NO$$

$$2\,NO + O_2 = 2\,NO_2$$

Depending on conditions of the reaction conducting, it is possible to intensify formation of these ions and considerably decrease the same of others.

The smell of the ozone is what we feel in air after a heavy rainfall with thunder and lightning. The ozone is present in coniferous forests, especially cone forests. At pine tar decomposition, some ozone is formed. Also, the ozone is formed in air in sunny weather on snowy fields. At people, these rather small quantities of ozone cause euphoria. In lower layers of air, the ozone is dispersed and its content is not high. However, in some places, we feel especially well. For example, in the air of the spa town Sochi (Russia), the ozone concentration is changed in the range from 11 to 72µg/m^3. The ozone top content in the air of a coniferous forest reaches 84÷100 µg/m^3. Also, there is an increased content of ozone in locations being close to waterfalls (Gubernsky, 1991).

The ozone can exist in three aggregative states: gas, liquid and hard substance. In ordinary conditions, the ozone stays in the gaseous state and has a strong smell. The ozone basic properties are given in the Table 2 (Filippov, 2008).

The high density of gaseous ozone together with its good diffusion ensures its fast distribution on a reaction volume. Another important factor is its high – as compared to oxygen – solubility in water. Also, important feature is the ozone ability to be decomposed spontaneously both in air and in water. This fact gets a special significance at ozone application as a disinfectant. The ozone decomposition takes place – and even in a quite intensive manner – at its contact with some substances. Active initiators of ozone decomposition are metal oxides (Filippov, 2008).

The ozone ecological friendliness and its easy obtainment in necessary amounts with aid of modern ozone generators allow hoping for its increasingly wide use in enterprises of agrarian production and food industry. Apart from the traditional spheres of application such as decontamination of drinking water and effluents, the gaseous ozone use has been started for the following purposes:

- Sanitary treatment of industrial and domestic premises
- Treatment of ready products
- Elimination of having bad smell and bacteria-bearing blowouts from chambers and production facilities
- Disinfection-aimed treatment of equipment, inventory, containers, etc.
- Creation of ozonized atmosphere in chambers and cargo boxes of vehicles for prevention of bacterial contamination at transportation and for foodstuffs realization times elongation.

Table 2. Basic physical properties of ozone

Indicator		Value	Measurement Unit
1. Molecular weight		48	°C
2. Boiling temperature at 760 mm of mercury		-111,9	°C
3. Melting temperature at 760 mm of mercury		-192,5	°C
4. Critical temperature		-12,1	atm.
5. Critical pressure		54,6	g/cm^3
6. Critical density		0,437	g/l
7. Density at 298 K		2,144	cal/mole grade
8. Thermal capacity at 298 oK		9,406	cm^2/s
9. Diffusion ratio at 300 oK and 760 mm of mercury		0,157	kcal/mole
10. Heat of dissolution in water at:			
	0 °C	5,9	kcal/ mole
	18 °C	3,4	
11. Ionization potential		12,8 – 13,5	eV
12. Entropy		57,1	cal/mole grade

However, the ozone technologies application is restrained by an absence of special technical procedures of ozone obtainment suitable for use in agrarian production. The use of the existing ozone generators requires added complexity of technological schemes, high qualification of operational personnel and increased requirements to machinery safety at operation (GOST).

Because of its activeness, the ozone can to become a dangerous factor for health of humans and animals in a case that its acceptable content limit would be exceeded in the air. However, in the nature, this scenario has never takes place. At present time at effectiveness evaluation of this or that disinfectant, there is used so called C·T criterion, i.e. product of reagent concentration by time of exposure. It would be correct to say that Action (inactivation) = Concentration · Action Time. In the Table 3, for a comparison, values of C·T criterion are represented for various microorganisms and disinfectant agents.

Separately, it is necessary to consider an interaction between ozone and biological objects at the cellular level. In opinion of some researchers, their interaction mechanism depends on ozone insufficiency or redundancy. Biological cells are protected against external action by their membranes and in a case of exceeding of a certain threshold of action, membrane integrity fails, which results in cell death. Upon that the running destruction is expressed in fast accumulation of lipids peroxidation and in intensive formation of protein-protein and protein-lipid covalent linking. As a result, the induced by ozone structural modification of membranes results in discontinuity of its barrier function and in particular in its penetrability growth for exo- and endogenous organic compounds (Chelnokov et al., 1984; Konev, et al., 1982).

Studying of ozone interaction with erythrocytes was conducted both in our country and abroad; but the molecular mechanisms of its cytotoxicity stay unclear in many ways. The spectrum of induced by ozone effects is very large, which founds its explanation in the multiplicity of its ways of interaction with a cell. An opinion exists that the ozone action mechanism consists in destruction of microorganisms by way of inactivation of bacterial proteins, i.e. by way of ozone diffusion through cell membrane into cytoplasm and damage of vital centers (Freeman and Mudd,1981).

The experimental data obtained at erythrocytes exposed to ozone action showed that an oxidation starts at the lowest doses of ozone. At its lower doses, the least destruction is observed of hemoglobin contained inside of erythrocytes in relation to total protein amount. With a dose growth, the hemoglobin relative destruction is not increased (Konev et al., 1982). Ozone oxidizes actively free amino acids and ones being part of protein composition, sensitivity of which declines in the following order: methionine – tryptophane – tyrosine – histidine – cystine – phenylamine (Chelnokov et al., 1984). Also sensitive to the ozone are cellular lipids, in composition of which there are unsaturated fatty acids. At the same time, the ozone does not affect penetrability of membranes consisting of lipids represented by the desaturated fatty acids (Goldstein and Balchum (1967).

*Table 3. Value of C·T criterion for various microorganisms (99% inactivation atu 5-25 °C. C*T criterion (mg/l·min)) (Butko & Frolov, 2001)*

Kind of Microorganisms	Ozone pH:6÷7	Free Chlorine pH:6÷7	Chloramin pH:8÷9	Chlorine Dioxide pH:6÷7
E-coli	0.02	0.03-0.05	95-180	0.4-0.75
Polio viruses	0.1-0.2	1.1-2.5	770-3470	0.2-6.7
Rotaviruses	0.006-0.06	0.01-0.05	3810-6480	0.2-2.1
Gardialambdacysts	0.5-0.6	47-150	-	-
Gardiamuricysts	1.8-2.0	30-630	1400	7.2-18.5

Also, the ozone reacts with nucleic acids. In 1963, Dubinina put forward the assumption that the formed at ozone decomposition atomic oxygen causes formation of hydroxyl radicals, which are exactly those compounds that react with nucleic acid. Later on, this assumption was confirmed. In secondary reactions, the hydroxyl radicals lead to formation of hydro-peroxide radicals, which react even to a greater extent with the nucleic acids. Especially active is run of the reaction with pyrimidine bases and, first of all, with thiamine, then with cytosine and uracyl. The destructed nucleic acids are digested intensively by the survived cells, which can explain the high growth of cells of microorganisms after ozonation at relevant treatment modes (Trukhacheva et al., 1992). The ozone reactions with the nucleic acids confirm its mutagenetic. The scientific observations made in the zone of Chernobyl Nuclear Plant provides with a lot of interesting material on ozone role in vital cycle arrangement. It is well-known that the radiation action results in ozone formation in the air. The ozone action together with the radioactive precipitation occurred in the zone resulted in death of microorganisms and hence to cessation of decay decomposition processes. As a result of this event, their vital cycle stopped forever – the conserved remainders give no material for a new life. Where the action turned out to be less intensive, mutations had place.

In a number of experimental works conducted by the authors, at analysis of the processed material, there was observed a faster growth of microorganisms. Here from certain requirements follow: the ozone for objects treatment is necessary to be used in a close controlled volume and along with this, ozone action modes must guarantee the inactivation of microorganisms.

Based on studying of the processes of objects oxidation by the ozone, it is possible to say that the mechanism of microorganisms' action on cells looks like as follows:

- Primary processes of oxidative destruction at cell exposure to ozone are localized in cells plasmatic membranes.
- As a result of ozonolysis of plasmatic membrane components, relatively stable products form; those are secondary oxidants able to enter into further reactions after the ozone action stoppage.
- Apart of damage of functionally active proteins, the ozone toxic action can be connected with deterioration of conditions of their functioning being a result of membrane structure modifications.

The ozone is distinguished with high activeness and universal nature of its action. The ozone can be used in all 3D-technologies such as disinfection, desinsection and deratization. The unlimited availability of the raw material (which is the atmospheric air), the comparatively low cost, the simplicity of obtainment, the ecological friendliness – all these factors promote expansion of ozone application (Borodin et al., 1989). However, it is necessary to keep in mind the ozone action duality on biological structures; see Figure 3.

Figure 3. Ozone action on biological objects

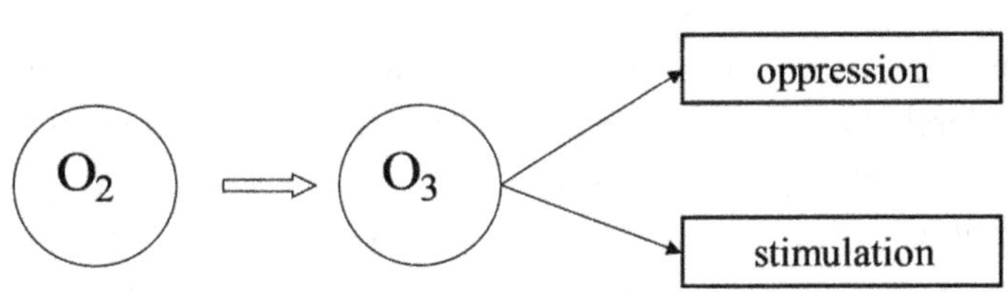

The ozone bactericidal action is based on ozonolysis, i.e. rapture of organic compounds by atomic oxygen forming at ozone decomposition. In fact, a microorganism's membrane, its external cover turns to be broken out. The mechanism itself of the ozone action on a cell is quite sophisticated and runs in several steps.

According to some data, to small specific doses of ozone, the initial fragment of the authors' kinetic curve corresponds. That is interval of slow death, after which cells fast inactivation is observed. The obtained results were interpreted by the authors as the result of cells competition for ozone molecules (Grits, Fomichev, 1985).

A certain difference is observed between various kinds of microorganisms on their resistance ability to the ozone action. In quite short time, incitants of angina and diphtheria as well as various fungi perish. As a rule, most stable are microbes covered with a membrane such as tubercle bacilli and microbe spores. An effectiveness of ozone sterilizing action depends on its concentration, exposure time, temperature, humidity, kind of a microorganism, pH factor and initial microorganisms' content in the would-be decontaminated air (Peretyagin, 1995). The ozone action effectiveness in relation to microorganisms is fair to represent in a form of a row: – viruses – bacteria – fungi – mushrooms – spores. In the Table 3, the comparative data were given on ozone action effectiveness on various kinds of microorganisms obtained by various authors. However, not always it is clear, at what conditions of ozone action, the data was obtained.

Storage of vegetables and fruits is an important aspect of foodstuffs supply to the country population. Annually in Russia, approximately 10-15 mln tons of vegetables and 2-5 mln tons of fruits are produced. Along with this, population's providing with own products reaches maximum 80% speaking of vegetables and 30% of fruits. In part, this is due to products losses on every step, from growing up to a trading bar.

The issue of effectiveness increase of vegetables and fruits storage has a complex nature. It is necessary to take into account a lot of factors, which can result in products losses as well as extensive consumption of electric energy. Also, one should take into consideration the modern methods of biological and chemical treatment of fruits, cost and characteristics of the modern equipment intended for microclimate maintenance, tools of parameters control and of process automation and use of modern kinds of containers.

The main amount of consumed electric energy falls on treatment of fruits, microclimate maintenance and illumination. In the conditions of the constant growth of tariffs, the electric energy consumption amount makes a considerable impact on products cost. In the case of application of the ecologically clean and energy-effective technologies based on the ozone, it is possible to cut considerably both losses of vegetable products and expenses on its storage.

BACKGROUND

Modern systems for storing fruits and vegetables differ from their predecessors in full automation (Polshina, 2018)

The use of biopreparations is the most efficient, environmentally friendly and cost-effective way of storing plant materials (Pershakova, Lisovoi, Kupin, Panasenko, Viktorova, 2016).

In this regard, the analysis of international scientific experience in ensuring the safety of vegetables using biologics is relevant. Consumers of biotechnology products are predominantly highly developed countries: the USA, Canada, Japan, the European Union, China, India, Brazil.

However, the United States is a leader in the field of biotechnology, accounting for about 40% of the global market. Thus, in the United States, scientists have identified strains of the Metschnikowia species, in particular Metschnikowia pulcherrima, which have antagonistic activity against Penicillium expansum, to fight blue mold in apples. These strains are particularly active during refrigerated storage (Pat. US 6991930, 2006).

To prevent vegetables from spoiling, Chinese scientists have patented a surface treatment method using phenethyl alcohol extracted from a culture of bacteria Loeckera apiculata (Pat. CN 102204588, 2011).

Modern methods of processing materials during storage, such as processing chemicals, artificial cooling, electromagnetic radiation and others, prevent microbial spoilage, but at the same time, destroy the biological barrier that prevents the penetration of pathogenic microorganisms, reducing the natural immunity of plant materials (Pershakova, Lisovoi, Kupin, Panasenko, Viktorova, 2016).

The most progressive way of long-term storage of vegetables is storage in a regulated gas environment (CA). In Italy, Germany, Belgium, Holland, England, Great Britain (Bishop D, 1990, 1994), crops of vegetables and fruits are stored in a controlled atmosphere.Controlled atmosphere storage has continued to be used on an increasing scale (Thompson, 2010, 2016; Khanbari, and Thompson, 1996).

There is considerable evidence in the literature that storing vegetables in CA storage can affect their subsequent shelf or marketable life (Thompson, 2010, Wills, Pitakserikul, and Scott. 1982, Burdon et al., 2008, Thompson, 1996; Gunes, Liu and Watkins, 2002).

Studies on ozonation of equipment and fruit surfaces with a high concentration of gaseous ozone show inactivation of spores of green mold, blue mold and acid rot (Penicillium digitatum, Penicillium italicum and Geotrichum citri-auranatii, respectively) were conducted for 1 hour at 200 μmol mol-1 (5 ° C, 95% relative humidity) (Smilanick, 2003). Oranges, grapes and tomatoes infected with green mold (Penicillium digitatum) and gray mold (Botrytis cinerea) were stored in "clean air" or in an atmosphere enriched with ozone (concentrations up to 200 mg / m^3). The development of injuries and the growth of spores decreased markedly in fruits stored in an atmosphere enriched with ozone (Tzortzakis, 2015). However, high levels of ozone can damage treated products and have significant toxicological effects on human health and safety.

The ozone treatment can be an effective method of fight with potato tuber moth and other infestants; also, it can be used as one of alternative of methyl bromide (Ibrahim, Al-Ahmadi, 2014). The traditional methods of fight have a lot of shortages and among them low penetration of pesticides into a damaged place, environment contamination and dangers for human health.

Also, the ozone can be applied for conducting of deratization, extermination of such gnawing animals as rats and mice. The fight with rats and mice is an actual issue for the countries all around the world. The rodents cause an essential harm by eating part of collected yield. But the loss is aggravated from other sources, too, and among them there are constructions destruction, damage of heat isolation and electric wires. Besides, the rats destroy a great amount of offspring in rabbits and poultry farms and induce a stress state at adult animals, which results in products amount and quality lowering. Per day, a rat eats food in amount of up to 10% of its mass, which makes up to 10-18 kg annually. As for mice, this value reaches 3.5 kg of foodstuff. The total losses in the world make up to 33 mln tons of wheat and rice (Butko, Frolov, 2001). Upon that, with their excrements, those rodents contaminate 10 times more foodstuff than they eat. Also, they are dangerous in that relation that they can carry up to 200 various human diseases and more than 40 diseases of agrarian animals. Among them there such dangerous ones as pox, aphthous fever, erysipelas, dysentery, dermatological toxicosis, helminths, ectoparasites and so on.

At present time for fight with the rodents, various methods are used with diverse principle of action: mechanical, physical, chemical and biological. But quite soon after the deratization conducting, the rodents' number is restored quite quickly and that is why the farmers are forced to repeat the treatment again and again, which leads to increase of material expenses. That is why the ozone application for this purpose attracts interest (Butko et al. 1995; Pershin,1993).

The essence of the technology consists in the following procedure application. In the zone of rodents dwelling, an increased concentration of ozone is created, for example, not less than 0.015 mg/m^3. This causes a fear feeling at animals and they leave this zone. At greater concentrations of the ozone, some morphological changes in their organisms take place, namely their muscles get to be flabby and they lose their appetite. At even greater concentration of ozone (it is especially true for mice), pulmonary edema is developed at them and they die. (Butko & Frolov, 2001)

Also, the ozone affects cockroaches. The cockroaches communicate with each other with aid of pheromones. The pheromones are biologically active substances excreted by animals into environment and making a specific influence on behavior, physiological and emotional state or metabolism of other animals of the same type. As a rule, pheromones are produced by specialized glands. The pheromones biological action is performed via chemo-receptors. The pheromones do not belong to a certain one class of chemical compounds. Their structure is various. The pheromones can be represented by single chemical compounds but more often their biological action is cumulative and is rendered by several components. As a rule, for the pheromones biological action, species specificity is typical. As pheromones, different kinds of animals or insects use different chemical substances or mixtures with different components combination. Mainly, the pheromones perform the chemical communication between animals of the same species.

For a number of insects (many kinds of bed bugs, flies, cockroaches and some beetles), an accumulation is typical of a great number of individual living beings on a limited area. This phenomenon ensured by so called aggregation pheromones is best of all studied at such forest infestants as bark beetles. There are pheromones of "alarm"; they cause the reaction of escapement, hiding behavior or, on contrary, – as it is observed at social insects – an aggressive reaction and collective enemy attacking. Usually, these pheromones are relatively simple and more volatile in comparison with sexual pheromones. (At domesticated bee as one of components of alarm pheromone, isoamyl acetate serves). The ozone reacts actively with the pheromones represented mainly by organic substances and thus the contained in them information disappears. A premise processed by the ozone gets unsuitable for cockroaches and they leave it.

The case of interest is the "ventafresh", the technology of agrarian products treatment before their placement in storage. The technology consists in washing of, for example, potatoes in ozonized water with further use of UV-radiation and then storage in slightly ozonized aerial environment. Before realization, the product is sorted based on bulbs size, washed with the ozonized water, processed with ultraviolet rays, dried and packed. By the author, also economical indicators of such treatment are given for the storehouse by volume of 6000 tons. At natural storage of potatoes, the expenses made 900 thousand Swiss francs and upon that, the losses were quite noticeable: 10 - 12%. At refrigerating storage, the expenses reached 1 mln Swiss francs, while the losses 6%. At use of the "ventafresh" technology, the expenses made 740 thousand Swiss francs and the losses less than 4%. The potatoes cost made 0.5 Swiss francs per a kilogram (Steffen, 2007).

Also, in Russia, Ukraine, Kazakhstan and Republic Belarus, studies are performed and already practical results have been obtained on the ozone application for pre-sowing treatment of grain, which promotes yield enhancement of cereal crops, ecological friendliness of both the treatment process and final products (Vasilchuk & Epstein, 2007; Shesterin, 2004; Avdeeva et al., 2008).

The analysis of the state-of-the-art of technologies for vegetable produce storage has shown that the use of ozone generators still is sporadic and experimental. In part, this is due to insufficient knowledge of the process of ozone treatment of object in closed space. The purpose of this paper is to study conditions affecting ozone disinfection process and to develop methods of calculation for an ozonation plant.

MOTIVATION

Disinfection is usually carried out in disinfection chambers (DC) or closed premises. Ozone is supplied to volume to be processed by ozone-air mixture (OAM) injection and required concentration is reached through its dilution in the volume (Butko, Volkhov, Pershin, & Frolov, 2010). In particular cases, especially when low concentration ozone is used, it is supplied by diffusion (Trotskaya, 2005; Rezgo, 2011).

The most widespread scheme of ozone use for treatment of some objects in a closed chamber is disinfection with the use of ozone generators with barrier type of discharge. As a power source for barrier-type ozonizers are often used amplitude-modulated AC power supply (Zhang et al., 2018). A promising technology for ozone generation is a pulsed dielectric barrier discharge (Wei et al., 2018). Treatment is carried out according to the following scheme: air is supplied by a compressor for drying and then fed to a processing chamber. Concentrated ozone-air mixture is supplied to a disinfection chamber and then ozone concentration is decreased due to natural diffusion and gravity sedimentation.

The disadvantage of this method is the danger of ozone leakage due to excessive pressure and the need to pre-clean the air from dust and moisture.

DISINFECTION WITH OZONE GENERATOR OF CORONA DISCHARGE

Let us consider the most often used scheme of ozone application for treatment of some objects in a close chamber with aid of a barrier ozone generator. Here only one variant is possible, namely as follows. By an air compressor, objects are conveyed for drying and then they enter into the treatment chamber. In the area of agrarian production, this chamber is called disinfection chamber (DC).

The ozone-air mixture (OAM) arrives into DC in its concentrated form and then due to a natural diffusion and gravitational precipitation, the ozone concentration falls.

For the treatment intensification and more uniform distribution of ozone in the DC, it is recommended to install an additional ceiling fan (Krivopishin, 1988). After the required exposure, the ozone supply is stopped, the DC is ventilated and the treatment objects are removed. It should be said that at the present time, the emission of the remained ozone with the concentration 3,3μg/m^3 into the environment is forbidden. It is evident that the DC work in this mode is not optimal. One has to maintain the ozone influx until the exposure time expiration and to pump the ozone-air mixture (OAM) into the disinfection chamber, which results in pressure increase in it and hence in possibility of ozone leakage into adjacent premises.

When using a corona discharge ozone generator, the picture of treatment is changed. Corona discharge ozone generators have extremely low hydraulic resistance due to large discharge gap, so for OAM supply in recirculation mode, axial ventilators can be used, as they have higher capacity and lesser dimensions, weight and energy consumption in comparison with other flow boosters – pumps and compressors.

In this case processing scheme may be presented as in Fig. 4.

Let's consider working modes of ozone generator from point of view of ways of ozone supplying to a processed object. Bringing ozone to an object of treatment can be accomplished with use of following three methods: 1) pumping; 2) due to diffusion of the ozone itself from an ozone generator; 3) re-circulation. As it is seen from the following experimentally obtained functions represented in the Fig.5, the re-circulation mode is quickest one for reaching a required value of ozone concentration.

Let us compare various scenarios of OAM supply for processing objects in the chamber in order to reveal the most efficient ones (Baidukin, Pershin & Fedorov, 1988).

The following assumptions are made:

- Microorganisms reside on the surface of an object to be processed in a thin layer;
- Under the effect of convection flow microorganisms can break away from the partitioning and processed surfaces and be suspended alongside with dust papers;
- Under the effect of ozone microorganisms pass into inactivated form, and oxidation of obtained organic matter may be prolonged up to its transformation into inorganic compounds.

Figure 4. Operation scheme of a disinfection chamber when using a corona discharge ozone generator: 1 – ozone generator; 2 – ventilator; 3 – disinfection chamber

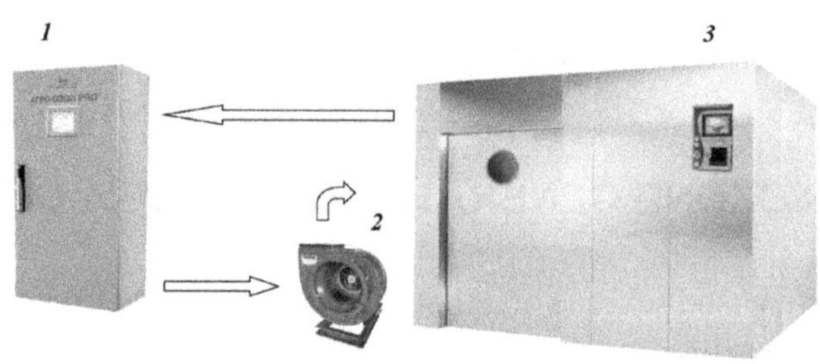

Figure 5. Dependence of ozone concentration change on time at use of ozone generators of the same productivity: 1 –re-circulation; 2 – pumping; 3 – diffusion

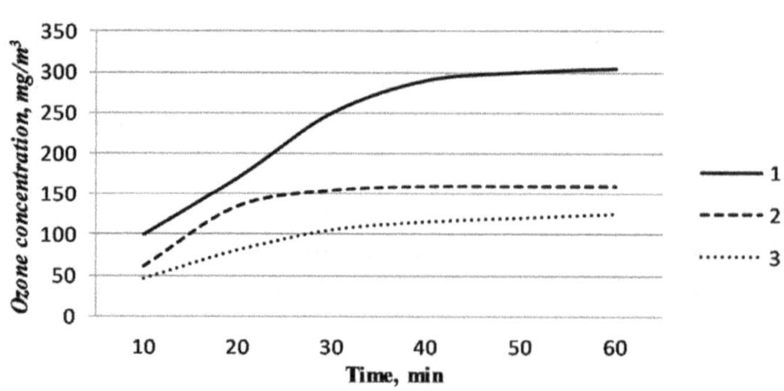

General condition of disinfection in closed volume:

$$\frac{q_{in} + q_p + q_y}{q_H} < 1 \tag{1}$$

wherein q_{in} is initial content of microorganisms in the volume; q_p – growth of microorganism's content in the treatment process (reproduction and carrying from outside); q_y – reduction of microorganism content due to ozone oxidation.

For ozonation under normal conditions:

$$q_y = q_n \tag{2}$$

wherein: q_n is reduction of microorganism content on surfaces limiting the volume, and in the air.

In the case of using an ozone generator operating in circulation mode -

$$q_y = q_o + q_{in} \tag{3}$$

wherein: q_o is reduction of microorganism content due to flow passing directly through the ozone generator.

We consider that air after passing through the ozone generator is practically sterile and does not content microorganisms. This is quite possible, as while passing the discharge zone, microorganisms are affected not only by ozone, but by high-gradient electric field and intensive UV-radiation, accompanying discharge. These effects are used to control microorganisms (Vasiljak, Kostjuchenko, & Kol'cov, 2001). Moreover, the ozone generator discharge chamber also serves as an electrofilter wherein dust papers and microorganisms are settled (Vozmilov, Astaf'ev, & Matveev, 2007).

In this case

$$q_o = A \cdot q_н \tag{4}$$

wherein: $q_н$ is total microorganisms' content in the volume; A – coefficient considering share of microorganisms being in suspension that have passed through the ozone generator; B – coefficient considering share of microorganisms residing on the surfaces limiting the volume A+B =*1*.

$$q_y = A\, q_н + B\, q_n \tag{5}$$

In case of ozone expansion in the volume due to diffusion, the mass transfer equation is as follows:

$$M = B_y\, F\, (y - y_{ph}) \tag{6}$$

wherein: B_y is mass transfer coefficient; F – phase boundary; y – ozone concentration in the volume; y_{ph} – ozone concentration at phase interface boundary.

Considering convection component, mass transfer equation is as follows:

$$M = B_x \cdot F\left(y - y_{ph}\right)\frac{y}{y_{av}} \tag{7}$$

wherein: y_{ph} is average logarithmic ozone concentration in boundary layer.

If $y >> y_{ph}$, expression (7) becomes:

$$M = 2\, B_x\, F(y\text{-}y_{ph}) \tag{8}$$

Therefore, presence of convective flows increases mass transfer coefficient nearly twice which considerably affects processing modes (Borisov, 2010).

CONDITIONS AFFECTING OZONE DISINFECTION PROCESS

While processing objects in the volume, ozone concentration depends on dynamic balance of the two basic processes – supply of ozone-air mixture and ozone decomposition. Ozone decomposition proceeds directly in the volume, on the surfaces limiting this volume, on objects to be processed – in this case microorganism inactivation and organic matter transfer into non-organic form occur. Moreover, sorption processes are possible, as ozone easily diffuses into porous materials. In the presence of water vapors, ozone decomposition proceeds in accordance with monomolecular scheme described by the formula (Samojlovich, Gibalov, & Kozlov, 1989):

$$\frac{dC}{dt} = -KC \tag{9}$$

wherein: C is ozone concentration, mg/m^3; t – time, hour; K – coefficient of ozone consumption rate.

Under permanent ozone concentration in the volume, the material balance equation for the ozone-air recirculation mode is as follows (Evdoseeva, Pershin, & Fedorov, 1989):

$$\frac{dC}{dt} = N\left(C - C_0\right) - KC \tag{10}$$

wherein: N – circulation rate, hour^{-1}, $N = \frac{W}{V}$; C_o, C – ozone circulation at the ozone generator input and output, g/ m^3; W – capacity of the ozone generator ventilator, m^3/hour; V – the chamber volume, m^3.

Under boundary conditions $C = 0$ и $t = 0$, the equation takes the following form:

$$C = C_0 \frac{N}{N + K}\left(1 - e^{-t(N+K)}\right) \tag{11}$$

The resulting expression makes it possible to calculate current ozone concentration at a time at known other values or at measured values of ozone concentration, to determine the value *K*.

Capacity of ozonation equipment capable of providing required modes of objects processing, depends on the method of technical implementation of gas exchange process in the "disinfection chamber–ozone generator" system.

In the Fig.6, the dependence is shown of speed ratio of ozone flow on gases interchange ratio in the chamber at re-circulation and ordinary pumping.

For empty premises of large volume, this value *K* can also be assumed, and in the presence of additional equipment or objects made of relatively inert materials (carton boxes, various containers) K=*10÷20*. In this case, required ozone generator capacity can be determined from the semi-empirical formula derived from the expression (11).

$$G = \frac{KCV\left(1 - K_3\right)}{T} \tag{12}$$

wherein: *G* is ozone generator capacity, g/hour; *V* – volume occupied by an object to be processed, m^3; K_3 – coefficient of fullness, C – ozone concentration, mg/m^3.

However, the represented dependence is suitable for use at a relatively short period of work time of an ozone generator as it does not take into account the ozone spontaneous decomposition in a volume, which lasts not longer than 15-25 minutes. The following formula is proposed for calculation of ozone concentration change with time based on the assumption that the ozone decomposition is similar to decomposition of radioactive substances (Chetvergov et al., 2005). In this paper for obtainment of time-related characteristics of ozone decomposition, it was proposed to use the substance's half-life period.

The value of ozone concentration C in a close premise in a case of ozone generator functioning in it at t→∞ will be equal to:

$$C = \frac{\beta Q M\left(O_2\right)}{\left[\left(1 + \beta - \alpha\right)Q + \lambda M\right]V} \tag{13}$$

Figure 6. Dependence of speed ratio of ozone flow on gases interchange ratio in the chamber by 0.42 m3 in volume at re-circulation of ozone-air mixture (1) and pumping (2)

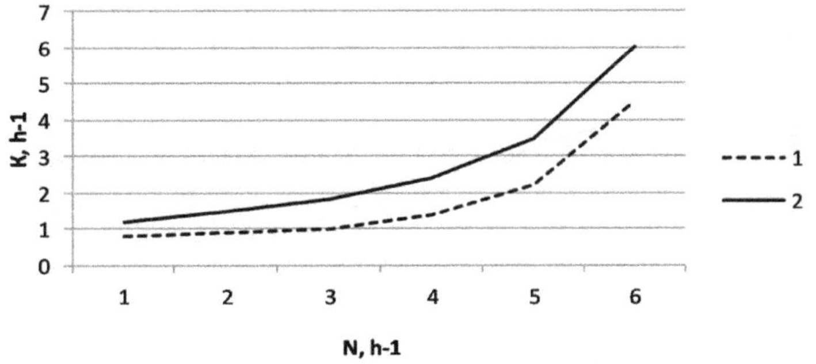

wherein: Mo (O_2), kg – mass content of oxygen in initial aerial environment; M = t•Q = 1•2.4 = 2.4 g – total gas mass in premise; T, c – half-life period of ozone in premise; λ =0.693/T, 1/h - constant of ozone decomposition in normal conditions; α – parameter specifying decomposition extent of ozone, which got repeatedly into the zone of the ozone generator discharge.

Indeed, at ozone generator work in a closed volume, the noticeable discrepancies were obtained between expected and experimental values of the concentration (Pershin et al., 2012). In the Fig.7, the obtained results are represented.

If to assume that in a premise volume V, an ideal gas mixing takes place, the following ratio is true:

$$\frac{m(O_2)}{Q} = \frac{M(O_2)}{M} \tag{14}$$

From the air physical constants, it is known that at T = 00C, the weight of 1 m^3 of air is equal to m = 1.2929 kg; while at T = 20°C, the weight of 1 m^3 of air is equal to m = 1.2047 kg.

The oxygen content in the air is 21%, and hence:

$M(O_2) = V \cdot m \cdot 0{,}21 = 10 \cdot 1{,}2047 \cdot 0.21 = 2.53 \text{ kg} = 2530 \text{ g}$

Based on the formula (1), we express λ. On substituting, we get $\lambda = 4.449$. T=0.693/ λ = 0.15 h ≈ 9 minutes.

In the Fig. 8, the comparison is given between the expected dependence of ozone concentration change at the value $\alpha = 0.7$ (solid line) and the experimental ozone concentration change in the premise by 9.4 m^3 in volume with working ozone generator inside of it (dashed line).

As it is seen from the Fig. 8, at this value of the parameter α, the experimental and the calculated dependencies show a quite satisfactory coincidence.

So, the proposed dependence for determination of ozone generator productivity for this volume of a disinfection chamber will look like as follows:

$$G = 0{,}7\ K\ C\ V\ (1 - K_3)\ /T \tag{15}$$

Figure 7. Expected and experimental dependencies of ozone concentration change in time

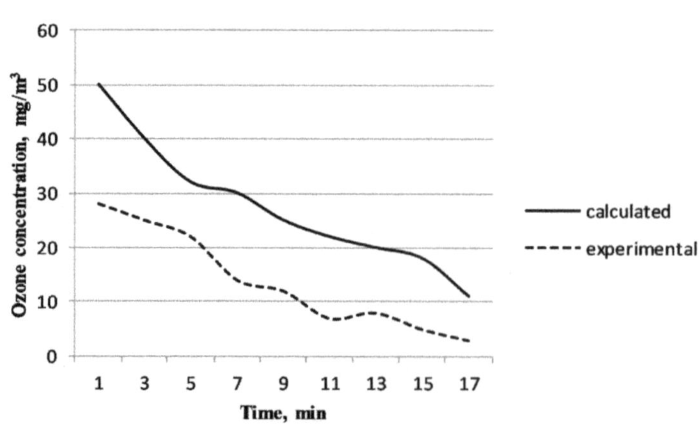

Figure 8. Change of ozone concentration in premise depending on ozone generator work time

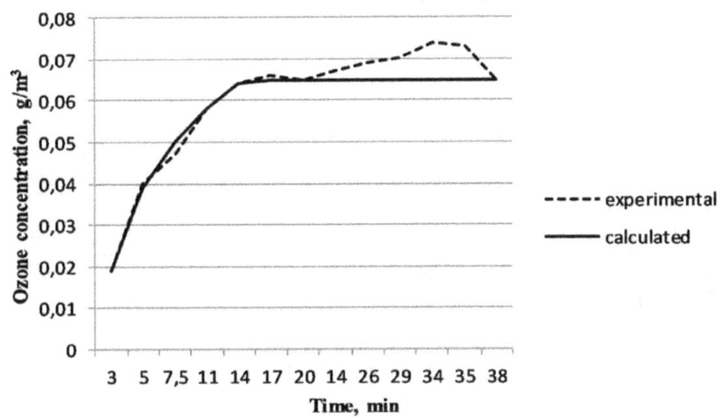

The Fig. 9 represents the experimentally obtained dependence of the ozone concentration change with time in a closed incubator. The time of ozone generator work is equal to 0.5 hours, its productivity in relation to the ozone is 1.4 g/h, while the fan productivity is 140 m^3/h. The volume of a standard incubator is 11 m^3. The ozone treatment was performed after a wet cleaning procedure.

The calculation on the formula 9 provides with a little bit different value 56 мg/m^3. However, here the air increased humidity is not taken into account inside of the incubator after its washing. The air increased humidity results in ozone generator productivity fall and more accelerated ozone decomposition. (Filippov et al., 1987).

It is forbidden to throw away an ozone-air mixture (OAM) with the ozone concentration more than 3.3 μg/m^3 into the atmosphere. However, in practice, a DC is ventilated immediately after ozone generator work completion, so OAM is thrown away with its concentration making tens and hundreds мg/m^3. That is why the treatment mode was checked up with the ozone accelerated decomposition, which was fulfilled with enforced OAM convection or under action of UV-radiation with the wave length 390-410 nm. In the Fig.10, the obtained dependence is represented. Ozone concentrations close to zero values were not obtainable with aid of the ozonometer AF-2 due to its resolving power. As one can judge from the represented results, the action of UV-radiation or convection accelerates the ozone decomposition

Figure 9. Dependence of ozone concentration change on time in incubator (ozone generator OP – 3, work time – 30 minutes)

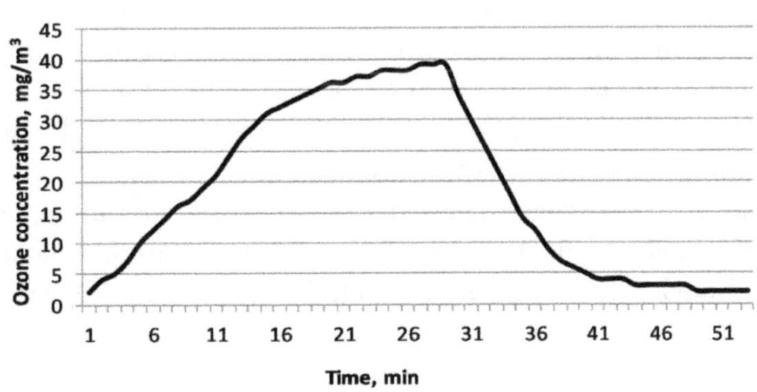

Figure 10. Ozone concentration change with time under action of UV-radiation and additional convection in comparison with natural decomposition

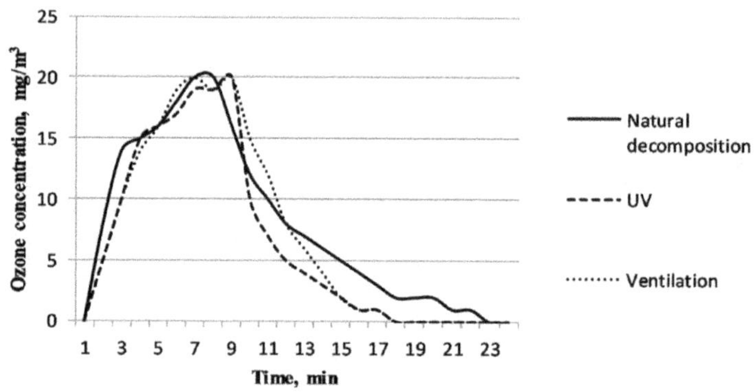

process but is not able to solve the problem. The remaining ozone would pass through the decomposition process during a long time until the required values reaching. The way-out of the obstruction is the application of ozone catalytic decomposition on Goptalume catalysts developed in M.V. Lomonosov Moscow State University (Tkachenko et al., 2001).

STORAGE AND DISINFECTION PROCESS CONTROL SYSTEM

For microclimate control in vegetable storehouses there exist automated control systems. Automation of the process of air ozonation makes it possible to control ozonation process to the specified accuracy, while providing its total safety. One of such automated control systems has been developed and successfully implemented by the AgroEngineering company (Vladimir) (Kljuev, & Zvjagincev, 2016) with the use of Russian controllers and other devices produced by the OVEN company (Fig. 12).

The automated control system of the complex "Agro – 7 microclimate" includes: touch panel controller OVEN CPK207-web, fan control unit (OVEN MBY8 or MB110-xx), power switching unit, converter unit for sensors (OVEN MBA8 or MB110-8A), corona discharge ozonizers. The wide capabilities of the complex are provided by the controller with the function of remote monitoring of the storage process. The controller allows you to manage the equipment in several independent sections or several storages. With the help of active bookmarks, it is easy to move from one screen to another.

You can create a special program for managing distribution channels, depending on the load, mode and temperature of the product. In addition to the standard set, for the convenience of the operator in the complex, additional functions are provided, for example, the function of calling help with the recommendation of the order of actions of the operator in each specific case. Access to the control system (local or remote) is protected. Each operator creates its own access password. The most important control screen settings are hidden by a special password. The main energy consumption falls on the ventilation system and heating. The controller program has created a special algorithm for adaptive control of the operation of EC fans, which adjusts their power depending on the current state of the microclimate in the storage, the selected mode, and the product's humidity temperature. On average, in the auto save mode, a potato storage area of 500 m2 requires about 7–12 kW per day. One or several fans and ozonizers are controlled by the MBY8 module with analog outputs (0 ... 10 V). In each ventilation chamber, two DACs (0 ... 10 V) are installed to control the fans and the damper actuators. The accuracy of the air dampers is determined by the software PID controller. The management of the numerous dampers of the distribution channels can be arranged from the operator's screen. Temperature and humidity control. DTC125 type resistance thermometers are used as temperature sensors in the main channel, and solid pin DTC sensors with a three-wire switch-on circuit are used to monitor the product temperature. The number of sensors is determined by the storage design. Each sensor can be used to calculate the average temperature in the control program, emergency sensors are automatically excluded from this calculation. The automated control system is designed for controlling medium in vegetable storehouses of bulk and container types. It provides optimal storage conditions at minimal electricity consumption. The system allows to control all the necessary technological equipment: damper, ventilators, valves, ozone generators, dosing units, etc. The system can be managed from various devices: from PC to smartphones. A wide range of the system possibilities is provided by the OVEN controller with the function of remote monitoring of storage process.

For each type of fruits and vegetables in the recipe software "View Manager" user parameter settings can be made and storage modes can be set up. In the event log all emergency situations are registered. Data archiving makes it possible to view storage conditions for a selected period of time. Up to 16 operating parameters are constantly archived at 1-hour interval. Files can be transferred to standard programs for viewing, data processing and graphing. The size of the archive is determined by the capacity of the flash drive, for example, the archive for 72 days (16 parameters / hour) takes less than 1.5 MB. The archive allows you to confirm the conditions of the product during the entire storage period and objectively judge the quality of different varieties and types of vegetables. If necessary, the customer is issued a product storage passport. To account for electricity consumption in storage, meters can be installed, for example, Schneider Electric (iEM3255) or a three-phase multimeter OVEN ME110-220.3M. Power and current are archived along with other system parameters.

For the ozone treatment of premises of various intention and technological equipment – in particular disinfection chambers of small volume – mobile ozone generators OP-4 have been developed and produced. The further ozone generators OS-5 and OS-10 are intended for equipping of disinfection chambers and made in wall-mounted version. A remote-control panel is installed outside of the chamber. On the panel, the following regulators are located: power switch, controller and indicator of ozone generator work mode. For safe conducting of the treatment onto the chamber door, stop limit switch is installed. The ozone generator switching-on is possible only in the case of tightly closed door of the chamber. The installation works in the mode of re-circulation. At the ozone generator switching-on, the highly effective fans allow increasing the ozone concentration in a short time. The high motion speed of the ozone-air mixture promotes its more effective influx and penetration to treated surfaces. The comparison was conducted with the earlier used ozone generator "Ozone 2M". After the treatment in the end of the ozone generator work time (1 hour), i.e. just before the ozone generator switching-off, the ozone concentration in the loaded-up chamber by about 20 m^3 in volume made 220 - 250 мg/m^3 in the case of the ozone generator "Ozone 2M" and 160 – 170 мg/m^3 in the case of the ozone generator OSC–5, though the productivity of the latter is 7 times less and consumed power 15 times less.

Figure 11. Diagram of the automated system for control of microclimate and disinfection in a vegetable storehouse

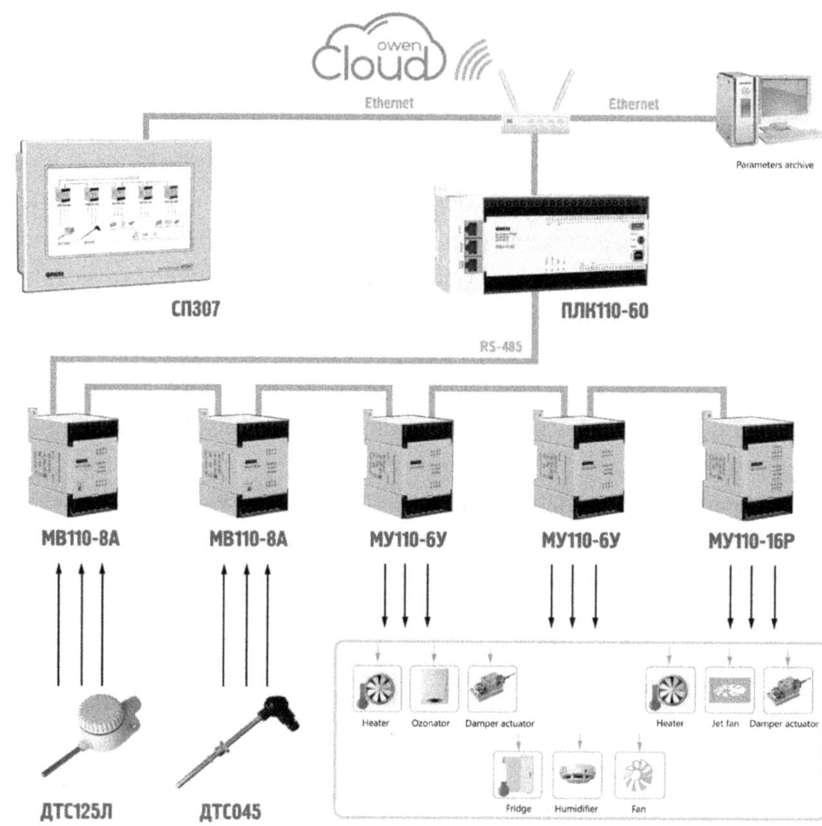

EXPERIMENTAL CONFIRMATION

The described automated system "Agro-7 Microclimate" with ozone disinfection has been installed in two vegetable storehouses in the Moscow region, where obtained analytical dependences have been tested experimentally.

Table 1 presents data on ozone efficiency at 21st day after ozonation. Empty chamber of a vegetable storehouse with volume V = 300 m^3 was processed. Ozone concentration amounted to 9 mg/m^3, exposure – 8 hours, with triple processing a day later (1 – 6).

Microbiological data is obtained with the use of standard methods –rinse sampling from the surfaces and inoculation in the Petri dish in a thermostat. To generate ozone, in these experiments ozone generators with corona discharge were used with capacity: ozone – 1,2 g/hour; the ventilator – 140 m^3/hour. The ozone generator electric mode was controlled with the use of the above-mentioned control system.

The main indicators of the economic assessment of the use of ozone are the increase in yield due to the combined stimulating and disinfecting effects, as well as the annual actual effect obtained in the economy in the form of net present value (NPV). Net present value is defined as the sum of current effects for the entire billing period, reduced to the initial step (year, quarter, month), or as the excess of the integral results over the integral costs.

Table 4. Results of ozonation of an empty chamber for fruits and vegetables storage

Experiment	Microflora Species					
	Selerotinia Selerotio-rum	**Alternaria Radicicola**	**Penicilium glaucum**	**Botrytis cinerea**	**Mucor sp.**	**Associated Fungal Species**
	The Chamber Floor Surface					
Control Processing	100 75	85 85	100 35	100 0	100 100	- -
	The Wall at the 2 m Height					
Control Processing	100 25	100 0	100 0	100 0	- -	- -

Table 5. Data for calculating the economic efficiency of the use of the ozonating unit

Indicators	Value Indicator
Number of units, pcs	2
The cost of materials, rub / pc	1495
The cost of components, rub / pc	15395
The complexity of the staff, pers. hours	14
The total tariff rate of staff, rub 8	8
Cost of chemical disinfectant, rub 12386	12386
Cost of 1 m3 of water, rub 8	8
Power of the designed installation, kWh 1.15	1,15
The cost of 1 kWh of electricity 1.56	3,56
Rate of return (bank lending rate), % 13	13
Forecast inflation rate, % 5	5

Thus, the evaluation of the effectiveness of the use of ozone-air mixture for one year showed that the net discounted annual income would be 1,532 thousand rubles.

CONCLUSION

Ozonation of fruits and vegetables increases their storage life while reducing losses. Ozone concentration in the vegetable storehouse volume depends on dynamic balance of the two basic processes – supply of air-ozone mixture and ozone decomposition. Experimental test results have shown that ozone effect is long-term and efficiency of sterilization of surface contaminated with microflora amounts to 100%.

The research findings can be used, with corresponding correction, in processing premises of large volume - livestock houses, storage areas, warehouses, etc.

Further research will be directed to the study of ozone adsorption by various porous materials to refine the mathematical model of ozone decomposition in the volume of the processed room.

The proposed ozonation method can be considered from the point of view of the global optimization method - the Monte-Carlo method (Gazizov, 2017), as the most common and most well-known random search algorithm, which consists of a uniform random "throwing" points into the search space. In the theory of global optimization, this algorithm is used as a reference for theoretical or numerical comparison of algorithms and as an integral part of some global random search algorithms. This method will be considered in further research.

REFERENCES

Avdeeva V.N., Starodubtseva G.P., Lyubaya S.I. (2008). Predposevnaya obrabotka semyan pshenitsy ozonom [Preseeding treatment of wheat seeds with ozone]. *Agrarian Science*, *5*, 19-20. (in Russian)

Bajdukin, J.A., Pershin, A.F., & Fedorov, A.V. (1988). Jelektroozonirujushhaja ustanovka dlja dezinfekcionnyh kamer pticefabrik [Electrozoniruyuschaya installation for disinfection chambers of poultry farms] *Nauka – APK [Science agribusiness]*, *2*, 42-46. (in Russian)

Belov V.V., Matkin A.Ju., & Semenova M.A. (2018). Problemy hranenija ovoshhej v hozjajstvah [Problems of storage of vegetables in farms]. (in Russian)

Bishop, D. (1990). Controlled atmosphere storage. In C. J. V. Dellino (Ed.), *Cold and Chilled Storage Technology* (pp. 66–98). London, UK: Blackie.

Bishop, D. J. (1994). Application of new techniques to CA storage. *Proceedings of the International Institute of Refrigeration International Symposium*, Istanbul, Turkey, June 8–10 (pp. 323-329). Academic Press.

Borisov, G. S. (2010). *Osnovnye processy i apparaty himicheskoj tehnologii* [*Basic processes and apparatuses of chemical technology*]. Moscow: Al'jans. (in Russian)

Borodin I.F., Pershin A.F., Evdoseeva A.Yu., Fedorov A.V. (1989). Perspektivy ispol'zovaniya koronnogo razryada v sel'skokhozyaystvennykh elektroozoniruyushchikh ustanovkakh *[Prospects for the use of corona discharge in agricultural electro-zoning installations].* Collection of scientific papers MIISP. P.3-9. (in Russian)

Burdon, J., Lallu, N., Haynes, G., McDermott, K., & Billing, D. (2008). The effect of delays in establishment of a static or dynamic controlled atmosphere on the quality of 'Hass' avocado fruit. *Postharvest Biology and Technology*, *49*(1), 61–68. doi:10.1016/j.postharvbio.2008.01.002

Butko, M.P. & Borisov, S.G. (1995). Patent 2608.2.03. Russia.

Butko, M. P., & Frolov, V. S. (2001). Ozon: sintez i yego primeneniye *[Ozone: synthesis and its application].* Moscow. [in Russian]. *GNU VNIIVSGE, 2001*, 357s.

Butko M.P., Volohov B.V., Pershin A.F., & Frolov V.S. (2010). Primenenie ozona na predprijatijah pishhevoj promyshlennosti [The use of ozone in the food industry] *Pererabotka i hranenie sel'hozsyr'ja [Processing and storage of agricultural products]*, 3, 8-11. (in Russian)

Chelnokov, A. A., & Trukhacheva, T. V. (1984). Opyt i perspektivy kompleksnogo ispol'zovaniya ozona v mikrobiologicheskoy promyshlennosti [Experience and prospects for the integrated use of ozone in the microbiological industry]. Moscow: ONTIDI microbioprom. (in Russian)

Chetvergov, N. A., Bezrukikh, N. S., & Bezrukikh, E. G. (2005). *Raschot dinamicheskikh kharakteristik kontsentratsii ozona v vozdukhe pomeshcheniy pri rabote v nikh ozone generatorov razlichnoy proizvoditel'nosti* [Calculation of the dynamic characteristics of the concentration of ozone in the indoor air when ozone generators of various capacities are working in them]. (pp. 111–115). Moscow: Technical Physics. (in Russian)

Daus, A., Horev, B., Dvir, O., Ish-shalom, S., & Lichter, A. (2011). The efficacy of ultrasonic fumigation for disinfestation of storage facilities against postharvest pathogens. *Postharvest Biology and Technology, Amsterdam*, *62*(1), 310–313. doi:10.1016/j.postharvbio.2011.06.012

Evdoseeva A.J., Pershin A.F., & Fedorov A.V. (1989). Issledovanie rezhimov raboty ozonirujushhej ustanovki s dezinfekcionnoj kameroj [Study of operating modes of the ozonating unit with a disinfection chamber]. *Jelektrotehnologii v sel'skohozjajstvennom proizvodstve [Electrotechnology in agricultural production]*, 73-78. (in Russian)

Filippov, Yu. V., Voblikova, V. A., & Panteleev, V. I. (2008). Elektrosintez ozona [Electrosynthesis of Ozone] Moscow State University. (in Russian)

Freeman, J. E., & Mudd, J. B. (1981). Reaction of ozone with sulfhydryls of human erythrocytes. *Archives of Biochemistry and Biophysics*, *208*(1), 212–220. doi:10.1016/0003-9861(81)90142-9 PMID:7259179

Gazizov, T. T. (2017). *Metody global'noy optimizatsii: uchebnoye posobiye [Global optimization methods: a tutorial]*. Tomsk: In-Spectrum. (in Russian)

Goldstein, B. D., & Balchum, O. J. (1967). Effect of ozone on lipid peroxidation in the red blood cell. *Proceedings of the Society for Experimental Biology and Medicine*, *126*(2), 356–358.

GOST 12.1.007-76 Occupational safety standards system. Harmful substances. Classification and general safety requirements. (n.d.). (in Russian)

Grits I.V., Fomichev A.V. (1985). Zavisimost' stepeni antibakterial'nogo i antifagovogo deystviya ozona ot kontsentratsii kletok i fagovykh chastits v pitatel'noy srede [Dependence of the degree of antibacterial and antiphage action of ozone on the concentration of cells and phage particles in the nutrient medium]. *Microbiology*, *3*, 410-413. (in Russian)

Gubernsky, Yu. D., & Litskevich, V. K. (1991). Zhilishche dlya cheloveka [Residence for a man Stroyizdd]. Academic Press. (in Russian)

Gunes, G., Liu, R. H., & Watkins, C. B. (2002). Controlled-atmosphere effects on postharvest quality and antioxidant activity of cranberry fruits. *Journal of Agricultural and Food Chemistry*, *50*(21), 5932–5938.

Ibrahim R.A., Al-Ahmadi S.S. (2014) Utilization of Ozone to Control Potato Tuber Moth, Phthorimaea operculella (Lepidoptera: Gelechiidae), in Storage. *African Entomology*, *22*(2), 330-336. doi:10.4001/003.022.0209

Khanbari, O. S., & Thompson, A. K. (1996). Effect of controlled atmosphere, temperature and cultivar on sprouting and processing quality of stored potatoes. *Potato Research*, *39*(4), 523–531. doi:10.1007/BF02358471

Kljuev, S., & Zvjagincev, V. (2016). Kompleksnaja sistema upravlenija dlja ovoshhehranilishh [Integrated management system for vegetable stores]. *Avtomatizacija i Proizvodstvo [Automation and production]*, *1*(46), 16–17.

Konev, S. V., Matus, V. K., Melnikova, A. M., & Rudenok, A. N. (1982). Deystviye ozona na membranozavisimyye funktsii drozhzhevykh kletok kandidoza [The effect of ozone on the membrane-dependent functions of yeast cells of candidiasis]. *Microbiology Today*, *51*(2), 220–228. (in Russian)

Krivopishin, I. P. (1988). *Ozon v promyshlennom ptitsevodstve* [Ozone in industrial poultry farming]. Moscow: Rosagropromizdat. (in Russian)

Long, L. C. P., & Cheng, Y. (2011). Patent CN102204588, China, IPC A23B 7/154.

Ozkan, R., Smilanick, J. L., & Karabulut, O. A. (2011). Toxicity of ozone gas to conidia of Penicillium digitatum, Penicillium italicum, and Botrytis cinerea and control of gray mold on table grapes. *Postharvest Biology and Technology*, *60*(1), 47–51. doi:10.1016/j.postharvbio.2010.12.004

Peretyagin, S. P. (1995). Ozon v biologii i meditsine [Ozone in biology and medicine]. In Abstracts of the 2nd All-Russian Scientific and Practical Conference. Academic Press. (in Russian)

Pershin, A. F., Bogdanov, K. V., & Smirnov, A. A. (2012). Povysheniye mer bezopasnosti pri ekspluatatsii dezinfektsionnykh kamer s primeneniyem ozona [Improving security measures in the operation of ozone disinfection chambers]. *Proceedings of the 8th International Scientific and Technical Conference "Energy Saving and Energy Supply in Agriculture"* (pp. 317-321). Academic Press. (in Russian)

Pershin, A. F., Kazev, Yu. R., & Fedorov, A. V. (1993). RF patent N°2002413 IPC A 01 M 19/00 B.I. No. 39-49.

Pershin, A. F., Smirnov, A. A., & Bogdanov, K. V. (2012). K voprosu bezopasnoy ekspluatatsii dezinfektsionnykh kamer s primeneniyem ozona. [On the issue of safe operation of disinfection chambers with the use of ozone]. Innovation in agriculture, 1. 50-54. (in Russian)

Polshina, V. G. (2018). Sovremennaya tekhnologiya khraneniya plodoovoshchnoy produktsii [Modern technology of storage of fruits and vegetables]. (in Russian). Retrieved from http://web.snauka.ru/issues/2018/01/85522

Rezgo, G.J. (2011). Ozonirovanie kak innovacionnyj metod hranenija polukopchjonnyh kolbas [Ozonation as an innovative method of storage of smoked sausages]. *Tovaroved prodovol'stvennyh tovarov [Merchandise production goods]*, *2*, 35 - 39. (in Russian)

Rice, R. G. (1999). Ozone in the Unites States of America – States of the art. *Ozone Science and Engineering*, *21*(2), 99–118. doi:10.1080/01919519908547244

Samojlovich V.G., Gibalov V.I., & Kozlov K.V. (n.d.). Fizicheskaja himija bar'ernogo razrjada [*Physical chemistry of the barrier discharge*] Moscow State University. (in Russian)

Shesterin, I. V. (2004). Vliyaniye ozona i protraviteley na posevnyye kachestva i ozdorovleniye yarovoy pshenitsy [Influence of ozone and disinfectants on sowing qualities and improvement of spring wheat]. [Dissertation abstract]. (in Russian)

Smilanick, J. L. (2003). Use of ozone in storage and packing facilities. *Proc. 99th Meeting* (pp. 131-139). Washington State Horticultural Association.

Steffen, H. (2007). Fruit and vegetable disinfection at SAMRO by means of ozone and UV Radiation. *Proceedings of the 18th World Ozone Congress*. Academic Press.

Thompson, A. K. (2010). *Controlled atmosphere storage of fruits and vegetables* (2nd ed.). Oxford, UK: CAB International. doi:10.1079/9781845936464.0000

Thompson, A. K. (2016). Controlled Atmosphere Storage. In *Fruit and Vegetable Storage*. Cham: Springer. doi:10.1007/978-3-319-23591-2_2

Tkachenko, S. N., Golosman, E. Z., & Lunin, V. V. (2001). Tsementosoderzhashchiye katalizatory ochistki gazov ot ozona [Cement-containing catalysts for ozone gas purification]. *Catalysis in Industry*, (2), 52–55. (in Russian)

Trockaja T.P. (2005) Osnovnye napravlenija isp. ozona v mjasomolochnoj promyshlennosti [The main directions isp. ozone in meat and dairy industry]. In *Conference materials: Modern technologies and complexes of technical means in agricultural production*. Academic Press. (in Russian)

Trukhacheva, T. V., Gavrilov, V. B., Malama, G. A., & Astakhov, V. A. (1992). Kineticheskiye zakonomernosti gibeli mikroorganizmov pod deystviyem ozona [Kinetic regularities of the death of microorganisms under the action of ozone]. *Microbiology Today*, *61*(4), 660–665. (in Russian)

Tzortzakis, N. (2015). Gaseous ozone-enrichment for the preservation of fresh produce. *Acta Horticulturae*, (1079), 273–278. doi:10.17660/ActaHortic.2015.1079.32

Vasilchuk N.S. & Epstein V.A. (2007) Vliyaniye predposevnoy obrabotki semyan sistemnymi protravitelyami i ozonom na nachal'nyye rostovyye protsessy i produktivnost' ozimoy pshenitsy [Effect of presowing treatment of seeds with systemic disinfectants and ozone on the initial growth processes and productivity of winter wheat]. *Vestnik SSAU*, *2*, 70 -71. (in Russian)

Vasiljak, L. M., Kostjuchenko, S. V., & Kol'cov, G. V. (2008). Primenenie impul'snogo i nepreryvnogo uf-izluchenija dlja obezzarazhivanija vody i vozduha [The use of pulsed and continuous UV radiation for disinfecting water and air] [Sanitary engineering]. *Santehnika*, *3*, 75–78.

Vozmilov, A.G., Astaf'ev, D.V., & Matveev, S.D. (2007). Primenenie ozona v tehnologicheskih processah pticevodstva i kriterii sravnitel'noj ocenki ozone generatorov [The use of ozone in poultry production processes and criteria for comparative evaluation of ozone generators]. *Mehanizacija i jelektrifikacija sel'skogo hozjajstva [Mechanization and electrification of agriculture]*, *3*, 13-16. (in Russian)

Wei L.S.; Liang X.; Zhang Y.F. (2018). Numerical investigation on the effect of gas parameters on ozone generation in pulsed dielectric barrier discharge. Plasma science & technology, 20(12). doi:10.1088/2058-6272/aadca6

Wills, R. B. H., Pitakserikul, S., & Scott, K. J. (1982). Effects of pre-storage in low O2 or high CO2 concentrations on delaying the ripening of bananas. *Australian Journal of Agricultural Research*, *33*, 1029–1036. doi:10.1071/AR9821029

Wojciech, J. (2006). US Patent 6991930, USA, IPC C12N 1/14 A01N 63/04.

Zhang, Y. F., Wei, L., Liang, S. X., Deng, H. Z., & Šimek, M. (2018). Characteristics of the Discharge and Ozone Generation in Oxygen-Fed Coaxial DBD Using an Amplitude-Modulated AC Power Supply. *Plasma Chemistry and Plasma Processing*, *38*(6), 1199–1208. doi:10.100711090-018-9922-2

Chapter 3
Large-Scale Renewable Energy Monitoring and Forecast Based on Intelligent Data Analysis

Mehmet Baris Ozkan
Energy Institute, TÜBİTAK Marmara Research Center, Turkey

Dilek Küçük
Energy Institute, TÜBİTAK Marmara Research Center, Turkey

Serkan Buhan
Energy Institute, TÜBİTAK Marmara Research Center, Turkey

Turan Demirci
Energy Institute, TÜBİTAK Marmara Research Center, Turkey

Pinar Karagoz
Computer Engineering Deptartment, Middle East Technical University, Turkey

ABSTRACT

Intelligent data analysis techniques such as data mining or statistical/machine learning algorithms are applied to diverse domains, including energy informatics. These techniques have been successfully employed in order to solve different problems within the energy domain, particularly forecasting problems such as renewable energy and energy consumption forecasts. This chapter elaborates the use of intelligent data analysis techniques for the facilitation of renewable energy monitoring and forecast. First, a review of the literature is presented on systems and forecasting approaches applied to the renewable energy domain. Next, a generic and large-scale renewable energy monitoring and forecast system based on intelligent data analysis is described. Finally, a genuine implementation of this system for wind energy is presented as a case study, together with its performance analysis results. This chapter stands as a significant reference for renewable energy informatics, considering the provided conceptual and applied system descriptions, heavily based on smart computing techniques.

DOI: 10.4018/978-1-7998-1216-6.ch003

Copyright © 2020, IGI Global. Copying or distributing in print or electronic forms without written permission of IGI Global is prohibited.

INTRODUCTION

Applications of intelligent data analysis methods such as machine learning algorithms are commonly observed in diverse domains including energy, health, education, and media. Considering the energy domain, such intelligent methods are used in order to improve energy forecast, management, planning, and optimization, among others. For instance, machine learning algorithms such as linear regression, artificial neural networks (ANNs), and support vector machines (SVMs) are known to perform well for tasks such as forecasting load or electricity consumption as well as renewable energy resources such as wind, solar, hydro, and wave energy (Ahmad et al., 2014; Mocanu et al., 2016).

In this chapter, we first describe a generic and large-scale renewable energy monitoring and forecast system which employs various intelligent data analysis techniques. Monitoring different renewable energy plants is achieved through deployment of meteorological sensors and power analyzers at these plants, and collecting data from them into a central database. On the other hand, forecasting renewable energy resources and electrical power generated from them is achieved by training several different machine learning algorithms on this database which also includes meteorological forecasts obtained from related meteorological services.

This generic system is realized for wind energy and the latest version of this large-scale wind power monitoring and forecast system is described as a case study. The system makes use of different machine learning algorithms to perform wind power forecasts required for operational and trading purposes. Wind energy is a ubiquitous renewable energy resource with limited disadvantages compared to other energy resources and hence its share in overall energy production is consistently increasing worldwide. High-performance wind energy forecasts are useful both for the operation of the overall electricity grid (by the transmission system operators) and for optimized trading of the produced energy (by the plant owners). The presented system is built within the course of Wind Power Monitoring and Forecast Center (Rüzgar Gücü İzleme ve Tahmin Merkezi (RİTM) in Turkish) project is developed for General Directorate of Renewable Energy of Turkey[1]. The system provides facilities for almost real-time monitoring of the wind power plants as it automatically collects high-resolution and synchronized energy production data from these plants. It also automatically collects mesoscale numerical weather predictions from international and national weather services, to be used by the wind power forecasting modules. The implemented system performs wind power forecasts employing singleton physical, statistical, and machine learning algorithms, as well as hybrid methods combining these algorithms. In addition to producing very short-term and short-term forecasts for each plant, the system is able to produce regional forecasts. The system performs both point forecasts and probabilistic (interval) forecasts. Descriptions and comparative evaluation results of the employed algorithms for all of these forecast types are provided. The operational system currently covers more than 150 wind power plants in Turkey with a total installed capacity of about 6.7 GW.

As the presented system for wind energy is an implementation of a generic one for renewable energy plants, the system can be extended to monitor and forecast the energy production of hydro-power and solar power plants by applying the existing implemented infrastructure for wind energy forecasting using singleton and combined machine learning algorithms. We provide discussions on this system extension for monitoring and forecast of different renewable energy resources. Other important directions of future work based on the presented intelligent system architecture are also provided at the end of the chapter.

Below listed are the main objectives of this chapter:

- To present a generic and large-scale system architecture for centralized monitoring and forecast of renewable energy resources such as wind, solar and hydro-power where the system is heavily based on intelligent data analysis techniques to produce highly accurate renewable energy forecast,
- To describe a genuine implementation in detail and application of this generic architecture for wind energy in country-wide settings, and to provide performance evaluation results of this application,
- To provide pointers for the ways of extension of the application to cover other renewable energy resources, so that a fully-applied version of the generic system can be achieved,
- To provide a convenient review of the literature on systems and intelligent data analysis methods for renewable energy monitoring and forecasting so that the readers can put the presented generic and applied systems into perspective,
- To list recommendations of more recent smart computing techniques to the interested readers, including deep learning methods for renewable energy forecast.

The rest of the chapter is organized as follows: In the next section, literature review on state-of-the-art methods and systems for the purposes of renewable energy monitoring and forecast is provided. The following section presents the generic and large-scale system architecture for centralized monitoring and forecast of the renewables. Then, the case study, which presents the implementation of the generic system for wind energy in Turkey, is described. Finally, the future research directions and the conclusion remarks are presented a summary of the chapter's main points.

BACKGROUND

The instantaneous monitoring and prediction of the power from the energy sources are the most important two issues for the institutions that manage the energy flow in the country. For a good planning, the data collected from the field can be stored securely on the central servers, and system operators can instantly monitor the reliability of the systems by using the user software. Apart from instant monitoring, the most important second pillar of a reliable system is reliable forecast. The integration of energy sources such as wind, solar and water, which depends on weather conditions, to the system should be carried out with reliable day-ahead forecasts. In this section, general structures of energy monitoring and prediction centers in the literature and algorithmic approaches used for forecast applications are examined.

Monitoring and Forecast Centers

In (Erinmez et al., 2002), the authors present the details of the Geomagnetically Induced Current (GIC) monitoring and forecast system which is used by The National Grid Company who is the operator of the high-voltage electric power systems in England and Wales. The main purpose of this system is, by using the global solar weather forecasts, to determine the power output swings, voltage dip and transformer failures in the transmission system. Forecast time horizon in the system is determined as long term (up to a year ahead), medium term (up to a month ahead), short-term (up to a week ahead) and real time (1 hour ahead). According to the system performance results, most useful forecasts are obtained for the 72, 24 and 1 hours ahead and the system is correctly forecast the two major solar storm events and alarm the system operator.

The Special Regime Control Centre (CECRE) is located in Spain for the purpose of the integration of the renewable energy sources to the transmission line (de la Torre et al., 2012; Rodriguez et al., 2008). This center is used for monitoring and controlling wind power generation in Spain. The system uses active and reactive power, voltage, connectivity, temperature and wind speed values as the inputs. By using these inputs, CECRE calculates the wind power which must be integrated to the energy flow at any given time automatically. Additionally, in the center, voltage dips and generation trips are estimated and this information is used in the congestion management. According to the regulation, generators with an installed power of 10 MW must be directly monitored and integrated to CECRE system.

The communication technologies and standards of a smart grid system are discussed by the authors in (Gungor et al., 2011). Smart grid systems are the extended version of the traditional electric power grid infrastructure which is integrated with the renewable and alternative energy sources through automated control and modern communication technologies. In such systems, it is important to determine the safety of the data and the standard of the data format. In the paper, the authors discuss the main properties of the communication standards for several parts of the system, such as powerline networking, home area network device communication measurement and control, inter-control and interoperability center communications.

In (Cheung et al., 2010), the authors propose a smart dispatch framework for regional transmission organizations and transmission system operators in order to control the large power grids. Unlike traditional methods, it is intended that by using this framework, the system operator can make real-time decisions based on changing system conditions such as load, generation, interchanges in the electricity grid. According to their experimental results, the proposed dispatch framework provides better holistic and forward-looking view of system conditions.

The researchers in (Mahoney et al., 2012) propose a wind power forecasting framework which is used by The National Center for Atmospheric Research (NCAR) for optimizing the grid integration. In the system, the researchers provide both deterministic mesoscale wind power forecasts and probabilistic wind power forecasts, which are calibrated by analogue Kalman filter and quantile regression. Produced forecasts are used for not only real-time grid integration, but also day-ahead trading in the energy market. For the data inputs, Variational Doppler Radar Assimilation System (VDRAS), which is developed by NCAR, is used and the system obtains the data from by lidars, Doppler radars and surface networks.

Forecast Approaches

In central monitoring systems, forecast algorithms have an important role for the transmission system operators to plan an effective and accurate energy management. By using meteorological weather forecasts and the power data collected from the power plant sites, various power estimates can be made by using these algorithms for renewable energy sources in the monitoring and forecast center (Giebel et al., 2011; Gellert et al., 2019). We can classify the power forecast models needed for renewable energy sources as follows (Monteiro et al., 2009):

- Very short-term power forecasting
- Short-term power forecasting
- Probabilistic power forecasting
- Regional power forecasting

Very short-term power forecasting models generally deal with 0-6 hours range with high forecast resolution and these forecasts are usually used in real-time grid managements. In (Blonbou, 2011), the author proposes a very short-term wind power forecasting algorithm by using neural networks and adaptive Bayesian learning. The algorithm uses a three-outputs-neural network with 5 hidden layers. A wind speed vector contains 30 minutes moving average wind speed values are used as the input of this network. Performance of the algorithm is tested with 10 min, 20 min and 30 min intervals separately and model performs better than the persistence model for all intervals. In (Golestaneh et al., 2016), the authors discuss a very short- term forecasting algorithm which is adapted to the solar energy sources. They use an Extreme Learning Machine (ELM) as fast regression model and the model is trained with both point and probabilistic forecasts of solar power generation. The model is tested with two different data sets, which contain one-minute resolution and highly fluctuating patterns. According to the performance results of the model, it detects the sharp densities in the data up to one-hour lead times effectively. They also compare the seasonal performance of the proposed method and they conclude that better results are obtained in winter season while worst results are obtained in the summer season.

Short-term power forecasting algorithms generally deal with day-ahead power forecasts such as up to 48 or 72 hours. These kinds of forecasts are more crucial for the plant owners since they are used in the energy market. The authors of Fleten and Kristoffersen (2008) develop a multi-stage mixed-integer linear stochastic based short- term power production plan for hydropower plants. Such day ahead planning is important to determine how much water must be allocated among the reservoirs for the management of the grid. They test the proposed model in a hydropower plant which is located in Norway with a 33.9 MW generation capacity. According to experimental results, by employing the expectation-based objective function criterion in short-term planning reduces the risks significantly and in the long-term hydropower producers have a natural hedge. For the short-term forecasts of the power generation obtained from the wind sources, the authors in (Osório et al., 2015) propose a new algorithm which combines adaptive neuro-fuzzy inference system with particle swarm optimization. Proposed model is tested in 24 hour day-ahead interval and compared with the other methodologies in the literature such as persistence, neural networks and ARIMA. According to experimental results, proposed approach shows better performance compared other models and lowest forecast error rates obtained in the fall season while highest rates obtained for the winter season.

Probabilistic power forecast models deal with the uncertainty analysis in the power sources. In these models, forecasts are presented with some confidence interval and generally these forecasts are used in the load decision by the transmission system operator. The authors use quantile regression on the point wind forecasts in order to produce probabilistic wind power forecasts in their work (Nielsen et al., 2006). For the 25% and 75% quantiles, probabilistic wind power forecasts are generated by using the existing wind power point forecasts. The quantiles are modelled as the sum of non-linear smooth functions of variables forecasted by the meteorological model. According to experimental results in 500 out of 11000 cases the upper and lower band of the confidence intervals are broken by the real production data. In (Owens et al., 2017), the authors propose a probabilistic forecast method that can be applied on solar wind and geomagnetic forecasting. The model is based on analog ensemble approach. In the model, the most similar day to the current conditions from the past historical meteorological forecast data is searched. Then by considering an ensemble of past analogues, the forecasts are inherently probabilistic and provide a measure of the forecast uncertainty. According to the experimental findings, best forecasts are obtained by using the previous 6-12 hours observations.

Regional forecast models are crucial for the maintenance planning. Generally, for large-scale systems, it is hard to obtain real generation data for all plants in the country and these forecast models are used for producing a power forecast for a specific region or whole country. In the methodologies, some reference plants are selected and the whole region is modelled by using these reference plants. The authors of Davò et al. (2016) propose a regional forecasting model for wind and solar power by using post-processing techniques and Principal Component Analysis (PCA). For the wind power, the model is tested for the wind power plants located in Sicily, and for the solar irradiance, it is tested on the state of Oklahoma. Proposed model first takes gridded forecast data from Numerical Weather Prediction (NWP) models over the geographic region then PCA is applied on the NWP data for reducing the size of the gridded forecast data. In the final step, they apply Analog Ensemble (AnEn) technique and NN to produce point forecasts. The researchers in (Lobo and Sanchez, 2012) develop a regional wind power forecast algorithm based smoothing techniques and this model is integrated to Spanish Peninsular System. The model makes direct aggregate wind power forecasts in a specified region without using any sub-area level. The method is based on non-parametric techniques for local modelling and this approach shows better performance compared to classic Kernel methods. They compare the proposed model with another forecast tool, namely SIPREOLICO. According to their results, computing time of the proposed approach is lower than the compared forecast tool and it shows considerably smaller prediction errors than the competitors.

A GENERIC AND LARGE-SCALE RENEWABLE ENERGY MONITORING AND FORECAST SYSTEM

In this section, the details of a generic and large-scale renewable energy monitoring and forecast system (henceforth, REMFS) are presented. The architecture of this system proposal is presented in Figure 1. The main capabilities of the system are listed below and they are described in the following subsections.

- Data collection facilities
- Forecast facilities
- Data presentation facilities

Data Collection Facilities

The main data sources of REMFS are (1) systems, sensors and devices of renewable energy plants including (but not limited to) wind, solar, hydro-power, and wave energy plants, and (2) national and international weather services.

As depicted in Figure 1, there are mainly three types of data that are retrieved from the plant sites in an automatic and periodic manner:

1. Measurements of energy and/or power analyzers deployed at the plants which correspond to actual generations of the plants.
2. Measurements of meteorological sensors such as wind speed and direction for wind plants, solar radiation for solar plants, and precipitation and temperature for hydro-power plants, and from the corresponding meteorological sensors for other types of plants.

3. On/off status of the generating units of the plants from their SCADA (Supervisory Control and Data Acquisition) systems, whenever applicable.

The above-listed types of data are retrieved by the Data Collection Services (DCSs) of the overall system and stored in a database at Renewable Energy Monitoring and Forecast Center (REMFC).

Apart from the data obtained directly from the plants, numerical weather predictions (NWPs) produced by weather services are known to be significant features for learning-based energy and power forecasting for renewable energy plants. Within the proposed architecture, NWPs at different ranges (such as very short-range, short-range, medium-range, etc.) can be automatically obtained and stored in the REMFC's database, again with the corresponding services (DCSs). It should be noted that, national weather services usually provide high-resolution NWPs for the country or region, with a large number of grid points for the particular country/region, while international weather services such as ECMWF (European Centre for Medium-Range Weather Forecasts)[2] and NOAA (National Oceanic and Atmospheric Administration)[3] usually provide NWPs with low resolution with less number of grid points for the country/region under consideration. As a final note on data collection, there may be several means of data collection: some data may be presented by the related providers through file sharing, while others may be provided through Web services or Inter-Process Communication (IPC) methods.

Figure 1. The architecture of the renewable energy monitoring and forecast system

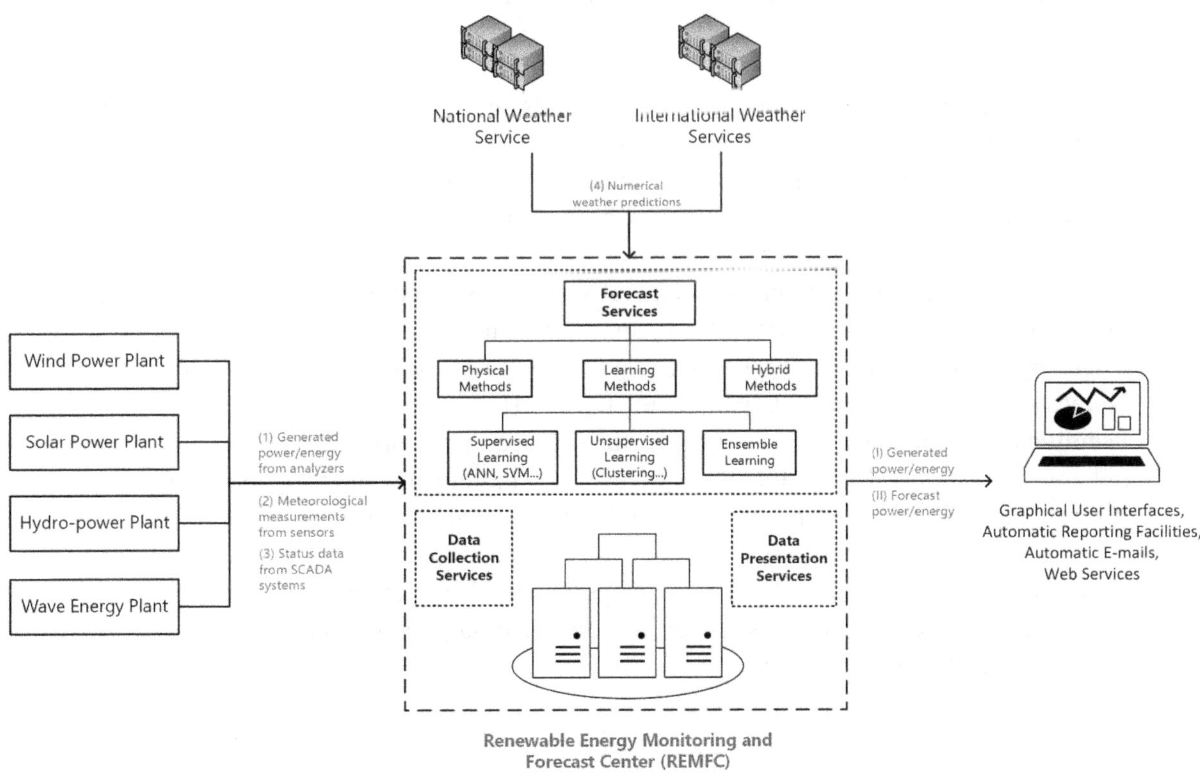

Forecast Facilities

As mentioned earlier in this chapter, power and energy production forecasts of renewable energy resources are particularly important for two main reasons:

1. The organization within the country that is responsible for the operation of the electrical grid (which is usually called transmission system operator (TSO) or grid operator) requires energy forecasts from the renewable energy plants to balance electricity demand and supply, and to perform this task in an optimized way in order not to waste the potential of natural resources.
2. The corresponding power/energy forecasts are also useful for the plant owners themselves to optimize their daily and maintenance operations and to optimize their gain in the energy market.

Hence, considerable research effort has been devoted to renewable energy forecast in a high-performance fashion, as reviewed in the previous section.

As demonstrated in Figure 1, Forecast Services of the overall system, REMFS, can be considered in three classes first: physical methods, learning methods, and hybrid methods. Physical methods are usually based on the geographical properties of the plant site and on related methods such as Computational Fluid Dynamics (CFD) to transform NWPs to power/energy generation forecasts. The learning methods will further be elaborated below and hybrid methods (in our system proposal) correspond to those methods that produce combined methods using physical and learning methods.

Learning (or, machine learning) methods used for forecasting purposes can be classified into three groups:

1. **Supervised Learning:** Supervised learning methods such as artificial neural networks (ANNs), Support Vector Machines (SVMs), and proprietary statistical methods are known to perform favorably for forecasting renewable energy production. Similarly, regression methods such as linear regression are known to be successful at very short-term (very short-range) forecasting of the renewables (Oquab et al., 2015; Miyato et al., 2018).
2. **Unsupervised Learning:** Common unsupervised learning methods such as k-means clustering, Principal Component Analysis (PCA), and Singular Value Decomposition (SVD) are known to be particularly useful within the preprocessing stages of the other learning-based/statistical methods of forecasting (Lei et al., 2016; Ghahramani et al., 2018).
3. **Ensemble Learning:** Ensemble learning methods that combine more than one learning method (a singleton method) are also used for renewable energy monitoring and they commonly outperform the singleton methods when they perform individually (Wang et al., 2015).

All of the above-mentioned intelligent data analysis methods are included in REMFS, and both the individual outputs of singleton methods and/or the combined outputs of hybrid methods and ensemble learning methods can be presented to the system users.

Data Presentation Facilities

REMFS has a number of Graphical User Interface (GUI) applications, automatic and periodic reports, automatic e-mails, and Web services as its data presentation modules. These modules enable interested parties (system users) to access both the actual generations and generation forecasts stored in the system center. The system center hosts a suite of software called Data Presentation Services to interface with and supply data to these modules.

Details of these modules are provided below:

1. **Desktop Applications:** REMFS includes a suite of desktop, Web-based, and mobile GUI applications to help its users access the generation and forecast data. Desktop applications usually facilitate detailed analysis of the underlying data, including detailed long-range reports, long-range queries for historical data (similar to a historian application), and graph, table, and map-based visualizations.
2. **Web-Based Applications:** These applications will facilitate access to the generation and forecast data with some limitations on the time range of queries and reports. Yet, these applications are easier to maintain due to centralized installation at the Web server, compared to desktop applications, which require updates to the computers of the users.
3. **Mobile Applications:** These applications will enable ubiquitous access to the data stored at REMFC. However, similar to Web-based applications, there may be some limitations on the time range of queries and reports within mobile applications.
4. **Automatic Report Generation and Periodic E-mails:** The system also has facilities to produce general and plant-based automatic reports which include the error rates of previous forecasts and new forecasts. These reports can be periodically sent to system operators and plant owners through automatic e-mails.
5. **Web Services:** The data collected and produced within REMFC can also be supplied to related stakeholders like the grid operator and plant owners through Web services which also include an authentication mechanism.

WIND POWER MONITORING AND FORECAST SYSTEM: A CASE STUDY

System Overview

Wind Power Monitoring and Forecast Center for Turkey (RİTM) project was started in 2010 by the Scientific and Technological Research Council of Turkey (TÜBİTAK) (Terciyanli et al., 2013; Terciyanli et al., 2014). Aim of this project is to forecast the electrical power to be generated from the wind, using the meteorological data received from Numerical Weather Prediction (NWP) model outputs and the data received from the power quality analyzers. Project is launched for the General Directorate of Renewable Energy of Turkey (YEGM). Project is made up of two phases and first phase is conducted 14 pilots Wind Power Plant (WPP) selected from the different regions of the country (Özkan et al., 2013). In the second phase of the project, all power plants in the country are started to be included in the system after the forecast performance criteria were met for the pilot plants used in the first phase. By the end of 2018, approximately 150 wind farms with a total installed capacity of 6700 MW can be monitored instantly and daily power estimates are made for these power plants in the center. Power forecasts produced in the

Figure 2. General architecture of the RİTM system

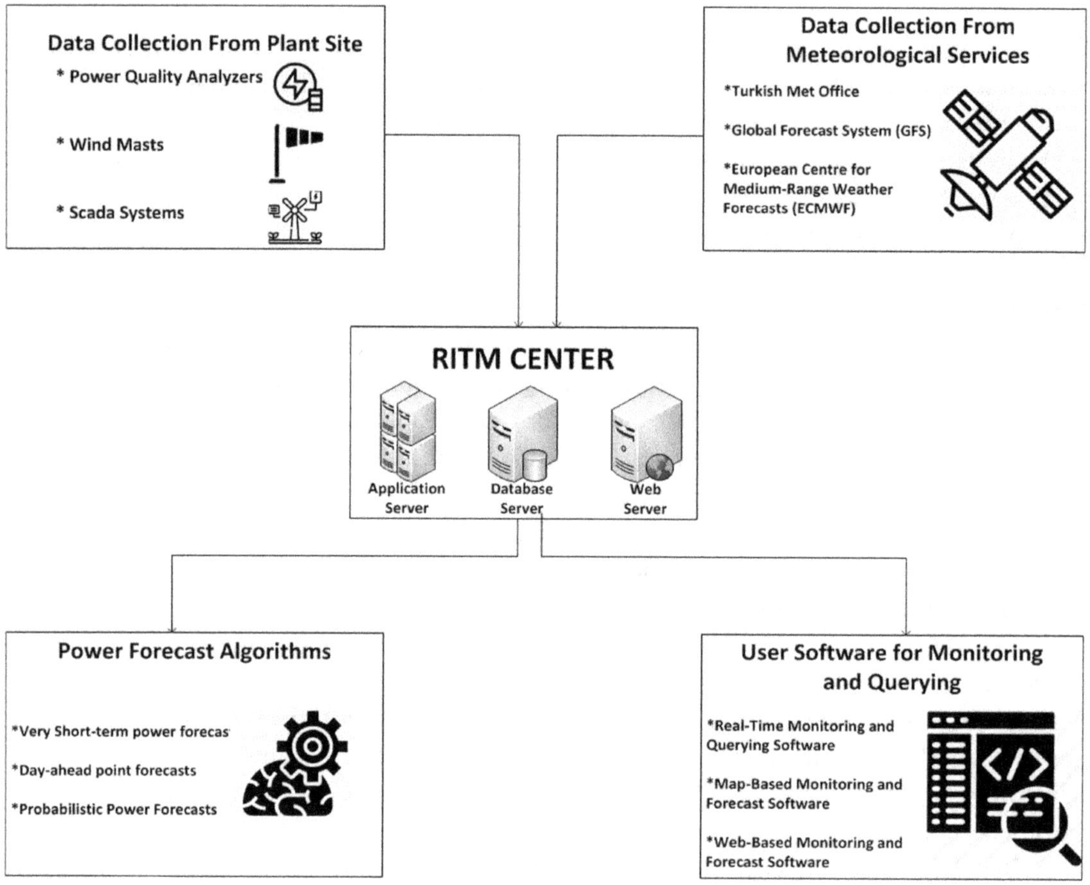

system are used by the Transmission System Operators (TSOs) in grid management and planning while they are used by the power plant owners in the energy market. For a better planning they need reliable power forecasts for all kind of energy sources. Also a reliable wind power forecast system is inevitable for the plant owners. WPP owners attend the energy markets with their day-ahead and very short-term power forecasts and according to the accuracy of these forecasts they get profits or pay penalty. This system solves these problems for the wind energy sources.

System Modules

General architecture of RİTM system is shown in Figure 2. In this structure, inputs required for power forecasts are provided from the plant site and the meteorological stations and these values are stored in the central servers. Then, the power prediction algorithms developed by using these inputs are run on the servers and the estimation-power values can be displayed by the user software.

Data Collection Modules

In order to produce power estimations, the most important component of the system is to collect the parameters related to the wind in the center. For this purpose, estimation inputs from the plant site and meteorological sources are collected in the system with many different methods and the details of these components are summarized in the following subsections.

Data Collection from Plant Site

The inputs collected from the plant site consist of atmospheric information, such as wind speed and direction in the field, real production power value and SCADA data on turbine basis. The details of the system components used to collect the data of these inputs are presented below.

Power Quality Analyzers

Wind power quality analyzers collect current and voltage data from two feeders for three phases, carry out real-time analyzes and have GPS (Global Positioning System) synchronization feature for the simultaneous measurement. With the IP based network interface on the wind power analyzer, communication is possible with RİTM over any kind of IP based network. Sampling frequency of each channel is 25.6 kHz, which means that each cycle for a signal at 50 Hz basic frequency is represented with 512 samples (Demirci et al., 2011). Measurement accuracy class of the wind power analyzers according to IEC 61000-4-30 is A. According to IEC-61000-4-30 standard, voltage sags, swells and interruptions are accepted as the power quality events. When these events are captured, wind power analyzers connect to the RİTM instantly and send 3-second long event raw data. With these devices, the following power quality parameters are sent to the center in 3-second resolution and 10-minute average:

- Voltage and Current
- Active, Reactive and Apparent Power
- Power Factor
- Harmonics and Inter Harmonics
- Unbalance
- Frequency
- Flicker
- Sag, Swell and Interrupt events

Power values are the main parameters used to train the statistical algorithms and test the forecast performances. Other parameters are generally used by the system operator to monitor power quality events at the plant site.

Wind Masts

Wind masts are the components that measure and send the wind parameters at the plant site. With the help of these components, wind speed, wind direction, air temperature, humidity and pressure values are measured by the sensors on the mast and these values are stored on the memory card in the logger device. Logger device samples speed sensors, direction sensors, heat sensor, pressure sensor, humidity sensor and battery data on a 1-second cycle time. Then data sent to data logger software which is running in RİTM Web Database with the help of GPS module. Measurements are taken with 15-second and 10-minute interval with based on the IEC 61400-12-1 standard (Terciyanli et al., 2014).

SCADA Systems

SCADA is the system that enables all technical data of plants kept in a central database. Collected data that is downloaded from SCADA system is used for many different purposes. SCADA subsystem is established for transferring various turbine data from plants to YEGM servers and using this data in forecasting system. Generated power quantity measured from the turbine, wind speed, wind direction and status data is stored in one text file on the SCADA system by an operator every 10 minutes in a file in the plant file server. Then, with the help of an application running on RİTM server, these files are transferred to the center with SFTP protocol and values are stored in the database for using in the forecast algorithms.

Data Collection from Meteorological Services

Numerical Weather Predictions (NWPs) are the main input of the power forecast algorithms. Due to their core functionality, these inputs are included in the system from three different sources, namely Turkish Met Office (MGM), Global Forecast System (GFS) and European Centre for Medium-Range Weather Forecast (ECMWF). As shown in Figure 3, GFS and ECMWF are served as the initial input of the Weather Research and Forecasting (WRF) model, while MGM is served as the initial input of ALADIN[4] model. For the GFS and ECMWF sources WRF model run at RİTM center after initial data are download from the weather sources. ALADIN model is run at the servers located in the Turkish Met Office and outputs are taken to RİTM center. These forecasts cover all country and the grid resolution is 6 km for GFS and ECMWF and 4 km for MGM forecasts (Terciyanli et al., 2014).

Figure 3. Numerical weather forecasts in RİTM system

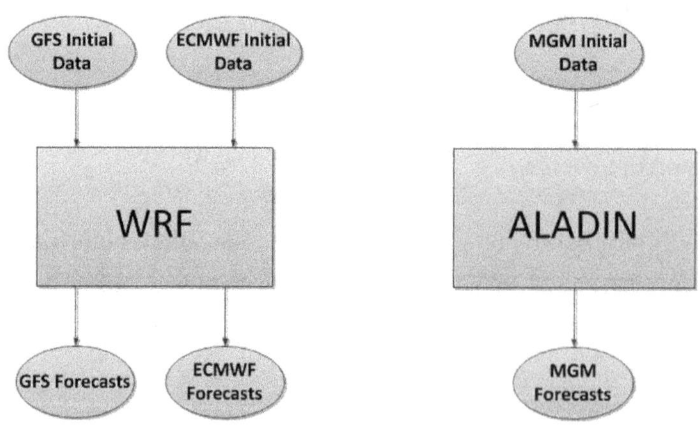

RİTM Center

RİTM Center is located within YEGM building. It includes a study room and a server room. Through the video surveillance system in the study room, researchers can monitor all wind farms in the country which are integrated to the system.

The servers are located in the server room and the tasks of these servers are as follows:

- 2 Database servers with one is served as a backup: The job of the servers is keeping forecast related data obtained from the system components.
- 2 Application servers with one is served as backup: The job of the servers is running the power forecast algorithms and giving service to user software applications.
- 2 NWP servers: These servers run the WRF model for the GFS and ECMWF sources.
- 1 Web server: This server provides services for RİTM web page and Web-Based Monitoring and Forecast Software.

Power Forecast Modules

Power forecast algorithms running at the center generate power estimates for various problems by using inputs from the plant site area and meteorological stations. The details of these power forecast problems and developed algorithms are presented in the following subsections.

Very Short-Term Power Forecast Algorithm

Very short-term power estimates have an important role in the instant planning of the system for the Transmission System Operator (TSO). In very short time forecasting models, autoregressive or self-autoregressive models, which depend on the trend of WPP production, are used in the center. These models use the last production of WPP or old forecasts to find trends and seasonality and determine forecast values for future by moving these. The forecasts are updated by each hour within which forecasts for the next 6 hours are made using the data received online. In the system, two approaches, namely Autoregressive Model (AR) and Autoregressive Exogenous Model (ARX), are used together. For the forecasts of the first 2 hours AR is used while ARX is used for the next 4 hours (Buhan et al., 2013).

The mathematical formulations of AR and ARX models are presented in Equation 1 and Equation 2, respectively. In the formulas, y_t is the power production at time t, a_0 and a_i are the regression coefficients, p is the order of the autoregression model and e_t is the error term at time t. For the second ARX formula, u_t represents the power forecasts obtained from the day-ahead forecast models at time t, β_0 and β_i are the regression coefficients, and q is the order of the exogenous part. Both of these models use real production measurements obtained by power quality analyzers and power forecasts are updated every hour.

$$y_t = a_0 + \sum_{i=1}^{p} a_i y_{t-i} + e_t \tag{1}$$

$$y_t = a_0 + \sum_{i=1}^{p} a_i y_{t-i} + \beta_0 + \sum_{i=1}^{q} \beta_i u_{t-i} + e_t \quad (2)$$

Day-Ahead Point Forecast Algorithms

Day ahead forecast algorithms constitute an important part of the system and the owners of the plants take part in the intraday market by using these forecasts. For this purpose, power forecast are produced by 3 statistical methods and final forecasts are presented as a combination of these models.

SHWIP

Statistical Hybrid Wind Power Forecast Model (SHWIP) is based on combining dynamic clustering with linear regression (Özkan and Karagoz, 2015). Model classifies the weather events dynamically according to the most important weather parameters namely wind speed, wind direction, temperature and pressure by using Principal Component Analysis (PCA) and linear regression on the NWP data. The general architecture of the SHWIP is presented in Figure 4.

In the training phase, by using the historical NWP and power data most similar weather clusters are determined. For the training period 90 day is used since it covers one season and the number of clusters is dynamically determined for each WPP independently. In the next step of the model, by using the power curve of the WPP, 48-hour NWP test data and clusters determined in the training phase, 48-hour interim power forecasts are calculated for every 3 NWP sources. Finally according to most recent forecast performances, interim forecasts are combined to a final power forecasts with a combination algorithm.

ANN

The developed ANN based forecast model consists of the best grid selection stage, the multi-ANN forecast stage and the final Model Output Statistics (MOS) correction stage. A three-layer with one hidden layer, feedforward ANN structure is selected to handle wind-to-power forecasts. In the three layer structure of the ANN, input layer has two neurons (for wind speed and direction features), hidden layer has a varying number of neurons between 6 and 35 (with respect to wind characteristics and training data size), and the output layer has one neuron (for wind-electric power output). Sigmoid activation function is selected for both the hidden layer and the output layer. The batch-type resilient backpropagation algorithm, which is adaptive to a large training data size and fast enough for use in forecast problems, is found to be the best one in terms of forecast accuracies (Buhan and Çadırcı, 2015; Buhan et al.,2016).

Figure 4. General structure of SHWIP model

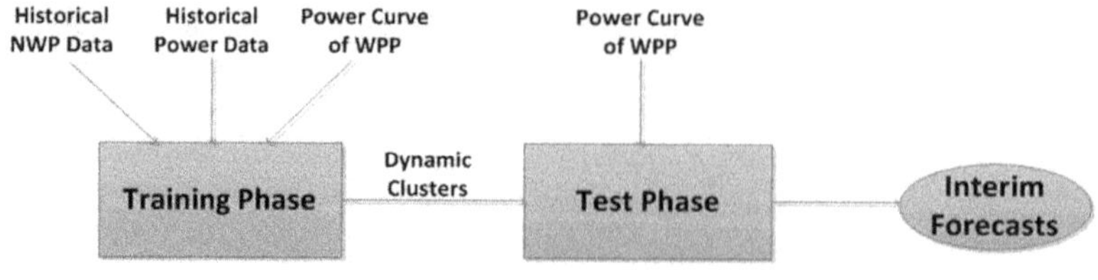

In order to select the best grid for each of the multiple NWP input data, a new approach is adopted by using the associated data of the nearest nine grids to each WPP. While training the model, wind speed and wind direction data is used as input, and the generated power is used as the output. The mean square error (MSE) is used for obtaining the parameters of the ANN model with a batch-type resilient back-propagation algorithm. The MSE calculation is as given in Equation 3.

$$MSE = \frac{1}{L} + \sum_{u=1}^{L} \left(P_u - R_u\right)^2 \tag{3}$$

In this equation, L is the number of samples, P_u is the uth forecast, and R_u is the uth observation. The parameters of ANN structures are either stored or used in 48-hour ahead real-time power forecasts. The proposed model is then trained with a proper data size, for tuning parameters of the ANN structures and forecasting of power generation of each WPP up to 48-hour ahead.

SVM

The developed SVM wind-electric power forecast model consists of the best grid selection stage, the multi-SVM forecast stage, and the final MOS correction stage. For power forecast, SVM regression is used under RBF kernel function. The kernel function is selected after validation experiments with various kernel functions such as the linear, polynomial, RBF, and sigmoid functions in the forecast models and the best performance result is obtained under the RBF kernel. The kernel parameters of the SVM model are obtained by an iterative process in order to reach minimum MSE at the training stage. In training of the model wind speed and wind direction data is used as the input while generated power is used as the output (Buhan and Çadırcı, 2015; Buhan et al.,2016).

The radial base kernel function (RBF) is defined as given in Equation 4.

$$K\left(z_j, z_i\right) = \exp\left(-\lambda \left\|z_i - z_i\right\|^2\right) \tag{4}$$

In this equation, z_j and z_i are the jth and i^{th} training samples, and λ is the kernel parameter with a positive value. Similar to ANN model, best grid selection for each of the multiple NWP input data is adopted by using the associated data of the nearest nine grid points to each WPP. The parameters of the SVM model are used in 48-hour ahead real-time wind-electric power forecasts.

Probabilistic Forecast Algorithm

Probabilistic forecasts, unlike point forecasts, can give a range or distribution as forecast output. As a result of the literature review carried out under the project, normal distribution is found to estimate the wind forecasts best. Therefore, in the process of creation of local probabilistic forecast, "*Gaussian*" models are used (Özkan et al., 2017). The most important advantage of this model is that, local effects can be reflected in the model with different coefficients. Local Gaussian model basically aims to find mean value and standard deviation of the distribution. For this purpose, a minimization problem for squares of error rates is solved. By using the point forecasts and boundary coefficients for 5-25-75-95 percentiles, for every plant, probabilistic power forecasts are generated in the system daily. For these forecasts, it is important for system planning to keep production value within confidence intervals.

User Interface Modules

In order for the system users to view power forecast results obtained from the models, user interface software is required. In order to fulfill these needs, three main graphical user interface modules have been developed within the scope of the project as follows:

- **Real-Time Monitoring and Querying Module:** This module is used for tracking the real-time production and query forecast. In addition, the SCADA data of the power plant and the measurements obtained from the wind mast can also be investigated in detail with this module.
- **Map-Based Monitoring and Forecast Module:** With this module, the locations of the power plants, connection statuses and the placement of the turbines are displayed dynamically on the map. A view of this module is presented in Figure 5. The sample screenshot belongs to the Turkish version of the software used in the center. There is also an English version of the software.
- **Web-Based Monitoring and Forecast Module:** With this module, the forecast data is presented to users from the web environment. Through this module, plant owners can query production and forecast data of their own wind farms for the date ranges they select.

Figure 5. Map-based monitoring and forecast user interface module

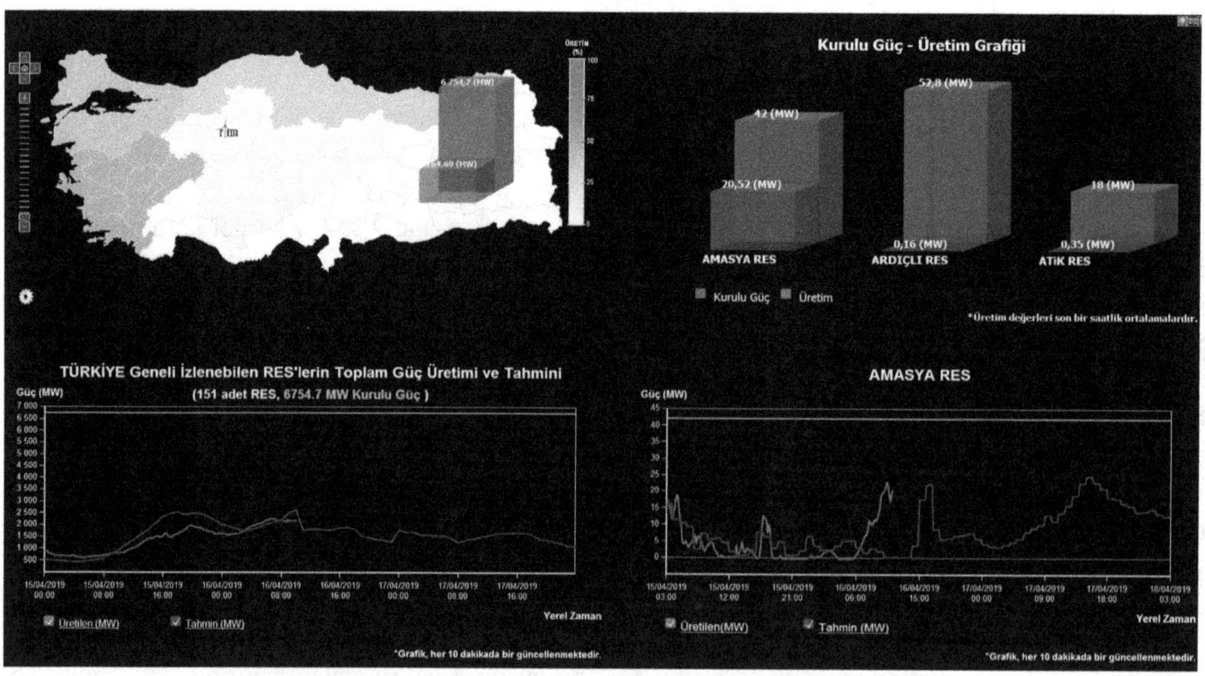

Experimental Forecast Results and Discussions

In this section, the performance of the power forecast methods developed within the scope of the RİTM project are examined. The Normalized Mean Absolute Error (NMAE) ratio, whose formula is given in Equation 5, is used as performance criteria (Kariniotakis et al., 2004). In this formula, y_i represents the power forecast at i^{th} hour, x_i is the corresponding real power at time i, C is the total installed capacity of the plant or country and N is the total number of data used in the calculation. All test results are obtained by using the power forecasts obtained in year 2018.

$$NMAE = \frac{\frac{\sum_{i=1}^{N} \left| y_i - x_i \right|}{N}}{C} \times 100\% \tag{5}$$

Very Short-Term Power Forecast Evaluation

The performance of very short-term power forecast algorithms in the RİTM system (AR, ARX) have been investigated for the country forecasts along with the persistence model.

According to the experimental results presented in Table 1, persistence model shows better performance for the time intervals less than 40 minutes in average (Buhan et al., 2013). Similarly, between 40 minutes and 2 hours, AR model performance is better than ARX model. For long time intervals, after 2 hours, adding the power forecasts of the day-ahead models as the exogenous input of the ARX model improves the forecast performances for this time interval. According to the results of the model performances, system operator can use AR model outputs for 0-2 hour planning and ARX model outputs for 2-6 hours grid management.

Table 1. Very short-term average power forecast errors for different time intervals for Turkey

Ahead Time (Minute)	Persistence NMAE (%)	AR NMAE (%)	ARX NMAE (%)
10	0.48	0.65	0.72
20	0.61	0.82	0.92
30	0.81	0.94	1.07
40	1.31	1.02	1.25
50	1.55	1.15	1.46
60	1.63	1.25	1.79
120	2.81	2.40	2.83
180	4.13	3.79	3.45
240	5.89	4.85	4.12
300	7.54	6.58	5.74
350	9.43	8.72	7.18

Day-Ahead Power Forecast Evaluation

The day-ahead power forecasts obtained from SHWIP, ANN and SVM models are combined with the combination algorithm and published as the final power forecasts (Terciyanli et al., 2014). The average annual error rates of these forecasts, also by seasonal, are presented in Table 2 for 10 pilot WPPs and the whole country. Installed capacities of the pilot WPPs vary between 100 MW and 240 MW. Due to agreement between plants their names are represented as WPP in the results. These 10 pilot WPPs are the biggest plants in the system in terms of installed power capacity. According to the results, the biggest error rates are obtained in winter and the lowest error rates obtained in summer season for all WPPs and Turkey. The main factor in creating these seasonal differences is the accuracy of weather forecasts. In the winter months, because weather events are more unstable, the accuracy of weather forecasts is declining, which directly affects the accuracy of power forecasts. On the other hand, because weather events are more stable in summer months, meteorological forecasts are more consistent at the same rate. This is the main factor in improving the performance of wind power forecasts in the summer.

Error distributions for WPP1 and Turkey are presented in Figure 6. The error rates for both centers according to this histogram plot are generally in the range of -10% and 10%, which indicates that the power estimates are stable and reliable for the system operator and plant owners. From this distribution graph, we can also see that the positive and negative bias rates in the forecasts are similar both for WPP1 and Turkey.

Probabilistic Power Forecast Evaluation

For testing the performance of the probabilistic forecasts we have used the Gaussian distribution metric. According to this distribution for a reliable model, at least 90% percentage of the real power must be in 5-95 percentile and 70% of the power must be in 25-75 confidence intervals. Probabilistic power forecast results for Turkey is presented in Table 3.

Table 2. Day-ahead average power forecast errors

Power Plant	Winter NMAE (%)	Spring NMAE (%)	Summer NMAE (%)	Fall NMAE (%)	1 Year NMAE (%)
WPP1	10.97	9.15	7.41	9.8	9.09
WPP2	13.7	11.88	10.09	10.49	11.35
WPP3	14.69	12.4	10.24	10.4	11.92
WPP4	12.17	11.91	9.15	10.56	10.81
WPP5	11.2	9.72	9.58	12.02	10.4
WPP6	13.22	11.63	10.32	11.60	11.56
WPP7	12.3	11.29	11.31	13.4	12.17
WPP8	11.88	10.72	7.78	10.77	10.2
WPP9	10.76	11.55	9.92	11.72	10.97
WPP10	12.52	12.79	9.14	11.25	11.34
TURKEY	4.52	4.65	3.49	3.91	4.16

Figure 6. Error distributions for WPP1 and for Turkey

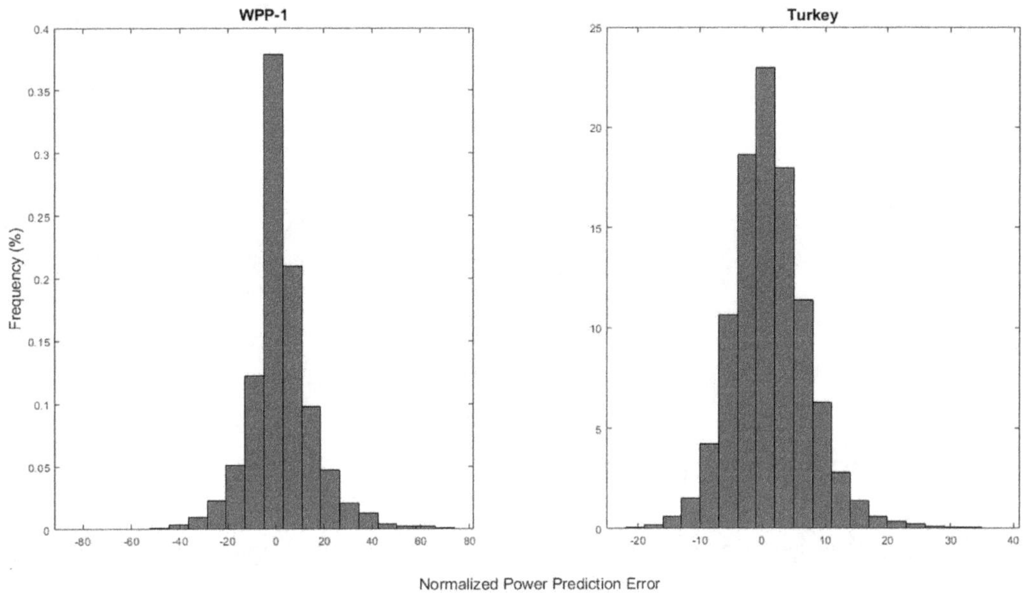

Table 3. Probabilistic power forecast results

Power Plant	5-95 Percentile (%)	25-75 Percentile (%)
WPP1	92.39	66.19
WPP2	81.21	55.34
WPP3	90.43	66.56
WPP4	87.51	72.43
WPP5	91.42	71.22
WPP6	92.71	73.94
WPP7	85.42	70.12
WPP8	89.84	68.63
WPP9	87.83	70.74
WPP10	92.93	72.73
TURKEY	89.61	69.40

According to results in Table 3, except for WPP2 in all of the WPPs and whole country nearly 90% of the real power values are in 5-95 confidence intervals and 70% of the real power in the 25-75 confidence intervals (Özkan et al., 2017). WPP2 plant site is larger than the other power plants and the distribution of turbines varies in this plant. This issue may be the reason the lower success rate of probabilistic power forecasts for WPP2. Additionally, a sample time series chart is presented in Figure 7 for Turkey. In this figure, the green line represents the real production data and gray intervals denote 25-75 percentile and 5-95 percentile interval forecasts.

Figure 7. Sample probabilistic wind power forecasts for Turkey

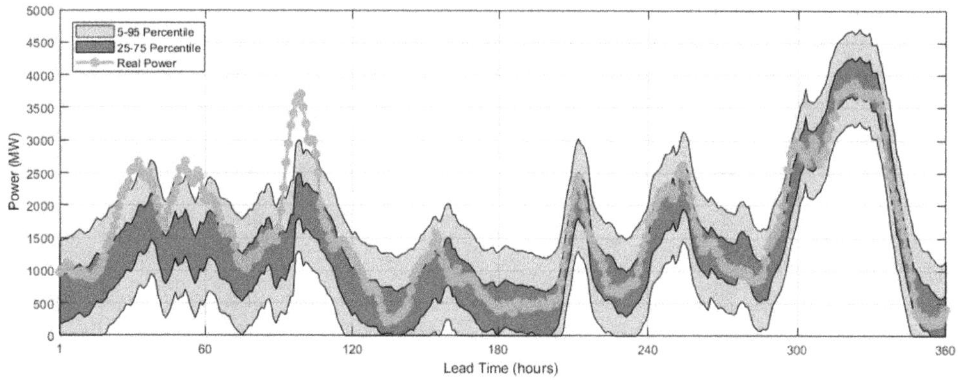

FUTURE RESEARCH DIRECTIONS

Future work based on the generic system proposal (REMFS) and its implementation (RİTM system) are listed below:

- Based on the generic architecture given in Figure 1, RİTM system can be extended with the addition of necessary hardware (sensors, analyzers, etc.) and software components to cover other renewable energy plants, such as solar and hydro-power plants. After this stage, a further significant extension will be to include optimization modules to the overall system to better operate some of these plants, particularly hydro-power plants.
- Forecasting methods of REMFS can be extended to cover other traditional machine learning methods. For instance, methods such as Adaptive Neuro-Fuzzy Inference System (ANFIS) are known to be applicable to renewable energy forecasting (Mohandes et al., 2011). Such learning methods can also be tested and included in the set of supervised learning services of REMFS.
- Deep learning methods such as Recurrent Neural Networks (RNNs), Convolutional Neural Networks (CNN) and their different variants are known to be successful in several application domains including computer vision and speech recognition. Recent studies show that such deep neural networks are also applicable to renewable energy forecasting (Hu et al., 2016; Yu et al., 2019). A plausible line of future work is to include deep learning methods in the forecast services of the overall generic system. After this objective is attained, better performance rates can be attained, and further comparisons with traditional machine learning methods can be performed.

CONCLUSION

Renewable energy forecast is an important research topic both for efficient operation and planning of the electrical grid by the grid operator organization and for economically viable operations in the energy market by the renewable plant owners. The aim of this chapter is to present the contributions of intelligent data analysis techniques to produce high-performance forecasts for renewables. Hence, a generic and centralized system architecture for monitoring and forecast of the renewable energy resources is presented first, and next, a genuine and large-scale application of the generic system for wind energy is described. This latter application currently collects data from more than 150 wind power plants in Turkey and forecasts are automatically produced for these plants which correspond to about 99% of all wind power plants in the country. Future research directions based on the presented content is also provided at the end of the chapter, which basically include: (1) the extension of the country-scale application to cover other renewable energy types such as solar, hydro-power, and biomass, (2) forecast experiments with other state-of-the art singleton methods such as those based on fuzzy logic and deep neural networks, in addition to ensemble methods combining different classifiers employed so far.

ACKNOWLEDGMENT

This work is conducted within the scope of RİTM project (with number 5122807), which is conducted by Energy Institute of TÜBİTAK Marmara Research Center. We would like thank all of the partners of the project, particularly to Renewable Energy General Directorate of Turkey (YEGM).

REFERENCES

Ahmad, A. S., Hassan, M. Y., Abdullah, M. P., Rahman, H. A., Hussin, F., Abdullah, H., & Saidur, R. (2014). A review on applications of ANN and SVM for building electrical energy consumption forecasting. *Renewable & Sustainable Energy Reviews*, *33*(1), 102–109. doi:10.1016/j.rser.2014.01.069

Blonbou, R. (2011). Very short-term wind power forecasting with neural networks and adaptive Bayesian learning. *Renewable Energy*, *36*(3), 1118–1124. doi:10.1016/j.renene.2010.08.026

Buhan, S., & Çadırcı, I. (2015). Multistage wind-electric power forecast by using a combination of advanced statistical methods. *IEEE Transactions on Industrial Informatics*, *11*(5), 1231–1242. doi:10.1109/TII.2015.2431642

Buhan, S., Özkazanç, Y., & Çadırcı, I. (2016). Wind pattern recognition and reference wind mast data correlations with NWP for improved wind-electric power forecasts. *IEEE Transactions on Industrial Informatics*, *12*(3), 991–1004. doi:10.1109/TII.2016.2543004

Buhan, S., Terciyanli, E., Özkan, M. B., Küçük, D., Haliloğlu, A. B., Demirci, T., ... Ünsal, H. (2013, May). Verification of a very short term wind power forecasting algorithm for Turkish transmission grid. *Proceedings of the 4th International Conference on Power Engineering, Energy and Electrical Drives* (pp. 1217-1221). IEEE. 10.1109/PowerEng.2013.6635786

Cheung, K., Wang, X., Chiu, B. C., Xiao, Y., & Rios-Zalapa, R. (2010, January). Generation dispatch in a smart grid environment. In *2010 Innovative Smart Grid Technologies (ISGT)* (pp. 1-6). IEEE.

Davò, F., Alessandrini, S., Sperati, S., Delle Monache, L., Airoldi, D., & Vespucci, M. T. (2016). Post-processing techniques and principal component analysis for regional wind power and solar irradiance forecasting. *Solar Energy*, *134*, 327–338. doi:10.1016/j.solener.2016.04.049

de la Torre, M., Juberias, G., Dominguez, T., & Rivas, R. (2012, July). The CECRE: Supervision and control of wind and solar photovoltaic generation in Spain. *Proceedings of the 2012 IEEE Power and Energy Society General Meeting* (pp. 1-6). IEEE.

Demirci, T., Kalaycıoglu, A., Küçük, D., Salor, Ö., Güder, M., Pakhuylu, S., ... Bilgen, S. (2011). Nationwide real-time monitoring system for electrical quantities and power quality of the electricity transmission system. *IET Generation, Transmission & Distribution*, *5*(5), 540–550. doi:10.1049/iet-gtd.2010.0483

Erinmez, I. A., Kappenman, J. G., & Radasky, W. A. (2002). Management of the geomagnetically induced current risks on the national grid company's electric power transmission system. *Journal of Atmospheric and Solar-Terrestrial Physics*, *64*(5-6), 743–756. doi:10.1016/S1364-6826(02)00036-6

Fleten, S. E., & Kristoffersen, T. K. (2008). Short-term hydropower production planning by stochastic programming. *Computers & Operations Research*, *35*(8), 2656–2671. doi:10.1016/j.cor.2006.12.022

Gellert, A., Florea, A., Fiore, U., Palmieri, F., & Zanetti, P. (2019). A study on forecasting electricity production and consumption in smart cities and factories. *International Journal of Information Management*.

Ghahramani, A., Castro, G., Karvigh, S. A., & Becerik-Gerber, B. (2018). Towards unsupervised learning of thermal comfort using infrared thermography. *Applied Energy*, *211*, 41–49.

Giebel, G., Brownsword, R., Kariniotakis, G., Denhard, M., & Draxl, C. (2011). The state-of-the-art in short-term prediction of wind power: A literature overview. ANEMOS.

Golestaneh, F., Pinson, P., & Gooi, H. B. (2016). Very short-term nonparametric probabilistic forecasting of renewable energy generation—With application to solar energy. *IEEE Transactions on Power Systems*, *31*(5), 3850–3863. doi:10.1109/TPWRS.2015.2502423

Gungor, V. C., Sahin, D., Kocak, T., Ergut, S., Buccella, C., Cecati, C., & Hancke, G. P. (2011). Smart grid technologies: Communication technologies and standards. *IEEE Transactions on Industrial Informatics*, *7*(4), 529–539. doi:10.1109/TII.2011.2166794

Hu, Q., Zhang, R., & Zhou, Y. (2016). Transfer learning for short-term wind speed prediction with deep neural networks. *Renewable Energy*, *85*, 83–95. doi:10.1016/j.renene.2015.06.034

Kariniotakis, G., Martí, I., Casas, D., Pinson, P., Nielsen, T. S., Madsen, H., . . . Brownsword, R. (2004, November). What performance can be expected by short-term wind power prediction models depending on site characteristics. *Proceedings of the European Wind Energy Conference EWEC* (pp. 22-25). Academic Press.

Lei, Y., Jia, F., Lin, J., Xing, S., & Ding, S. X. (2016). An intelligent fault diagnosis method using unsupervised feature learning towards mechanical big data. *IEEE Transactions on Industrial Electronics*, *63*(5), 3137–3147. doi:10.1109/TIE.2016.2519325

Lobo, M. G., & Sanchez, I. (2012). Regional wind power forecasting based on smoothing techniques, with application to the Spanish peninsular system. *IEEE Transactions on Power Systems*, *27*(4), 1990–1997. doi:10.1109/TPWRS.2012.2189418

Mahoney, W. P., Parks, K., Wiener, G., Liu, Y., Myers, W. L., Sun, J., ... Haupt, S. E. (2012). A wind power forecasting system to optimize grid integration. *IEEE Transactions on Sustainable Energy*, *3*(4), 670–682. doi:10.1109/TSTE.2012.2201758

Miyato, T., Maeda, S. I., Ishii, S., & Koyama, M. (2018). Virtual adversarial training: A regularization method for supervised and semi-supervised learning. *IEEE Transactions on Pattern Analysis and Machine Intelligence*. PMID:30040630

Mocanu, E., Nguyen, P. H., Gibescu, M., & Kling, W. L. (2016). Deep learning for estimating building energy consumption. Sustainable Energy. *Grids and Networks*, *6*, 91–99.

Mohandes, M., Rehman, S., & Rahman, S. M. (2011). Estimation of wind speed profile using adaptive neuro-fuzzy inference system (ANFIS). *Applied Energy*, *88*(11), 4024–4032. doi:10.1016/j.apenergy.2011.04.015

Monteiro, C., Bessa, R., Miranda, V., Botterud, A., Wang, J., & Conzelmann, G. (2009). *Wind power forecasting: State-of-the-art 2009*. Argonne National Lab.

Nielsen, H. A., Madsen, H., & Nielsen, T. S. (2006). Using quantile regression to extend an existing wind power forecasting system with probabilistic forecasts. Wind Energy. *An International Journal for Progress and Applications in Wind Power Conversion Technology*, *9*(12), 95–108.

Oquab, M., Bottou, L., Laptev, I., & Sivic, J. (2015). Is object localization for free-weakly-supervised learning with convolutional neural networks. *Proceedings of the IEEE Conference on Computer Vision and Pattern Recognition* (pp. 685-694). IEEE. 10.1109/CVPR.2015.7298668

Osório, G. J., Matias, J. C. O., & Catalão, J. P. S. (2015). Short-term wind power forecasting using adaptive neuro-fuzzy inference system combined with evolutionary particle swarm optimization, wavelet transform and mutual information. *Renewable Energy*, *75*, 301–307. doi:10.1016/j.renene.2014.09.058

Owens, M. J., Riley, P., & Horbury, T. S. (2017). Probabilistic solar wind and geomagnetic forecasting using an analogue ensemble or "similar day" approach. *Solar Physics*, *292*(5), 69. doi:10.100711207-017-1090-7

Özkan, M. B., Guvengir, U., Küçük, D., Secen, A. U., Buhan, S., Demirci, T., ... Karagoz, P. (2017, September). Probabilistic Wind Power Forecasting by Using Quantile Regression Analysis. *Proceedings of the International Workshop on Data Analytics for Renewable Energy Integration* (pp. 72-82). Springer. 10.1007/978-3-319-71643-5_7

Özkan, M. B., & Karagoz, P. (2015). A novel wind power forecast model: Statistical hybrid wind power forecast technique (SHWIP). *IEEE Transactions on Industrial Informatics*, *11*(2), 375–387.

Özkan, M. B., Terciyanli, E., Küçük, D., Buhan, S., Demirci, T., Yıldız, C., & Günindi, M. (2013). Verification of a real-time wind power monitoring and forecast system for Turkey. *Proceedings of the IET Renewable Power Generation Conference*, Beijing, China. Academic Press.

Rodriguez, J. M., Alonso, O., Duvison, M., & Domingez, T. (2008, July). The integration of renewable energy and the system operation: The Special Regime Control Centre (CECRE) in Spain. *Proceedings of the 2008 IEEE Power and Energy Society General Meeting-Conversion and Delivery of Electrical Energy in the 21st Century* (pp. 1-6). IEEE.

Terciyanli, E., Demirci, T., Küçük, D., Buhan, S., Özkan, M. B., Köksoy, C. E., ... Günindi, M. (2013, May). The architecture of a large-scale wind power monitoring and forecast system. *Proceedings of the 4th International Conference on Power Engineering, Energy and Electrical Drives* (pp. 1162-1167). IEEE. 10.1109/PowerEng.2013.6635775

Terciyanli, E., Demirci, T., Küçük, D., Sarac, M., Cadirci, I., & Ermis, M. (2014). Enhanced nationwide wind-electric power monitoring and forecast system. *IEEE Transactions on Industrial Informatics*, *10*(2), 1171–1184. doi:10.1109/TII.2013.2294157

Wang, S., Minku, L. L., & Yao, X. (2015). Resampling-based ensemble methods for online class imbalance learning. *IEEE Transactions on Knowledge and Data Engineering*, *27*(5), 1356–1368. doi:10.1109/TKDE.2014.2345380

Yu, R., Gao, J., Yu, M., Lu, W., Xu, T., Zhao, M., ... Zhang, Z. (2019). LSTM-EFG for wind power forecasting based on sequential correlation features. *Future Generation Computer Systems*, *93*, 33–42. doi:10.1016/j.future.2018.09.054

ADDITIONAL READING

Collotta, M., & Pau, G. (2017). An innovative approach for forecasting of energy requirements to improve a smart home management system based on BLE. *IEEE Transactions on Green Communications and Networking*, *1*(1), 112–120. doi:10.1109/TGCN.2017.2671407

Du, P., Wang, J., Yang, W., & Niu, T. (2018). Multi-step ahead forecasting in electrical power system using a hybrid forecasting system. *Renewable Energy*, *122*, 533–550. doi:10.1016/j.renene.2018.01.113

Hosseini, S. M., Aslani, A., Naaranoja, M., & Hafeznia, H. (2017). Analysis of Energy System in Sweden Based on Time series Forecasting and Regression Analysis. *International Journal of Energy Optimization and Engineering*, *6*(3), 97–113. doi:10.4018/IJEOE.2017070105

Olivares, D. E., Lara, J. D., Cañizares, C. A., & Kazerani, M. (2015). Stochastic-predictive energy management system for isolated microgrids. *IEEE Transactions on Smart Grid*, *6*(6), 2681–2693. doi:10.1109/TSG.2015.2469631

Vergara, G., Carrasco, J. J., Martínez-Gómez, J., Domínguez, M., Gámez, J. A., & Soria-Olivas, E. (2019). Global and Local Clustering-Based Regression Models to Forecast Power Consumption in Buildings. In Architecture and Design: Breakthroughs in Research and Practice (pp. 506-536). Hershey, PA: IGI Global. doi:10.4018/978-1-5225-7314-2.ch018

KEY TERMS AND DEFINITIONS

SCADA: Supervisory Control and Data Acquisition System.
ANN: Artificial Neural Networks.
SVM: Support Vector Machines.
WRF: Weather Research and Forecasting Model.
ALADIN: Air Limit Adaptation Dynamic Development International.
ARMA: Autoregressive Moving Average.
ARMAX: Autoregressive Moving Average with Exogenous.
GFS: Global Forecast System.
ECMWF: European Center for Medium range Weather Forecasting.

ENDNOTES

1 http://www.ritm.gov.tr/root/index_eng.php
2 https://www.ecmwf.int/
3 https://www.noaa.gov/
4 http://www.umr-cnrm.fr/aladin/

Chapter 4
Optimization of Sectionalization Parameters of Distributive Electric Networks

Kozyrskyi Volodymyr
National University of Life and Environmental Sciences of Ukraine, Ukraine

Gai Oleksandr
National University of Life and Environmental Sciences of Ukraine, Ukraine

Sinyavsky Oleksandr
National University of Life and Environmental Sciences of Ukraine, Ukraine

Vitaliy Savchenko
National University of Life and Environmental Sciences of Ukraine, Ukraine

Makarevich Svitlana
https://orcid.org/0000-0002-9601-5156
National University of Life and Environmental Sciences of Ukraine, Ukraine

ABSTRACT

The reliability of electrical consumers depends on the presence of sectioning devices in electrical network distribution. Determining the number and installation sites of sectioning machines is an important optimization problem, based on a comparison of the cost of installing devices and reducing the damage to consumers from an undersupply of electricity. A mathematical model is proposed to determine the cost-effectiveness of installing partitioning switching devices for networks with a means of increasing reliability. This allows for the choice when installing switching devices not only meets an economic criterion (in the form of reducing damage to the power supply system as a whole), but also takes into account the change in the duration of downtime of local consumers.

DOI: 10.4018/978-1-7998-1216-6.ch004

Copyright © 2020, IGI Global. Copying or distributing in print or electronic forms without written permission of IGI Global is prohibited.

INTRODUCTION

According to the concepts of optimization (Leshchinskaya, 2003; Ishizaka, & Nemery, 2013; Brown, 2009) and the implementation of measures to improve the operation of the systems providing electric energy to the consumer, the research should be based on a comprehensive study of the object and its characteristic processes, i.e. on a systems approach. The following postulates are the basis of the systems approach: 1) the system should be considered as a whole, and not as a simple set of its components; 2) the optimization goals should be based on clearly defined decision objectives; 3) the optimization model must take into account all the defining properties of the system, which are necessary for drawing up a fairly complete picture of the system behavior.

From the point of view of the system approach, according to the above postulates, it is necessary to decompose the system for providing electrical energy to the consumer into its components in order to carry out optimal planning and management of its further development.

BACKGROUND

The system for providing electrical energy to the consumer consists of the following subsystems: power supply subsystems, representing a set of electrical installations designed to provide consumers with electrical energy (power lines, transformer substations, installed capacities of consumers' electrical installations); locations of electrical installations, which determines the nature of weather conditions and the landscape of the territory; measures aimed at improving the reliability of power supply to consumers. One of the main tasks of the system for providing electrical energy to the consumer is to supply the consumer with the goods in the required volume and of a given quality, which is done by increasing the reliability of the power supply.

A large number of scientific papers (Prakhovnik, Popov, & Nakhodov, 2010; Zorin, Tislenko, Kleppel', & Adler, 1984; Begovic, & Lambert, 2002; Tislenko, 1991; Pruss, & Tislenko, 1989) are devoted to the analysis of the effectiveness of using one or another method of improving reliability. In modern scientific research (Buynyy, Podolnyi, & Zorin, 2004; Javadian, & Massaeli, 2011), it has been shown that one of the most effective methods for increasing the reliability of power supply for an agricultural consumer is network partitioning using circuit breakers, in particular reclosers, with simultaneous organization of localized and selective operation of relay protection.

We will use the methodology of the systems approach to describe the object of study and identify factors that have a significant impact on the efficiency of partitioning.

According to the principle of operation of the distribution network, the following factors will affect the total annual outage: network design, power line length, which, due to its length, causes about 80% of damage to the distribution network 10 kV (Tislenko, Kholmskiy, & Gurbich, 1975); the number of transformer substations (the second object in terms of the number of damage to distribution networks); weather conditions characteristic of a given territory (there are a number of studies (Makoklyuyev, Pavlikov, Vladimirov, & Fefelova, 2003; Vaccaro, Popov, Villacci, & Terzija, 2011) showing the existence of a dependence of the amount and time of damage on weather conditions characteristic of a given territory — temperature, precipitation, humidity, thunderstorm intensity); landscape characteristics that affect the speed of restoration of power supply (Pruss, & Tislenko, 1989) due to the varying degree of difficulty in moving when searching for a place of damage; the degree of depreciation of fixed assets, which determines the change in the probability of damage to the latter, as well as the presence of already established means of increasing reliability.

The magnitude of the total annual damage is influenced by the nature and capacity of the connected load (Bernardon, Sperandio, et al, 2004), due to changes in the under-supply of electrical energy and the degree of severity of possible consequences due to interruptions in the power supply. The efficiency of the installation is also determined by the economic criterion in the form of the installation price, the payback period and the amount of depreciation.

Thus, in order to improve the reliability of the distribution network of the rural region, it is advisable to use partitioning by automatic switching devices. When compiling a mathematical model for finding the optimal location and installation scheme of a sectioning switching device, it is required to evaluate the impact of a possible change in the source data on the optimization result.

MAIN FOCUS OF THE CHAPTER

Mathematical Model of Switching Devices Location With Radial Distribution Network

It is known (Dehghani, & Dashti, 2011) that the optimization of reliability as applied to the power grids of rural regions consists in the choice of the composition, volume, location and sequence of inputting the means of improving reliability, the use of which in electrical networks allows you to change the characteristics of network reliability and power supply to consumers. At the same time, optimization of reliability comes down to three tasks: ensuring the maximum level of reliability in the presence of a limited resource; ensuring minimal resource consumption at a given level of reliability; ensuring the level of reliability corresponding to the minimum costs associated with unreliability (including resource costs and damage from power supply disruptions). In all three optimization problems, reliability is understood as the numerical value of the adopted indicators characterizing reliability. The resource is specified in the form of the physical volume of the means for improving reliability or capital investments (reduced costs) in the means for increasing reliability.

Let us form a mathematical model (Brown, 2009) of placing a means of improving reliability, in particular, installing sectional switching devices, in order to optimize the reliability of distribution networks in rural regions. Based on the fundamental principles of this definition, we construct a mathematical model. In this case, we base the methodology described in (Javadian, & Haghifam, 2010), taking into account the characteristics of a radial line, this is a transmission line, consumers of which receive power from one supply center.

When creating a mathematical model that describes the processes in distribution networks, it is necessary to take into account a number of features that distinguish these networks from higher voltage networks. These features include (Pruss, & Tislenko, 1989; Dehghani, & Dashti, 2011) a lower degree of redundancy and automation compared to networks of other purposes. At the same time, if as a complex indicator of reliability we apply the mathematical expectation of the total undersupply of electricity, which is equal to the mathematical expectation of the undersupply caused by the downtime of one, two, three, etc. elements, which is typical for higher voltage networks, since the latter have all the main elements reserved, then for rural distribution grids the expected total electricity undershoot will be approximately equal to the expected electricity undershoot caused by a single element being idle, namely, according to statistical information, in 97 out of 100 cases of failure, damage is attributable to elements belonging to the distribution network (transmission line with components taken into account (structure round wire, insulators, material supports), switching devices installed in the lines, transformer substations). The low level of automation leads to the fact that all the basic operations in the network are carried out by operational-field teams.

We will construct a mathematical model for the placement of partitioning switching devices in distribution networks of rural regions, taking into account the above features. When composing the model, we accept the fact that the process of functioning of the distribution network obeys the laws (Zorin, Tislenko, Kleppel, & Adler, 1984), which determines the flow of events that satisfies the properties of stationarity, lack of consequences and ordinary-ness.

It is accepted (Zorin, Tislenko, Kleppel, & Adler, 1984) that the properties of the simplest flow of failures are possessed by power supply system elements that are in the normal period of work and have sufficient power reserve. With the simplest flow of failures, the flow parameter and failure rate do not depend on time and are equal to each other: $\omega(t) = \lambda(t) = \omega = \lambda = const$, andom failures are distributed according to the Poisson law (a = λt), and the intervals between successive failures are distributed exponentially. It is known that partitioning is effective if the effect of partitioning. *Eff* more than the annual reduced costs for installation and maintenance of partitioning points *C* (Javadian, & Haghifam, 2010):

$$Eff > C \tag{1}$$

In the general case, installing a device on the network that optimizes network operation results in several types of effects (Pruss, & Tislenko, 1989). The installation of a partitioning switching device is characterized by the following types of effects: 1) the main effect achieved by using a switching device for its intended purpose is to reduce the value of the disconnected load; 2) the additional effect is achieved through the use of a disconnector for a non-intended purpose, for example, as a damage detection device; 3) the negative effect is due to possible damage to the switching device itself, which leads to an emergency outage of the distribution network or possible non-selective operations.

The feasibility of the introduction of the switching apparatus will be evaluated by the main effect. To do this, we will develop a mathematical model that allows us to evaluate the effectiveness of a particular scheme and the location of the switching apparatus.

According to the methods (Pruss, & Tislenko, 1989; Bertling, 2002; Lagland, & Kauhaniemi, 2007), the damage (Inj) from the undersupply of electric energy is determined by the formula:

$$Inj = P_{\Sigma} t y \tag{2}$$

P_{Σ} –total power of the disconnected consumers, kW;
y – specific damage from interruptions of power supply, \$/(kW hour);
t –off time, hour.

Off time (Bertling, 2002) is defined as:

$$t = \alpha L \tag{3}$$

α –specific annual duration of shutdown of distribution lines, hours/(year km);
L –length of the disconnected line, km.

The specific annual duration of disabling the distribution line (Lagland, & Kauhaniemi, 2007), is determined by the formula:

$$\alpha = \omega_0 \tau + \lambda \nu_0 \eta \tag{4}$$

ω_0, v_0 – the frequency of emergency and planned outages, respectively;
τ, η –duration of emergency and planned outages, respectively;
λ –coefficient taking into account the lesser severity of planned outages.

The total power of disconnected consumers (Javadian, & Haghifam, 2010) is determined by:

$$P_\Sigma = \sum_i k_{0i} S_i \cos \phi_i \tag{5}$$

k_0 –coefficient of simultaneity;
S_i –installed capacity of electrical installations of consumers, kVA;
$cos\varphi$ – power factor that determines the nature of the load connected to the transformer substation.

Used in the mathematical model, the indicators of the reliability of power supply to consumers are determined for the system (Tislenko, 1991; Pruss, & Tislenko, 1989; Dehghani, & Dashti, 2011) consisting of two subsystems - the power supply system and the consumer

The damage from emergency situations for the year for the line without means of increasing reliability is:

$$Inj_{no\ cd} = P_\Sigma y\ \alpha L \tag{6}$$

In the case of placing the switching device at the point of automatic partitioning (*PAP*) in the scheme of the distribution network (Fig. 1), the damage from undersupply of electrical energy will be determined by the formula (7).

When forming the expression (7), the following assumptions were used:

- If damage to the distribution network occurred before the installation of the switching device, then all consumers connected to this line will be disconnected in an emergency, and the location of the damage search will be limited to the area l_1;

Figure 1. Plot distribution network

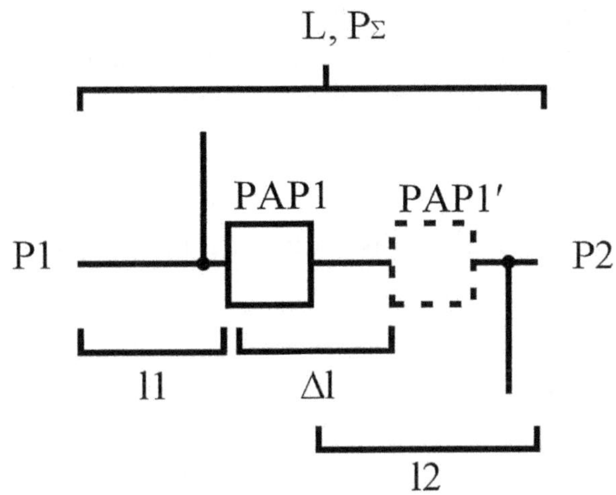

- If the damage occurred beyond the installation point of the Switching apparatus, then the consumer disconnection zone will be determined by the power P2 of the consumers beyond the switching point of the switching apparatus and the length l2.

$$Inj_{ce} = P_\Sigma y\alpha' l_1 + P_2 y\alpha' l_2 \quad (7)$$

Accordingly, the effect of installing of the switching device will be as follows:

$$Eff = Inj_{no\ cd} - Inj_{cd} = y\alpha(P_\Sigma(L - l_1) - P_2 l_2) \quad (8)$$

Prove that the placement of point of automatic partitioning in the open circuit of the distribution network is the place in the immediate vicinity of the right of the branch. Assume two installation locations are possible (Figure 1), PAP 1 and PAP 1'.

At the same time we get:

$$\begin{aligned} &Inj_{PAP\,1} = P_\Sigma y\alpha l_1 + (P_\Sigma - P_1)y\alpha \cdot l_2 = y\alpha[P_\Sigma(l_1 + l_2) - P_1 l_2], \\ &Inj_{PAP1'} = P_\Sigma y\alpha(l_1 + \Delta l) + (P_\Sigma - P_1)y\alpha \cdot (l_2 - \Delta l) = y\alpha \cdot [P_\Sigma l_1 + l_2) - P_1 l_2 - \Delta l)], \\ &Inj_{PAP1} < Inj_{PAP1'} \end{aligned} \quad (9)$$

Thus, the damage caused by the lack of electricity to consumers in the first case will be less than in the second, and accordingly the efficiency of the first measure will be higher:

$$Eff_{PAP\,1} > Eff_{PAP\,1'} \quad (10)$$

Consider a sequential partitioning scheme (Figure 2). In this, it is necessary to check the effectiveness of the PAP2 installation, if PAP1 is already installed.

Damage when installing the second switching device in a sequential scheme will be:

$$Inj_{PAP2} = P_{\Sigma} y\alpha l_1 + \left(P_{\Sigma} - P_1\right) y\alpha' l_2 + \left(P_{\Sigma} - P_1 - P_2\right) y\alpha' l_3 \quad (11)$$

Summarize the resulting formula, in the case of the possibility of installing the *i*-switching device in a sequential scheme:

$$Inj_{PAP\ i} = \sum_{i} \left(P_{\Sigma} - \sum_{0}^{i-1} P_n^{posl} \right) y\alpha' l_i, \ P_0 = 0 \quad (12)$$

P_n^{posl} –total zone power limited by the switching device from two sides;
l_i –the total length of the zone bounded by the switching device.

We obtain mathematical expressions for determining the damage and the efficiency of the installation of the switching apparatus for various partitioning schemes.

The effectiveness of the installation of the *i*-th switching device in a sequential scheme is determined from the expression:

$$Eff_{PAP\,i} = Inj_{no\ \mathrm{cd}} - Inj_{PAP\,i} > \sum_{i-1} Z_{PAP\ n} \quad (13)$$

Damage when installing PAP3 in parallel scheme:

$$Inj_{PAP\ 3} = P_{\Sigma} y\alpha (L - l_4) + P_4 y\alpha l_4 \quad (14)$$

Figure 2. Plot distribution network

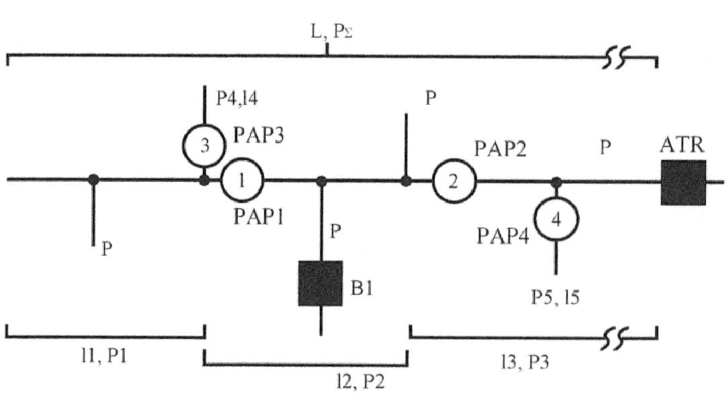

Damage when installing PAP4 in parallel scheme:

$$Inj_{PAP\,4} = P_{\Sigma} y\alpha \left(L - l_4 - l_5 \right) + P_4 y\alpha l_4 + P_5 y_p \alpha l_5 \quad (15)$$

Respectively damage at installation n-th PAP in parallel scheme:

$$Inj_{PAP\,n} = P_{\Sigma} y\alpha \left(L - \sum_{1}^{n} l_n^{paral} \right) + \sum_{1}^{n} P_i^{paral} y\alpha l_i^{paral} \quad (16)$$

P_i^{paral} –active power consumers on branch;
l_i^{paral} –branch length

The efficiency of the installation of the i-th switching apparatus in parallel is determined by the expression (13).

Consider a mixed circuit arrangement of switching apparatus. As was shown in (Kozyrskyi, & Guy, 2004), at first, all switching devices are installed according to a parallel scheme, for which more costs are necessary, and then taking into account the already installed switching devices, the switching device is arranged according to a sequential layout.

The damage in case of an accident in the line in the case of installation of PAP1 is determined by the formula (for the case when PAP3 and PAP4 are installed):

The efficiency of the installation of the i-th switching apparatus in parallel is determined by the expression (13).

Consider a mixed circuit arrangement of switching apparatus. As was shown in (Kozyrskyi, & Guy, 2004), at first, all switching devices are installed according to a parallel scheme, for which more costs are necessary, and then taking into account the already installed switching devices, the switching device is arranged according to a sequential layout.

The damage in case of an accident in the line in the case of installation of PAP1 is determined by the formula (for the case when PAP 3 and PAP 4 are installed):

$$Inj_{PAP\,1} = P_{\Sigma} y\alpha (l_1 - l_4) + P_4 y\alpha l_4 + P_5 y\alpha l_5 + (P_{\Sigma} - P_1) y\alpha (L - l_1 - l_5). \quad (17)$$

A similar expression for calculating damage in case of installation of PAP2 (when PAP1, PAP3, PAP4 are already installed):

$$Inj_{PAP\,2} = P_{\Sigma} y\alpha (l_1 - l_4) + P_4 y\alpha l_4 + P_5 y\alpha l_5 + (P_{\Sigma} - P_1) y\alpha l_5 + (P_{\Sigma} - P_1 - P_2) y\alpha (l_3 - l_5). \quad (18)$$

Thus, the generalized damage formula for the arrangement of a switching device according to a mixed scheme will be:

$$Inj_{PAP\,i} = \sum_{i=1}^{a\_paral} P_i^{paral} y\alpha\, l_i^{paral} + \sum_{j=1}^{a\_seq+1} \left(l_j^{seq} - \sum_{k=1}^{m} l_k^{paral} \right) \left(P_\Sigma - \sum_{n=0}^{a\_seq} P_n \right) y\alpha \tag{19}$$

l_i^{paral}, l_j^{seq} –the length of the zone, respectively, parallel and sequential partitioning;

P_i^{paral}, P_j^{seq} –power of the zone of parallel and sequential partitioning;

a_paral, a_seq – the number of devices included in the parallel and serial partitioning schemes, taking into account the checked;

m – the number of parallel partitioning devices that are included in the zone of the switching device included in a serial circuit.

The efficiency of the installation of the i-oro switching device according to a mixed scheme is determined by the formula (13).

Using the approach described above to determine the effect of installing additional partitioning switching devices, we will determine the optimal locations and layout of switching devices for the distribution network of rural regions shown in Figure 3. We will search for the optimal solution, based on the method of mathematical programming with Boolean variables.

We write for the network diagram shown in Figure 3, the objective function that determines the efficiency of the installation of partitioning switching devices:

$$f := Eff_1 x_1 + Eff_2 x_2 + \ldots + Eff_{15} x_{15} + Eff_{16} x_{16} \tag{20}$$

x_i –the indicator that determines the installation location of the partitioning switching device, which is a Boolean variable;

Eff_i –coefficient determining the effect of the installation of a switching device, taking into account the earlier installed switching devices.

To solve problems of integer programming with Boolean variables (Taha, 2017), in addition to the objective function, it is necessary to define a zone of feasible solutions, in the form of constraints.

Figure 3. Scheme of the distribution network

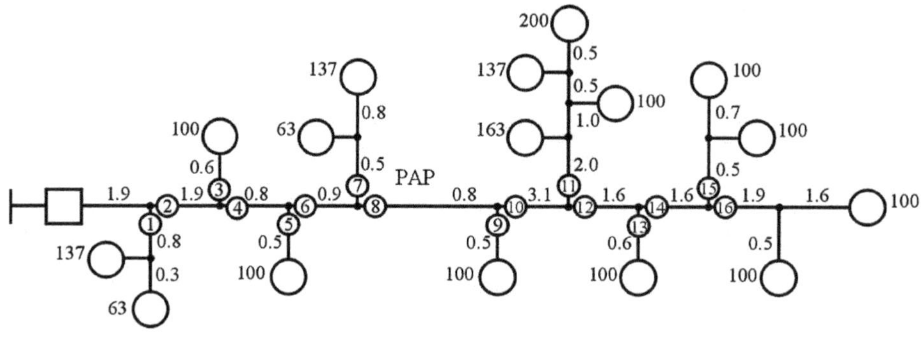

In our case, the following restrictions apply:

1. restriction on the range of decisions made:

$x_i,\ where\ i = (1 \div 16)$–integers, Boolean variables.;

2. economic constraint based on the condition of partitioning efficiency of the rural distribution network, from which it follows that partitioning is effective if the total partitioning effect is greater than the total annual reduced costs:

$$Eff_1 x_1 + Eff_2 x_2 + \ldots + Eff_{15} x_{15} + Eff_{16} x_{16} \geq Z(x_1 + x_2 + \ldots + x_{16}) \tag{21}$$

Z -annual reduced costs for the installation of a partitioned switching device;

3. the limit on the number of installed spacecraft needed to eliminate the interdependence of the coefficients *Eff*. According to (Kozyrskyi, & Guy, 2004), the possible number of installed switching devices ranges from zero to three, therefore:

$$\sum\nolimits_1^{16} x_i = (1 \div 3) \tag{22}$$

Based on the last limitation, the task of determining the optimal location of the switching device is divided into three subtasks, for which, respectively, the number of switching devices is one, two or three. Among the options considered, choose the one that is characterized by the maximum effect, but it is within the scope of the restrictions.

Thus, the mathematical formulation of the problem takes the form:

maximize

$$f = Eff_1 x_1 + Eff_2 x_2 + \ldots + Eff_{15} x_{15} + Eff_{16} x_{16} \tag{23}$$

under restrictions

$$\begin{cases} Eff_1 x_1 + Eff_2 x_2 + \ldots + Eff_{15} x_{15} + Eff_{16} x_{16} - Z(x_1 + x_2 + \ldots + x_{16}) \geq 0, \\ \sum\nolimits_1^{16} x_i = (1 \div 3), \\ 0 \leq x_i \leq 1, \\ x_i \in N, N - \text{set of natural numbers.} \end{cases} \tag{24}$$

To determine the partitioning efficiency factors (Eff_i) and the subsequent optimization of the number and locations of the switching apparatus, an algorithm is used, which is based on the division of a common task into three independent, particular tasks. Each of the tasks has a different priority, which is determined by the possible number of installed switching devices, and, accordingly, by the total effect of partitioning. In the case of the possibility of obtaining a solution to a problem with a higher priority, the solution of problems with a lower priority is omitted. This is possible due to the evidence obtained earlier, according to which for an open distribution network, the total effect of installing several switching devices does not depend on the place of the beginning of optimization calculations (or on the main switch on the line, or on the end section of the trunk) the communicative law of addition is fulfilled.

Let us give an example of finding the optimal number and layout of the switching apparatus for the power supply system shown in Fig. 2. In the case of testing the possibility and subsequent search for the installation locations of three switching devices, the optimization task will take the following form:

maximize

$$\begin{aligned} f = {} & 678.0x_1 + 769.9x_2 + 674.3x_3 + 798.3x_4 + 672.7x_5 + 825.8x_6 + 715.6x_7 + \\ & +859.0x_8 + 699.0x_9 + 853.7x_{10} + 859.0x_{11} + 859.0x_{12} + 705.9x_{13} + \\ & +826.6x_{14} + 742.365x_{15} + 777.2x_{16} \end{aligned} \tag{25}$$

under restriction

$$\begin{cases} 678.0x_1 + 769.9x_2 + 674.3x_3 + 798.3x_4 + 672.7x_5 + 825.8x_6 + 715.6x_7 + \\ +859.0x_8 + 699.0x_9 + 853.7x_{10} + 859.0x_{11} + 859.0x_{12} + 705.9x_{13} + \\ +826.6x_{14} + 742.365x_{15} + 777.2x_{16} - 1201.4(x_1 + x_2 + \ldots + x_{16}) \geq 0, \\ \sum_1^{16} x_i = 3, \\ 0 \leq x_i \leq 1, \\ x_i \in N, N - \text{set of natural numbers} \end{cases} \tag{26}$$

The result of the search for optimal partitioning places of the distribution network of rural regions, which consists in maximizing the objective function (25) with the condition of fulfilling the restrictions (26), concludes that it is not advisable to install three sectional switching devices for this type of power supply circuit Fig.4. The search for a solution to the problem of mathematical programming with Boolean variables is based on the recommendations given in (Chowdhury, & Koval, 2009).

In the case of testing the possibility and subsequent search for the installation locations of two switching devices for the circuit shown in Fig. 3, the optimization problem will take the following form:

maximize

$$\begin{aligned} f = {} & 910.6x_1 + 905.8x_2 + 800.7x_3 + 905.8x_4 + 795.2x_5 + 947.1x_6 + 919.2x_7 + \\ & +996.9x_8 + 749.0x_9 + 989.0x_{10} + 978.7x_{11} + 996.9x_{12} + 749.0x_{13} + \\ & +948.3x_{14} + 803.7x_{15} + 874.2x_{16} \end{aligned} \tag{27}$$

under restrictions

$$\begin{cases} 910.6x_1 + 905.8x_2 + 800.7x_3 + 905.8x_4 + 795.2x_5 + 947.1x_6 + 919.2x_7 + \\ +996.9x_8 + 749.0x_9 + 989.0x_{10} + 978.7x_{11} + 996.9x_{12} + 749.0x_{13} + \\ +948.3x_{14} + 803.7x_{15} + 874.2x_{16} - 1201.4(x_1 + x_2 + ... + x_{16}) \geq 0, \\ \sum_1^{16} x_i = 2, \\ 0 \leq x_i \leq 1, \\ x_i \in N, N - \text{set of natural numbers.} \end{cases} \tag{28}$$

The result of the solution of the task is the conclusion about the inexpediency of installing two switching devices.

In the case of checking the possibility and subsequent search for the installation locations of one switching device for the circuit shown in fig. 2, the optimization problem will take the following form:

maximize

$$f = 213.8x_1 + 563.8x_2 + 123.9x_3 + 754.5x_4 + 103.3x_5 + 942.8x_6 + 252.7x_7 + \\ +1253.9x_8 + 103.3x_9 + 1343.8x_{10} + 583.2x_{11} + 1374.2x_{12} + 123.9x_{13} + \\ +1105.7x_{14} + 233.3x_{15} + 719.3x_{16} \tag{29}$$

under restrictions

$$\begin{cases} 213.8x_1 + 563.8x_2 + 123.9x_3 + 754.5x_4 + 103.3x_5 + 942.8x_6 + 252.7x_7 + \\ +1253.9x_8 + 103.3x_9 + 1343.8x_{10} + 583.2x_{11} + 1374.2x_{12} + 123.9x_{13} + \\ +1105.7x_{14} + 233.3x_{15} + 719.3x_{16} - 1201.4(x_1 + x_2 + ... + x_{16}) \geq 0, \\ \sum_1^{16} x_i = 1, \\ 0 \leq x_i \leq 1, \\ x_i \in N, N - \text{set of natural numbers} \end{cases} \tag{30}$$

The result of solving the task is the conclusion about the possibility of installing one switching device. At the same time, according to the obtained results, the optimal place for the installation of the switching apparatus is the location number 12 (Fig. 2), which implies sequential sectioning. In the case of the installation of a switching device in the location number 12, the effect of the installation will be 274.8, with a reduced installation cost of $ 240.3.

Thus, an approach has been disclosed and a model has been obtained that allows determining the optimal location and layout of automatic partitioning switching devices for radial networks, which proves the effectiveness of using partitioning in order to increase the reliability of electricity supply to consumers in rural regions. It is shown that the solution of the optimization problem of determining the number, location and installation scheme of partitioning switching devices for a typical distribution network scheme allows us to estimate the efficiency of installation of partitioning switching devices.

MATHEMATICAL MODEL OF SWITCHING DEVICES LOCATION WITH RING AND LOCAL BACKUP DISTRIBUTION NETWORK SCHEMES

We construct mathematical models to determine the optimal installation location of switching devices for a ring circuit and a circuit with a local reserve.

Consider a sequential partitioning scheme. In this case, according to Figure 3 it is necessary to check the effectiveness of the PAP2 installation, if PAP1 is already installed, if there is an ATR (automatic transfer reserve) and / or local reserve connected to the B1 switch in the circuit.

According to (13) the damage when installing the second switching device in a sequential scheme will be:

$$Inj_{PAP2} = P_1 y\alpha l_1 + P_2 y\alpha l_2 + P_3 y'\alpha' l_3 \tag{31}$$

Let us generalize the obtained formula for the case of checking the possibility of installing the i-th switching device by the sequential scheme:

$$Inj_{PAP\ i} = \sum_i P_{\Sigma i} y\alpha_i' l_{\Sigma i}. \tag{32}$$

$P_{\Sigma i}$ –power of the *i*-th allocated by the relevant partitioning switching devices;
$l_{\Sigma i}$ – total length, taking into account the lengths of highways and branches of the i-th zone.

We will show that the optimal place for a sequential arrangement of partitioning switching devices in a circular distribution network can be places both at the end and at the beginning of the main section.

Suppose that there are two possible locations for the installation of a switching device, PAP 1 and PAP 1'. At the same time we get:

$$\begin{aligned} Inj_{PAP\ 1} &= P_1 y\alpha l_1 + P_2 y\alpha l_2 = y\alpha[P_1 \cdot l_1 + P_2 l_2], \\ Inj_{PAP\ 1'} &= P_1 y\alpha(l_1 + \Delta l) + P_2 y\alpha(l_2 - \Delta l) = y\alpha[P_1 l_1 + P_2 l_2 - \Delta l(P_1 - P_2)], \end{aligned} \tag{33}$$

Until $P_1 < P_2$, it is more efficient to install the switching device in the PAP1' place, otherwise in the PAP1. Accordingly, when searching for the optimal installation location of a switching device using a sequential or mixed scheme, if there is an ATR point in the scheme, it is necessary to check whether the switching device can be installed both at the beginning and at the end of the trunk section.

Damage during installation of this PAP in a parallel scheme for a looped distribution network:

$$Inj_{PAP\,i} = P_\Sigma y\alpha\left(L - \sum_i l_i\right) + \sum_i P_i y\alpha_i l_i \tag{34}$$

If there is a local reserve, the damage when installing the i-th PAP in a parallel scheme:

$$Inj_{PAP\ i} = P_{\Sigma} y\alpha \left(L - \sum_i l_i^{paral} \right) + \sum_i P_i^{paral} \cdot y\alpha l_i^{paral} - \sum_m P_m^{paral} y\alpha l_m^{paral} \tag{35}$$

P_i^{paral} –active power consumers on the branch;
l_i^{paral} –branch length;
m – number of lines with local reserve and parallel sectioning.

In the presence of a local reserve and the ringed nature of the distribution network, the damage caused by installing the i-th PAP in a parallel scheme:

$$Inj_{PAPi} = P_{\Sigma} y\alpha \left(L - \sum_i l_i \right) + \sum_i P_i y\alpha_i l_i - \sum_m P_m y\alpha_m l_m \tag{36}$$

m – number of lines with local reserve and parallel sectioning .

Damage during installation of the i-th PAP according to a mixed scheme for a looped-down distribution network:

$$Inj_{PAPi} = \sum_j^{n_{paral}} y\alpha_j l_j P_j + \sum_i^{n_{seq}+1} P_i \alpha_i y_p \left(l_i - \sum_{k=1}^{m} l_k^{paral} \right) \tag{37}$$

In the presence of a local reserve, the damage when installing the i-th PAP according to a mixed scheme:

$$\begin{aligned} Inj_{PAP\ i} = & \sum_{i=1}^{a\_paral} P_i^{paral} y_p \alpha\, l_i^{paral} - \sum_{r=1} P_r^{paral} y_p \alpha\, l_r^{paral} + \\ & + \sum_{j=1}^{a\_seq+1} (l_j^{seq} - \sum_{k=1}^{m} l_k^{paral})(P_{\Sigma} - \sum_{n=0}^{a\_seq} P_n) y\alpha \end{aligned} \tag{38}$$

l_i^{paral}, l_j^{seql} –the length of the zone, respectively, parallel and sequential partitioning;
P_i^{paral}, P_j^{seq} –power of the zone of parallel and sequential partitioning;
a_paral, a_seq – the number of devices included in the parallel and serial partitioning schemes, taking into account the checked;
m – the number of parallel partitioning devices included in the zone of the switching device included in a sequential scheme;
r – number of lines with local reserve and parallel sectioning.

In the presence of a local reserve and the ringed nature of the distribution network, the damage caused by installing the i-th PAP according to a mixed scheme:

$$Inj_{PAP\,i} = \sum_{j}^{n_{paral}} y\alpha_j l^{paral}{}_j P^{paral}{}_j - \sum_{m}^{n_{paral}} y\alpha_m l^{paral}{}_m P^{paral}{}_m + \\ + \sum_{i}^{n_{seq}+1} P^{seq}{}_i \alpha_i y (l^{seq}{}_i - \sum_{k=1}^{m} l_k^{paral}) \tag{39}$$

m –number of lines with local reserve and parallel sectioning;
l_i^{paral}, l_j^{seq} – the length of the zone, respectively, parallel and sequential partitioning;
P_i^{paral}, $P_j^{seq\pi}$ –power of the zone of parallel and sequential partitioning.

Thus, models have been obtained that allow determining the optimal location and layout of automatic partitioning switching devices for both looped circuits and circuits with a local reserve.

MATHEMATICAL MODEL OF SWITCHING DEVICES IN THE ENERGY MARKET

The interaction between producers and consumers of services in all spheres of modern society occurs according to the rules of market relations. In order to regulate the producer-consumer relationship, there is a need to create methods that determine the amount of possible damage to the parties and the way to compensate for it (Vasil'yev, & Guk, 2000).

In order to regulate the relationship in the system "power supply company - consumer", mathematical models have been developed that take into account such relationships (Buynyy, Podolnyi, & Zorin, 2004; Perova, 1998). Analysis of existing information reflecting market relations in the energy industry indicates the need to take into account not only the damage caused to the electricity supplying company (as the amount of lost profit proportional to downtime), but also the individual consumer, which is required to be reflected when developing their own models that regulate such relations.

In (Kozyrskiy, Kuz'menko, & Guy, 2005), a mathematical model was developed, which allows to unambiguously determine the optimal number and installation location of partitioning switching devices. The damage (2) and the effect of the installation (13) are determined as follows:

$$Inj_{no\ cd} = P_{\Sigma} y\alpha L. \tag{40}$$

The disadvantage of the point of sectioning lines (PSL) deployment model (Kozyrskiy, Kuz'menko, & Guy, 2005) is the lack of consideration of possible reliability enhancing tools installed on the line. We construct a refined model (Kozyrskiy, Kuz'menko, & Guy, 2005), based on the theory of models and methods for calculating reliability. It is recommended to determine the load idle time per sustained damage (for distribution networks in rural regions) according to (Pruss, & Tislenko, 1989; Dehghani, & Dashti, 2011):

$$\tau_i = \tau_{1i} + \tau_{2i} + \tau_{3ii} + \tau_{4ii} + \tau_{5i} + \tau_{6i} \tag{41}$$

τ_i –recovery time of normal power supply to consumers in case of damage to the *i*- th area;

τ_{1i} –the duration of the stage from the moment of disconnecting the line to the beginning of the search for the damaged section of the line, taking into account the crossing and the trial head switch of the line;

τ_{2i} –the duration of the search stage of the damaged area, taking into account the possible analysis of information from consumers of electrical energy by the dispatch service;

τ_{3i} –the duration of the stage of localization of the damaged area;

τ_{4i} –the duration of the stage of switching loads, which can be powered when the damaged section is removed from the circuit, by existing means of increasing reliability.;

τ_{5i} –the duration of the stage of bypassing the damaged area to identify the location of the damage.;

τ_{6i} –the duration of the repair phase and the inclusion of the line in normal mode.

Consequently, the installation of PSL significantly reduces the total time spent on restoring the normal power supply mode, which is due to the reduction in the duration of the following stages: τ_{2i}, τ_{3i}, τ_{4i}, τ_{5i}.

Studies of distribution networks in rural areas have shown that the equipment of power lines with linear disconnectors is 0.2 units/km. This fact will be used in analyzing the effectiveness of the PSL installation in the distribution networks of rural regions.

According to statistical studies (Pruss, & Tislenko, 1989; Javadian, & Haghifam, 2010), the repair time per one sustained damage, based on the formula (42) and tabl 1:

$$\tau_6 = \frac{\omega_s}{\omega}\tau_s + \frac{\omega_w}{\omega}\tau_w + \frac{\omega_{im}}{\omega}\tau_{im} + \frac{\omega_d}{\omega}\tau_d + \frac{\omega_o}{\omega}\tau_o. \tag{42}$$

Thus, the average repair time will be – τ = 2,3 h.

We construct a mathematical model to find the optimal partitioning scheme, based on the distribution network diagram shown in Fig. 4. To do this, we will determine the magnitude of damage from under-supply of electrical energy to consumers and the duration of downtime for each of the consumers during the operation of the distribution network of rural regions in nominal mode and subject to the installation of additional means of improving reliability - airline partitioning points. The scheme (Fig. 4) is divided

Table 1. Reliability indicators of rural distribution networks

The Nature of the Damage	Specific Frequency of Sustained Failures Due to Damage (year-km)$^{-1}$		Average Repair Time, h, When Damaged	
	Symbol	Numerical Designation	Symbol	Numerical Designation
Supports	ω_s	0.040	τ_s	2.8
Wires	ω_w	0.035	τ_w	2.2
Insulation	ω_{in}	0.100	τ_{in}	1.0
Difficult types	ω_d	0.025	τ_d	4.9
Other reasons	ω_o	0.050	τ_o	3.1

Figure 4. Distribution network diagram with installed line disconnectors

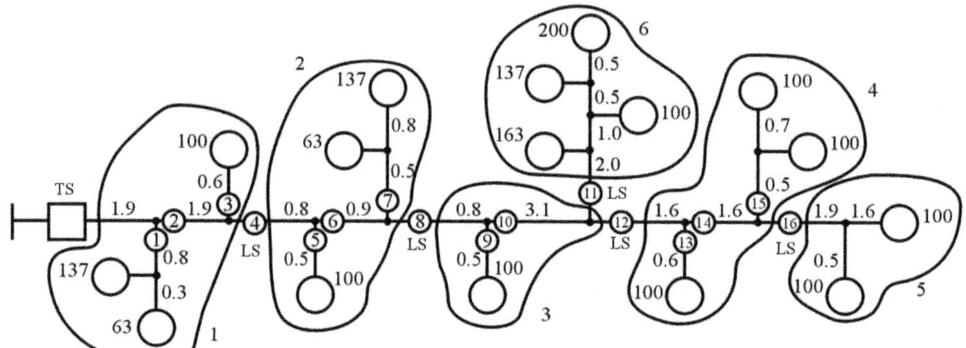

into 6 zones, the boundaries of which are determined by linear disconnectors installed in the network, the zone is such a part of the distribution network scheme that is separated from the entire circuit by switching by installed switching devices and without containing a switching device. switching devices installed in a distribution network do not belong to any of the zones and are zone boundaries. Note that the zones are numbered as follows: first, all zones limited by switching devices of the sequential type (from first to fifth), then parallel (sixth zone) are numbered.

The first item in the search for a solution will be to obtain a mathematical model for the operation of a network with previously established means of increasing reliability. When conducting a mathematical experiment, we assume that the telecontrol of all switching devices already installed and newly installed belonging to the distribution network under study (Fig. 4) is absent.

Mathematical expression to calculate the duration of the search of the damaged area (τ_i) according to figure 5:

$$\tau_2 = \begin{pmatrix} \frac{k_{kp}}{v} \cdot 2 \cdot L_{m1}; \frac{k_{kp}}{v} \cdot 2 \cdot \left(L_{m1} + L_{m1} + L_{m2}\right); \ldots; \frac{k_{kp}}{v} \cdot 2 \cdot \left(2 \cdot \sum_{1}^{i-1} L_{mj} + L_{mi}\right); \\ \frac{k_{kp}}{v} \cdot 2 \cdot \sum_{1}^{i} L_{mj}; \ldots; \frac{k_{kp}}{v} \cdot 2 \cdot \left(2 \cdot \sum_{k}^{q_n} L_{mk}\right) \end{pmatrix} \tag{43}$$

i, n –the number of linear disconnectors of serial and parallel type, respectively;
q – the number of sections of the highway to the m-th disconnector parallel type;
L_i, L_{mi} –;the total length and the length of the line of the i-th section, respectively;
L_Σ – power line length;
k_{kp} –the curvature of the roads in relation to the air line connecting the end points of the brigade's crossing route (for Chernigov distribution networks, k_{kp}=0.4);
v – brigade speed (v=40 km/h).

Mathematical expression for calculating the duration of the localized area (τ_l), which consists in separating the damaged area from those parts of the overhead line, the power of which can be restored from the main or backup sources:

$$\tau_3 = \left(\frac{k_{kp}}{v} \cdot 0; \frac{k_{kp}}{v} \cdot L_{m1}; \ldots; \frac{k_{kp}}{v} \cdot \sum_{j=1}^{i-2} L_{mj}; \frac{k_{kp}}{v} \cdot \sum_{j=1}^{i-1} L_{mj} \right) \tag{44}$$

Search Duration Calculation Results (τ_2) and localization of the damaged area (τ_3) in relation to the scheme of the network depicted in Figure 4, are given in Table 2

The mathematical expression for calculating the duration of the inclusion of loads that can be powered when the damaged section is removed from the circuit by existing means of improving reliability (τ_4), which consists in restoring power supply to those sections of the line that can be switched on when the damaged section is removed from the circuit (Fig. 4):

$$\tau_4 = \begin{cases} \tau_3, & \text{if there is no provision for the reserve} \\ \tau_3 + L_{M\Sigma} \dfrac{k_{kp}}{v}, & \text{with manual activation of the reserve or ATR} \end{cases} \tag{45}$$

The results of the calculation of the duration of the stage of inclusion of loads that can be powered when the damaged section is removed from the circuit, by the existing means of improving reliability as applied to the circuit of the network shown in Figure 5, are given in Table 3.

Mathematical expression for calculating the duration of the stage of bypassing the damaged area to identify the location of the damage (τ_5) (Pruss, & Tislenko, 1989):

$$\tau_{5i} = \frac{L_i}{2v_x} + \frac{L_i - L_{mi}}{2\alpha v_x}, \tag{46}$$

Table 2. Duration of search and localization of the damaged area

Zones	1	2	3	4	5	6
τ_2, h	0.076	0.186	0.563	0.814	0	0.374
τ_3, h	0	0.039	0.056	0.094	0	0

Table 3. Duration of the stage of inclusion of loads

Zone	1	2	3	4	5	6
τ_{4i}, h without reserve	0	0.039	0.056	0.094	0	0
τ_{4i}, h with manual input reserve	0.146	0.184	0.217	0.240	0.146	0
τ_{4i}, h with automatic input reserve	0.146	0.184	0.217	0.240	0.146	0

L_i, L_{mi} –the total length and the length of the line considered section, respectively;
v_x- average crawl speed (v_x=km/h);
αv_x – return speed with branches (movement without inspection), $\alpha v_x = 7.5$ km/h.

For the distribution network shown in Figure 4, the damage from the undersupply of electrical energy for the network diagram without a reserve:

$$\begin{aligned} Inj^1 &= P_\Sigma y L_1(\omega_0(\tau_1^{'} + \tau_1^{''}) + \lambda\nu_0\eta) + yL_2(\omega_0(\tau_2^{'}P_\Sigma + \tau_2^{''}(P_\Sigma - P_1)) + (P_\Sigma - P_1)\lambda\nu_0\eta) + \ldots + \\ &+ yL_4(\omega_0(\tau_4^{'}P_\Sigma + \tau_4^{''}(P_\Sigma - P_1 - P_2 - P_3 - P_6)) + (P_\Sigma - P_1 - P_2 - P_3 - P_6)\lambda\nu_0\eta) + \ldots + \\ &+ yL_6(\omega_0(\tau_6^{'}P_\Sigma + \tau_6^{''}P_6) + P_6\lambda\nu_0\eta) + \ldots = \sum_i^{seq+1}\left(y_p L_i(\omega_0(\tau_i^{'}P_\Sigma + \tau_i^{''}(P_\Sigma - \sum_j^{i-1} P_j - \sum_k^{i-1} P_k^{paral})) + \right. \\ &\left. + (P_\Sigma - \sum_j^{i-1} P_j - \sum_k^{i-1} P_k^{paral})\lambda\nu_0\eta) \right) + \sum_k^{paral} yL_k(\omega_0(\tau_i^{'}P_\Sigma + \tau_i^{''}P_k) + P_k\lambda\nu_0\eta), \end{aligned} \tag{47}$$

$\tau_i^{'} = \tau_{1i} + \tau_{2i} + \tau_{3i} + \tau_{4i}$, $\tau_i^{''} = \tau_{5i} + \tau_{6i}$; $\sum_k^{i-1} P_k^{paral}$ – the sum of the active powers of the zones allocated by parallel-type linear disconnectors.

After calculations, we get: $Inj^1 = 5519.3$ \$/year.

For subscribers located in different zones, the shortage time for the scheme (Figure 4) without a reserve:

$$\begin{aligned} t_i^{iseq} &= \omega_0\left(\sum_{j=1}^{seq+paral+1} L_j\tau_j^{'} + \sum_{k=1}^{i} L_k\tau_k^{''}\right) + \lambda\nu_0\eta\sum_{k=1}^{i} L_k; \\ t_i^{paral} &= \omega_0\left(\sum_{j=1}^{seq+paral+1} L_j\tau_j^{'} + \sum_{k=1}^{ZPi} L_k\tau_k^{''} + L_i\tau_i^{''}\right) + \lambda\nu_0\eta\left(\sum_{k=1}^{ZP_i} L_k + L_i\right) \end{aligned} \tag{48}$$

ZP_i –the number of zones allocated by linear disconnectors of sequential type, located before the i-th zone, including the one that includes the i-th zone.

The results of the calculation of the duration of the detour of the damaged area to identify the location of the damage (τ_5) and the calculation of the undersupply time in relation to the network diagram (Fig. 5) without the installed BP are shown in Table 4

Table 4. Duration of the stage of bypassing the damaged area and the time of the short supply to consumers connected to certain zones

Zone	1	2	3	4	5	6
τ_5, h	0.66	0.47	0.47	0.62	0.43	0.4
t, h/year	24.37	29.60	36.16	43.81	49.74	42.06

Knowing the duration of downtime of individual consumers, we can determine the total damage for the entire electrical distribution network

$$Inj^1\_t = y\sum_{i}^{paral+seq+1} P_i t_i = 5519.3 \text{ \$/year},$$

that is $Inj^1\_t = Inj^1$, which proves the validity of the approach used.

For the distribution network shown in Figure 4, we will determine the damage from the undersupply of electrical energy when using manual and automatic switching on reserve:

$$\begin{aligned} Inj^1 &= yL_1(\omega_0(P_\Sigma\tau_1^{'} + P_1\tau_1^{''}) + P_1\lambda\nu_0\eta) + \ldots + yL_6(\omega_0(P_\Sigma\tau_6^{'} + P_6\tau_6^{''}) + P_6\lambda\nu_0\eta) = \\ &= \sum_{i}^{seq+1} yL_i(\omega_0(P_\Sigma\tau_i^{'} + P_i\tau_i^{''}) + P_i\lambda\nu_0\eta) + \sum_{k}^{paral} yL_k(\omega_0(P_\Sigma\tau_k^{'} + P_k\tau_k^{''}) + P_k\lambda\nu_0\eta). \end{aligned} \tag{49}$$

After calculations, we obtain numerical expressions of damage to the line with automatic and manual switching on reserve, respectively:

$$Inj^1ARR = Inj^2MTR = 3714.1 \text{ \$/year}.$$

For subscribers located in different zones, the time of the short supply for the distribution network scheme (Fig. 4) with automatic and manual BP is determined by the:

$$\begin{aligned} t_i^{seq} &= \omega_0\left(\sum_{j=1}^{paral+seq+1} L_j\tau_j^{'} + L_i\tau_i^{''}\right) + L_i\lambda\nu_0\eta; \\ t_i^{paral} &= \omega_0\left(\sum_{j=1}^{paral+seq+1} L_j\tau_j^{'} + L_i\tau_i^{''} + L_m\tau_m^{''}\right) + \left(L_i + L_m\right)\lambda\nu_0\eta \end{aligned} \tag{50}$$

m – is the number of the zone.

The results of the calculation of the time outflow in relation to the network diagram (Fig. 5) with automatic and manual switching on reserve, are given in Table 4.

Thus, the obtained mathematical dependencies make it possible to estimate the damage caused to the power supplying organization due to the disconnection of the entire transmission line, as well as the time and amount of undersupply for an individual consumer for cases of the presence and absence of

Table 5. Time of undersupply of electricity to consumers, for networks with redundancy

Zone	1	2	3	4	5	6
t_{MTR}, h/year with manual input reserve	25.21	21.96	23.30	24.39	22.67	29.20
T_{ATR}, h/year with automatic input reserve	25.21	21.96	23.30	24.39	22.67	29.20

ATR. The analysis of the obtained results shows the inexpediency of the installation of automatic input reserve, when used as a means of search and localization of damage to linear disconnectors.

The second item of the search for a solution will be the receipt of a mathematical model when installed in the distribution network of rural regions.

When installing PAP, the scheme shown in Figure 4 is divided into several partition zones. In particular, when installing one PAP to the coordinates recommended by research (Kozyrskiy, Kuz'menko, & Guy, 2005) - to the location at number 12 in Figure 4, into two zones, therefore, the first of the zones includes LS 1-3, 6, and the second 4.5. We will evaluate the effectiveness of the PAP installation as an additional means of increasing reliability, analyzing the change in the length of downtime for the entire line as a whole, and for individual consumers in particular, for this purpose we will use the dependency (41).

Duration of search for a damaged area (τ_2):

$$\tau_2^{PAP} = \left(\frac{k_{kp}}{v} \cdot 2 \cdot \begin{cases} \sum_{i=1}^{ZC_1} L_{mi}, \\ \sum_{i=ZC_1}^{ZC_2} L_{mi}, \\ \ldots, \\ \sum_{i=ZCn-1}^{ZCn} L_{mi}. \end{cases} \right) \tag{51}$$

ZC_i – the number of zones allocated to LS in that sectioning zone.

The duration of the stage of localization of the damaged area ():

$$\tau_3^{PAP} = \left(\frac{k_{kp}}{v} \cdot \begin{cases} \sum_{i=1}^{ZC_1-1} L_{mi}, \\ \sum_{i=ZC_1-1}^{ZC_2-1} L_{mi}, \\ \ldots \\ \sum_{i=ZC_{n-1}-1}^{ZC_n-1} L_{mi} \end{cases} \right) \tag{52}$$

The results of the calculation of the duration of the search (τ_2) and localization of the damaged area (τ_3) in relation to the network diagram shown in Figure 4 are shown in Table 6

The duration of the stage of inclusion of loads:

$$\tau_4 = \begin{cases} \tau_3, & \text{without reserve} \\ \tau_3 + L_{ZC_i} \dfrac{k_{kp}}{v}, & \text{if there is an automatic switch on reserve} \\ \tau_3 + \left(L_{m\Sigma} - L_{3C_{i-1}} \right) \dfrac{k_{kp}}{v}, & \text{with manual switch on reserve} \end{cases} \tag{53}$$

Table 6. Duration of search and localization of a damaged area for a network with an installed switching device

Zone	1	2	3	4	5	6
τ_2, h	0.076	0.186	0.374	0.064	0	0
τ_3, h	0	0.039	0.056	0	0	0

Table 7. Duration of the stage of switching loads for the network with installed switching device.

Zone	1	2	3	4	5	6
τ2, h without reserve	0	0.039	0.056	0	0	0
τ3, h with manual input reserve	0.146	0.183	0.200	0.051	0	0
τ4, h with automatic input reserve	0.094	0.133	0.150	0.051	0	0

The results of the calculation of the duration of the stage of inclusion of loads (τ_4) in relation to the network diagram shown in Figure 4, are given in Table 7

The duration of the stage of bypassing the damaged area to identify the location of damage (τ_5) (Lagland, & K. Kauhaniemi, 2007):

$$\tau_{5i} = \frac{L_i}{2v_x} + \frac{L_i - L_{mi}}{2\alpha v_x}, \tag{54}$$

L_i, L_{mi} – is the total length and the length of the line of the considered section, respectively;
v_x -is the average crawl speed (km/h);
αv_x –is the speed at return from a branch (movement without inspection), αv_x =5 km/h.

The results of the calculation of the duration of the detour of the damaged area to identify the location of the damage (τ_5) in relation to the network diagram shown in Figure 4 are given in Table 8

For the distribution network (Figure 3), the damage is due to the undersupply of electrical energy for the PSL network scheme, but without a reserve.:

$$\begin{aligned} Inj^{seq} &= P_\Sigma y L_1(\omega_0(\tau_1^{'} + \tau_1^{''}) + \lambda\nu_0\eta) + yL_2(\omega_0(\tau_2^{'}P_\Sigma + \tau_2^{''}(P_\Sigma - P_1)) + (P_\Sigma - P_1)\lambda\nu_0\eta) + \ldots + \\ &+ yL_4(\omega_0(P_4 + P_5)(\tau_4^{'} + \tau_4^{''}) + (P_4 + P_5)\lambda\nu_0\eta) + yL_5(\omega_0((P_4 + P_5)\tau_5^{'} + \tau_4^{''}P_5) + P_5\lambda\nu_0\eta) + \\ &+ yL_6(\omega_0(\tau_6^{'}P_\Sigma + \tau_6^{''}P_6) + P_6\lambda\nu_0\eta) = Inj1 + Inj2, \end{aligned} \tag{55}$$

Table 8. Duration of the stage of bypassing the damaged section for the network with the switching device installed

Zone	1	2	3	4	5	6
τ_5, h	0.66	0.47	0.47	0.62	0.43	0.4

$\tau_i^{'} = \tau_{1i} + \tau_{2i} + \tau_{3i} + \tau_{4i}$, $\tau_i^{''} = \tau_{5i} + \tau_{6i}$; $\sum_k^{i-1} P_k^{paral}$ -is the sum of the active powers of the zones allocated by parallel-type linear disconnectors.

$$Inj1 = \sum_{i=1}^{KPS}\left(yL_i(\omega_0(\tau_i^{'}P_\Sigma + \tau_i^{''}(P_\Sigma - \sum_j^{i-1} P_j - \sum_k^{i-1} P_k^{paral})) + (P_\Sigma - \sum_j^{i-1} P_j - \sum_k^{i-1} P_k^{paral})\lambda\nu_0\eta)\right) +$$
$$+\sum_k^{paral} yL_k(\omega_0(\tau_i^{'}P_\Sigma + \tau_i^{''}P_k) + P_k\lambda\nu_0\eta),$$

$$Inj2 = \sum_{i=KPS}^{seq+1}\left(yL_i(\omega_0(\tau_i^{'}(P_\Sigma - \sum_{j=1}^{KPS} P_j) + \tau_i^{''}(P_\Sigma - \sum_{j=1}^{KPS} P_j - \sum_{j=KPS}^{i-1} P_j - \sum_k^{i-1} P_k^{paral})) + \right.$$
$$\left. +(P_\Sigma - \sum_{j=1}^{KPS} P_j - \sum_{j=KPS}^{i-1} P_j - \sum_k^{i-1} P_k^{paral})\lambda\nu_0\eta)\right) + \sum_k^{paral} yL_k(\omega_0(\tau_i^{'}(P_\Sigma - \sum_{j=1}^{KPS} P_j) + \tau_i^{''}P_k) + P_k\lambda\nu_0\eta). \quad (56)$$

Making the calculations, we get: $Inj^{nocлe} = 4781.13$ \$/h.

For subscribers located in different zones, the time for the outflow for the scheme (Fig. 7) without the installed BP:

$$t_i^{seq} = \begin{cases} \omega_0(\sum_{j=1}^{KPS} L_j\tau_j^{'} + \sum_{j=1}^{paral\_1} L_j\tau_j^{'} + \sum_{k=1}^{i} L_k\tau_k^{''}) + \lambda\nu_0\eta\sum_{k=1}^{i} L_k, 1 \le i \le KPS \\ \omega_0(\sum_{j=1}^{seq+1} L_j\tau_j^{'} + \sum_{j=1}^{paral\_1+paral\_2} L_j\tau_j^{'} + \sum_{k=1}^{i} L_k\tau_k^{''}) + \lambda\nu_0\eta\sum_{k=1}^{i} L_k, KPS+1 \le i \le seq+1, \end{cases}$$

$$t_i^{seq} = \begin{cases} \omega_0(\sum_{j=1}^{KPS} L_j\tau_j^{'} + \sum_{k=1}^{ZPi} L_k\tau_k^{''} + L_i\tau_i^{''}) + \lambda\nu_0\eta(\sum_{k=1}^{ZP_i} L_k + L_i), 1 \le i \le KPS \\ \omega_0(\sum_{j=1}^{nocL+1} L_j\tau_j^{'} + \sum_{k=1}^{ZPi} L_k\tau_k^{''} + L_i\tau_i^{''}) + \lambda\nu_0\eta(\sum_{k=1}^{ZP_i} L_k + L_i), KPS+1 \le i \le seq+1, \end{cases} \quad (57)$$

ZP_i - is the number of zones allocated by linear disconnectors of the sequential type, located up to the i-th zone, including the one that includes i-th zone.

For the distribution network shown in figure 3, the loss of electric power undershoot for the network with manual and automatic BP:

$$Inj^{seq} = yL_1(\omega_0(\tau_1^{'}\sum_{i=1}^{ZS_1} P_i + P_1\tau_1^{''}) + P_1\lambda\nu_0\eta) + ... + yL_6(\omega_0(\tau_6^{'}\sum_{i=1}^{ZS_1} P_i + P_6\tau_6^{''}) + P_6\lambda\nu_0\eta) =$$
$$= \sum_{j=1}^{n}\sum_{i=1+ZS_{j-1}}^{ZS_j} yL_i(\omega_0(\tau_i^{'}\sum_{k=1+ZS_{j-1}}^{ZS_j} P_k + P_i\tau_i^{''}) + P_i\lambda\nu_0\eta). \quad (58)$$

ZS_1, ZS_2, …ZS_{n+1} –is the number of zones created by PSL;
n – number of zones formed by PSL.

Making calculations, we obtain a quantitative expression of damage to the line with automatic and manual BP, respectively $Inj^{seq}ATR = 2502.73$ \$/год; $Inj^{seq}MTR = 2520.57$ \$/год.

For subscribers located in different zones, the duration of the shortfall for the distribution network diagram (Figure 3) with automatic and with manual input reserve is determined on the basis of expressions.:

$$t_i^{seq} = \omega_0 (\sum_{j=1+ZS_{k-1}}^{ZS_k} L_j \tau_j^{'} + L_i \tau_i^{''}) + L_i \lambda \nu_0 \eta ;$$
$$t_i^{seq} = \omega_0 (\sum_{j=1+ZS_{k-1}}^{ZS_k} L_j \tau_j^{'} + L_i \tau_i^{''} + L_m \tau_m^{''}) + (L_i + L_m) \lambda \nu_0 \eta \quad (59)$$

m –is the number of the zone,
k –is the number of the zone allocated to the PSL, which includes the i-th zone under study

When installing the PSL in place 12 (Fig. 4) during the computational experiment, the results were obtained, summarized in Table 9. To find the effect of installing an additional partitioning switching device, in a network with installed means of improving reliability, it is necessary to search for the difference in the amount of damage obtained by mathematical modeling of the network, for cases without and with a partitioning switching device: $Eff = Inj^1 - Inj^2$.

When installing the PSL in place 8 (Figure 4), in the course of a mathematical experiment, the results were obtained, summarized in Table 10.

Analyzing the results of tables 9 and 10, it can be concluded that for different installation sites of partitioning switching devices with almost the same level of damage for the power system as a whole, for individual consumers, the idle time varies widely, which indicates the need to choose the optimal option according to two criteria - maximization efficiency for the power supply company and minimizing downtime for both all consumers in general and individual consumers in particular with various degrees of equipment distribution network. Conducting a comparative assessment of the values, we conclude that the optimal installation location of sectional switching devices, both in the analysis of specific parameters and calculated time values based on the power supply restoration model (Pruss, & Tislenko, 1989; Dehghani, & Dashti, 2011) for damages in distribution networks of rural regions, will be similar, which allows use a more simplified model to find the optimal number and locations for the installation of partitioned switching devices.

Table 9. Electricity undersupply time for consumers for networks with installed switching devices.

Zone	1	2	3	4	5	6	Damage, \$/year	Effect, \$/year
t_1, h/year without reserve	18.04	23.26	29.83	42.06	47.99	36.73	23905.66	3690.90
t_2, h/year with manual input reserve	18.52	15.28	16.62	12.29	10.57	22.52	12602.86	5967.43
t_4, h/year with automatic input reserve	18.35	15.11	16.45	12.29	10.57	22.35	12513.64	6056.65

Table 10. Electricity undersupply time for consumers for networks with installed switching devices

Zone	1	2	3	4	5	6	Damage, $/ year	Effect, $/ year
t_1, h/year without reserve	13.07	18.30	34.29	41.94	47.87	40.19	23941.80	3654.76
t_2, h/year with manual input reserve	13.40	10.16	16.21	17.29	15.57	22.11	12252.93	6317.36
t_4, h/year with automatic input reserve	13.15	9.90	16.21	17.29	15.57	22.11	12191.88	6378.41

FUTURE RESEARCH DIRECTIONS

Further research should consider the problem of optimizing the number and location of sectioning devices in electrical networks in the presence of distributed sources of electricity (small solar, wind and other power plants) in the context of incomplete source information.

CONCLUSION

1. A systematic analysis of the problem to be solved was carried out and possible factors influencing the increase in reliability of distribution networks in rural regions were identified due to the installation of partitioned switching devices.
2. The possibility of building a model for optimizing the number and installation locations of partitioning switching devices, with a different level of generalization of the processes occurring during the installation of additional means of improving reliability, and accordingly with a different level of complexity and accuracy of the final result, is shown.
3. An approach is revealed and a mathematical model is obtained, which allows determining the optimal location and layout of partitioning switching devices for radial networks. An example of finding the optimal scheme and partitioning location is shown, based on the method of mathematical programming with Boolean variables.
4. The approach is revealed and a mathematical model is obtained that allows determining the optimal location and layout of partitioning switching devices for networks with automatic transfer switch and networks with local reserve. The substantiation of the most effective locations for the installation of switching devices within the same trunk section has been carried out.
5. A mathematical model is proposed and an example of a solution for determining the economic efficiency of installing partitioned switching devices for networks with installed means of improving reliability is shown. The possibility of using, when choosing the place of installation of sectional switching devices, is indicated not only an economic criterion (in the form of reducing damage to the power supply system as a whole), but also a criterion that takes into account the change in the idle time of certain consumers that have a higher level of significance, which is caused by the contract between the power supply company and consumer.

ACKNOWLEDGMENT

The authors are deeply grateful to the editors of the book for the Prof. Valeriy Kharchenko and Pandian Vasant and the reviewers for valuable comments that were taken into account in the preparation and finalization of the chapter.

REFERENCES

Bernardon, D. P., Sperandio, M., Garcia, V. J., Russi, L. N., Canha, J., Abaide, A. R., & Daza, E. F. B. (2011). Methodology for allocation of remotely controlled switches in distribution networks based on a fuzzy multi-criteria decision making algorithm. Electric Power Systems Research, 81(2), 414-420. doi:10.1016/j.epsr.2010.10.010

Bertling, L. (2002). Reliability Centred Maintenance for Electric Power Distribution Systems. Thesis submitted in accordance with academic requirements for the Degree of Doctor of Philosophy, Stockholm: Royal Institute of Technology (KTH).

Brown, R. E. (2009). *Electric Power Distribution Reliability*. CRC Press.

Buynyy, R. A., Podolnyi, S. V., & Zorin, V. V. (2004). Modyfikovanyi metod optymizatsii nadiinosti rozpodilchykh merezh 10 kV [Modified method reliability optimization of distribution 4. Begovic, M., Lambert, F.C. (2002). Case Studies: Recloser Positioning and Reliability Indices.

Chowdhury, A., & Koval, D. (2009). *Power distribution system reliability practical methods and applications*. IEEE.

Dehghani, N., & Dashti, R. (2011). Optimization of Recloser Placement to Improve Reliability by Genetic Algorithm. Energy and Power Engineering, 3, 508-512. doi:10.4236/epe.2011.34061

Ishizaka, A., & Nemery, Ph. (2013). *Multi-criteria decision analysis: methods and software*. New Delhi: Times by Aptara Inc. doi:10.1002/9781118644898

Javadian, S. A. M., & Haghifam, M.-R. (2010). Optimal placement of protective devices in distribution networks based on risk analysis. Proceedings of the IEEE/PES Transmission and Distribution Conference and Exposition: Latin America (T&D-LA), Sao Paulo, November 8-10 (pp. 147-154). IEEE. doi:10.1109/TDC-LA.2010.5762875

Javadian, S. A. M., & Massaeli, M. (2011). Impact of distributed generation on distribution system's reliability considering recloser-fuse miscoordination - A practical case study. *Indian Journal of Science and Technology*, *4*(10), 1279–1284.

Kozyrskiy, V. V., Kuz'menko, B. V., & Guy, A. V. (2005). Podkhod k opredeleniyu mest razmeshcheniya sektsionnykh kommutatsionnykh apparatov [Approach to determining of the location of sectional switching devices]. *Elektryfikatsiia ta avtomatyzatsiia silskoho hospodarstva*, *1*, 32-44.

Kozyrskiy, V. V., Kuz'menko, B. V., & Guy A. V. (2005). Matematicheskaya model' razmeshcheniya kommutatsionnykh apparatov v usloviyakh rynochnykh otnosheniy mezhdu elektrosnabzhayushchey kompaniyey i potrebitelem [Mathematical model of switching devices in the conditions of market relations between the power supply company and the consumer]. *Elektrifikatsiya ta avtomatizatsiya sil's'kogo gospodarstva*, *3*, 40-52.

Kozyrskyi, V. V. & Guy, O. V. (2004). Vybir optymalnoi kilkosti sektsionuiuchy prystroiv dlia rozpodilnykh merezh napruhoiu 10 kV [Choosing of the optimal number of partitioning devices for 10 kV distribution networks]. *Elektryfikatsiia ta avtomatyzatsiia silskoho hospodarstva*, *2*, 12-20.

Lagland, H., & Kauhaniemi, K. (2007). Outage Cost Comparison of Different Medium Voltage Networks. *Proceedings of the CIRED 19th International Conference on Electricity Distribution*, Vienna, May 21-24. Academic Press.

Leshchinskaya, T. B. (2003). Primeneniye metodov mnogokriterial'nogo vybora pri optimizatsii sistem elektrosnabzheniya sel'skikh rayonov [Application of multicriteria choice methods for optimizing of rural power supply system]. *Elektrichestvo*, *1*, 14-22.

Makoklyuyev, B. I., Pavlikov, V. S., Vladimirov, A. I., & Fefelova, G. I. (2003). Vliyaniye metereologicheskikh faktorov na elektropotrebleniye energoob"yedineniy [The influence of meteorological factors on the power consumption of energy interconnections]. *Energetik*, *6*, 45-47.

Perova, M.B. (1998). Regulirovaniye otnosheniy mezhdu sel'skimi potrebitelyami i energosnabzhayushchimi organizatsiyami [Regulation of relations between rural consumers and energy supplying organizations]. *Mekhanizatsiya i elektrifikatsiya sel'skogo khozyaystva*, *4*, 18-19.

2. Prakhovnik, A. V., Popov, V. A., & Nakhodov, V. F. (2010). Risk and uncertainty assessment in the development of strategy for forming the potential of dispatchable generation in the national power system of Ukraine. International Journal Global Energy Issues, 34(1-4), 132-138. doi:10.1504/IJGEI.2010.037400

Pruss, V. L., & Tislenko, V. V. (1989). *Povysheniye nadezhnosti sel'skikh elektricheskikh setey* [The reliability improving of rural electrical networks]. Leningrad: Energoatomizdat.

Taha, H. A. (2017). *Operations Research: An Introduction.* Pearson.

Tislenko, V. V. (1991). Modeli i metody upravleniya nadezhnost'yu sistem elektrosnabzheniya sel'skokhozyaystvennogo naznacheniya [Models and methods for managing of the reliability of agricultural power supply system]. Kyiv: Instytut elektrodinamiki AN Ukrainy.

Tislenko, V. V., Kholmskiy, D. V., & Gurbich, R. F. (1975). Issledovaniye zavisimosti pokazateley nadezhnosti elektrosnabzheniya potrebiteley ot dliny vozdushnykh liniy elektroperedachi raspredelitel'noy seti [Study of the reliability indicators dependence of electricity supply on the length of the air-distribution network of power lines]. *Energetika i elektrifikatsiya*, *2*, 15-17.

Transmission and Distribution Committee. (2003). IEEE Guide for Electric Power Distribution Reliability Indices. IEEE Std 1366.

Vaccaro, A., Popov, M., Villacci, D., & Terzija, V. (2011). An integrated framework for smart microgrids modeling, monitoring, control, communication, and verification. *Proceedings of the IEEE*, *99*(1), 119–132. doi:10.1109/JPROC.2010.2081651

Vasil'yev, A.P. & Guk, Yu.B. (2000). Obespecheniye nadezhnosti, bezopasnosti i energosberezheniya v usloviyakh energeticheskogo rynka [Ensuring reliability, safety and energy saving in the conditions of the energy market]. *Energetika i elektrifikatsiya*, *3*, 45-53.

Zorin, V. V., Popov, V. A., & Ekel', P. E. (1988). Modeli optimizatsii nadezhnosti raspredelitel'nykh elektricheskikh setey [Optimization models of reliability of electrical distribution networks]. *Energetika i elektrifikatsiya*, *3*, 46-49.

Zorin, V. V., Tislenko, V. V., Kleppel', F., & Adler, G. (1984). Nadezhnost' sistem elektrosnabzheniya [Reliability of power supply system]. Kyiv: Vishcha shkola.

KEY TERMS AND DEFINITIONS

Means of Improving Reliability: A set of technical devices that contribute to improving the reliability of power supply to electricity consumers.

Recloser: A device for automatic control and protection of overhead power lines based on vacuum circuit breakers under the control of a specialized microprocessor.

Sectioning of Overhead Power Lines: Segmentation of the power line into areas using switching devices.

The Point of Sectioning the Lines: An electrical device containing a switching device, a system of relay protection and automation, is installed in a sealed enclosure on a power line support.

Chapter 5
Application of the Software System of Finite Element Analysis for the Simulation and Design Optimization of Solar Photovoltaic Thermal Modules

Vladimir Panchenko
https://orcid.org/0000-0002-4689-843X
Russian University of Transport, Russia

Sergey Chirskiy
https://orcid.org/0000-0001-8196-3068
Bauman Moscow State Technical University, Russia

Valeriy Vladimirovich Kharchenko
https://orcid.org/0000-0003-3725-2976
Federal Scientific Agroengineering Center VIM, Russia

ABSTRACT

The chapter discusses the simulation of thermal operating conditions and the optimization of the design of solar photovoltaic thermal modules. As a realization of the developed method, two photovoltaic thermal modules with one-sided solar cells with one-sided heat removal and two-sided solar cells with two-sided heat removal are presented. The components of the developed models of solar modules must be optimized on the basis of the required indicators of the thermal mode of operation of the modules. For this task, a method has been developed for visualizing thermal processes using the Ansys system of finite element analysis, which has been used to research thermal modes of operation and to optimize the design of the modules created. With the help of the developed method, the temperature fields of the module components, coolant velocity and its flow lines in the developed models of a planar photovoltaic thermal roofing panel and a concentrator photovoltaic thermal two-sided module are visualized.

DOI: 10.4018/978-1-7998-1216-6.ch005

Copyright © 2020, IGI Global. Copying or distributing in print or electronic forms without written permission of IGI Global is prohibited.

INTRODUCTION

The principle of creating a single device that provides simultaneous generation of electrical and thermal energy is the creation of a photovoltaic thermal module (PVTM) (Kharchenko, Panchenko, Tikhonov & Vasant, 2018), when the structure of such a module is provided by placing solar cells on a heated absorbing surface of a flat solar collector. The absorber in this design performs a dual function – firstly, it cools the photovoltaic panel, removing excess energy that is not involved in the generation of electricity, thereby increasing its efficiency, and secondly, it produces thermal energy (figure 1 at the top). Such photovoltaic thermal modules can be used both for autonomous and parallel with the power supply network of facilities (Gapeeva & Zhilenko, 2018; Solovyov, 2018; Sokut & Murovskaya, 2017).

BACKGROUND

Publications in which the well-known collectors (Bekirov & Karkach, 2018; Mitrofanova, 2016) and photovoltaic solar cells (Bekirov & Asanov, 2017; Asanov, Bekirov & Voskresenskaya, 2014) are widespread. The history of the development of such a class of devices as solar photovoltaic thermal modules goes back to the work of Wolf (Wolf, 1976), where it was shown that such a combined system is a promising goal for further development. In the works (Zharkov, 2014; Kern & Russell, 1978; Hendrie, 1979), the authors studied the modules that heat water and air, identified key concepts and main

Figure 1. The principle of creating a photovoltaic thermal module: on the left a photovoltaic module, in the middle a solar collector, on the right a photovoltaic thermal module (at the top) and the most common PVTM designs (below)
(Kharchenko, Panchenko, Tikhonov & Vasant, 2018; Kharchenko, 2014)

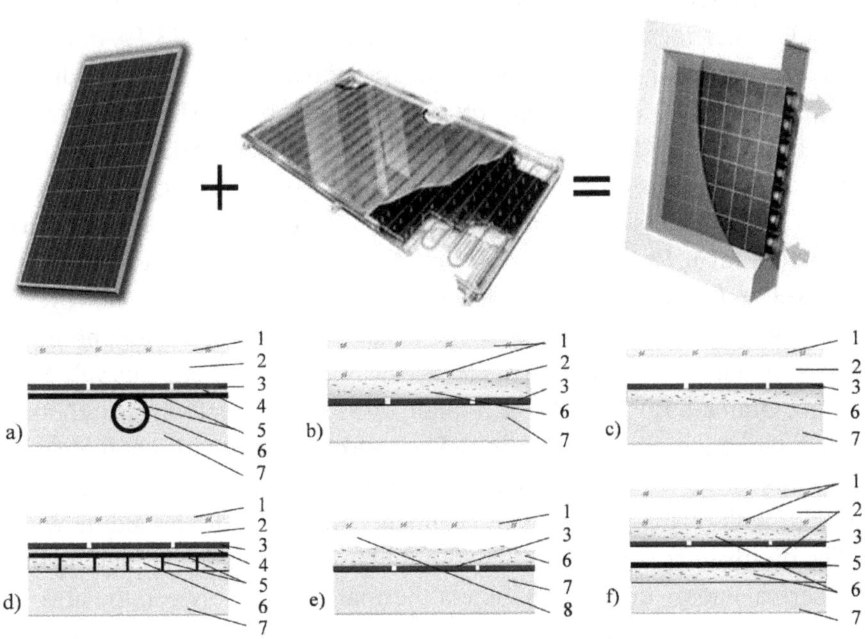

priorities for the development of cogeneration systems of this type. Studies (Al Harbi, Eugenio & Al Zahrani, 1998; Kamilov, Muminov & Tursunov, 2008) showed that high ambient temperature leads to a decrease in photoelectric efficiency, and thermal efficiency in these conditions remains high. In the works (Sandnes & Rekstad, 2002; Bosanac, Sørensen, Katic, Sørensen, Nielsen & Badran; Kharchenko, Nikitin, Tikhonov & Gusarov, 2013), it was shown that increasing the efficiency of thermal conversion of solar radiation implies a high operating temperature, which at the same time reduces the efficiency of photoelectric conversion. It is not surprising that in the reviewed papers attention is paid to the development of the optimal design of the heat generating part of the PVTM, since the optimal design of such a module will ensure high efficiency of the photovoltaic installation and high thermal output. In the study (Garg & Agarwal, 1995), the authors considered a PVTM with forced circulation of the refrigerant with optimization of the operating modes of the circulation pump. In the work with thermosyphon PVTM (Garg & Agarwal, 1994), it was noted that the average efficiency of solar cells does not depend on the area occupied by them on the absorbing surface of the solar module, however, thermal efficiency significantly depends on this factor. Moreover, an important parameter affecting the operation of the installation is the amount of water in the storage tank of the thermosyphon system. In the study (Prakash, 1994), an analysis of transients that occur in a flat PVTM designed for heating air and water, where the efficiency of PVTM with air was significantly lower than the efficiency of PVTM with water was carried out. In the work (Bergene & Lovvik, 1995), the effect of the ratio of the width to the diameter of the pipe in PVTM was considered, where the total efficiency of the module was 60–80%. In the Netherlands (De Vrie, 1998), stationary PVTMs of various configurations (with single glazing, double, without glazing) were investigated, where the best performance was demonstrated by PVTM with single glazing. Analysis of the PVTM exergy of the module (Fujisawa & Tani, 1997) showed that the density of the output radiation of an open structure is slightly higher than that of a single-glazed structure. In an experimental study (Hosseini R., Hosseini N. & Khorasanizadeh, 2011), it was shown that the overall efficiency of a combined module is greater than that of a conventional solar panel. Confirmatory experiments (Rawat, Debbarma, Mehrotra et al., 2014) were carried out under full-scale conditions with a fixed water flow rate of 0,002 kg/s and different water temperatures at the inlet of the PVTM. Daily thermal efficiency was 50,1% and the overall efficiency of the developed module exceeded 73%. An increase in electrical efficiency with the use of cooling was also shown in (Buonomano, Calise & Vicidimini, 2016). In the work (Norton & Edmonds, 1991) with the year-round operation of PVTM with antifreeze, it was shown that in summertime there is a slight decrease in efficiency. Improving the design of the PVTM (Ibrahim, Jin, Daghigh, Salleh, Othman, Ruslan et al., 2009), where the absorber is made in the form of a rectangle in cross section, it is possible to manufacture an absorber in the form of a V-shaped triangle, which is presented in the work (Othman, Ruslan, Sopian & Jin, 2009), thereby reducing heat losses and improving heat removal. In (Sevela & Olesen, 2013) a PVTM with a tubular heat exchanger is presented, where the maximum efficiency of a liquid solar collector in the installation was 48% with photoelectric converters turned off, and with simultaneous production of electrical and thermal energy, its value decreased to 42%. In the work (Ibrahim, Othman, Ruslan, Mat & Sopian, 2011) a flat photoelectric thermal collector was considered, in which the radiation receiver is a light-absorbing plate with photoelectric elements, and under this plate are tubes with a circulating coolant. Studies (Dubey & Tay, 2012) have shown greater efficiency of a PVTM with rectangular coolant channels compared with a module with a tubular sheet heat exchanger. The disadvantages of PVTMs with tubular heat exchangers include the low efficiency of the transfer of thermal energy due to insufficient thermal contact between the absorber and the substrate of the photovoltaic cells. To solve this problem, a PVTM is presented, in which the photoreceiver is mechanically pressed against the thermal collector without the aid of any mounting glue. Compared to a photoelectric thermal collector with a tubular radiator, this solution provides better thermal contact between the photoreceiver and the heat exchanger, which increases the efficiency of solar energy conversion, but this solution is problematic due to the fragility of the photoelectric converters and the need for a sealant to maintain high electrical characteristics.

Today, one of the most advanced designs in terms of optimization is the modules of Solimpeks (www.solimpeks.com) and Sunsystem (www.sunsystem.bg), which produce PVTMs with a tubular heat exchanger. However, even commercially manufactured modules are distinguished by high material consumption, mass, and, accordingly, cost. In addition, the modules are performed in two versions for various purposes – maximum electrical efficiency and maximum thermal efficiency, which makes one doubt the universality of this development, since in any case one component of the power supply is reduced. It is also a question of the service life of the electrical part of the module at the level of the rated power, the level of which falls over the years.

PVTMs can be classified according to various criteria, such as: type of solar cells used (monocrystalline, polycrystalline or amorphous silicon, thin films, etc.), glazed or unglazed module, type of cooling fluid (water, glycol or air), presence of a concentrator etc. In practice, the most common air and water PVTMs, which can be used to supply hot water. PVTM with water are created on the basis of a solar collector, which is complemented by solar elements placed on the surface of the absorber. PVTMs with concentrators are also of interest because they help reduce the number of expensive solar cells and increase their efficiency. The best results shows PVTM with single glazing, where the density of the output radiation is lower than that of an open structure.

From the above review and studies (Kharchenko, Panchenko, Tikhonov & Vasant, 2018; Kharchenko, 2014) it is possible to depict generalizing designs of PVTMs (figure 1 below), where a) PVTM with an absorber of the "metal sheet-tube" type; b) PVTM with an absorber in the fluid channel above the photovoltaic panel; c) PVTM with an absorber in the form of a channel for the liquid under the photovoltaic panel; d) PVTM absorber with rectangular channels; e) PVTM with an absorber in the channel, partially filled with liquid and located above the photovoltaic panel; f) PVTM with double absorber and air gap. The above PVTM designs have common elements: 1 – transparent coating; 2 – air; 3 – photovoltaic panel; 4 – heat-conducting electrical insulating fixing (adhesive) material; 5 – heat absorber; 6 – coolant (liquid); 7 – insulation material; 8 – evaporating liquid.

PVTM with a metal sheet-tube absorber is a traditional solar collector, on the working surface of which a photovoltaic panel is placed (figure 1a below) (one of the most common types of PVTM, this type of module can be called a thermal photoelectric collector). With additional transparent heat insulating coating increases the reflection of solar radiation, reduced heat loss from the working side of the solar cell, which leads to deterioration of electricity generation. A PVTM with a channel-shaped absorber is a channel with a liquid (water) adjacent to a photovoltaic panel in front or behind. If the channel with the liquid is in front of the working surface of the solar cell, then the liquid absorbs part of the solar radiation that passes through it, and thus is part of the absorber. The advantage of this design compared with a metal sheet-tube design is uniform dispersion of solar energy. When using this type of PVTM, it is desirable that the liquid absorbs as much sunlight as possible with the wavelength of that part of the spectrum that solar cells do not convert into electricity, thanks to which the solar cell will work under conditions close to optimal due to the removal of unused energy by the liquid layer (which imposes restrictions on the type of fluid). The choice of a suitable liquid for such a channel is a very difficult task, because the liquid absorbs part of the working spectrum of solar cells and their electrical efficiency decreases by about 4% when using water in such PVTM (Bakker, Zondag & Helden, 2002). The PVTM design may have another configuration with a heat removal channel located behind the photovoltaic panel (figure 1c below), that is, adjacent to the back side of the solar cell. In the figure 1d below, the bottom of the absorber consists of a set of vertical rectangular channels that are located behind the photovoltaic panel. This structure forms ribs-absorbers, which increase the surface area to remove heat from the liquid, thereby increasing the heat sink.

As a result of the analysis of the literature and the research conducted, it can be concluded that the receivers of solar PVTMs of planar and concentrator constructions may have different designs, but it should be noted that the structural components used in such modules have a fundamentally similar structure. The main task of modern research is to find the optimal design construction, at which the overall efficiency of the module will be maximum. In the search papers (Florschuetz, 1979; Kalogirou, 2001; Dubey & Tiwari, 2008; Ji, Lu, Chow, He & Pei, 2007; Fontenault & Gutierrez-Miravete; Kiflemariam, Zevallos, Almas & Lin; Karthick & Helen Catherine) authors present various mathematical models of the operation of solar PVTMs. However, the presented models are not illustrated by three-dimensional models of the modules being developed, their design features and design recommendations for their production. Also, in the reviewed papers, the visualization of the processes occurring in such modules is poorly implemented; no confirmation of the investigations using finite element analysis systems has been identified. At the same time, along with mathematical models and graphs, there is a need a holistic view of the processes occurring in such complex design modules, using visualizations of thermal fields, coolant velocity and flow lines, which will allow to fully analyze the proposed calculations and designs.

In order to implement such search tasks, it is necessary to create a design method for modules that would allow creating such solar PVTMs of various designs (one-sided, two-sided, planar, concentrator) in the computer-aided design system (AutoCad, Compass 3D (www.kompas.ru), etc.) in the form of three-dimensional models. To clarify the parameters of three-dimensional models (material, dimensions, etc.), it is also necessary to use the method of finite element analysis to optimize the design taking into account those or other requirements of the developer (maximum electrogeneration, maximum heat generation, maximum overall efficiency). The use of such a method is implemented in software systems of finite element analysis, for example, Ansys (www.ansys.com), where the designer has the opportunity to study thermal processes in detail and visualize the thermal processes that occur in a module under various conditions.

MODELING OF SOLAR PHOTOVOLTAIC THERMAL MODULES

As a tool for creating two-dimensional and three-dimensional models of solar PVTMs, it should be noted computer-aided design systems, for example, the Ascon software complex – KOMPAS 3D, which for several decades has established itself as a worthy domestic three-dimensional design tool with full compliance with the uniform standard of Russian Federation design documentation. KOMPAS 3D – family of computer-aided design systems with the possibility of registration of design and construction documentation according to standards. Programs of this family automatically generate associative types of three-dimensional models.

In the considered method of creating models of solar PVTMs, designs of modules of stationary and mobile power generation are developed, the main differences of which are the number of illuminated sides of solar cells (single and bilateral) and types of heat removal (front, rear and bilateral) (figure 2 at the top). The geometrical parameters included in the component models are optimized in the Ansys finite-element analysis software package. The developed method is used to create the main types of solar PVTMs that are used in planar and concentrator systems (Panchenko, Kharchenko & Vasant, 2019).

The first type of solar PVTM for use in a concentrator system is a solar module with a two-sided ray-sensing surface ("Model 4" in figure 2 below). The number of components used in this type of module is limited by the need to ensure the transparency of both beam-sensing sides of the module in the spectrum of solar radiation in which the solar cell generates electricity. The second type of solar PVTM for use in a planar system is a solar module with a single-sided beam-receiving side ("Models 1, 2 and 3" in figure 2 below). In this type of module, the number of components used can be expanded and the design is complicated. Both types of receivers of PVTMs are also subdivided according to the type of cooling of the radiation-receiving side of the receiver (heat removal) – front, rear, two-sided. Depending on the radiation-receiving sides and the type of heat removal, the module is created according to one of the four models (figure 2 below), which include the various components shown in figure 2 at the top.

The developed method of creating models of solar PVTMs allows to create models of photoreceiver with one-sided solar cells and facial heat removal ("Model 1" in figure 2 below); one-sided solar cells and rear heat removal ("Model 2" in figure 2 below); one-sided solar cells and two-sided heat removal ("Model 3" in figure 2 below); bilateral solar cells (Strebkov, Polyakov & Panchenko, 2013; Panchenko, Strebkov, Polyakov & Arbuzov, 2015) and bilateral heat removal ("Model 4" in figure 2 below).

Created three-dimensional models of solar PVTMs based on the developed method should be investigated in the system of finite element analysis (for example, Ansys) to visualize the thermal processes occurring in these models (figure 3), then optimize the layer-by-layer module design. Ansys is a universal software system of finite element analysis, which is widely used in the field of automated engineering calculations, solving tasks of fluid and gas mechanics, heat transfer and heat exchange, as well as a wide range of other specialized tasks. In the developed method, the Ansys software system allows analyzing the three-dimensional display of thermal processes occurring in the modules, the flow and speed of the cooling agent, the temperature distribution in the layers of the components, as well as the optimization of the design parameters of the components included in these modules.

The developed method allows obtaining an array of temperature values of the components used, which are analyzed and optimized on the basis of the required output parameters of the module (maximum electrogeneration, maximum heat generation, maximum overall efficiency). Simulation of the thermal state in the system of finite element analysis with obtaining an array of temperatures allows obtaining a more detailed three-dimensional picture of the thermal state of the module, as opposed to two-dimensional analytical modeling, that allows comparing the thermal state of all components and each separately for different parameters.

As an example of the implementation of the thermal process modeling method, the modules developed on the basis of the Model 2 in the form of a photovoltaic thermal roofing panel and the Model 4 in the form of a bilateral photovoltaic thermal module are presented.

Simulation of thermal processes and optimization of the layered structure of a solar photovoltaic thermal module with front illuminating and rear cooling ("Model 2")

As an illustration of the application of the developed method of creating three-dimensional models of solar PVTMs, a model of solar photovoltaic thermal roofing panel (Panchenko, 2018; Strebkov, Kirsanov & Panchenko; Panchenko, 2019) (figure 4 on the left) was developed, consisting of 11 components. To create an assembly in the computer-aided design system KOMPAS 3D, 11 components have been cre-

Figure 2. Scheme of creating three-dimensional models of solar photovoltaic thermal modules with different ray-sensing sides and types of heat removal (4 models of modules) (at the top) and created models (below)

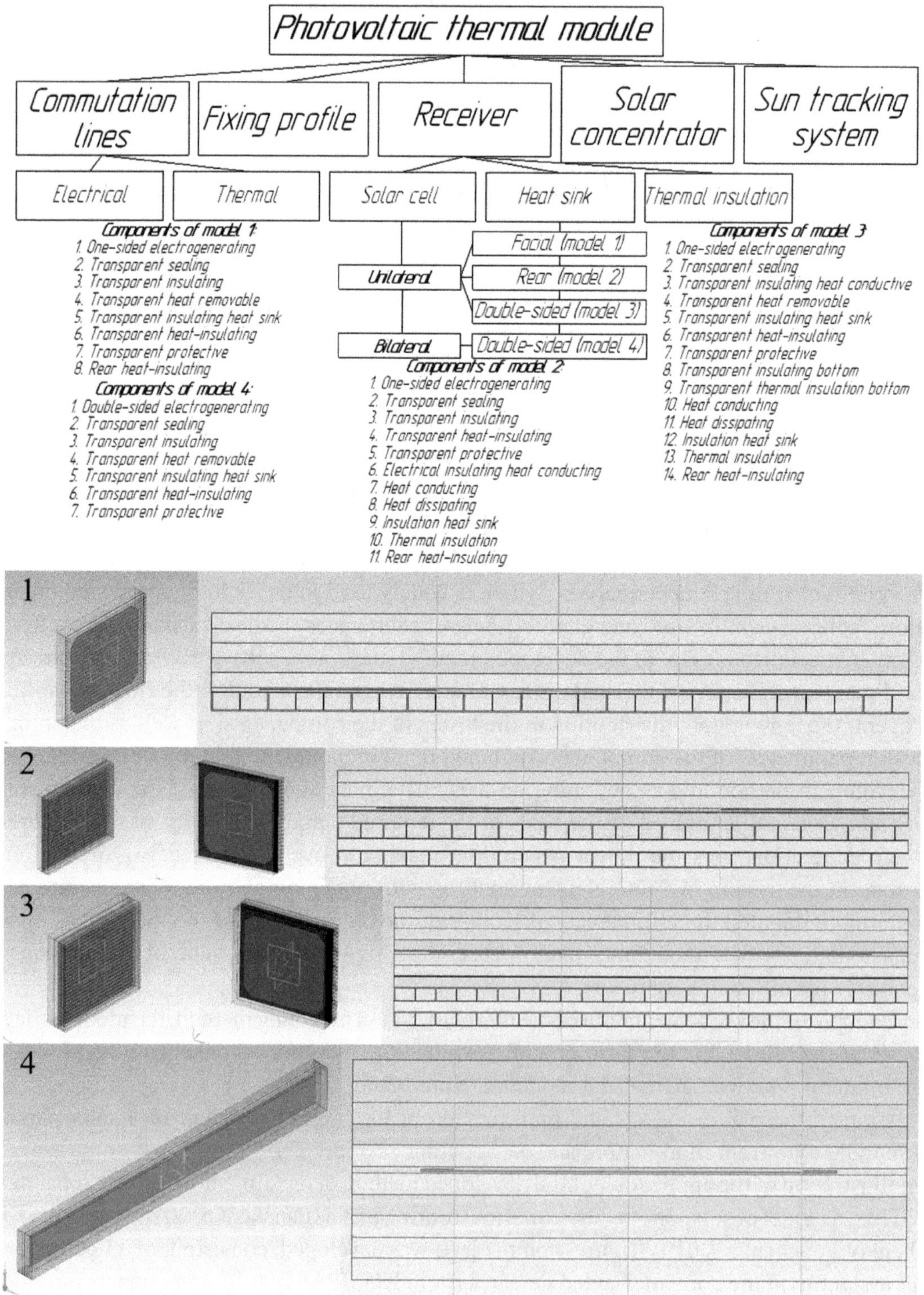

Figure 3. Method of visualization of thermal modes of operation of models of solar photovoltaic thermal modules with a layered structure of components

1. Justification of the overall dimensions of the module	11. Configuring the three-dimensinal model, the matrial properties of each of the layers (components) and boundary conditions in the module "CFX"	21. Setting the end adiabatic surfaces of the module
2. Development of two-dimensional sketches in computer-aided design	12. Disabling "Constant Domain Physics" setting to provide fluid domains with different properties	22. Setting th boundary condition of the type "Inlet", which describes the entry of the cooling agent
3. Creating a three-dimensional solid model in computer-aided design	13. Creation and configuration of all materials used in the studied structure	23. Setting te boundary condition of the type "Opening", describing the place of free expiration
4. Saving a three-dimensional solid model in Parasolid format	14. Specifying in "Basic Settings" the geometric bodies entering the domain, detrmining the phase state of the substance of the domain and the choice of material	24. Selection of interface type "Interface" and the inclusion of heat exchange through it
5. Creatig a project in the CFX module of the ANSYS Workbench package	15. Configure in the "Solid Models" way the domain interacts with other domains	25. Setting the heat source
6. Opening of a three-dimensional model in the module "Geometry" ANSYS Workbench	6. Select from the list of "Heat Transfer", "Option" item "Thermal Energy", which allows the domain to exchange heat with other domains	26. Setting the maximum number of iterations in "Solver Control" in "Max. Iterations"
7. Creating a calculation using "Import External Geometry File ..."	17. In the "Iniialization" tab, configure the initial state of the system (initial temperature)	27. Running a calculation in the "Solution" section of the "CFX" module
8. Setup and construction of a finite element mesh in the "Mesh" module	18. In the "Solver Control" tab, the solver setting	28. View results in the "Results" section of the "CFX" module
9. Select the type of CFD task to be solved and the CFX solver used	19. Configuring an interface of the type "Symmetry"	29. Displays plots of velocity, pressure, temperature, etc.
10. Dimensioning finite element sizes using partitions of different sides of layers	20. Setting the outer top and bottom surface of the module	30. Construction of streamlines of liquids and gases

ated that perform various functions as individual parts that make up the module being developed. The main structural element is the housing (mounting profile), to which other components are attached. The cooling agent is water that washes black aluminum radiator. Solar cells are sealed using a two-component polysiloxane compound (Poulek, Strebkov, Persic & Libra 2012; Strebkov, Persits & Panchenko, 2014), a thin, transparent film and a black tape around the perimeter. The heat-insulating area is an air gap that is limited to the optically transparent glass from the front surface of the module. The created three-dimensional model is associated with a two-dimensional drawing for obtaining design documentation and the subsequent development of the manufacturing technology of an experimental module (figure 4 on the right).

Solar planar photovoltaic thermal roofing panel (the example of "Models 2" in figure 2 at the top and in figure 2 below) consists of 11 different components (figure 2 at the top and table 1), which have different physical and geometric properties and which are entered as input data in the simulation of thermal processes in the system of finite element analysis.

When sequentially creating and assembling (example of component assembly is "Model 2" in figure 2 below) of the presented components in table 1 with the specified component thicknesses (which are subsequently optimized in the finite element analysis system), an assembly unit is formed in the CAD system (figure 4 on the left), with which can get design documentation. The length and width of the presented components was determined based on the size of standard solar cells, and can be optimized in the system of finite element analysis. In the figure 4 below, the layer-by-layer structure of the developed module is visible (the base "Model 2" in the figure 2 below). The thicknesses of the various components are subsequently optimized for the production of the prototype (figure 4 on the right).

In the process of modeling thermal modes of operation according to the developed method in the first stage, the narrow middle part of the roofing panel with a coolant flow rate (water) of 0,5; 5 and 50 g/s, where the gas gap was optimized from 3 to 9 mm (transparent heat-insulating component) and coolant flow rate (heat dissipating component) (figure 2 at the top – "Components of model 2") (table 2).

After optimization of the gas gap (9 mm) and the coolant flow rate (0,5 g/s), half of the symmetric model of the roofing panel (the coolant flow rate corresponding to the model is 1,7 g/s) was tested, where the temperature fields of the model components in the longitudinal section were visualized (figure 5), coolant flow lines (water) (figure 6), temperature of the heat dissipating and electrogenerating components (figure 7), as well as the temperature of the module components in the inlet and outlet zones of

Figure 4. Three-dimensional model of the assembly unit of a planar photovoltaic thermal roofing panel (left), an optimized prototype (right) and its layer-by-layer structure ("Model 2") (below)

Table 1. Thickness of layers (components), materials and their properties (the transparent side of the module is located on top)

Component Name	Thick., mm	Material	Therm. Conduct., W/m·K	Dens., kg/m^3	For Liquids		Heat Cap., J/ kg·K	Therm. Exp. Coeff., 1/K
					Kin. Vis., Pa·s	Dyn. Vis., m^2/s		
1. Transparent protective *glass*	4	Glass	0,937	2530			750	$8{,}9 \cdot 10^{-6}$
2. Transparent heat-insulating *air*	9,5	Air Nitrogen at 0 °C	0,0244 $2{,}43 \cdot 10^{-2}$	1,293 1,21	$17{,}2 \cdot 10^{-6}$ $16{,}7 \cdot 10^{-6}$	$13{,}28 \cdot 10^{-6}$ $13{,}78 \cdot 10^{-6}$	1005 1051	
3. Transparent sealing *germ*	0,1	Polyethylene PET Glass	0,14 0,937	1330 2530			1030 750	$60 \cdot 10^{-6}$ $8{,}6 \cdot 10^{-6}$
4.1. Transparent insulating	0,4	Butyl tape	0,12	920			1950	$110 \cdot 10^{-6}$
4.2. Transparent insulating	0,23	Polysiloxane PMS Ethylenevinylacetate EVA Epoxy resin Epoxy	0,167 0,35 0,59	950 931 1200			1175 1400 950	$100 \cdot 10^{-6}$ $180 \cdot 10^{-6}$ $55 \cdot 10^{-6}$
5. One-sided electrogenerating *PVe*	0,2	Silicon	148	2330			714	$2{,}54 \cdot 10^{-6}$
6. Electrical insulating heat-conducting *insul*	0,2	Polysiloxane PMS Ethylenevinylacetate EVA Epoxy resin Epoxy	0,167 0,35 0,59	950 931 1200			1175 1400 950	$100 \cdot 10^{-6}$ $180 \cdot 10^{-6}$ $55 \cdot 10^{-6}$
7. Heat conducting *HEx*	0,3	Copper Aluminum Polyethylene	385 230 0,14	8900 2700 1330			383 897 1030	$16{,}6 \cdot 10^{-6}$ $22{,}2 \cdot 10^{-6}$ $60 \cdot 10^{-6}$
8. Heat dissipating *cool*	3	Water Air Freon (gas)	0,569 0,0244 0,0117	1000 1,293 4,39	$1788 \cdot 10^{-6}$ $17{,}2 \cdot 10^{-6}$ $11{,}2 \cdot 10^{-6}$	$1{,}789 \cdot 10^{-6}$ $13{,}28 \cdot 10^{-6}$ $0{,}778 \cdot 10^{-6}$	4182 1005 867	
9. Insulating heat sink *frame 1*	3	Polyethylene ABS PLA	0,14 0,2 0,13	1330 1040 1220			1030 1800 1800	$60 \cdot 10^{-6}$ $90 \cdot 10^{-6}$ $68 \cdot 10^{-6}$
10. Thermal insulation *gap*	4	Air	0,0244	1,293	$17{,}2 \cdot 10^{-6}$	$13{,}28 \cdot 10^{-6}$	1005	
11. Rear heat-insulating *frame 2*	3	Polyethylene ABS PLA	0,14 0,2 0,13	1330 1040 1220			1030 1800 1800	$60 \cdot 10^{-6}$ $90 \cdot 10^{-6}$ $68 \cdot 10^{-6}$

the coolant (figure 8). Based on the presented drawings, an array of temperatures of all components was created (table 2) for the subsequent optimization of their thicknesses; the quality of the radiator washer (heat conducting component, table 1) and the uniformity of coolant flow lines (heat dissipating component, table 1) were analyzed at various water consumption; the locations of overheating of the solar cells (one-sided electrogenerating component, table 1) and underheating of the heat carrier, and the place of its stagnation in the housing were analyzed. The presented drawings (the result of three-dimensional modeling of the thermal state) carry the necessary information for analyzing and optimizing the thermal state of the components of the module and the quality of its cooling.

Table 2. Simulation results at various values of the gas gap (transparent heat- insulating component) and coolant flow rate (heat dissipating component)

Gas gap size (transparent heat-insulating component), mm	**3**			**5**			**7**			**9**		
Mass flow rate of the cooling agent (heat dissipating component), g/s	0,5	5	50	0,5	5	50	0,5	5	50	**0,5**	5	50
Glass temperature at the coolant inlet (transparent protective component), °C	22	20	20	21	20	20	21	20	20	21	20	20
Glass temperature at the coolant outlet, °C	24	21	21	24	21	21	25	21	21	22	21	20
Change criteria, % (the priority of the fourth category – the optimum minimum)	9	5	5	14	5	5	19	5	5	5	5	0
The temperature of the gas inside the gas gap at the entrance of the coolant (transparent heat-insulating component),°C	22	21	20	22	21	20	22	21	20	22	21	20
The temperature of the gas inside the gas gap at the coolant outlet, °C	26	22	21	26	22	21	26	22	21	25	22	21
Change criteria, % (the priority of the fifth category – the optimum minimum)	18	5	5	18	5	5	18	5	5	14	5	5
The temperature of the solar cells at the entrance of the coolant (one-sided electro-generating component), °C	23	22	21	23	22	21	23	22	21	23	22	21
The temperature of the solar cells at the coolant outlet, °C	28	23	22	28	23	22	28	23	22	28	23	22
Change criteria, % (the priority of the **second** category – the optimum minimum)	22	5	5	22	5	5	22	5	5	22	5	5
The temperature of the rear heat insulation at the inlet of the coolant (rear heat-insulating component), °C	20	20	20	20	20	20	20	20	20	20	20	20
The temperature of the rear heat insulation at the outlet of the coolant, °C	22	20	20	23	20	20	23	20	20	22	20	20
Change criteria, % (the priority of the third category – the optimum minimum)	10	0	0	15	0	0	15	0	0	10	0	0
Coolant temperature at the outlet (heat dissipating component), °C	26	20	20	26	20	20	26	20	20	26	20	20
Change criteria, % (the priority of the first (main) category – optimum maximum)	30	0	0	30	0	0	30	0	0	30	0	0
Changes in the total criteria (maximum priority)	-14,9			-22,4			- 25,9			**-9,7**		

It should be noted that the optimization of the gas gap and the coolant flow rate took place according to the criteria of a change of the coolant temperature (priority of the first category). The remaining minor criteria for which no optimization took place were of secondary importance, but also had priority depending on their contribution to the heating of the coolant (the second category – the coefficient of 0,9, the third – 0,8 the fourth category – 0,7 and the fifth – 0,6) (table 2). Also for the change of criteria,

Figure 5. Temperature fields of the model by components in longitudinal section

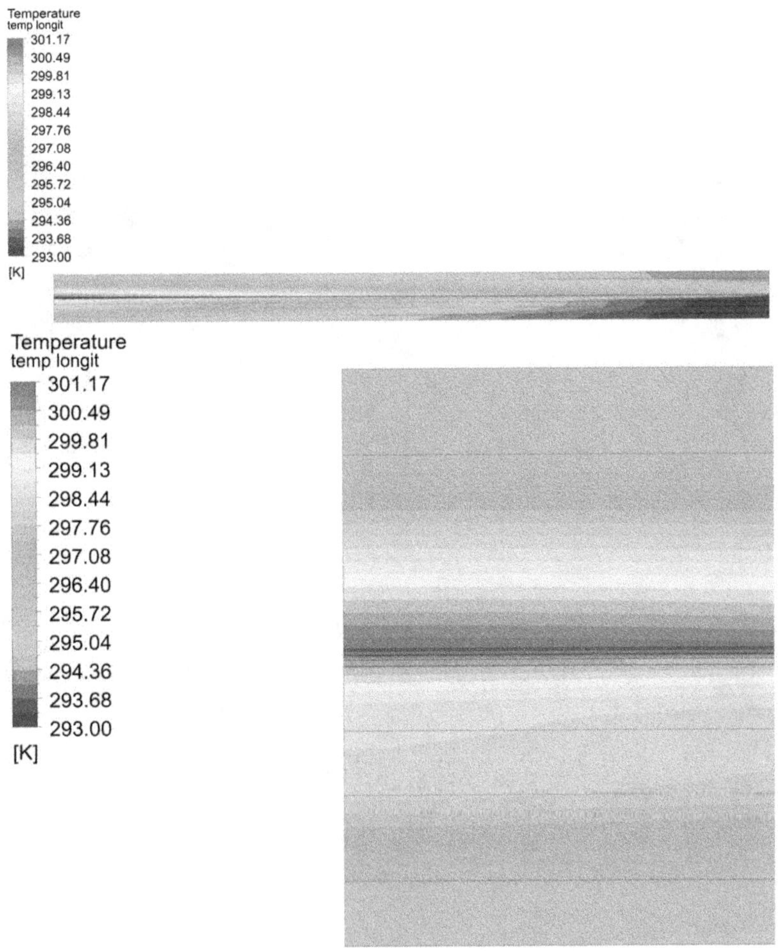

Figure 6. Flow lines of heat carrier (heat dissipating component) (water)

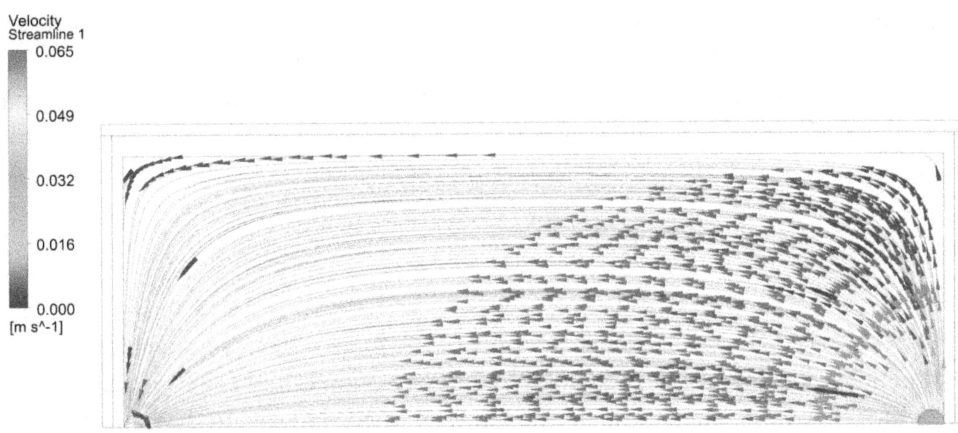

Figure 7. Temperatures of heat carrier (heat dissipating component) (water) and solar cells (electro-generating component) (top view)

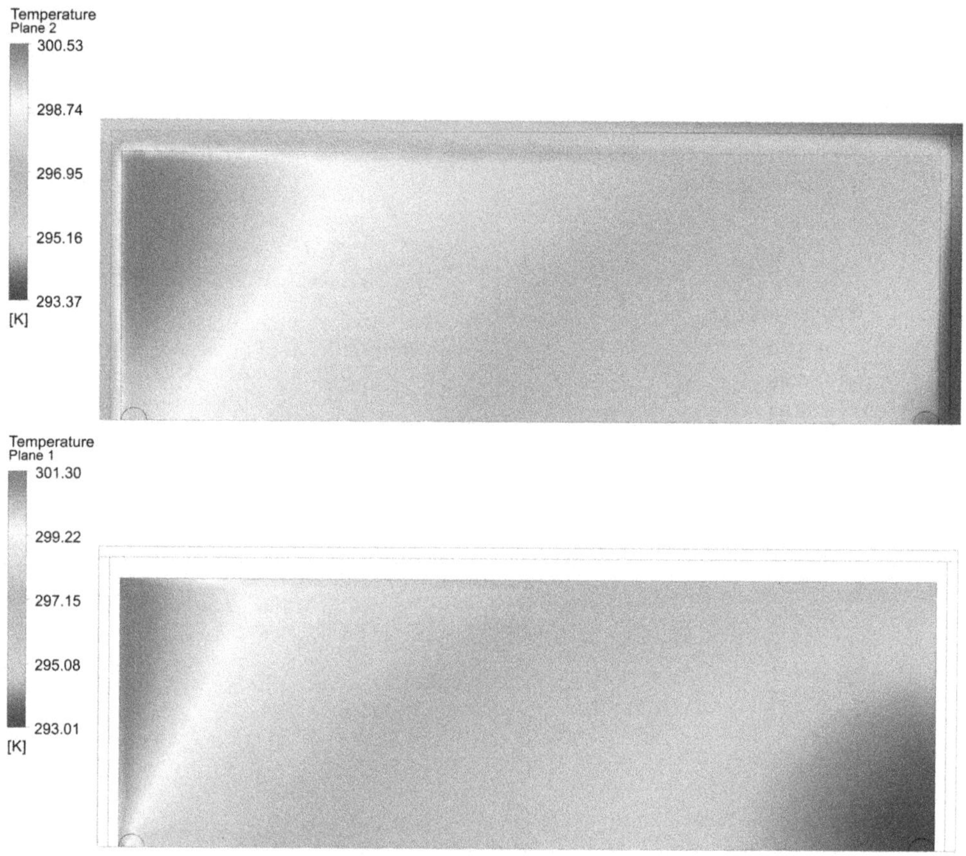

table 2 shows the achievement of their desired values in terms of finding the optimum – minimum or maximum, which is also taken into account when summing up the secondary criteria, that is, the smaller the change, the better the heating of the heat carrier. If the change in the value of the secondary criteria reduces the main criteria (the priority of the first category – heating of the coolant, relative to the value of which a decision is made on the appropriateness of a particular module design), then when summed, the change in such a criterion is given with a negative sign. With the same change in the main criteria (coolant temperature), the change in the total criterion is calculated, which includes changes in all the minor criteria and preference is given to the maximum value of the total criterion:

$$\sum K = 1 \cdot K_1 - 0,9 \cdot K_2 - 0,8 \cdot K_3 - 0,7 \cdot K_4 - 0,6 \cdot K_5$$

Thus, the total criteria reflects the change not only of the main criteria parameter (change in coolant temperature), but also takes into account changes in the remaining minor criteria, that is, changes in the temperatures of the other components, which also plays an important role in the optimization process in view of the level of their contribution to the thermal state of the module and heat carrier in particular.

Figure 8. Temperatures of the components of the module in the zone of inlet and outlet of the heat carrier (heat dissipating component) (water)

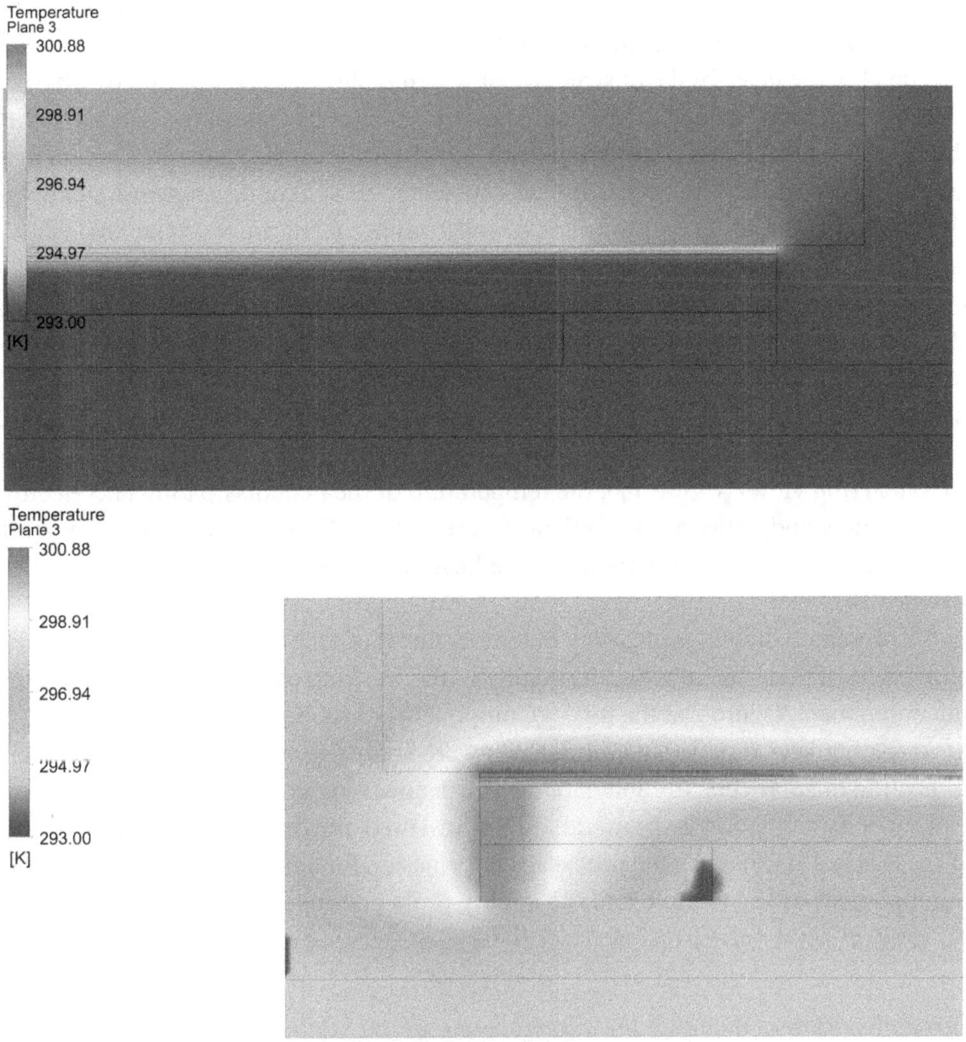

As a result of the optimization, a gas gap with a value of 9 mm and a coolant flow rate of 0,5 g/s for the selected model section (table 2) were selected. However, the number of models studied can be expanded with an increase in the ranges of the geometric characteristics of other components and the coolant flow rate.

Based on preliminary tests the area of solar cells 0,09 m^2; short circuit current 6,2 A; no-load voltage 4,1 V; electric power 20 W; panel operation temperature -50...+90 °C; volume of coolant in the panel 0,3 l; coolant flow rate ...0.1 - 1... l/min (adjustable); inlet temperature 10-30 °C; coolant outlet temperature 35-55 °C.

Simulation of thermal processes and optimization of the layered structure of a solar photovoltaic thermal module with double-sided illuminating and double-sided cooling ("Model 4")

The receiver of a double-sided concentrator folding solar photovoltaic thermal module (the example of "Model 4") consists of 7 different components (figure 2 at the top – "Components of model 4", figure 2 below – "Model 4"), the layer-by-layer structure of which is shown in figure 9 below (basic "Model 4" in figure 2 below). The high-voltage double-sided matrix solar cells (Strebkov, Polyakov & Panchenko, 2013; Panchenko, Strebkov, Polyakov & Arbuzov, 2015) (figure 9 on the right) are located in the focus of the concentrator solar module. The simulation was subjected to a bilateral photovoltaic thermal module located in the focus of the folding concentrator of solar radiation (figure 9 on the left).

When modeling the thermal state of the module, the effect of the concentration of solar radiation (3, 6 and 9 times), the gas gap (3, 5 and 7 mm), as well as the coolant flow (0,5; 5 and 50 g/s) on the thermal state of the coolant was considered (table 3). In the simulation, the temperature of solar cells should not exceed 60 °C due to a decrease in their efficiency.

As a result of modeling in the system of finite element analysis, the temperature fields of the components of the model are presented in longitudinal section (figure 10), the lines and flow velocity of the heat carrier (water) (top view) (figure 11), the temperature of the heat dissipating and electrogenerating components at the inlet and outlet of the coolant (figure 12), the flow rate of the coolant at the inlet and outlet (figure 13), as well as the temperature of the heat dissipating and electrogenerating components (top view) (figure 14).

Based on the presented drawings, an array of temperatures of all components was created for the subsequent optimization of their thicknesses; the quality of the washing of the transparent radiator (transparent insulating component, figure 2 at the top – "Components of model 4") and the uniformity of coolant flow lines (transparent heat removable component) were analyzed at various water flow; analyzed the place of overheating of solar cells (double-sided electrogenerating component) and underheating of the coolant, the place of its stagnation in the module. The presented drawings (the result of three-dimensional modeling of the thermal state) carry the necessary information for analyzing and optimizing the thermal state of the components of the module and the quality of its cooling.

Figure 9. Three-dimensional model of the assembly unit of the bilateral concentrator folding photovoltaic thermal module (on the left), its layer-by-layer structure ("Model 4") (below) and its high-voltage solar cells (on the right)

Table 3. Simulation results for various values of concentration, gas gap and coolant flow rate

The concentration of solar radiation, value	**3**			**6**						**9**					
Gas gap size (transparent heat-insulating component), mm	3	5	7	3		5		7		3		5		7	
Mass flow rate of the cooling agent (heat dissipating component), g/s	5	0,5	5	0,5	5	0,5	5	0,5	5	0,5	5	0,5	5	0,5	5
Glass temperature at the coolant inlet (transparent protective component),°C	20	20	20	20	20	20	20	20	20	20	20	20	20	20	20
Glass temperature at the coolant outlet, °C	20	23	20	29	20	29	20	29	20	34	20	34	20	34	20
Change criteria, % (the priority of the third category – the optimum minimum)	0	15	0	45	0	45	0	45	0	70	0	70	0	70	0
The temperature of the gas inside the gas gap at the entrance of the coolant (transparent heat-insulating component), °C	20	20	20	20	20	20	20	20	20	20	20	20	20	20	20
The temperature of the gas inside the gas gap at the coolant outlet, °C	20	24	20	35	20	34	20	34	20	39	20	38	20	37	20
Change criteria, % (the priority of the fourth category – the optimum minimum)	0	20	0	75	0	70	0	70	0	95	0	90	0	85	0
The temperature of the solar cells at the entrance of the coolant (double-sided electrogenerating component), °C	22	23	22	26	26	26	25	26	26	28	29	28	29	28	29
The temperature of the solar cells at the coolant outlet, °C	26	33	26	44	32	44	32	44	32	56	39	56	39	56	39
Change criteria, % (the priority of the second category – the optimum minimum)	18	44	18	69	23	69	28	69	23	100	35	100	35	100	35
Coolant temperature at the outlet (heat dissipating component), °C	21	29	21	38	22	38	23	38	21	44	23	45	23	44	23
Change criteria, % (the priority of the first (main) category – optimum maximum)	5	45	5	90	10	90	15	90	5	120	15	125	15	120	15
Changes in the total criteria (maximum priority)				-60,6		-57,1		-57,1		-92,5		-84		-80,5	

Figure 10. Temperature fields of the components of the model in longitudinal section

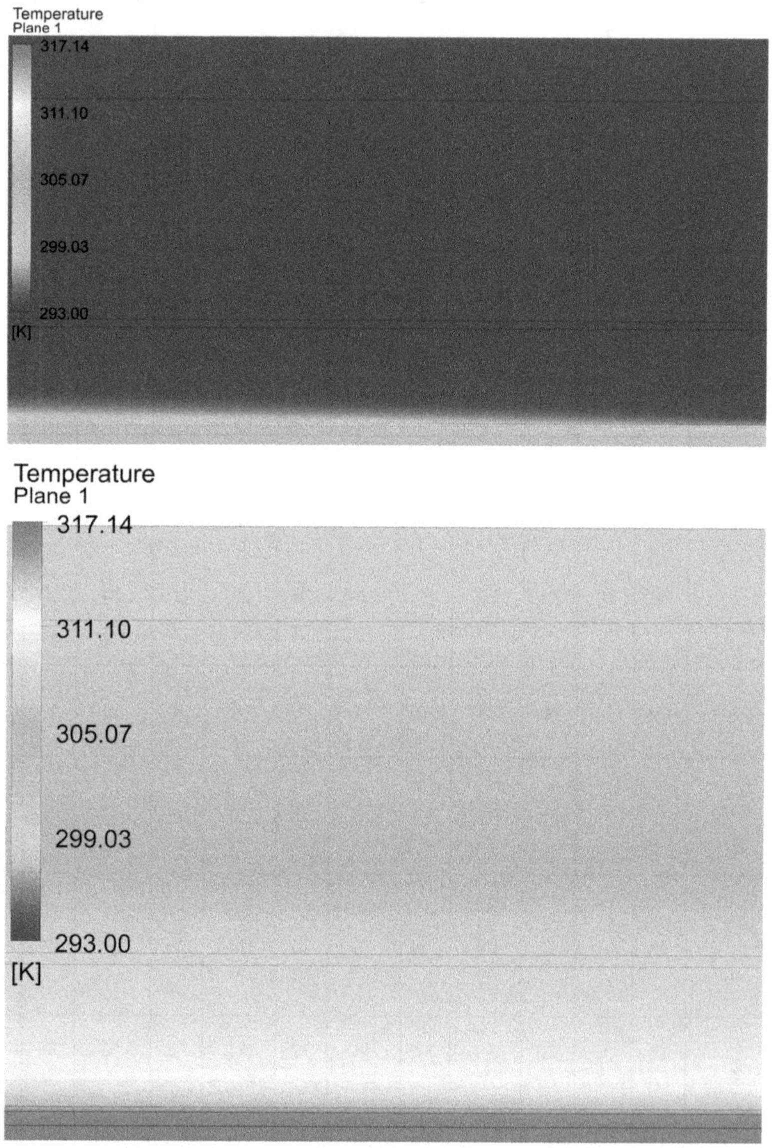

As a result of analyzing the temperature array of the "Model 4" components obtained in the process of modeling the thermal state of the three-dimensional model using the developed method (figure 3), optimization was performed according to the criteria of change in the coolant temperature. With the same change in the main criteria (coolant temperature), the change in the total criteria is calculated (optimization occurred like "Model 2"). As a result of the design optimization, a gap of 5 mm was chosen, the concentration was 6 times and the coolant flow rate was 0,5 g/s for the selected part of the module.

Figure 11. Lines and flow velocity of the coolant (water) (top view)

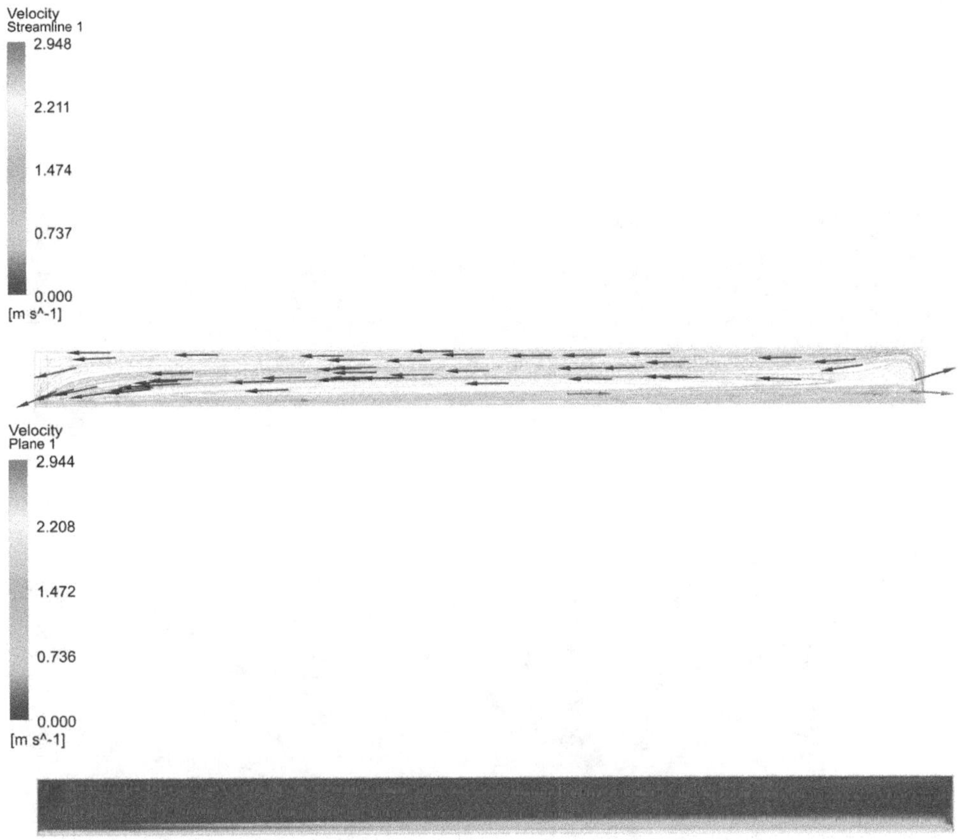

CONCLUSION

Three-dimensional modeling of solar photovoltaic thermal modules according to the developed method, implemented in the computer-aided design system, allows creating modules of various designs for the subsequent determination of their thermal modes of operation. The developed method of thermal calculation of solar photovoltaic thermal modules using visualization of processes in the system of finite element analysis allows judging the feasibility of the created model design and optimizing its layered structure and thermal modes of operation. The developed method allows visualizing and analyzing the thermal mode of the created three-dimensional model, the velocity of the heat carrier and the flow line of the heat carrier of the radiator of the solar photovoltaic thermal module. With the help of the developed method, modeling and optimization of the layered structure of solar photovoltaic thermal modules of planar and concentrator construction was carried out. The modules developed are based on one-sided solar cells with rear cooling and double-sided solar cells with front and rear cooling. For each model, the temperature fields of the layer-by-layer components of which each model consists are shown, visualization of coolant flow lines inside the water cooling radiator, coolant flow rates both along the entire length of the model and at the coolant inlet and outlet, as well as temperatures along the entire length of the heat dissipating (water) and the electrogenerating component (solar cells). Such a visualization of the thermal modes of the module, the nature and quality of cooling allows optimization of the geometrical

Figure 12. Temperatures of the heat dissipating and electrogenerating components at the inlet and outlet of the coolant

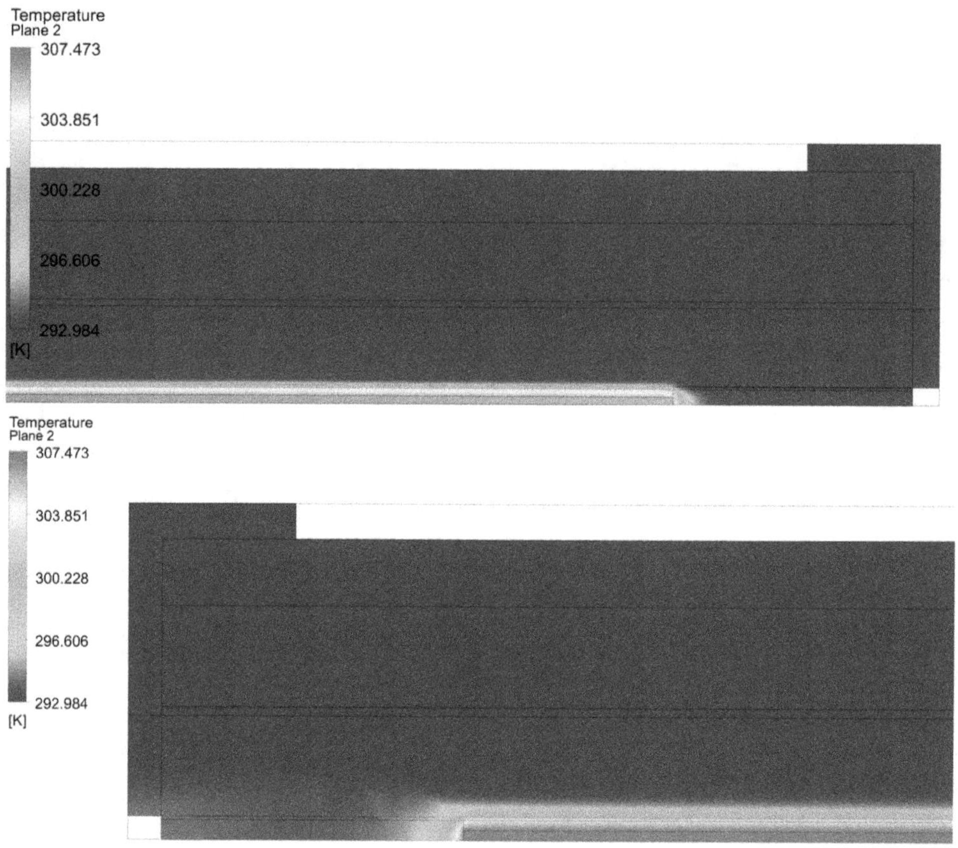

characteristics of each component of the module depending on the selected criteria (coolant temperature at the output of the module). Research and optimization of the structures of solar photovoltaic thermal modules make it possible to achieve maximum efficiency of solar modules due to the summation of electrical and thermal energy in one module, that is not possible to achieve photovoltaic modules and solar collectors separately.

ACKNOWLEDGMENT

The researches were carried out on the basis of financing the state assignments of the All-Russian Institute for Electrification of Agriculture and the Federal Scientific Agroengineering Center VIM; on the basis of funding of the grant "Young lecturer of RUT" of the Russian University of Transport; on the basis of funding of the Scholarship of the President of the Russian Federation for young scientists and graduate students engaged in advanced research and development in priority areas of modernization of the Russian economy, direction of modernization: Energy efficiency and energy saving, including the development of new fuels, subject of scientific research: Development and research of solar photovoltaic thermal modules of planar and concentrator structures for stationary and mobile power generation.

Figure 13. The flow rate of the coolant at the inlet and outlet

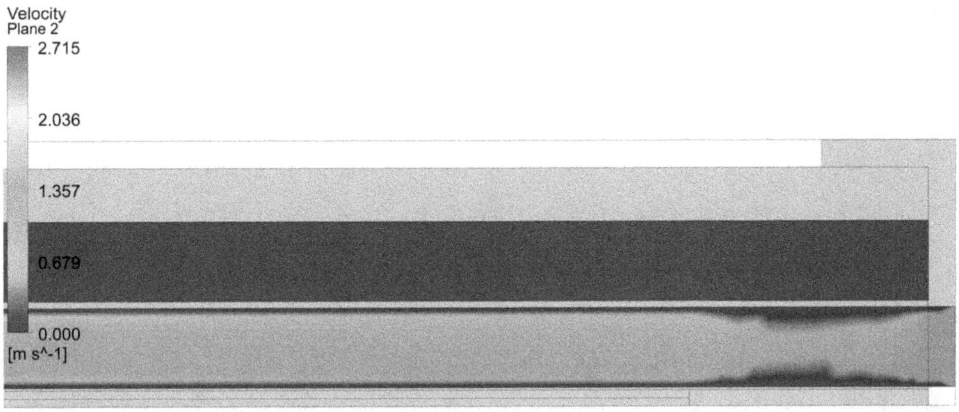

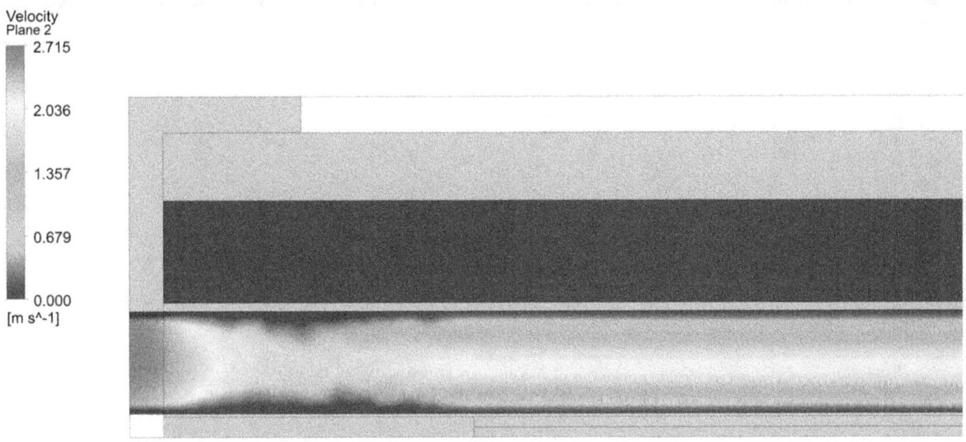

Figure 14. Temperatures of the heat dissipating and electrogenerating components (top view)

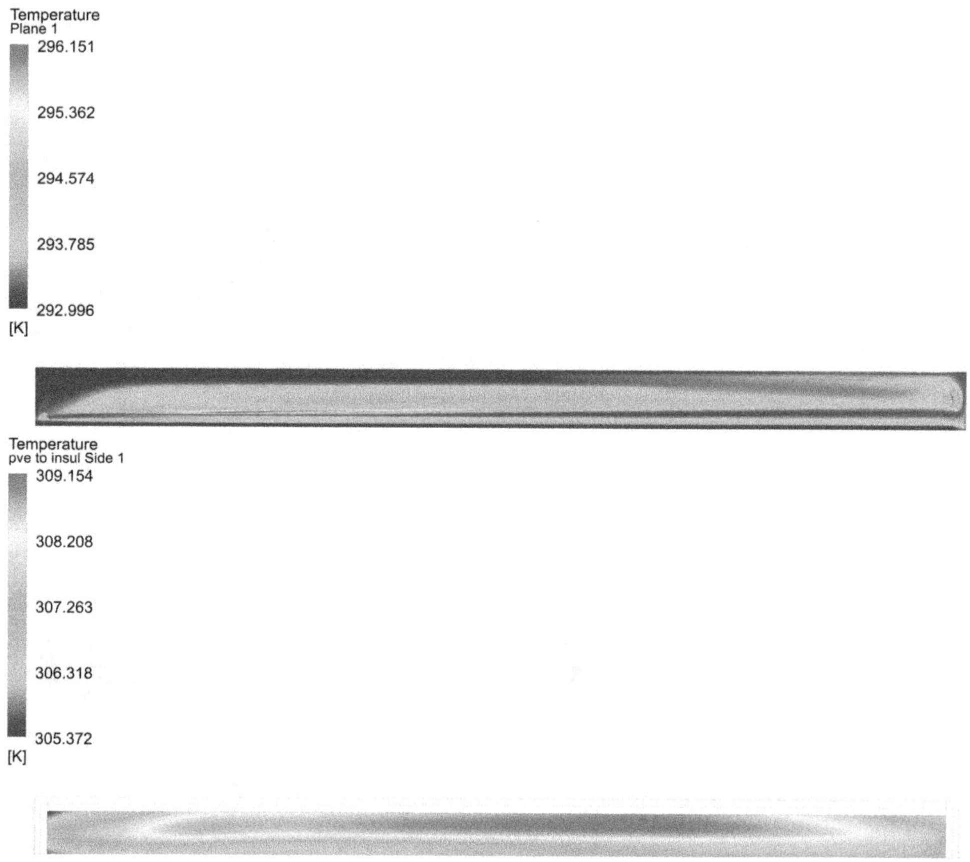

REFERENCES

Al Harbi, Y., Eugenio, N. N., & Al Zahrani, S. (1998). Photovoltaic-thermal solar energy experiment in Saudi Arabia. *Renewable Energy*, *15*(1-4), 483–486. doi:10.1016/S0960-1481(98)00209-2

ANSYS - Engineering Simulation & 3D Design Software. (n.d.). Retrieved from http://www.ansys.com/

Asanov, M.M., Bekirov, E.A. & Voskresenskaya, S.N. (2014). Snizhenie vliyaniya nagreva poverhnosti fotoehlementa na ehffektivnost' ego raboty (Reducing the influence of the heating of the photocell surface on the efficiency of its work). *Stroitel'stvo i tekhnogennaya bezopasnost' [Construction and Technogenic Safety]*, 51, 92-97. (in Russian)

Bakker, M., Zondag, H. A., & van Helden, W. G. J. (2002). Demonstration on a dual flow photovoltaic/thermal combined panel. *Proceedings of PV in Europe-From PV Technology to Energy Solutions Conference and Exhibition*, Rome, Italy, October 7-11. Academic Press.

Bekirov, E.A. & Asanov, M.M. (2017). Analiz sistem vodyanogo ohlazhdeniya fotoehlektricheskih panelej (Analysis of water cooling systems for photovoltaic panels). *Stroitel'stvo i tekhnogennaya bezopasnost' [Construction and Technogenic Safety]*, *6*(58), 55-59. (in Russian)

Bekirov, E.A. & Karkach, D.V. (2018). Dvumernaya model' teplovyh processov v solnechnom kollektore i ee ehksperimental'naya proverka (Two-dimensional model of thermal processes in a solar collector and its experimental verification). *Stroitel'stvo i tekhnogennaya bezopasnost' [Construction and Technogenic Safety]*, 10 (62), 191-201. (in Russian)

Bergene, T., & Lovvik, O. M. (1995). Model calculations on a flat-plate solar heat collector with integrated solar cells. *Solar Energy*, *55*(6), 453–462. doi:10.1016/0038-092X(95)00072-Y

Bosanac, M., Sørensen, B., Katic, I., Sørensen, H., Nielsen, B., & Badran, J. (n.d.). Photovoltaic/Thermal Solar Collectors and Their Potential in Denmark. Retrieved from https://www.ecn.nl/fileadmin/ecn/units/egon/pvt/pdf/EFP1713_00-0014.pdf

Buonomano, A., Calise, F., & Vicidimini, M. (2016). Design, Simulation and Experimental Investigation of a Solar System Based on PV Panels and PVT Collectors. *Energies*, *9*(7), 497. doi:10.3390/en9070497

De Vrie, D. W. (1998). *Design of a photovoltaic/thermal combi-panel* [PhD thesis]. Eindhoven Technical University.

Rawat, P., Debbarma, M., Mehrotra, S., & Sudhakar, K. (2014). Design, development and experimental investigation of solar photovoltaic/thermal (PV/T) water collector system. *International Journal of Science. Environmental Technology*, *3*(3), 1173–1183.

Dubey, S., & Tay, A. A. O. (2012). Experimental Study of the Performance of Two Different Types of Photovoltaic Thermal (PVT) Modules under Singapore Climatic Conditions. *Journal of Fundamentals of Renewable Energy and Applications*, *2*, 1–6. doi:10.4303/jfrea/R120313

Dubey, S., & Tiwari, G. (2008). Thermal modeling of a combined system of photovoltaic thermal (PV/T) solar water heater. *Solar Energy*, *82*(7), 602–612. doi:10.1016/j.solener.2008.02.005

Florschuetz, L. W. (1979). Extension of the Hottel–Whiller model to the analysis of combined photovoltaic/thermal flat plate collectors. *Solar Energy*, *22*(4), 361–366. doi:10.1016/0038-092X(79)90190-7

Fontenault, B. J., & Gutierrez-Miravete, E. (n.d.). Modeling a Combined Photovoltaic-Thermal Solar Panel. Retrieved from http://www.comsol.com/paper/download/150905/gutierrezmiravete_paper.pdf

Fujisawa, T., & Tani, T. (1997). Annual exergy evaluation on photovoltaic-thermal hybrid collector. *Solar Energy Materials and Solar Cells*, *47*(1-4), 135–148. doi:10.1016/S0927-0248(97)00034-2

Gapeeva, N.A. & Zhilenko, O.B. (2018). Avtonomnoe teplosnabzhenie vysotnyh zdanij (Autonomous heat supply of high-rise buildings). *Stroitel'stvo i tekhnogennaya bezopasnost' [Construction and Technogenic Safety]*, *10*(62), 77-89. (in Russian)

Garg, H. P., & Agarwal, R. K. (1994). Study of a photovoltaic-thermal system-thermosyphonic solar water heater combined with solar cells. *Energy Conversion and Management*, *35*(7), 605–620. doi:10.1016/0196-8904(94)90044-2

Garg, H. P., & Agarwal, R. K. (1995). Some aspects of a PV/T collector/forced circulation flat plate solar water heater with solar cells. *Energy Conversion and Management*, *36*(2), 87–99. doi:10.1016/0196-8904(94)00046-3

Hendrie, S. D. (1979). Evaluation of combined photovoltaic/thermal collectors. *Proceedings of the ISES international congress* (Vol. 3, pp. 1865-1869). Academic Press.

Hosseini, R., Hosseini, N., & Khorasanizadeh, H. (2011). An Experimental study of combining a Photovoltaic System with a heating System. *Proceedings of the World Renewable Energy Congress*, Linkoping, Sweden (pp. 2993-3000). Academic Press. 10.3384/ecp110572993

Ibrahim, A., & Jin, G.L. (2009). Hybrid photovoltaic thermal (PV/T) air and water based solar collectors suitable for building integrated applications. *American Journal of Environmental Sciences*, 618–624.

Ibrahim, A., Othman, M. Y., Ruslan, M. H., Mat, S., & Sopian, K. (2011). Recent advances in flat plate photovoltaic/thermal (PV/T) solar collectors. *Renewable & Sustainable Energy Reviews*, *15*(1), 352–365. doi:10.1016/j.rser.2010.09.024

Ji, J., Lu, J., Chow, T., He, W., & Pei, G. (2007). A sensitivity study of a hybrid photovoltaic/thermal water-heating system with natural circulation. *Applied Energy*, *84*(2), 222–237. doi:10.1016/j.apenergy.2006.04.009

Kalogirou, S. A. (2001). Use of TRYNSYS for modeling and simulation of a hybrid PV-thermal solar system for Cyprus. *Renewable Energy*, *23*(2), 247–260. doi:10.1016/S0960-1481(00)00176-2

Kamilov, A. G., Muminov, R. A., & Tursunov, M. N. (2008). Ocenka jeffektivnosti sistemy solnechnogo jelementa i kollektora v uslovih zharkogo klimata [Evaluation of system effectiveness solar cell and collector in a hot climate]. *Geliotehnika*, *2*, 32–35. (in Russian)

Karthick, S. P., & Helen Catherine, R. L. (n.d.). Modeling and Analysis of Solar Photovoltaic Thermal Collector. Retrieved from http://www.rroij.com/open-access/modelling-and-analysis-of-solar-photovoltaicthermal-collector.pdf

Kern, J. E. C., & Russell, M. C. (1978). Combined photovoltaic and thermal hybrid collector. *Proceedings of the 13th IEEE photovoltaic specialists* (pp. 1153-1157). IEEE.

Kharchenko, V., Nikitin, B., Tikhonov, P., & Gusarov, V. (2013). Investigation of experimental flat PV thermal module parametres in natural conditions. *Conference Proceeding - 5th International Conference, TAE 2013: Trends in Agricultural Engineering*, Prague, Czech Republic, September 2-3 (pp. 51-55).

Kharchenko, V., Panchenko, V., Tikhonov, P., & Vasant, P. (2018). Cogenerative PV Thermal Modules of Different Design for Autonomous Heat and Electricity Supply. In Handbook of Research on Renewable Energy and Electric Resources for Sustainable Rural Development (pp. 86-119). Hershey, PA: IGI Global. doi:10.4018/978-1-5225-3867-7.ch004

Kharchenko, V. V. (2014). PVThermal modules. *Proceedings of XI International Annual Conference "Renewable and small energy"* (pp. 248-257). Academic Press.

Kiflemariam, R., Zevallos, F., Almas, M., & Lin, C. (n.d.). Enhanced Transient Modeling of Hybrid Photovoltaic Air (PVT) Module. Retrieved from https://www.comsol.com/paper/download/194257/kiflemariam_paper.pdf

KOMPAS-3D v17. (n.d.). Retrieved from http://kompas.ru/

Mitrofanova, S.A. (2016). Vliyanie vida otrazhayushchih ehlementov ploskih solnechnyh kollektorov na obluchennost' priemnika [The influence of the type of reflecting elements of flat solar collectors on the irradiation of the receiver]. *Stroitel'stvo i tekhnogennaya bezopasnost' [Construction and Technologic Safety]*, *4*(56), 25-28. (in Russian).

Norton, B., & Edmonds, J. E. J. (1991). Aqueous propylene–glycol concentrations for the freeze protection of thermosyphon solar energy water heaters. *Solar Energy*, *47*(5), 375–382. doi:10.1016/0038-092X(91)90031-Q

Othman, M. Y. H., Ruslan, H., Sopian, K., & Jin, G. L. (2009). Performance study of photovoltaic thermal (PV/T) solar collector with V-grooved absorber plate. *Sains Malaysiana*, 537–541.

Panchenko, V., Kharchenko, V., & Vasant, P. (2019). Modeling of Solar Photovoltaic Thermal Modules. *Intelligent Computing & Optimization. Advances in Intelligent Systems and Computing*, *866*, 108–116. doi:10.1007/978-3-030-00979-3_11

Panchenko, V. A. (2018). Solar Roof Panels for Electric and Thermal Generation. *Applied Solar Energy*, *54*(5), 350–353. doi:10.3103/S0003701X18050146

Panchenko, V.A. (2019). Patent No. 188073. Russia.

Panchenko, V.A., Strebkov, D.S., Polyakov, V.I. & Arbuzov, Y.D. (2015). Vysokovol'tnye solnechnye moduli s naprjazheniem 1000 V (High-voltage solar modules with a voltage of 1000 V). *Al'ternativnaya energetika i ekologiya [Alternative energy and ecology]*, *19*(183), 76-81. (in Russian)

Poulek, V., Strebkov, D. S., Persic, I. S., & Libra, M. (2012). Towards 50 years lifetime of PV panels laminated with silicone gel technology. *Solar Energy*, *86*(10), 3103–3108. doi:10.1016/j.solener.2012.07.013

Prakash, J. (1994). Transient analysis of a photovoltaic thermal solar collector for cogeneration of electricity & hot air/water. *Energy Conversion and Management, 35*(11), 967–972. doi:10.1016/0196-8904(94)90027-2

Sandnes, B., & Rekstad, J. (2002). A photovoltaic/thermal (PV/T) collector with a polymer absorber plate: Experimental study and analytic model. *Solar Energy, 72*(1), 63–73. doi:10.1016/S0038-092X(01)00091-3

Sevela, P. & Olesen, B.W. (2013). Development and Benefits of Using PVT Compared to PV. *Sustainable Building Technologies,* 90-97.

Sokut, L.D & Murovskaya, A.S. (2017). Perspektivy razvitiya sistem ehlektrosnabzheniya za schet podklyucheniya vetrovyh i solnechnyh ehlektrostancij s nakopitelyami ehnergii v obshchuyu ehnergosistemu [Prospects for the development of power supply systems due to the connection of wind and solar power plants with energy storage in a common power system]. *Stroitel'stvo i tekhnogennaya bezopasnost' [Construction and Technogenic Safety],* 7 (59), 113-123. (in Russian)

Solimpeks. Volther Hybrid PV-T Panels. (n.d.). Retrieved from http://www.solimpeks.com

Solovyov, A.K. (2018). EHkonomiya ehnergii pri ehkspluatacii zdanij i passivnye sistemy ispol'zovaniya solnechnoj ehnergii [Energy savings in the operation of buildings and passive systems of solar energy use]. *Stroitel'stvo i tekhnogennaya bezopasnost' [Construction and Technogenic Safety], 10*(62), 179-191. (in Russian)

Strebkov, D. S., Kirsanov, A. I., & Panchenko, V. A. (n.d.). Patent for invention 17773387.0-1230PCT/IB2017051470. RU/28.03.16/RUA2016111201. Russia.

Strebkov, D.S., Persits, I.S. & Panchenko, V.A. (2014). Solnechnye moduli s uvelichennym srokom sluzhby [Solar modules with extended service life]. *Innovacii v sel'skom hozyajstve [Innovations in agriculture],* 3(8), 154-158. (in Russian)

Strebkov, D.S., Polyakov, V.I. & Panchenko, V.A. (2013). Issledovanie vysokovol'tnyh solnechnyh kremnievyh modulej [Investigation of high-voltage solar silicon modules]. *Al'ternativnaya energetika i ekologiya [Alternative Energy and Ecology], 6*(2), 36-42. (in Russian)

Sunsystems. (n.d.). Retrieved from http://www.sunsystem.bg/en/fotovoltaika/PV-T/

Wolf, M. (1976). Performance analysis of combined heating and photovoltaic power systems for residences. *Energy Conversion, 16*(1-2), 79–90. doi:10.1016/0013-7480(76)90018-8

Zharkov, S. V. (2014). Assessment and enhancement of the energy supply system efficiency with emphasis on the cogeneration and renewable as main direction for fuel saving. *International Journal of Energy Optimization and Engineering, 3*(4), 1–20. doi:10.4018/ijeoe.2014100101

KEY TERMS AND DEFINITIONS

Computer-Aided Design System: An automated system that implements an information technology for performing design functions, is an organizational and technical system designed to automate the design process, consisting of personnel and a set of technical, software and other automation tools for its activities.

Cooling Radiator: A technical device used for removing heat for a liquid or gaseous coolant from a cooled object in order to cool the object and heat the coolant.

Finite Element Analysis System: A software package based on a numerical method for solving partial differential equations, as well as integral equations arising in solving problems of applied physics, which is widely used to solve problems of deformable solid mechanics, heat exchange, hydrodynamics and electrodynamics.

Modeling: The study of objects on their models; building and studying models of real-life objects, processes or phenomena in order to obtain explanations of these phenomena, as well as to predict phenomena that interest the researcher.

Optimization: The process of maximizing profitable characteristics, ratios and minimizing costs.

Photovoltaic Thermal Module: A solar module, the design of which consists of photovoltaic solar cells for electrical conversion of solar radiation and heat absorber for their cooling and heat transfer to the coolant.

Receiving Surface: The surface of the photovoltaic module/solar collector/photovoltaic thermal module, which receives solar radiation.

Solar Battery: A combination of photoelectric converters (photocells) – semiconductor devices that directly convert solar energy into direct electric current.

Solar Collector: A device designed to heat the coolant (water, air, freon, antifreeze) using solar radiation.

Solar Concentrator: A technical device designed to focus solar radiation into a focal spot with an increase in the concentration of solar radiation.

Solar Radiation Intensity: The density of solar radiation (energy illumination), coming per unit area of the photoelectric module.

Chapter 6
Gainful Utilization of Excess Heat From Power Transformers

Alexander Vinogradov
Federal Scientific Agroengineering Center VIM, Russia

Anatoly Sopov
Oryol State Agrarian University, Russia

Vadim Bolshev
https://orcid.org/0000-0002-5787-8581
Federal Scientific Agroengineering Center VIM, Russia

Alina Vinogradova
Federal Scientific Agroengineering Center VIM, Russia

ABSTRACT

The study analyzes the various methods of gainful utilization of excess heat from power transformers. The ways to reduce heat loss inside the tank power transformer are found. The potential amount of heat emitted by power transformers of different capacities is calculated. New ways of combining the functions of electric transformation and heating in a power transformer are described. A system has been developed to use the excess heat of power transformers in the agricultural power systems. There are the structural and schematic diagrams of the system and a method for calculating its main elements. An improved design of the power transformer cooling system has been developed to combine the functions of electric transformation and heating. Experiments to verify the effectiveness of decisions are described. A feasibility study of the implementation of the developed system was carried out.

DOI: 10.4018/978-1-7998-1216-6.ch006

Copyright © 2020, IGI Global. Copying or distributing in print or electronic forms without written permission of IGI Global is prohibited.

INTRODUCTION

Nowadays, much more attention is paid to energy saving issues, various options for reducing energy costs are increasingly being sought. New ways of generating electric and thermal energy are emerging due to the development of renewable energy sources and distributed energy (Kharchenko, Gusarov & Bolshev, 2019) as well as methods of managing such systems (Vinogradov et al., 2019) reducing costs for the operation of power supply systems (Vinogradov et al., 2018).

Power transformers are characterized by high efficiency reaching 99% and high reliability as they do not contain moving parts. At first glance, the loss of electricity by 1-3% may seem insignificant. But when the power of a transformer is measured in tens of megavoltampers, the energy loss reaches hundreds of kilowatts. Since these losses are allocated in the form of heat, it makes sense to take it for further use.

Combining the functions of voltage transformation and electrical heating in a power transformer is a very promising idea for heating not only electrical installations, but also agricultural facilities. These heat consumers are often located away from central heating systems. Therefore, it is required to install autonomous heating systems, the work of which is accompanied by considerable financial expenses for electricity or other resources. The power transformer can be installed at power plants, at step-down substations, at consumer power centers and directly at electrical consumers where there are facilities that need heating (Baishev & Toropov, 2018).

High-power transformers (from 25 MVA) always acted as heat sources in the works of predecessors since the amount of released heat is so great that it was possible to get enough thermal energy without resorting to significant changes in the design of transformers when high percentage of heat loss. Therefore, there is a need to develop an improved design of the transformer cooling system allowing to maximize the targeted heat extraction for the needs of heat supply. This will make it possible to use power transformers of much lower power (from 1000 kVA) as heat sources. It will expand the area of application of the gainful utilization of the excess heat from power transformers by tens of times. So, the aim of the work is to develop methods and means of gainful utilization of excess heat from power transformers with improved design of their cooling system.

BACKGROUND

When creating and operating any system, the first question is its efficiency. Although a power transformer has almost maximum efficiency, yet a small part of the electric power is lost in its active elements (magnetic core, winding) and is released in the form of excess heat. And if this thermal energy is used to heat any objects? This theme was studied in many works. For example, Lavrentiev (2018) describes the system of the use of transformer heat for the purpose of heating substation premises of 110–400 kV on the basis of an external low-water heat exchanger, a thermal pump and a control cabinet. Kjell (2013) offers a system allowing to profitably use the excess energy of a power transformer. This is achieved by introducing an external heat exchanger, a heat pump and a heat accumulator. At the same time, Novikov and Taybakhtin (2015) described a system that provides for the collection of air transformer emitted by radiators and its transfer to a heated room. A number of options for the taking and using heat from power transformers for heat supply at power substations (PS) are also proposed in works of Gridin & Petrenko (2013), Mostofizadeh and Kulick (1998) and Cortes & Rivera (2010). These methods can be divided into the following groups:

1. With direct supply of heated transformer oil to the heating system
2. With heating water of the heating system in an oil-water heat exchanger
3. With subsequent heating of the heating system water by the heat pump
4. With air heating in the room by an oil-air heat exchanger
5. With direct removal of heated air from the cooling radiators into the room
6. With air heating in the room with a water-air heat exchanger

The use of one or another scheme depends on the type, quantity and power of installed transformers, their load, required thermal performance, type of heat carrier used in the system and remoteness of consumers. So, the scheme with heating of the water of the heating system in an oil-water heat exchanger has been used mainly only at 110-400 kV PS. A circuit with subsequent heating of the heating system water by a heat pump is also applied at 110-400 kV substations. The use of circuits with air heating in an oil-air heat exchanger room and with direct removal of heated air from cooling radiators is possible at 110-220 kV substations with transformers with a capacity of 16-160 MVA, but the heat output in these variants is very low compared to other schemes due to the low heat dissipation properties of air (Table 1).

The use of the same circuit with air heating in a water-to-air heat exchanger (where water heated in an oil-water heat exchanger is supplied through an insulated pipe through a pump) is practical only at 400kV PS with 250 MVA transformers where heated objects are far from heat sources.

Thus, the use of all these developments is possible only for the premises heating of high voltage level PS. At the same time, heat losses from more common transformers of lower voltage levels remain outside the scope of existing methods for the gainful utilization of excess heat from transformers. But in the places of their installation are often objects that require heating.

THEORETICAL UNDERPINNING OF THE POTENTIAL UTILIZATION OF THE EXCESS HEAT FROM POWER TRANSFORMERS

The average temperature of the upper layers of oil in a transformer varies depending on the modes of its operation under the load conditions and the season. According to Russian standart GOST R 52719-2007, the maximum temperature limit of the windings above the ambient temperature is 65-70 °C depending on the mode of oil circulation. Excess oil temperature is allowed within 55-60 °C. At the nominal load of the transformer, the temperature of the upper oil layers should not exceed +95 °C for oil transformers with natural oil circulation.

Table 1. Heat-removing properties of gases and liquids (Semenov, 2015)

Cooling Medium	Pressure,MPa	Physical Properties in Fractions of Air		
		Thermal Conductivity	Density	Heat Transfer
Air	0,100	1,0	1,0	1,0
Transformer oil	0,100	5,3	848,0	21,0
Water	0,100	23,0	1000,0	50,0

Energy losses in the steady state mode of the transformer operation are turning into heat and are transferred from the heated oil through the tank wall to the surrounding air. At the same time, part of the thermal energy from the outer surface of the tank is dissipated due to radiant heat exchange.

The total flow of thermal energy depends on the load of the transformer and in any mode of its operation can be approximately determined by the formula (1) (Borzov & Shabalina, 2002):

$$Q_O = \Delta P_{PT} = \Delta P_{NLL} + \beta^2 \Delta P_{SCL} \tag{1}$$

where Q_O is the heat flow from the tank surface to air due to heat transfer and radiation, W

ΔP_{PT} is the total power loss in the power transformer, W

ΔP_{NLL} is the no-load losses of a transformer, W

ΔP_{SCL} is the short-circuit losses of a transformer, W

β is transformer load factor

Transformer load factor is determined by the formula 2

$$\beta = I_{PT} / I_{PT.nom} \tag{2}$$

where $I_{PT}, I_{PT.nom}$ are currents in the windings of the transformer in the nominal mode of its operation and, accordingly, in different one from this mode, A.

Obviously, at the nominal load of the transformer $\beta = 1$

For example, let's take a building with an area of 150 m^2 and a ceiling height of 3 m as a consumer of excess heat from a power transformer.

Calculation of heat consumption for heating is necessary to create a comfortable stay in the premises.

Determine the minimum heating power for heating the building (formula 3) (Rossii, 2003):

$$Q = \left(k * V * \Delta T\right) / 860 \tag{3}$$

where Q is power consumption of the room, kW / h;

k is the average coefficient of heat loss by the building (in our case, 1.3);

V is the total volume of the entire heated room, in cubic meters. (this room has a volume of 450 m^3);

ΔT is the maximum temperature difference between the street and inside the room (according to Russian construction rules and regulations 41-01-2003, the estimated indoor temperature is 18 °C, and the external one is -26 °C, the temperature difference is 44 °C); 860 is the value required to convert the resulting kcal to kW (Gosstroi Rossii, 2003).

Calculate the power consumption of the room:

$$Q = \frac{1,3 * 450 * 44}{860} = 29,9\,\text{kWh}$$

Thus, a heat source capable of producing up to 30 kW of heat energy is needed to cover the heat demand of a room. Next step is determining the kind of power transformers for this.

Firstly, consider the amount of heat loss of power transformers of different capacities.

For this, the heat losses of the power transformer of a complete transformer substation installed in the power centers of consumers or for individual consumers are determined. The heat losses of several types of transformers will be found from technical characteristics shown in Table 2

The amount of excess heat generated by:

TMG-400/10/0.4: $Q_O = 0,75 + 0,7^2 * 5,5 = 3,5\,\text{kWh}$

TMG-1000/10/0.4: $Q_O = 1.4 + 0,9^2 * 10,6 = 10\ \text{kWh}$

TMN-4000/35: $Q_O = 5,3 + 0,7^2 * 34 = 22\,\text{kWh}$

TDTN-40000/110: $Q_O = 30 + 0,5^2 * 200 = 80\,\text{kWh}$

This leads to the conclusion that transformers with power up to 2500 kVA inclusively lose a very small amount of thermal energy. This volume is only enough for heating a small object located near the PS. The transformers with a capacity of 4,000 kVA to 16,000 kVA emit already enough energy to heat any room or building near a power transformer. The power transformers with a capacity of 25MVA and more are already capable of covering the heat load on the heating of all the premises of the substation.

It can be concluded that transformers with a capacity of more than 25 MVA are most advantageous as sources of heat in electrical substations. But the lower limit of the power of transformers will drop significantly if it is possible to increase the efficiency to removal and transfer heat energy of a power transformer. Before the development of a more advanced system of using the excess heat of power transformers, it is necessary to define the difficulties.

Table 2. Technical characteristics of several types of power transformers (Silovye transformatory, 2018)

Type	Rated Power, kVA	Rated Voltage, kV		Power Loss, kW		Tank Size, mm		
		HV	LV	ΔP_{NLL}	ΔP_{SCL}	Length	Width	Height
TMG-400/10/0.4-U1	400	10	0,4	0,75	5,5	1280	800	1180
TMG-1000/10/0.4-U1	1000	10,5	0,4	1,4	10,6	1380	780	1250
TMN-4000/35	4000	35	10,5;6,3	5,3	34	2290	1390	1960
TDTN-40000/110	40000	115	38,5;6,6	30	200	4610	3370	4860

During the operation of a transformer in its active materials energy losses occur. They are released as heat and the windings and the magnetic system of the transformer begin to heat up gradually. When the temperature rises, temperature difference arises between the active elements of the power transformer and the environment, i.e. transformer oil and air. It causes heat exchange between them. As a result, part of the heat generated in the windings and the magnetic system goes to heat them while the second part is discharged into the environment. The heat accumulation in the tank gradually decreases and the heat transfer increases with increasing temperature. As a result, with long-term preservation of the load mode, the temperature stops rising and all the heat released by the transformer is released into the environment (El Wakil, Chereches, & Padet, 2006; Mufuta & Van den Bulck, 2000; Swift, Molinski, & Lehn, 2001).

The heat flow in the oil transformer goes through a difficult path that can be divided into the following areas (Tikhomirov, 2015):

1. From the inner points of the material of the winding or the magnetic system to their outer surfaces, washed with oil. Here the heat transfer occurs by conduction
2. From the outer surface of the winding or magnetic system to the surrounding oil
3. From active elements to the inner surface of tank walls by oil. Here the heat is transferred by the convection current of the oil
4. From the oil to the inner surface of the tank walls
5. From the outer surface of the tank walls to the surrounding air. Here the heat transfer occurs by radiation and convection

Transformer power increases with increasing current density. The permissible current density is determined by the balance of the amount of heat released in the winding wire (I^2R) and the amount of heat released by the surface of the winding to the external environment. The thermal conductivity of the winding is determined by the thermal conductivity of the wire insulation and the interlayer air. Thermal conductivity can be increased by reducing the total thickness of all types of insulation, using insulation materials with high electrical and mechanical strength. The presence of gaskets in the winding impairs heat transfer (Tikhomirov, 2015).

Currently, most transformers use paper insulation windings impregnated with varnish or resin. Thermal conductivity of such insulation is at the level of 0,17 W/m*°C. Despite the fact that the thermal conductivity of copper is infinitely large compared with the thermal conductivity of all dielectrics, a slight increase in the thermal conductivity of the insulation plays a big role. Among various types of insulation materials, insulation of bakelite varnish has the highest level of thermal conductivity (Table 3).

Thermal conductivity of insulating materials depends on the technology of their processing. Its thermal conductivity increases during the impregnation of insulation, moisture and when compacted by external pressure.

Thus, one of the ways to reduce heat losses in a power transformer is winding insulation using Bakelite varnish or electrical insulating varnished fabrics giving high thermal conductivity and heat transfer.

Transformer oil plays a huge role in the diversion of heat from the active elements of the transformer (Bertrand & Laurichesse, 2007). Heat is extracted both by the oil flowing in the oil channels inside the winding and by the free-flowing oil from the outer winding of the rod.

The temperature of the oil washing all the windings is not constant and increases as it moves upward in the winding channels what leads to unequal temperature distribution in the axial direction of the winding.

Table 3. Specific thermal conductivity of insulating materials (Tikhomirov, 2015)

Material	λ, W/m*⁰C
Dry cable paper	0,12
Paper cable in oil	0,17
Paper cable, varnished	0,17
Insulating cardboard	0,17
Electrical insulating varnish	0,25
Getinax	0,17—0,175
Textolite	0,146—0,162
Glass fiber	0,178—0,182
Bakelite varnish and other varnishes	0,3
Oil in the absence of convection	0,1

The oil heated at the surface of the transformer windings rises in the upper part of its tank and comes into contact with the walls of the tank and then goes down giving them some of its heat. If there are waves, pipes or specially attached radiators (coolers) on the tank walls, some of the oil goes down, washing their inner surface. The cooled oil again approaches the windings and the convection current of the oil inside the tank is closed.

A possible way for improved heat extraction is to replace transformer oil with another filler with better thermal characteristics (Perrier & Beroual, 2009). Currently, an alternative to transformer oil are liquid dielectrics Midel 7131, Sofeksil TSZH and others. The characteristics of Sofexil are shown in Table 4.

In addition to heat removal due to losses in the transformer, transformer oil also plays the role of an insulating medium with high dielectric strength where the core and windings are immersed. This allows transformers to be more compact and cheaper. In addition, the oil removes heat from the elements much better than air. Sofexil performs all these functions more efficiently (breakdown voltage at a gap of 2.5 mm is 35 kV, flash point is 300 °C, pour point is 75 °C). The use of Sofexil allow reducing the total volume of fluid in the transformer and its size. In addition, by reducing the volume of dielectric fluid and lower heat capacity relative to transformer oil, heat loss to heat this volume is reduced as well, thereby accelerating the onset of heat transfer to the heating circuit. The disadvantage of such a liquid is only a higher price relative to the oil.

Table 4. Comparison of dielectric fluid Sofexil TSZH and transformer oil ("Silikonovaia transformatornaia zhidkost Sofeksil-TSZH", 2016)

	Heat Conductivity, W/m*C	Heat Capacity, kJ/kg*C
Sofexil TSZH	0,156	1,51
Transformer oil	0,107	1,85

The greatest heat losses in transformers occur from the surface of the tank. The metal tank has excellent thermal conductivity (for steel - 64 W/m*°C). It is good for transformer cooling, but thermal energy is wasted which contradicts the concept of our development. In this regard, one of the solutions to prevent the release of heat into the environment is the manufacture of a transformer tank from a material with low thermal conductivity.

Fiberglass were chosen as an alternative. Fiberglass has a low thermal conductivity, steel strength, durability, biological and chemical resistance and is not subject to rotting (Bobovich, 2014). Products made of it can operate in a wide range of temperatures and voltages. It also possesses fire retardancy and in case of fire it does not emit potent dioxin gas unlike polyvinyl chloride.

The thermal conductivity of this material is only 0,75 W/m*°C (Table 5), so almost all the heat will remain inside the tank. In addition, fiberglass will significantly reduce the weight of the transformer since its density is 4 times lower than steel with the same level of strength.

Thus, the manufacture of fiberglass tank provides many advantages, and minimizes the loss of heat from its release into the atmosphere. This is also facilitated by the absence of external radiators for cooling since heat exchangers for heat extraction will be located inside the tank.

DEVELOPMENT OF TECHNICAL MEANS FOR GAINFUL UTILIZATION OF EXCESS HEAT FROM POWER TRANSFORMERS

The structural scheme of the installation for gainful utilization of excess heat from power transformers is shown in Figure 1.

According to Figure 1, 1 is power transformer; 2 is plate heat exchanger; 3 is multi way valve; 4 is heat pump; 5 is heat consumer; 6 is heat accumulator; 7 is analyzing complex;

In the active elements (winding and magnetic core) of the power transformer 1 there is a loss of energy in the form of heat which is removed by means of transformer oil. Further, this transformer oil enters the plate heat exchanger 2. It transfers thermal energy to another coolant and returns to the active elements.

Table 5. Comparative characteristics of fiberglass, steel and aluminum alloys

	Steel	Aluminum Alloys	Fiberglass
Density (kg/m3)	7800	2640-2800	1800-1900
Elastic modulus, HPA	210	70-71	55
Strength (for metals, yield strength) in tension, MPa	240	50-440	1700
Heat conductivity at 20 degrees Celsius, W/m*°C	64	105-200	0,75
Specific volume electrical resistance, Ohm*m	Conductor	Conductor	$1,0x10^{10}$
Resistance to chemically aggressive media, salt solutions	Not resistant	Subject to electrochemical corrosion	Resistant
Operating costs	Maintenance work required at least 1-2 times a year	Maintenance work required at least once a year	Restoration of color coloration as its intensity decreases
The possibility of implementing architectural design solutions	Requires expensive equipment reconstruction	Requires expensive equipment reconstruction	Production of low-cost tooling is required

Figure 1. The structural scheme of the developed installation for gainful utilization of excess heat from power transformers

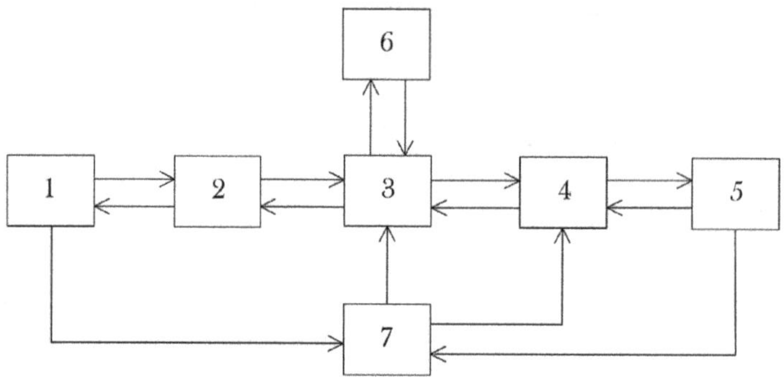

The heat carrier from the heat exchanger through the pipeline approaches the multi-way valve 3. Here two variants of events are possible. In winter, when this installation serves as a heating source for the substation building, the valve directs the coolant to the heat pump 4. In this case, the connection with the heat accumulator 6 is blocked. In the heat pump, the low-grade heat energy of the excess heat of the power transformer increases its potential, is transferred to the heat carrier of the heating system and goes through the pipeline to the heat consumer 5. During the period of severe frost or low load of the transformer, communication with the heat accumulator 6 can be opened to obtain additional thermal energy.

In the summer, when there is no need for heating, the multi-way valve directs the coolant to the heat accumulator 6, where the excess heat is partially utilized and partially stored. At the same time communication with the heat pump is blocked.

In rare cases when the transformer is disconnected, the multi-way valve closes the pipeline to the heat exchanger in the transformer and circulates the coolant between the heat accumulator and the heat pump. Thereby continuing to support the heating of the heat consumer.

One of the most important elements of the system is the analyzing complex 7. It collects data from temperature sensors installed in the power transformer and the heat consumer, and controls the multi-way valve and the heat pump maintaining medium temperature values within the normal range.

The functional scheme of the developed installation is shown in Figure 2.

This system serves to remove heat from the active elements of the transformer to the heating system or to the heat accumulator. To this end, two plate-type heat exchangers 2 are placed in a high-capacity transformer tank 1 filled with transformer oil. This type of heat exchanger is characterized by a permissible coolant temperature from -300 °C to 1800 °C and pressure up to 1.6 MPa; it has a large heat exchange surface with small dimensions, which is excellent for the presented installation (Bannykh, 2012). They are positioned longitudinally on both sides of the tank. In this case, the heat exchangers produce heated oil in the upper part of the volume and withdraw it to the bottom. it ensures uniform cooling of the entire volume of oil.

In the heat exchangers, the heat of the transformer oil is transferred to the coolant in the pipeline 8 connected to the cooling devices. Pipeline 8 is made of polyethylene pipes. It is planned to be insulated by foamed polyethylene foam in the winter period. The heat carrier of the pipeline 8 can be represented by a brine (a mixture of water and alcohol). The brine has excellent anti-corrosion properties and does not freeze at low temperatures. It is important if the transformer can be stopped during the cold period of the year. In addition, the brine has excellent heat dissipation properties.

Figure 2. The functional scheme of the developed installation for gainful utilization of excess heat from power transformers

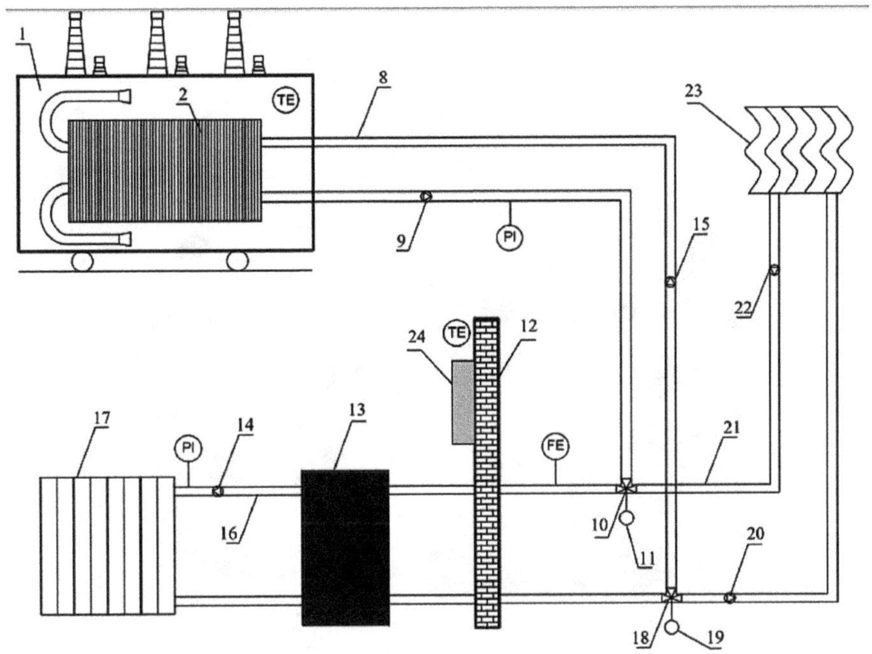

This coolant is pumped through circulation pumps 9 and 15 to the evaporation zone 3 of the heat pump 13 located in a heated building 12. His working body (it can be freon) boils wasting heat from the phase transition heat generated during the operation of the transformer 1. The working fluid moves in a closed pipeline 5. From the evaporator 3, the vapor of the working fluid of the heat pump 13 is compressed by the compressor 6. In the condensate zone 7 of the heat pump 13, the steam of its working fluid condenses with returning the heat of the phase transition to the heating medium of the heating system, and then returns to the evaporator through the expansion valve 4 (Figure 3).

The heat carrier (water) of the heating system of the building moves through the pipeline 16 to the heat consumer 17. This pipeline can be made of plastic aluminum-reinforced pipes. The movement of the coolant is provided by a circulation pump 14.

A steel pipeline 21 with thermal insulation is introduced into the scheme in order to provide transformer cooling during the period when heating is not required. Its inputs have two multi-way valves 10 and 18 regulating the flow of heat carrier into this pipeline and to the heat pump. Valves are controlled by electric motors 11, 19. Pipeline 21 communicates with the heat accumulator 23. The heat accumulator is represented by a mixture of granite and sand. In the period of extreme cold, the heat accumulated in the ground through the pipeline 21 can be directed to the heat pump 19. Circulation of brine in the pipeline is provided by two circulation pumps 20 and 22.

The following parameters should be monitored for stable system operation: the oil temperature in the tank (at the inlet to the oil cooler should not exceed + 70 °C (Krasnik, 2011), the air temperature in the room, the pressure in the pipelines and the coolant flow. This data is collected using temperature sensors installed in the transformer tank and the room as well as using pressure gauges and flow meters

Figure 3. The principle of the heat pump of the developed installation for gainful utilization of excess heat from power transformers

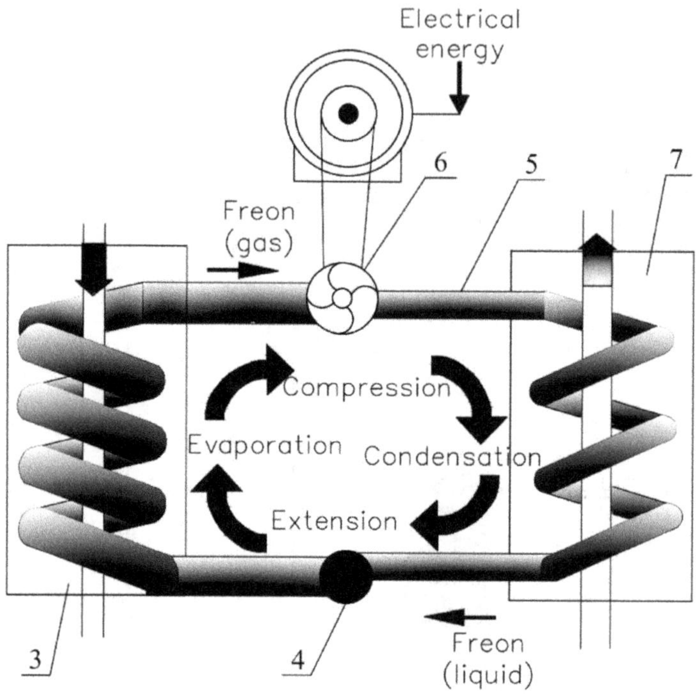

installed on the pipelines. All information from these devices is collected in the analyzing complex 24. This complex determines the stable operation of the entire system by controlling the actuators (circulation pumps, heat pumps and multi-way valves).

Thus, this installation scheme for using excess heat from power transformers successfully provides a constant cooling of transformer oil at any time of the year and also gives heat to nearby buildings during the cold period.

The proposed installation has automatic control performed by the controller in accordance with the specified programs. Management is characterized by three modes of operation of the system: cooling, heating and cooling, heating (Figure 4).

The cooling mode of operation corresponds to the period of time when the temperature outside is above + 7 °C and when the temperature in the heated room is above + 18 °C. In this case, the controller turns off the heat pump, acts on the motors of multi-way valves and activates the circulation pumps 1, 3 and 4. A result of this the coolant is directed between the transformer and the heat accumulator. In this mode, the heat collected from the transformer is stored in the heat accumulator.

The heating and cooling modes turn on when the temperature in the room drops below + 18 °C and works until the temperature reaches + 24 °C. In this case, the controller turns on the heat pump, turns off the circulation pumps 3 and 4, acts on the motors of multi-way valves and activates the circulation pumps 1, 2 and 5. The coolant is directed between the transformer and the heat pump. In this mode, heat is supplied to the heat pump, which transmits it to the heating circuit of the room increasing its energy potential. When the outside temperature drops below + 7 °C, this mode is activated on a permanent basis and ceases to depend on the room temperature until the outdoor temperature rises above + 7 °C.

Figure 4. Scheme of the control system of the developed installation for gainful utilization of excess heat from power transformers

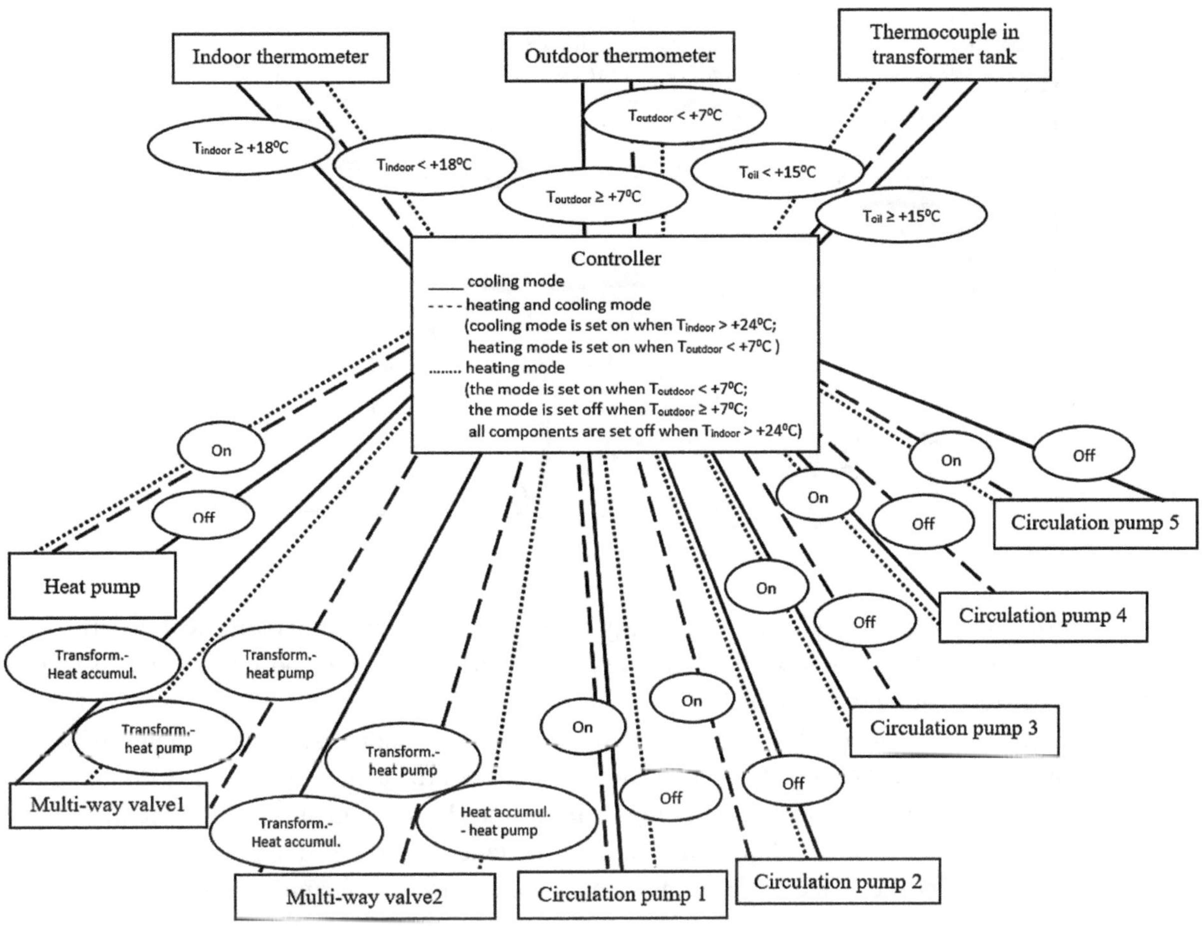

The system switches to the heating mode by lowering the oil temperature below + 15 °C. (for example, when the power transformer is disconnected). In this case, the controller turns off the circulation pumps 1 and 2, acts on the multi-way valve and turns on the circulation pumps 3, 4 and 5. The coolant is directed tbetween the heat pump and the heat accumulator. This mode only serves to heat the substation building and acts until the transformer is put into operation and the oil temperature rises. The dependence on temperature is the same as in the heating and cooling mode.

DESIGN SOLUTIONS FOR THE DEVICES USING THE EXCESS HEAT OF POWER TRANSFORMERS

Since our system is planned to be installed at other power facilities with power transformers of lower power, we considered a pool for breeding fish of 10 m^3 as a heat consumer (Figure 5). In hot summer weather, little energy is needed for heating water, but the pool water requires a significant amount of

Figure 5. The scheme of heating the pool with a heat pump ("Kak vybrat teploobmennik", n.d.)

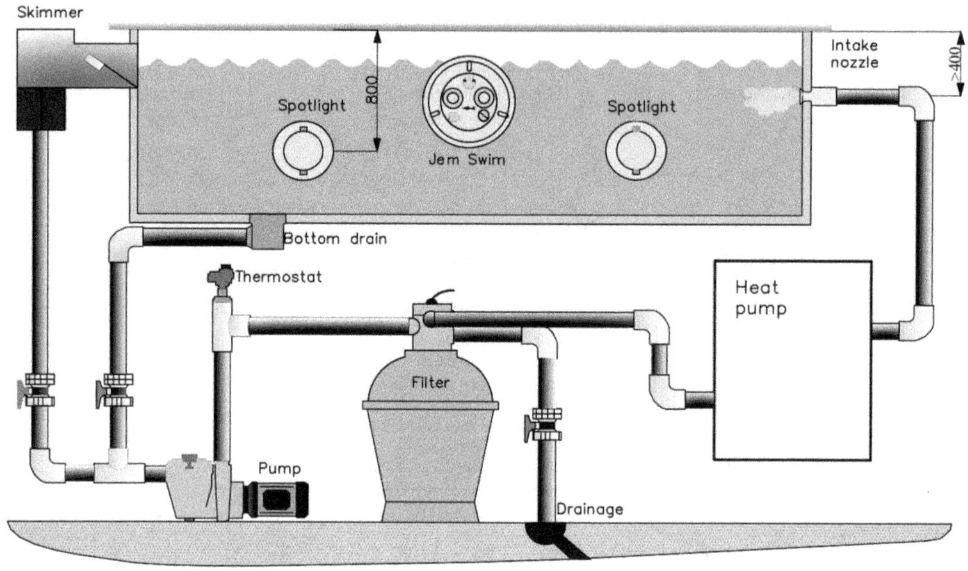

heat in the remaining colder months. It is necessary to ensure the water temperature within +18 - +22 °C for the breeding of sturgeon. The manufacturers of equipment recommend using the 13-kW heat exchanger power for heating the pool of 10 m^3.

We choose the TMG-1000/10/0.4 power transformer as a heat source since the amount of generated excess heat is 10 kWh. The previously considered solutions are taken to reduce the heat loss in the transformer: insulation of the transformer windings is performed by impregnation with Bakelite varnish, Sofexil TSZH is used as a dielectric liquid, the transformer tank is made of fiberglass.

Find the amount of heat emitted by the transformer tank into the air (formula 4) (Abramov, 2016).:

$$Q = Q_e * S * \Theta_{tank\&air}\text{, kWh} \quad (4)$$

where Q_e is heat emission in air by radiation from a surface unit, (0.91 W/(m2*°C) for fiberglass) (Bobovich, 2014);

S is surface area of heat exchange tank, m2;

$\Theta_{tank\&air}$ is temperature difference between the tank wall and air, °C.

It is necessary to find out the size of the transformer tank to determine the heat loss. The volume of the tank of an ordinary oil transformer is approximately 1,36 m^3 including 0,36 m^3 occupied by active elements and 1m^3 by oil. Since Sofexil has higher insulating properties and the transformer tank is made of a material with high electrical resistance (1*1013 Ohm), the volume of dielectric liquid will be less by 20%, i.e. 0,8 m^3. So, the required tank volume will be 1,16 m^3, i.e. existing tank sizes should be reduced by 5%. The surface area of heat transfer tank will be 6,87 m^2.

Now we determine the heat loss from the surface of the tank:

$$Q = 0,91 * 6,87 * 40 = 0,25 \text{ kWh}$$

Thus, the heat exchangers of the cooling system extract approximately 9,75 kWh of heat.

Since the transformer tank is made of heat-insulating material and its volume is very small, a double-circuit spiral copper heat exchanger was chosen as the heat exchanger. This type of heat exchanger provides a more uniform cooling of the dielectric fluid and the active elements of the transformer. In addition, this heat exchanger takes up less space.

The calculation of the spiral heat exchanger is as follows (Barulin, Isaev & Sokolsky, 2009):

Temperature scheme of heat carriers:

75°C → 55°C,

10°C → 30°C,

Δt_{max}=65°C,

Δt_{min}=25°C.

The average temperature difference (formula 5):

$$\Delta t_{1av} = (\Delta t_{max} - \Delta t_{min}) \,/\, \ln(\Delta t_{max} - \Delta t_{min}) \quad (5)$$

$$\Delta t_{1av} = \left(65 - 25\right) / \ln\left(65 / 25\right) = 41,86\ °\text{C}$$

The average temperature of the solution (formula 6):

$$\Delta t_{2av} = t_{max} - \Delta t_{1av} \quad (6)$$

$$\Delta t_{2av} = 65 - 41,86 = 23,14\ °\text{C}.$$

The thermophysical properties of Sofexil at an average temperature are: thermal conductivity λ_2 = 0,156 W/(m*K); density p_2 = 942 kg/m^3; dynamic viscosity coefficient μ_2=5 MPa*s; Prandtl criterion Pr_2 = 83,6; heat capacity c_2 = 1510 J/(kg*K) (Silikonovaia transformatornaia zhidkost Sofeksil-TSZH, 2016).

The thermophysical properties of a 40% solution of ethyl alcohol with water as an intermediate coolant at an average temperature are: thermal conductivity λ_2 = 0,41 W/(m*K); density p_2 = 935,2 kg/m^3; dynamic viscosity coefficient μ_2 = 2,71 mPa*s; Prandtl criterion Pr_2 = 26; heat capacity c_2=3961 J/(kg*K) (Levin, 2004).

The velocity of the solution is W_2=1,5 m/s, the circulation rate of the dielectric fluid is W_1=0,1 m/s. The outer diameter of the heat exchanger channel is taken d_1 = 0,024 m, the internal diameter of the channel is d_2 = 0,02 m.

The Reynolds criterion for Sofexil will be found by formula 7:

$$R_{e1} = W_1 d_1 p_1 / \mu_1 \tag{7}$$

$$R_{e1} = 0,1 * 0,024 * 942 / \left(5 * 10^{-3}\right) = 452$$

The Reynolds criterion for solution will be found by formula 8:

$$R_{e2} = W_2 d_2 p_2 / \mu_2 \tag{8}$$

$$R_{e2} = 1,5 * 0,02 * 935,2 / \left(2,71 * 10^{-3}\right) = 10352$$

The criterion equation 9 is recommended to use for spiral heat exchangers with a value of Re2>2000.

$$Nu = 0,021 * \mathrm{Re}^{0.8} * \mathrm{Pr}^{0.43} * \left(\frac{\mathrm{Pr}}{\mathrm{Pr}_{st}}\right)^{0,25} \tag{9}$$

The first approximation $\left(\frac{\mathrm{Pr}}{\mathrm{Pr}_{st}}\right)^{0,25} = 1$

$$Nu_1 = 0.021 * 452^{0.8} * 83.6^{0.43} = 18,74$$

$$Nu_2 = 0.021 * 10352^{0.8} * 26^{0.43} = 138,92$$

The heat transfer coefficient from the dielectric fluid to the wall is determined by the formula 10:

$$\alpha_1 = Nu\lambda_1 / d_1 \tag{10}$$

$$\alpha_1 = 18,74 * 0,156 / 0,024 = 121,81 \text{ W/(m}^2\text{S)}$$

The heat transfer coefficient from the wall to the solution (formula 11):

$$\alpha_2 = Nu\lambda_2 / d_2 \tag{11}$$

$$\alpha_2 = 138,92 * 0,41 / 0,02 = 2847,86 \text{ W/(m}^2\text{S)}$$

The amount of thermal resistance of the wall (formula 12):

$$\sum r = r_{pol.Sof.} + \delta_{cop.} / \lambda_{cop.} + r_{pol.sol.} \tag{12}$$

where $r_{pol.Sof.}$ and $r_{pol.sol.}$ are thermal resistances of pollution from Sofexil TSZH and solution of water with ethyl alcohol;

$\delta_{cop.}$ = 2 mm is heat exchanger wall thickness;

$\lambda_{cop.}$ = 394 W/(m*k) is the thermal conductivity of copper.

$\sum r = 0,00017 + 0,002 / 394 + 0,00017 = 0,000345$ (m²K)/W

Heat transfer coefficient (formula 13):

$$K = 1/\left(1/\alpha_1 + \sum r + 1/\alpha_1\right) \tag{13}$$

$K = 1/\left(1/121,81 + 0,000345 + 1/2847,86\right) = 112,3$ W/(m²K)

The wall temperature on the side of the solution (formula 14):

$$t_{cop.2} = t_{2av.} + K * \Delta t_{av.} / \alpha_2 \tag{14}$$

$t_{cop.2} = 23,14 + 112,3 * 41,86 / 121,81 = 61,7$ °C

The temperature of the wall on the side of the dielectric fluid (formula 15):

$$t_{cop.1} = t_{1av.} + K * \Delta t_{av.} / \alpha_1 \tag{15}$$

$t_{cop.1} = 65 - 112,3 * 41,86 / 2847,86 = 63,35$ °C

Heat transfer coefficients from the wall to the solution and from Sofexil to the wall are refined according to the formulas 16 and 17:

$$a_2^* = \alpha_2 \left(\Pr / \Pr_{cop.2}\right)^{0,25} \tag{16}$$

where $\Pr_{cop.2} = 9$ is Prandtl criterion for the solution at 61 ° C.

$$a_2^* = 2847,86 * \left(26 / 9\right)^{0,25} = 3712,8 \text{ W/(m}^2\text{K)}$$

$$a_1^* = \alpha_1 \left(\Pr / \Pr_{cop.1}\right)^{0,25} \quad (17)$$

where $\Pr_{cop.2} = 83,5$ is Prandtl criterion for Sofexil at 63 °C.

$$a_1^* = 121,82 * \left(83,6 / 87,2\right)^{0,25} = 121 \text{ W/(m}^2\text{K)}$$

Heat transfer coefficient is refined as follows:

$$K = 1 / \left(1 / 121 + 0,000345 + 1 / 3712,8\right) = 113 \text{ W/(m}^2\text{K)}$$

Required heat exchange surface is defined by formula 18:

$$F = Q / K * \Delta t_{av.} \quad (18)$$

$$F = 12000 / 113 * 41,86 = 2,5 \text{ m}^2$$

The actual surface is taken more than the calculated up to 15%. Therefore, we accept the standard spiral heat exchanger with a surface F_d=3 m^2. Thus, a spiral heat exchanger 40 m long is required for cooling a 1000 kVA power transformer at full load. A coil of such length should consist of 12 turns. To reduce the length of the heat exchanger, it is possible to increase the surface area of the coil by installing copper plates along its entire length.

If the heat loss during transportation is taken equal to 5%, the amount of thermal energy at the consumer will decrease to 9,3 kWh. Since the minimum power of thermal energy for heating the pool is 13 kW, the thermal energy at the maximum thermal load at the consumer will not be enough. Therefore, a heat pump must be installed near the pool to increase the transformer heat potential.

In accordance with the power consumption of the 13-kW pool, it was chosen WI14ME heat pump with the following technical characteristics: thermal output is 13,4 kWh; consumed electric power is 2,87 kW; heat capacity coefficient (COP) is 3,8; brine/coolant supply temperature is W10/W45 (“Teplovye nasosy”, n.d.).

Due to the fact that the temperature of the ground has a much smaller amplitude of oscillations relative to the open air, it was decided to deepen the heat accumulator into the ground, i.e. to use a heat accumulator based on a mixture of sand and granite.

In our case, the heat accumulator will be represented by a certain isolated volume of granite crushed stone of mark M1100 mixed with fine river sand. To do this, it is necessary to dig a ditch, isolate the walls and the bottom of the ditch with polystyrene foam. Next, to lay a mixture of granite rubble and river sand in layers in a ditch, between layers of which lay a horizontal ground heat exchanger. After the top of the heat accumulator also heat insulated and covered with a thin layer of soil to protect against direct solar heating.

Taking 1 day for possible tripping of the power transformer, we determine the required volume of the heat accumulator. To ensure the heat load of the pool (13 kWh), it is necessary to use the heat accumulator able to store 230 kW of heat.

The volume of granite rubble and sand can be calculated by the formula 19 and 20 (Nagornaya, 2007):

$$V = Q / c * \rho * \Delta t \tag{19}$$

where V is the volume of the heat accumulator, m^3;

Q is the amount of heat transferred, J;

c is specific heat capacity, kJ/(kg*°C);

ρ is the density of the heat accumulator substance, kg/m^3;

Δt is the temperature difference, °C.

$$V_s = 0,28 * V_g \tag{20}$$

where V_s is the volume of sand in a heat accumulator;

V_g is the volume of crushed granite in a heat accumulator.

$$V = V_g = 828 * 10^6 / 15 * \left(\left(0,92 * 10^3 * \left(2700 * 0,72\right)\right) + 0,28 * \left(0,88 * 10^3 * 1600\right)\right) = 26,5\ \mathrm{m}^3$$

That is, it is necessary to fill the ditch with the size of 9m*3m*1m with a mixture of sand and crushed granite. Such a volume of rubble with sand can be an excellent foundation for the installation of a power transformer. It will not only serve as a heat accumulator, but will also serve as a reliable and stable foundation for a power transformer.

The choice of crushed granite as a heat accumulator is associated with a high level of heat removal (Table 6).

The calculation the horizontal ground heat exchanger will be done as follows. Firstly, it is necessary to find the removed power of the contours of the ground heat exchanger of the heat pump according to the formula 21 ("Gruntovye teploobmenniki", n.d.):

$$P_e = P_n * \left(1 - 1 / COP\right) \tag{21}$$

where P_e is the removable capacity of the soil heat exchanger circuits;

Table 6. Heat removal of the soil heat exchanger with various types of rocks ("Gruntovye teploobmenniki", n.d.)

Type	Heat Removal, W/m
Dry sedimentary rocks	20
Dry sand, gravel	25
Wet sand	35-40
Stony soil saturated with water, wet clay	40-50
Limestone	50-60
Granite	60-80

P_n is the rated power of the heat pump;

COP is the conversion factor.

$$P_e = 13 * \left(1 - 1 / 3,8\right) = 9,6 \text{ kW}$$

The required length of the horizontal heat exchanger of the heat pump is equal to the ratio of the required power to the removed power of one meter of pipe (formula 22):

$$L = P_e / q \qquad (22)$$

where q is the heat removal from the horizontal heat exchanger.

$$L = 9,6 / \left(0,08 + 0,035\right) = 83 \text{ m}$$

To find out what area such a collector will occupy, it is necessary to multiply the value obtained by the pipe laying step (a step is taken equal to 0.7 m) (Krasnik, 2011).

$$S = 83 * 0,7 = 58 \text{ m}^2$$

Thus, it is necessary to lay a horizontal ground heat exchanger with an area of 58m^2 to ensure the operation of the system using heat accumulator in its volume.

According to the results of the study, the main differences of the proposed system from the existing developments was designated in Table 7.

Table 7. The main differences of the proposed system from the existing developments

Developer	Heat Extraction Method	Intermediate Heat Carrier Type	Heat Pump Presence	Backup Heat Source Type	Heat Accumulation Presence	Degree of Tank Design Change	Availability of Automated Control System
The proposed system	Double-circuit spiral copper heat exchanger “oil-water”, placed in the transformer tank	40% ethylene glycol water solution	Yes	Ground battery heat	Yes	New fiberglass tank filled with Sofeksil dielectric liquid	Yes
Kjell (2013).	Outdoor plate heat exchanger “oil-water”	Mixture of alcohol and water	Yes	Ground battery heat	Yes	No cooling radiators filled with oil	No
Novikov Taybakhtin (2015)	Extraction of heated radiator air using a mounted casing	Air	No	Electric heating or central heating	No	Radiator casing filled with oil	Yes
Lavrentiev (2018)	External heat exchanger “oil-water”	Water	Yes	Boiler	No	No cooling radiators filled with oil	Yes
Timonin et al. (2013)	External heat exchanger “oil-water”	Air	No	Electric heating or central heating	No	No cooling radiators filled with oil	Yes
Tsoi et al. (2013)	External heat exchanger “oil-water”	Ethylene glycol water solution	Yes	Electric heating or central heating	No	No cooling radiators filled with oil	Yes

EXPERIMENTS

The goal of the experiments was to study the heat losses in the transformer tank using different materials of its execution as well as the intensity of heat transfer between the hot oil in the tank and the coolant with different cooling methods.

The first experiment to study heat losses in metal and fiberglass tanks consisted of two stages.

At the first stage, we heated the oil in a metal tank. For this we fixed a bimetallic thermometer on the tank surface. Next, we poured 10 liters of transformer oil into the tank and placed the electric heater inside sealing the remaining openings. Then the heater was connected to the network. When the oil temperature in the tank reaches 50°C (the oil temperature in the transformer tank at medium load), we began to take readings of the oil temperature in the tank and the surface temperature of the tank every 6 minutes. The temperature of the ambient air during the experiment was 18°C. After the experiment, we obtained the following results (Table 8).

Heat loss from the surface is determined using formulas 23 and 24:

$$q_{em} = k * \sqrt[4]{\Theta_{tank\&air}} \quad (23)$$

Table 8. Measurement results when the oil is heated in a metal tank

Time	Oil Temperature, °C	Surface Temperature, °C	Heat Loss, W
11:59	50	28	18,92
12:05	75	43	59,48
12:12	98	53	90,58
12:19	115	62	120,57
12:25	131	72	155,75

where q_{em} is the heat emission in air by radiation from a unit of surface, W/(m^2*°C);

k is heat transfer coefficient (for steel – 2,8; for fiberglass – 0,91) (Bobovich, 2014);

$\Theta_{tank\&air}$ is temperature difference between the tank wall and air, °C.

$$Q = q_{em} * S * \Theta_{tank\&air} \quad (24)$$

where S is the surface area of heat exchange tank, m^2.

At the second stage, we heated the oil in a fiberglass tank. To do this, we perform the same manipulations as in the first stage. When the temperature of the oil in the tank reaches 50°C, we also began to take readings every 6 minutes. After the experiment, we obtained the following results (Table 9).

It should also be noted that the oil got cooled for 1 hour in the fiberglass tank to 36°C, in the metal to 48°C.

Comparing the results obtained when testing a metal and fiberglass tank, it can be concluded that the fiberglass tank is much more efficient in terms of heat preserving. In addition, fiberglass showed the necessary resistance to chemical exposure to oil and high thermal and mechanical strength.

The second experiment was carried out to study heat transfer between heated oil and cooling water depending on the location of the heat exchanger at two stages.

In the first stage, the heat exchange between oil and water was considered when the heat exchanger is placed in a tank with heated oil and cold water circulates through it. To do this, we poured 10 l of oil into the tank with the heat exchanger and 10 l of water into the other tank. Next, the electric heater began to raise the temperature of the oil. When the oil temperature reached 107°C, we turned on the pump for circulation of cold water to start the heat exchange. The water temperature was 12°C. And then

Table 9. Results of measurements when the oil is heated in a fiberglass tank

Time	Oil Temperature, °C	Surface Temperature, °C	Heat Loss, W
12:45	50	25	3,94
12:51	75	34	11,06
12:58	98	42	18,37
13:04	115	50	26,32
13:10	131	56	32,63

every 6 minutes we measured the temperature of the oil and water. The temperature of the ambient air during the experiment was 14°C. According to the results of measurements, we obtained the following results (Table 10)

There is a cooling oil and heat transfer to the coolant.

In the second stage, we changed the liquid in some places. The volumes of liquids in the tanks are the same as in the first stage. The pump was turned on at an oil temperature of 105°C. Water temperature at the beginning of circulation was 27°C. The starting water temperature was higher than in the first stage because colder water can lead to a strong increase in the viscosity of the oil inside the heat exchanger. There are two detrimental things. First, the load on the circulating pump can increase and lower circulation rate can worsen the heat transfer. Secondly, a viscous oil has less heat transfer. It also reduces heat transfer. Measurements were also carried out every 6 minutes. As a result, we obtained the following results (Table 11).

The results of the second stage show that despite the circulation of oil through the heat exchanger, its temperature continues to grow, albeit more slowly. The water also heats up less. Thus, the placement of the heat exchanger in the tank with oil is characterized by a much higher heat exchange efficiency, and, consequently, contributes to better cooling of the oil.

FEASIBILITY STUDY FOR THE INTRODUCTION OF DEVELOPED DEVICES

Capital investments for the implementation of this system consist of the cost of materials and equipment, the cost of their delivery and installation. They are defined by formula 25: (Nagornaya, 2007)

$$K = C + M + T \tag{25}$$

where K are capital investments; C is the cost of components;

M are installation mounting costs (15 ... 25% of C);

T are transportation costs (10 ... 15% of C).

Table 10. Measurement results when placing the heat exchanger in the tank with oil

Time	Oil Temperature, °C	Water Temperature, °C
18:03	107	12
18:09	113	14
18:15	101	27
18:21	97	31
18:27	95	36

Table 11. Measurement results when placing the heat exchanger in the tank with water

Time	Oil Temperature, °C	Water Temperature, °C
18:47	105	27
18:53	104	32
18:59	105	36
19:05	107	34
19:11	111	35

Table 12. Estimated estimate of the developed installation

Name	Unit Cost, dollars	Total Cost, dollars
Fiberglass transformer tank, 1,2 m^3	686,4275	686,4275
Double-circuit copper spiral heat exchanger 3m^2, 1pcs	187,2075	187,2075
Dielectric liquid Sofexil TSZH, 0,8 m^3	4875,195	3900,156
Heat pump WI 14ME	1981,279	1981,279
Three-way valve ESBE VRG 131, 2 pcs	68,01872	136,0374
Electric rotary ESBE ARA661, 2 pcs	147,4259	294,8518
Crushed stone granite, 26,5 m^3	8,736349	231,5133
Circular pump Caliber NTs-32/6, 3 pcs	30,57722	91,73167
Polyethylene pipe PND 20mm, 80 m	0,280811	22,4649
Pipe insulation 22/6, 80 m	0,156006	12,4805
Stainless steel pipe 20mm, 83 m	1,092044	90,63963
Thermal insulation Penofol 10 mm, 24 m^2	1,544462	37,05148
Controller OVEN PLC100 220.R-M	119,6568	119,6568
Sand, 7,5 m3	5,460218	40,95164
Other installation products	0	78,00312
TOTAL	0	7910,452

The specification of the basic configuration of such a transformer is presented in Table 12.

The total cost of components according to estimates is 7910 $.

Transportation costs for the delivery of components (Formula 26) (Nagornaya, 2007):

$$T = 0{,}1 * C \tag{26}$$

$$T = 0{,}1 * 7910 = 791\,\$$$

Installation mounting costs (Formula 27)

$$M = 0{,}15 * C \tag{27}$$

$$M = 0{,}15 * 7910 = 1185\,\$$$

Total capital investments will be:

$$K = 7910 + 1185 + 791 = 9874{,}99\,\$$$

The cost of operating automation equipment is determined by the formula 28:

$$O = D + RM + E + OT \tag{28}$$

where O is operational costs;

D is depreciation;

RM is the cost of repair and maintenance of the system;

E is cost of electricity consumed;

OT are other direct costs.

Cost for depreciation is determined by the formula 29:

$$D = R_D * K \tag{29}$$

where R_D is the rate of deductions for depreciation (we take 5%).

$$D = 0,05 * 9874,99 = 493,75\,\$$$

Cost of repair and maintenance of the system is determined by the formula 30

$$RM = R_{RM} * K \tag{30}$$

where R_{RM} is the rate of deductions for repair and maintenance of the system (we take 3%).

$$RM = 0,05 * 9874,99 = 296,24\,\$$$

The cost of electricity consumed by the equipment is determined by the formula 31:

$$E = T_E * \left(\frac{\sum P * T * K_{load} * K_{sim.}}{K_{loss} * \eta} \right) \tag{31}$$

where T_E is the electricity tariff, \$/kWh (we accept tariff 0,064 \$/kWh) (Nagornaya, 2007);

$\sum P$ is total power of electricity consumers, kW;

T is the fund of the equipment operation time, h (amounts to 4380 hours);

K_{load} is load factor (0,7-0,8);

$K_{sim.}$ is the coefficient of simultaneity (0,9-1.0);

K_{loss} is loss coefficient in the network (0,95);

η is efficiency (0,9).

The total power of the installation is determined by the formula 32:

$$\sum P = P_{HP} + \sum P_{CP} \tag{32}$$

where P_{HP} is heat pump power, kW;

$\sum P_{CP}$ is total power of circulation pumps, kW.

$\sum P = 2,5 + 0,64 = 3,14\,\text{kW}$

$$E = 0,064 * \left(\frac{3,14 * 4380 * 0,8 * 0,95}{0,95 * 0,9} \right) = 780,92\,\$$$

Other direct costs are determined depending on the value of all direct costs (Formula 33):

$$OT = R_{OT} * \left(D + RM + E \right) \tag{33}$$

where R_{OT} is the rate of deductions for other costs (we take 5%).

$$OT = 0,05 * \left(493,75 + 296,24 + 780,92 \right) = 78,54\ \$$$

Total operating costs will be:

$$O = 493,75 + 296,24 + 780,92 + 78,54 = 1649,44\ \$$$

Thus, the cost of installation work during the year is 1649,44 $.

The developed power transformer with an improved design of the cooling system to combine the functions of transformation and electric heating gives a good economic effect.

A metal tank in a 1000 kVA power transformer weighs about 280 kg and costs about 3739,20 $. The fiberglass tank will weigh about 100 kg, and it costs only about 685,52 $.

There is 1m^3 of transformer oil in the usual metal transformer for 1000kVA. The oil volume cost about 1090,60 $. In the developed transformer there is 0.8 m^3 of Sofexil liquid which cost about 3895 $.

Usually, a 10 m^3 pool is heated with a 13-kW water heater for 311,60 $ and a 13-kW flow heat exchanger for 140,22 $. In addition, a pipeline and a circulation pump of 0,1 kW are needed with a total cost of about 46,74 $. The total cost of traditional equipment for heating the pool in this case will be 498,56 $. And if to add it up with the cost of a metal tank and transformer oil, the total cost of the equipment will be 5328,36 $. Taking into account the transportation costs and installation costs of the system, the total investment will result in 6660,45 $.

The cost of electricity consumption by such a heating system for the year according to the formula 30:

$$E = 0,064 * \left(\frac{13,1 * 4380 * 0,8 * 0,95}{0,95 * 0,9} \right) = 3257,95\,\$$$

The power consumption of such a power transformer with the combination of functions with an electric power of 3,14 kW will be 780,92 \$. Thus, the annual energy savings will be 2477,03 \$.

The total operating costs of the traditional system taking into account depreciation the cost of maintenance and repair and other expenses per year will amount to 3980,32 \$. It turns out that the operating costs of the developed transformer is 2330,88 \$ per year less. If the traditional transformer oil is as a dielectric fluid in such a transformer, the annual savings will be 2648,24 \$.

The developed installation also allows reducing wear of the active elements of the transformer. It leads to a reduction in the cost of capital repairs, an increase in the service life of the transformer and a decrease in the undersupply of electricity. Numerically, this effect is not taken into account in this paper due to insufficient data.

The payback period of capital investments in the developed installation is determined in two versions according to the formula 34:

1. Using Sofexil TSZH as dielectric fluid:

$$T_{payback} = K \,/\, S_{annual} \tag{34}$$

where $T_{payback}$ is payback period, years.;

S_{annual} is annual savings, \$.

$$T_{payback} = 9855,99 \,/\, 2326,37 = 4,3 \text{ years}$$

The coefficient of economic efficiency of investment is determined by Formula 35:

$$K_{econ.efficiency} = 1 \,/\, T_{payback} \geq K_{econ.ef.normative} \tag{35}$$

where $K_{econ.efficiency}$ is the coefficient of economic efficiency of capital investments; $K_{econ.ef.normative}$ is the normative coefficient of economic efficiency of additional investments (According to Nagornaya (2007) for electrical installations $K_{econ.ef.normative} = 0,12$)

$$K_{econ.efficiency} = 1 \,/\, 4,3 = 0,23 \geq 0,12$$

2. Using transformer oil as dielectric fluid:

$$T_{payback} = 6085,12 \,/\, 2643,13 = 2,3 \text{ years}$$

$$K_{econ.efficiency} = 1 \,/\, 2,3 = 0,43 \geq 0,12$$

It can be concluded that the introduction of a 1000kVA power transformer with the combination of transformation and electrical heating functions using Sofexil as dielectric liquid is economically very effective ($T_{payback} = 4,3$ years). But, in spite of better cooling, better insulation, fire safety and other advantages of this dielectric compared with transformer oil the use of Sofexil is noticeably inferior to the use of conventional transformer oil in terms of economic efficiency ($T_{payback} = 2,3$ years) (Table 13).

Thus, the introduction of a 1000 kVA power transformer with an improved design of the cooling system to combine the functions of transformation and electric heating shows high economic efficiency. Therefore, the developed system is very promising for reducing the costs of enterprises especially when using more powerful transformers and greater heating needs.

FUTURE RESEARCH DIRECTIONS

The presented work reveals the principle of operation and interaction of the developed means and methods of gainful utilization of excess heat from power transformers and the advantages relative to other analogues. Further development of this work will be aimed at solving a number of tasks related to the search for new constructive and technical solutions to improve the efficiency of the system and reduce its cost.

First of all, a software control algorithm for all elements of the system should be developed depending on the input parameters for the controller. All the system operation options up to emergency and extreme modes should be worked out in order to create a ready-made software product suitable for monitoring and automatic control for such utilization systems of power transformer heat.

Table 13. Economic efficiency of the introduction of a 1000kVA power transformer with a developed design

Indicators	Using Sofexil TSZH	Using Transformer Oil
Capital investment, dollars	633826	391326
Total cost of components, dollars	507061	313061
Operating costs, rub, dollars	105870	85500
Annual savings, dollars	149606	169976
Payback period, years	4,3	2,3
The coefficient of economic efficiency of capital investments	0,23	0,43

The next task is the search for a new type of heat accumulator, which would save more thermal energy, have a large heat output per unit length of the heat exchanger, be with a smaller volume and easy to install. Also, an important task is to develop a new design of a power transformer tank with an integrated heat exchange unit. Much attention should be paid to the subject of a detailed study of the operation of each element of the system in actual operating conditions.

CONCLUSION

A number of conclusions can be drawn based on the research:

1. Power transformers emit a sufficient amount of heat to cover the thermal loads of the substation; the gainful utilization of excess heat is associated with certain difficulties, the solution of which requires careful study.
2. The ways to reduce heat loss in a power transformer are winding insulation using bakelite varnish or other varnishes, using Sofexil TSZH silicone fluid as an alternative to transformer oil and manufacturing a transformer tank made of fiberglass;
3. The scheme for extracting excess heat from transformers using heat pumps looks more preferable for heating the substation buildings in comparison with ones without heat pumps. It is due to the fact that a heat pump increases the potential of the heat obtained to the level necessary for its own needs;
4. Transformer load schedule and average load factor should be paid attention to when calculating heat generation from active elements of a transformer. The greater the power of the transformer is, the smaller required volume of the heat accumulator with an equal heated area is; The developed design of the power transformer cooling system for combining the functions of electric transformation and heating opens up wide possibilities for use.
5. It is much more efficient to make a transformer tank of fiberglass. This material retains heat much better inside the tank while having reliability, durability, resistance to chemical influences. At the same time, such a tank has a strong electrical resistance reducing the energy loss for magnetization; Heat exchange between hot oil and coolant is best accomplished by placing a spiral tube heat exchanger inside the transformer tank.
6. The introduction of a power transformer with an improved design of the cooling system to combine the functions of electric transformation and heating is also characterized by high economic efficiency.

REFERENCES

Abramov, Yu. M. (2016). *Elektricheskie stantsii i podstantsii [Electrical stations and substations]*. Russia: Don State Technical University. Retrieved from: https://studfiles.net/preview/6023617/

Baishev, A. V., & Toropov, A. S. (2018). Assessment of the possibility of using waste heat of power transformers for heating and hot water supply of electrical substations [Otsenka vozmozhnosti ispolzovaniia sbrosnogo tepla silovykh transformatorov dlia otopleniia i goriachego vodosnabzheniia elektricheskikh podstantsii]. *Bulletin of the Khakass State University*, *24*, 5–7.

Bannykh, O.P. (2012). *Osnovnye konstruktsii i teplovoi raschet teploobmennikov* [The basic design and thermal calculation of heat exchangers]. Russia: SPbNUU ITMO.

Barulin, E.P., Isaev, V.N., & Sokolsky, A.I. (2009). *Raschety teploizoliruiushchego oborudovaniia* [Calculations of insulating equipment]. Ivanovo, Russia: Ivan state chemical-tehnol. un-t.

Bertrand, Y., & Laurichesse, D. (2007, November). Comparison of the oxidation stabilities of vegetable based and mineral insulating oils. *Proceedings of MATPOST conference*. Academic Press.

Bobovich, B. B. (2014). Polimernye konstruktsionnye materialy [Polymeric construction materials]. Moscow, Russia: FORUM: INFRA-M.

Borzov, V. P., & Shabalina, L. N. (2002). Sbornik zadach po teplotekhnike [Collection of tasks for heat engineering]. Kostroma, Russia: KSAA.

Cortes, E., & Rivera, W. (2010). Exergetic and exergoeconomic optimization of a cogeneration pulp and paper mill plant including the use of a heat transformer. *Energy*, *35*(3), 1289–1299. doi:10.1016/j.energy.2009.11.011

El Wakil, N., Chereches, N. C., & Padet, J. (2006). Numerical study of heat transfer and fluid flow in a power transformer. *International Journal of Thermal Sciences*, *45*(6), 615–626. doi:10.1016/j.ijthermalsci.2005.09.002

Gosstroi Rossii. (2003). SNiP 41-01-2003. Otoplenie, ventiliatsiia i konditsionirovanie [SNiP 41-01-2003 Heating, ventilation and air conditioning].

Gridin, S. V., & Petrenko, A. F. (2013). Energoeffektivnost sposobov utilizatsii otrabotannogo tepla sistem okhlazhdeniia silovykh transformatorov [Energy Efficiency of Methods for Utilization of Waste Heat from Cooling Systems of Power Transformers]. *Energy Saving*, *7*, 9–13.

Gruntovye teploobmenniki dlia geotermalnogo teplovogo nasosa [Ground heat exchangers for a geothermal heat pump]. (n.d.). Retrieved from http://solarsoul.net/gruntovye-teploobmenniki-dlya-geotermalnogo-teplovogo-nasosa

Kak vybrat teploobmennik dlia basseina? [How to choose a heat exchanger for the pool?]. (n.d.). Retrieved from http://udobnovdome.ru/kak-vybrat-teploobmennik-dlya-bassejna/

Kharchenko, V., Gusarov, V., & Bolshev, V. (2019). Reliable Electricity Generation in RES-Based Microgrids. In *Handbook of Research on Smart Power System Operation and Control* (pp. 162–187). IGI Global; doi:10.4018/978-1-5225-8030-0.ch006

Kjell, A. (2013). R.F. Patent No. 2234755. Patent Service of the Russian Federation.

Krasnik, V. V. (2011). *Ekspluatatsiia elektricheskikh podstantsii i raspredelitelnykh ustroistv* [Operation of electrical substations and switchgear]. Russia: NTS ENAS.

Lavrentiev, S. V. (2018). Primenenie teplovykh nasosov dlia utilizatsii teplovykh poter v transformatornykh podstantsiiakh [Application of heat pumps for utilization of heat losses in transformer substations]. *Proceedings of the X International Student Scientific Conference "Student Scientific Forum"*. Retrieved from https://scienceforum.ru/2018/article/2018003793

Levin, B.D. (2004). *Protsessy i apparaty pishchevykh proizvodstv. Teplovye protsessy [Processes and equipment for food production. Thermal processes]*. Krasnoyarsk, Russia: Krasnoyars. state Agrarian Un.

Mostofizadeh, C., & Kulick, C. (1998). Use of a new type of heat transformer in process industry. *Applied Thermal Engineering*, *18*(9-10), 857–874. doi:10.1016/S1359-4311(97)00115-4

Mufuta, J. M., & Van den Bulck, E. (2000). Modelling of the mixed convection in the windings of a disc-type power transformer. *Applied Thermal Engineering*, *20*(5), 417–437. doi:10.1016/S1359-4311(99)00034-4

Nagornaya, N. V. (2007). *Ekonomika energetiki* [*Energy Economics*]. Vladivostok, Russia: FESTU Publishing House.

Novikov, D. A., & Taybakhtin, E. S. (2015). *Ispolzovanie tepla silovykh transformatorov dlia otopleniia proizvodstvennykh pomeshchenii podstantsii s tseliu snizheniia raskhoda elektroenergii [The use of heat power transformers for heating the industrial premises of substations in order to reduce power consumption]*. Retrieved from: http://gridology.ru/projects/453/documents

NTTS FSK EES [STC FGC UES]. (n.d.). Retrieved from http://www.ntc-power.ru/media/.pdf

Perrier, C., & Beroual, A. (2009). Experimental investigations on insulating liquids for power transformers: Mineral, ester, and silicone oils. *IEEE Electrical Insulation Magazine*, *25*(6), 6–13. doi:10.1109/MEI.2009.5313705

Semenov, Yu. P. (2015). *Teplotekhnika* [*Thermal Engineering*]. Russia: Infra-M.

Silikonovaia transformatornaia zhidkost Sofeksil-TSZH Rekomendatsii po primeneniiu [Sofaxil-TSZh Silicone Transformer Liquid. Recommendations for use]. (2016). Moscow, Russia.

Silovye transformatory. Katalog [Power transformers. Catalog]. (2018). Retrieved from: http://silovoy-transformator.ru/110kv/tdtn-40000-110-u1.htm

Swift, G., Molinski, T. S., & Lehn, W. (2001). A fundamental approach to transformer thermal modeling. I. Theory and equivalent circuit. *IEEE Transactions on Power Delivery*, *16*(2), 171–175. doi:10.1109/61.915478

Teplovye nasosy [Heat pumps]. (n.d.). Retrieved from http://www.nibe-evan.ru/nibedocuments/17337/M11540-1.pdf

Tikhomirov, P. M. (2015). *Raschet transformatorov* [*Calculation of transformers*]. Russia: Tyumen State Oil and Gas University. Retrieved from https://studfiles.net/preview/5332633/page:76/

Timonin, Yu. N., Gorelov, Yu. I., & Stepanov, V. M. (2013). *R.F. Patent No. 128776*. Patent Service of the Russian Federation.

Tsoi, A. D., Kopyrin, V. S., & Sosnovsky, O. G. (2013). *U. Patent No.* 2505. Patent Service of Ukraine.

Vinogradov, A., Vasiliev, A., Bolshev, V., Semenov, A., & Borodin, M. (2018). Time factor for determination of power supply system efficiency of rural consumers. In Handbook of Research on Renewable Energy and Electric Resources for Sustainable Rural Development (pp. 394–420). Hershey, PA: IGI Global. doi:10.4018/978-1-5225-3867-7.ch017

Vinogradov, A., Vasiliev, A., Bolshev, V., Vinogradova, A., Kudinova, T., Sorokin, N., & Hruntovich, N. (2019). Methods of Reducing the Power Supply Outage Time of Rural Consumers. In Renewable Energy and Power Supply Challenges for Rural Regions (pp. 370–392). Hershey, PA: IGI Global. doi:10.4018/978-1-5225-9179-5.ch015

KEY TERMS AND DEFINITIONS

Heat Accumulator: A device for accumulating heat for the purpose of its further use.

Heat Exchanger: is a technical device where heat transfer occurs between two media having different temperatures.

Heat Pump: A device for the transfer of thermal energy from a source of low-grade thermal energy to the consumer.

Heat Supply: A system for distributing heat generated in a centralized location through a system of insulated pipes for residential and commercial heating requirements.

Heat Transfer: The physical process of transferring heat energy from a hotter body to a less hot one either directly or through a separating partition of any material.

Power Supply System: A set of sources and systems for the conversion, transmission and distribution of electrical energy.

Power Transformer: A static electromagnetic device that has two (or more) windings connected inductively and is intended to be converted one AC system to another AC system by electromagnetic induction.

Chapter 7
To Methodology Research of Microwave:
Convective Processing of Frain

Alexey N. Vasilyev
https://orcid.org/0000-0002-7988-2338
Federal Agricultural Research Center VIM, Russia

Alexey Alexeyevich Vasilyev
Federal Agricultural Research Center VIM, Russia

Dmitry Budnikov
https://orcid.org/0000-0002-8729-1300
Federal Agricultural Research Center VIM, Russia

Aleksandr Belov
https://orcid.org/0000-0002-9216-9852
Federal Agricultural Research Center VIM, Russia

ABSTRACT

In the presented materials, a system of principles and approaches for studying processes of microwave-convective grain treatment is discussed. The essence of this research is described starting from the initial phase of designing structural diagrams of technological processes. Based on the explicitly-defined process target, a functional-physical analysis has been performed that made it possible to specify the most essential factors of process. The morphological analysis of the processes was carried out in several stages. At each stage, the specification of the goal and excluded variants of factors that at this stage, the development is impractical to implement. It is advisable to carry out a check of the effectiveness of the options selected from the morphological table using modeling. To this end, developed a mathematical and computer models of the process. The results of modeling the drying of grain in the microwave - convective zone confirmed the feasibility and efficiency of using the obtained model.

DOI: 10.4018/978-1-7998-1216-6.ch007

Copyright © 2020, IGI Global. Copying or distributing in print or electronic forms without written permission of IGI Global is prohibited.

INTRODUCTION

In technological processes of grain post-harvest handling, those involving thermal treatment belong to leaders, in terms of energy consumption. However, it is impossible to avoid these processes since they are commonly applied in grain drying and sanitation techniques. That is why reduction of energy intensity of these processes is a problem of prime importance.

Application of microwave electromagnetic fields is regarded as one of the effective methods for energy consumption reduction in grain drying and sanitation technologies, in which case grain dielectric properties are employed for its heating. It makes it possible to avoid energy loss associated with heat consumed by heat-carrier and removed from a technological cycle. In case of grain convective drying, gradients of temperature and moisture content are oppositely directional, within grain seeds, which adversely affects process effectiveness. When microwave heating is used internal layers of seeds are heated to higher temperatures owing to their higher moisture content. Therefore, gradients of moisture content and temperature are unidirectional. That is why energy consumption for grain drying is lower compared to conventional convective drying techniques (Ospanov, Vasilyev et al., 2017; Budnikov, & Vasiliev, 2018).

Studies in the field of microwave technologies application in grain postharvest handling, particularly for grain drying, have been carried out for many years (Horynski, 1999; Kupfer, 1999; Barbosa-Canovas, et.al. 2001; Jumah, Raghavan, 2001). Nevertheless, implementation of these technologies in agricultural production has not yet gained desirable scales. It can be explained by the need to employ more complicated equipment compared to conventional convective drying methods. But, in a greater degree, it is associated with the absence of required nomenclature of reliable equipment and machinery. However, such machinery can not be designed without detailed and comprehensive study of convective heat-and-moisture exchange processes in grain layer under the effect of microwave fields. For this purpose, an adequate research methodology has to be developed.

RESEARCH METHOD

Methodology, in its practical sense, is a system of principles and approaches for research activities that have to be adhered by researchers in the course of obtaining and development of knowledge, in the frames of a specified problem (Novikov, & Novikov, 2013; Lukashevich, 2001, Ruzavin, 1975). The following system of principles and operation sequence was applied in the course of studying the process of microwave-convective treatment of grain:

1. Development of process block diagram.
2. Development of the target statement for technological process.
3. Defining controlled parameters, control actions, initial conditions and status options for the object under control.
4. Application of functional-physical analysis to controlled parameters, control actions, initial conditions and status options of the controlled object.
5. Morphological analysis of the process under development and selection of appropriate variants of initial conditions, controlled parameters and control actions. Adjusting the purpose of the process.
6. Development of mathematical model for the process under study.

   a. Description of dynamic properties of the subject of research.
   b. Specifying initial and boundary conditions.
7. Method selection for mathematical model solution. Solving the problem.
8. Software selection for dynamic programming of the process under development.
9. Implementation of mathematical model with the use of computer. Program checkout. Model-based analysis of the process under study.
10. Optimization procedure. Application of computer simulation technique for optimization of parameters of the process under study or for that of design parameters of equipment used to implement this process.
11. Planning of experimental studies designed to check validity of model-based analysis results.
12. Development of experimental equipment and carrying out experimental studies.
13. Experimental results analysis.
14. Adjustment of the mathematical model of the process.

In accordance with the described sequence, it is advisable, at the first stage, to design block diagram for technological process under study. For the process of microwave-convective grain treatment, this diagram has the following layout (see Figure 1).

W is moisture content of grain upon completion of the process, Θ is grain temperature a upon completion of the process, $\Delta\Theta$ is temperature growth rate, W_0 is initial moisture content of grain, Ku is parameter characterizing specific properties of particular crop spices under processing, S is status parameter of grain layer, Z is grain contamination degree, Pc is intensity of microwave field, T is air temperature, V is air velocity, F is air relative humidity (F) or moisture content (D).

Let us define the target of microwave-convective grain treatment. First of all, initial data have to be specified. Microwave-convective grain treatment process may be aimed at various targets including the following ones: grain drying to a required value of moisture content, grain sanitation, grain micronization, grain pre-sowing treatment. We will consider the process grain sanitation. Let us take freshly harvested grain delivered to threshing floor as the product subject to processing. It may be wheat, barley, triticale, etc. Moisture content of grain delivered to a threshing floor depends on harvest time and on weather con-

Figure 1. Block diagram of the process of grain microwave-convective treatment

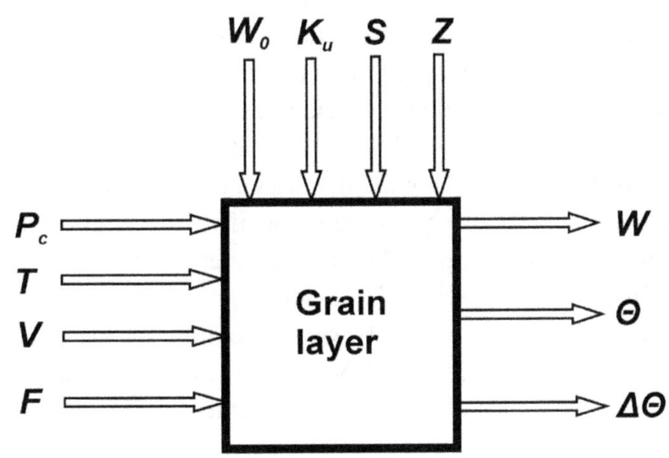

ditions. The wetter is grain delivered from field the faster is the process of microorganisms' development contributing to its contamination. At the same time, moisture content of grain affects substantially the rate and amplitude of its heating in microwave fields. That is why this parameter is essentially important for grain processing. In our studies, we assume that grain has standard values of moisture content.

Grain delivered to threshing floor has to be cleaned. Depending on selected grain treatment technology, grain preliminary clearing from extraneous objects may be included in the processing cycle, as well as final cleaning combined with fractions separation operations. Normally, grain subject to drying has to be preliminary cleaned in order to enhance process effectiveness and to reduce its energy consumption rate.

Applied soil handling agro-technologies affect substantially grain contamination with microorganisms and their metabolic byproducts, i. e. mycotoxins. Therefore, sanitation process has to solve the problem of reducing grain contamination to a required degree.

Taking the foregoing into consideration, the following target of the microwave-convective grain layer handling can be defined:

To perform sanitation of freshly harvested grain of various crop spices with standard moisture content, that has passed preliminary cleaning, in order to neutralize microorganisms and their metabolic products to the level stipulated in accordance with the requirements of standards or with those applied to grain to be treated in certain processing lines.

For more detailed description of technological process under study, it is advisable to apply the method of functional physical analysis (FPA) (Feofanov, 2005; Polovinkin, 1988).

Let us carry out functional-physical analysis for input parameters of the process, in accordance with the specified target.

INITIAL CONDITIONS

$\mathbf{F_1}$ is initial, standard, moisture content of grain (W_0). Moisture content, as a factor, is important since it affects substantially dielectric properties of grain. The higher is moisture content the higher is dielectric permeability of grain and, therefore, microwave field absorption which leads to the growth of grain heating rate. In accordance with the specified target, moisture content has to have standard value. However, standard values of moisture content considerably differ for seeds of various crop spices (Vasiliev, Budnikov et al., 2018; Zverev, & Zvereva, 2007). For seeds of each grain crop (Ku), its own value of normal moisture content is stipulated. Thus, for such crop species as wheat, rye, barley and oat value 14% is considered as the standard one, for moisture content. For sunseeds, castor beans, and dredge corn, these values are 9%, 8% and 16%, respectively. For example, relative permittivity of hard winter wheat with moisture content of 11.5% is 0.29 (Norena, Caciano & Lescano-Anadon, 2016; Mishra, Pooja, & Mishra, Bharat. 2017; Bansal, Namita et.al.2016), while that of wheat with moisture content of 14.6% is 0.4. Since hygroscopic property of free-flowing materials, i.e. their ability to sorb water vapors from the air, depends on their granular structure grain crops specific properties affect considerably moisture absorption ability (Egorov, 1985; Schinabeck, Weigler, & Floeter, 2018).

Distribution of water inside grain seeds is also essentially important for grain microwave-assisted processing. It depends on particular crop specificity, weather conditions in the harvest time and on period of grain complete ripeness. Studies (Aksjonov, 2004; Vasil'ev, Budnikov, & Vasil'ev, 2014) show that freshly harvested wheat grain features maximal moisture content in the center of seeds where germ is located. Moisture distribution in the direction towards the seed surface may change depending on harvest time and weather conditions. However, the main character of water distribution remains one and the same; i.e. the highest moisture content is in the central part of the seed while the layers adjacent to the surface contain less moisture. For such moisture distribution profile, it is important to select microwave field modes of grain treatment so that temperature in the area of germ does not exceed 55 °C.

Therefore, for factor F_1, initial standard moisture content of grain and options for its variation are defined by specific properties of particular crop spices and weather conditions, independently on the processing method.

Factor F_2 describes the grain layer status. In equipment designed for postharvest handling, grain layer may be in either static (N) or dynamic (D) state. Depending on the selection of layer mode, design and structure parameters of equipment for grain processing vary essentially which may have a strong effect on amount of materials per unit, energy intensity, production rate and grain processing modes. Thus, in case of microwave-assisted treatment of fixed grain layer, high production rates can be only attained in processing lines of large size. At the same time, grain layer thickness shall not exceed a certain value (normally, 15 cm to 18 cm) in order to ensure uniform heating of the entire grain layer, in microwave field. Otherwise, critical nonuniformity of temperature distribution may occur over elementary layers of grain which will cause incompliance with the requirements for stipulated grain processing modes.

For moving grain layers (D), layer thickness is critically limited, along with the layer velocity, since these parameters stringently define technological conditions of grain processing in microwave fields.

Apart from layer dynamic or static status, it is important to specify the properties of its structure that include: dense layer (P), loose layer (R), fluidized layer (G) and vortex layer (V). Layer structure depends on the flow rate of air blowing through it. Block diagram of the process (see Figure 1) contains control parameter V (air velocity), That is why, while describing the process of microwave-convective treatment, there is every reason to consider all options for grain layer status, including both moving and static layer modes. Finally, we will deal with the following eight status options for grain layer: NP is static dense layer, NR is static loose layer, NG is static fluidized layer, NV is static vortex layer, DP is moving dense layer, DR is moving loose layer, DG is moving fluidized layer, DV is moving vortex layer.

Therefore, factor F_2 describing grain layer status is directly linked with control parameter V, i.e. velocity of air blowing through grain layer. For further analysis, we assume that, within a particular option of grain layer status, air velocity variations shall not cross the limits of the range that defines grain layer initial state. It has to be taken into account that in currently applied structures of grain processing lines only one grain layer status is used, as a rule.

It has to be noted, as well, that processes of grain microwave-convective treatment may, in a general case, include changeover of grain layer status depending on particular technological mode.

We have studied two factors of input disturbance. The other two are grain crop spices parameter (Ku) and that of contamination degree (Z). These factors define, to a great extent, options for factor F_2 variation. Depending on particular crop spices, its standard moisture content and grain layer processing status have to be specified. In accordance with the target of microwave-convective processing, grain has to pass preliminary cleaning procedure. That is why the value of factor Z remains fixed.

Let us consider factors that describe control actions.

F_3 is intensity of microwave impact (Pc). Use of this factor that specifies control effect of microwave field in the process of grain convective handling makes it possible to control such process parameters as grain temperature Θ and grain temperature growth rate $\Delta\Theta$. Various options to implement the effect of this factor have to be considered:

- Microwave field intensity remains constant (PcC)
- Microwave field intensity is controlled smoothly, depending on grain temperature ($Pc\Theta$)
- Microwave field intensity is controlled smoothly, depending on grain temperature growth rate $\left(Pc\Delta\Theta\right)$
- Microwave field intensity is controlled on positional basis, depending on grain temperature $\left(PcPz\Theta\right)$
- Microwave field intensity is controlled on positional basis, depending on grain temperature growth rate $\left(PcPz\Theta\right)$
- Microwave field intensity is controlled depending on grain layer status ($PcS1$)...($PcS8$).

In the above options, specific features of processing lines design are not, absolutely, taken into consideration that may substantially affect variations and control of microwave field impact on grain layer. These specific features can be analyzed as separate factors when technology of grain microwave-convective processing has been distinctly defined.

The next control factor is temperature T of drying agent (factor F_4). In order to analyze variation options for this factor, let us, again, refer to the target of the process. Air temperature does not apply directly to the target but it can be used for more accurate control of temperature Θ or that of heating rate $\Delta\Theta$, in the process of grain treatment. Based on this principle, we can consider variation options for factor F_4:

- Ambient air convection is used to heat grain layer (Ta),
- Ambient air is heated/cooled to a certain temperature value (Tk) that has to have grain layer upon process completion,
- Air is heated/cooled up to the maximum value of grain temperature (Tm),
- In the process of grain treatment, air temperature (Tv) is controlled in the range from Tm to Tk.

Factor F_5 defines air velocity. This factor may essentially affect control parameters Θ and $\Delta\Theta$ thus changing the grain layer status. It has been noted above that grain layer status is defined by design parameters of processing equipment, grain crop specific properties, as well as by the purpose of grain processing. Therefore, let us assume that air velocity remains within a range that does not change grain layer status. We will consider the following possible variation options for factor F_5:

- (Vc) remains constant during the entire processing cycle,
- Air velocity ($Vshf$) is controlled depending on microwave field intensity,
- Air velocity (Vt) is controlled depending on air temperature,
- Air velocity ($V\theta$) is controlled depending on grain temperature,
- Air velocity (Vt) is controlled depending on grain processing time.

The next control factor is relative humidity (F) or moisture content (d) of air blowing through the grain layer. This factor does not apply to the target of grain microwave-convective processing. We will study its possible effect on control parameters. Process of microwave-convective grain treatment, no matter whether it is a short-term operation, results in moisture transfer within grain seeds producing slight drying effect. That is why required values of grain moisture content upon processing cycle completion can be achieved by controlling moisture content of air entering the grain layer during the process. At the same time, there is a possibility to control grain surface temperature by moistening it locally with the use of wet air blowing through the layer and so on. Let us describe variation options for factor F_6 characterizing air humidity:

- In the process of grain treatment, air having ambient value of moisture content (Fa) is used
- In the process of grain treatment, air having constant value of moisture content (Fc) is used
- In the process of grain treatment, wet air (Fm) is used
- Moisture content of air (Fv) blowing through the grain layer is controlled depending on the process stage
- Air moisture content (Ft) is controlled depending on grain temperature

We will consider all factors that were shown on the block diagram for grain microwave-convective processing. However, it has to be noted that the list of considered variation options for particular factors is far from being complete and they may be extended, in case of more comprehensive analysis of the process. Nevertheless, even this combination of factors with their variation options enables to develop morphological table of the process (Odrin, 1989; Wang, Xuefeng et al., 2017).

Morphological Analysis

Simple calculations show that morphological table comprises 5760 variants for implementing grain microwave-convective processing method (Table 1). Evidently, purely-formal approach is not applicable to our case.

On the basis of thorough specification of process targets, comprehensive analysis of each variation option for each factor has to be made with the account of limitations for parameters of equipment and energy intensity of the process. For example, grain layer status defines, to a great extent, size and weight characteristics of plants for grain microwave-convective processing. Evidently, requirements for production rate of equipment have to be considered, as well. Processing lines with fixed dense grain layer take much larger room compared with flow-through plants designed for moving grain layer. Weight-size

Table 1. Morphological table for the process of grain microwave-convective treatment aimed at its decontamination from microorganisms

Factors	Variation Options							
	1	2	3	4	5	6	7	8
F_2	*NP*	*NR*	*Na*	*NV*	*DP*	*DR*	*DA*	*DV*
F_3	PcC	$Pc\Theta$	$Pc\Delta\Theta$	$PcPz\Theta$	$PcPz\Delta\Theta$	PcS	*	*
F_4	Ta	Tk	Tm	Tv	*	*	*	*
F_5	Vc	$Vshf$	Vt	$V\Theta$	$V\Delta\Theta$	Vt	*	*
F_6	Fa	Fc	Fm		Ft	*	*	*

parameters of flow-through mode plants and those designed for processing fixed dense grain layers are comparable for rated production rate of about 5 t/h. In case that technological cycle under development requires a plant of much larger performance (for instance, 50 t/h to 100 t/h) weight and size values have to be increased substantially, and grain layer status may drastically affect these parameters for equipment of the two types.

Additional functions of technological equipment are also of great importance. For example, in postharvest handling lines, forced-aeration bins are often applied. These devices are multifunctional since they can be used for seed grain drying, grain postharvest storage and pre-drying, prior to the final drying in grain-dryers for long-term storage. It can be regarded as an advantage of processing plants for fixed grain layer. As far as the process of microwave-convective grain process is concerned, multi-functionality of equipment may present in a wider range of targets such as, for example, pre-drying of wet grain aimed to extend its safe storage period, grain drying to s moisture content, sanitation processing in grain storage barns, grain micronization for fodder preparation, etc. (Vasil'ev, 2016; Smith, Deandrae et al., 2018; Jafari, Hassan et al., 2017).

While considering options for magnetron output power control (factor F_3), functional parameters of test-and-measurement equipment, as well as cost indicators of control and supervision system, have to be taken into account. Thus, the option of microwave output power control depending on current values of grain temperature ($P_c\Theta$) requires accurate monitoring the temperature of grain. It has been noted above that moisture distribution within grain seeds is not uniform which leads to more intensive heating of germ. Therefore, the question arises: how to control germ temperature in a moving grain layer in the case when moisture distribution within seeds changes during processing?

Stepless control of microwave radiation is a source of additional technical problems since the range of magnetron output power change by input voltage variation is not wide enough while other control methods are associated with relatively high investments (Ospanov, Vasiliev, Budnikov et al., 2016).

Previous studies show that morphological table for grain microwave-convective processing has to be composed on the basis of analysis made with the account of all technological targets, design requirements and energy consumption limitations. For this purpose, the target of the process has to be checked for possible errors. At this stage, the requirements for the plant designed to implement the process have to be considered, to their maximum extent.

Extended target can be specified as follows: Sanitation and pre-drying, as an auxiliary effect, of freshly harvested grain of any crop spices, having passed postharvest cleaning in technological cycle of postharvest handling, with the possibility of temporary storage grain postharvest cleaning area so that grain, upon completion of processing, can be transported, in correspondence with the following options: into forced-aeration bin for temporary storage, into grain dryer, to threshing floor for grain batch formation, to storage house for long-term storage, to feed processing area for compound feed preparation. Plant production rate is 10 t/h.

The extended target enables comprehensive analysis of morphological table. Since clear requirements for processing rate (10 t/h) and size of equipment (installation in area for grain postharvest cleaning) have been specified selection of processing plant for moving grain layer is the most appropriate option.

Based on the analysis of factor F_3, we will exclude options involving microwave field intensity control from the table in order to reduce costs related to implementation of control and supervision system.

Options for factor F_4 will remain without change.

As far as factor F_5 is concerned, let us exclude the option involving air velocity control depending on microwave radiation intensity since it is identical to option ($V\Theta$). Option Vt, has to be excluded, as well, because duration of processing will depend on initial parameters of grain that may vary considerably. Therefore, developing a universal control program $V = f\left(t\right)$ may appear a complicated problem.

Factor F_6 has to be excluded from the table for the following reasons. Air humidity and temperature are interrelated with each other. That is why control of humidity with the purpose of its reduction results in air temperature growth which is not acceptable, in certain cases.

Morphological table composed with the account of our brief analysis has the following content.

This comprehensive morphological table (Table 2) contains only 256 options for implementation of grain microwave-convective treatment method. It may be a subject of certain interest in terms of novelty of options but not many of them may appear effective in practice.

Therefore, it is reasonable to pursue comprehensive analysis of this table in order to exclude more variation options of certain factors or factors as such. For this purpose, targets of the process have to be specified to a deeper extent. However, it is important to realize that the problem of option selection remains of rather complicated ones even if we exclude maximum number of variations.

Let us assume that, as a result of manifold reduction of morphological table, we obtained a matrix of size 2×2. It means that only four options are implementable. Which of these options is the most viable? Which has to be preferred? This question is not so easy to answer without additional studies related to each of these options. The cheapest way to do this is to apply mathematical model-based analysis of the process, and results of model-based analysis will yield more information for further studies.

MODELING OF HEAT AND MOISTURE EXCHANGE IN THE GRAIN LAYER UNDER MICROWAVE CONVECTION INFLUENCE

Mathematical description of the of heat and moisture exchange processes in the elements of the processed products and the layer of these products allows to determine the operating modes of the electrical equipment and features of the design.

Table 2. Comprehensive morphological table for grain microwave-convective a processing

Factors	Variation Options			
	1	2	3	4
F_2	*DP*	*DR*	*DA*	*DV*
F_3	PcC	$PcPz\Theta$	$PcPz\Delta\Theta$	PcS
F_4	Ta	Tk	Tm	Tv
F_5	Vc	Vt	$V\Theta$	$V\Delta\Theta$

With microwave influence on grain, the system of heat and moisture transfer differential equations during microwave heating has the following form (Lykov, 1963; Lykov, Maksimov, 1957; Maksimov, 1956):

$$\frac{\partial \theta}{\partial \tau} = a\frac{\partial^2 \theta}{\partial x^2} + \frac{r'}{c_z \varepsilon} \cdot \frac{\partial W}{\partial \tau} + \frac{Q_v}{c_z \rho_z} \tag{1}$$

$$\frac{\partial W}{\partial \tau} = a_{m_2}\frac{\partial^2 W}{\partial x^2} + a_{m_2}\delta_2 \cdot \frac{\partial^2 \theta}{\partial x^2} + \varepsilon\frac{\partial W}{\partial \tau} \tag{2}$$

$$\frac{\partial P}{\partial \tau} = a_p\frac{\partial^2 P}{\partial x^2} + \frac{\varepsilon}{c_v} \cdot \frac{\partial W}{\partial \tau} \tag{3}$$

where a - coefficient of thermal diffusivity, m^2/s; ε - coefficient of liquid phase transformation -vapor, p.u.; c_z – specific heat of grain, kJ/kg °C; r' – specific heat of vaporization, kJ / kg; Q_v – specific power dissipated in the dielectric under the influence of the microwave field, W/m^3; ρ_z – density of grain dry matter, kg/m^3; a_{m_2} – coefficient of fluid diffusion, m^2/s; δ_2 - relative coefficient of thermal diffusion, p.u.; P - excess pressure in the sample, Pa; c_v - body capacity relative to moist air, $c_v = \frac{1}{P_n}$, $\frac{1}{Pa}$; P_n - vapor pressure of the material at a given moisture content, Pa; θ - temperature of grain, $°C$; a_p – coefficient of convective vapor diffusion, m^2/s.

Heat and moisture exchange with convective influence are described by the following system of differential equations (Aniskin, 1972; Baum, 1983; Vasilyev, Rudenko, 2009; Vasiliev, Vasiliev, Budnikov et al., 2019):

$$\frac{\partial T}{\partial \tau} + 3600V\frac{\partial T}{\partial x} = -\frac{\gamma_z c_z}{\gamma_v c_v \varepsilon} \cdot \frac{\partial \theta}{\partial \tau} - \frac{\gamma_z r'}{\varepsilon \gamma_v c_v \cdot 100} \cdot \frac{\partial W}{\partial \tau} \tag{4}$$

$$\frac{\partial T}{\partial \tau} + 3600V\frac{\partial T}{\partial x} = -\frac{\alpha_q s_v}{\varepsilon \gamma_v c_v} \cdot \left(T - \theta\right) \tag{5}$$

$$\frac{\partial W}{\partial \tau} = -\frac{\varepsilon \gamma_v}{10^3 \gamma_z} \cdot \left(\frac{\partial D}{\partial \tau} + 3600V \cdot \frac{\partial D}{\partial x}\right) \tag{6}$$

$$\frac{\partial W}{\partial \tau} = -K\left(W - W_p\right) \tag{7}$$

where T - temperature of the drying agent, °C; D-humidity content of the drying agent, g/kg; W - grain moisture,%; W_p - equilibrium humidity of grain,%; K - coefficient of drying 1/h

$$K = 7,4897 - 0,1022T - 0,6438W + 0,0134T \cdot V$$
$$+0,0148T \cdot W + 0,0029V \cdot W - 0,0026T^2 + 0,0071W^2$$

(Vasilyev, Severinov, 2015; Vasiliev,Vasiliev, Budnikov, & Samarin 2019; Vasilyev, Vasilyev, & Budnikov, 2018); θ - grain temperature, °C; V-speed of the drying agent, m; c_v-air thermal capacity, kJ/kg °C; ε - porosity of the grain layer; s_v - specific surface of seeds, 1/m; α_q - coefficient of heat transfer, kJ/kg s °C; γ_z - bulk weight of grain, kg/m³; γ_v - air absolute weight, kg/m³; τ - time, h.

The process of microwave convection drying is as follows. At the initial stage, grain is exposed only to the microwave field. At this time, grain warms up. The central part of the grain is heated to a higher degree, since it has a high humidity relative to the outer layers. This allows us to create a temperature gradient directed from the center to the surface of the grain. There is a partial conversion of moisture in the grain into vapor and its movement through the capillaries is closer to the grain surface. The process of grain heating in microwave field proceeds quickly and depends on the moisture content of grain. The higher humidity is, the faster the heating rate. So, within 2-3 minutes grain can be heated to a temperature of more than 90 °C. To prevent grain heating higher than 55 °C, it is necessary to turn off the supply of microwave field to the grain layer in time.

In the second stage, the drying agent is supplied into the grain layer. Moisture is removed from the grain layer and the grain is cooled.

Then (the third stage), the microwave field is supplied again into the cereal layer together with drying agent supply. Then (the fourth main drying stage), the process is cyclic supplying of microwave field into the grain layer, in order to intensify the transfer of moisture from the center to the grain surface, and constant supply of the drying agent.

Cooling of dried grain to the required moisture is carried out by supplying the atmospheric air to grain layer.

It shows that in the process of microwave convection drying, the heat and-moisture exchange in the grain layer at different stages are described by different systems of differential equations. At the first stage - system (1) - (3). At the second stage - combined (1) – (7), At the third stage – (4) – (7). At the main stage of drying - interchange (1 - 7) / (4 - 7). At the final stage, the grain cooling, - (4) - (7).

At the first stage of microwave convection drying, intensive heating of grain takes place and moisture in it moves from the center to surface. There is also a certain transfer of moisture from grain to intergrain space. Since the drying agent is not supplied into the grain layer, moisture is not removed, the moisture content of the grain layer does not decrease. Therefore, equations (2) and (3) can be neglected in the first stage of drying. We use only the following equation for describing heating of grain in the microwave field:

$$\frac{\partial \theta}{\partial \tau} = a\frac{\partial^2 \theta}{\partial x^2} + \frac{r'}{c_z \varepsilon} \cdot \frac{\partial W}{\partial \tau} + \frac{Q_v}{c_z \rho_z} \qquad (8)$$

At the next stage microwave convection drying is carried out. Assuming that convective heat and moisture exchange is carried out from the entire grain surface, the equations for changing the grain temperature will be presented in the following form:

$$\frac{\partial \theta}{\partial \tau} = a\frac{\partial^2 \theta}{\partial x^2} + \frac{r'}{c_z \varepsilon} \cdot \frac{\partial W}{\partial \tau} + \frac{Q_v}{c_z \rho_z} - \frac{\alpha_q s_v}{\varepsilon \gamma_v c_v} \cdot \left(T - \theta\right) \tag{9}$$

and the remaining equations remain as before:

$$\frac{\partial T}{\partial \tau} + 3600V\frac{\partial T}{\partial x} = -\frac{{}^{3}{}_{z}c_{z}}{{}^{3}{}_{v}c_{v}\mu} \cdot \frac{\partial \theta}{\partial \tau} - \frac{{}^{3}{}_{z}r'}{\mu^{3}{}_{v}c_{v} \cdot 100} \cdot \frac{\partial W}{\partial \tau} \tag{10}$$

$$\frac{\partial W}{\partial \tau} = -\frac{\varepsilon \gamma_v}{10^3 \gamma_z} \cdot \left(\frac{\partial D}{\partial \tau} + 3600V \cdot \frac{\partial D}{\partial x}\right) \tag{11}$$

$$\frac{\partial W}{\partial \tau} = -K_e\left(W - W_p\right) \tag{12}$$

It should be noted, however, that the drying coefficient K_e is calculated by a different equation than for convective drying, taking into account microwave field parameters in the grain layer.

An interchange of equation (1) with the system of equations (10), (11), (9), (12) describes the microwave convection drying process of grain. At the final stage of grain cooling, the system of equations (4) - (7) is applied. In microwave convection drying of grain, the unevenness of moisture removal depends on uneven propagation of the field in the grain layer and on uneven propagation of the drying agent. In order to specify the parameters of microwave field and the speed of the drying agent, a step-type calculation method is used (Metodicheskie rekomendacii…, 1977; Vasilyev, Vasilyev, 2018; Vasilyev, Severinov, 2014). To do this, a dense grain layer is represented as a set of elementary layers. As an elementary layer is one grain. With this assumption in mind, in equations (8) and (10), (11), (9), the spatial coordinate x can be replaced by the grain equivalent diameter d_{eq}.

The essence of the step-type method is that the parameters of the drying agent passing through the elementary layer become input parameters for the next elementary layer. When passing through the grain layer, the air becomes saturated with moisture, its temperature changes and, as a consequence, the equilibrium humidity. Therefore, at the input of each elementary layer, it is necessary to calculate the moisture content and equilibrium humidity. For this, the known dependencies (Metodicheskie rekomendacii…, 1977) are used:

$$W_r = \left[\frac{\ln\left(1 - F\right)}{5.4710^{-6}\left(T + 273\right)}\right]^{0,435} \tag{13}$$

$$F = \frac{745D}{\left(622 + D\right)e^{\left(0.622+\frac{7.5T}{238+T}\right)}} \tag{14}$$

The mathematical model of the grain layer microwave convection drying can be represented as a set of elementary layer mathematical models. It was assumed that the thickness of the dried grain layer is composed of equivalent diameters d_{eq} of the successively arranged grain. The number of elementary layer models will be equal to the number of grain that fits within the given thickness.

To develop a computer model of the elementary layer, an attempt is made to represent the differential equations (1), (4) - (9), in the form of transfer functions.

Equations (1), (4) - (8) contain partial derivatives. Let us use the method of the Laplace integral transformation for the intermediate solution. In this case, transformation of partial derivatives is carried out according to the following rule (Martinson, & Malov, 2011):

For example, for the drying agent temperature T, which varies both in time and in thickness of the grain layer, the transformation is carried out as follows. If $T = T\left(x,\tau\right)$ the Laplace transformation is carried with variable $\tau\left(\tau \geq 0\right)$ then

$$L\left[T\left(x,\tau\right)\right] \equiv T\left(x,p\right)\int_{0}^{\infty} e^{-p\tau}T\left(x,\tau\right)d\tau ,$$

it is possible (by integration by parts) to establish the relation:

$$L\left[\frac{\partial T}{\partial \tau}\right] = pT\left(x,p\right) - T\left(x,0\right),$$

$$L\left[\frac{\partial T}{\partial \tau}\right] = \frac{\partial T}{\partial x}\left(x,p\right),$$

where $T\left(x,0\right)$ is an initial temperature value.

For the equation (4), we perform the Laplace transformation procedure according to A:

$$\begin{gathered} pT\left(x,p\right) - T\left(x,0\right) + 3600V\frac{\partial T}{\partial d_{eq}} \\ = -\frac{\gamma_z c_z}{\varepsilon \gamma_v c_v}\left(p\theta\left(x,p\right) - \theta\left(x,0\right)\right) - \frac{\gamma_z r'}{100\gamma_v c_v \varepsilon}\left(pW\left(x,p\right) - W\left(x,0\right)\right) \end{gathered};$$

$$pT(x,p) - T(x,0) + 3600V\frac{\partial T}{d_{eq}}$$
$$= \frac{\gamma_z c_z}{\gamma_v c_v \varepsilon}\theta(x,0) - \frac{\gamma_z c_z}{\varepsilon \gamma_v c_v} p\theta(x,p) + \frac{\gamma_z r'}{100\gamma_v c_v \varepsilon} W(x,0) - \frac{\gamma_z r'}{100\gamma_v c_v \varepsilon} pW(x,p)$$, or

$$pT(x,p) - T(x,0) + 3600V\frac{\partial T}{d_{eq}} = A_1\theta(x,0) - A_1 p\theta(x,p) + A_2 W(x,0) - A_2 pW(x,p),$$

where $A_1 = \frac{\gamma_z c_z}{\varepsilon \gamma_v c_v}$, $A_2 = \frac{\gamma_z r'}{100\gamma_v c_v \varepsilon}$.

Writing the equation for partial derivative and replacing coordinate *x* by the grain equivalent diameter d_{giv}, we obtain the following expression:

$$3600V\frac{\partial T}{\partial d_{tr}} = A_1\theta(d_{eq},0) - A_1 p\theta(d_{eq},p) + A_2 W(d_{eq},0)$$

$$-A_2 pW(d_{eq},p) - pT(d_{eq},p) + T(d_{eq},0) \tag{15}$$

Using the superposition principle, let us write differential equation only with variable *T*:

$$3600V\frac{dT}{dd_{eq}} = pT(d_{eq},p) + T(d_{eq},0).$$

In addition, *p* plays the role of a variable. We solve differential equation with respect to d_{eq}, writing it as boundary condition $T(d_{eq},0) = T_0$.

Solving the equation using MATLAB package (D'yakonov, 2008a; D'yakonov, 2008b) we obtain the following result:

$$T(p) = \frac{T_0 - e^{-p\tau_1}T_0 + pT_0 e^{-p\tau_1}}{p} \tag{16}$$

where $\tau_1 = \frac{d_{eq}}{3600V}$ is the delay time in one grain layer at the speed of drying agent *V*.

To obtain in operator form the dependence of the coolant temperature image T (p) on the output of the elementary layer from the initial air temperature T_0 (p), grain temperature $\theta_0(p)$, and its humidity $W_0(p)$, we substitute these terms into equation (12). A similar operation can be performed for the crent temperature values $\theta(p)$ and grain moisture $W(p)$. As a result, the following equation is obtained in the operator form:

$$T(p) = T_0 \frac{1}{p}\left(1 + pe^{-p\tau_1} - e^{-p\tau_1}\right) - A_1\theta(p)p + A_1\theta_0 - A_2W(p)p + A_2W_0 \quad (17)$$

According to the existing method (Sharma, Dzh, & Sinh, 2002), to find the equation describing the dependence $T = f(\theta, W, \theta_0, W_0, T_0)$ it is necessary to perform the inverse Laplace transformation of equation (17). We have tested (Vasilyev, Ospanov, Budnikov, et al., 2015; Vasilyev, Severinov, & Makarova, 2016) a technique that allows to perform numerical modeling of the process, presenting this dependence as transfer functions. The functional diagram realizing this dependence is shown in Figure 2.

We can perform a similar procedure for equations (1), (5), (6), (9). From equation (5), we find the dependence for grain temperature of the elementary layer θ under convection drying, from equation (1), the dependence for grain temperature under the action of microwave field, from equation (9), the dependence for grain temperature in the microwave convection influence on grain. To find the dependence of moisture content D for the drying agent on its output from the grain layer, we use equation (6).

Figure 2. Functional diagram for the implementation of the intergrain space temperature dependence on the parameters of grain and air

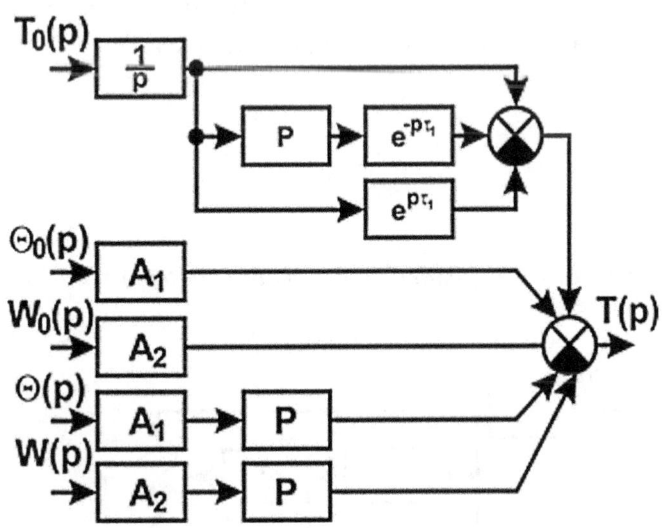

Representing $\frac{\varepsilon\gamma_v}{10^3\gamma_z} = A_3$ and performing the Laplace transformation for (6), we obtain the following relationship:

$$3600V\frac{dD}{dd_{giv}} = -D(p)p + D_0 - \frac{1}{A_3}pW(p) + \frac{1}{A_3}W_0$$

We solve this equation using the superposition principle in d_{giv} only for D.

$$D(p) = \frac{D_0 - e^{-p\tau_1}D_0 + pD_0e^{-p\tau_1}}{p} \tag{18}$$

Substituting the moisture term of grain in (18), we obtain operator form the coolant moisture content image dependence *D*(p) at the output of elementary layer from the moisture content image at the input to the grain layer D_0, the initial grain moisture content of W_0, from the image of the current grain moisture $W(p)$:

$$D(p) = D_0\frac{1}{p}\left(1 + pe^{-p\tau_1} - e^{-p\tau_1}\right) - \frac{1}{A_3}W(p)p + \frac{1}{A_3}W_0 \tag{19}$$

The functional diagram realizing this dependence in the transfer functions is shown in Figure 3.

In order to obtain the equation for the change in the grain temperature under microwave influence, it is necessary to solve equation (1); and for microwave convection influence it is equation (9). We use the principle of superposition and solve heat equation for the sphere:

$$\frac{d\theta}{d\tau} = \left(\frac{d^2\theta}{dr^2} + \frac{2}{r}\frac{d\theta}{dr}\right) \tag{20}$$

Figure 3. Functional diagram for the dependence of moisture content on the grain layer output from the input parameters

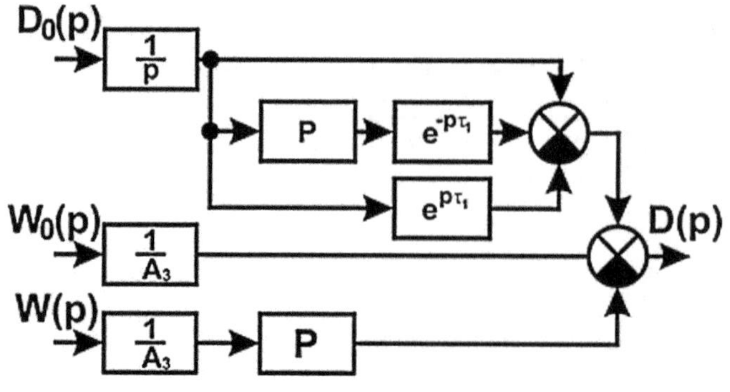

where r – grain equivalent radius.

The initial conditions determine the temperature at the center and on the surface of grain at the beginning of drying process: $\left.\frac{\partial \theta}{\partial r_0}\right|_{\tau=0} = \left.\frac{\partial \theta}{\partial r_1}\right|_{\tau=0} = \theta_0$, $r_0 = 0$, $r_1 = r$. As boundary conditions, let us write: heat transfer in the center of the grain - $\frac{\partial \theta}{\partial r_0} = 0$, on the surface –

$$\frac{\partial \theta}{\partial r_1} = -A_4 \cdot \left(\theta - T\right), at\, x = r.$$

where $A_4 = \frac{\alpha_q s_v}{\varepsilon \gamma_v c_v}$.

The grain is presented in the form of a sphere, from the surface of which the heat is discharged into the environment at a temperature of T = const. During the entire heating / cooling period, the heat transfer coefficient is the same for all areas of the surface and remains constant ($\alpha = const$).

Solving equation (20) and subordinating it to the initial and boundary conditions, in the criterial form we obtain (Pikulin, Pohozhaev, 2004):

$$\theta_{rel} = \sum_{n=1}^{n=\infty} A\left(\mu_n\right) U\left(\mu_n R\right) | \exp\left(-\mu_n^2 Fo\right) \tag{21}$$

where $R = r / r1$ is the dimensionless radius; $\theta_{rel} = \left(\theta - T\right) / \left(\theta_0 - T\right)$ - dimensionless excess temperature; μ_n - the root of the characteristic equation $tg\mu = -\mu / \left(Bi - 1\right)$, which has an infinite number of roots; Fo - Fourier criterion, $\text{Fo} = \frac{a\tau}{d_{giv}^2}$.

The dimensionless excess temperatures at the center ($R = 0$) and on the surface ($R = 1$) of the sphere can be determined by the formulas:

$$\theta_{R=0} = N_1\left(Bi\right) \exp\left(-\mu_1^2 Fo\right) \tag{22}$$

$$\theta_{R=1} = P_1\left(Bi\right) \exp\left(-\mu_1^2 Fo\right) \tag{23}$$

In this equation, $Bi = \frac{\alpha_q d_{giv}}{\lambda}$ dimensionless complex called Biot number representing ratio of the internal thermal resistance of thermal conductivity to the external thermal resistance of heat transfer. The numerical values of the range $\mu_1, N_1\, and\, P_1$ depending on the number Bi are given in Table 3.

Table 3. The numerical values of the range μ_1, N_1 and P_1 for calculation of the cooling (heating) of the sphere (Karminskij, 2005)

Bi	μ_1	N_1	P_1	Bi	μ_1	N_1	P_1
0.00	0.0000	1.000	1.000	1.00	1.5708	1.273	0.811
0.01	0.1730	1.003	0.998	1.20	1.6887	1.320	0.776
0.02	0.2445	1.006	0.996	1.40	1.7906	1.364	0.743
0.04	0.3450	1.012	0.992	1.60	1.8798	1.405	0.712
0.06	0.4217	1.018	0.988	2.00	2.0288	1.479	0.654
0.10	0.5423	1.029	0.979	3.00	2.2889	1.623	0.534
0.15	0.6609	1.044	0.970	4.00	2.4557	1.720	0.444
0.20	0.7593	1.059	0.960	5.00	2.5704	1.787	0.376
0.30	0.9208	1.091	0.943	10.00	2.8363	1.925	0.204
0.40	1.0528	1.116	0.921	21.0	2.9930	1.980	0.0979
0.50	1.1656	1.144	0.902	51.0	3.0801	1.996	0.0398
0.60	1.2644	1.171	0.883	81.0	3.1028	1.998	0.0250
0.80	1.4320	1.224	0.846	101.0	3.1105	1.999	0.0200

The heat transfer coefficient α_q can be determined using the dependence for the Reynolds criterion

$$\mathrm{Re} = \frac{Vd_{giv}}{\nu},$$

where V - speed of the drying agent, m/s; d_{giv} - characteristic dimension of particle, m; ν - climatic viscosity of the gas, m2/s; Re - Reynolds criterion, which is the ratio of inertia forces to the viscous forces. $Nu = \frac{\alpha_q d_{giv}}{\lambda}$, and for a dense grain layer $Nu = 0,075\,\mathrm{Re}$ (Pticyn, 1964), or $\alpha_q = 0,075\frac{V\lambda}{\nu}$, where Nu - Nusselt criterion (dimensionless heat transfer coefficient).

For $W = 18\%$ and $V = 0,7 m / s$ $\alpha_q = 1,629\frac{kJ}{m^2 c \cdot °C}$, $Bi = 16,492$,

$Fo = 0,004\tau$. Then equation (23) can be written in the following form:

$$\theta_{R=1} = P_1\left(Bi\right)\exp\left(-\mu_1^2 Fo\right) = 2,332e^{-0,0343\tau}.$$

$\theta_{rel} = \left(\theta - T\right) / \left(\theta_0 - T\right)$, so, the temperature of the single grain with convective heating can be found from the following equation:

$$\theta = \left(\theta_0 - T\right) \cdot 2,332e^{-0,0343\tau} + T \text{ .}$$

In general, the dynamic properties of the temperature of the single grain in convective heat transfer can be represented by a first-order aperiodic link $W_\theta\left(p\right) = \frac{1}{T_1 p + 1}$, with a time constant of T_1. Then, in operator form, equation (1) will be represented as:

$$\theta\left(p\right) = \left(\theta_0 - T\left(p\right)\right)\frac{1}{T_1 p + 1} + \theta_0 + A_5 W\left(p\right)p + A_5 W_0 + \frac{1}{p}A_6 Q_v \tag{24}$$

where $A_5 = \frac{r'}{c_z \varepsilon}$, $A_6 = \frac{1}{c_z \rho_z}$.

As a result of these transformations, a complete system of equations and transfer functions has been obtained that allow us to calculate heat and moisture exchange in the elementary grain layer under convection and microwave convection processing. Let us write it again:

$$W_r = \left[\frac{\ln\left(1 - F\right)}{5.4710^{-6}\left(T + 273\right)}\right]^{0,435} \tag{25}$$

$$F = \frac{745D}{\left(622 + D\right)e^{\left(0.622 + \frac{7.5T}{238 + T}\right)}} \tag{26}$$

$$T\left(p\right) = T_0 \frac{1}{p}\left(1 + pe^{-p\tau_1} - e^{-p\tau_1}\right) - A_1\theta\left(p\right)p + A_1\theta_0 - A_2 W\left(p\right)p + A_2 W_0 \tag{27}$$

$$D\left(p\right) = D_0 \frac{1}{p}\left(1 + pe^{-p\tau_1} - e^{-p\tau_1}\right) - \frac{1}{A_3}W\left(p\right)p + \frac{1}{A_3}W_0 \tag{28}$$

$$\theta\left(p\right) = \left(\theta_0 - T\left(p\right)\right)\frac{1}{T_1 p + 1} + \theta_0 + A_5 W\left(p\right)p + A_5 W_0 + \frac{1}{p}A_6 Q_v \tag{29}$$

$$W\left(p\right) = W_p \frac{K}{p + K} \tag{30}$$

With the use of the computer modeling system, a numerical solution of the equations can be found with obtaining the values of grain parameters and drying agent. Simulink software package was used for modeling (D'yakonov, 2008b).

Computer modeling also uses a step-type method. The computer model, for the previously wrote system of equations, is represented in the form of typical automation locks (Figure 4).

This model describes the processes of heat and moisture exchange in the single grain with microwave convection drying. The initial values of grain temperature and humidity are set using the "Constant" blocks "Q0" and "W0", respectively. This approach to the construction of the model allows you to set

Figure 4. Computer model (Simulink) of heat and moisture exchange in the single grain with microwave convection influence

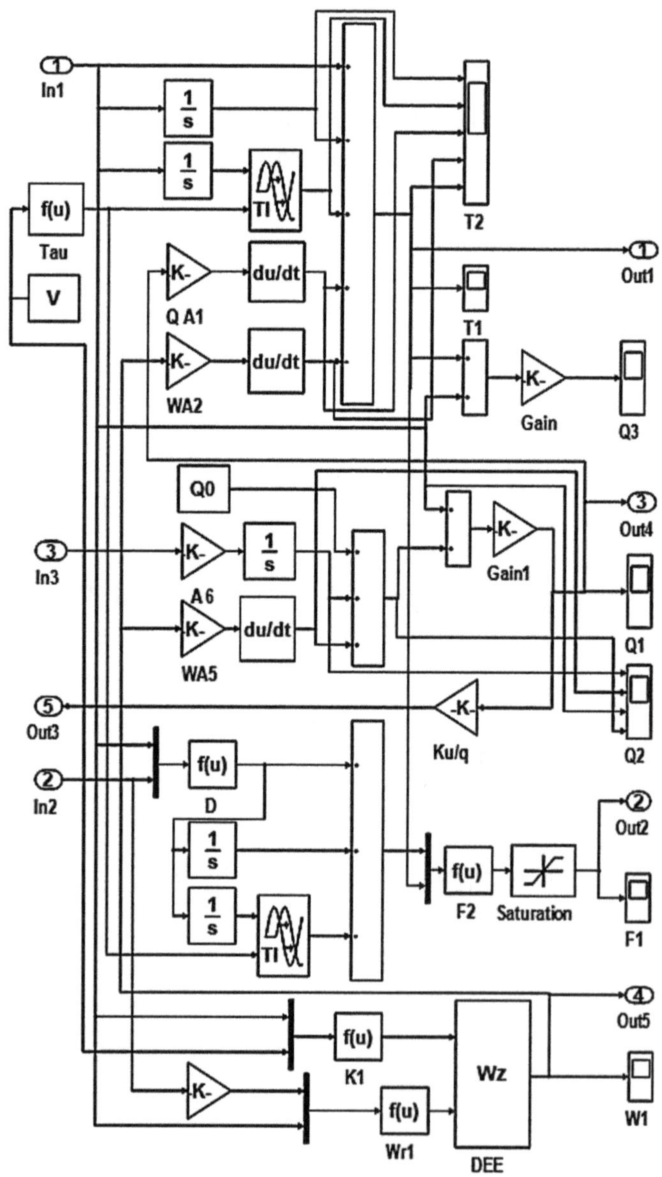

initial parameters for each grain layer separately. Also, we may take into account the uneven distribution of the drying agent over the grain layer, which is reflected in its speed. Block "V" is used to set the speed of the drying agent in the grain layer. The temperature of drying agent T1 at the output of the single grain layer is the input temperature for the subsequent grain layer in the course of air movement. Also, the relative air humidity F1 at the output of a single grain layer is the input to the subsequent grain layer. This construction of the model makes it possible to determine the specific power of microwave field absorbed by the grain at a given point in the grain layer. Taking into account the fact that propagation of microwave field is uneven in the grain layer, such opportunity makes the model more flexible in use.

When constructing a computer model of the grain layer microwave convection drying, the modeling results of microwave field propagation in the microwave active zone were taken into account. It was assumed that the grain layer thickness, which is subjected to microwave convection influence, is 15 cm. This is comparable to the depth of microwave field penetration into the wheat grain layer of the conditioning humidity (14%). It was assumed that the grain layer can be divided into sections of 5 cm thick, within which microwave field is evenly propagated.

Simulink model of the grain layer with 15 cm thickness, divided into three layers of 5 cm, is shown in Figure 5.

The grain layer is represented by three blocks-layers "Grain layer", in each layer the microwave field has a different intensity, by analogy with the real grain layer. The drying agent successively passes through all the layers. Its input parameters are set using blocks *Tvh* and *Fvh*.

With the help of the Relay block, you control the on and off of the magnetrons. The magnetrons are switched off when the temperature of the grain, at the point of microwave field maximum power, reaches 55 °C. The oscillographs display graphs of grain temperature change in each layer (block Q), grain moisture in the layer (block W), relative humidity at the output of each grain layer (Fvih), temperature of the drying agent at the output of each grain layer (Tvih) (Vasilyev et al., 2016; Vasilyev, Ospanov, Dautkanova, Budnikov et al., 2016).

Since the specific energy absorbed by the grain varies depending on the moisture content of the grain and the magnetron distance, a two-factor experiment was conducted to determine this dependence. The results of the experiment are partially shown in Table 4.

Figure 5. Simulink model of microwave convection drying of grain in a dense layer

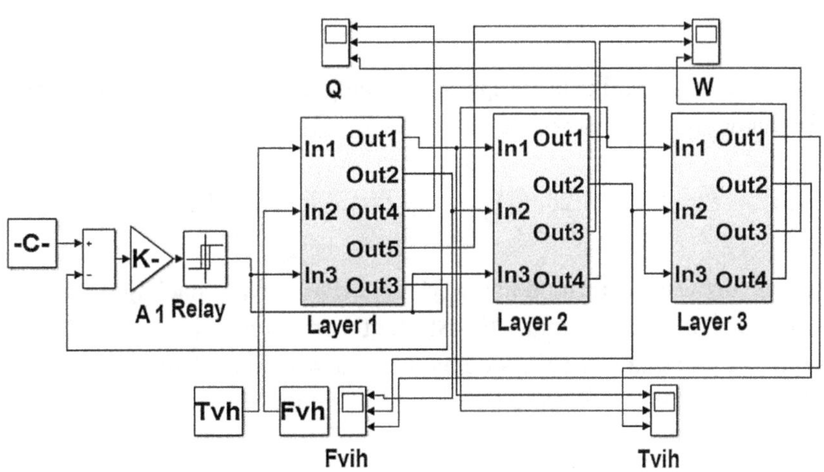

Table 4. Experimental data on the influence of grain moisture and the distance to magnetron on the specific energy absorbed by grain

Grain Moisture	Axial Distance to the Magnetron	Specific Energy, Absorbed by Grain at the Control Point
W, %	L, cm	Qv, kJ/m³
21.3	3	3180.3
21.3	6	1458.3
21.3	9	773.7
17.8	3	3146.1
17.8	6	1646.2
17.8	9	703.6
16.9	3	3580.0
16.9	6	1962.0
16.9	9	913.9
15.9	3	3090.4
15.9	6	1916.2
15.9	9	886.6
15	3	3452.0
15	3	3153.3
15	6	1632.5
15	9	861.4
15	6	1578.7
15	9	929.1
13.9	3	3757.6
13.9	6	1572.2
13.9	9	911.4
12	3	3235.4
12	6	1644.8
12	9	1148.9

Using the obtained experimental data, a regression analysis was performed and the following regression model was obtained:

$$Qv = 5778 - 14,92W - 846,2L + 0,2332W^2 - 3,727W \cdot L + 42,74L^2 \quad (31)$$

The regression analysis was performed using MATLAB software package.

The model coefficients are significant by the Student's criterion. The model adequacy was tested by the F-ratio test. The model accuracy is estimated by the following indicators: *SSE*: 1.409e+06; *R-square*: 0.9731; *Adjusted R-square*: 0.9698; *RMSE*: 185.4, which characterizes a good level of reliability.

Using the regression equation, the response surface (see Figure 6) was constructed to change the specific energy.

The regression equation is used in the computer model, which allows to model the change in the specific energy of microwave field in the grain layer during drying.

When developing the design of microwave convection zones, the direction of drying agent's movement, its temperature, the uniformity of distribution in the grain layer is important. Proceeding from this, modeling of the microwave convection grain drying process was carried out. Figure 7 shows a schematic diagram of the grain layers location, the propagation of the microwave field and the movement of the drying agent.

The drying agent first passes through the grain layer 1 with this movement option. Since this layer is closest to the source of microwave field, it must have a higher temperature than Layer 2 and Layer 3. Therefore, when air passes through Layer 1 → Layer 2 → Layer 3, it must be heated from the grain in Layer 1. Then, its temperature can be decreased or kept at a constant level. It all depends on the specific modes of grain drying. Figure 8 shows the results of modeling the change in grain temperature according to the layers.

The simulation was carried out with the following initial parameters: $W_0 = 22\%$; $F_0 = 65\%$; $T_0 = 20$ °C; $V = 0,5\frac{m}{c}$.

The air temperature change in the grain layer is shown in Figure 9.

The results of modeling show that the drying agent temperature at the output of the grain layer "follows" the grain heating temperature. As the temperature of the drying agent rises, its moisture absorbing capacity increases. A large amount of moisture is carried out by air from the grain layer.

The change in the relative humidity of the air at the output of the grain layers is shown in Figure 10.

The results of change modeling in moisture content of the grain are shown in Figure 11.

Figure 6. Response surface $Q_v = f(L,W)$

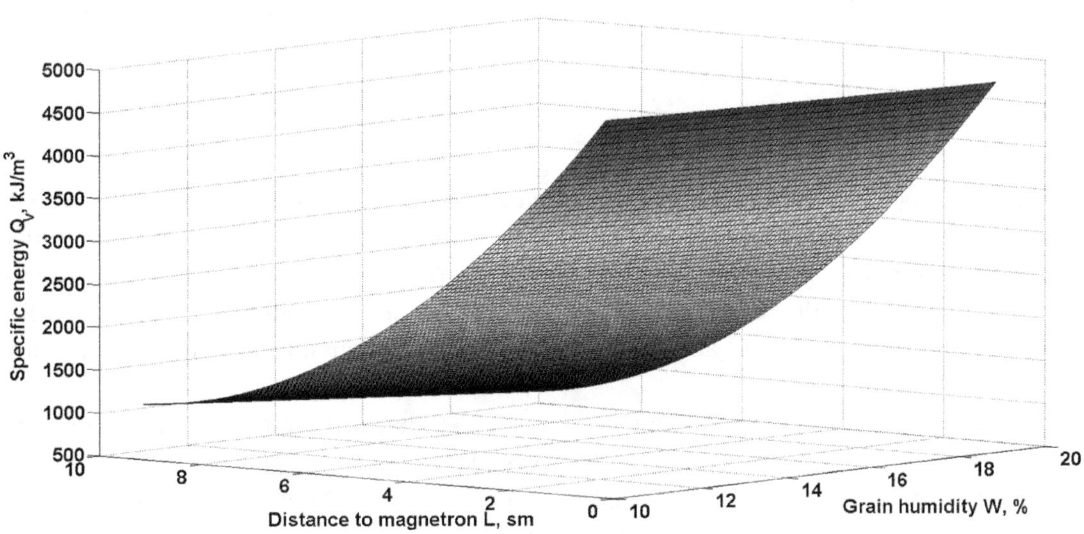

Figure 7. Scheme of the grain layers and propagation of the microwave field and the drying agent (option 1)

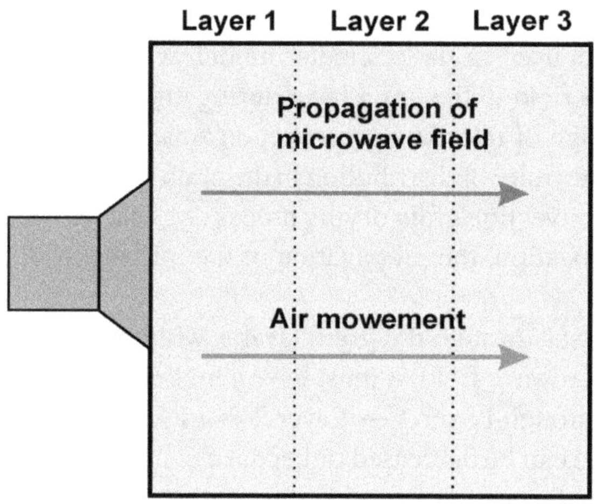

Figure 8. Change in grain temperature according to the layers in the microwave convection process

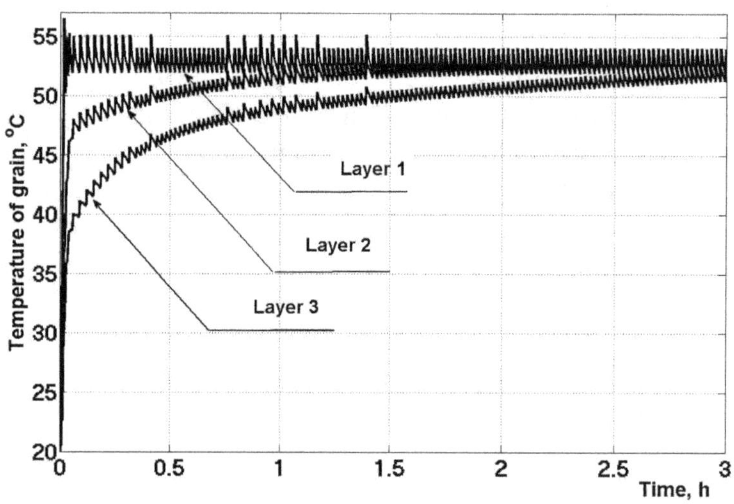

The results of modeling show that the drying rate differs significantly according to the grain layers. Moreover, the maximum specific power of the microwave field, absorbed by the grain layer, does not provide the maximum drying speed at all. If the first layer does not dry up to a conditioned moisture content of 14% for 4 hours, the second layer dries in 1.5 hours, and the third in 1.3 hours. At the same time, an average moisture content of all layers reaches 14% in 1.7 hours. Uneven moisture according to the layers is 3%.

Based on the results of the modeling, it can be concluded that such system of air distribution gives a sufficiently large uneven drying of the grain across the layers.

However, in many microwave convection grain dryers, before reaching the grain layer, atmospheric air is sent to cool the magnetrons power units and the magnetrons themselves. As a result, the atmospheric air is heated and is already warmed up to dry the grain. The objective was to model the drying

Figure 9. Temperature change of the drying agent at the output of the layers under microwave convection processing

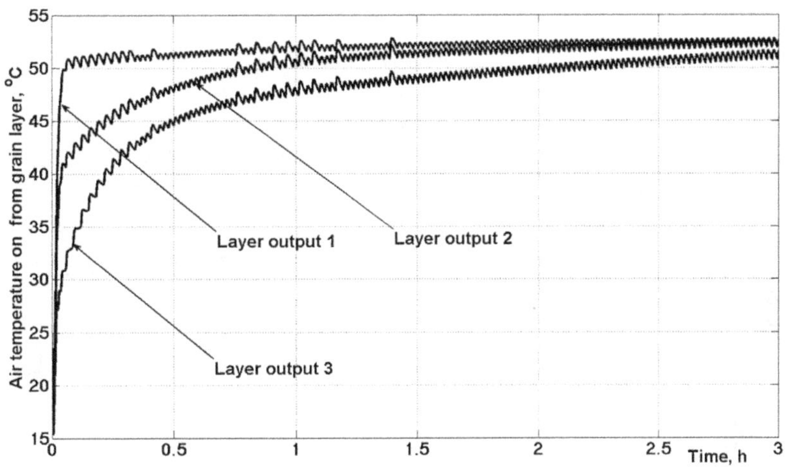

Figure 10. Change in air humidity at the output of the cereal layers

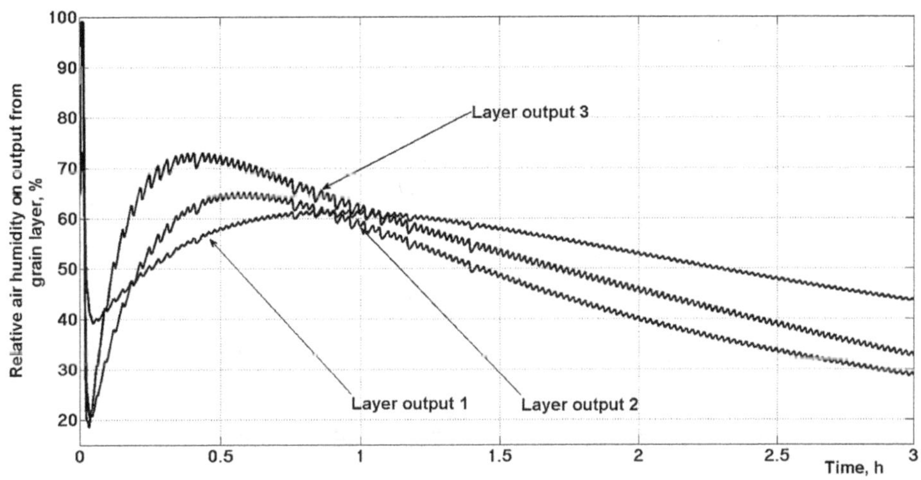

process, when heated air is fed into the grain layer. The initial parameters of the drying agent are: $T_0 = 40$ °C; $F_0 = 15\%$; $V = 0,5\frac{m}{c}$. The results of modeling are shown in Figure 12.

The results show that, in contrast to the previous option, grain dries more quickly in the first layer for 0.9 hours. Then, for 1.3 hours, the grain dries in the second layer. And only after 1.6 hours the grain dries in the third layer. The average moisture content of the grain layers will reach 14% in 1.25 hours, with an uneven drying over the layers of 2%. It turns out that the heated air supply into the grain layer accelerates the drying process, reduces the uneven drying of the grain.

The scheme for supplying the drying agent to the grain layer is possible, in which it first passes a grain layer far from the magnetron (Layer 3), then Layer 2 and Layer 1 (Figure 13)

Figure 11. Curves of grain drying by layers

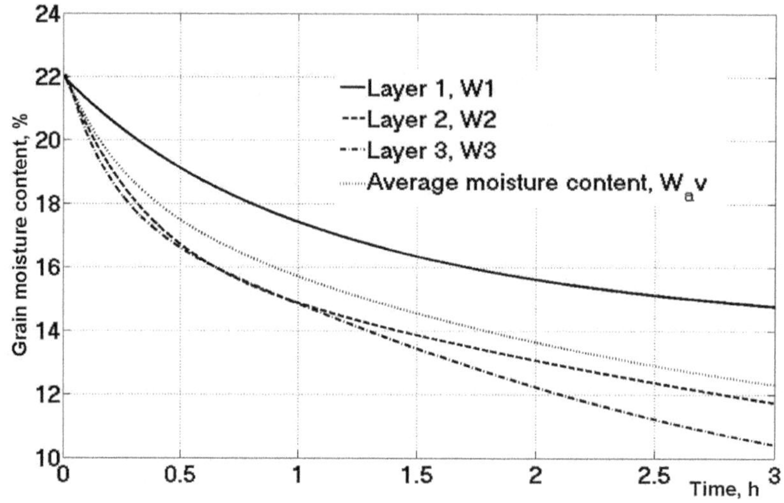

Figure 12. Curves of grain drying with heated air

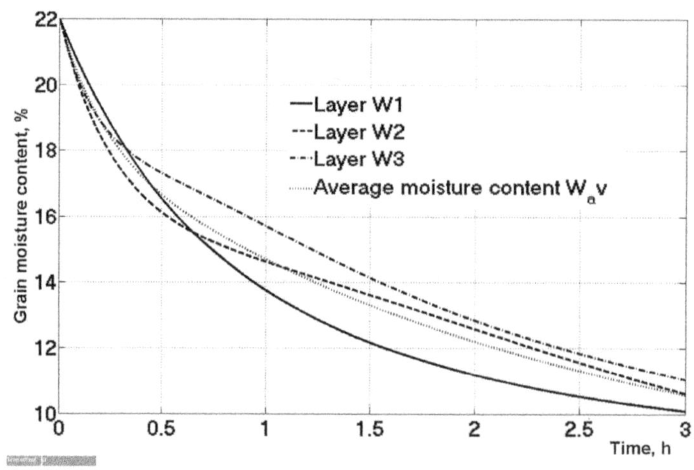

To determine the change in the parameters of the grain layers and the drying agent under microwave convection processing, this process was modeled (Figure 14). The modeling was carried out with the following initial parameters: $W_0 = 22\%$; $F_0 = 65\%$; $T_0 = 20$ °C; $V = 0,5\frac{m}{c}$.

The results of modeling show that, unlike the previous air flow scheme, the grain temperature in the second and third layer does not approach the grain temperature in the first layer. This is due to the fact that the grain in the third and second layers is actively cooled by the drying agent, and the power of the microwave field is insufficient to heat the grain. Similar pattern is observed for the temperature changes of the drying agent at the grain layer output (Figure 15).

During the modeling drying curves were also obtained (Figure 16).

Figure 13. The scheme of the grain layers and propagation of microwave field and the drying agent (option 2)

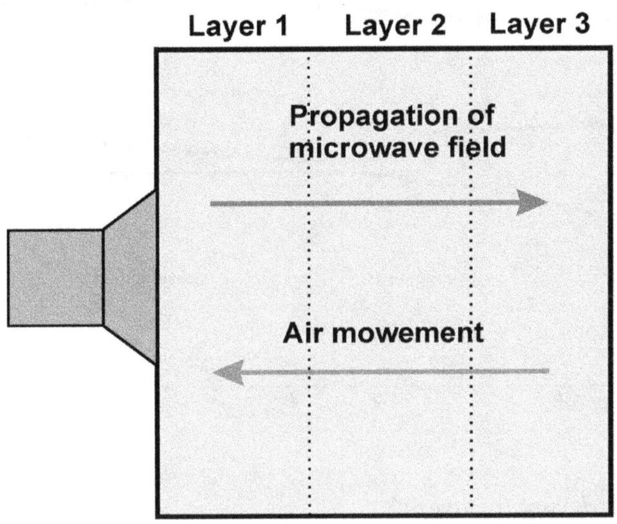

Figure 14. Change in the grain layers temperature

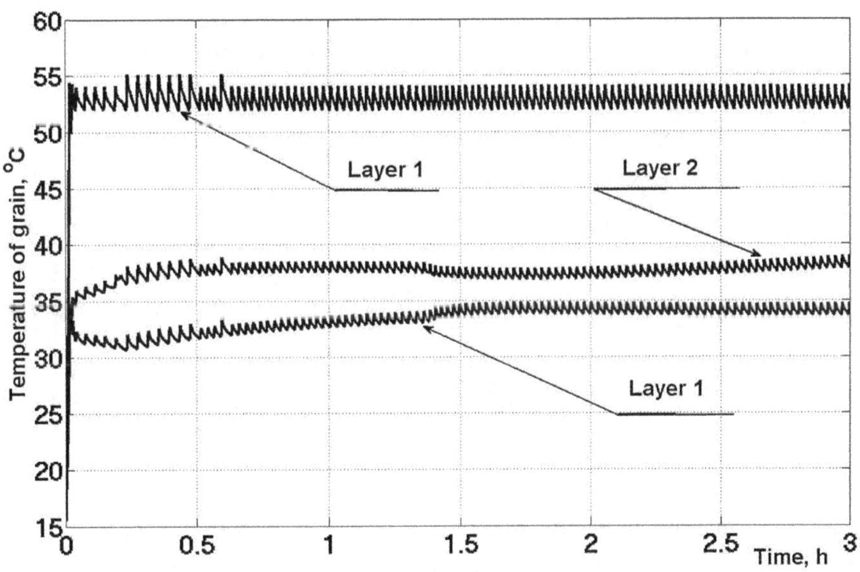

The obtained drying curves allow us to say that the dynamics of change in moisture across layers is significantly different from that observed in another coolant supply scheme. The first, in 1.4 hours, Layer 1 dries up. Then in 2.5 hours Layer 3 dries up, and only after 3.8 hours Layer 2 will reach moisture content of 14%. At the same time, an average moisture content by grain layers will reach 14% in 4 hours This situation became possible because Layer 1 began to moisten, picking up the moisture and drying agent that pass-through Layers 2 and 3.

Figure 15. Temperature change of the drying agent at the grain layers output

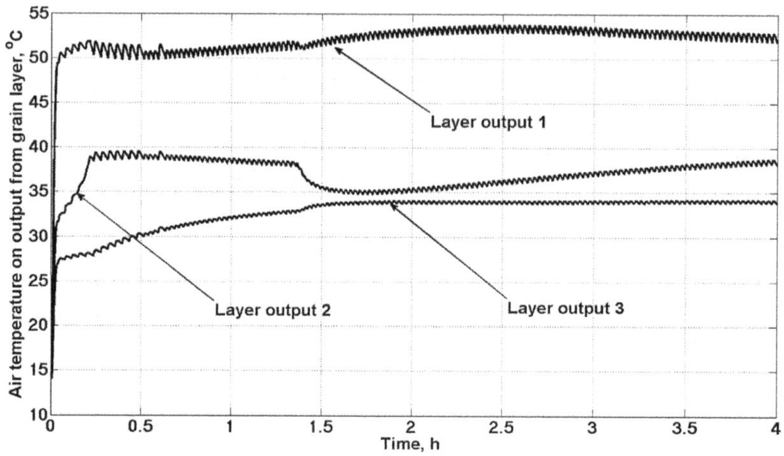

Figure 16. Curves of grain drying by layers

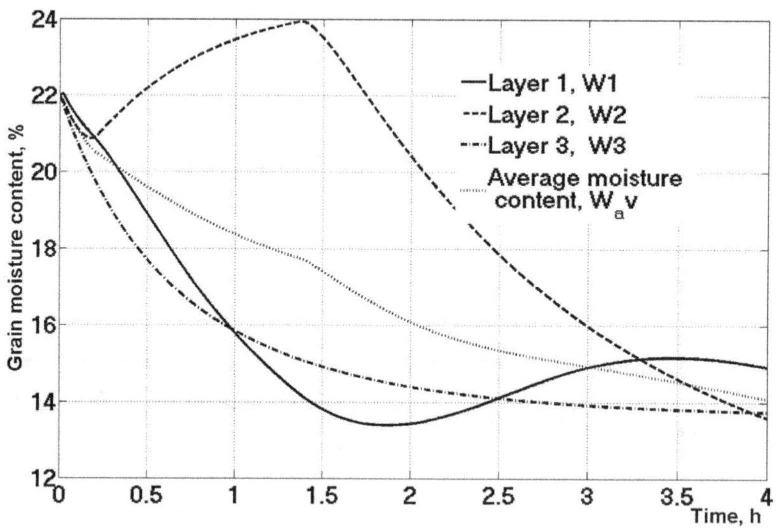

By analogy with previous machine experiments, a model was carried out for the case when the drying agent was preheated before supplying into the grain layer. The results of modeling are shown in Figure 17.

The initial parameters of the drying agent are: $T_0 = 40$ °C; $F_0 = 15\%$; $V = 0,5\,\frac{m}{c}$. The supply of preheated air in Layer 3 leads to its quicker drying, in 0.8 hours. Then Layer 1 dries up for 1.5 hours. Layer 2 dries to conditioned moisture for 2.6 hours. An average moisture content by the layers reaches the standard value for 1.74 hours. The unevenness of the grain moisture content along the layer, at the period when an average moisture reaches the conditioning value, is 9%. Unlike the previous option of the air supply, increasing temperature of the drying agent leads to an increase in drying speed, but it significantly increases the non-uniformity of the grain drying.

Figure 17. Curves of grain drying with heated air

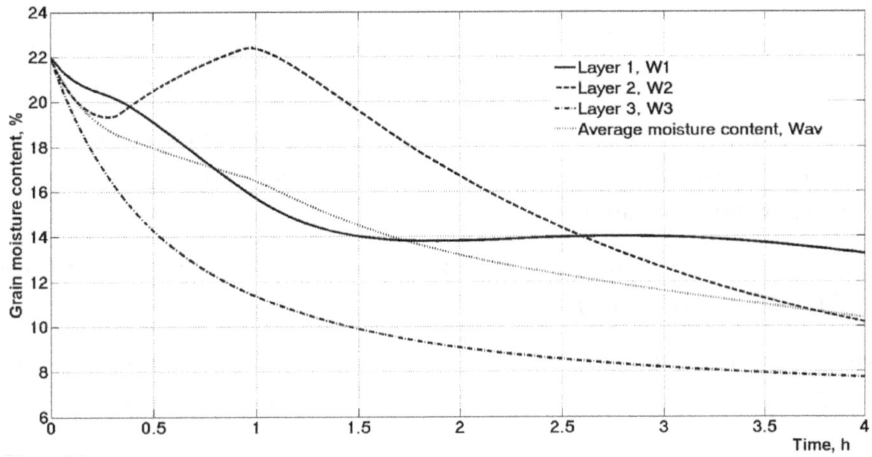

Based on the results of grain drying modeling in the microwave convection zone it should be noted:

- more effective in terms of drying time and uniformity of grain over the layer is the design of microwave convection zone that ensures the drying agent movement in the layer of grain from the layers that are heated more by microwave field to the less heated layers;
- preheating the drying agent leads to an increase in the drying speed of the grain and a decrease in the non-uniformity of the grain drying along the layer.

The obtained simulation results are intended to introduce certainty when choosing options for combining the factors of the morphological table 2.

CONCLUSION

The described principles and approaches that have been used to study the process of microwave-convective treatment of grain are well in line with traditions of system-based studies of technical objects.

Application of functional-physical analysis to studying the process of grain microwave-convective treatment makes it possible to discover all specific features of process and to substantiate options for its improvement.

Morphological analysis of processes has to be performed in several stages. At each of these stages, targets of processes have to be specified, and options of factors have to be excluded that are not expedient to be implemented at this particular stage.

The results of carried out multiple stage morphological analysis are options of technological processes whose effectiveness has to be proved with the use of model-based analysis and experimental research.

The presence of a computer model, with sufficient accuracy reflecting the processes in the microwave - convective zone, allows you to consciously choose the options for combining the factors of the morphological table.

REFERENCES

Aksjonov, S.I. (2004). *Voda i ejo rol' v reguljacii biologicheskih processov [Water and its role in the regulation of biological processes].* Moskva: Institut komp'juternyh issledovanij. (in Russian)

Aniskin, V. I., & Rybaruk, V. A. (1972). *Teoriya i tekhnologiya sushki i vremennoj konservacii zerna aktivnym ventilirovaniem [Theory and technology of drying and temporary preservation of grain by active gates]*. Moskva: Kolos. (in Russian)

Bansal, N., Dhaliwal, A. S., & Mann, K. S. (2016). Dielectric characterization of rapeseed (Brassica napus L.) from 10 to 3000 MHz. *Biosystems Engineering, 143*, 1–8.

Barbosa-Canovas, G. V., Vega-Mercado, H., & Gongora-Nieto, M. M. (2001). Dehydration of foods. *Past, present and future Conference: 2nd Inter-American Drying Conference (IADC),* Boca del Rio, Mexico (pp. 43-55). Academic Press.

Baum, A. E., & Rezchikov, V. A. (1983). *Sushka zerna* [Drying grain]. Moskva: Kolos. (in Russian)

Budnikov, D. A., & Vasiliev, A. N. (2018). *The Use of Microwave Energy at Thermal Treatment of Grain Crops.* In V. Kharchenko & P. Vasant (Eds.), *Handbook of Research on Renewable Energy and Electric Resources for Sustainable Rural Development* (pp. 475–499). Hershey, PA: IGI Global. doi:10.4018/978-1-5225-3867-7.ch020

D'yakonov, V. P. (2008a). *MATLAB 7.*/R2006/2007 (tutorial).* Moskva: DMK-Press. (in Russian)

D'yakonov, V. P. (2008b). *SIMULINK 5/6/7 (tutorial).* Moskva: DMK-Press. (in Russian)

Egorov, G.A. (1985). *Tehnologicheskie svojstva zerna [Technological properties of grain].* Moscow: Agropromizdat. (in Russian)

Feofanov, J.A. (2005). *Funkcional'nyj analiz tehnicheskih obektov [Functional analysis of technical objects].* Moscow: MGTU MAMI. (in Russian)

Horynski, M. (1999). The influence of electric field and mechanical stresses on the electric properties of the grain. *Proceedings of the International Conference on Trends in Agricultural Engineering Location,* Prague, Czech Republic (pp. 35-39). Academic Press.

Jafari, H., Kalantari, D., & Azadbakht, M. (2017). Semiindustrial continuous band microwave dryer for energy and exergy analyses, mathematical modeling of paddy drying and it's qualitative study. *ENERGY Vol, 138*, 1016–1029. doi:10.1016/j.energy.2017.07.111

Jumah, RY, Raghavan, GSV. (2001). Analysis of heat and mass transfer during combined microwave-convective spouted-beddrying. *Drying technology, 19*(3-4), 485-506.

Karminskij, V. D. (2005). Tekhnicheskaya termodinamika i teploperedacha [Technical thermodynamics and heat transfer]. Moscow: Marshrut. (in Russian)

Kupfer, K. (1999). Drying oven method, infrared- and microwave drying as reference methods for determination of material moisture. *Technisches messen, 66(6)*, 227–237.

Lukashevich, V.K. (2001). *Osnovy metodologii nauchnyh issledovanij [Fundamentals of research methodology].* Jelajda. (in Russian)

Lykov, A. V. (1963). Teoriya teplo- i massoperenosa [Theory of heat and mass transfer]. Moscow: Gosehnergnoizdat. (in Russian)

Lykov, A. V., & Maksimov, G. A. (1957). Issledovanie processa sushki v pole vysokoj chastoty: Teplo-i massoobmen v kapillyarno-poristyh telah [Heat and mass transfer in capillary-porous bodies]. Moscow: Gosehnergoizdat. (in Russian)

Maksimov, G. A. (1956). Issledovanie processov teplo- i massoobmena pri vnutrennem istochnike: avtoref.

Martinson, L. K., & Malov, Y. U. I. (2011). Differencial'nye uravneniya matematicheskoj fiziki [Differential equations of mathematical physics]. Moscow: Izd-vo MGTU im. N.Eh. Baumana. (in Russian)

Metodicheskie rekomendacii po matematicheskomu modelirovaniyu processa sushki i ohla-zhdeniya zerna v ustanovkah plotnogo sloya [Guidelines for mathematical modeling of the drying and cooling of grain in dense layer installations]. (1977). Moscow: VIEHSKH. (in Russian)

Mishra, P. & Mishra, B. (2017). Dielectric Study of pure and insecticide treated rice seeds (ORYZA SATIVA L.). *Journal of science and arts, 4,* 823-828.

Norena, C.P.Z. & Lescano-Anadon, C.E. (2016). Dielectric Properties of Importance in Operations of Post-harvest of Sorghum. *International journal of food engineering, 13*(4).

Novikov, A.M., & Novikov D.A. (2013). *Metodologija [Methodology].* Moscow: SIN-TEG. (in Russian)

Odrin, V.M. (1989). *Metod morfologicheskogo analiza tehnicheskih system [Method of morphological analysis of technical systems].* Moscow: VNIIPI. (in Russian)

Ospanov A.B., Vasiliev A.N., Budnikov D.A., Karmanov D.K., & Dautkanova D.R. (2016). Changing Parameters of the Microwave Field in the Grain. *Layer Journal of Engineering and Applied Sciences ll,* (Special Issue 1), 2915-2919.

Ospanov, A. B., Vasilyev, A. N., Budnikov, D. A., & Karmanov, D. K. (2017). Improvement of grain drying and disinfection process in the microwave field. Almaty: Nur-Print.

Pikulin, V. P., & Pohozhaev, S. I. (2004). Prakticheskij kurs po uravneniyam matematicheskoj fiziki [Practical course on the equations of mathematical physics 2nd ed.]. Moscow: MCNMO. (in Russian)

Polovinkin, A.I. (1988). *Osnovy inzhenernogo tvorchestva [Fundamentals of engineering creativity].* Moscow: Mashinostroenie.

Pticyn, S. Yu. (1964). Znachenie temperatury sredy pri sushke i uvlazhnenie zerna. *[Proceedings of VIM]* (Vol. 34, pp. 5-28). VIM.

Ruzavin, G.I. (1975). *Metody nauchnogo issledovanija [Research Methods].* Moscow: Mysl'. (in Russian)

Schinabeck, T. M., Weigler, F., & Floeter, E. (2018). Variability in determination of the single kernel moisture content of grain by means of TD-NMR spectroscopy. *Quality assurance and safety of crops & foods, 10*(1), 75-82.

Sharma, D.N. & Sinh K. (2002). Uravneniya v chastnyh proizvodnyh dlya inzhenerov [Partial differential equations for engineers]. Moscow: Tekhnosfera. (in Russian)

Smith, D. L., Atungulu, G. G., Sadaka, S., & Rogers, S. (2018). Implications of micro-wave drying using 915MHz frequency on rice physico-chemical properties. *Cereal Chemistry*, *95*(2), 211–225. doi:10.1002/cche.10012

Vasiliev, A. N., Budnikov, D. A., Gracheva, N. N., & Smirnov, A. A. (2018). Increasing efficiency of grain drying with the use of electroactivated air and heater control. In V. Kharchenko & P. Vasant (Eds.), *Handbook of Research on Renewable Energy and Electric Resources for Sustainable Rural Development* (pp. 255–282). Hershey, PA: IGI Global; doi:10.4018/978-1-5225-3867-7.ch011

Vasiliev, A. N., Vasiliev, A. A., Budnikov, D., & Samarin, G. (2019). Analysis of experimental research results obtained for grain drying with electrically activated air. doi:10.4018/978-1-5225-7573-3.ch009

Vasiliev, A. N., Vasiliev, A. A., Budnikov, D., Tikhomirov, D., & Ospanov, A. B. (2019). Control and Optimized Management of Grain Drying in Forced-Aerated Bins Advanced Agro-Engineering Technologies for Rural Business Development. doi:10.4018 / 978-1-5225-7573-3.ch008

Vasilyev, A. N. (2016). Research Methodology for Microwave Convective Grain Processing. *Innovation in agriculture, 3*(18), 143-154. Retrieved from http://ej.viesh.ru/wp-content/uploads/2016/03/insel16.pdf

Vasilyev, A.N., Budnikov, D.A., & Vasilyev A.A. (2014). Vlijanie osobennostej uborki zerna na rezhimy ego obezzarazhivanija v pole SVCh [Influence of features of grain harvesting on modes of its disinfection in the microwave field]. In *Vestnik VNIIMZh /Serija: Mehanizacija, avtomatizacija i mashinnye tehnologii v zhivotnovodstve* (pp. 82-85). Moscow: GNU VNIIMZh. (in Russian)

Vasilyev, A. N., Ospanov, A. B., Budnikov, D. A., Vasilyev, A. A., & Karmanov, D. K. (2015). Kriterii optimal'nogo upravleniya sushkoj zerna. Mezhdunarodnaya agroinzheneriya [International Agroengineering], 2, 12-17. (in Russian)

Vasilyev, A.N., Ospanov A.B., Dautkanova D.R., Budnikov D.A., Karmanov D.K., Shalginbaev D.B., & Vasilyev A.A. (2016). Primenenie diehlektricheskogo nagreva dlya konvektivnoj sushki zerna/*Vestnik ETU, 2*, 16-20. (in Russian)

Vasilyev, A.N., Ospanov A.B., Dautkanova D.R., Budnikov D.A., Karmanov D.K., Shalginbaev D.B., Vasilyev A.A., & Moshanov K.A. (2016). Modelirovanie mikrovolnovogo oblucheniya v processah teplovoj obrabotki zernovyh kul'tur. *Issledovanie i rezul'taty [Research and Results], 4.* (in Russian)

Vasilyev, A. N., & Rudenko, N. B. (2009). *Model' teplovlagoobmena pri recirkulyacii zerna. In Vestnik agrarnoj nauki Dona* (pp. 37–41). ACHGAA. (in Russian)

Vasilyev, A. N., & Severinov, O. V. (2014). Matematicheskaya model' teplo- i vlagoobmena v ehlementarnom sloe pri sushke zerna aktivnym ventilirovaniem. *AgroEhkoInfo,* (2). (in Russian) Retrieved from http://agroecoinfo.narod.ru/journal/STATYI/2014/2/st_13.doc

Vasilyev, A. N., & Severinov, O. V. (2015). K raschyotu teplo-i vlagoobmena v plotnom sloe zerna. *Teoreticheskij i nauchno-prakticheskij zhurnal Innovacii v sel'skom hozyajstve [Theoretical, scientific and practical journal "Innovations in Agriculture"]. 1*(11). (in Russian) Retrieved from http://ej.viesh.ru/journal/2015-01/

Vasilyev, A. N., Severinov, O. V., & Makarova, Yu. M. (2016). Razrabotka komp'yuternoj modeli teplo-i vlagoobmena v plotnom sloe zerna/ *Vestnik NGEHI (tekhnicheskie nauki), 4*(59), 63-71. (in Russian) Retrieved from http://elibrary.ru/item.asp?id=26083913

Vasilyev, A. N., & Vasilyev, A. A. (2018). Control of parameters of grain drying process in dense layer. *Proceedings of the International Russian Automation Conference RusAutoCon 2018.* IEEE. doi:10.1109/RUSAUTOCON.2018.8501729

Vasilyev, A. N., Vasilyev, A. A., & Budnikov, D. A. (2018). The Functional Dependencies of the Drying Coefficient for the Use in Modeling of Heat and Moisture-Exchange Processes. In P. Vasant & I. Zelinka & G.W. Weber (Eds.), *Intelligent Computing & Optimization ICO 2018.* Springer. doi:10.1007/978-3-030-00979-3_24

Wang, X., Ma, P., Huang, Y., Guo, J., Zhu, D., Porter, A. L., & Wang, Z. (2017). Combining SAO semantic analysis and morphology analysis to identify technology opportunities. *Scientometrics, 111*(1), 3–24. doi:10.100711192-017-2260-y

Zverev, S.V., & Zvereva N.S (2007). Fizicheskie svojstva zerna i produktov ego pererabotki [Physical properties of grain and its products]. Moscow: DeLi print. (in Russian)

Chapter 8
Obtaining and Using Biogas in Agricultural Production

Sergey Nikolaevich Devyanin
Russian State Agrarian University – Moscow Timiryazev Agricultural Academy, Russia

Vladimir Anatolyevich Markov
Moscow State Technical University Named After N.E. Bauman, Russia

Andrei Alexandrovich Savastenko
Moscow State Technical University, Russia

ABSTRACT

In recent years, an increased interest has been shown in fuels derived from renewable energy resources of animal and vegetable origin, raw materials of which are practically unlimited. The use of biofuels makes it possible not only to replace petroleum motor fuels with alternative ones, but also to improve exhaust gas toxicity indicators (as a rule, there is a marked decrease in emissions of toxic exhaust gas components). This work is devoted to the study of the possibilities of solving these problems in order to adapt diesel to biogas, a theoretical study of ways to create an experimental model of a power plant for generating electricity from biogas, a theoretical study on the processes occurring in the main components of the plant being developed and the creation of this plant.

DOI: 10.4018/978-1-7998-1216-6.ch008

Copyright © 2020, IGI Global. Copying or distributing in print or electronic forms without written permission of IGI Global is prohibited.

INTRODUCTION

Energy supply of agricultural production underlies the production culture, living standard in the countryside, productivity and efficiency of performed work. Solid fuel, gas and liquid fuel remain among the main energy carriers despite the active introduction of electric power engineering into rural life. Proportion of these fuels is more than 80% in the industry, and the approximate ratio between them is 3:3:2, respectively.

Biological factors of production play a key role in the rural energy sector. And they form a unique agro-energy system together with man-made means and objects of labor. This system is not only consuming and transforming energy resources, but also producing energy. Increasing the proportion of energy supply from bio-resources remains an important task, especially for agricultural production from waste. Recently, many scientists have been paying a lot of attention to this issue, which is described in detail in the book edited by Dahlquist (2013).

Resource conservation is a set of measures to find reserves on the basis of reducing waste and losses. It is a waste-free use of resources, the maximum elimination of all types of losses, maybe the fuller involvement of secondary material and energy resources into economic circulation.

The functioning feature of the agricultural industry is that biological objects (soil, plants, and animals) act as objects of energy technologies. This affects the characteristics of energy consumption and distribution, as well as possible energy sources. The energy intensity of the products produced is a factor in its competitiveness. The increase in agricultural products by 1% entails an increase in consumption of energy recourses by 2-3%.

The use of alternative fuels is necessary to reduce dependence on expensive energy resources and eliminate the instability of their supply (especially electricity). Biogas is obtained from crop and livestock waste, which is used for heating and hot water production, as well as power generation. The work of Weiland (2010) is devoted to this question.

In recent years, an increased interest has been shown in fuels derived from renewable energy resources of animal and vegetable origin, the raw materials of which are practically unlimited. The use of biofuels makes it possible not only to replace petroleum motor fuels with alternative ones, but also to improve exhaust gas toxicity indicators (as a rule, there is a marked decrease in emissions of toxic exhaust gas components). First of all, this refers to the exhaust opacity and emissions of other products of fuel incomplete combustion, which are reduced by 1.5…2 times when using biofuels.

At the same time, biofuels have physicochemical properties that differ from traditional petroleum diesel fuels. Therefore, when transferring engines that were originally adapted to work on diesel fuel to biofuels, a number of problems arise associated with the organization of work processes, primarily the processes of fuel feeding, mixture formation and combustion. This is much more relevant to gaseous fuels (in particular, to biogas), since these fuels are in a different state of aggregation compared to diesel fuel.

This work is devoted to the study of the possibilities of solving these problems in order to adapt diesel to biogas, a theoretical study of ways to create an experimental model of a power plant for generating electricity from biogas, a theoretical study on the processes occurring in the main components of the plant being developed and the creation of this plant.

Biogas as a Waste Conversion Product

Intensification of agricultural production leads to an increase in a variety of organic waste and their concentration. As pointed out by Shakhov (2011) in their work, collateral crop production and animal waste are classified as agricultural waste and are characterized by specific properties:

- In their natural form, they have low quality characteristics, which makes it possible to classify them as low-grade fuel;
- Low density and significant changes in the calorific value with changes in humidity are inherent in crop waste.
- Animal waste has a high natural humidity, which prevents their direct use without prior drying;
- A common problem in the energy use of agricultural waste is their territorial distribution, which makes it difficult to create large energy processing centers.

Most of the waste is generated from crop production by-products, which accounts for 83%. Livestock waste is generated about 5 times less every year. The recoverable energy potential of agricultural waste is approximately 24% of potential resources in the economically developed countries of Western Europe and America, including about 21% for crop waste and 42% for animal waste.

As pointed out by Shakhov (2010) in his article, the concentration of livestock of animals and poultry on farms, complexes and poultry farms contributes to the daily accumulation in a relatively small area of a significant amount of their waste products in the form of manure and litter, which require timely removal to meet the demands of ecological purity of the environment. The average daily manure output on a dairy farm with a livestock of 200 heads is more than 10 tons, on complexes it is from 30 to 60 tons. The manure yield can be from 3 to 5 kg based on 1 kg of milk.

Disposal, storage, disinfection of manure and litter by traditional methods require a large land area, fuel and energy costs. Only handling and transportation work requires from 25 to 40% of all costs on a livestock farm or complex. And about half of them are accounted for the manure removal system.

The amount of waste can reach up to 30% of the main products. Disposal of this waste requires additional energy consumption and leads to a deterioration of the ecological situation. The main source of raw materials in biogas technology is waste in list of which may include waste from agricultural production.

In the process of processing waste using biogas technology, highly efficient organic fertilizer is obtained, which will always be in demand in agricultural production and does not require additional costs for the purchase of expensive mineral fertilizers, whose efficiency is 2 times lower than that obtained from the bioreactor.

Raw Materials for Processing into Biogas

It is recommended to distinguish between the gross, economic and industrial-technological (energy) potential of bioenergy, based on the results of research and generalization of existing methods for assessing the potential of renewable energy, when assessing the bioenergy potential of agricultural waste.

The gross energy potential of organic agricultural waste is the total yield of crop and livestock waste for all categories of farms. At the same time, in crop production, gross potential is determined on the basis of the gross collection and yield of main agricultural crops.

The economic potential is a part of the gross energy potential that can be realized at large agricultural enterprises, since biological waste from agricultural production in personal subsidiary farms is used, as a rule, as fertilizer in the farms themselves.

The industrial and technological energy potential of the waste will be a part of the economic potential that can be directly realized for the production of bioenergy.

In the analysis of the resources of the agricultural sector, done by the authors (Revyakin, Tabashnikov, Samoylenko, & Dragaytsev, 2011) in their scientific publication notes the great potential of unused resources. The bioenergy potential assessment of organic waste from agricultural production and energy consumption showed that its level is about 80 million tons of standard fuel in Russia that considers:

- Crop waste is about 70%
- Animal waste is more than 25%
- Waste poultry is less than 5%

Table 1 shows the average values of crop waste of some crops and their specific energy potential during processing into biogas.

The level of specific energy potential ranges from 7,500 to 9,200 MJ per ton of crop waste when it is processed into biogas. Direct combustion in boilers and furnaces becomes inappropriate because plant waste has low density and significant changes in calorific value with changes in humidity.

In contrast, livestock and poultry waste has a high natural moisture content, which prevents their direct use without prior drying. However, the development of large livestock farms and poultry farms contributes to the daily accumulation in a relatively small area of a significant amount of their livelihoods in the form of manure and litter, which require timely removal to meet the demands of the environment ecological purity. Table 2 presents the average values of livestock and poultry waste and their specific energy potential during processing to biogas.

The level of specific energy potential ranges from 1,800 to 6,300 MJ per ton of livestock waste when they are recycled into biogas, which is about 4 times lower than the values for crop waste due to the greater moisture content of animal waste.

Most biogas plants operate in mesophilic mode, i.e. at a temperature of from 25 to 40 °C. This mode is recognized as optimal for medium latitudes of their operation. For the production of biogas you can use any degradable organic substances. Most biogas plants are focused on working with liquid cattle manure and less often with pigs. In more detail with biogas plants and processes occurring in them can be found in the work of Vedenev and Vedeneva (2011). But the main substrate is co-fermented with so-

Table 1. The ratio of waste and the gross collection of crops and the calorific value of biogas during waste processing

Type of Organic Waste	Waste Yield, as a Percentage of Gross Production	The Specific Value of the Energy Potential, GJ per Ton of Waste
Cereal straw	100	7,5
Sugar beet, potato and vegetable tops	25,0	9,2
Oilseed waste	280	7,7
Feed crops (average)	–	8,9

Table 2. Average daily yield of waste in livestock and the specific value of their energy potential

Livestock Industry	Average Daily Yield of Waste by Period (kg per head)		The Specific Value of the Energy Potential GJ per Ton of Waste
	Winter (stall)	Summer (grazing)	
Dairy cattle breeding (main herd)	46	15	2,0
Beef cattle breeding (raising and feeding of cattle)	25	8	1,8
Pig breeding	4	4	2,3
Small cattle (sheep breeding)	4	-	1,7
Poultry farming:			
laying hens broilers	0,2 0,115	0,2 0,115	6,3 6,3

called cosubstrates since the content of the total mass of carbohydrates, fat and protein in liquid manure is relatively low. As a result, the production of biogas is increased. As the authors point out Moset, Xavier, and Moller (2015) the process optimization is needed.

Crop residues of agricultural production and biomass can serve as co-substrates obtained from deposits and virgin lands, as well as cultivated plants, specially grown for the production of biogas. This is indicated by the authors in the article (Sondergaard, Ioannis, Fotidis, Kovalovszki, & Angelidaki, 2015). The release of biogas from livestock waste is also affected by the types of feed, the method of keeping animals and removing excreta, the duration of manure storage. All types of plant biomass are suitable for co-fermentation, but between them there are significant differences in the yield and composition of biogas, the effectiveness of their use for these purposes. This is indicated by the authors in the article (Jankowski, Dubis, Budzyński, Bórawski, & Bułkowska, (2016). These indicators are influenced by soil and climatic conditions, growing technology, fertilizers, harvesting time, method and shelf life. An example of such influence is discussed in detail in the article (McEniry, Allen, Murphy, & O'Kiely, 2014).

Not only the quantity of biogas output changes, but also its qualitative composition depending on the raw materials used, the thermal state of the bioreactor, the presence of catalysts and process inhibitors. A great influence on the composition of biogas has a type of raw materials, which can be used in a biogas plant in the proportions formed randomly. Table 3 shows the averaged amount of methane by volume in biogas and its output during the processing of various raw materials.

Analysis of the presented results shows that the methane content in BG can vary from 50 to 70%, which significantly complicates the process of its use as a motor fuel.

The Use of Biogas for Heat Engineering Purposes in Agricultural Production

Biogas as an energy carrier can be used to solve a large number of tasks, both for domestic purposes, and in the production, storage and primary processing of products both instead of the used natural gas and liquefied petroleum gas, and in the newly formed production territories. The heat supply in boiler plants of residential, public and industrial rooms is important task for the regions of Russia.

Table 3. The output of biogas produced from various raw materials

Methane Content in Biogas (volumetric), %	Biogas Output l/kg (in Terms of Dry Matter)	Raw Material
	Crop Waste	
70	630	Grass
62	615	Chaff
60	420	Potato haulm
60	300	Sunflower husks
59	360	Flax straw
58	340	Wheat straw
53	420	Corn stalks
	Livestock and Poultry Waste	
60	250	Cattle dung
62	400	Bird droppings
	Timber Industry Complex Waste	
59	220	Tree foliage
69	370	Pine needle
	Municipal Solid Waste and Sewage	
60	200	Sewage
65	180	Industrial waste
70	570	Sludge of wastewater
60	270	Fecal sediments
50	600	Household waste

Replacing solid fuels with biogas can improve the sanitary and hygienic level, the culture of production and the productivity increasing of stoves in domestic, public and industrial premises.

The energy-intensive process of grown agricultural products drying, such as oilseeds, part of cereals, roots, fruits, various herbs, etc., can be solved with the help of biogas. An example of such use of biogas is described in the article (Ho, Roberts, & Lucas, 2015).

The tasks of energy supply are also solved in the processes of storage and preparation of seed material, cultivation, harvesting, storage and processing of potatoes, as well as the preparation of feed, pasteurization, cooling and deep processing of milk and cheese making in animal husbandry.

In addition, biogas can also be used to heat barns, pigsties, poultry houses and other production premises.

Fertile land is often located in remote from the central energy supply in the manufacturing of agricultural products, which requires autonomous power supply to remote deep-seated areas.

The biogas using as a motor fuel allows the operation of mobile agricultural machinery (tractors, combines, loaders, self-propelled chassis, cars and other equipment). This significantly reduces the need for the delivery of traditional oil fuel and the dependence of the work execution timeliness and, consequently, increases the production efficiency.

A comparison of the physicochemical properties of various gases used in heat engineering equipment is presented in Table 4. These gases are methane (CH_4), propane (C_3H_8), butane (C_4H_{10}). It also shows properties for biogas consisting of 60% methane and 40% carbon dioxide (CO_2). The presented gases differ both in chemical composition (C, H, O), molecular weight M_m, and in terms of the lower heating value of volumetric Q_{low} and mass H_u. Gas density ρ_o and adiabatic index k_o are given for normal conditions.

The stoichiometric coefficients are shown for the gases presented in Table 4. It shows the required amount of air for the complete combustion of the gas in mass expression l_0 and volumetric one L_0. The ratio of the calorific value of the gas to the stoichiometric coefficient for both the volumetric values Q_{low}/L_o and the mass values H_u/l_o varies little (less than 3%). Therefore, it suffices to change only the gas supply and not to change the air supply system in order to preserve the thermal power of the equipment.

The mass flow rate of gas G_g in jets of heat engineering devices is determined by the expression:

$$G_g = \mu \cdot f_{jet} \cdot \sqrt{\frac{2 \cdot k}{(k-1)} \cdot \left[\left(\frac{p_>}{p_A}\right)^{\frac{2}{k}} - \left(\frac{p_>}{p_A}\right)^{\frac{k+1}{k}} \right] \cdot \left(\frac{p_>}{p_A}\right)^{-\frac{1}{k}} \cdot (p_A - p_>) \cdot \rho_>} ,$$

where μ is consumption coefficient and f_{jet} is jet section; k is adiabatic index for gas; p_c is compressed gas inlet pressure; ρ_o, p_o are gas and pressure densities at the outlet of the hole.

The analysis of the presented dependence of the gas flow shows that when the characteristics of the gas change, it changes. This leads to the deviation of the thermal capacity of the equipment, which can cause on the one hand a lack of thermal performance, and on the other - the failure of equipment and fire hazard.

Table 4. Physical and chemical properties of various gases

Gas Characteristic	Unit of Measure	Gas			
		CH_4	C_3H_8	C_4H_{10}	$Biogas_{60/40}$
Carbon proportion, C	-	0,75	0,818	0,828	0,441
Hydrogen share, H	-	0,25	0,182	0,172	0,088
Oxygen share, O	-	0	0	0	0,471
Molar mass, M_m	kmole/kg	16	44	58	27,2
Calorific value, Q_{low}	MJ/m^3	35,8	91,0	119	21,5
Calorific value, H_u	MJ/kg	50,1	46,3	45,8	17,7
Density, ρ_o	kg/m^3	0,72	1,96	2,59	1,21
Adiabatic index, k_o	-	1,31	1,14	1,10	1,31
Stoichiometric coefficient, l_0	kg/kg	17,42	15,82	15,59	6,14
Stoichiometric coefficient, L_o	m^3/m^3	9,67	24,09	31,29	5,78
Q_{low}/L_o	MJ/m^3	3,7	3,78	3,8	3,72
H_u/l_o	MJ/kg	2,88	2,93	2,94	2,88
Wobbe's number W	MJ/m^3	48,1	73,8	84,0	22,1

To preserve the thermal performance of the equipment, it is customary to change the jets with a different flow area f_{jet}, the gas pressure in front of the jet p_c, or control the gas flow through the electrically controlled nozzles by frequency regulation.

The operation of gas equipment occurs at a low pressure of the supplied gas (3...5kPa) and the effect of the adiabatic index on the gas flow rate can be neglected. Thermal power of the heat engineering equipment will be determined by the product of the calorific value of the gas and its consumption. Changing these products for different gases will show how many times the heat output changes from one gas to another, if nothing changes in the design of the equipment.

The necessity of changing the flow rate when switching from one gas to another is usually estimated by Wobbe number W, which is determined by the expression:

$$W = \frac{Q_{low}}{\sqrt{\frac{\rho_o}{\rho_{air}}}}$$

where ρ_o, ρ_{air} are gas and air densities, respectively.

If the Wobbe index W is different when switching to another gas, then the gas consumption should be changed in the same time by any of the methods listed above.

Many manufacturers of heat engineering equipment now offer boilers operating on both liquid fuel and various gases, including biogas. As an example, Table 5 presents the model range of *ATTSU* boilers operating on biogas.

Biogas as Fuel for Internal Combustion Engines

The biogas using as a motor fuel is the most difficult process, since it is connected with the need to meet the high requirements of standards for motor fuels and ensure the preservation of the reliability and service life of the machines.

Table 5. Model range of ATTSU boilers

Model	Power, кW	Fuel*	Heat Carrier
HH SERIES	1500…15000	LF, NG, B	steam
RL SERIES	38…3800	LF, NG, B	steam
HHAS	1500…15000	LF, NG, B	water
RLAS	38…3800	LF, NG, B	water
S	200…3500	LF, NG, B	water
EAGLE KHW	300…5000	LF, NG, B	water
EAGLE KDHW	300…3000	LF, NG, B	water
YGNIS	15…450	LF, NG, B	water
TPB	15…90	NG, B	water

* - LF, NG, and B are liquid fuel, natural gas and biogas, respectively

Biogas is a mixture of various gases and during the processing of agricultural waste there are among them:

- Methane in the amount of 45 ... 75%;
- Carbon dioxide (27 ... 44%);
- Nitrogen (0 ... 3%);
- Oxygen (0…3%)
- Hydrogen (0.01 ... 0.002%);
- Carbon monoxide (0.01 ... 0.002%);
- Hydrogen sulfide (0 ... 1%).

The component composition of biogas determines its physicochemical properties. The main characteristics of the biogas components are given in Table 6.

If the amount of methane in biogas is desirable to have more, because the energy value of the biogas depends on this, the presence of carbon dioxide is ballast. It reduces the specific energy characteristics of the biogas and complicates the process of managing the power supply system of the internal combustion engine. The variability of methane content in biogas occurs mainly due to changes in the amount of CO_2 in biogas. This leads to instability of the generated power of thermal power plants.

The change in the net calorific value of biogas (H_u) and the relative change in air-fuel ratio (α') of the combustible mixture is shown in Figure 1 (while maintaining the ratio between the amount of air and gas, as for $\alpha = 1$ on methane, from the volume content of methane).

The presented data shows that with a decrease in the volume concentration of methane in the combustible mixture to 30% and maintaining the amount of air supplied, the excess-air ratio decreases by more than 7 times. It will significantly complicate the process of its efficient combustion and will require the development of special internal combustion engine systems. The net calorific value will change 1.7

Table 6. Physical and chemical properties of biogas components

Indexes	CH_4	CO_2	N_2	O_2	H_2	CO	H_2S
Molecular mass, g/kmole	16,04	44,01	28,01	32,00	2,02	28,01	34,08
Molecular volume, m³/kmole	22,38	22,26	22,40	22,39	22,43	17,61	22,14
Density at 0 °C, and 101,3 kPa, kg/m³	0,717	1,977	1,251	1,429	0,090	1,25	1,536
Relative density by air	0,554	1,529	0,968	1,105	0,070	0,967	1,188
Net calorific value, MJ/m³	35,76	-	-	-	10,83	12,68	23,65
Superior calorific value, MJ/m³	40,1	-	-	-	12,80	12,68	25,85
Theoretically necessary amount of air for gas combustion, m³/m³	9,52	-	-	-	0,5	2,38	2,38
Flammability limits at the content of gases mixed with air, % (volumetric)							
lower upper	5,0 15,0	- -	- -	- -	4,0 75,0	12,5 74,0	4,3 45,5
Self-ignition temperature of the mixture with air, °C	545	-	-	-	510	610	290

times for the considered biogas compositions with a volume concentration of methane from 50 to 70%. It also complicates the organization of an effective combustion process and will require the creation of a special system for monitoring the concentration of methane in biogas.

A low proportion of methane and the presence of harmful impurities in biogas are undesirable. The heat value of biogas is reduced with a lower proportion of methane and, as a result, the power and efficiency of a thermal installation decrease. This is indicated in the book (Simon, & Johan, 2011). However, the use of biogas enrichment systems is advisable only in large biogas plants, where their share in the cost of the installation is not salt is large.

The peculiarity of the gaseous biofuels using is that they are in the same aggregative state with an oxidizing agent (air). This facilitates the organization of mixing and allows you to get a homogeneous air-fuel mixture, to ensure its complete combustion. Fuel is prevented from entering the walls of the cylinder and into the engine oil, which occurs when using liquid fuels. This reduces wear on engine parts and extends engine oil life. But, as a rule, it is necessary to make additional design changes related to the necessity of organization the engine's workflow when converting internal combustion engines to gaseous fuels in their design. This engine operation provides the required quality indicators for the self-ignition processes and combustion of these fuels. This question is discussed in more detail in the monograph (Markov, Devyanin, Zykov, & Gaidar, 2016).

The Efficiency of Biogas Using as a Motor Fuel

The effectiveness evaluation of the biogas using in the internal combustion engine was made according to the tests results on motor stands of engines operating in the gas-diesel cycle. The impact on power, economic and environmental indicators was considered in the process of research. These indexes are methane concentrations in biogas and using of methyl ester vegetable (rapeseed) oil in ignition dose instead of diesel fuel.

Figure 1. The change in the net calorific value of biogas (H_u) and the relative change in the air-fuel ratio (α') from the proportion of methane in biogas (C_{CH4}).

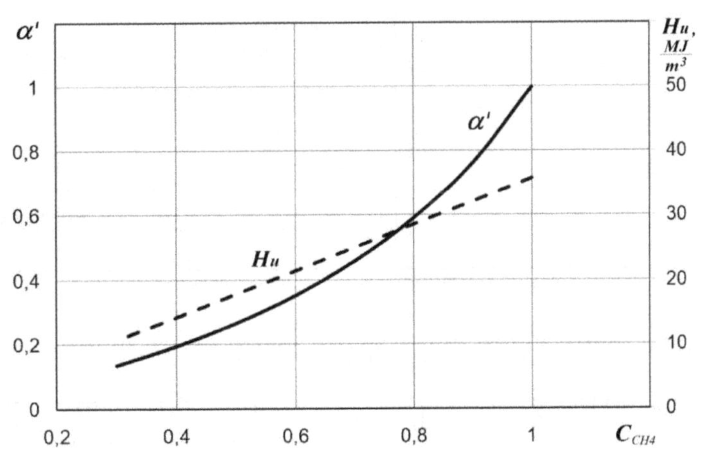

The Operation of the Gas-Diesel Engine on Biogas of Different Quality

Evaluation of engine performance on biogas is carried out using Lister-Petter's *1C 8/11* diesel engine, some of whose technical characteristics are presented in Table 7. This type of engine is widely used to drive pump and electric power plants.

The diesel engine was studied on the modes of the regulatory branch with the installation advance angle of fuel injection $\theta = 31^{\circ}$ of the crankshaft rotation to the top dead center. These modes correspond to the operation of a diesel engine on a generator drive or pumping station. The proportion of biogas in the fuel was set to the maximum possible, while maintaining the stability of the engine. Studies were conducted on biogas with methane concentration C_{CH4} = 60.0; 72.8; 77.4 and 84.8%. Biogas with the specified methane content was stored in four steel tanks. The values of biogas heat combustion were calculated by the volume content of methane and were given in Table 8.

The main modernization of the diesel engine was in the equipment of the mixer built into the engine intake manifold as in the article (Ortiz-Canavate, Hills, Chancellor, 1981). This allowed using the engine both in the basic version and gas-diesel modification. In this regard, the details and design of the engine remained unchanged, which simplifies the obtaining of gaseous modification. These engine conversion solutions and other possible conversion options are described in detail in the books (Vasiliev, Zolotarevsky, & Ksenofontov, 1992; Erokhov, & Karunin, 2005).

Table 7. Some parameters of the 1C 8/11 diesel engine

Parameters	Value
Engine type	four-stroke, diesel
Number of cylinders	1
Cylinder diameter *D*, mm	80
Piston stroke *S*, mm	110
Working volume of cylinder V_h, l	0,553
Compression ratio ε	16,5
Nominal rotating speed *n*, rpm	1380
Nominal power N_e, kW	4,5
Type of combustion chamber	Semi-split combustion chamber in the piston
Valve control	Valve type with overhead valves
Cooling system	Water, forced
Lubrication system	Forced, with spatter
Fuel (feed) system	Fuel injection pump assembly with centrifugal generator
Weight, kg	150

Table 8. The heat of combustion of biogas with different CH_4 content

Methane Content in Biogas, %	60,0	72,8	77,4	84,8
Net Calorific Value, MJ/kg	18,52	22,47	24,01	26,17

Engine tests were carried out on the load and speed modes of its operation.

The calculation of the effective efficiency η_e of a diesel engine was carried out when processing the results of experimental studies, taking into account the different calorific value of the fuels used by the expression:

$$\eta_e = \frac{3,6Ne}{H_{Umix}\left(G_{DF} + G_{gas}\right)},$$

where Hu_{mix} is calorific value of supplied fuel conditional mixture, MJ/kg N_e is effective engine output, kW. The combustion heat of the mixture supplied to the combustion chamber of fuels was determined by the ratio of hourly consumption of diesel fuel G_{DF} and biogas G_{gas} [kg/h] and their calorific value by ratio given below:

$$H_{U_{mix}} = \frac{G_{DF}}{G_{\Sigma}} \cdot H_{U_{DF}} + \frac{G_{gas}}{G_{\Sigma}} \cdot H_{Ugas},$$

where H_{uDF}, H_{ugas} are calorific values of diesel fuel and biogas correspondingly, MJ/kg.

The results of processed data are presented as graphs in Figures 2…8. Figure 2 shows the dependence of the effective power N_e of the investigated diesel engine on its speed limit (crankshaft rotation speed n) when operating on diesel and biogas. The highest power ($N_e = 4.5$ kW) is achieved using diesel fuel at the maximum load mode (at $n = 1380$rpm). There is a decrease in power when switching to biogas. Its maximum reduction is 21.2% when using biogas with the content of $C_{CH4} = 60.0\%$, and the minimum is 4.1% when using biogas with the content of $C_{CH4} = 84.8\%$. Further (when the load is reduced to 70%), the power spread sharply decreases when using biogas with different methane content and then does not exceed 3%.

The power received during the diesel engine operation on diesel fuel is 12.5% higher, on average, in the load range from 70% to 35% than when operating on biogas. The difference in power values does not exceed 3% when using all types of investigated fuels in modes with low load. The reasons for the decrease in power during engine operation with gas-diesel cycle are lower heat of combustion of biogas compared to diesel fuel. And also there is a reduction in filling of the cylinder with fresh charge due to changes in the intake system (mixer installation) and replacement of a part of the intake air supplied by biogas.

Thus, increasing the methane concentration of CH_4 in biogas provides a greater value of the output power of the gas-diesel engine. The most noticeable effect of methane concentration on the output power occurs at high loads (Figure 3). Thus, an increase in the methane concentration in biogas (C_{CH4}) from 60.0 to 84.8% leads to an increase in the maximum power N_e from 3.6 to 4.3 kW (by 16%) at the maximum load at the rotational frequency of $n = 1380$ rpm. An increase in methane concentration in biogas (C_{CH4}) from 60.0 to 84.8% leads to an increase in power from 3.5 to 3.8 kW (8%) in the mode with a rotation frequency of $n = 1420$ rpm and a load of about 90% of the full load. An increase in the methane concentration (C_{CH4}) from 60.0 to 84.8% leads to an increase in power from 2.7 to 2.8 kW, i.e. by 3.6% in the mode with a rotation frequency of n = 1500 rpm, and with a load of about 65% of the total. The change in methane concentration in biogas does not have a significant impact on the effective performance of the diesel at low loads.

Figure 2. The dependence of the engine effective power N_e on the rotational speed of the crankshaft: 1 – diesel fuel (DF); 2 – C_{CH4}=60,0%; 3 – C_{CH4}=72,8%; 4 – C_{CH4}=77,8%; 5 – C_{CH4}=84,8%.

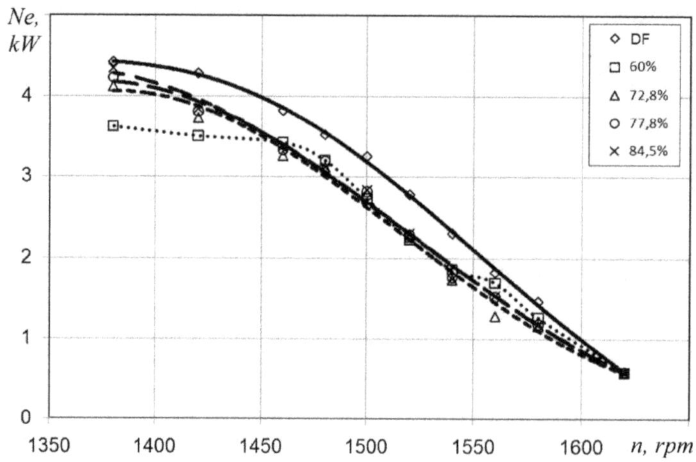

Figure 3. The dependence of the engine effective power N_e on the concentration of methane in biogas C_{CH4}: 1 – n=1380 rpm; 2 – n=1420 rpm; 3 – n=1500 rpm; 4 – n =1620 rpm

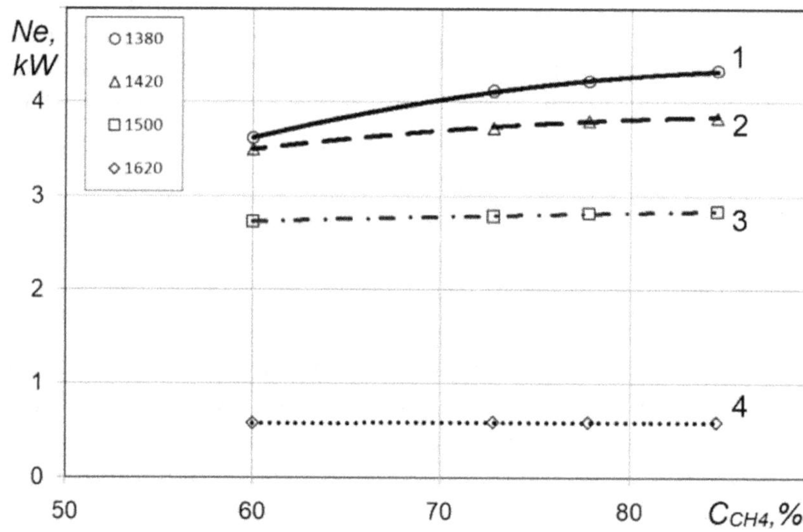

The main parameter characterizing the efficiency of the diesel engine is the effective efficiency η_e. It shows how fully the heat obtained during the combustion of a particular fuel is converted into mechanical work, given to the consumer. In contrast to the specific fuel consumption g_e, the effective efficiency allows you to compare the parameters of engines operating on different types of fuel.

Figure 4 shows the effect of methane concentration in biogas (C_{CH4}) on the effective efficiency of the gas-diesel engine η_e in various modes of the regulatory characteristics. As follows from the presented data, an increase in the concentration of methane in biogas (C_{CH4}) leads to a decrease in the effective efficiency η_e at high loads and does not have a noticeable effect on η_e at low loads. So, an increase in methane concentration in biogas (C_{CH4}) from 60.0 to 84.8% leads to a decrease in the effective efficiency

Figure 4. The dependence of the diesel effective efficiency η_e on the concentration of methane in biogas C_{CH4} in the modes of the regulatory characteristics at different rotating speeds n: 1 – n = 1380 rpm; 2 – n = 1420 rpm; 3 – n = 1500 rpm; 4 – n = 1620 rpm.

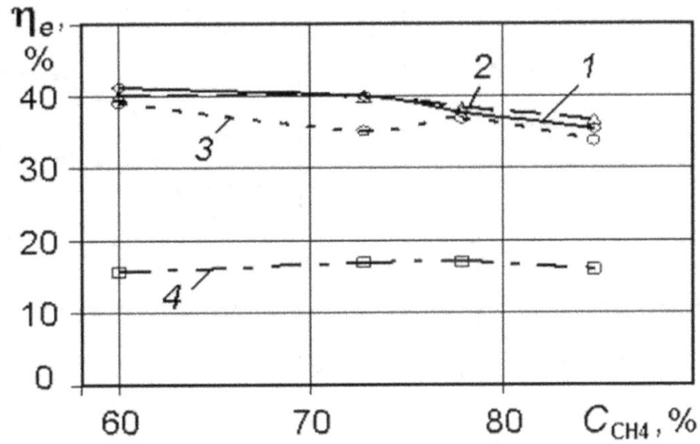

η_e from 41.2 to 35.4% in the regime with rotation frequency of $n = 1380$ rpm and the maximum load. An increase in methane concentration of (C_{CH4}) from 60.0 to 84.8% leads to a decrease in the efficiency η_e from 40.1 to 36.7% in the mode with a rotation frequency of $n = 1420$ rpm and the load of about 90% of the total. An increase in methane concentration (C_{CH4}) from 60.0 to 84.8%, on the contrary, leads to an increase in the efficiency η_e from 15.6 to 16.0% with a rotation frequency of $n = 1620$ rpm and a load of about 15% of the total. The maximum values of the effective efficiency η_e are noted for the modes with high and medium loads at rotational frequencies of $n < 1500$ rpm.

Figure 5 shows the data on the effective efficiency of the gas-diesel engine η_e for the conditions of its work on the diesel and gas-diesel cycles at various concentrations of methane in biogas C_{CH4}. The biogas supply into combustion chamber with methane content $C_{CH4} = 60.0\%$ allows you to raise the maximum efficiency η_e from 27.9 to 41.2% (Figure 5, characteristics 1 and 2).

A significant increase in efficiency η_e is probably associated with a significant improvement in the process of mixing when using biogas instead of diesel fuel. The process is typical for small-sized diesel engines with low injection pressure.

The conducted experimental studies have shown that the biogas supply into combustion chamber and the increase in methane concentration in biogas leads to a slight decrease in the coefficient of charge (Figure 6, *a*). The fall of the delivery ratio is due to the replacement of a certain amount of air with the corresponding volume of biogas supplied during the intake stroke. The engine coefficient of charge η_v changed relatively little for all values of the methane content in biogas and for all engine operating conditions during the experimental studies.

In experimental studies, a more pronounced dependence was noted of the excess-air coefficient α on the concentration of methane in biogas C_{CH4} (Figure 6b).

It is necessary to make the maximum replacement of diesel fuel when diesel engine is operating on biogas. But it is necessary to ensure the stable engine work. This goal was achieved by gradually changing of biogas consumption (increasing its shares) while reducing the supply of diesel fuel in the regulatory branch for each diesel operation mode. Figure 7 shows the results of such an investigation, which show the areas of stable and unstable gas-diesel operation at various concentrations of methane in biogas C_{CH4} and the change in the share of substituted diesel fuel G_{gas} when the engine is running on biogas.

Figure 5. Dependence of the effective efficiency of the diesel engine η_e on the load (effective power N_e) at different concentrations of methane in biogas C_{CH4}: 1 – operation on diesel fuel; 2 – operation on biogas with C_{CH4} = 60,0%; 3 - C_{CH4} = 72,8%; 4 - C_{CH4} = 77,8%; 5 - C_{CH4} = 84,8%

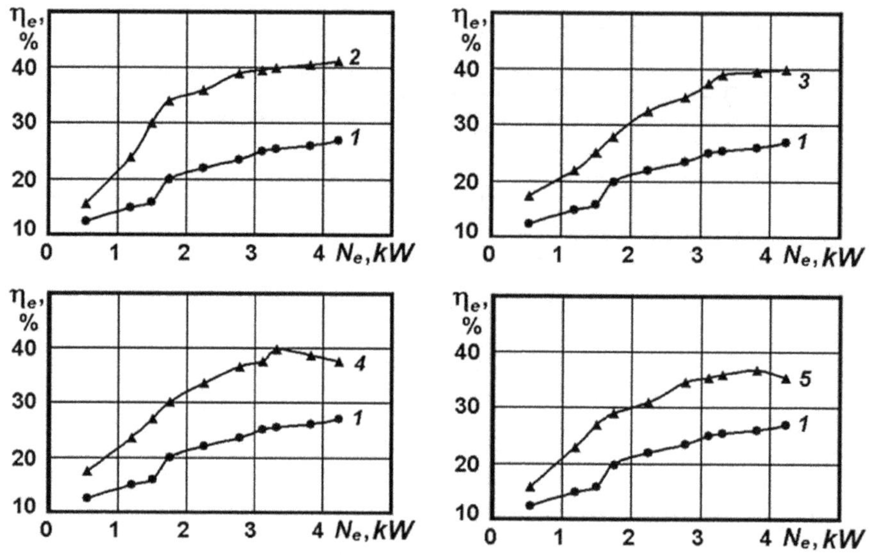

Figure 6. The dependence of the coefficient of charge η_v (a) and the excess-air coefficient α (b) on the concentration of methane in biogas C_{CH4} in the modes of the regulatory characteristics at different rotational frequencies n: 1 – n = 1380 rpm; 2 – n = 1420 rpm; 3 – n = 1500 rpm; 4 – n = 1620 rpm

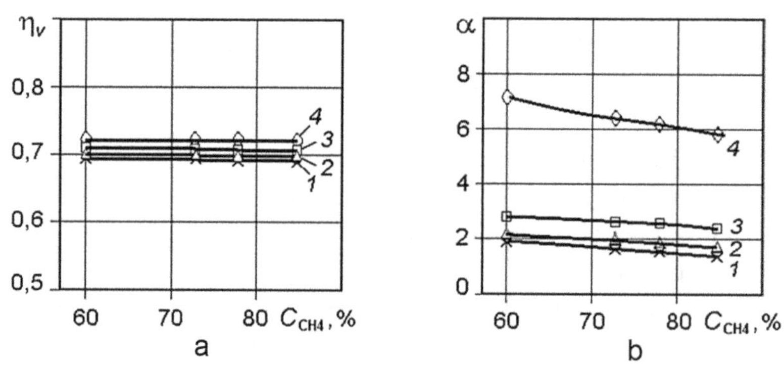

The curves shown are the limiting values for the maximum proportion of biogas in the supplied fuel. It should be noted that the diesel mode (rotational speed and load) does not have a significant effect on the fraction of the replaced diesel fuel, which ensures the stable operation of the diesel engine. All curves for different modes of operation fit into the data spread area of 5%.

From the data presented in Figure 7, it follows that in the zone of stable engine operation, an increase in methane concentration in biogas C_{CH4} allows increasing the proportion of biogas in the supplied fuel G_g and, therefore, reducing the amount of diesel fuel supplied. The amount of ignition diesel fuel supplied decreases on average from 30 to 20%, i.e. 1.3 times, with an increase in C_{CH4} from 60 to 70%. A further increase in C_{CH4} leads to a smaller effect on the replacement of diesel fuel. This phenomenon is probably due to the peculiarities of ignition and combustion of the ignition dose of diesel fuel. Its value cannot be reduced below 10...15% when only methane is supplied to the diesel during the gas-diesel process.

The results of the studies confirm the dependence of the diesel indicators on the concentration of methane in biogas C_{CH4}. Therefore, it may be possible to optimize the composition of biogas to increase the effectiveness of its use. With such optimization, the main engine indicators should be taken into account, not allowing the transition beyond their limit values.

Thus, it is necessary to ensure maximum engine efficiency (maximum values of effective efficiency η_e) when optimizing the composition of biogas of the studied diesel engine with limitations imposed on the temperature of the exhaust gases T_r. It should be noted that the efficiency of the combustion process rises with increasing load (with decreasing rotational speed n on the modes of the regulatory characteristics). It is accompanied with a simultaneous increase in the effective efficiency of the diesel η_e and the exhaust gases temperature T_r (Figure 8b). Moreover, if the limiting exhaust gases temperature reaches

Figure 7. The stable zone of the gas-diesel operation in the coordinates: the share of fuel oil replaced by biogas G_{gas} is the concentration of methane in biogas C_{CH4} at different rotational frequencies: 1 – n = 1380 rpm; 2 – n = 1420 rpm; 3 – n = 1500 rpm; 4 – n = 1620 rpm

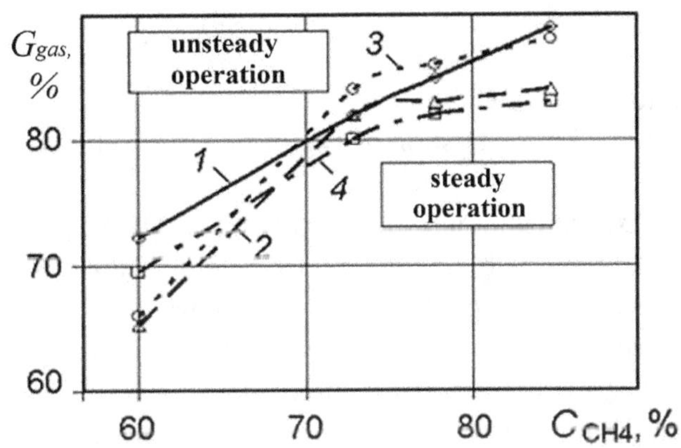

Figure 8. Exhaust gases temperature dependency T_r in the regulatory characteristics modes of the diesel engine from methane content in biogas C_{CH4} (a) and effective efficiency η_e (b) at different rotation speeds n and methane concentrations in biogas C_{CH4}: 1 – n = 1380 rpm; 2 – n = 1420 rpm; 3 – n = 1500 rpm; 4 – n = 1620 rpm; 5 – C_{CH4} = 60,0%; 6 – C_{CH4} = 72,8%; 7 – C_{CH4} = 77,8%; 8 – C_{CH4} = 84,8%

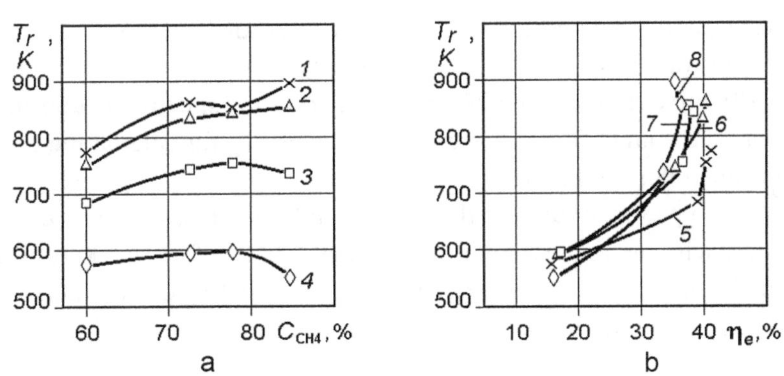

the level of $T_r = 800$ K during the operation of the researched engine for the diesel cycle (only on diesel fuel), then, as noted above, the maximum exhaust temperature increases to $T_r = 897$ K when the engine is running through the gas-diesel cycle (on biogas with the ignition dose of diesel fuel). This can lead to overheating of parts of the diesel engine combustion chamber and, first of all, its exhaust valves and engine failure. Therefore, a contradiction arises between the desire to increase the diesel engine effective efficiency η_e and the need to limit the exhaust gases temperature T_r. Analysis of the data in Figure 8 shows that the limitation of the exhaust gases temperature at the level of $T_r = 800$ K, which is characteristic of the base diesel engine, is achieved only when using biogas with methane content $C_{CH4} = 60.0\%$. This restriction is not enforced when using biogas with methane content $C_{CH4} = 72.8$; 77.8 and 84.8%

Gas-Diesel Engine Operation Without Petroleum Fuel

The complete exclusion of petroleum fuel from the process is possible by replacing it with an alternative one having similar characteristics in motor physical and chemical properties. As such a role, vegetable oil methyl ester can be used, having the following characteristics:

- Density of the liquid phase at 20 °C is equal 877 kg/m3
- Net calorific value is 37.8 MJ/kg
- Cetane number is 48
- Auto-ignition temperature is 230 °C
- The amount of air required for combustion of 1kg of rapeseed oil methyl ester is equal 12.7 kg

The main element of the experimental installation was diesel D-243 produced by the Minsk Motor Plant (MMZ). The diesel is converted to biogas ignited from the ignition dose of the rapeseed oil methyl ester. A diesel-generator set commercially produced at the MMZ was used when creating the experimental stand. The fuel injection pump assembly was developed to convert this engine to gaseous fuels. And it is equipped with a regulation system of gas and ignition-fuel supply. This system includes a regulator of liquid fuel supply (rapeseed oil methyl ester), a biogas supply regulator and an air-gas mixer installed in the gas-diesel engine intake system.

The un-supercharged tested diesel of the type *4C 11/12.5* had a nominal power of $N_e = 36$kW at a crankshaft rotating speed of $n = 1500$rpm. In this engine, as noted above, the main fuel was biogas, fed through the engine intake system into its cylinders. Ignition of biogas in the combustion chamber occurred from the ignition dose of the rapeseed oil methyl ester injected into the cylinders by a regular fuel delivery system.

The researches were carried out under normal climatic conditions at ambient temperature (20 ± 10) °C. Relative air humidity was from 45 to 80% and atmospheric pressure was from 630 to 800 mm Hg. (from 84 to 107 kPa). Effective engine output N_e, crankshaft rotating speed n, air consumptions, liquid and gaseous consumptions, and some additional parameters necessary for calculating engine fuel economy indicators were determined during the experiments.

The diesel *4C 11/12.5* type was studied on the motor stand at the maximum modes of regulatory characteristic. It was formed with the regulator at the crankshaft speed of $n = 1500$ rpm. These regimes are the standard modes of the diesel-generator set that produces alternating electric current. During the tests, the fuel injection advance angle was set equal to $\theta = 13^{o}$ of the crankshaft rotation to the top dead center.

The dual-fuel gas-diesel cycle is implemented for the purpose of reliable ignition of biogas in conditions of the diesel engine combustion chamber. It is caused by the worse combustibility of biogas compared to petroleum diesel fuel (their auto-ignition temperatures are 650 and 250 °C, respectively, and cetane numbers are 0 and 45 units).When converting the diesel to biogas, the developed regulator of the fuel injection pump assembly maintained an approximate constancy of the hourly consumption of the ignition rapeseed oil methyl ester at the level of 1.93 kg/h (Figure 9). It means that the approximate constancy of the ignition dose of the rapeseed oil methyl ester was at the level of q_c = 12.2 mm³.

Only the ignition rapeseed oil methyl ester was supplied into the engine in the entire range of load modes at idle speed (at N_e = 0 kW and n = 1515 rpm) because of the approximate constancy of the hourly consumption of the rapeseed oil methyl ester at the level of G_{ROME} =1.93 kg/h. Biogas consumption was G_{gas} = 0 and it raised with the load increase. The biogas feed was maximal and equal to G_{biog} = 9.00 kg/h in the full load mode (at N_e = 33.7 kW and n = 1434 rpm) and the hourly consumption of the rapeseed oil methyl ester was $G_{diesel\ fuel}$ = 1.92 kg/h.

The values of the hourly air flow rates G_{air}, diesel fuel $G_{diesel\ fuel}$, the ignition dose of rapeseed oil methyl ester G_{ROME} and biogas G_{gas} are presented in Figure 9. According to these data the values were determined of the excess-air coefficient α for the diesel and gas-diesel cycles of the studied diesel engine. In this case, the generally accepted formulas are used to calculate this coefficient, the derivation of which can be considered in more detail in the articles (Devyanin, Chumakov, & Markov, 2012; Markov, Devyanin, & Shimchenko, 2014). For the diesel cycle, the coefficient of excess air α was determined by the expression:

$$\alpha = \frac{G_{air}}{l_{o\,diesel\,fuel} G_{diesel\,fuel}}$$

and the following relationship was used during researches of the gas-diesel cycle to calculate the coefficient of excess air:

$$\alpha = \frac{G_{air}}{l_{o\,ROME} G_{ROME} + l_{o\,gas} G_{gas}}$$

where G_{air}, $G_{diesel\ fuel}$, G_{ROME} and G_{gas} are hourly consumptions of air, diesel fuel, rapeseed oil methyl ester and biogas kg/g; $l_o = L_o \cdot \mu_{air}$ is the amount of air required for combustion of 1kg of fuel, kg of air/kg of fuel (for diesel fuel $l_{o\ diesel\ fuel}$ = 14.3;for rapeseed oil methyl ester $l_{o\ ROME}$ = 12.7; for biogas $l_{o\ biogas}$ = 12,0); μ_{air} =28,93 that is molecular mass of air. Amount of air L_o [kg of air/kg of fuel] required for combustion of 1 kg of fuel was determined by the following formula:

$$L_o = \frac{1}{0,21}(C/12 + H\ /\ 4 - O/32)$$

where C, H and O are mass fractions of carbon, hydrogen and oxygen in the fuel.

Figure 9. Dependence of hourly air flow rates G_{air}, diesel fuel $G_{diesel\ fuel}$ (G_{df}) rapeseed oil methyl ester G_{ROME} and natural gas G_{biogas} (G_{bg}), excess-air coefficient α from the effective power N_e of the diesel engine 4C 11/12.5 operating at the modes of the nominal regulatory characteristic with n_{nom} = 1500 rpm: 1 - diesel cycle (work on diesel fuel); 2 - gas-diesel cycle (operation on biogas with the ignition dose of rapeseed oil methyl ester)

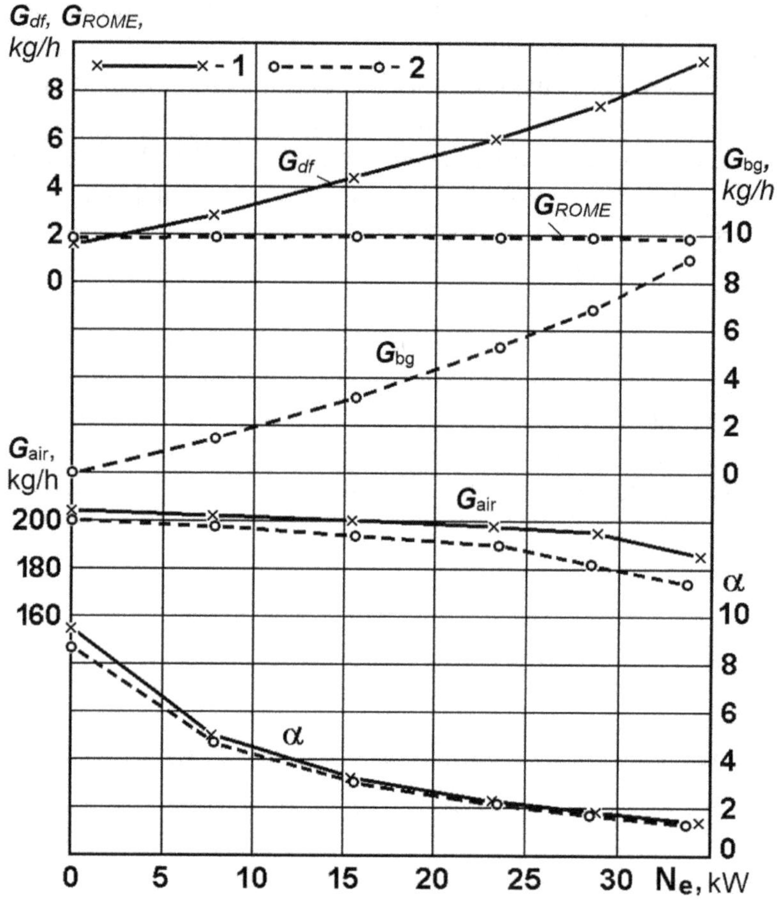

The indicators of fuel efficiency were determined of the researched diesel engine on the basis of consumptions values of petroleum diesel fuel, rapeseed oil methyl ester and biogas, as well as the power indicators obtained during the tests. In this case, the total specific fuel consumption of the diesel $g_{e\ total}$ was carried out taking into account the hourly diesel fuel consumption $G_{diesel\ fuel}$, rapeseed oil methyl ester G_{ROME} and biogas G_{gas} obtained in the experiments, as well as the effective engine power N_e according to the formula:

$$g_{e\,total} = \frac{G_{ROME} + G_{gas}}{N_e}$$

The data presented in Figure 10 suggests that the total specific effective fuel consumption was generally higher than in the pure diesel cycle with the implementation of the gas-diesel cycle. This is due to the lower calorific value of biogas and the ignition rapeseed oil methyl ester compared to petroleum diesel fuel. Net calorific value of these fuels is 42.5 MJ/kg for petroleum diesel fuel; 34.8 MJ/kg for biogas and 37.8 MJ/kg for rapeseed oil methyl ester.

It should be noted that the considered indicator of the engine fuel efficiency $g_{e\,total}$ does not take into account the different calorific value of the fuels used. Therefore, it is preferable to use the effective efficiency of the diesel engine η_e for a comparative assessment of the efficiency of the combustion process in the diesel and gas-diesel cycles. The formula for determining it includes the net calorific value of these fuels, as well as their mixtures.

When implementing the gas-diesel cycle, the total combustion heat of the mixture $H_{U\Sigma}$ of supplied fuels to the combustion chamber was determined taking into account the hourly consumptions of the rapeseed oil methyl ester G_{ROME} and biogas G_{gas} [kg/h] and their calorific value by the ratio:

$$H_{U_\Sigma} = \frac{G_{gas}}{G_{fuel\,\Sigma}} \cdot H_{U\ gas} + \frac{G_{ROME}}{G_{fuel\,\Sigma}} \cdot H_{U_{ROME}}$$

where $G_{fuel\,\Sigma}$ is the total hourly consumption of biogas and rapeseed oil methyl ester kg/h; $H_{U\ gas}$ and $H_{U\ ROME}$ are net calorific value of these fuels, MJ/kg (the net calorific value is the amount of heat that is released during complete combustion, without taking into account the condensation heat of water vapor contained in the working mixture).

The calculation of the effective efficiency η_e of the diesel engine was carried out, when processing the results of experimental studies, taking into account the different calorific value of the fuels used by the ratio:

$$\eta_e = \frac{3,6\ N_e}{H_{U_\Sigma}\left(G_{gas} + G_{ROME}\right)}$$

or by the following formula:

$$\eta_e = \frac{3600}{H_{U_\Sigma} g_{e\,total}}.$$

The values of the effective efficiency ηe calculated in accordance with the presented methodology are shown in Figure 10. The effective efficiency of the diesel η_e tends to zero in idle (with $N_e = 0$). As the load increases, the effective efficiency η_e increases and reaches its maximum with an effective power of $N_e = 23\ldots28$ kW. The highest effective engine efficiency of the engine ($\eta_e = 0.331$) corresponds to the gas-diesel cycle mode with an effective power of $N_e = 28.5$ kW at $n = 1440$ rpm.

The obtained experimental data on the exhaust gases smokiness of the studied diesel type *4C 11/12.5* (Figure 10) show that the use of gaseous fuel (biogas) leads to a noticeable decrease in carbon emission C (reduce exhaust smokiness). There are no high-molecular hydrocarbons in biogas and a better quality

Figure 10. Dependence of the total specific effective fuel consumption $g_{e\ total}$, effective efficiency η_e and exhaust smokiness on the Hartridge scale K_X from effective power N_e of 4C 11/12.5 diesel engine operating at the modes of the limit regulatory characteristics at n_{nom} = 1500 rpm: 1 - diesel cycle (work on diesel fuel); 2 - gas-diesel cycle (operation on biogas with the ignition dose of rapeseed oil methyl ester)

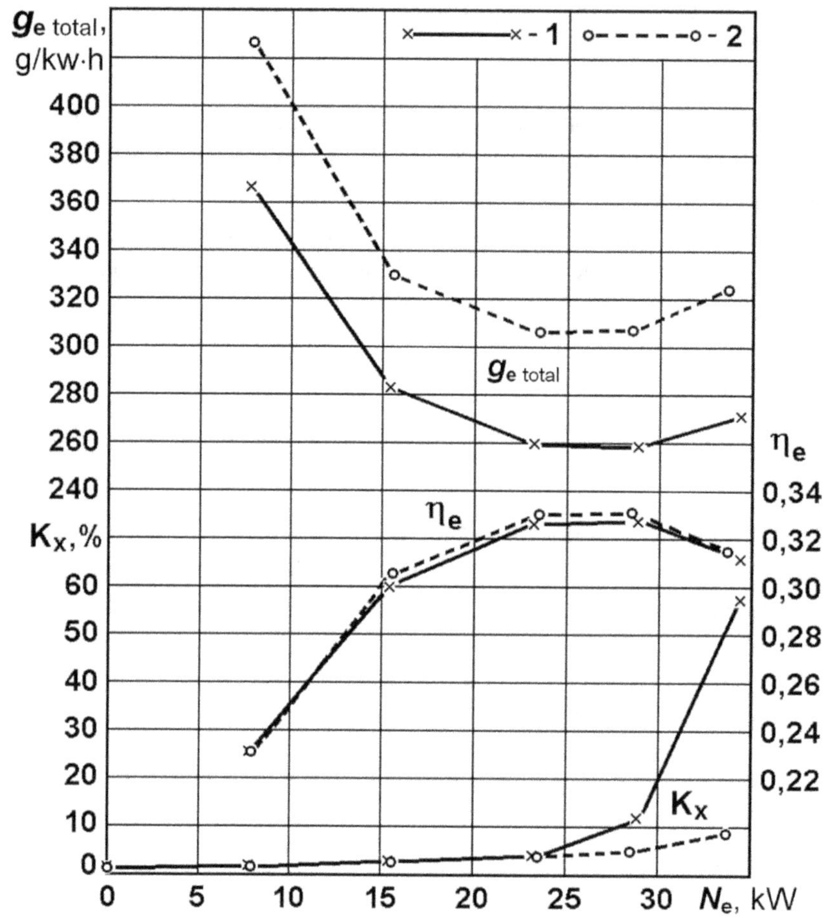

of the mixture formation process occurs. A lack of oxygen in the working mixture begins to affect in modes with high loads. Therefore, a significant decrease in the exhaust gases smokiness is marked in high load modes in the gas-diesel cycle, compared to the diesel cycle. Similar data are provided in the article (Shaik, Sudheer, & Vijaya, 2013).

The greatest exhaust smokiness is noted in the diesel cycle K_X = 57.3% on the Hartridge scale and in the gas-diesel cycle K_X = 8.7% on the Hartridge scale, in the full load mode (with N_e = 33.7…34.4 kW). Thus, in this mode, the transition from the diesel cycle to the gas-diesel cycle leads to a decrease in the exhaust gases smokiness by about 6.6 times.

This transition from the diesel to the gas-diesel cycle also affects the content of other components in the exhaust gas (Figure 11). These indicators are also largely dependent on the engine load from its effective power N_e.

Figure 11. Dependence of the volume concentrations in exhaust gases of oxygen C_{CO2}, carbon dioxide C_{NOx}, nitrogen oxides C_{CO}, carbon monoxide C_{CO} and light unburned hydrocarbons C_{CHx} on the effective power N_e of the 4C 11/12.5 diesel engine operating at the modes of the limiting regulatory characteristic at n = 1500 rpm: 1 - diesel cycle (work on diesel fuel); 2 - gas-diesel cycle (operation on biogas with the ignition dose of rapeseed oil methyl ester)

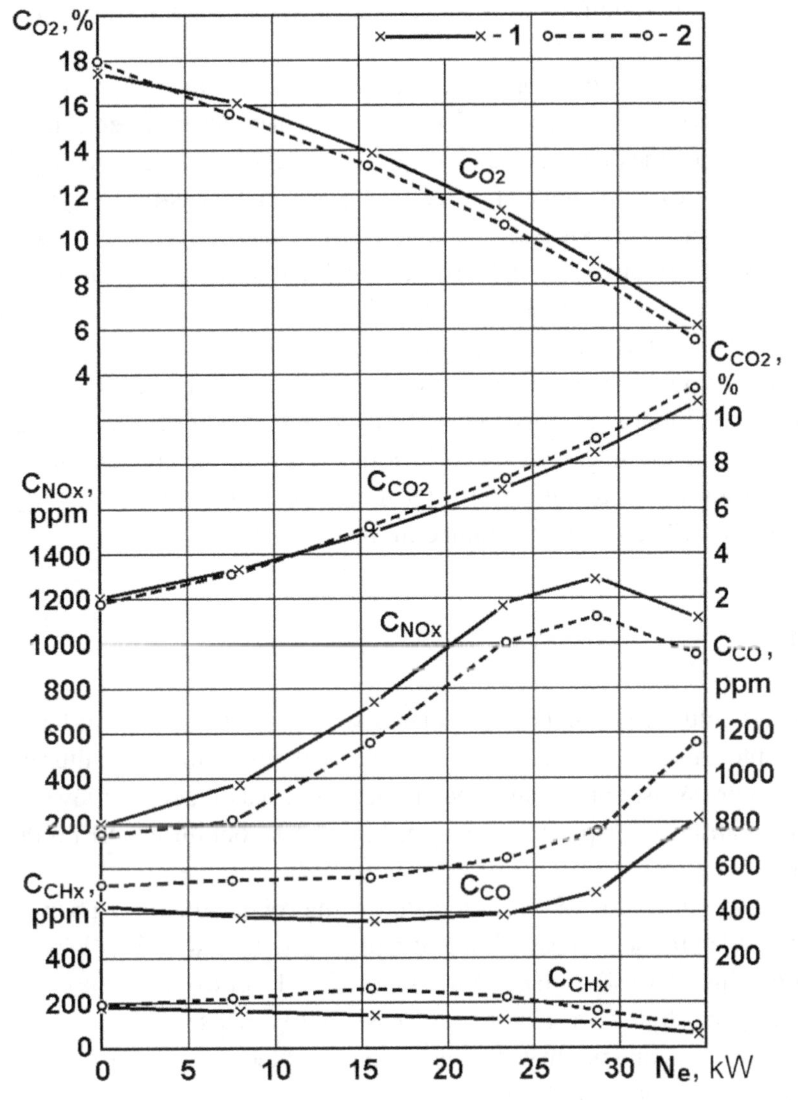

The obtained experimental data on the diesel *4C 11/12.5*-type show that with increasing engine load (before the load corresponding to N_e = 28...29 kW), an increase in the exhaust gas concentration of the main gaseous toxic component (nitrogen oxides C_{NOx}) is observed (Figure 11). In the gas-diesel cycle, an increase in the load from N_e = 0 kW (idling mode) to N_e = 28.7 kW (maximum power mode) leads to an increase in the content of nitrogen oxides C_{NOx} in the exhaust gases from 151 to 1120 ppm. The concentration of nitrogen oxides in the exhaust gases increases from 196 to 1284 ppm in the same range of the load increase in the diesel cycle. In both cycles, a further increase in load leads to a slight decrease in C_{NOx}. In general, it is necessary to note the lower content of nitrogen oxides in the exhaust gases in the gas-diesel cycle compared to the diesel cycle. This is due to the better quality of the mixing process and more homogeneous working mixture in this cycle, because gaseous fuel (biogas) and oxidizer (air) are in the same aggregative state (gaseous), which facilitates the mixing of these components.

It should be noted that an increased content in the exhaust gases of incomplete combustion products (carbon monoxide CO and light unburned hydrocarbons CH_x) is characteristic for the gas-diesel cycle of the studied diesel engine (Figure 11).

In general, the analysis of the emission characteristics of the exhaust gases toxic components of the researched diesel engine showed that the marked improvement in its environmental performance was noted when it was transferred from the diesel cycle to the gas-diesel cycle. There was a significant reduction in emissions of two major toxic components of the exhaust gases in the gas-diesel cycle. These are nitrogen oxides and soot (exhaust gases smokiness). At the same time, there was an increase in emissions of two other standardized toxic components of exhaust gases which are carbon monoxide CO and unburned hydrocarbons CH_x. Wherein, the toxicological significance of carbon monoxide CO, nitrogen oxides NO_x, unburned hydrocarbons CH_x and soot (solid particles) is estimated as a ratio of 1:41; 1:3; 16:200, which is noted by the authors of the book (Markov, Bashirov, & Gabitov, 2002). It should also be noted that the increase in carbon monoxide emissions of CO and unburned hydrocarbons CH_x during the tests of 4C 11/12.5 diesel engine in the gas-diesel cycle compared with the conventional diesel cycle can be easily eliminated by installing a catalytic converter in the exhaust system of an engine that effectively cleans exhaust gases from CO and CH_x contained in them.

CONCLUSION

The production of agricultural products is associated with the formation of a large amount of waste. Processing of them into a useful product improves the efficiency of the agricultural sector.

The use of agricultural waste processing technology in biogas reactors allows not only to improve the environmental situation by disposing of waste, but also to obtain useful products in the form of fertilizers and biogas.

The main share of waste (up to 70%) is associated with crop production. The main disadvantage of this waste is their dispersal in the process of cleaning products. This situation does not allow organizing the centralized processing of waste in large enterprises. The level of energy potential of crop waste is from 7,500 to 9,200 MJ per ton of waste when processed into biogas.

Waste of modern livestock and poultry farming is concentrated in livestock farms and poultry houses, which simplifies the process of waste collection and recycling in the bioreactor, compared with crop waste. The level of specific energy potential of these wastes ranges from 1,800 to 6,300 MJ per ton, which is about 4 times lower than the values for crop waste. It is recommended to add plant residues of agricultural production to increase the yield of biogas from the bioreactor, that is, to mix both types of waste.

The placement of plants for the production of biogas is most efficient in the places of formation of organic waste, i.e. at the enterprises of agriculture, the food industry, treatment facilities, landfills for municipal solid waste, etc.

The biogas produced from the bioreactor is a mixture of various gases, the main share of which is methane and carbon dioxide. The methane content in biogas can vary from 50 to 70%, which depends on the raw materials used, the thermal state of the bioreactor, the presence of catalysts and process inhibitors.

The problem solving of heat supply with the help of biogas is simplified by the presence of the developed production of gas heating equipment operating on natural and liquefied petroleum gases. The creation of new production areas and the development of new structures will not be required for the production of equipment operating on biogas, and it will suffice to limit the modernization of the existing one. It is customary to change the flow area of gas jets to preserve the thermal performance of this equipment, or the gas pressure before the jet, or to control the gas flow through the electrically controlled nozzles by frequency control.

Now many manufacturers of heat engineering equipment offer boilers operating on both liquid fuel and various gases, including biogas.

The use of biogas as a motor fuel is the most complex process, since it is connected with the need to meet the high requirements of standards for motor fuels and ensure the preservation of the reliability and service life of machines. For the considered biogas compositions with a volume concentration of methane from 50 to 70% of the excess-air coefficient will change 1.7 times. It complicates the organization of the efficient combustion process, which is easier to organize in gas-diesel engines.

Tests of the 1C 8/11 engine showed the effect of methane concentration in biogas on its power and economic indexes with the igniting of diesel fuel. The decrease in methane concentration in biogas C_{CH4} from 84.8 to 60.0% leads to decrease in the maximum power N_e from 4.3 to 3.6 kW (by 16%) and the increase in the effective efficiency η_e from 35.4 to 41.2% at maximum load at the rotating speed of $n =$ 1380 rpm.

Reducing the ignition dose of diesel fuel is limited with the beginning of an unstable engine operation. The minimum ignition dose of diesel fuel depends on the concentration of methane in biogas. The decrease in methane concentration in biogas C_{CH4} from 84.8 to 60.0% leads to the increase of 2.2 times of the minimum permissible ignition dose at maximum load and the rotational frequency of $n = 1380$ rpm.

The use of biogas instead of diesel fuel has significantly increased the effective efficiency. It is associated with a significant improvement in the process of mixing in small-sized diesel engines with low injection pressure. The effective efficiency reached 37.8% instead of 27.9% on diesel fuel, during the use of biogas with a methane concentration of $C_{CH4} = 84.8\%$ at maximum power mode.

The complete exclusion of petroleum fuel from the process was tested on the 4C 11/12.5 diesel engine as part of the diesel generator, using a feed of methyl ester of vegetable oil as the ignition portion. Tests have shown that the diesel engine is fully operational without petroleum fuel in the gas-diesel process using biogas while maintaining the nominal power.

Replacing the diesel process with the gas-diesel and biogas supply does not degrade the effective efficiency of the engine and has a positive effect on emissions reduction of soot and nitrogen oxides. At the same time, there was an increase in emissions of two other standardized toxic components of exhaust gases. These are carbon monoxide CO and unburned hydrocarbons CH_x.

The conducted complex of experimental studies confirmed the possibility of using biogas as an alternative fuel for diesel engines. The use of biogas allows solving the issue of obtaining energy resources for their local use not only partially, but also deciding a number of problems facing the power industry in general:

- Partially or completely replace of traditional petroleum fuels and, thereby, save oil resources for future generations;
- Meet the needs of the population in electricity and domestic gas, significantly reducing the negative impact of internal combustion engines on the environment;
- Use of alternative fuels in existing engines without a significant change in their design;
- Create of energy-autonomous agricultural enterprises;
- Recycle (burn in the engine) of the greenhouse gases entering the atmosphere during recycling of household and agricultural waste.

REFERENCES

Dahlquist, E. (Ed.). (2013). *Technologies for Converting Biomass to Useful Energy: Combustion, Gasification, Pyrolysis, Torrefaction and Fermentation.* Sweden: Taylor & Francis. doi:10.1201/b14561

Devyanin, S.N., Chumakov, V.L., & Markov, V.A. (2012). Biogas - alternative fuel for diesel engines. *Alternative Fuel Transport, 2*, 68-73.

Erokhov, V. I., & Karunin, A. L. (2005). Gas-diesel cars (design, calculation, operation). Moscow: Graf-Press.

Ho, T. B., Roberts, T. K., & Lucas, S. (2015). Small-scale household biogas digesters as a viable option for energy recovery and global warming mitigation – Vietnam case study. *Journal of Agricultural Science and Technology A, 5*, 387–395.

Jankowski, K. J., Dubis, B., Budzyński, W. S., Bórawski, P., & Bułkowska, K. (2016). Energy efficiency of crops grown for biogas production in a large-scale farm in Poland. *Energy, 109*, 277–286. doi:10.1016/j.energy.2016.04.087

Markov, V. A., Bashirov, R. M., & Gabitov, I. I. (2002). *Toxicity of exhaust gases of diesel engines.* Moscow: Moscow State Technical University.

Markov, V.A., Devyanin, S.N., & Shimchenko, S.P. (2014). Experimental uses of a diesel generator set operating on biogas. *Truck, 4*, 41-47.

Markov, V. A., Devyanin, S. N., Zykov, S. A., & Gaidar, S. M. (2016). Biofuels for internal combustion engines. Moscow: SIC Engineer.

McEniry, J., Allen, E., Murphy, J. D., & O'Kiely, P. (2014). Grass for biogas production: The impact of silage fermentation characteristics on methane yield in two contrasting biomethane potential test systems. *Renewable Energy, 63*, 524–530. doi:10.1016/j.renene.2013.09.052

Ortiz-Canavate, J., Hills, D. J., & Chancellor, W. J. (1981). Diesel Engine Modification to Operate on Biogas. *Transactions of the ASAE. American Society of Agricultural Engineers, 24*(4), 808–813. doi:10.13031/2013.34342

Revyakin, E. L., Tabashnikov, A. T., Samoylenko, E. M., & Dragaytsev, V. I. (2011). *Resource-saving technologies: state, prospects, efficiency.* Moscow: RosInformagrotekh.

Shaik, M. H., Sudheer, P. K., & Vijaya, K. R. (2013). Biogas–Diesel Dual Fuel Engine Exhaust Gas Emissions. *International Journal of Advanced Research in Engineering and Technology*, *4*(3), 211–216.

Shakhov, A.B. (2010). Ecological and economic basis of energy conversion of agricultural waste. *Bulletin of the Federal State Educational Establishment of Higher Professional Education*, *5*(44), 82-86.

Shakhov, A. V. (2011). *Organizational and economic basis for the implementation of bioenergy potential of agricultural production: monograph*. Moscow: MSAU.

Simon, J., & Johan, W. (2011). Cryogenic biogas upgrading using plate heat exchangers.

Sondergaard, M. M., Ioannis, A., Fotidis, I. A., Kovalovszki, A., & Angelidaki, I. (2015). Anaerobic co-digestion of agricultural byproducts with manure for enhanced biogas production. *Energy & Fuels*, *29*(12), 8088–8094. doi:10.1021/acs.energyfuels.5b02373

Vasiliev, Yu. N., Zolotarevsky, L. S., & Ksenofontov, S. I. (1992). *Gas and Gas-Diesel Engines*. Moscow: RAO Gazprom.

Vedenev, A. G., & Vedeneva, T. A. (2011). *Guide to biogas technology*. Bishkek, Kyrgyzstan: DEMI.

Weiland, P. (2010). Biogas production: Current state and perspectives. *Applied Microbiology and Biotechnology*, *85*(4), 849–860. doi:10.100700253-009-2246-7 PMID:19777226

Chapter 9
Modeling of Electrohydraulic Technology in Agriculture

Jorge Vinna Sabrejos
The Antenor Orrego Private University, Peru

Alexander Anatolievich Belov
https://orcid.org/0000-0002-9216-9852
Federal Scientific Agroengineering Center VIM, Russia

Alexey N. Vasilyev
https://orcid.org/0000-0002-7988-2338
Federal Scientific Agroengineering Center VIM, Russia

Victor Nikolayevich Toporkov
Federal Scientific Agroengineering Center VIM, Russia

Andrey Anatolievich Musenko
Federal Scientific Agroengineering Center VIM, Russia

ABSTRACT

The authors reveal the relevance of modeling electrohydraulic technology in terms of determining the parameters that cannot be measured by existing methods and technical means. The purpose of the chapter is modeling of electrohydraulic technology in agriculture. By modeling, the authors are developing an installation for obtaining nutrient solutions under irrigation conditions in order to increase the yield in greenhouse vegetable combines while ensuring the environmental friendliness of the process. The authors used the methods of research, analysis and calculation of technical means and control systems applied to high-voltage electro-technological installations of agricultural use. Mathematical patterns are established and a logically based relationship is established between the changes in electrical conductivity of water as a result of electrohydraulic processing.

DOI: 10.4018/978-1-7998-1216-6.ch009

Copyright © 2020, IGI Global. Copying or distributing in print or electronic forms without written permission of IGI Global is prohibited.

INTRODUCTION

Technologies for the production of various materials with the desired properties using the electrohydraulic (EH) effect have been developed for a long time (Yutkin, 1986). The advantages of these technologies are the high efficiency of the processes due to the possibility of the formation of materials of a given structure from the source material (peat, water, etc.), which gives a significant technical and economic effect and obtaining an environmentally product with minimal negative impact on the environment.

Currently, consumers' attention is drawn to organic food. Environmentally product is a product that is made from natural raw materials according to modern technology, which ensures minimal ingress of foreign bodies into the product, does not contain any impurities, and is obtained without the use of mineral fertilizers, pesticides and in the absence of man-made impacts. The products of farms that use predominantly organic fertilizers of natural origin, and sell their products in markets and supermarkets are in great demand. However, due to the limited nature of such fertilizers and the duration of the processes of their formation, the cost of organic food is still quite high.

To improve the quality and economic indicators of agricultural enterprises is possible only by using cheap raw materials. Currently, there is an increasing interest in the application of electrohydraulic technologies in agriculture and national economy. This is due to the fact that in recent years there has been a tendency in the Russian Federation to reduce the fertility of the soil. The soil cover is most susceptible to pollution, thereby losing the ability to restore properties, as well as the reproduction of fertility. According to minimal estimates, about 30 million hectares of productive land have been lost in Russia over the past 15 years.

The applied doses of mineral and organic fertilizers do not compensate for the loss (at harvest) of the nutrient yield in the soil. It can also be used effectively in greenhouses and electrohydraulic sterilization of soil with simultaneous fertilizer. Electrohydraulic processing of peat, including crushing, sequential settling and drying, also has great prospects. To study and evaluate the technological and technical features of the processes during electrohydraulic processing, a theoretical base is needed, which is not sufficiently represented in modern literature sources. The solution of these problems contributes to the modeling of technology and technical means for its implementation.

BACKGROUND

To increase the gross production of greenhouse products, it is important to provide plants with all the necessary nutrients that are found in the soil (especially in peat) in large quantities, but in a form inaccessible to plants. In various countries, original technologies are being developed that allow nutrients to be made available to plants in the form and, in most cases, exclude the addition of other fertilizers, restoring lost fertility to the soil (Feng et al., 2016).

The advantage of these and other technologies using high-voltage pulsed discharges is the high efficiency of the processes, the absence of a negative impact on the environment. This technology is designed to solve the problem of environmental safety. During electrohydraulic processing, most of the soil is ground to particles with a diameter of 0.002 mm. The size of the resulting surface becomes larger than even the most highly dispersed silty fractions of ordinary soil. This contributes to the cost-effective transfer to the solution of the nitrogen and phosphorus, trace elements contained in the soil and in the air, which, in a soluble state, are easily absorbed by plants. Under natural conditions, this takes years, and with EH-effect it occurs almost instantly. For a sharp increase in the nitrogen content released from the soil and air after electrohydraulic treatment, it is advisable to sow the soil with strains of various nitrifying and ammonifying bacteria using the "bacterial explosion" capabilities. The authors propose to apply electrohydraulic treatment of water and its solutions, using the main advantages of the results of tillage. In electrohydraulic technologies, the principle of "electrohydraulic effect" is realized, when electrical energy passes into mechanical energy when an electrical discharge appears (Bian et al., 2018). In order for a discharge to appear, a current pulse with a sufficiently steep front up to 2×10^{11} A / s is needed and with an absolute value of current up to 250 kA. In addition, the dignity of these technologies is high efficiency due to the direct (without intermediate links) conversion of electrical energy into mechanical, high efficiency of the process. The electrohydraulic effect is accompanied by physical and chemical processes. For microseconds, the EH-impact power in the discharge channel reaches hundreds of megawatts. In the treated environment, cavitation is created around the discharge zone, high and ultrahigh pressures occur, capable of grinding not only solid materials, but even breaking chemical bonds in the molecules, the fragments of which re-join, partially forming new substances. Only a few stages of discharge appear: the appearance of a conductive channel between the electrodes, the release of energy in the discharge channel, the final stage, all electrical processes end (Vasiliev et al., 2018).

For EH-processing of peat pulp and other solutions, electrohydraulic crushers are commonly used, which work in industrial fertilizer production lines. For small-scale farms, personal subsidiary farms and greenhouses where a small amount of fertilizer is required, it is more economical to purchase energy means of small capacity (2.0-2.5 kV A) and several EH-devices with the help of which is possible to carry out a set of works in: production of fertilizers from peat pulp, water, disinfection of wastewater, crushing of organic materials, processing of plant and animal feed (Lee et al., 2018). The authors propose to partially or completely abandon the application of chemical fertilizers, replacing them with treated EH-activated water, which will be a nutrient solution and partly fertilizer. The main problem for watering plants is to obtain a nutrient solution with certain parameters. Water is part of the plant body; it dissolves mineral salts that enter the plants through the root system. The need for water varies depending on the species, variety, phase of plant development. To obtain a nutrient solution, it is necessary to take into account the chemical composition of water.

It was established experimentally that during electrohydraulic processing of soil pulp from it, about 30 chemical elements available for plant nutrition turn into a water-soluble state. The content of ammonia nitrogen increases by 1.4-4.5 times, and water-soluble organic matter by 1.5-5 times, depending on the type of peat. The most appropriate use of treated peat - as an additive to greenhouse soils. Energy costs for electrohydraulic crushing and peat concentration do not exceed 50-60 kW per 1 ton. Using the EH-treatment, 9 kg of nitrogen, 4 kg of potash, 0.5 kg of phosphate fertilizers and about thirty different trace elements can be obtained from one cubic meter of peat.

To get the maximum yield you need to make from a few grams to tens of grams of EH-treated soil per plant. In many cases it is more profitable to process not the soil, but the water for irrigation. EH-water treatment allows to increase the number of nitrogen compounds dissolved in it up to 300 times. Such water can serve as an excellent, environmentally friendly nitrogen fertilizer. It promotes active leaching from the soil and the conversion of chemical elements into digestible forms for plants. Biologically active nitrogen dissolved in irrigation water is converted into oxides, thereby increasing the content of NO_2 and NO_3 ions. And the OH anions prevailing in the discharge zone are converted into hydrogen peroxide, which decomposes into H_2O and O–. Atomic oxygen actively oxidizes insoluble and difficultly soluble salts in the fertile soil layer (Ruma et al., 2018).

The use of EH-treatment of plant materials significantly increases the excretion of biologically active substances, trace elements, vitamins, and other substances important for the body of animals. These substances, enclosed in a solid hard-to-digestible shell, are poorly supplied to destruction by traditional mechanical methods. EH-treatment destroys this shell and releases the nutrients in it. Another reason for the possibility of using EH-treatment for disinfecting water and solutions.

There is no technical equipment in Russia and the world with electrohydraulic installations for use in obtaining nutrient solutions under irrigation (irrigation) conditions in order to increase yields in greenhouse vegetable complexes. Similar technologies and technical means for obtaining nutrient solutions under irrigation (irrigation) conditions in order to increase the yield in greenhouse vegetable complexes (complexes) are not used.

To create an energy tool and EH-devices, it is necessary to conduct research on the required parameters and their operating modes, ensuring the least energy-consuming process and cost reduction.

The authors reveal the relevance of modeling in determining parameters that cannot be measured by existing methods and technical means. It is proposed to simulate both the technology and the means for its translation into a practical plane. In addition, the modeling of technological processes serves as a means of reducing the cost of design and the project itself, which is certainly important to achieve acceptable economic performance in the implementation of the developed theoretical ideas in practice.

The purpose of the chapter is modeling of electrohydraulic technology in agriculture. By modeling, the authors are developing an installation for carrying out comprehensive studies on the development of new energy-efficient agrotechnologies for the use of a specially formed pulsed high-voltage electrical discharge inside a fluid volume in technological processes of agricultural production, in particular, to obtain fertilizers from solutions. We used the methods of research, analysis and calculation of technical means and control systems applied to high-voltage electrotechnological installations of agricultural use.

MODELING OF ELECTROHYDRAULIC TECHNOLOGY AND TECHNICAL MEANS

The technical device for the implementation of electrohydraulic technologies is modeled in the graphic editor Kompas-3D, taking into account the observance of the dimensions and technical features of the design to withstand the necessary technological modes of electrohydraulic processing of agricultural raw materials. Thus, the authors justify the benefits of using a hemispherical electrode developed by them. The authors have developed an electrohydraulic installation, which relates to crop production in agriculture and can be used to obtain nutrient water solutions, which are proposed to be used as fertilizers in conditions of irrigation of greenhouse vegetable crops in personal subsidiary and peasant (farmer) farms.

During the development of the installation, both previously published works and the results of new studies (Liu et al., 2017) were taken into account. Analysis of publications and patent search showed that the existing technical means for the implementation of the EH-effect have drawbacks, since they are unregulated in length and power of the spark discharge with power losses in the discharge channel. A device for grinding, mixing and emulsifying materials or cleaning products using electrohydraulic shocks, made in the form of a vessel and equipped with a system of electrodes placed in a liquid developed by L. A. Yutkin, is known (Yutkin, 1986). A disadvantage of the known device is the short length of the spark discharge of 5-20 mm at high generated voltages of 15-50 kV and, as a result, the low effect of electrohydraulic shock. In addition to this device, taken as an analogue, there are many others that are similar in functional characteristics, but have common negative qualities.

The closest in technical essence to the proposed by the authors of the invention is an electrohydraulic device, which consists of a flat negative electrode immersed in a conductive liquid and a positive rod-shaped electrode, to which a metal plate is electrically connected (the author of the invention is L. A. Yutkin).

The disadvantage of this known device, taken as a prototype, is that it is not possible to increase the effect of electrohydraulic shock, since the regulation of the length of the spark discharge is carried out only in the direction of decreasing it by bringing the plate closer to the negative electrode, which entails an increase in power loss and as a result, a reduction in the length of the spark discharge in a liquid.

The objective of the proposed installation is to increase the effect of electrohydraulic shock to obtain nutrient water solutions used as fertilizer for agrotechnical purposes, carrying out irrigation in crop production, allowing varying technological regimes from periodic to continuous depending on the conditions of the irrigation process.

As a result of the use of the present invention, the effect of electrohydraulic shock is increased and it becomes possible to reduce the dissipated energy on the negative electrode plane by providing localization, concentration and compaction of the electromagnetic field in the plane of the tip of the hemispherical negative electrode.

The above technical result is achieved by the fact that the proposed installation for electrohydraulic treatment of aqueous solutions contains a working body in the form of a cylinder in which the negative electrode and the positive rod-shaped electrode are immersed in a conductive liquid. According to the invention, the negative electrode is made in the form of a cup-shaped hemispherical dish-shaped disk with a tip the plane of which is localized and concentrated electromagnetic field. In this case, the electrodes are fixed in a cylindrical container and fixed in the unions with cap nuts. It should be noted that the inlet and outlet nozzles for supplying and discharging the treated aqueous solution are made in the container. At the lower and upper ends of the container of the working body of the installation, the bottom cover of metal is mounted, and there is also the top cover of organic glass.

It was established experimentally that in liquids with ionic conductivity, a discharge occurs only at very small distances between the electrodes and the pressures arising from the discharge channel in the liquid are insignificant. It was possible to solve the problem of a sharp increase in the action of a spark discharge in ion-conducting liquids by introducing an air-induced spark gap into the electrical circuit. To enhance the EH-effect in an ion-conducting fluid and efficiently convert electrical energy into mechanical energy, it is necessary to increase the surface of the spark discharge channel through which energy is transferred to the fluid.

The essence of the proposed technical means is illustrated by the drawing, which shows the general scheme of the working body of the installation for electrohydraulic processing of aqueous solutions (longitudinal section) in accordance with figure 1. The working body of the installation for electrohydraulic processing of aqueous solutions contains a cylindrical container 1, a positive electrode 2, a hemispherical negative electrode 3, tip 4, fittings 5, the outlet 6, the bottom cover 7, cap nuts 8, the inlet 9, the top cover 10.

The working body of the installation contains a cylindrical container 1 for a set of treated aqueous solution. At the ends of the container, the bottom cover 7 made of metal and the top cover 10 made of organic glass are installed for visual assessment, observation and control of the electrohydraulic treatment process. At the top of the container 1 there is an inlet for supplying the aqueous solution 9, and at the bottom of the outlet for discharging the treated aqueous solution 6, which ensures the process manufacturability. The device has fittings 5 for fixing electrodes 2 and 3 with cap nuts 8. Positive 2

Figure 1. The working body of the installation for electrohydraulic treatment of solutions

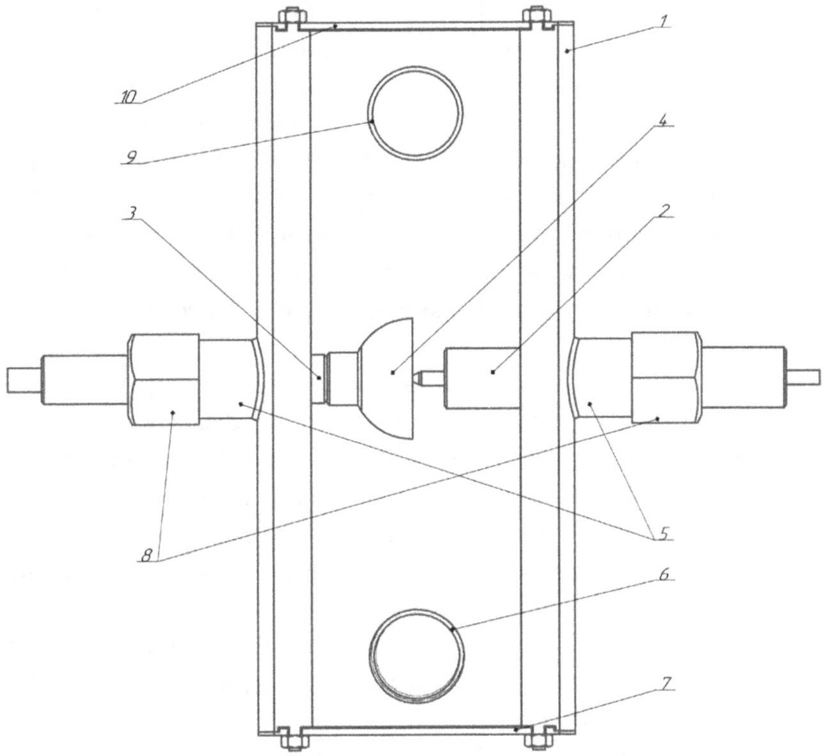

and hemispherical negative 3 electrodes are made of a copper core with an insulator, the material of which must be a dielectric to mitigate the impact arising during the breakdown of the discharge gap. The hemispherical negative electrode 3 is made in the form of a cup-shaped dish-shaped disk, while the electromagnetic field is localized and concentrated in the plane of the end cap 4. The negative electrode is considered to be the anode, and the positive electrode is the cathode. The positive electrode is a pointed rod with a diameter of 12 mm threaded in an insulating (except for the front pointed end) shell. The area of its contact with water should be minimal so that the storage capacitor of the discharge circuit is not discharged to the water before the main discharge channel has been formed, and a significant part of the source energy would not be useless to flow through the water, thereby reducing the efficiency of the process. The mutual arrangement or gap of the negative and positive electrodes is adjusted depending on the required power and energy.

The design of the working body is determined by the technological process performed by the installation. The working body of the installation for electrohydraulic processing solutions works as follows.

The aqueous solution fills the volume of the container or chamber so that the air area always remains necessary to compensate for the action of the shock waves that occur during the breakdown and move in the working volume of the chamber perpendicular to the axis of the channel discharge. The container 1 through the inlet 9 is filled with the treated water solution to a level that must be higher than the tip 4 of the negative electrode 3 by more than 50 mm and above the positive electrode 2 by more than 75 mm, depending on the processing modes for the formation of electrohydraulic shocks. The working body is made subject to the assumption that it is possible to vary the technological regimes from periodic to continuous depending on the conditions of the irrigation process. The hemispherical negative electrode 3, is made in the form of a cup-shaped dish-shaped disk in the form of a hemisphere with a tip 4. The basis for the execution of the negative electrode 3 is hemispherical: during the formation of an electric discharge the electromagnetic field is maximally localized and concentrated in the plane of the tip with minimal power loss. The surface area of the tip 4 of the hemispherical negative electrode 3 is 1700 mm^2. The electrodes 2 and 3 are held, embedded and sealed in fittings 5 with cap nuts 8 in order to ensure tightness. The positive electrode 2 is supplied with a voltage of 15-35 kV. Hemispherical negative electrode 3 is located at a distance of 5-15 mm from the tip of the positive rod-shaped electrode. This distance forms the working gap of the spark discharge. The bottom cover 7 and the top cover 10 are sealed. The formation of an electrical discharge is visually assessed and controlled through the top cover 10, the material of which is organic glass.

In the water in which the electric discharge is made, there are only two types of ions. Negative H–, which play the main role in the development and existence of the discharge. The greater their concentration, the further the streamer germinates. This increases the length of the discharge channel and, accordingly, its surface and the mechanical efficiency of the discharge.

Positive H+ ions do not take in the growth of streamers and is useless for the development of the discharge. Therefore, to increase the efficiency of the discharge, it is necessary to create conditions under which negative H– ions would be formed in the working gap and hinder the formation of positive H+ ions. Such conditions can be created by changing the areas in contact with the water of the electrodes.

It is necessary to minimize the surface (in contact with water) of the positive electrode (tip with a diameter of 2-3 mm) and significantly increase the active surface of the negative electrode to sizes limited by the size of the working chamber (in a laboratory setting, the diameter of the negative electrode of the working element should be 20-60 mm). This creates a significant field asymmetry between the electrodes. As a result, the number of negative H– ions sharply increases, and a negative space charge arises from these ions between the electrodes, which transfer their electrons to the streamer channel, which grows over considerable distances.

When using the proposed working body installation for electrohydraulic treatment of aqueous solutions is achieved electrohydraulic effect, that is, oscillations and movement of the aqueous solution, which gets accelerated as a result of the formation of cavitation cavities in the event of ultra-high hydraulic pressures in the discharge zone. The minimum capacity is calculated from the fact that the developed working body processes 5 liters of an aqueous solution in 1 minute.

The process of electrohydraulic discharge is divided into 5 stages. Table 1 presents the generalized and average data.

Table 1. Pressure zone (stage) the formation of electrohydraulic discharge

Pressure Zone (stage)	Physical and Chemical Phenomena
I	Spark discharge.
II	The material breaks down into dispersed parts. The fluid acquires the properties of a solid elastic body.
III	Observed a riveting of metals. The fluid is in a solid elastic state.
IV	There is an emission of particles due to a powerful pushing action. The fluid is in a state of elastic body.
V	The pressure decreases with distance from the center of the discharge. Large volumes of fluid moving.

EH-technologies use high-voltage pulsed power sources with capacitive storage. The discharge circuit of an RC generator includes connected in series: capacitance - C, inductance - L, consisting of load inductance, connecting wires and internal drive inductance and resistance - R, which includes load resistance, connecting wires and internal resistance of the drive.

If the parameters of the discharge circuit are linear, then the transition process can be described by known equations. The type of discharge depends on the coefficient γ, which depends on the active, reactive resistance and capacitance. If ($\gamma < 1$), then the discharge will be oscillatory, ($\gamma > 1$) - aperiodic and at ($\gamma = 1$) - critical. Aperiodic discharge is characteristic mainly of electrohydraulic technologies (Vasilyev, Vasilyev, & Budnikov, 2019).

Energy release during the electrohydraulic effect occurs on the active resistance of the circuit, which is close to critical. The main factors affecting the occurrence of the electrohydraulic effect are the amplitude, the steepness of the front, the shape and duration of the electrical impulse.

The electrical parameters necessary for EH-processing of various organic structures were established experimentally.

It was established experimentally that when processing peat pulp, the following operating modes of installations are possible:

- Hard – U > 50 kV; C < 0.1 μF;
- Medium – 20 kV < U < 50 kV; 0.1 μF < C < 1 μF;
- Soft – U < 20 kV; C > 1.0 μF.

Set the duration of the discharge course at different processing modes in accordance with table 2.

Electrohydraulic treatment of water requires little energy and the necessary electrical parameters are within the following limits: U = 20-60 kV; C = 0.05-0.2 μF;

When crushing organic materials and processing plant and animal feed, it is necessary: U = 20-70 kV and C = 0.1-0.6 microfarads.

The efficiency of EH-processing is influenced not only by the absolute value of the electrical parameters, but also by their interrelation. The magnitude of the voltage determines the length of the discharge channel, the capacitance - the diameter of the channel. An increase in capacitance, at a constant voltage, increases the pulse energy and the amplitude of the current. With an increase in voltage, the discharge is shorter; an increase in the capacitance leads to an increase in the discharge duration. By reducing the active and inductive resistance of the circuit increases the efficiency of the discharge.

Table 2. Discharge duration

No. p/p	Stage	Duration on Modes, μs		
		Hard	Medium	Soft
1	Predigit	0.000003-0.3	0.03-30.0	3.0-300.0
2	Front-line	0.000005-0.05	0.005-5.0	0.5-50.0
3	Rear front	0.000015-0.15	0.015-15.0	1.5-150.0
4	Subsequent half-waves	0.00005-0.5	0.05-50.0	5.0-500.0
5	Reaction attenuation	From a few seconds to hours, days		

The EH-effect will be the more effective, the greater the voltage level of the pulse current generator and the energy introduced into the working gap, the steeper the pulse front, the shorter the pulse duration. Therefore, the main tasks of creating a laboratory installation will be:

- The optimal scheme of a high-voltage pulsed current source, providing all the necessary EH-processing modes;
- Stable (against erosion) electrode design;
- Independent adjustment of output parameters.

The authors are developing an electrical circuit installation for electrohydraulic treatment of aqueous solutions, for example, the impact on water from various sources in accordance with figure 2.

Power is supplied to the circuit using a KM magnetic starter, which closes KM1.1 and KM1.2 contacts. To close the contacts KM1.1 and KM1.2, you must use a control system that consists of a QF circuit breaker that acts as a limiting device for opening the circuit when an emergency current pass. In the control system there is a "start" button SB1, which sends a signal to the magnetic starter, which closes the contacts KM 1.1 and KM 1.2. In turn, the button "stop" SB2, serves as a control element that opens the contacts KM1.1 and KM1.2, if necessary. Also, the control system includes a lock SB3, locking the contacts KM1.1 and KM1.2 in a certain position.

The design parameters of the installation include: the composition of the installation, ease of operation, etc. Installation for electrohydraulic treatment of solutions should have a block design, be easy to maintain, assemble and dismantle, reliable in operation, electrically safe and convenient for the work of the staff in the process of maintenance and current, planned, unscheduled repairs. The installation should

Figure 2. Electrical schematic diagram of the device for electrohydraulic treatment of solutions

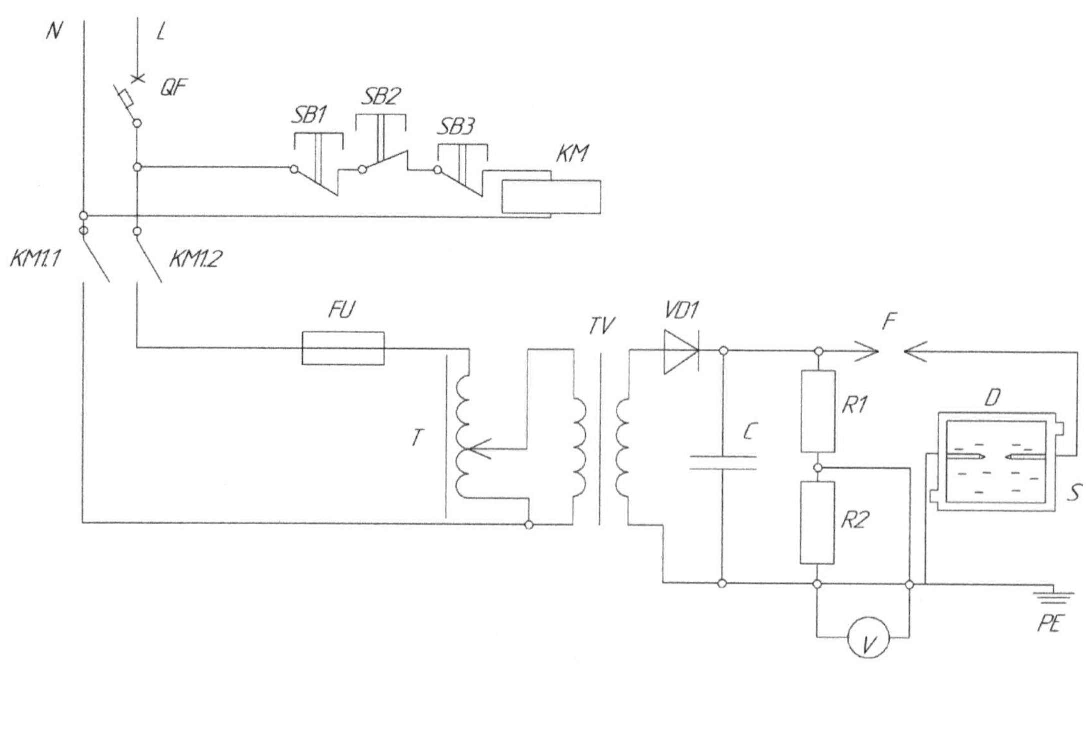

provide the range of electrical parameters necessary for the study, it should provide for the protection of personnel from high voltage and noise by placing the equipment in a metal and grounded housing and the working body in a special soundproof chamber.

The installation of the "electrohydraulic discharger" includes a case in which the following elements and components are installed.

TV: Is a single-phase step-up transformer for converting electrical energy. The transformer has two windings of low 220 V and high voltage 50 kV. Allowed to use a high-voltage power supply based on a TVI-50/70 transformer with a rectifier.
VD1: Rectifying diode is needed to convert AC to DC and discharge-charge capacitor.
C: High-voltage pulse capacitor stores electrical energy, which is released in the spark gap.
R1, R2: Resistive divider lowers the voltage to connect a voltmeter.
V: Voltmeter is designed to measure the voltage in an electrical circuit.
PE: A device that protects a person from electric shock.
S: In the spark gap a spark discharge is formed, due to which a water hammer occurs.
D: A device (tank or discharger) is necessary for the formation of an electrohydraulic shock, that is, impulses with a steep front; it contains electrodes of copper.
F: Forming gap is needed to formation the voltage on the secondary winding.

In addition, the installation must have controls that allow you to change the installation parameters when conducting research.

Before starting work, it is necessary to verify the absence of voltage at the installation and inspect the installation for defects and faults. After performing pre-launch maintenance, it is necessary to withstand a number of measures and conditions for safe operation of the installation for electrohydraulic treatment of solutions:

1. The QF circuit breaker is turned to the on position;
2. Is started by pressing the button SB2, which activates the magnetic starter KM, which closes the contacts KM1.1 and KM1.2;
3. The SB3 lock is moved to the working position;
4. As a result of the supply voltage, the TV transformer starts to convert energy;
5. Passing through the rectifying diode VD1 alternating current is converted to direct current;
6. When a constant voltage is applied to the plates, charge accumulates;
7. The source of the accumulation and release of discharge energy in this circuit is the capacitor; after the capacitor C is fully charged, it will begin to discharge into a circuit; when a certain potential is reached, an air gap F is punched; the energy stored in the capacitor transfers to a spark gap concentrated in the working device (reservoir) liquid D, thereby an electric pulse appears having a sufficient discharge power;
8. In the working device (tank) between the electrodes there is a discharge, accompanied by the occurrence of an arc;
9. Claps are formed, accompanied by a splash and movement of water.

The energy losses that are spent on the formation of the discharge channel negatively affect the result of the process. Acceptable to get rid of these losses, add pulse energy, but it affects the service life of the electrode elements, namely the insulation of the electrodes. This process of capacitance and voltage is resumed at a frequency that depends on the selected power of the transformer.

The main technical characteristics of the installation for the electrohydraulic treatment of solutions are shown in table 3.

The advantage over analogs.

The advantages of an electrohydraulic installation, for obtaining nutrient solutions under irrigation conditions to increase yields in greenhouse vegetable plants, with respect to irrigation (irrigation) untreated nutrient solutions, are the technology and technical solution to increase yields, reduce production costs, efficient use of capital investments and environmental protection.

The forecast for the needs of the proposed development.

In Eastern Europe countries, the greenhouse vegetable complex of year-round use of the 3rd light (middle) zone in block or hangar greenhouses is required in average indicators of winter-spring, autumn and transitional periods per irrigation by sprinkling 20 l / m^2 with a sprinkling intensity of no more than 1 l per 1 m^2 / min at the rate of - cucumber 5 l / m^2 and tomato 15 l / m^2. The frequency of irrigation (irrigation) is taken averaged - 2 times a week. Total for every 1000 m^2 of area required 2080000 liters of nutrient solution per year. For a greenhouse plant of small production capacity with a greenhouse area of 3000 m^2, 6240000 liters of nutrient solution per year is required. Thus, the required number of electrohydraulic units with a design capacity is three. Considering the approximate total area of greenhouses in greenhouses - more than 2000 hectares, the forecast for the needs of production of a serial batch of an electrohydraulic unit is 10 pcs.

RESULTS OF ELECTRICAL CONDUCTIVITY STUDIES

As a result of modeling the developed electrohydraulic technology, it becomes possible to conduct experimental studies. Electrohydraulic processing in a liquid medium is carried out to achieve different results and, as a result, it is used quite widely in various industries and in agriculture (Imai, Sakaguchi, & Shirafuji, 2018). This disinfection of wastewater, increasing the concentration of nitrogen compounds absorbed by plants and others (Zheng, 2017). In addition, this technology is used in the recovery of heat exchangers for descaling (Nussupbekov et al., 2017). Electrohydraulic water treatment is accompanied by the following electrophysical characteristics of the process: high and ultra-high impulse hydraulic

Table 3. Technical characteristics of electrohydraulic installation

No. p/p	Specifications	Electrohydraulic Setting
1	Productivity, l / h	1000
2	Voltage of the secondary winding of the transformer TVI-50, kV	50
3	Capacity of capacitor bank, μF	0.2
4	Installed power, kW	2
5	Rated voltage, V	220/380
6	Overall dimensions, mm	1000x1000x1000

pressures; mechanical resonance phenomena; intense pulsed light, heat, ultraviolet radiation; multiple ionization of compounds and elements contained in a liquid. To study these effects, quite expensive instrumentation technical means of measurement are required. However, all the above factors are interrelated. Therefore, the authors propose a method of integrated assessment of electrophysical characteristics based on studies of the dynamics of changes in specific electrical conductivity (hereinafter referred to as electrical conductivity) by identifying patterns according to which a change in electrical conductivity will correspond to a change in a number of electrophysical parameters. This will serve to interpret the variation of technological regimes and possible constructive improvements of the installation in order to increase the achievable EH-effects (Aksenov et al., 2018).

Experimental studies are conducted to identify changes in the electrical conductivity of water as a result of EH-treatment. Applied equipment and methods for measuring the electrical conductivity of water: appendix to certificate No. 58191 on the approval of the type of measuring instruments using an ATLANT 1200 conductometric multichannel liquid analyzer.

The selected instrument for measuring electrical conductivity meets the requirements of functionality, compactness and portability. The grounds for the choice of a technical measuring instrument: the technical characteristics and the working range of measurements of the selected conductometer satisfies the expected conditions for conducting the experiment on EH-effects in the part of the range of change in conductivity from $50 \cdot 10^{-6}$ to $2 \cdot 10^{-3}$ S / cm; enters the State Register (No. 60144-15); easy to use, easy to maintain, sufficient accuracy of readings (absolute measurement error).

The electrical conductivity of water characterizes its ability to conduct electrical current. In this work, the term water refers to both distilled and tap and lake water. However, tap and lake waters are not chemically pure; in this connection, the definition of aqueous solutions will be more accurate. Aqueous solutions are created either industrially, in which case they can be considered artificial, or the surrounding nature, then they can be considered waters from sources of natural origin. Aqueous solutions are used for various purposes to carry out different processes. Therefore, aqueous solutions can be considered working. In addition to water molecules, they contain ions of mineral salts and organic substances, which mainly determine the electrical conductivity of water and working aqueous solutions. The level of the dissolved salts in water and substances of organic origin can be judged on the possibility of creating an electrohydraulic effect. Electrical discharges during EH-processing are a consequence of the circuit closure and the flow of electric current between the electrodes. Consequently, the electrical conductivity of water and aqueous solutions is the electrophysical factor that can serve as a characteristic of all the basic properties of the EH-treatment process (Zhu et al., 2014).

For experimental studies of changes in the electrical conductivity of water as a result of EH-treatment, an EH-installation with a spark gap was used. The EH-installation includes a sinusoidal voltage source of 220 V. The transformer increases the input voltage, which is stored in capacitors. There is an autotransformer for regulating the voltage value and correct supplying it to the load. Upon reaching a voltage sufficient to break the forming gap and the working gap, a spark is generated.

Marked design parameters of the installation. The system of electrodes used is the needle, the disk. The electrodes are immersed in the test sample of water, on the side of supplying voltage to them, insulation is made to prevent breakdown on the metal case of the container.

Three technological modes of EH-impact on water were selected using the installation. The selection bases are the minimum applied voltage to the capacitive-inductive load. At this stage of the experiments, the inductance of the circuit and the capacitance are constant values due to the design features of the installation. The parameters of technological regimes were selected taking into account the sifting experiment conducted using the methods of Plackett-Berman and random balance, which revealed significant and insignificant factors influencing the process of electrohydraulic processing.

The first technological mode is provided by the following parameters: voltage - 15 kV; storage capacitor capacitance - 0.1 microfarad; inductance - 20 μH; pulse energy - 11.3 kJ; air gap - 10 mm; working gap - 8 mm. The pulse energy is taken by calculation, it is considered equal to the accumulated energy on the plates of the capacitor battery of the installation.

In the second technological mode, the following parameters are accepted: voltage - 38 kV; storage capacitor capacitance - 0.1 microfarad; inductance - 20 μH; pulse energy - 72.2 kJ; air gap - 13 mm; working gap - 10 mm.

The third process mode is characterized by the following technological parameters: voltage - 50 kV; storage capacitor capacitance - 0.1 microfarad; inductance - 20 μH; pulse energy - 125.0 kJ; air gap - 20 mm; working gap - 20 mm.

The air and working gaps were set according to experimental observations regarding the formation of discharge pulses with a power sufficient to achieve EH-shocks. The criteria for achieving the EH-strikes and, accordingly, ensuring the EH-effect for the processed medium were visual and acoustic characteristics. These are claps, the mechanical movement of fluid flows directed from the discharge channel, the generation of a spark discharge, in the foreign literature called plasma. In this context, the analogy is carried out with the breakdown of an air gap with a strength of 30 kV / cm. The air, which is considered an insulator, is transformed into an ionized gas. That is, one can consider this phenomenon as a transition of a substance from one state of aggregation to another.

Initial conditions: the temperature of the investigated water samples was 18 °C; loading volume of the investigated water samples 2.5 liters.

Water samples were taken of three types: plumbing, distilled and lake. The basis of the choice of these particular varieties of water is their possible use as nutrient solutions in vegetable production.

The electrical conductivity of distilled water is minimal relative to tap and lake water. Samples of distilled water were obtained by reverse osmosis, the initial value of electrical conductivity 108 μS / cm. A sample of tap water was taken from the city water supply of the South-Eastern Administrative District of the city of Moscow, the initial value of electrical conductivity was 355 μS / cm. A sample of lake water was taken from an open water body in the Moscow region, the initial value of electrical conductivity was 258 μS / cm.

It was experimentally confirmed that due to the EH-effect from the physical properties, the electrical conductivity changes in accordance with the dynamics in figures 3, 4, 5.

The electrical conductivity of the water samples was measured directly in the working device (tank), where the exposure took place, in the full volume of the water sample by immersing the electrical conductivity sensor of the ATLANT 1200 measuring instrument in water no later than 1 minute after the EH-treatment. The temperature was measured in the total volume of the treated water after mixing.

Diagrams constructed from the results of experimental studies contain curves that are approximated by exponential functions and are described by the following empirical expressions, shown with an approximation confidence value:

Figure 3. Dynamics of the specific conductivity of water at a voltage of 15 kV

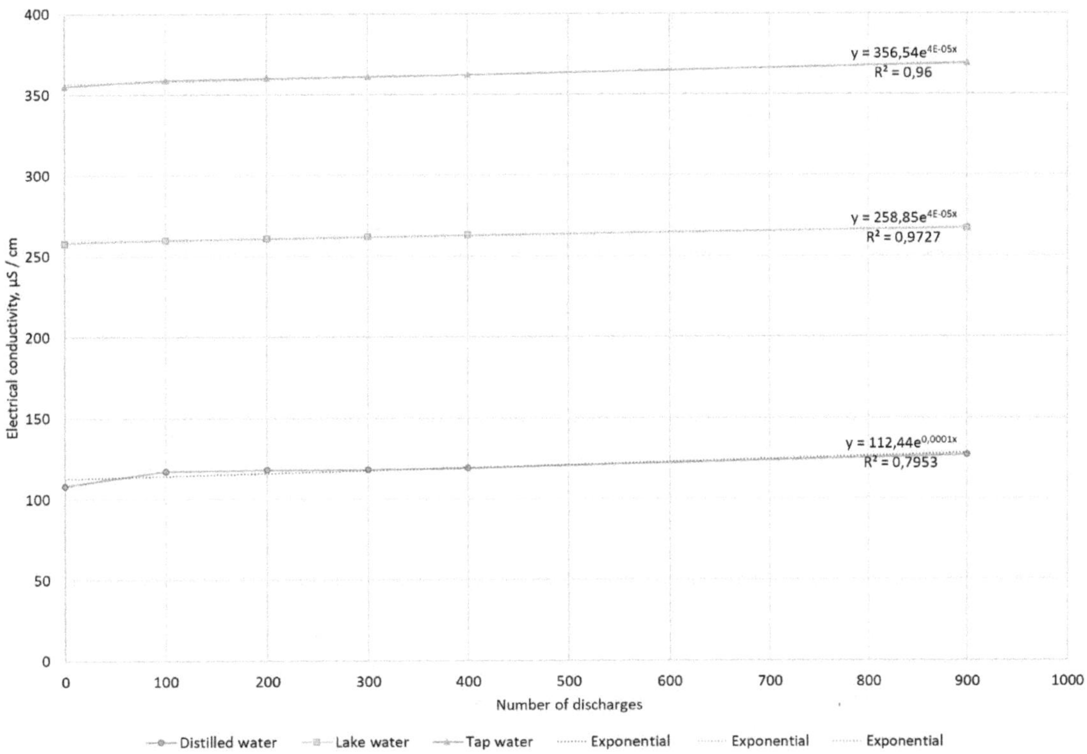

Figure 4. Dynamics of the specific conductivity of water at a voltage of 38 kV

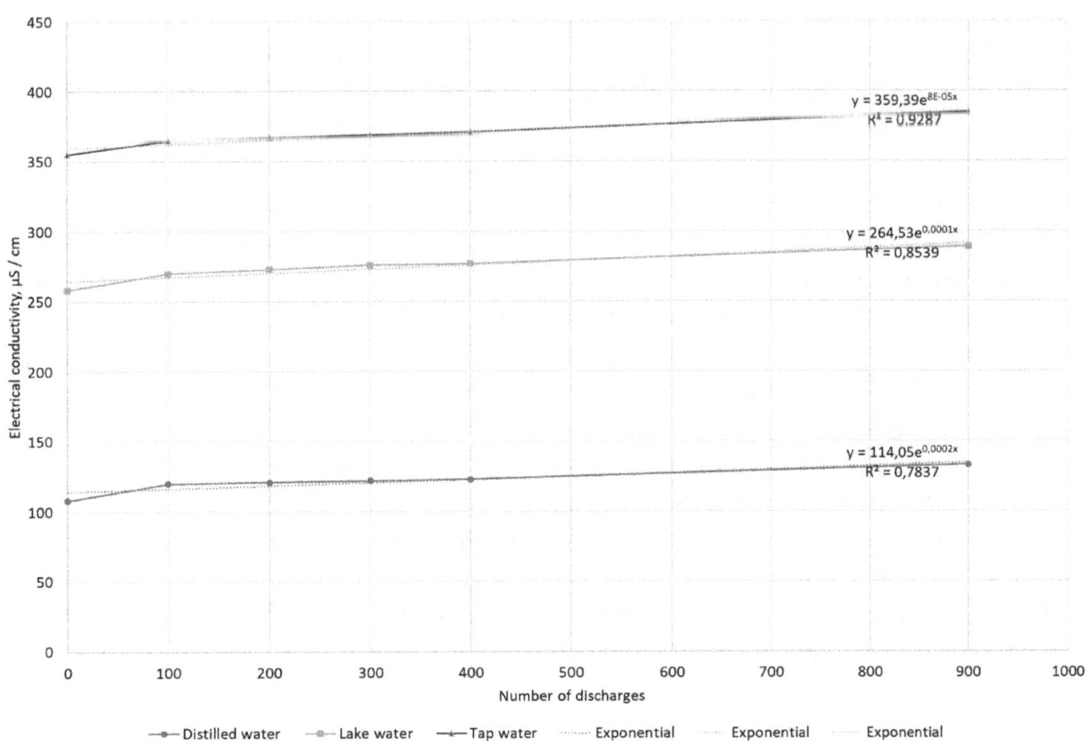

Figure 5. Dynamics of the specific conductivity of water at a voltage of 50 kV

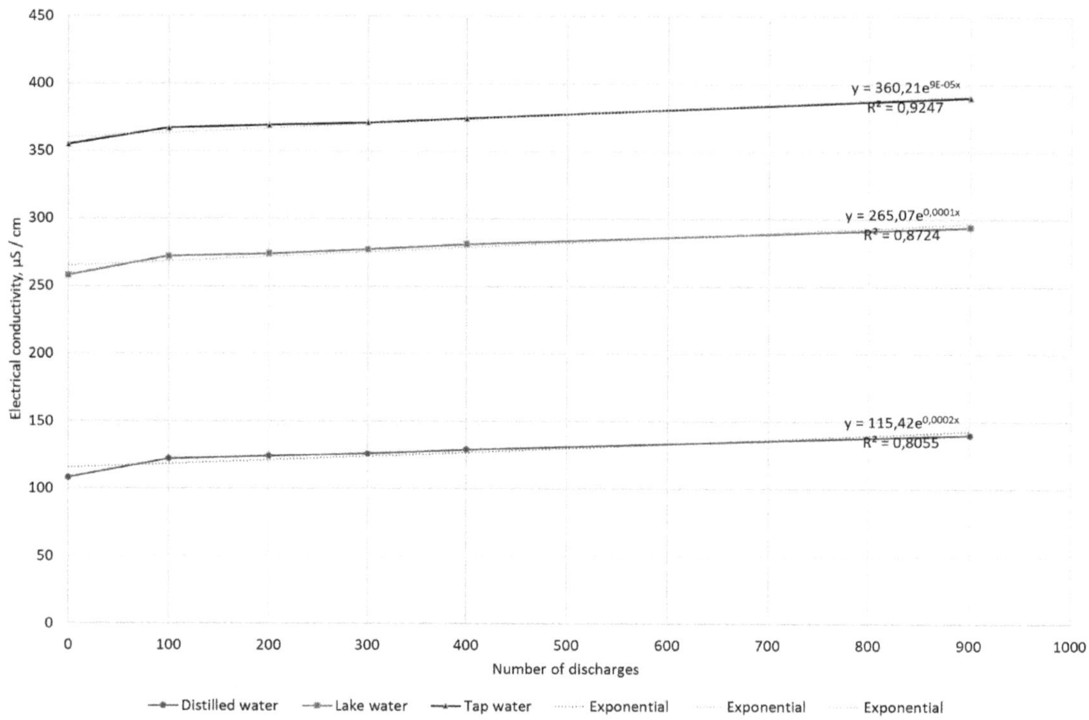

- At a pulse energy of 11.3 kJ for distilled water, $y = 112.44\ e^{0.0001x}$, $R^2 = 0.9753$; for lake water, $y = 258.85\ e^{4E\text{-}0.5x}$, $R^2 = 0.9727$; for tap water, $y = 356.54\ e^{4E\text{-}0.5x}$, $R^2 = 0.96$;
- At a pulse energy of 72.2 kJ for distilled water, $y = 114.05\ e^{0.0002x}$, $R^2 = 0.7837$; for lake water, $y = 264.53\ e^{0.0001x}$, $R^2 = 0.8539$; for tap water, $y = 359.39\ e^{8E\text{-}0.5x}$, $R^2 = 0.9287$;
- At a pulse energy of 125.0 kJ for distilled water, $y = 115.42\ e^{0.0002x}$, $R^2 = 0.8055$; for lake water, $y = 265.07\ e^{0.0001x}$, $R^2 = 0.8724$; for tap water, $y = 360.21\ e^{9E\text{-}0.5x}$, $R^2 = 0.9247$.

A positive dynamic was revealed, generalizing the increase in the electrical conductivity of the water of the samples under study with an increase in the number of electrical discharges from 100 to 900, regardless of the pulse energy of each technological mode. The choice of the number of spark discharges is due to the fact that at 900 discharges the minimum result of the EH-processing is achieved. This corresponds to the disinfection of water and an increase in the concentration of nitrogen compounds, that is, their conversion into easily digestible forms. This is confirmed by microbiological analyzes of water samples and studies of changes in the concentration of nitrogen in them (Belov, Toporkov, & Vasilyev, 2018).

A comparative analysis of the results of each of the three experiments showed that the percent increase in electrical conductivity is the higher, the higher the voltage source. At a voltage of 15 kV, the increase in electrical conductivity in percentage ratio is: 8.3% for distilled water, 3.5% for lake water, 3.9% for tap water. At a voltage of 38 kV, the increase in electrical conductivity in percentage is: 23.1% for distilled water, 12.0% for lake water, 8.5% for tap water. And at a voltage of 50 kV, the increase in electrical conductivity in percentage ratio is: 29.6% for distilled water, 14.0% for lake water, 9.9% for tap water.

The second process mode is considered to be the most preferable for further experimental studies from the point of view of safety, energy intensity and performance, while ensuring the parameters of which achieve a pulse energy of 72.2 kJ. When operating at a voltage of 38 kV, a larger volume of water covered by the discharge is provided than at 15 kV, but smaller than at 50 kV (the third process mode). The energy intensity of the EH-processing process increases simultaneously with the increase in the source voltage, therefore, when working with the parameters of the second process mode, the relative energy efficiency takes average intermediate values.

The change in the electrical conductivity of tap, distilled and lake water towards an increase in the process of EH-exposure is due to the formation of water bubbles and the structuring of water in terms of polarization of charged particles, that is, a change in their movement and activity. Increasing the temperature of all types of water samples after EH-exposure is not more than 5 °C after 900 discharges, which will have no effect on the results of the EH-treatment in the preparation of the nutrient solution from water. Therefore, temperature is a minor factor that can be neglected. An increase in the electric conductivity of water leads to a decrease in the resistance to the flow of electric current, therefore, the generation of electric discharges will be carried out with less required energy. Consequently, the more discharges are given, the less they are required in dynamics for sparking.

The results of the experiments performed show the requirements for changes in the design of some units of the EH-installation to reduce inductance and increase the capacitance. Thus, the mode of generating the maximum specific energy per unit time will be simulated, which is necessary to create the maximum steepness of the pulse front in order to ensure sufficient conditions for the EH-process to proceed. The results of experimental studies at various values of the accumulated energy suggest that these energy values affect the electrical conductivity, therefore, affect the result of the EH-treatment in terms of the formation of a nutrient solution from water.

FUTURE RESEARCH DIRECTIONS

In the course of solving the tasks and obtaining the results, the authors identified the shortcomings in the work done. This made it possible to outline certain plans for future research. It is planned to investigate experimentally and theoretically changes in the electrophysical characteristics, chemical composition of water and aqueous solutions as a result of varying the various technological regimes.

CONCLUSION

In conclusion, the authors noted:

1. Conducted modeling of electrohydraulic technology in agriculture. The developed device for the realization of electrohydraulic action on aqueous solutions with the purpose of activation is substantiated. The arguments in favor of using a hemispherical tip of the negative electrode are the required lower voltages when forming a discharge of the same length. The design parameters of the device as a model in a graphical editor for testing technological modes are shown. The developed technology and technical solution for its implementation, in particular with the help of modeling, provide increased yields, lower production costs, efficient use of capital investments and environ-

mental protection by obtaining nutrient solutions in irrigation conditions in greenhouse vegetable combines. These factors are advantages in relation to irrigation (irrigation) of unprocessed nutrient solutions, which is currently used in world practice.

2. Mathematical patterns are established and a logically based relationship is established between changes in electrical conductivity as a result of electrohydraulic processing. A change in the electrical conductivity of samples of tap, distilled and lake water in the direction of an increase in the process of electrohydraulic action is caused by the formation of water bubbles and the structuring of water in terms of polarization of charged particles, that is, a change in their movement and activity. A scientific suggestion has been made that electrical conductivity can serve as a criterion for evaluating the process of electrohydraulic processing with respect to transferring an aqueous working solution to a nutrient one.

REFERENCES

Aksenov, A. G., Izmaylov, A. I., Dorokhov, A. S., & Sibirev, A. V. (2018). Onion bulbs orientation during aligned planting of seed-onion using vibration-pneumatic planting device. *INMATEH - Agricultural Engineering*, *55*(2), 63–70.

Belov, A. A., Toporkov, V. N., & Vasilyev, A. N. (2018). Planning and conducting screening experiment to study the production of fertilizers in the electrohydraulic treatment of solutions. *International Technical and Economic Journal*, *5*, 22–28.

Bian, D. C., Yan, D., Zhao, J. C., & Niu, S. Q. (2018). Experimental Study of Pulsed Discharge Underwater Shock-Related Properties in Pressurized Liquid Water. *Advances in Materials Science and Engineering*.

Feng, J. W., Jiang, L., Zhu, D., Su, K. Z., Zhao, D. Y., Zhang, J. B., & Zheng, Z. (2016). Dielectric barrier discharge plasma induced degradation of aqueous atrazine. *Environmental Science and Pollution Research International*, *23*(9), 9204–9214. doi:10.100711356-016-6148-9 PMID:26832879

Imai, S., Sakaguchi, Y., & Shirafuji, T. (2018). Simultaneous generation of acidic and alkaline water using atmospheric air plasma formed in water. *Japanese Journal of Applied Physics*, *57*(1).

Lee, S.J., Ma, S.H., Hong, Y.C., & Choi, M.C. (2018). Effects of pulsed and continuous wave discharges of underwater plasma on Escherichia coli. *Separation and Purification Technology*, *193*, 351–357. doi:10.1016/j.seppur.2017.10.040

Liu, Y., Li, Z.Y., Li, X.D., Zhou, G.Y., Li, H., Zhang, Q., & Lin, F.C. (2017). Energy Transfer Efficiency Improvement of Liquid Pulsed Current Discharge by Plasma Channel Length Regulation Method. *IEEE Transactions on Plasma Science*, *45*(12 part 2), 3231-3239.

Nussupbekov, B. R., Khassenov, A. K., Karabekova, D. Z., Stoev, M., Beysenbek, A. Z., & Kazankap, B. I. (2017). New technology for the recovery of heat exchangers. *Bulletin of the University of Karaganda-Physics*, *1*(85), 67–72. doi:10.31489/2017Ph1/67-72

Ruma, H., Sakugawa, T., & Akiyama, H. (2018). The Role of Pulse Voltage Amplitude on Chemical Processes Induced by Streamer Discharge at Water Surface. *Catalysts*, *8*(5).

Vasiliev, A. N., Ershova, I. G., Belov, A. A., Timofeev, V. N., Uhanova, V. Y., Sokolov, A. V., & Smirnov, A. A. (2018). Energy-saving system development based on heat pump. *Amazonia Investiga*, *7*(17), 219–227.

Vasilyev, A. N., Vasilyev, A. A., & Budnikov, D. A. (2019). The Functional Dependencies of the Drying Coefficient for the Use in Modeling of Heat and Moisture-Exchange Processes. *Advances in Intelligent Systems and Computing*, *866*, 239–245. doi:10.1007/978-3-030-00979-3_24

Yutkin, L. A. (1986). Electrohydraulic effect and its application in industry/Leningrad. *Engineering*.

Zheng, JS. (2017). Inactivation of Staphylococcus aureus in water by pulsed spark discharge. *Scientific Reports*, *7*.

Zhu, L., He, Z. H., Gao, Z. W., Tan, F. L., Yue, X. G., & Chang, J. S. (2014). Research on the influence of conductivity to pulsed arc electrohydraulic discharge in water. *Journal of Electrostatics*, *72*(1), 53–58. doi:10.1016/j.elstat.2013.11.004

Chapter 10
Placement Plants on the Field Area by Seeding Machines:
Methodical Aspects Assessment Rationality

Pavel Lavrukhin
FSBEI HE, Don State Agrarian University, Russia

Sergey Senkevich
https://orcid.org/0000-0001-6354-7220
Federal State Budgetary Scientific Institution, Federal Scientific Agroengineering Center VIM, Russia

Pavel Ivanov
FSBEI HE, Don State Agrarian University, Russia

ABSTRACT

Any harvest is primarily the realization of the potential productivity of plants in certain conditions, depending on the parameters and operating modes of the sowing machine. The result of the operation of mechanized sowing will be the parameters of the nutrition area of each plant is the shape and numerical value of these area. The resulting schemes of plant placement on the field correspond to certain characteristics and parameters of the areas of plant nutrition, determining the availability of all elements of mineral nutrition, photosynthetically active radiation, and the relationship of units of agrophytocenosis. This is one of the decisive conditions for the harvest. The development of theoretical ideas about the uniformity of plant placement and the use of appropriate criteria and characteristics will create the necessary understanding of the role of the sowing machines in crop production during cultivated plants in the framework of mechanized technology.

DOI: 10.4018/978-1-7998-1216-6.ch010

Copyright © 2020, IGI Global. Copying or distributing in print or electronic forms without written permission of IGI Global is prohibited.

INTRODUCTION

In the next decade, humanity will enter a new stage of technological development, which is characterized by a focus on the use of high technologies. In crop production, these technologies will be characterized by high productivity, efficiency, creation of comfortable working conditions, but first of all they will overcome the main drawback of previous technologies – the lag in the growth of crop yields from the growth of costs for their production. Modern technologies make it possible to realize 30-40% of the potential productivity of plants and for this indicator came to the limit of saturation of their efficiency. In high technologies the level of potential realization should be not less than 70% (Krasnoshchekov, 2002; O'Leary et al., 2018).

One of the ways to solve the problem of increasing the efficiency of crop production is the creation of new varieties and hybrids, as well as genetically modified plants. It is believed that the use of GM plants has its advantages: they are resistant to diseases and weather ripen faster, stored longer, independently produce insecticides against pests (De Vita, 2017; Ewbank, 2002). On the other hand, experts identify three main threats posed by the use of GM plantsisa threat to the human organism, a threat to the environment, global risks. The noted inconsistency in the use of GM plants limits the scope of their application.

A great contribution to the development of agricultural technologies is made by the creation of new varieties. The development of new varieties and the production of hybrids is not a quick process, so it is not possible to achieve the required increase in crop production efficiency at a rapid pace. It takes 10 to 12 years to create a new variety.

These ways make possible increase the biological potential of cultivated plants, but do not create conditions for its implementation. All achievements of geneticists, breeders, biologists, agronomists, agrochemists, and soil scientists can be implemented in wide practice only with the use of developed machines complexes that make up the mechanized technologies base.

The solution problem of increasing the efficiency of crop production involves the creation of high technologies. All technologies of crop production should provide plants with optimal conditions for the formation of the crop. One of the main roles in the formation of such conditions belongs to the sowing machine, the modes and parameters of which determine the size and shape of the nutrition area of each of the plants in the sowing. Thus, the parameters and characteristics of the processes performed by sowing machines should meet the biological requirements of cultivated plants to yield conditions. The main direction should be the creation of a new generation of machines, rather than modernization and improvement of the produced ones. Development of new equipment and technologies are possible only on the basis of new and improvement of existing methods of research, testing and quality management of sowing machines for intensive and high technologies in crop production (Bespamyatnova, 2010; Bespamyatnova, 2002). In turn, the development of such methods should be based on the latest achievements of engineering science and practice, which are based on solutions aimed at maximum satisfaction of the vital needs of plants for the most complete realization of their biological potential yield.

GENERAL CONTENTS

Sowing Machines Constructs Developing and Seeds Placement Methods

Sowing machines modern constructions development largely repeats the becoming those ways that emerged during the formation previous technological structures and reflect the urgent issues faced by farmers in ancient times. The history of sowing machines began around 1500 BC when the Sumerians began to use something similar to a primitive seeder with one seed pipe. However, the Sumerian seeder was never known in Europe. Since the II century BC in China there was an iron seeder which had several seed pipes, and its application allowed adjusting the system of food production allowing to feed the population of the country for many centuries and to provide social progress. In Europe, the first agricultural machine, which had all the sufficient signs of a seeder, was patented by a man named Camillo Torello in 1566. Later, in 1701, the design of this seeder was modified by the Englishman Jethro Tall. The seeder turned out to be too expensive and not very reliable, and it is for this reason that the widespread use of seeders in Europe began closer to the middle of the XIX century (Lyubushko, 2001). The seeders arranged the seeds in rows. With the coulter seeders creation, important tasks of seeding into the moist soil layer and guaranteed germination even in severe weather and soil conditions were solved. Modern seeders place seeds in rows or ribbons, which are due to the design of the coulter groups, while the seeds are located at some intervals in a row, and the coulters locates the seeds rows with the constant aisles. The width of the row spacing selects from the conditions of the Coulter possibility in the soil, skips serving crops equipment, personnel, or other considerations. The number of seeds sown per unit area of the field is changes by loading the row with the number of seeds, i.e. by changing the interval between the seeds in the row. This leads to the fact that the concept of uniform distribution of plants degenerates into the position of equality of intervals between plants in a row, beyond which the methodology, in consequence, does not go. Thus, the interval between seeds became the basic concept for determining the uniformity of plant placement by area. It is this interval that underlies the modern methodology of engineering and agrobiological sciences, as the doctrine of the organization of theoretical and practical activities of scientists in the field of crop production by plants. It is this basic concept that has led to the modern choice of ways of development of mechanization of crop production technologies and the level of development of technologies themselves.

There is a paradoxical situation when the size of the nutrition area is recognized, but its form is ignored. The reason for the discrepancy is insufficient axiomatation in the description of the phenomenon. The shape of the area is as axiomatic as its size. This divergence of concepts generates a number of existing facts and phenomena that cannot be explained from the standpoint of the applied logic fixed in the fundamental standards and methods of research of the influence of parameters and modes of operation of sowing machines on the productivity of cultivated plants. It is here that a situation is created when the conditions for the formation of the crop limit the level of realization of the potential productivity of plants.

It follows from the above that in order to create a new generation of sowing machines that meet the requirements for the creation of high technologies of crop production, it is necessary to clarify the basis of methodical constructions. Unified methodological base of engineering and agrobiological sciences Improvements will provide further development of methodological support and development of new principles of mechanized technologies will ensure the transition to a new level of technological structure of the industry.

Modern Researches Level Uniformity Plants Placement in the Field Area

Considering the issues of mechanized placement of plants on the field, first of all, it should be noted that the research was carried out both within the biological sciences and in the field of mechanization of sowing. Such issues as the size of the row spacing, not much attention was paid. The reasons for the size of the aisles can be called a large number, but such considerations as providing plants with optimal nutrition area, as a rule, are in the last places. The analysis of this situation with the size of the row spacing sends to the very beginning of research activities in the agronomic science of seeding. To highlight the mutual influence joint moments research two directions on the modern result of development, recall that in the scientific literature are known descriptions seeding methods and planting crops since the works of ancient authors. For example, in the well-known epochal work of I.I. Sinyagin "Plant nutrition areas", it is reported about the works of Mark Portia Katona and Pliny the elder about the methods of planting and seeding rates of a large number of cultivated crops. Further works of a large number of the European and Russian agronomists by definition of norms of seeding of seeds of various cultures followed. They include such names as Christian Richards, J. Liebig, C. Wollny, and V. I. Edelstein. In particular, V. I. Edelstein believed that not only the size, but also the configuration of the feeding area, which he associated with the shape of the plant (the projection of the Bush, rosettes, etc.), was essential. His experiments showed the advantage of the square shape of the nutrition area. Further, for plants with a low nutrition area size, they were made the conclusions about the difficulty of applying machinery to care for the crops and for mechanized harvesting. Therefore, for such crops it was recommended to increase the spacing, reducing the distance between the seeds in rows. This approach to the placement of plants on the field area has survived to our time and in the works of modern scientists. Works, which would consider the optimization of the values of row spacing or other ways of placing plants over an area of a very limited number. Basically, the elements of such studies were conducted by engineering specialties scientists. Agronomic science accepts the current settings, seeding machines, including the row spacing width as traditional well-established fact. Few exceptions to this rule can be found in the analysis of agrotechnical requirements or advertising brochures of machines for sowing vegetable crops. There you can find ways to set the schemes of plant placement either as requirements for the organization of the seeding process, or as other recommendations. The scheme plants placement is created by a sowing machine-seeder, but the placement of seeds over the field area in rows, the distances between plants in which there are fewer rows at times, remains the same. With the total area provided to the plants, the plants are placed so that the nutrition area configuration every plant is a rectangle with such a sides ratio when the length of one side exceeds the length other in several times. Scientists engineering specialties question row width was interested only from the point of view mechanization sowing possibility or doing other technology requirements. The scheme coulters arrangement determines the sowing machine technological reliability, the possibility inter-row treatments performing, the track size for the passage of sprayers and fertilizer spreaders, the operation of dosing systems and transfer mechanisms, etc. Special studies aimed at determining the reaction of plants to changes in the size and configuration of the plant nutrition area in the scientific literature practically does not occur. There is no holistic view of the rules and requirements for the placement of plants on the field area.

Proposed Methods for Assessing Uniformity Plants Placement on the Field Area

In engineering studies, the plants nutrition area configuration has always been considered indirectly as some geometric figure. To perfection the view seed uniformity placement it was proposed to consider their mutual position in the form of squares, in the corners of equilateral triangles (triangular arrangement) (Bakhmutov, 1980;Machnev, 2012), the location in the vertices and centers of hexagons (hexagonal arrangement), and it was proposed to recreate the real obtained shape of the nutrition area in the form of an irregular polygon. As a criterion for assessing the effectiveness of the proposed scientific views and technical solutions, a comparative assessment level of agrophytocenosis yield formed taking into account the new view, but while no systematic observations of the reaction of plants were carried out.

To establish the optimal plant stand density, seeding rate and other seeding parameters, many studies have been carried out in a variety of soil and climatic conditions (Banaj, 2017; Poncet, 2018; Horbe, 2016), but General guidelines and recommendations on seeding rate, the size of the feeding area are still indicative. Often recommendations for the purpose of seeding rates for different regions are contradictory (Bakhmutov, 1980). The main reasons for this situation, in our opinion, can be called the lack of a unified methodology and criteria for assessing the placement of plants in the area of sowing, regardless crop type. Attempts of theoretical understanding of the issues of uniformity of plant placement on the field area, as one of the most important methodological aspects, led to the creation of a number of techniques, which were based on attempts to assess the relative position of plants using statistical methods. At the time, these methods have had a significant impact on the development of agricultural machinery, technological methods, and in some cases, and the emergence of new technologies. Let us briefly consider some of the most typical of them.

In the 50-ies of the last century it was proposed to determine the uniformity of plant placement at cross and rowlesssowing by applying a frame of 500×500 mm with a grid of squares, the side of which is 50 mm. Has been counted the numbers of empty squares with one plant, then with two plants, then with three, etc.After counting the number of squares with different numbers of plants, was determined the calculated parameters as indicators of the statistical series or simply stated the fact of the number of cells containing a certain number of plants. This method does not allow comparing the results of experiments with changes in the number of plants, for example, when changing the seeding rate or when changing distance between rows. The size of the sides of the square of 5 cm is obtained as the square root of the quotient from the division of the number of square centimeters in one hectare by the seeding rate equal 4 million seeds ($\sqrt{\frac{10^4 m^2 / ha}{4 \times 10^6 \, pieces / ha}}$ = 25sm^2/pieces).

An example is the methodic in which the number of cells of 5×5 cm size containing 1, 2, 3 was determined plants, etc. Then determined their percentage of the total area and the result was summarized. Thus, a share of the field area rationally used by plants was obtained (Semenov, 1959).The overall uniformity of plant distribution was calculated as an indicator:

$$C = \left(Z_1 + \frac{Z_2}{2} + \frac{Z_3}{3} + \ldots + \frac{Z_n}{n} \right) \times \frac{100\%}{\sum_{i=0}^{i=n} Z_i} \tag{1}$$

where Z1, Z2, Z3, …, Zn are the numbers of squares with 1, 2, 3, …n plants.

Note that the seeding rate is variable value, and the distribution was calculated for the seeding rate of 4 million seeds and for others seeding rates the application of the method is limited. To assess other types of plant placement method is not applicable at all.

Seeding rates of cereals may differ in 2-3 times, respectively, the size of the cell must also change. The method is unsuitable for the assessment of placement uniformity on the accounting area of the cultures sown with aisles more than 25 cm. It is not clear how to conduct sampling in the survey of crops sown with wide ribbons. Moreover, no single estimated figure and the limits of permissible deviations, although the implementation of the planting will vary with several factors. Apparently for these reasons, the method of evaluation is not widespread, and later regulations grades were canceled.

The Department of "Agricultural machines" of Orenburg agricultural Institute proposed a method for determining the uniformity of the distribution of seeds, in which the criterion of uniformity took the average nearest distance between plants, determined by measurements of the shortest distances between neighboring plants (Kovrikov, 1995).

$$y = \frac{\sum_{i=1}^{n} y_i}{n-1} \qquad (2)$$

where

y_i is the i-th value of the shortest distance between two plants;
n is the number of plants in the area.

This method was used in determining the rational modes of operation of the coulter rowless seeding. In the future, this technique has been developed in terms of application of the method to row crops. In order to carry out appropriate assessments coefficient was adopted, which is the ratio of the average actual distance between plants to the optimum, with their ideal placement. With visible progress, the method leaves a number of questions. So, the highest estimate of uniformity can be obtained only with the "staggered" arrangement of the sown seeds, since the ideal placement of plants, the authors provided for the placement of seeds at the corners of the squares. Note that in this representation, the concept of ideal seed placement should correspond to the hexagonal placement of plants (at the corners of equilateral triangles that make up the hexagon). Also, with this method, it is impossible to take into account a number of biological features of cultivated crops, such as plant architectonics or artificial increase in the density of crops. It is not possible to take into account the shape of the nutrition area and its distortion, and, therefore, objectively set the parameters of the sowing machine in terms of seed placement uniformity will be difficult. For these reasons, the authors themselves admit that an increase in the uniformity coefficient does not always lead to an increase in yield. These facts indicate the limitations of the method. The method also did not get its further distribution and was not accepted by specialists of agrobiological directions.

There are a number of techniques in this direction, which are some intermediate options, or corresponding to individual elements of the methods considered. The uniformity of plant distribution in the evaluation of the work of corn seeders is estimated by varying the intervals between adjacent seeds in a row. Statistical methods provide only general information about the quality of sowing without taking into account the most important indicators of the development of agrophytocenosis and the parameters of its implementation.

Also known is a method for assessing the placement of plants using the coefficient of uniformity of plant placement by area, which is the ratio of the actual average distance between plants to the average distance between plants at their ideal location, i.e. (Grishchenko, 1963).

$$\eta = b_f / b_{opt} \tag{3}$$

where b_f and b_{opt} is the average distance between plants, respectively, under ideal and actual placement of plants.

The merits of this method include the introduction of the concept of ideal placement, which is already dependent on the actual seeding rate. But the method can also be considered special for cases seeding with throwing.

At the Department of "Tractors and cars" Azov-Black Sea Institute was developed a method of determining the conditional individual area of plant nutrition on the results of direct sample measurements of the coordinates of the plants. To determine the individual area of plant nutrition, a number of tolerances were made, as a result of which the boundary of such area of nutrition for all plants has been established. The boundary between adjacent plants was established as normal to the middle of the segment connecting adjacent plants. Its equation –

$$y_{ij} = a_{ij} + b_{ij}x, \tag{4}$$

where a_{ij}, b_{ij} are coefficients:

$$b_{ij} = (x_i - x_j)/(y_j - y_i);$$

$$a_{ij} = y_{ijcp} - b_{ij}x_{ijcp};$$

i, j-sequence numbers of neighboring plants;
x_{ijcp}, y_{ijcp}- average values of the coordinates of neighboring plants:

$$x_{ijcp} = (x_i + x_j)/2; \; y_{ijcp} = (y_i + y_j)/2$$

The authors understand that the number of configurations of feeding areas is unlimited, because they are individual for each plant. Therefore, it is proposed to establish the basic configuration of areas in the form of a set of regular shapes – circle, square, triangle, rectangles with different aspect ratios, pentagon hexagon. As a quantitative assessment of the configuration, they offer the polar moment of inertia of the nutrition area - is the sum of the products of the area of elementary sites per square of their removal to the center, which is the stem of the plant. This method of evaluation will be very time-consuming, and the in formativeness of the method is questionable. It is unlikely that the method can be applied in General practice. The number of base nutrition areas values will also be large, and the values of moments of inertia will have to rank and determine the boundaries of ranks. In addition, the essence of the coefficient determining problem the is reduced to the definition of a certain configuration of the nutrition area and its parametric evaluation, but need to required evaluate the mutual location of plants. In this part, the method is practically useless.

Summarizing the results of attempts to create methods for assessing the uniformity of plant placement, the following can be noted. Engineers are trying to obtain parametric values of the areas of plant nutrition, as these configurations must determine the parameters and modes of sowing machines. The main drawback of such methods can be considered as mixing into a single whole the size and shape of the area and the desire to obtain some universal criterion. This criterion should simultaneously show both the employment of the field area by plants, the size of the nutrition area, and the configuration of the area. This approach provides a mixture in a single whole a series of assessments that in our opinion cannot be rational.

A special place is occupied by methods for determining the quality of seeders of row crops. The task of assessing the relative position of plants in the area in this direction was not carried out. All estimates are reduced to the calculation of the statistical series of the sample values of the intervals between plants along the length of the seeded row. The width between rows was taken to be the standard defined in the framework of the standard zonal technology.

Thus, the absence of criteria for describing the mutual location of plants is a factor limiting the further development of the designs of sowing machines, since there is no connection between the parameters and operating modes of seeders with the need for plants in the living space.

In a more detailed consideration of methodical approaches to determining the uniformity of plant placement on the acreage, the following important aspects attract attention:

- In any of the cases of theoretical understanding of the concept of uniformity, as well as the accuracy of sowing, always the beginning of the study is a kind of geometric scheme, when the area of nutrition of one plant is taken as a circle, square or equilateral triangle, the position equality of intervals between plants in a row having been accepts, beyond which the researchers do not exit;
- In the future, during the construction of distribution models, the category of the shape of the nutrition area is often lost or mixing with the concept of its magnitude;
- Most of the methodological approaches relate to special cases of sowing - either to crops of row crops, or to grain crops, etc.;
- Criterias for evaluating the uniformity of the plants placement on the field area is either absent or there are only a certain descriptions (the width of the strip of sieving of the seeds, distance between the outermost plants along the width of the tape, etc.).
- Also, attempts are made to estimate the values of plant placement in the row as statistical - the arithmetic mean of distances between plants, the quadratic mean of distances between neighboring plants, the coefficient of variation, etc.;
- The results of the experiments are evaluated using agrotechnical indicators, rather than developed criteria.

Generalizations allow us to draw the following conclusions:

1. Mechanistic and statistical approaches are not capable of giving a complete solution to the problem of describing the location of plants over the area of the field and do not define the criteria for the influence of technological parameters of sowing machines on biological processes occurring in agrophytocenoses.
2. In this approach, the task of creating sowing machines has received some development, but sufficient clarity about the mechanisms and conditions for the implementation of biological potential is not made. Also not developed the idea of how should be appointed by the seeding rate and what is the scheme seeds location must be implemented.
3. It is necessary to proceed to the analysis of not only the agrotechnical indicators of sowing machines, but also the biological laws of plant development in conditions that are provided to the entire community of cultivated plants.
4. To a greater or lesser extent, all the considered types of seeding machines designs were created in the pursuit of equality of distances between seeds or plants. This statement of the problem was the basis of the methodology for creating not only machine designs, but also regulatory documents, test standards, research methods, creating varieties and hybrids, i.e. the whole complex of issues related to the sowing of seeds and the crop technologies content.

Standard of Ideal Distribution and the Actual Placement of the Seeds

The necessary direction of improving the scientific and methodological support of specialties that determine the technological development of the industry can be determined by the development of new concepts that reflect the relationship between the placement of plants on the field area by sowing machines and the biological requirements of plants to the vital space. The working process of seed distribution over the field area in the longitudinal and transverse directions can be represented as the interaction of two systems with distributed parameters (Lavrukhin, 2003).

$$N = f(Q)/U(q, b) \quad (5)$$

where

Q is the seeding rate;
q and b intervals between seeds in longitudinal and transverse directions.

Based on this view, it is necessary to obtain an assessment of the mutual location of the set of seeds in the soil, forming some aggregating of seeds. The assessment should be based on the needs of plants in the vital space.

In this task, the size of the nutrition area per plant is taken as the basic unit:

$$|\Delta_i| = S/Q \quad (6)$$

where:

Δ_i – the size of the specific or elemental area of nutrition;
S – is the unit of area for which the seeding rate is set;
Q – actual seeding rate for this field, pcs.

We will assume that this area unit is the optimal vital space for the unit of agrophytocenosis – plants. This assumption follows from the main task of the seeding operation – uniform placement of plants on the field area. The resulting area unit is a characteristic that depends on the seeding rate or technical specifications for sowing. In this case, the shape of the nutrition area can be taken in the configuration of the square, then it is possible to determine the placement step of the seeding rate:

$$\delta = \sqrt{\Delta i} \tag{7}$$

Step making – the value distance between neighboring plants, or the magnitude of the sides of the elementary square of the area unit in the configuration of a square. In the General case, the step of introduction should be divided into the distance in the longitudinal and transverse directions, δ_x and δ_y (as arguments of the function $U(q, b)$).

At this, it is assumed that only the position of germinating seeds will be considered. This assumption allows to combine the concept of "seed" and the concept of "plant", recognizing them as equivalent. Seeding of a non-germinating seed at single-grain seeding will be considered as an error, and further excluded from the General data set and taken into account separately. At the same time, we believe that the recommended seeding rates are optimized from agronomically appropriate values.

The sown area of the field can be represented as a certain area divided into elementary of nutrition areas. In accordance with the given seeding rate, thousands pieces/ha, each elementary nutrition area site will have only one plant. The area of the field thus divided into elementary nutrition areas containing only one plant will represent the distribution standard of the seeding rate. The standard reflects the ideal distribution of the seeding rate.

Unlike all previous methodical approaches, the value of the specific nutrition area and seeding step are not constant, but depend on the seeding rate. These values are determined for each case of placement of the seeding rate on the field area and thus take into account the biological needs of plants in the nutrition area, set for reasons of agronomic feasibility.

The formulation of the concept of ideal plant placement will have the form – the ideal placement of plants is such their mutual location on the field area, when one specific area of nutrition accounts for only one plant, located in the center of this elementary site. This formulation of the ideal placement allows you to display the placement of plants in the figure, even without observing its scale and makes it possible to perform a table recording of the placement of plants on a fragment of the field area.

A tabular record of the plants number on each field site, made from the calculation when one plant will be placed to each nutrition area will be a table E(m×n), each element of which is equal to one, showing the case of ideal plant placement. The size of such a table E (m×n) can be arbitrary and will depend on the specific conditions determined by the objectives of the study. Such a table of recording the number of plants E (m×n)will have the form shown in figure 1.

Figure 1. Example of record of the table of ideal placement of plants

1	1	1	1	1	1	1	1
1	1	1	1	1	1	1	1
1	1	1	1	1	1	1	1
1	1	1	1	1	1	1	1
1	1	1	1	1	1	1	1
1	1	1	1	1	1	1	1

The standard must be created for every execution of sowing, to meet the uniform requirements for comparing different constructions of seeders through the implementation of the variety of options for placement of plants and comparing with a single type of ideal distribution because each planting will have its own individual characteristics depending on a large number of specific conditions for its implementation.

The actual planting will be characterized by some mutual arrangement of seeds in the field area – the scheme of sowing to establish the differences between the actual seeding from the standard, it is necessary to recreate the mapping of plant placement at this seeding rate in the form of a grid of reference placement superimposed on the actual seeding. The area of the field is to be divided into elementary areas, the dimensions of which correspond to the calculated values of the specific area Δ and the seeding stepδ. To compare the standard with the actual sowing, record the number of plants in each cell of the grid, observing the order of mutual arrangement of specific areas as in table. Cells of the record table of plants number should be coordinated, i.e. have the certain place in columns and lines. Each elementary site will contain a number of plants. As a result of the table record of the number of plants in each cell, will be obtain a table of size $m\times n$, each cell of which will contain a number corresponding to the number of plants in the cell of the coordinate grid. Such a table showing the real distribution of plants in the field area on the grid cells A (m×n), will be a "digitized" or parametric display of the relative position of plants on the field fragment of the same size as the standard. Since with the help of such a table the relationship between the actual placement of plants and its abstract display is carried out, such a table can be called a commutative table. Figure 2 shows an example of a tabular record of the plants number (commutative table), which can be obtained in real crops.

This operation will allow presenting the compared values in the same dimension and scale. The size of the elementary nutrition areas - the size which depends on the appointed seeding rate, and, in each case, will have the specific value. This measure is necessary to compare the placement of plants or seeds by different machines, different seeding rates and other features relating to the organization of the seeding operation, depending on the requirements of crops to the size and shape of the nutrition area. In conclusion, will show all the assumptions and consequences of these assumptions.

Figure 2. Example a tabular record of the plants number on the area units (commutative table)

	1		2	1	3	1	
	1		1	1	2	1	2
	1	1	1	2	2	1	3
					2		
1		1			3		1
1	1	1	1	2	1	1	2

1. Only the position of germinating seeds is considered. Under this assumption, the position of the seed and the position of the plant are equivalent. Sowing of non-germinating seeds in single-grain sowing should be considered as an error.
2. We will assume that the size of the specific (elementary) area of food is the optimal size of the vital space for the unit of agrophytocenosis – one plant. The optimal value of the vital space of the plant is obtained from the considerations of agronomic expediency in the appointment of the seeding rate.
3. The shape of the elementary nutrition area is taken in the square configuration. The assumption should be checked for inaccuracies in the application of other principle schemes of placing plants on the field area.
4. The sown area is represented as a certain area divided into elementary nutrition areas.
5. The part of the field area divided into elementary nutrition areas when each elementary area has only one plant, will consider as a standard with ideal placement of plants.
   The standard is created for each specific seeding rate, serves to compare with the actual placement of plants.
6. A sequential table record of the number of plants contained in the elementary sites of any part of the field and made in accordance with the cell number of the table, is a parametric display of the location of plants on this field fragment.
7. The commutative table will be called the record of the number of plants in the elementary areas of the field fragment performed in described way.
8. At this stage, we strive to solve the problem of representation of the mutual arrangement of plants, which is the sum of the multidimensional process of realization of many functions. The resulting record of the number of plants reflects the sum of interactions, but not some functional relationship between the final position of the seed and the work of dosing, transporting and sealing systems. To establish such functional relationships, it is necessary to conduct independent studies of the processes of seed placement over the field area with a set of large numerical statistics, which is an independent goal that is not within the scope of the tasks and goals of these studies.

Evaluations of Host Plants on the Area of the Field

Statement of the problem it is necessary to evaluate the placement of plants as a result of the seeding operation implemented by the sowing machine and expressed by the commutative table A(m×n), compared it with the table record of the ideal distribution (standard) E(m×n). At the same time, the design, purpose of the sowing machine and the seeding rate should not affect the evaluation results. To do this, it is necessary to establish criteria for differences and characteristics of the received crops placements. Since in each case the deviation of the actual placement of plants from their ideal placement, expressed by the standard created by the same rules, will be estimated, it is possible to assess the differences between one placement and another. This assessment will allow to compare the work of all types of sowing machines, regardless of the purpose of sowing, seeding rates, methods of operation, and other specific conditions.

First of all, we introduce the notation of the values obtained in the calculation of the numerical characteristics of the commutative table. The size of the table record the plants number A(m×n) and the table recording the number of plants in ideal sowing E(m×n), corresponds to the number of nutrition areas at the site.

$$S_2 = m \times n \quad (8)$$

or

$$S_2 = \sum_{i=1}^{m} \sum_{j=1}^{n} e(i, j) \quad (9)$$

As a result of the sowing machine and under the influence of a number of natural factors in the selected area will be a number of plants. The number of plants in the surveyed area S_1 will be equal to the sum of the numbers of all plants in the cells of the commutative table.

The sum of S_1 is determined from the expression:

$$S1 = \sum_{i=1}^{m} \sum_{j=1}^{n} a(i, j) \quad (10)$$

It is obvious that for sowing not corresponding to the ideal, the some part of the nutrition areas will not contain plants, then the number of specific areas of nutrition containing at least one plant is denoted by M.

Seeding characteristics will be obtained as a certain ratio of the number of plants in the cells of the table. Compliance with the specified seeding rate and the actual in this specific area of the field can be obtained as the ratio of the sums S_1 elements of commutative table A(m×n) among the specific nutrition areas of sowing S_2.

$$K_b = S_1/S_2 \quad (11)$$

The resulting ratio of the amounts gives a coefficient showing the possible completeness of the filling area according to the selected seeding rate (the coefficient of possible completion). For the standard, this coefficient will be equal to 1, and any deviation from this value is a qualitative difference between the performed sowing and the ideal placement of plants.

Herewith:

$K_b > 1 \pm N_o$ - means exceeding the actual seeding rate with a permissible deviation from the specified seeding parameters N_o;

$K_b < 1 \pm N_o$ means an understated actual seeding rate with an acceptable deviation from the specified seeding parameters N_o;

The permissible deviation of the N seeding parameters takes into account the value of deviations that affect the size and configuration of the nutrition area: the deviation of the actual seeding rate from the set, the uneven seeding of seeds along the seed pipelines, the unevenness of the total seeding. For indicators of seeding quality, the deviation value is the ability to adjust the seeding machine to a given seeding rate and the ability to steadily perform the function of dosing, both in time and when comparing the work of dosing devices along the width of the machine. In a broader sense, this criterion can serve as a measure of variability in the number of plants in certain areas of the field. Indeed, the values of the coefficients will be different in different parts of the field. Then a number of values of the coefficients of K_b, obtained by randomized scheme of their calculation, will show the magnitude of the values of the number of plants on the field as heterogeneity of the density of standing. Such a task can also be set, for example, when assessing the stability of sowing function the influence on the number of shoots of soil cultivation, or other factors.

The effectiveness of the sowing machine in terms of the quality of the operation of the distribution of the seeding rate over the occupied area can be defined as the ratio of the number of occupied specific nutrition areas is M to their total number in the allocated area. We show this ratio by the actual fill factor:

$$K_f = M/S_2 \tag{12}$$

This criterion for the ideal placement case is also 1. For crops carried out with the placement of plants in rows, the coefficient shows the proportion of the number of elementary sites containing plants from the total number of elementary areas of nutrition determined by the seeding rate. For the parameters of the sowing machine, the deviation of this coefficient from the unit means a deviation from the most appropriate value of the row spacing, which can be considered as a criterion for the quality of the placement of the seeding rate in the transverse direction. In General, the closer the value of the criterion to one, the higher the quality of the distribution of plants in the transverse direction.

Also, here can be considered the number of specific plots with a given number of seeds or the number of specific plots containing a given number of plants is x.

To do this, the number of MX cells of the commutative table is calculated $A(m \times n)$, the elements of which are equal to either a given value x, or on variants of filling of specific areas containing $x_1 = b$ plants, $x_2 = c$ plants, $x_3 = d$ plants, etc. or the total number of cells containing these plant groups:

$$Mx = \sum_{i=1}^{m} \sum_{j=1}^{n} x(i,j) \tag{13}$$

or for different *x* values

$$\begin{cases} Mb = \sum_{i=1}^{m} \sum_{j=1}^{n} b(i,j) & PRIx_1 = b \\ Mc = \sum_{i=1}^{m} \sum_{j=1}^{n} c(i,j) & PRIx_2 = c \\ \cdots\cdots\cdots & \cdots\cdots\cdots \\ Md = \sum_{i=1}^{m} \sum_{j=1}^{n} d(i,j) & PRIx_3 = d \end{cases} \tag{14}$$

Taking into account options of filling of specific areas we have:

$$K_f = \begin{cases} K_{fb} = Mb / S_{2b} \\ Kfc = Mc / S_{2c} \\ \cdots\cdots\cdots \\ Kfd = Md / S_{2d} \end{cases} \tag{15}$$

The characteristic allowing comparing the conditions created to plants at their placement on the allocated site is the ratio of the number of plants S_1 to the number of the occupied elementary areas *M* on a field fragment. Modern seeding machines, performing seeding in rows, to change the seeding rate, by changing the number of seeds in a row, increasing or decreasing the distance between the seeds. Then part of the elementary nutrition areas will contain all the seeds sown, and the other part will not be loaded. This characteristic shows the load of the occupied areas of food and shows the average number of plants per one loaded specific area of food.

$$H_{yd} = S_1 / M \tag{16}$$

The characteristic showing completeness of use of the allocated fragment of the area of the field can be received from the following judgments. The inverse value of the H_{yd} gives the average actual value of the area of nutrition of one plant is $\overline{\Delta_f}$ (the actual average value of the nutrition area).

$$\frac{\Delta i}{H_{yd}} = \frac{\Delta i}{\frac{S_1}{M}} = \frac{M \Delta i}{S_1} = \overline{\Delta_f} \tag{17}$$

Loading the elementary areas sowing machine, in fact, reduces the value of the actual nutrition areas on the one hand, and on the other hand part of the field area for plants is not be available. The product of the value of the average actual area of nutrition of one plant and the factor of actual filling of the surveyed area of K_f will show the share of rationally used area on the surveyed area. This will take into account the accuracy of the distribution plants scheme created by the sowing machine on the surveyed area according to the formulation of the basic requirements for sowing, sowing quality of seeds, dosing quality and other factors affecting the final result of the work. The semantic load of the obtained criterion is the coefficient of use of the area of the K_n field is the proportion of rationally used area at a given seeding rate.

Generally

$$\frac{1}{H_{yd}} \times K_f = K_n \tag{18}$$

or taking into account a number of options shown

$$K_n = \begin{cases} K_{nb} = \dfrac{1}{H_{yd}} \times K_{fb} \\ K_{nc} = \dfrac{1}{H_{yg}} \times K_{fc} \\ \cdots\cdots\cdots\cdots \\ K_{nd} = \dfrac{1}{H_{yd}} \times K_{fd} \end{cases} \tag{19}$$

Another record of this indicator can also be made, which is more convenient for practical calculations:

$$K_n = \frac{1}{\frac{S_1}{M}} \times \frac{M}{S_2} = \frac{M^2}{S_1 \times S_2} \tag{20}$$

or for the filling of specific areas

$$K_n = \begin{cases} K_{nb} = \dfrac{M_b^2}{S_1 \times S_{2b}} \\ K_{nc} = \dfrac{M_c^2}{S_1 \times S_{2c}} \\ \cdots\cdots\cdots\cdots \\ K_{nd} = \dfrac{M_d^2}{S_1 \times S_d} \end{cases} \tag{21}$$

We will assume that outside the limits of the average actual size of the area of nutrition, plants enter into competition with other cultivated plants, i.e. the remaining specific areas where plants compete with each other or cannot use them to obtain mineral nutrition, are used irrationally. Further, the plants, developing with the help of their adaptation mechanisms, will master the free area and thereby change the ratio K_n. In addition, as a result of intraspecific competition and the impact of other adverse factors, part of the cultivated plants will die (fall out), which will also change the value of the characteristic.

Therefore K_n, the obtained results of germination will differ from the characteristics of other ages of agrophytocenosis, and the dynamics and magnitude of its changes taking into account the permissible deviations of size and shape will show the specific values of the optimal plant placement for these environmental conditions and varieties. Thus, the representation of crops in the form of models numeric fields as an ideal distribution of individual plants or groups with a given number of plants and compare them with the real picture of the placement of plants allows by far in absolute terms to determine the resulting unambiguously on absolute indicators to define the received deviations of the areas of food of plants and to draw conclusions about completeness of rational use of the area of the field.

The Quality of Plants Distribution on Field Area

The obtained relations characterize the placement of plants on the area, and their semantic load allows formulating the concept of quality of distribution of plants on the area of the field. The quality of plants distribution on field area this approximation of the characteristics and the criteria's of the planting real performance to the standard of the ideal distribution of this seeding rate for a particular crop. There is no doubt that the deviation has its tolerances, when the yield of agrophytocenosis is not damaged by reducing the quality of the operation on the placement of plants. These tolerances should form the basis of the performance requirements of the processes reproduced by the seeders. The main processes of sowing machines that affect the quality are:

- The accuracy of the machine settings on the seeding rate and the preservation of these settings in time for the whole machine and individual dosing devices (stability of the total seeding and stability of individual devices);
- Uniformity of the indicator of placement of seeds or plants in the longitudinal direction, characterized by a deviation from a given seeding between individual machines or coulters;
- Method of accommodating seeds of the working bodies of seeders ranks, belts, no ranks.

Evaluation of the Inaccuracy of the Quality Method Plant Placement

From the geometric point of view, the hexagonal arrangement of plants will be perfectly uniform, i.e. at the corners of equilateral triangles arranged so that they form regular hexahedral figures over the entire area of the field. We having been adopted the placement of plants in the centers of squares, as a display of the distribution functions of the seeding rate in two mutually perpendicular directions. But then there is a need to find the numerical differences (errors) between the schemes. Finding such differences is all the more necessary that the practical application method of the quality plant placement should provide for the creation of a standard in the form of systemically arranged grid cells, reflecting the size of specific areas of nutrition for this seeding rate.

Comparison of the plant placement scheme is feasible by applying one scheme to another so that the position of the plants from which the distances in the longitudinal and transverse directions are started to coincide.

The number of nutrition areas in the form of a squares S_{kb} and in the form of an equilateral triangle S_Δ is the same, and, consequently, their areas are also equal:

$$S_{kb} = S_\Delta \tag{22}$$

The area of a square S_{kb} with sides b is

$$b^2 = S_{kb} \tag{23}$$

the area of an equilateral triangle with side a is

$$S_{\Delta} = \frac{a^2 \sin 60^\circ}{2} \tag{24}$$

The ratio of the sides of a square and a triangle $\frac{b}{a}$ in the transverse direction, taking into account the equality of areas, is found as an infinite aperiodic fraction

$$\frac{b}{a} = \frac{\sqrt{S_{kb}}}{\sqrt{S_{\Delta}}} = \frac{\sqrt{b^2}}{\sqrt{\frac{2S_{\Delta}}{\sin 60^\circ}}} = \frac{\sqrt{b^2}}{\sqrt{\frac{2b^2}{\sin 60^\circ}}} = \frac{1}{\sqrt{2}}\sqrt{\frac{1}{\sin 60^\circ}} = 0.6095 \tag{25}$$

The ratio of the triangle height to the square height (in the direction of the OY axis) will also be an infinite aperiodic fraction:

$$\frac{h}{b} = \frac{a}{b}\sin 60^\circ \tag{26}$$

This suggests that the coincidence of the boundaries of the feeding areas in mutually perpendicular directions will not be at any values of the sides of the geometric shapes under consideration. In addition, since the side of the square will be less than the side of the triangle of equal area, the end of the segment forming the square will always lie within the segment forming the side of the triangle. Thus, within any area in the square configuration will lie at least one seed, common with the configuration of the area in the form of a triangle. The position is valid for longitudinal and transverse direction.

From other side, since there is a difference in the lengths of the sides square and triangle, then may be found a number of sides of the square and the sides of the triangle, at which the beginning and the end of the side of the square will lie within one side of the triangle (for the transverse direction), and one height of the triangle in the longitudinal direction.

The side of the square is smaller than the side of the triangle. Taking the side of the triangle as a unit, we get the difference of the sides in the shares of this unit (it should be noted that according to the condition of obtaining the value of the nutrition area, we will receive the value of such a unit for each seeding rate). The difference will be 1-0.609= 0.391 Then the multiplicity of the difference and the value of the triangle side will be

$$\frac{1}{0.391...} = 2.558.... \tag{27}$$

that is, each third side of the triangle in the transverse direction will contain the beginning and end of the side of the square.

The maximum difference between the schemes in the number of seeds in all specific areas in the transverse direction will be

$$\frac{1}{3}\times 100 = 33.3\ldots\% \tag{28}$$

In the longitudinal direction, the ratio of the side of the square and the height of the triangle is

$$\frac{b}{h} = \frac{b}{a\sin 60^{\circ}} = \frac{0,609}{\sin 60^{\circ}} = 0.703\ldots \tag{29}$$

Then

$$\frac{1}{1-0.703} = 3.37 \tag{30}$$

that is, on each fourth segment equal to the height of the triangle, there will be the beginning and the end of the side of the square or two seeds.

The maximum difference between the schemes in the number of seeds in all specific areas in the longitudinal direction will be

$$\frac{1}{4}\times 100 = 25\ \% \tag{31}$$

the maximum possible difference in the performance of the schemes will not exceed 0.33 ... ×0.25 = 0.083... or 8.3...%. This error is quite acceptable from an engineering point of view.

This approach can no longer be considered mechanistic, since the formation of the plant layout involves their biological need for a feeding area, expressed through the seeding rate. In fact, the question of the optimal shape of the nutrition area is beyond the scope of consideration. It deals with the biological characteristics of plants, such as plant architectonics, especially the structure of the root system, the reaction to sunlight, the need for such placement of plants, which will affect the required part of the crop.

An important point of the comparison is that the error can be considered acceptable, and this error does not contradict the requirement of comparability of schemes of crops and will be the same for all cases of comparison.

CONCLUSION AND FURTHER DIRECTIONS OF RESEARCH

1. The basic units for building an idea of the quality of plant placement should take the area of nutrition of one plant and the seeding step. Using these pieces of proposed construction actually performed and the standard of perfect placement of plants.
2. Only linear measures (m, cm, and mm) should be the units used to describe the placement of crops and determine the quality indicators of sowing machines work. Since only in this case it is possible to establish functional relationships between the quality of the sowing machine and the yield of agrophytocenosis, can be determined permissible errors in the location of seeds or plants, and requirements for the development of processes performed by sowing machines.
3. As the values obtained in the calculation of the numerical characteristics of the commutative table, for further calculations are taken:
    a. Number of plants in the survey area S_1;
    b. The number of specific nutrition areas on the observed part field $S_2 = m \times n$;
    c. Number of occupied nutrition grounds M.

As the main criteria for the description of uniformity of placement of plants at the mechanized sowing of seeds can be offered:

a. Coefficient of possible occupancy is K_b. The coefficient shows the compliance of the given seeding rate with the actually obtained number of plants, i.e. the deviations of the actual seeding and field germination of plants are taken into account (it is important to recall that the seeding rate is set as the total number of shoots per 1 ha);
b. Actual occupancy rate – K_f. Shows the proportion of specific areas of nutrition containing at least one plant to their total number in the surveyed area;
c. Specific load N_{yd}. It shows the average number of plants in the area occupied by plants and is an average characteristic of the local conditions of the unit of agrophytocenosis;
d. Use coefficient of the field are K_n. It is a characteristic of rational use of the entire area of the surveyed area by plants;
e. The average actual value of the area of nutrition of one plant is $\overline{\Delta_f}$ (the actual average value of the area of nutrition).

4. The proposed characteristics and criteria make it possible to correlate the technical parameters of the machine and the biological requirements of plants from a single point of view of the main tasks of the sowing operation is uniform distribution of plants over the area and providing plants with favorable conditions for the formation of the maximum yield. Propose a formulation of the quality of plant placement, as a correspondence to their reference placement.
5. Comparison of schemes of hexagonal placement of plants with the proposed placement showed that the maximum deviation is not more than 8.3% and is an acceptable value.

It is possible that further studies will show the need to use the grid patterns in the form of other configurations, for example, for plants with sharp differences in architectonics, but at this stage of research this representation is very consistent for most cultures. For these reasons, the shape of the nutrition area in the form of a square will acceptable for this method of describing the mutual arrangement of plants.

REFERENCES

Bakhmutov, V. A. (1980). Razmeshcheniye semyan po ploshchadi pri ryadovykh posevakh [seeds placement on the area at ordinary crops]. Mechanization and electrification of agriculture, 5, 24-27. (in Russian)

Banaj, A., Šumanovac, L., Hefer, G., Tadić, V., Banaj, Đ., & Kovačev, I. (2017). Yield of corn grain by sowing in twin rows with MaterMacc-2 planter. *Proceedings of the 45th International Symposium on Agricultural Engineering, Actual Tasks on Agricultural Engineering*, Opatija, Croatia, February 21-24 (pp. 141-152).

Bespamyatnova, N.M. (2002). *Nauchno-metodicheskie osnovi adaptatsii pochvoobrabativayushchih mashin [Scientific and methodological fundamentals of adaptation of soil cultivating and sowing machines]*. Rostov-on-Don, Russia: LLC "Terra".

Bespamyatnova N.M. (2010). *Vibratsii v tekhnologicheskikh protsessakh [Vibrations in technological processes]*. Rostov-on-Don, Russia: LLC "Terra".

De Vita, P., Colecchia, S. A., Pecorella, I., & Saia, S. (2017). Reduced inter-row distance improves yield and competition against weeds in a semi-dwarf durum wheat variety. *European Journal of Agronomy*, *85*, 69–77. doi:10.1016/j.eja.2017.02.003

Ewbank, N., & Buckby, L. (2002). Lynx engineering enters seed drill market with the seedline system. *Landwards*, *82*(1), 153–155.

Grishchenko, F. V. (1963). Osnovy teorii dvizheniya semyan i raspredeleniye ikh v pochve pri bezryadkovom poseve [Fundamentals of the theory of seeds movement and their distribution in the soil at sowing without rows]. *Ryazanskiy SKhI*, *10*, 31–33.

Hörbe, T. A. N., Amado, T. J. C., Reimche, G. B., Schwalbert, R. A., Santi, A. L., & Nienow, C. (2016). Optimization of Within-Row Plant Spacing Increases Nutritional Status and Corn Yield: A Comparative Study. *Agronomy Journal*, *108*(5), 1962–1971. doi:10.2134/agronj2016.03.0156

Kovrikov, I. T., Nogtikov, A. A., & Sazonov, S. N. (1995). Povysheniye Effektivnosti Ispolzovaniya Kombinirovannykh Pochvoobrabatyvayushche-posevnykh mashin [Increasing the efficiency of combined tillage and sowing machines using]. Sovershenstvovaniye tekhnologiy i tekhn. sredstv dlya mekhanizatsii s.-kh. protsessov, 82, 80-81.

Krasnoshchekov, N.V. (2002). O proyektirovanii tekhnologiy proizvodstva selsko-khozyaystvennoy produktsii [About design of production of agricultural products technologies*]. Prioritety mekhanizatsii rasteniyevodstva i zhivotnovodstva*, *138*, 11–20.

Lavrukhin, P.V. (2003). *Issledovaniye sovremennykh sposobov i skhem razmeshcheniya semyan po ploshchadi polya. [Study of modern methods and schemes of seed placement on the area of the field].* Otchet o NIR. FGOU VPO AChGAA, Zernograd, Russia.

Lyubushko, N.I., Zvolinskiy, V.N., (2001). Zernovyye seyalki na rubezhe XXI veka [Grain drills at the turn of the XXI century]. *Traktory i selskokhozyaystvennyye mashiny, 2*, 4-7.

Machnev, A.B. (2012). Soshnik s napravitelem-raspredelitelem semyan dlya poseva zernovykh kultur [Coulter with guide-distributor of seeds for sowing of crops] *Traktory i selkhozmashiny, 7*, 42-43.

O'Leary, G., Aggarwal, P., Calderini, D., Connor, D., Craufurd, P., Eigenbrode, S., ... Hatfield, J. (2018). Challenges and Responses to Ongoing and Projected Climate Change for Dryland Cereal Production Systems throughout the World. *Agronomy (Basel), 8*(4), 34. doi:10.3390/agronomy8040034

Poncet, A. M., Fulton, J. P., McDonald, T. P., Knappenberger, T., Shaw, J. N., & Bridges, R. W. (2018). Effect of Heterogeneous Field Conditions on Corn Seeding Depth Accuracy and Uniformity. *Applied Engineering in Agriculture, 34*(5), 819–830. doi:10.13031/aea.12238

Semenov, A. N. (1959). *Zernovyye seyalki* [*Grain drills*]. Moscow, Kiev: Mashgiz. Russia.

KEY TERMS AND DEFINITIONS

Estimates Of Plant Placement By Field Area: Are a number of indicators of differences between the reference location of plants from their actual placement, performed by the sowing machine.

Living Space: Is a part of the field area determined by agronomic expediency (by seeding rate), part of the air space, photosynthetically active radiation, moisture, elements of mineral nutrition, weed vegetation, diseases and pests, to on one cultivated plant.

Step Of Application: Is the value of the distance between adjacent plants specified by the seeding rate, or the value of the side of the elementary nutrition area in the square configuration: $\delta = \sqrt{(\Delta i)}$.

The Ideal Placement Of Plants: Is their mutual location on the area of the field, when one elementary area of nutrition accounts for only one plant, located in the center of this area.

The Interval Between Plants: Is the distance between one of the plants and the nearest plants surrounding it.

The Plant Placement Quality By Area: Is the value of the approximation of the characteristics and criteria for the placement of the actual performance of sowing, to the standard of the ideal distribution of this seeding rate for a particular culture, variety, hybrid or batch of seeds.

The Specific (Elementary) Area Of Food: Is the size of the area of the field falling on one cultivated plant and received as private from division: $|\Delta_i| = S/Q$, where: $\mathbf{\Delta}i$ – size of elementary or basic nutrition area; S - is the unit of area for which the seeding rate is set; Q - actual seeding rate for this area, pcs.

The Standard Distribution Of The Seeding Rate: Is the area of the field divided into elementary nutrition sites containing only one plant. The standard is created for each case of sowing.

Chapter 11
How to Manage the Effectiveness of Presowing Treatment of Seeds

Aleksey Vasilyev
https://orcid.org/0000-0002-7988-2338
Federal Agricultural Research Center VIM, Russia

Alexey Alexeyevich Vasilyev
Federal Agricultural Research Center VIM, Russia

Alim Dzhanibekov
Federal Agricultural Research Center VIM, Russia

Dmitry Budnikov
https://orcid.org/0000-0002-8729-1300
Federal State Budgetary Scientific Institution, Federal Scientific Agroengineering Center VIM, Russia

Gennadiy Samarin
https://orcid.org/0000-0002-4972-8647
Federal Agricultural Research Center VIM, Russia

ABSTRACT

Variants of seed response to external disturbances were studied to determine the principles that should be followed during pre-sowing treatment. The approach that involves analysis of seeds' adaptive response has been applied. The basic organizational rules have been specified for external disturbance application to seeds. The use of informational method for studying the biological response of seeds on external actions made it possible to find the rules that have to be followed when pre-sawing treatment is performed with the use of electrophysical excitation. The essence of these rules is that the following sequence of operations shall be carried out in the process of pre-sowing treatment: to define the lowest energy level of external disturbance corresponding to the seed's sensitivity threshold, to organize cyclic action of external factors on seeds, to vary the amplitude of external factor within cycles, to vary the mode of external disturbance. The field experiments were carried out for several years, during which the pre-sowing treatment of seeds was determined.

DOI: 10.4018/978-1-7998-1216-6.ch011

Copyright © 2020, IGI Global. Copying or distributing in print or electronic forms without written permission of IGI Global is prohibited.

INTRODUCTION

Presowing treatment of seeds belongs to important technological operations designed to improve quality of seeds providing pest control and enhancement of their sowing qualities. At the present stage of development of these technologies, methods of seeds exposure to various external electrophysical factors are in wide use making it possible to improve yields of agricultural crops at the expense of relatively low energy consumption for processing. This adds an incentive for further development of research in this field. Nevertheless, processing lines using these methods are not in sufficiently wide actual practical use, today. There exist a number of reasons for such situation of which the principal one is that seed is a biological object with all its specific reactions on various factors of external excitation inherent to each particular crop. Let us consider certain elements of effects that electrophysical external factors produce on properties of seeds in order to answer the questions that provide a subject of agricultural technologists' concern.

BIOLOGICAL ISSUES RELATED TO SAWING QUALITY OF SEEDS EXPOSED TO VARIOUS EXTERNAL FACTORS

Reviews of seeds presowing treatment methods show that there exists a general response pattern of living organisms, particularly that of seed, on external excitation. This response may appear in both retaining the initial status of seed (biological resistance) and self-adapting to new conditions.

Reactions of plant cells on various external factors, particularly, on temperature change, appear to be similar to those of other organisms (animals and humans) (Rubin and Riznichenko, 2014; Aleksandrov and Kislyuk, 1994; Gamburg, Borovskii, & Voinikov, 2017; Shapovalenko, Dorovskih, Korshunova, Shtarberg, Slastin, & Nevmyvako, 2011). Therefore, further on, we will involve data related not only to plants but also to other biological objects, in the course of our study.

In a number of works (Garkavil, 2015; Sel'eGans 1979) any external factor making its effect on plants is considered as stress that results in adaptation and development of immunity to it. In the course of immunity development in plants, response of specific types is a step-wise process. First, non-specific reactions on stress occur such as change in membrane permeability, hyperchromatism, change of pH value, drop of cell membrane potential leading to the change of cytoplasm structure. Due to the effect of these short-term responses intermediate reactions take place (acute phase proteins synthesis, inhibition of vegetation and photosynthesis). In Table 1 provided below, the sequence of non-specific stress reactions transformation into specific immunity response is presented.

Multi-year experience of studies in the field of life evolution shows that there exists a rather limited range of standard, uniform, reactions on an extensive variety of external excitation factors. Thus, all kinds of external factors (microwave fields, laser radiation, heating and so on) induce nearly similar stress in seeds. All reactions differ depending mainly on the penetration depth of particular external excitation (differently affecting different structural elements of seed), excitation intensity and time of exposure.

Stress is defined as response on a strong disturbance though there is no clear classification of disturbance sources, in their strength of effect. Results of fundamental studies of various kinds of stress applied to living organisms have been reviewed by Selye (Sel'eGans, 1979; Sel'eGans, 1960; Sel'eGans,

Table 1. Sequence of transformation of non-specific reactions on stress (Rubin A.B. 1999)

1. Non-Specific Response		
Change of membrane permeability	Change of cytoplasm structure	Change of pH value, drop of cell membrane potential
2. Intermittent Reactions		
Synthesis of acute phase proteins	Vegetation inhibition	Modification of photosynthesis, etc.
3. Specific Immunity Reactions		
Freezing tolerance (hormones, + sugars)	Salt resistance (- hormones, + polyamines)	Drought resistance (- cytokines, + abscinic acid)

It has to be noted that mechanisms of these transformations still remain unexplained.

1972). According to these studies, 6 hours after external stress stimulus application, the first stage of stress develops, the so called 'alarm reaction' that appears in activity inhibition (duration of this phase is 24 h to 48 h).

Alarm reaction is followed by resistibility stage, i. e. that of immunity formation. Selye found out that this stage is characterized by enhanced organism's immunity to negative external factors.

In the stage of resistibility, strong excitation gives way to protective inhibition, i. e. 'last-resort protection action', after I. P. Pavlov. Sensitivity to external stress excitation decreases, in conditions of protective inhibition.

Garkavi et all have shown (Garkavi, Kvakina, & Ukolova, 1997; Garkavi et al., 2003; Garkavi, et al., 2007; Garkavi et al., 2008) that, at least, three types of adoption reactions may develop in organism depending on the intensity (or/and time of exposure) of external excitation:

1. Response on weak external excitations, 'training' reaction,
2. Response on medium-intensity excitations (intermediate excitation level between weak and strong disturbances), activation reaction,
3. Response on strong excitations (extraordinary high external excitation), 'Selye stress'.

Organism has, first, to evaluate the level of disturbance and to define weather it is low. In other words, it has to 'measure' the intensity of disturbance and 'just for security reason' to 'mobilize' its own protection system. This is the first stage of organism's response on weak external disturbance, called 'stage of understanding'. Similarly, to the stress alarm reaction, this stage develops 6 hours after exposure to external disturbance, and its duration is 24 h to 48 though it completely differs from the first one, in its character (Garkavi et al., 1997).

Since protective inhibition that has developed on the 'understanding' stage suppresses sensitivity, organism ceases to respond on the acting disturbance if its intensity remains as low as it was when it started. With the help of protective inhibition mechanism, organism protects itself against multiple various weak disturbances repeating without change in their intensity since such disturbances can be ignored later on (Garkavi et al., 1997).

If we apply this assumption to the process of presowing treatment of seed it will explain why low levels of various external excitation factors do not affect its sawing quality. Therefore, it is important to define the 'lowest' disturbance level that will initiate the response of seeds that we exactly expect.

In order to support training reactions in organism we have to raise slightly disturbance level, in a goal-directed manner, so that its intensity grows oscillating attaining its minimum value that initiates generic reaction in organism.

In this case, the second stage of training reaction will develop called ‘rebuilding stage’. This process of enhancement of protection system activity by training does not progress linearly but develops in a cyclic manner (Garkavi et al., 1997).

Diagram of organism’s resistibility process, for training reaction, is shown in Figure 1.

Systematic sequence of weak but gradually growing, in their intensity, disturbances leads, in a long run, to high activity of protection systems and, therefore, organism gains immunity to destructive external factors not only owing to the development of protective inhibition but also due to enhanced activity of protection systems. It worth mentioning that independently on the origin of applied external ‘training’ factor we observe development of immunity to not only this particular external factor but also to disturbance sources of other nature (Garkavi et al., 2008).

Turning to the technology of presowing treatment of seeds by forced aeration, assumption can be made that desired effects that we have after such a long-term processing are achieved because air temperature and relative humidity change during a whole day. It is important that, as a rule, they change in a gradual manner i. e. this process has the character of systematic sequence of weak disturbances.

Studies carried out in (Garkavi et al., 2003) show that response on disturbances of medium intensity, whose amplitude is between their strong and weak values, stimulates activity of organism’s protection system. Such disturbances are not strong enough to launch processes of protection system inhibition in order to decrease organism’s response to them. At the same time, such disturbances shall not be too weak since they are applied for the sake of rather fast organism’s adaptations them and razing the activity of its protection system.

Activation is a multiple-stage process each stage of which is characterized by a certain level of organism’s non-specific resistibility. The first stage, called stage of initial activation, develops 6 hours after external disturbance application, and its duration is 24 h to 48 h. The character of this stage differs from both changes occurring in alarm reaction, during stress, and those that can be observed on the stage of evaluation, in training reaction (Garkavi et al., 2003).

Since, in stage of initial activation, sensitivity increases disturbance rate for the second reaction stage has to be slightly decreased. Nevertheless, this process has a cycle-wise character that is why we have to increase it in a certain proportion, after the rate of disturbance has been decreased, and then it has

Figure 1. Change of organism’s resistibility in training reaction
(Garkavi L.H. et al. 1997)

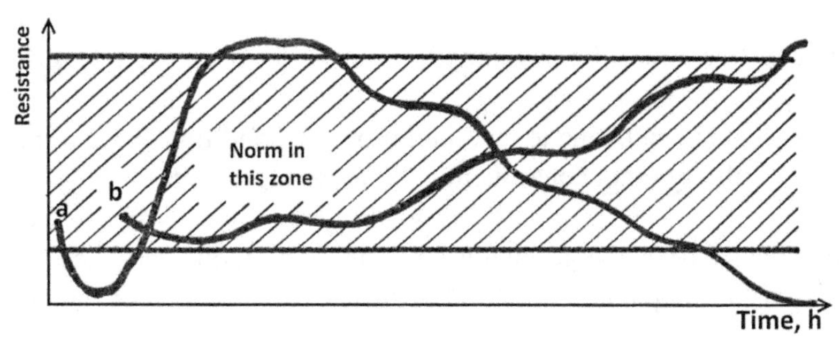

Figure 2. Change of organism's resistibility in activation reaction
(Garkavi et al., 2007)

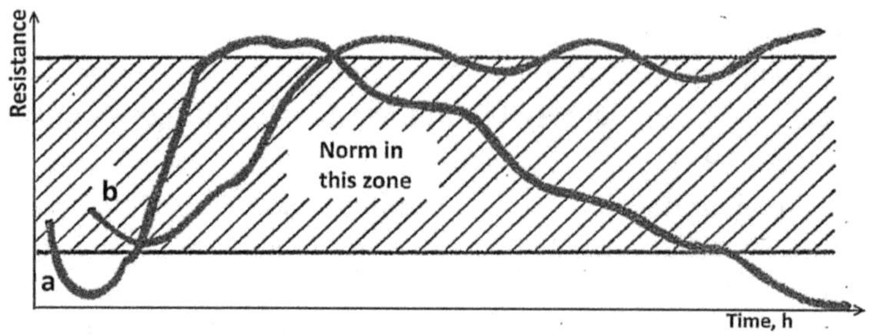

to be decreased again, and so on. The second stage of activation reaction develops in form of sequence of systematic disturbance of medium strength. It is called 'steady activation' stage because it features stability of organism's protection system activation (Garkavi et al., 2007). Diagram of organism's resistibility process for activation reaction is shown in Figure 2.

Based on description provided above, the assumption can be made that the maximum positive and stable effect can be achieved by application of similar processing modes, in the cause of seeds treatment by heated air.

The questions are: to what extent seed's resistibility level is linked to its sawing quality, what kinds and levels of disturbances have to be applied to make seeds enter the stage of 'steady activation'.

With regard to this problem, Garkavi (Garkavi et al., 1997) explains that the intensity of disturbance is a relative quantity which means that reactions develop as a response not on disturbance amplitude but on its gain (in relation to lower reactivity level) (Garkavi et al., 1997).

Periodicity in adoption reaction development provides organisms with an important mechanism making it possible, on the one hand, to adapt flexibly to even small change in surrounding media (starting from 10% to 20%), on the one hand, to extend the range of external actions within which organism retains its viability. In other words, 'periodical sequence of adoption reaction tetrads provides a real mechanism for multi-level homoeostasis regulation system' (Garkavi et al., 1997).

Let us consider status conditions of grain layer taking as example the change of grain temperature in the course of its treatment with the use of forced aeration.

Let us suppose that initial grain temperature is 20 °C. In order to transfer seeds in training mode it has to be heated to temperature $_{c}min = 22$ °C (10%). Value $_{c}max = 24$ °C (10%) is the maximum value, for this stage. For the stage of smooth activation, these values are $_{c}min = 26.4$ °C (10%) and $_{c}max = 28.8$ °C (20%). In high level activation mode, seeds may be exposed to temperatures $_{c}min =$ 31.7 °C (10%), $_{c}max = 34.6$ °C (20%).

Temperature limits of transitions, for these modes, will differ from those specified above, for other initial temperatures.

In view of this, conclusion can be made that the following conditions have to be adhered to for enhancement of seed's resistibility (and, consequently, their sawing quality):

- External disturbances have to be gradually growing, continuous and alternating (cyclic) so that adaptation reactions develop not on the level of structure but on that of functional change.

- In order to initiate activation reaction, external physical disturbance of certain strength is required that has to be then decreased or terminated. After that, the rate of external factor has to be decreased by 20% to 25% in relation to its initial value.
- Initial value of external disturbance has to minimal, for the range of adaptation reaction development, on lower reactivity levels. In this case, the highest reactivity can be attained.
- Combination of non-specific and specific reaction yields the highest effect of external physical factor application.

DATA-BASED APPROACH TO DESCRIPTION OF SEED REACTIONS DURING PRESOWING TREATMENT

In the sixties of the last century, Academician Shmalgauzen proposed to apply the data-based approach, in theoretical biology (Shmalgauzen, 1968). Nevertheless there exist a number of problems that, as a result of external and internal circumstances, has not been solved, up to the present moment. Close comparability between thermodynamic entropy and data on system status provides an instrument to use information as a criterion for system development evaluation, as well as to define trends of processes that occur in biosystems. First of all, let us define basic terms that wieldable to specify the required criterion.

According to C. Shannon (Shennon, 1963), amount of information that contains an event (status, object) *Y* related to event *X* equals to:

$$I\left(X,Y\right)=\log_2\left(p\left(X/Y\right)/p\left(X\right)\right) \tag{1}$$

where is probability of event *X* before event *Y* occurs, is probability of event *X* given that event *Y* has already occurred (conditional probability).

In case that information *Y* about event *X* is absolutely correct,

$$p\left(X/Y\right)=1 \text{ and } I\left(X,Y\right)=\log_2\left(1/p\left(X\right)\right) \tag{2}$$

We will use different terms for thermodynamic entropy *S* and data entropy *H*.

Suppose that information related to grain moisture content is less certain, for instance, either 13% or 14%. In this case, the sought-for amount of information can be found as follows:

$$I\left(X,Y\right)=\log_2\left(p\left(X/Y\right)/p\left(X\right)\right)=\log_2\left(\left(1/2\right)/\left(1/32\right)\right)=\log_2\left(16\right)=4\,\text{(bits)}.$$

Therefore, obtained information is the larger the higher is the data certainty and the lower is system uncertainty. The latter is defined by the presence and volume of conditional data (conditional entropy). For this case, the expression for calculation of Tran's information content has the following form, for these two events:

$$I(X,Y) = H(X) - H(X/Y) \quad (3)$$

where $H(X)$ is unconditional entropy (bit), $H(X/Y)$ is conditional entropy.

Based on provisions specified in (Golicin and Petrov, 2005) by Golitsyn and Petrov, extremum principle has been developed that defines many behavior types of organisms including those of seed, as single biological object, and grain layer, as a whole. We suppose that application of this principle will make it possible to evaluate the effectiveness of various methods of exposure seeds and grain layers to external factors with the purpose of both drying and presowing processing.

The essence of this principle is as follows (Golicin and Petrov, 2005). While describing the behavior of biological objects, three variables shall be taken account of. These are: result *R* of external action on biological object, external action *X* applied to biological object and organism reaction *Y* on external action (processes that develop within object after external factor action), influence of organism (system) on external factor.

In order to achieve optimal results of its reaction, biological object has to ensure maximum trans information content in relation to surrounding medium conditions and reactions:

$$I(X,Y) \to \max_{xy} \quad (4)$$

Indexes *x* and *y* mean that biological object (system) can proceed to this maximum by selecting both its own reactions and by stimulus choice. In respect of seeds and grain layer, the second option of events development is not applicable since seeds are intended to be exposed to external processing.

Principle of maximum information content can be represented in either of the two identical forms (Golicinand Petrov, 2005):

$$I(X,Y) = H(X) - H(X/Y) = H(Y) - H(Y/X) \to \max_{xy} \quad (5)$$

Achievable maximum of information content is a conditional one since, in real cases, there always exist limitations, particularly those of resources: $U(X,Y) \leq const$.

With the help of Langrangian multiplier and limitations imposed on resources, expression (1.4) can be represented in the following form (Golicin and Petrov, 2005):

$$I(X,Y) = H(X) - H(X/Y) - \lambda U(X,Y) \to \max \quad (6)$$

With the account of resources-related limitations, condition of maximum of information content can be substituted by the requirement of maximum effectiveness $L = I(X,Y)$.

As it follows from expression (7), effectiveness *L* changes mainly due to variation of conditional entropy $H(X,Y)$ (conditional probability of event). Other members of equation are subject to change to a narrower extent. That is why such changes can be neglected, in the first approximation.

Negative values of effectiveness apply to negative effects of reaction. For only conditional probability, its dependence can be expressed in the following form (Golicin and Petrov, 2005):

$$-L = \left\{ H\left(X / y_1\right) + \left[\lambda\left(U\left(x_1, y_1\right) - U\left(x_0, y_1\right)\right)\right] p\left(x_1 / y_1\right) \right\} p\left(y_1\right) + A \tag{7}$$

or

$$-L = \left\{ H\left(X / y_1\right) + Kp\left(x_1 / y_1\right) \right\} p\left(y_1\right) + A \tag{8}$$

where

$$H\left(X / y_1\right) = H\left(x_0 / y_1\right) + H\left(x_1 / y_1\right),$$

A is addend containing equation members that do not depend on $p\left(\frac{x_1}{y_1}\right)$, x_0 is stimulus affecting biological object and that reduces conditional entropy $p\left(\frac{x_1}{y_1}\right)$, x_1 is stimulus hat contributes to the growth of conditional entropy as a result of object exposure to external factors.

Organisms are capably to control conditional probability by either reduction or enhancement of flows x_0 and x_1. This is the form in which reactions of biological object become evident.

Change of reaction effectiveness may have graphical representation shown in Figure 3.

Graphical dependence $L = f\left(p\left(x_1 / y_1\right)\right)$ has the form of hill (addend $H\left(X / y_1\right)$) placed on slanting line (addend $Kp\left(x_1 / y_1\right)$.

Figure 3. Dependence of effectiveness of biological object's reaction on conditional entropy

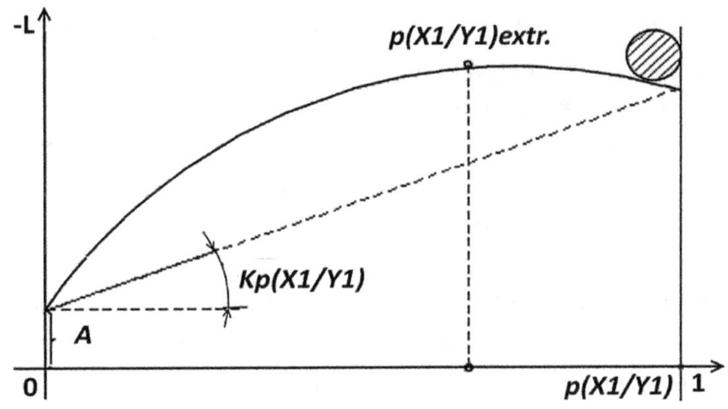

Let us describe graphically the status of biological system and its reaction on external disturbances in form of a ball on the surface of this hill. As it can be seen from Fig. 3, biological system has two points of steady-state conditions, $p\left(x_1 / y_1\right) = 0$ and $p\left(\frac{x_1}{y_1}\right) = 1$. Accordingly, there exist areas (attraction zones, areas of immunity), around each of these points. Biological object does not change its status once it is found in or has entered one of these areas.

Size of immunity area depends on the slope of hill basement (addend 5, in equation (8)). Conditional probability $p\left(x_1 / y_1\right)$ tends to reduce as a result of external disturbance, and biological object starts to 'stand against' occurring changes by means of internal forces. Such resistance proceeds while immunity resources last. Point $-L = f\left(p\left(x_1 / y_1\right)\right)$, on the surface of hill is that of maximum of function $p\left(x_1 / y_1\right)_{\lim}$. Once organism has reached this point it fails to 'resist' and 'slides down' into a new steady state. Since that moment, biological object will react on changes of an external disturbance by 'standing against' leaving this new steady state

Therefore, further change of effects produced by external action becomes impossible without change of type of external disturbance. Contrariwise, changing the type of external action helps to overcome seed's immunity to external disturbance, and it starts to 'resist' again. The sequence of effectiveness alternation has to be repeated. Terminating external action x_0 should, as well, initiate change of seed's reaction. If external conditions differ from external action conditions x_0 seed has to overcome the status that it entered due to the action of factor x_0 at the expense of internal resources (if any). Absence of external action x_0, is detected by biological object as external disturbance x_1.

From the standpoint of principle of maximum trans information, it turns out that processes of grain treatment by application of external factors leads to the extension of trans information. Maximum reaction effectiveness can be achieved for either alternating various external factors or in cycles of type 'disturbance–pause'. Effectiveness of reactions is the higher as lower is the energy level of external action. In case of either long-term external low-energy disturbance or short-term high-energy one, final effectiveness of seed's reaction and, therefore, results of presowing treatment will differ.

The above analysis makes it possible to conclude that principle of maximum trans information combination of adaptation reaction principle, that of minimum of energy consumption, as well as that of maximum and minimum entropy that biological object tends to 'adhere to', in its development and in reactions on external actions. Therefore, this principle can be considered as the basic one while designing methods of presowing treatment, evaluating existing pretreatment technologies and presowing processing of seeds. Application of maximum information principle made it possible to develop approaches (methods) for designing electricity-based technologies of grain presowing treatment (Vasil'ev and Kononenko, 2008; Vasilyev, 2013; Vasil'ev, 2014; Vasil'ev, Dzhanibekov, & Udincova, 2015; Vasiliev, Ospanov, Budnikov, Karmanov, Salginbayev, & Vasilyev, 2016):

- Definition of minimum energy levels of external disturbance, within the seed's sensitivity range,
- Organization of cyclic external disturbance,
- Varying the amplitude of external disturbance in cycles,
- Varying the type of external disturbance factor.

IMPLEMENTATION PROBLEMS OF METHODS FOR EFFECTIVENESS IMPROVEMENT OF GRAIN PRESOWING TREATMENT BY FORCED AERATION

Processing plants for grain forced aeration have essential advantages in terms of presowing treatment technology application. In many cases, there is a possibility to expose grain to forced aeration, during storage, which provides tools for presowing treatment 'on the spot' without resort to transportation. Besides, implementation of 'mild' presowing treatment modes is acceptable, in this situation, enabling to control both external disturbance action duration and its intensity level. Implementation of cyclic treatment modes is also ensured. In this case, high processing rate of treatment plants is easily retained.

In order to enhance grain drying processes, those of seeds sanitation or presowing treatment, ozone-air mixtures are actively used (Wozniak, Nowakowicz-Debek, Stepniowska, with co-authors. 2016; Wu, Jiangning, Doan, Huu, & Cuenca, Manuel Alvarez 2016). Equipment designed to produce ozone can be easily and organically fit into agricultural technological processes. Ozone concentration control is rather simple. Nevertheless, taking the principle of maximum trans information into consideration, the expediency of application of ozone and ozone-air mixtures instead of electrically activated air gives rise to doubt. We will define electrically activated air (EAA) as air that has passed through corona discharge field and that contains positive or negative ions, of sufficiently high density. It is assumed that EAA is capable to produce milder effect in comparison with ozone-air mixtures, in terms of energy level of external disturbance.

As a result of theoretical research carried out in (Vasil'ev and Udincova, 2000; Vasil'ev and Dzhanibekov), major factors have been specified that affect the effectiveness of treatment processes. In accordance with the maximum information principle, basic variants of presowing treatment with the use of forced aeration technologies have been defined.

Below partial morphological table is presented containing implementation options for presowing treatment in forced aeration processing plants (see Table 2).

Each of these factors is an external disturbance whose application initiates adaptation reactions development in seeds. It is assumed that particularly these reactions make their contribution to enhancement of seeds' sawing quality. However, the issues related to the optimal values of external disturbances' amplitude and time of exposure have to be defined. Table 2 contains options of various operation modes of processing equipment.

Table 2. Partial morphological table of technological factors applied for grain treatment with the use of heated air

Disturbance Factor	Factor Variation Options			
	1	2	3	4
F1 (processing cyclicity)	1B	NB	LT	
F2 (air parameters)	AA	RL	HC	
F3 (air velocity)	SC	SF	SW	ST
F4 (voltage polarity of electric activator)	UO	UP	UOP	UPO
F5 (corona current of electric activator)	IC	IV		

Let us consider possible variants in detail. FactorF1 defines processing duration and cyclicity (1B and NB correspond to single cycle and multiple cycles, respectively, LT is continuous treatment). Duration of treatment has to be defined based on experiments. It has to be taken in consideration that, according to hypothesis, external disturbance has to have a repeated character such as, for instance, 'heating–cooling–heating–cooling…'.

In the course of processing, air parameters can have any values in a wide range (AA) or they can follow a certain regular pattern (RL). Alternatively, air temperature can be controlled depending on its relative humidity (HC). Specific values of air parameters have to be defined, for each particular case. Factor 3 is air velocity that has four value options, in Table 2. In the input into grain layer, air velocity can either have constant value (SC) or it may vary depending on grain moisture content and air relative humidity (CF, CW), as well as on air temperature (ST). It has to be noted that air velocity changes as it flows through grain layer.

In order to define the desired variation ranges for factors F4 and F5, relevant model has to be run and series of field experiments have to be carried out. The objective of model experiment is studying the following issues:

- To define the minimum rate of seeds exposure to aero ions EAA capable to initiate the desired treatment effects in seeds.
- To make out whether the desired effects of grain treatment occur as a result of exposure to aero ions EAA of not only negative polarity but also of positive one.

Among a wide variety of processing equipment operation modes specified in Table 2, mode F1.3. – F2.2 – F3.1 is a standard one. Exactly this mode is implemented in currently applied forced aeration processing plants. Therefore, its effectiveness, in terms of seeds sawing quality enhancement, has to be, foremost, checked (proved).

EXPERIMENTAL DEFINITION OF GRAIN PRESOWING TREATMENT MODES FOR FORCED AERATION BINS

It was discovered, as the result of experimental modeling (Vasil'ev, 2009), that cyclic modes of grain treatment of type 'heating–cooling–heating–cooling' were the most effective ones. The best results were achieved for velocity of air passing through grain layer equal to $V_a = 0.3$ m/s, air temperature $T_a = 29$°C to 30°C and duration time $t_p = 4$ hours. The lying period was 2 days.

These results were obtained for grain treatment in experimental processing plant where layer thickness of grain under processing was 0.15m and 0.45m. In case that a thin grain layer is subject to processing, temperature of air and grain can be supposed to be uniform across the entire layer. In real forced aeration bins, air has to pass grain layer having a thickness of 1.2 m to 1.5 m.

Analysis of results of multi-factor experiments enabled to plot family of curves for two-dimensional section of response surface. Regression equations have been derived, for the following three factors: grain temperature increment $_c$ (in %), air velocity V_a (in m/s), half-cycle time of treatment t_p. These equations have the following forms:

For germinating energy:

$$EG = -27.5736 + 0.693_{c} + 62.256V_{a} + 20.736t_{p} + 0.0864_{c}V_{a}$$
$$+0.0392_{c}t_{p} - 4.08V_{a}t_{p} - 0.007_{c}^{2} - 69.44V_{a}^{2} - 9.8t_{p}^{2} \quad (9)$$

For germinating ability:

$$G = -23.905 + 0.598_{c} + 50.488V_{a} + 19.405t_{p} + 0.082_{c}V_{a}$$
$$+0.035_{c}t_{p} - 3.9V_{a}t_{p} - 0.0067_{c}^{2} - 56.32V_{a}^{2} - 9.8t_{p}^{2} \quad (10)$$

Counter graphs have been plotted, for equations (7) $G = f\left(_{c}, t_{p}\right)$ and $G = f\left(_{c}, V_{a}\right)$, see Figure 4 and Figure 5.

As it is seen from these graphs, the highest values of seeds' germinating ability, after their treatment by heated air, is achieved for grain temperature increment of 40% to 55%, in relation to its initial level, as well as for increments of velocity of air passing through grain layer and treatment duration in the ranges of 0.35 to 0.45 m/s and 1h to 1.25h, respectively. Let us assume that half-cycle treatment time that complies with the requirement of maximum germinating ability is 1.05h.

MORE PRECISE PARAMETER DEFINITION FOR GRAIN PRESOWING TREATMENT BY HEATED AIR IN FORCED AERATION BINS

In real processing plants designed for forced aeration, grain layer thickness exceeds 1 m. That is why an essential temperature drop $\left(_{c}\right)$ may occur. Therefore, output parameters of grain layer may not fall into the range of optimal treatment conditions even if electric heater provides optimal air temperature values

Figure 4. Counter graphs for dependence $G = f\left(_{c}, t_{p}\right)$

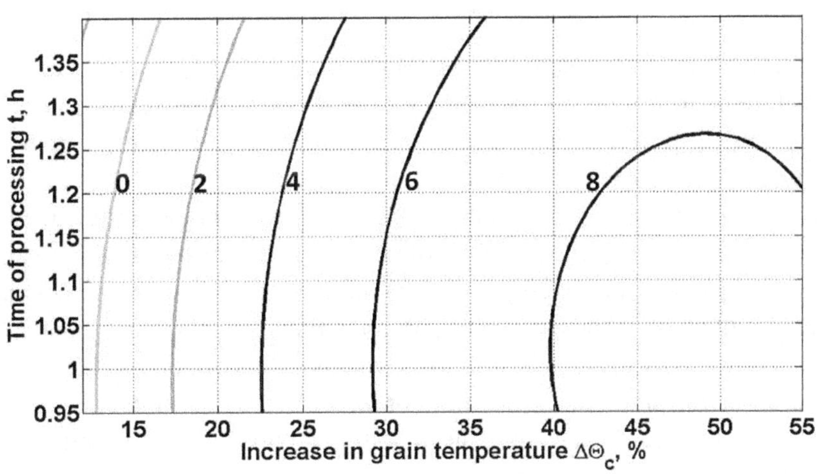

Figure 5. Counter graphs for dependence $G = f\left({}_{c}, V_{a}\right)$

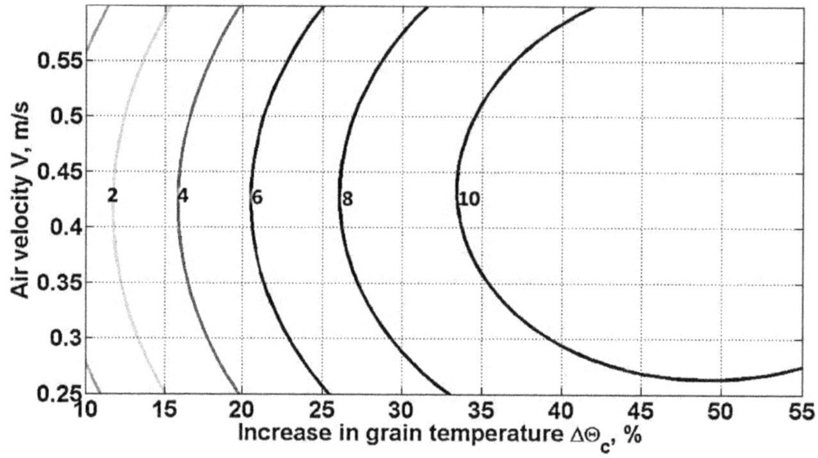

Figure 6. Frequency response function (a) and circuit phase response (b) for grain layers in forced aeration bin

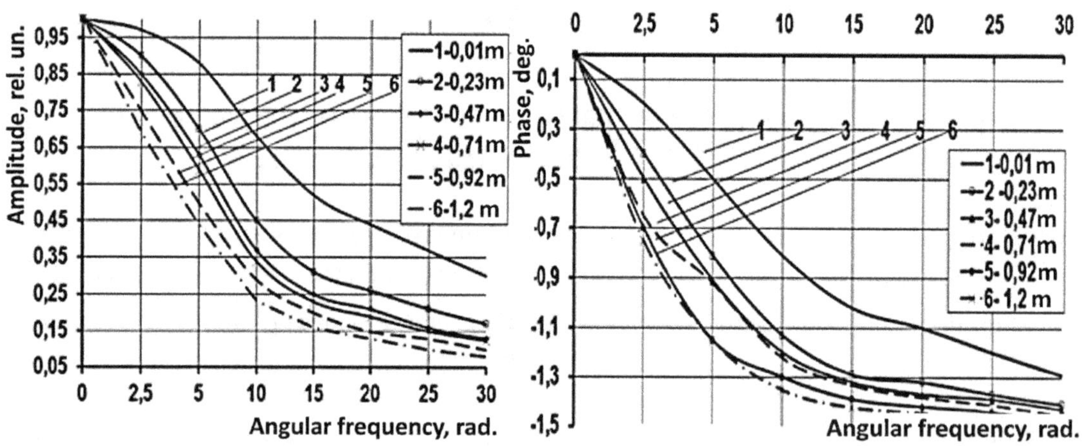

in its input (Vasil'ev, 2009). Thus, electric heater operation mode parameters (air temperature in the grain layer input and frequency of power switching on/off) have to be corrected to ensure optimal treatment conditions for seeds over the entire grain layer. In this regard, frequency response function and circuit phase response of grain layers were plotted (see Figure 6).

Analysis of frequency response function shows that, in terms of temperature change, grain layer is a low-frequency filter. In means that, in case of oscillating character of temperature variations of air fed into grain layer, the amplitude of grain temperature oscillation reduces as air passes the layer.

With the help of circuit phase response (see Fig.6), operation time of electric heater has been defined, for grain presowing treatment by heated air. It follows from analysis of circuit phase response that, in the input of grain layer, phase lag amounts to 0.69 rad., or 0.23h (14 min), for oscillation period of 2.1 h. It means that elementary grain layer located near the air output attains its maximum/minimum temperature value 14 min later than those in the air input.

Thus, we have to extend the half-period of temperature variation by 14 min which yields corrected values of $t_{pro} = 1.3$ hand $t = 2.6$ h, for half-period and entire period, respectively.

It has been found above that maximum possible effect of treatment can be achieved for grain temperature increment of 40% to 55%. Therefore, in stationary mode of forced aeration, calculated temperature of air fed into grain layer is $T_a = 1.5 \cdot {}_c in$.

However, it has to be noted that this formula is viable only for thin grain layers. In case of thick grain layer processing, the above air temperature is too low to heat elementary layers of grain located in the air output to the required desired temperature values.

Let us apply regression equation for germinating ability (7) and graphs for frequency response function (see Fig. 6) to define air temperature, in the input into the grain layer, corresponding to the maximum improvement of seed's sawing quality across the entire layer. Let us divide the grain layer into six thin elementary layers and let us calculate the mean seeds' germinating ability within grain layer with the use of formula:

$$G_{mean} = \frac{\left(G_1 + G_2 + G_3 + G_4 + G_5 + G_6\right)}{6} \qquad (11)$$

where G_i is germinating ability of the i-th elementary layer (i = 1, 2, 3, 4, 5, 6).

Grain temperature change (in %) can be defined as follows:

$$_c = \frac{_c - _c in}{_c in} 100\% \qquad (12)$$

With the account of frequency response function graphs, values of $_c$ have been derived, for each of the six elementary layers.

Mean value of germinating ability, for the entire (thick) grain layer can be found as:

$$G_{mean} = -141.615 + \frac{172.52 T_{ac} in - 49.178 T_a^2}{^2_{c\,in}} \qquad (13)$$

After differentiating the above expression for germinating ability mean value (13) with respect to variable T_a we find optimal air temperature in the input into the grain layer that insures maximum germinating ability across the entire layer:

$$\frac{dG_{mean}}{dT_a} = \left(\frac{172.52_c in - 98.35 T_a}{^2_{c\,in}}\right).$$

Table 3. Improvement of germinating energy (EG) and germinating ability (ΔB), for elementary layers (in accordance with regression equations)

Layer Number (thickness)	$_c$ (%)	V_a (m/s)	t_p (h)	EG(%)	G(%)
1 (0.01 m)	74.6	0.7	1.3	5.8	5.2
2 (0.23 m)	63.8	0.6	1.3	11.4	10
3 (0.47 m)	54.8	0.5	1.3	14	12.1
4 (0.71 m)	53.0	0.4	1.3	14.1	12.2
5 (0.95 m)	44.0	0.3	1.3	12.6	11
6 (1.2 m)	31.4	0.2	1.3	7.8	7

After setting derivative to zero we obtain:

$$T_a = 1.8\ _c in \quad (14)$$

For initial grain temperature $_c in$ = 20°C, air temperature has to be equal to 36 °C, in accordance with condition (14), in which case output layers of grain are heated to temperature 26.3 °C.

In accordance with regression equations (9) and (10), improvement of seeds' sawing quality is the following, for elementary layers (see Table 3).

Thus, the method of optimal presowing treatment of grain by heated air in forced aeration bins shall comply with the following conditions:

- Fan operates continuously during 5.2 h,
- Electric heater is switched on/off with periodicity of 1.3 h,
- Electric heater heats air to temperature value of $1.8\ _c in$.

DEFINITION OF MINIMUM RATE OF EXTERNAL DISTURBANCE IN CASE OF ELECTRICALLY ACTIVATED AIR APPLICATION FOR GRAIN PRESOWING TREATMENT

To define optimal technological modes of presowing grain treatment with the use of electrically activated air (EAA), field experiments were performed.

Super quality stock barley seeds of cultivar 'Victoriya' with moisture content of 14% were chosen as seed grain. Seeds that have not passed EAA treatment were used as reference samples. presowing treatment was carried out at air temperature T_a =20°C, and air relative humidity was 60% to 70%. Air flow velocity was $V_a = 0,2m / s$.

Experiments were made in cyclic processing modes, in accordance with the following schematic formulas:

N–A–N–A,
P–A–P–A,
N–P–N–P,
P–N–P–N.

The following designations are applied to components of above formulas: **N** and **P** are grain aeration operations with the use of EAA, for negative and positive polarity of electric activator voltage, respectively. **A** is aeration operation with electric activator switched off. Duration of aeration period with either switched on or switched off electric activator is 1 hour. Duration of single treatment operation is 4 hours. Corona current of electric activator is 175 μA. Density of aero ions for positive polarity is $24 \bullet 10^3$ cm^{-3}, while that of negative polarity is $28 \bullet 10^3 cm^{-3}$.

Finally, experiment formula had the following form:

- (–28) – A – (–28) – A,
- (+24) – A – (+24) – A,
- Reference mode,
- (–28) –(+24)–(–28)–(+24),
- (+24)–(–28)–(+24)–(–28),

where (–28) and (+24) is density of aero ions in air×10^3(cm^{-3}).

During sprouting of planted seeds on experimental plots, 10 to 14 days after planting, density of sprouts was defined. Analysis of green plants was made, in the tillering phase. Results of experimental studies are given in Table 4.

As it follows from Table 4, plants on the first and the fourth plots were leaders, in that phase of development.

However, situation had changed substantially by the phase of tillering. As it is seen from Table 4, plants on plot 2 were explicitly in advance of those grown from seeds treated in other modes, in relation to all indicators. Moreover, for certain parameters, these plants showed even multifold superiority.

Table 4. Results of sprout density test and analysis of green plants

Parameter	(–28) –A–(–28) –A	(+24) –A – (+24)–A	Reference	(–28) –(+24) –(–28)–(+24)	(+24) –(–28) –(+24)–(–28)
Sprout density	547	499	520	564	489
Number of culms	1.45	3.40	1.95	1.45	2.30
Number of leaves	5.5	10.9	6.5	5.7	7.7
Number of secondary roots	1.6	5.6	2.2	1.75	3.4
Plant length	19.0	26.3	19.3	20.6	20.2
Weight of 20 green plants	22.1	72.5	28.5	26.1	42.0
Weight of 20 dry plants	4.2	12.4	5.4	4.8	6.78

Table 5. Yield structure analysis

Parameter	(–28)–A–(–28)–A	(+24)–A–(+24)–A	Reference	(–28)–(+24)–(–28)–(+24)	(+24)–(–28)–(+24)–(–28)
Number of plants per m^2	531	474	497	516	472
Number of culms per m^2	1027	867	1046	784	988
Number of productive culms per m^2	923	786	814	745	885
Number of ears per plant (mean value)	1.74	1.66	1.64	1.44	1.88
Plant height, (sm)	73.9	78.2	77.7	74.2	82.5
Ears length (sm)	6.8	7.5	7	6.3	7.3
Weight of sheaf of 50 pc. (g)	185	175	125	108	158
Weight of grain per ear (g)	0.75	0.81	0.71	0.69	0.76
Number of seeds per ear (pc.)	14.9	16.2	15.5	13.2	15.5
Weight of 1000 seeds (g)	48.7	48.3	45.6	47.4	46.2
Biological productivity (t/ha)	6.92	6.37	5.76	5.16	6.74
Yield (t/ha)	5.86	5.79	5.67	5.14	6.23
Grain unit(g)	661	651	658	641	663

Thus, plants of the second sample outstripped those of reference group in number of secondary roots by a factor of nearly 2, and by a factor of nearly 3, in their weight. Good results were obtained for the fifth treatment mode. Plants grown on this plot outstripped those of reference group by a factor of 1.2 to 1.5, in all parameters that were checked.

Plants grown from seeds exposed to treatment modes 2 and 5 kept their superiority until harvesting. In fact, plants from the fifth plot outstripped those from other plots in number of ears per plant, plant height; biological productivity and effective crop yield (see Table 5).

Plants from the2nd plot had the maximum length of ear and, consequently, maximum number of seeds per ear. That is why yield for this plot exceeds that of reference one. Certain parameters of plants from the first plot were better compared to reference plot while values of other parameters were on the same level with the reference plot. At the same time, high sprout density and survival rate of plants, in the spring-summer period (531 plants of 547 initial sprouts) ensured maximum number of yielding culms per 1 m^2 owing to which productivity attained 5.86 t/ha. It is by 0.19 t/ha higher compared to reference plot. Practically all parameters of plants from plot 4 were lower than those of reference one. Plant productivity for this plot was substantially lower compared to the reference plot. That is why this mode of EAA presowing treatment is not recommended for practical application.

Therefore, cyclic modes **P–N–P–N** and **P–A–P–A** that gave considerable growth in yield owing to the change of plants' productive tilling capacity and that of number and weight of seeds per ear are of particular interest. In this case, mode P–N–P–N ensuring maximum yield is the most preferable one.

Field experiments were repeated a year late. Their results have proved the highest effectiveness of cyclic mode P–N–P–N for grain presowing treatment.

Field experiments were carried out for winter barley, as well. In these experiments, winter barley stock seeds 'Dobrynya' (germinating energy = 92%, germinating ability = 98%) were planted out. Experiment procedure was the same as for winter barley. Green plants analysis was performed two times i. e. just before the winter period and in spring time. The first analysis has showed that the best quality plants were yielded on plots where seeds were treated with the application of mode P–N–P–N. Plants cultivated on these plots exhibited high values of parameter that is essentially important for winter agricultural crops, namely, depth of tillering joint. It is explained by the fact that the wider is tillering joint the higher is temperature in its center and, consequently, the higher is plants' ability to survive during frosts. Thus, upon wintertime termination, plants on these plots were in better status compared to others, in practically all parameters. The best results were obtained with the application of mode P–N–P–N, both in the course of yield structure analysis and while harvesting. Productivity of this plot attained 8.55 t/ha which was 0.4 t/ha higher compared with the reference plot.

Therefore, based on results of field experiments, one can conclude that cyclic mode P–N–P–N (+24) – (–28)–(+24)–(–28) is the most preferable one.

FIELD EXPERIMENTS ON EVALUATION OF PLANTS' ADOPTABILITY TO WEATHER CONDITIONS

Field experiments make it possible not only to evaluate the effect of presowing treatment on laboratory germinating ability and germinating energy but also to perform monitoring of plants, in the process of their growth and yield formation. It may happen, in certain cases, that positive results obtained in laboratory conditions, after presowing treatment, either do not 'work' in conditions of field experiments or their positive effect on germinating ability does not always guaranty yield enhancement.

In field experiments, weather conditions play essentially important role. On the one hand, their changeability does not provide a possibility to perform experiments in identical conditions. At the same time, there exist various weather options enabling comparison of particular presowing grain treatment techniques' effectiveness in different environments. The less the results of particular grain presowing treatment method depend on weather conditions the higher is its effectiveness.

Such dependences were evaluated with the help of method of Eberhart and Russell (Eberhart S.A., Russel W.A. 1966). This analysis makes it possible to calculate plant adaptation power in relation to changing climatic factors.

The essence of the method of Eberhart and Russell can be reduced to the variance analysis of development results for plants grown from seeds exposed to presowing treatment, as well as those of their yield rate, in different years of cultivation. Research has to be performed in two stages. On the first stage, variance analysis has to be made. This stage enables to define relationships between weather conditions change and results of both plant development and its productivity.

On the second stage, indexes of environmental conditions are calculated for each year when field tests were carried out:

$$Is_i = \frac{\Sigma Pef_i}{n} - \frac{\Sigma\Sigma Pef_{ij}}{n\tau} \tag{15}$$

where Is_i is weather conditions index, ΣPef_i is sum of effectiveness indexes, for all presowing treatment modes, for the *i*-th year, n is number of presowing treatment options, $\Sigma\Sigma Pef_{ij}$ is sum for particular effectiveness index, for all options of presowing treatment, for all years when presowing treatment tests were carried out, τ is number of such years.

The next step of analysis is calculating regression coefficients for each option of presowing treatment mode:

$$b_i = \frac{\Sigma Pef_{ij} \cdot Is_i}{\Sigma Is_i^2} \quad (16)$$

Coefficient of linear regression b_i defines the extent to which cultivation conditions affect changeability of presowing treatment effectiveness. Both various plants' development parameters and their productivity may be selected to serve as indicators of effectiveness. For example, for winter wheat or barley cultivars, maximum development of root system is essentially important. It enables well-established plants to survive hard winter conditions. In this case, number of secondary roots can serve as indicator of effectiveness. If the purpose of plant cultivation is green feed, such as vetch and oat mix, length of green culms or weight of specified number of green plants can be chosen as effectiveness index. In case of breeding activities, ear size or number of seeds per ear may serve as index of effectiveness.

Numerical value of effectiveness index is characterized by that of regression coefficient and may be equal to 1, or be either smaller or greater than 1. The greater is the value of regression coefficient b_i compared to 1 the stronger is the effect of weather conditions on plant development. Contrariwise, the smaller is its value than 1 the weaker is plant's reaction on weather conditions change. If $b_i = 1$ it means that there exists complete correlation between effectiveness index change and that of plants cultivation conditions.

In order to evaluate the effectiveness of various presowing treatment mode options, three-year fielded pediments were carried out. In the course of experiments, seeds of spring barley having average values of germinating ability and germinating energy of 95% and 90%, respectively, were exposed to presowing treatment. Weight of biomass of green plants was chosen as effectiveness index. Experimental data results were analyzed in order to answer the following questions: which treatment type, among a specified variety of methods, yields the best effectiveness; which grain presowing treatment mode provides better plant's immunity to weather conditions change. Grain treatment before planting out was performed by three methods (Vasil'evand Dzhanibekov, 2014; Budnikov and Vasiliev, 2018; Vasiliev, Budnikov, Gracheva, & Smirnov, 2018; Kharchenko Valeriy 2018) (see Table 6).

Evaluation of change in yield of barley green plants biomass was performed, based on weight of sheaf. For each treatment method specified in Table 1, two-factor variance analysis was made. Treatment method (factor A) and year of experiment (factor B) were selected as factors. Weight of sheaves that was the basic subject of variance analysis is shown in Table 7.

Table 6. Modes of grain presowing treatment used in field experiments

Treatment Technique	Mode Number
Treatment by heated air in forced aeration processing plants	R1
Treatment by electrically activated air containing negative aero ions	R2
Treatment by microwave field with air cooling	R3
Treatment electrically by electrically activated air containing secondary air ions	R4

Table 7. Weight of sheaves of spring barley (g)

Factor A (Treatment Method)	Factor B (Year)	Test Number			Sum $\sum x$	Mean Value XV
		I	II	III		
Reference samples	1	112	102	86	300	100.0
	2	80	85	87	252	84.0
	3	118	106	130	354	118
$\sum p$		311	293	303	907	302.33
R1	1	146	115	127	388	129.33
	2	129	92	115	336	112.00
	3	136	148	126	410	136.67
$\sum p$		412.00	355.00	367.00	1136.00	379.67
R2	1	103	85	92	280	93.33
	2	102	78	93	273	91.00
	3	124	112	139	375	125.00
$\sum p$		329.00	275.00	324.00	928.00	309.33
R3	1	95	97	108	300	100.00
	2	94	108	72	274	91.33
	3	133	121	116	370	123.33
$\sum p$		322	326	296	944	314.67
R4	1	121	116	93	329	109.67
	2	92	79	107	277	92.33
	3	95	131	125	348	116.00
$\sum p$		308	326	325	959	319.67
$\sum\sum p$		1682.0	1575.00	1615.00	4874.00	1625.67

In Table 8, results of variance analysis are presented. Results of variance analysis have shown that there exists correlation between the factors of experiment, i. e. between treatment method A and year of experiment B. Calculated value of Fisher's criterion exceeds tabulated value, for these factors, $Ff > Ft$.

In Table 9, results of calculations for index of environmental conditions Is_i and coefficient of linear regression b_i, for sheaf weight, are presented.

Analysis of data given in Table 9 makes it possible to conclude that, in terms of weather conditions, the third year of experiment was the most favorable because the highest green weight of sheaf (136.67g) was reaped that year. It has to be noted that method R1 of presowing treatment is the most preferable, for greater amount of green biomass production. For this presowing treatment method, the average volume of produced green mass exceeded that for other methods even in year (2) when weather conditions were less favorable. Besides, presowing treatment of seeds by heated air provides better stability and poorer dependence of results on weather conditions change (regression coefficient is 0.85). The highest immunity to weather conditions effect exhibited plants grown from seeds treated by secondary aero ions (*R*4) in which case regression coefficient was as low as 0.78. Nevertheless, it has to be stated

Table 8. Results of two-factor variance analysis

Variability Source	Sum of Squares	Degrees of Freedom (dof)	Mean Square S2	Ff	Ft
Total	5085.66	20	–	–	–
Treatment method (A)	1531.74	6	255.29	8.68	3.0
Year (B)	3201.15	2	1600.58	54.45	3.89
Difference (error)	352.77	12	29.40	–	–

Table 9. Results of calculations for index of environment conditions and coefficient of linear regression

Factor A (Treatment Method)	Factor B (Year)			$\sum X_i$	X_i	b_i
	1	2	3			
Reference samples	100.0	84.0	118	302	100.67	1.22
R1	129.33	112.00	136.67	379.7	126.56	0.85
R2	93.33	91.00	125.00	309.3	103.11	1.13
R3	100	91.33	123.33	314.7	104.89	1.06
R4	109.7	92.33	116.00	319.0	106.00	0.78
$\sum Xj$	763.33	660.67	872.33	2296.3	109.35	
Xj	109.05	94.38	124.62			
Is_i	- 0.30	-14.97	15.27			

that this method is less effective as method R1, as far as weight of sheaf is concerned. The highest adoptability (flexibility) was exhibited by plants grown from seeds treated with the application of methods *R*2 and *R*3 in which case b_i was equal to 1.13 and 1.06, respectively. As far as method *R*3 is concerned, it showed the best stability, for three years of experiments.

Based on the above analysis, the following conclusions can be made:

- All methods of grain presowing treatment used in experiments made it possible to achieve positive effect that appeared in growth of green mass volumes compared to reference material,
- Presowing treatment method of seeds that involves secondary aero ions exhibited higher immunity of plants to the influence of weather conditions change,
- Taking results of field experiments into account, the most preferable, from the practical point of view, method of presowing treatment is that of heated air processing yielding the highest growth in green mass accompanied by considerable dependence of results on weather conditions.

FUTURE RESEARCH DIRECTIONS

The proposed approach to the management of seed pre-treatment allows the use of any kind of impact to improve the sowing qualities.

However, to implement the obtained processing modes, it is necessary to have the appropriate equipment. Even in the presence of such equipment, it is necessary to check whether it is possible to implement the found modes of presowing seed treatment. As a rule, the standard equipment used in technological processes requires improvement. Therefore, at the next stage of research, it is necessary to determine the required parameters of the plants, which provide the specified processing modes.

CONCLUSION

Theoretical research in the field of external disturbances effect on seeds in the process of grain presowing treatment made it possible to conclude that the following conditions have to be provided to improve resistibility of crop seeds (and, consequently, their sawing quality):

- External disturbance has to be gradually growing, continuous, alternating (cyclic) in order to ensure development of adaptation reactions on the level of functional change but not on that of structural ones,
- External physical actions of certain amplitudes have to be applied that have to be then reduced or cancelled, in order to initiate activation reactions. After that the amplitude of external action has to be increased by 20% to 25% compared to its initial value,
- Initial value of external excitation has to be possibly low to ensure adaptation reaction development on the lower levels of reactivity to ensure the highest possible response.
- The following optimal operation modes of electric equipment were defined, in the course of multifactor experiments on application of seeds presowing treatment by heated air in forced aeration bins:
- Fan operates continuously during 5.2h,
- Electric heater switches on/out with the half period of 1.3h,
- The electric heater controls the air temperature at a level 1.8 times higher than the initial grain temperature.

Two-year field experiments on the influence of electrically activated air on sawing quality of seeds made it possible to make it certain that the best results can obtained with the use of cyclic treatment modes when seeds are alternatively exposed to the effect of either negative or positive aero ions.

Three-year field experiments on the influence of weather conditions on results of presowing treatment with the use of various external electrophysical disturbances have shown that:

- All methods of grain presowing treatment used in experiments can yield positive effect that appears in growth of green mass volumes compared to reference plots;
- -Method of presowing treatment of seeds by secondary aero ions ensures the highest immunity to the influence of weather conditions;
- Results of field experiments show that, from the practical point of view, the most preferable, method of presowing treatment is that of heated air processing providing the highest growth in green mass accompanied by considerable dependence of results on weather conditions;

- The main purpose of the presented Chapter is to show the methodology for managing the efficiency of presowing processing. The presented method is universal. Presowing treatment regimes for each crop and seed variety should be determined individually.

REFERENCES

Aleksandrov, V. Y. A., & Kislyuk, I. M. (1994). Reakciya kletok na teplovoj shok: Fiziologicheskij aspekt. [Cell reaction to heat shock: physiological aspect]. *Citologiya. T.*, *36*(1), 5.

Budnikov, D. A., & Vasiliev, A. N. (2018). In V. Kharchenko & P. Vasant (Eds.), *The Use of Microwave Energy at Thermal Treatment of Grain Crops. In Handbook of Research on Renewable Energy and Electric Resources for Sustainable Rural Development* (pp. 475–499). Hershey, PA: IGI Global. doi:10.4018/978-1-5225-3867-7.ch020

Eberhart, S. A., & Russel, W. A. (1966). Stability parameters for comparing varieties. *Crop Science*, *6*(1), 36–40. doi:10.2135/cropsci1966.0011183X000600010011x

Gamburg, K. Z., Borovskii, G. B., & Voinikov, V. K. (2017). The study of the participation of heat shock proteins in the resistance to high and low temperatures with the use of thellungiella (thellungiellasalsuguinea) and transgenic lines of arabidopsis (arabidopsis thaliana). *Journal of Stress Physiology & Biochemistry*, *13*(1), 23–34.

Garkavi L.H., Kvakina E.B., Kuz'menko T.S., & Shihlyarova A.I. (2003). Antistressornye reakcii i aktivacionnaya terapiya. Reakciya aktivacii kak put' k zdorov'yu cherez process samoorganizacii [Antistress reactions and activation therapy activation reaction as a way to health through self-organization processes]. Ekaterinburg: Tom Chast' II.

Garkavi, L. H., Kvakina, E. B., & Ukolova, M. A. (1997). Adaptacionnye reakcii i rezistentnost' organizma [Adaptive reactions and body resistance.]. Izvestiya Rostovskoj-na-Donu gosudarstvennoj akademii stroitel'stva [News of the Rostov-on-Don State Academy of Civil Engineering].

Garkavi, L. H., Tolmachev, G. N., Mihajlov, N. Y. U., Esipov, Y. U. V., Benya, F. M., Zverinceva, M.M., … Plyaka, P. S. (2007). Adaptacionnye reakcii i urovn ireaktivnost i kak ehffektivnye diagnosticheskie pokazateli donozologicheskih sostoyanij [Adaptation reactions and levels of reactivity as effective diagnostic indicators of prenosological conditions.]. Vestnik Yuzhnogo nauchnogo centra RAN [Bulletin of the South Scientific Center of RAS], 3(1), 61-66.

Garkavi, L. H., Zhukova, G. V., Kapkina, N. N., Mashchenko, N. M., & Mihajlov, N. Y. U. (2008). Aktivacionnaya terapiya i kachestvo zhizni v processe reabilitacii [Activation therapy and quality of life in the process of rehabilitation]. Sibirskijonkologicheskijzhurnal [Siberian Oncology Journal], S2, 34-35.

Garkavi, L. Kh. (2015). Periodic system of unspecific adaptational reactions and its significance for theory and practice. Cardiometry, 7.

Golicin, G. A., & Petrov, V. M. (2005). Informaciya i biologicheskie principy optimal'nosti: Garmoniya i algebra zhivogo. [Information and biological principles of optimality: Harmony and living algebra.] Izd. 2-e, stereotipnoe. – M.: KomKniga, - 128 p.

Kharchenko, V., & Vasant, P. (Eds.). (2018). *Handbook of Research on Renewable Energy and Electric Resources for Sustainable Rural Development*. Hershey, PA: IGI Global.

Rakhym, U., Auyelbek, I., Tazhikhan, T., Taraba-yev, B., Maemerov, M., Iztayev, B., Dauletkeldi, Y., & Moldabekova, N. (2016). Influence of presowing ion-ozone cavitational processing and air ionization in the cultivation process on high-yielding and seed characteristics of grain crops. *IJPT*, *8*(2), 14317-14327

Rubin, A., & Riznichenko, G. (2014). *Mathematical biophysics*. Springer Series in Chemical Physics. T. XV. doi:10.1007/978-1-4614-8702-9

Rubin, A. B. (1999). Biofizika: V 2-h t. T. 2.: Biofizikakletochnyh processov (2-e izd.). Moscow: Knizhnyjdom Univer-sitet.

Sel'eGans. (1960) Ocherki ob adaptacionnom syndrome [Essays on Adaptation Syndrome]. Moscow: Gosudarstvennoe izdatel'stvo medicinskoj literatury.

Sel'eGans. (1972). Na urovne celogo organizma [At the level of the whole organism]. Moscow: Nauka.

Sel'eGans. (1979). *Stress bez distressa* [Stress without distress]. Moskva: Progress.

Shapovalenko, N. S., Dorovskih, V. A., Korshunova, N. V., Shtarberg, M. A., Slastin, S. S., & Nevmyvako, E. E. (2011). Vliyanie holodovogo stressa na intensivnost' perekisnogo okisleniya lipidov i antioksidantnuyu sistemu tkanej ehksperimental'nyh zhivotnyh [The effect of cold stress on the intensity of lipid peroxidation and the antioxidant system of experimental animal tissues]. Byulleten' fiziologii i patologii dyhaniya [Bulletin of physiology and pathology of respiration], 39, 22-25.

Shennon, K. (1963). Raboty po teorii informacii i kibernetike [Works on information theory and cybernetics]. Moscow: Inostrannaya literatura.

Shmal'gauzen, I. I. (1968). Kiberneticheskie voprosy biologii / Pod obshch. red. i s predisl. R. L. Berg i A. A. Lyapunova [Cybernetic questions of biology / Under total. ed. and with foreword. R. L. Berg and A. A. Lyapunov]. Novosibirsk, Nauka, sib. otdelenie.

Vasil'ev, A. N. (2009). Ehnergosberegayushchie ehlektrotekhnologii sushki i predposevnoj obrabotki zerna aktivnym ventilirovaniem [Energy-saving electrical technologies of drying and pre-sowing treatment of grain by active ventilation].

Vasil'ev A.N. (2014). Ispol'zovanie zapasyonnoj ehnergii biologicheskih ob"ektov v tekhnologicheskih processah APK s pomoshch'yu ehlektrotekhnologij [The use of stored energy of biological objects in the technological processes of the agro-industrial complex with the help of electrical technologies]. *Vestnik VIEHCH [Herald VIESH]*, *1*(14), 28-35.

Vasil'ev, A. N., & Dzhanibekov, A. K. (2014a). Kak obespechit' predposevnuyu obrabotku semyan v bunkerah aktivnogo ventilirovaniya s trebuemym kachestvom [How to ensure presowing treatment of seeds in bunkers of active ventilation with the required quality]. Moscow: FGBNU VIM.

Vasil'ev, A. N., & Dzhanibekov, A. K. (2014b). O vozmozhnosti ispol'zovaniya vtorichnyh ionov dlya predposevnoj obrabotki semyan. Teoreticheskij i nauchno-prakticheskij zhurnal [Innovacionnoe razvitie APK Rossii na baze intellektu-al'nyh mashinnyh tekhnologij. *Innovacii v sel'skomhozyajstve [Innovation in agriculture]*, *1*(6). Retrieved from http://ej.viesh.ru/wp-content/uploads/2014/04/insel6.pdf

Vasil'ev, A. N., Dzhanibekov, A. K., & Udincova, N. M. (2015). Principy predposevnoj obrabotki semyan [Principles of presowing seed treatment]. *Innovacii v sel'skomhozyajstve [Innovations in Agriculture]*, *4*(14), 29-33. Retrieved from http://ej.viesh.ru/wp-content/uploads/2016/02/insel14.pdf

Vasil'ev, A. N., & Kononenko, A. F. (2008). Ehlektrotekhnologiyai upravlenie v realizaciiadaptivnyhrezhimovpredposevnojobrabotkizernaaktivnymventilirovaniem [Electrotechnology and control in the implementation of adaptive modes of presowing treatment of grain by active ventilation]. Rostov-na-Donu: Terra Print.

Vasil'ev, A. N., Kononenko, A. F., & Ereshko, A. S. (2002). Predposevnaya obrabotka semyan yachmenya ehlektroaktivirovannym vozduhom [Presowing treatment of barley seeds with electroactivated air]. Materialy 41 Nauchno-tekhnicheskoj konferencii. Chelyabinsk [Materials 41 Scientific and Technical Conference. Chelyabinsk] (pp. 217-220). Academic Press.

Vasil'ev, A. N., & Udincova, N. M. (2000). Analizparametrovvozdushno-teplovojobrabotkidlyauluchsheniyaposevnyhkachestvsemyan [Analysis of the parameters of air-heat treatment to improve the sowing qualities of seeds]. Mekhanizaciya i ehlektrifikaciya sel'skogo hozyajstva [Mechanization and electrification of agriculture], 1, 9-10.

Vasiliev, A. N., Budnikov, D. A., Gracheva, N. N., & Smirnov, A. A. (2018). Increasing Efficiency of Grain Drying With the Use of Electroactivated Air and Heater Control. In *Handbook of Research on Renewable Energy and Electric Resources for Sustainable Rural Development*. Hershey, PA: IGI Global. doi:5225-3867-7.ch011 doi:10.4018/978-1-

Vasiliev, A. N., Ospanov, A. B., Budnikov, D. K., Karmanov, D. K., Salginbayev, D. B., & Vasilyev, A. A. (2016). Controlling reactions of biological objects of agricultural production with the use of electrotechnology. *International Journal of Pharmacy & Technology*, *8*(4), 26855–26869.

Vasilyev A.N. (2013). Elektrotechnologies as controlling influence on bio-logical objects in agricultural production. *Research in Agricultural Electric Engineering*, *3*(3), 96-102.

Wozniak, A., Nowakowicz-Debek, B., & Stepniowska, A., with co-authors. (2016). Effect of ozonation on microbiological and chemical traits of wheat grain. *Plant, Soil and Environment*, *62*(12), 552–557.

Wu, J., Doan, H., & Cuenca, M. A. (2006). Investigation of gaseous ozone as an anti-fungal fumigant for stored wheat. *Journal of Chemical Technology and Biotechnology (Oxford, Oxfordshire)*, *81*(7), 1288–1293.

Chapter 12
Lévy-Enhanced Swarm Intelligence for Optimizing a Multiobjective Biofuel Supply Chain

T. Ganesan
Royal Bank of Canada, Toronto, Canada

Pandian Vasant
Universiti Teknologi PETRONAS, Perak, Malaysia

ABSTRACT

Engineering systems are currently plagued by various complexities and uncertainties. Metaheuristics have emerged as an essential tool for effective engineering design and operations. Nevertheless, conventional metaheuristics still struggle to reach optimality in the face of highly complex engineering problems. Aiming to further boost the performance of conventional metaheuristics, strategies such as hybridization and various enhancements have been added into the existing solution methods. In this work, swarm intelligence techniques were employed to solve the real-world, large-scale biofuel supply chain problem. Additionally, the supply chain problem considered in this chapter is multiobjective (MO) in nature. Comparative analysis was then performed on the swarm techniques. To further enhance the search capability of the best solution method (GSA), the Lévy flight component from the Cuckoo Search (CS) algorithm was incorporated into the Gravitational Search Algorithm (GSA) technique; developing the novel Lévy-GSA technique. Measurement metrics were then utilized to analyze the results.

DOI: 10.4018/978-1-7998-1216-6.ch012

Copyright © 2020, IGI Global. Copying or distributing in print or electronic forms without written permission of IGI Global is prohibited.

INTRODUCTION

Optimization efforts in engineering are often plagued by various complexities and uncertainties. Hence the computational capability of metaheuristic algorithms is indispensable when tackling such problems (Ganesan et al., 2016a; Ganesan et al., 2018a; Yang, 2013). Among the complexities arising are: high levels of nonlinearity, many target objectives, non-convexity and multiple variables. The existence of all these complexities in a single problem could potentially overwhelm the solution method (conventional metaheuristics) - resulting in the failure of the method to reach the optimal solution. To further boost conventional metaheuristics, algorithms are hybridized as well as enhanced (Ganesan et al., 2015; Ganesan et al., 2018b; Hong et al., 2016; Dong et al., 2016). For dealing with optimization problems which have multiple objectives, various algorithms such as Non-Dominated Sorting Genetic Algorithm (NSGA-II) (Mousavi et al., 2016), Strength Pareto Evolutionary Algorithm (SPEA-2) (Zhao et al., 2016), weighted sum (Naidu et al., 2014) and the Normal-Boundary Intersection (NBI) (Ahmadi et al., 2015) have been developed. Scalarization approaches such as the weighted sum technique and NBI enable multiple objectives to be aggregated into a single-objective problem. Transforming the problem this way significantly reduces its complexity making it easier to solve.

Supply chain planning/optimization present various challenges to decision makers globally. This is due to its highly complicated nature as well as its large-scale structure. Over the years various state-of-the-art methods have been employed to model supply chains (Seuring, 2013; Brandenburg et al., 2014; Ahi and Searcy, 2013). Optimization techniques are then applied to such models to help with optimal decision making (Ogunbanwo et al., 2014; Mastrocinque et al., 2013).

Fuel supply chains have recently picked up interest by researchers in academia as well as decision makers in various industries (Ba et al., 2016; Yue et al., 2014; Ghaderi et al., 2016). Modeling and optimizing these large-scale fuel supply chains have deemed to be challenging. An example of such a model development could be seen in the work of Awudu and Zhang (2013). In that work, the authors proposed a stochastic planning model for a biofuel supply chain. The supply chain consisted of biomass suppliers, biofuel refinery plants and distribution centers. Targeted at optimizing profits, the Geometric Brownian Motion and Benders decomposition method with Monte Carlo simulation was applied. Besides benchmarking their stochastic model against a deterministic model, the authors performed sensitivity analysis on various parameters using their model. Another work towards modeling fuel supply chains was presented in Poudel et al. (2016). There a pre-disaster planning model was developed to strengthen the biofuel supply chain system. The pre-disaster model considered the linking and selection of facilities that reflected the post-disaster connectivity and biofuel-related costs. The model was developed using the generalized Bender's decomposition algorithm and validated using industrial data from Mississippi and Alabama, United States.

On the other hand, the optimization work in Lin et al., (2014) aimed to minimize the annual biomass-ethanol production costs. The supply chain problem solved in that work was a large-scale model consisting of biomass harvesting, stacking, in-field transportation, transportation, preprocessing, packing and storage, to ethanol production and distribution. Using the mixed integer programming technique, the authors managed to reduce the cost of production (biorefinery) by 62%. A similar work was carried out in Zhang et al. (2013), where the switchgrass-based bioethanol supply chain was modelled and optimized. The location of study was North Dakota, United States. Similar to Lin et al. (2014), the model was optimized using mixed integer linear programming - obtaining optimal usage of marginal land for switchgrass production to harvest bioethanol in a sustainable and economical manner. In Osmani and

Zhang (2014), a large-scale sustainable dual feedstock bioethanol supply chain was optimized in a stochastic environment. The optimization was carried out while considering the biomass supply, purchase price, demand and sales price. The optimization work used a mixed integer linear programming technique in tandem with a sample average approximation algorithm as a decomposition technique. Similar works using mixed integer linear/nonlinear programming approaches for optimizing fuel supply chain could be seen in Gao and You (2015) and Osmani and Zhang (2013). A work that does not employ a mixed integer approach for fuel supply chain optimization could be seen in Marufuzzaman et al. (2014). In that work, the solution method employed combined Lagrangian relaxation and L-shaped techniques. Their computational experiments provided key insights on carbon regulatory mechanisms and uncertainties in biofuel supply chains. A broad review on metaheuristic techniques implemented to bioenergy supply chains is given in Castillo-Villar (2014) and De Meyer et al. (2014).

Swarm intelligence has been employed extensively in various industrial and engineering scenarios (Ling et al., 2016; Gao et al., 2015; Divyesh et al., 2017; Ganesan et al., 2016b). Besides, swarm-based optimization techniques have also been actively used in various types of supply chain problems. For instance, in Yeh et al. (2016), optimization was done on a three-stage supply chain network while considering the deterioration effect. Minimizing the total cost, the authors employed a novel swarm algorithm called the Simplified Swarm Optimization technique. The results were then compared to the standard particle swarm optimization (PSO) and genetic algorithm techniques. In Chen et al. (2017), the closed loop supply chain MO problem for the solar cell industry was tackled. Using the multiobjective PSO (MOPSO), the Pareto frontier of the large-scale supply chain problem was effectively attained. Another interesting work involving the application of swarm intelligence in supply chains could be seen in Kumar et al. (2016). In that work, the vehicle routing problem (production and pollution) was solved using the self-learning PSO algorithm. Since the problem was MO in nature, the results obtained using the self-learning PSO was compared with the NSGA-II approach.

In addition to PSO-based algorithms, cuckoo search (CS) has also been effective in solving real-world supply chain problems. This can be seen in the work of Mattos et al. (2017) – where a series of metaheuristics as well as CS was applied to a real-world supply chain (consumer packaged goods industry). Rigorous comparative analysis was carried out on the performance as well as the results produced by the techniques employed in that work. Similarly, in Abdelsalam and Elassal (2014), CS was employed among other metaheuristics to solve the joint economic lot-sizing problem. The problem was for a multi-layered supply chain with multi-retailers and a single manufacturer/supplier. Upon comparing the computational results, CS technique was shown to favor the centralized policy as opposed to a decentralized one for safety stock. Another application of CS in supply chains is given in the work of Srivastav and Agrawal (2017). In that work CS was employed to solve a MO lot-size reorder point backorder inventory problem. The triple-objective problem consisted of the following targets (annual): total relevant cost, expected number of stocked out units and expected number of stocked out occasions. The CS technique managed to generate an efficient Pareto curve between cost and service levels for the decision-makers.

The primary objective of this paper is to solve the real-world, large-scale biofuel supply chain problem using swarm-based optimization techniques. As an enhancement, a Lévy flight mechanism was integrated into the conventional GSA approach to further boost its optimization performance. This paper is organized as follows: Section 2 presents the problem formulation of the MO biofuel supply chain model. In Section 3 the swarm intelligence (CS, GSA and PSO) as well as the proposed Lévy-GSA techniques are described while the computational results and analyses are given in Section 4. This chapter ends with research conclusions and the authors' recommendations for future works.

BIOFUEL SUPPLY CHAIN MODEL

The fuel supply chain model framework employed in this work was developed in Tan et al. (2017). In that work only two objective functions were considered: profit (Pr) and social welfare (SW). The environmental benefits objective was incorporated into the SW function. In this work, the environmental benefits were considered as isolated from the SW function - taken as an independent objective (denoted Env). Therefore, the fuel supply chain model in this chapter consists of three objective functions to be maximized followed by inequality constraints (see equations (6) - (15)). The objective functions are shown in equations (1) - (5):

$$\mathrm{Pr} = P(1-EC)\times\sum_t q_t - \left\{FCp + \sum_t\left[GC\cdot q_t + \sum_i\left(SC\cdot IQ_{i,t} + \sum_k SQ_{i,k,t}\cdot PP_i\right) + Y_{1t}\cdot extraY_1 + Y_{2t}\cdot extraY_2\right]\right\} \quad (1)$$

$$SW = ACS\cdot(1-EC)\sum_t q_t + GT - GS\cdot(1-EC)\sum_t q_t \quad (2)$$

$$Env = AC\cdot\left[CET\cdot(1-EC)\cdot\sum_t q_t - \left(CEcb - CEtp\right)\right] \quad (3)$$

such that,

$$CEcb = 2\sum_{i,k,t} PQ_{i,k,t}\cdot\sum_{i,k}\left[\frac{CEncb_{i,k}\cdot Dcb_{i,k}}{LCcb_{i,k}}\right] + \left(\sum_{i,k,t} PQ_{i,k,t}\cdot\sum_{i,k} CEicb_{i,k}Dcb_{i,k}\right) \quad (4)$$

$$CEtp = 2\sum_{i,k,t} SQ_{i,k,t}\cdot\sum_{i,k}\left[\frac{CEntb_{i,k}\cdot Dbp_{i,k}}{LCtp_{i,k}}\right] + \left(\sum_{i,k,t} PQ_{i,k,t}\cdot\sum_{i,k} CEicb_{i,k}Dcb_{i,k}\right) \quad (5)$$

The constraints for the biofuel supply chain model are given below:

$$q_t \le q_{\max} \quad (6)$$

$$IQ_{i,t} \ge SIlb_i \quad (7)$$

$$\sum_i IQ_{i,t} \le IQ_{\max} \quad (8)$$

$$Q_{\min} \le \sum_t q_t \le Q_{\max} \quad (9)$$

$$HV_{\min} \leq \sum_{i} HV_{i} \cdot BR_{i,t} \leq HV_{\max} \tag{10}$$

$$\sum_{i} SQ_{i,k,t} \geq SQ_{\min i,k} \tag{11}$$

$$\sum_{i} PQ_{i,k,t} \leq PQ_{\max} \tag{12}$$

$$\sum_{i} PQ_{i,k,t} \leq AQ_{\max,i,t} \tag{13}$$

$$WR_{i,k,t} \leq \left[\frac{1 - MCori_{i,t}}{1 - MC_{\max.i,t}}\right] \tag{14}$$

$$\sum_{i,t} SQ_{i,k,t} \cdot PP_{i,k,t} \geq \left[\begin{array}{l} FCb_{k} + \sum_{i,t} SQ_{i,k,t} \cdot \left(AP_{i,k,t} + \dfrac{TCcb_{i,k} \cdot Dcb_{ik}}{LCcb_{i,k}}\right) \dfrac{1}{WR_{i,k,t}} \\ + \sum_{i,t} SQ_{i,k,t} \cdot \dfrac{TCtp_{i,k} Dbp_{k}}{LCtp_{ik}} \end{array}\right] \cdot \left(1 + ER_{k}\right) \tag{15}$$

such that,

$$i \in [1,2],\ k \in [1,10],\ t \in [1,12] \tag{16}$$

The decision variables are: q_t, $IQ_{i,t}$, $SQ_{i,k,t}$, $PP_{i,k,t}$, $PQ_{i,k,t}$, and $BR_{i,t}$. Details on the parameter settings of the model used in this work are as specified in Tan et al. (2017). The detailed nomenclature of variables and parameters of the model is presented at the end of the paper.

SWARM INTELLIGENCE

Cuckoo Search

Similar to swarm intelligence techniques, CS is a population based stochastic search and optimization algorithm (Mareli and Twala, 2017; Joshi et al., 2017). CS was originally inspired by brood parasitism often found among some species of cuckoo birds. The parasitism occurs when the cuckoo birds lay their eggs in the nests of other bird species (non-cuckoo birds). In a CS algorithm, a search pattern corresponds to a nest - where an individual attributed to a search pattern is represented by a cuckoo egg. Thus, the cuckoo egg symbolizes a potential solution candidate to the optimization problem - which will be tested by the fitness function. Then the solution will be accounted for in subsequent iterations to see if it is fit (fulfilling the fitness criteria). Otherwise the solution (egg) will be discarded and another candidate solution would be considered. This way the solution to the optimization problem would be iteratively improved as the technique explores the search space. The stochastic generator of the CS technique is based on the heavy-tailed random walk probability distribution, Lévy flights. The equation for the iterative candidate solution *i* and iteration *t* for the CS algorithm is given as:

$$y_i^{t+1} = y_i^t + \beta \cdot Levy\left(\lambda\right) \tag{17}$$

such that the Lévy distribution is given as follows:

$$Levy\left(\lambda\right) = t^{-\lambda} \tag{18}$$

where *t* is the random variable, $\beta > 0$ is the relaxation factor (which is modified based on the problem at hand) and $\lambda \in (1, 3]$ is the Lévy flight step-size. With $t \geq 1$, λ is related to the fractal dimension and the Lévy distribution becomes a specific sort of Pareto distribution. The CS algorithm is based on a few fundamental philosophies. For instance, each cuckoo bird lays a single egg at one time and randomly places the egg in a selected nest. The second being: via fitness screening, the best egg (candidate solution) is carried forward into the next iteration. The worst solutions are discarded from further iterations. The nests represent the objective space (or the optimization problem landscape). The parameter setting for the CS technique used in this work is shown in Table 1 while its respective algorithm is given in Algorithm 1:

Algorithm 1: Cuckoo Search (CS)

- **Step 1**: Initialize algorithmic parameters; y_i, β, λ, N
- **Step 2**: Define parameters in the constraints and decision variable
- **Step 3**: Via Lévy flights randomly lay a cuckoo egg in a nest
- **Step 4**: Define fitness function based on solution selection criteria
- **Step 5**: Screen eggs and evaluate candidate solution
    - **IF:** fitness criteria is satisfied
    - Select candidate solution (egg) to be considered in the next iteration, $n+1$
    - **ELSE:** fitness criteria is not satisfied
    - Discard candidate solution (egg) from further iterations
- **Step 6**: Rank the best solutions obtained during fitness screening
- **Step 7**: If the fitness criterion is satisfied and $t = T_{max}$ halt and print solutions, else proceed to Step 3

Table 1. CS Settings

Parameters	Values
Total Number of Eggs (N)	20
Number of nests, nests	4
Lévy flight step-size, λ	1.5
Relaxation factor, β	0.8
Maximum number iteration, T_{max}	300

Gravitational Search Algorithm (GSA)

The GSA algorithm is a meta-heuristic algorithm inspired by concepts from Newtonian gravitation (Tan et al., 2017). This computational framework uses ideas from gravitation by considering the search agents as massive objects. Thus, the gravitational forces dominate the motion of these masses - where lighter masses interact by gravitating towards heavier masses. The solutions are ranked based on the mass of the object. The higher the mass, the better the solution candidate. The gravitational forces hence act as a communication channel between the masses. The position of the masses correlate to the solution space in the objective space while the masses characterize the fitness landscape. As the iterations progress, and more gravitational interactions occur, it is expected that the masses would eventually swarm and aggregate to their fittest positions; providing the optimal solution to the problem.

The GSA algorithm randomly initializes a distribution of masses, $m_i(t)$(search agents) in the objective space. Thus the initial positions for these masses, x_i^d are fixed. For a maximization problem, the most fit mass, $m^{best}(t)$ and the least fit mass, $m^{worst}(t)$ at time t are determined as follows:

$$m^{best}(t) = \max_{j\in[1,N]} m_j(t) \tag{19}$$

$$m^{worst}(t) = \min_{j\in[1,N]} m_j(t) \tag{20}$$

As for a minimization problem, the procedure is simply vice versa. The inertial mass, $m_j'(t)$ and gravitational masses, $M_j(t)$ are then computed based on the developed fitness map.

$$m_i'(t) = \frac{m_i(t) - m^{worst}(t)}{m^{best}(t) - m^{worst}(t)} \tag{21}$$

$$M_i(t) = \frac{m_i(t)}{\sum_{j=1}^{N} m_j(t)} \tag{22}$$

such that,

$$M_{ai} = M_{pi} = M_{ii} = M_i : i \in [1, N] \tag{23}$$

Consequently the gravitational constant, $G(t+1)$ and the Euclidean distance $R_{ij}(t)$ is computed as the following:

$$G(t+1) = G(t)\exp\left(\frac{-\alpha t}{T_{\max}}\right) \tag{24}$$

$$R_{ij}(t) = \sqrt{\left(x_i(t)\right)^2 - \left(x_j(t)\right)^2} \quad (25)$$

where α is an arbitrary constant and T_{max} is the maximum number of iterations, $x_i(t)$ and $x_j(t)$ are the positions of particle *i* and *j* at time *t*. The interaction forces at time *t*, $F_{ij}^d(t)$ for each of the masses are then computed:

$$F_{ij}^d(t) = G(t)\left(\frac{M_{pi}(t) \times M_{aj}(t)}{R_{ij}(t) + \varepsilon}\right)\left(x_j^d(t) - x_i^d(t)\right) \quad (26)$$

where ε is a low-valued parameter. The sum of forces acting on each mass *i* is defined stochastically:

$$F_i^d(t) = \sum_{\substack{j=1 \\ i \neq j}}^{N} rand(w_j) F_{ij}^d(t) : rand(w_j) \in [0,1] \quad (27)$$

where $rand(w_j)$ is a randomly assigned weight. The magnitude of acceleration of each of the masses, $a_i^d(t)$ is then as follows:

$$a_i^d(t) = \left(\frac{F_i^d(t)}{M_{ii}(t)}\right) \quad (28)$$

Once the particle acceleration is determined, the particle positions and velocities are then computed:

$$v_i^d(t+1) = rand(w_j) + v_i^d(t) + a_i^d(t) \quad (29)$$

$$x_i^d(t+1) = x_i^d(t(t) + v_i^d(t(t+1) \quad (30)$$

where $rand(w_j)$ is a randomly assigned weight. The computational iterations are then continued until all the masses are at their most fit positions in the fitness landscape. The program is then halted when the stopping criterion is respected. The computational procedures of the GSA algorithm are presented in Algorithm 2 and the parameter settings are given in Table 2:

Table 2. GSA Settings

Parameters	Values
Initial parameter (G_o)	100
Number of mass agents, *n*	40
Constant parameter, α	20
Constant parameter, ε	0.01

Algorithm 2: Gravitational Search Algorithm (GSA)

Step 1: Define no of particles, m_p, initial positions, $x_i(0)$ and define algorithm parameters $G(0)$, α.
Step 2: Randomly define the initial distribution of masses, $m(0)$ and/or update solutions
Step 3: Compute gravitational & inertial masses based on the fitness map
Step 4: Compute the gravitational constant, $G(t)$
Step 5: Compute distance between agents, $R_{ij}(t)$
Step 6: Compute total force, $F_i^d(t)$ and the acceleration $a_i^d(t)$ of each agent.
Step 7: Compute new velocity $v_i(t)$ and position $x_i(t)$ for each agent
Step 8: If the fitness criterion is satisfied and $t= T_{max}$, halt and print solutions else proceed to step 3

Particle Swarm Optimization

PSO has been employed in many real-world problems ranging across various fields (Shi et al., 2015; Pan and Das, 2016). PSO is based on the investigation of swarming or flocking behaviors among some species of organisms (such as; birds, fishes, etc.). In the spirit of evolutionary computing, PSO explores the objective space for optimal regions and then evaluates the found solutions with respect to a fitness condition. The optimal solutions obtained by this algorithm are achieved as a result of swarming particles moving through the objective space. The velocity and position updating rule plays a critical role in this technique. The velocities and the positions of each particle are updated as follows:

$$v_i(t+1) = wv_i(t) + c_1 r_1 [\hat{x}_i(t) - x_i(t)] + c_2 r_2 [g(t) - x_i(t)] \tag{31}$$

$$x_i(t+1) = x_i(t) + v_i(t+1) \tag{32}$$

where each particle is identified by the index i, $v_i(t)$ is the particle velocity and $x_i(t)$ is the particle position at iteration t. The expressions, $c_1 r_1 [\hat{x}_i(t) - x_i(t)]$ and $c_2 r_2 [g(t) - x_i(t)]$ represent the personal and social influence respectively. These terms influence the degree of exploration and exploitation of the PSO technique during the search. The parameters w, c_1 and c_2 are user- defined while r_1 and r_2 are randomly initialized. These parameters are typically constrained with the following ranges:

$$w \in [0,1.2],\ c_1 \in [0,2],\ c_2 \in [0,2],\ r_1 \in [0,1],\ r_2 \in [0,1] \tag{33}$$

The PSO program halts when all particles/candidate solutions have achieved their highest fitness during the iterations. The computational procedures of the PSO along with the initial parameters used in this work are shown in Algorithm 3 and Table 3 respectively.

Table 3. PSO Settings

Parameters	Values
Initial parameter (c_1, c_2, w)	(1, 1.2, 0.8)
Number of particles	6
initial social influence (s_1, s_2, s_3, s_4, s_5, s_6)	(1.1, 1.05,1.033, 1.025, 1.02, 1.017)
initial personal influence (p_1, p_2, p_3, p_4, p_5, p_6)	(3, 4, 5, 6, 7,8)

Algorithm 3: Particle Swarm Optimization (PSO) Algorithm

Step 1: Initialize no of particles, i and the algorithm parameters w, c_1, c_2, r_1, r_2, n_o

Step 2: Randomly initialize the positions $x_i(t)$ and velocities, $v_i(t)$

Step 3: Update solutions and/or compute individual and social influences

Step 4: Compute positions $x_i(t+1)$ and velocities $v_i(t+1)$ at next iteration

Step 5: If the swarm evolution time, $t > t_o+T_{max}$, update position x_i and velocity v_i and go to Step 3, else proceed to Step 6

Step 6: Evaluate fitness swarm.

Step 7: If fitness criterion is satisfied, halt program and print solutions, else go to step 3.

Lévy: Enhanced Gravitational Search Algorithm

Similar to the conventional GSA, the Lévy-enhanced GSA (Lévy-GSA) contains a stochastic generator. The stochastic generator of conventional metaheuristics function to randomly generate the initial population of agents. In conventional metaheuristics, the stochastic generator produces random variables based on a Gaussian distribution. For instance a random variable, $x \in X$ which is distributed with a mean, μ and variance, σ^2 is Gaussian or normally distributed when the probability distribution function is as follows:

$$G_X(x) = \frac{1}{\sigma\sqrt{2\pi}} \exp\left[-\frac{1}{2}\left(\frac{x-\mu}{\sigma}\right)^2\right] \tag{34}$$

The Gaussian distribution is widely applicable in various fields of studies for modelling real-valued random numbers - e.g. in studies that involve modeling Brownian motion (Sottinen and Tudor, 2006) and Monte Carlo simulations (Corney and Drummond, 2004). In this work, the standard normal distribution with $\mu = 0$ and $\sigma = 1$ is employed in the stochastic engine to generate random values in the conventional GSA and PSO techniques. On the other hand, the CS algorithm (in its original form) does not use a Gaussian generator. CS employs Lévy flight (a heavy-tailed random walk probability distribution) to generate random numbers (see equations (17) and (18)). To further boost the performance of the GSA approach, the Lévy flight-based stochastic generator was integrated into the GSA approach. Other components of the GSA were maintained and not augmented. As with conventional CS algorithm, the consequent iterations are randomized as well using Lévy flight. The algorithm and process flow for the Lévy-GSA is as in Algorithm 4 and Figure 1:

Algorithm 4: Lévy-GSA

Step 1: Define no of particles, m_i, define algorithm parameters $G(0)$, α and initial positions, $x_i(0)$

Step 2: Define the initial distribution of masses, $m(0)$ using Lévy flight/update current solutions and use Lévy flight to randomize the distribution of masses.

Step 3: Compute gravitational & inertial masses based on the fitness map

Step 4: Compute the gravitational constant, $G(t)$

Step 5: Compute distance between agents, $R_{ij}(t)$

Step 6: Compute total force, $F_i^d(t)$ and the acceleration $a_i^d(t)$ of each agent.

Step 7: Compute new velocity $v_i(t)$ and position $x_i(t)$ for each agent
Step 8: If the fitness criterion is satisfied and $t = T_{max}$, halt and print solutions else proceed to step 2

The symbols and abbreviations used in this section can be see in the nomenclature presented at the end of the chapter.

ANALYSIS

The weighted sum approach was employed to tackle the MO biofuel supply chain problem. All three objectives were aggregated to form a single weighted objective function - where all the weights are in the range of 0 to 1. Since the techniques employed in this work are stochastic by nature, the techniques were executed 3 times and the best solution was taken. The weights were varied into 28 values. Therefore, each technique in this work was executed 84 times. The individual solutions depicting the Pareto frontier in this work were ranked into three classes; best, median and worst. The ranking was done based on the

Figure 1. Process Flow of Lévy-GSA approach

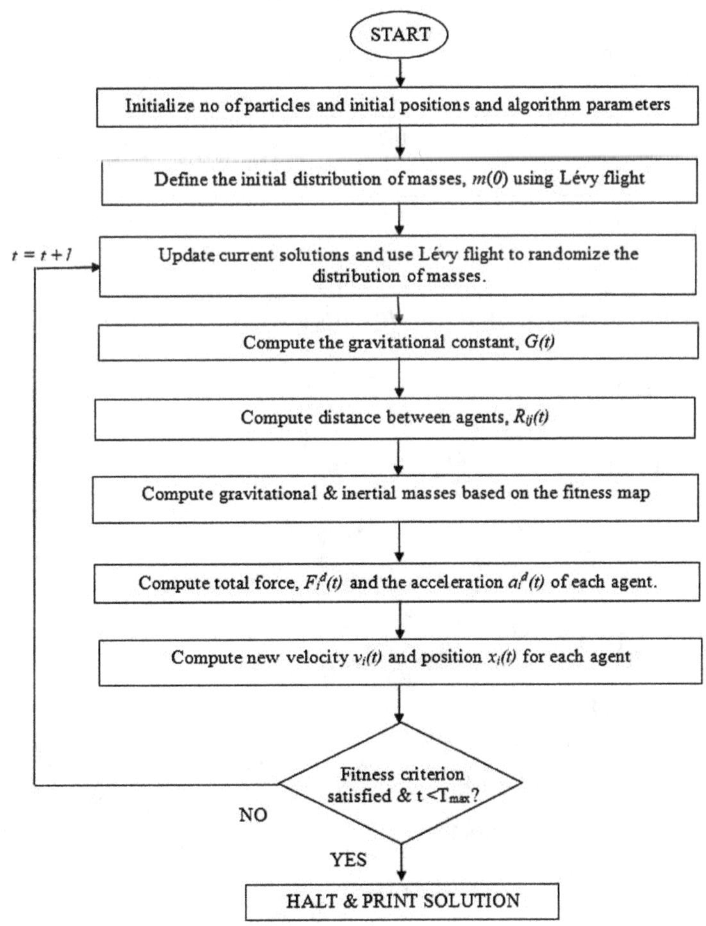

values obtained using the hypervolume indicator (HVI) (Jiang et al., 2015; Bringmann and Friedrich, 2013). The HVI is a useful metric which measures dominance levels of solutions to a MO optimization problem. The ranked solutions for the CS approach are given in Table 4:

The Pareto frontier obtained using the CS technique is depicted in Figure 2.

As for GSA approach, the ranked individual solutions as given in Table 5 while its associated Pareto frontier is depicted in Figure 3.

Table 6 and Figure 4 provides the individual solution ranking and the Pareto frontier obtained using the PSO algorithm.

As mentioned in the previous section, the GSA was improved by modifying its stochastic generator - utilizing a Lévy flight-based mechanism instead of the conventional Gaussian random number generator. The ranked individual solutions and the Pareto frontier of the Lévy-GSA is given in Table 7 and Figure 5:

Table 4. Ranked individual solutions for the CS technique

Description		Best	Median	Worst
weights	w_1	0.3	0.2	0.6
	w_2	0.2	0.6	0.3
	w_3	0.5	0.2	0.1
Objective functions	*PR*	513,600,000	511,207,000	513,417,000
	SW	1,018,670	1,021,820	1,020,790
	Env	142,326	101,600	70,408.1
Iterations	*t*	52	44	49
Metric	*HVI*	7307.18	5193.35	3594.96

Figure 2. Pareto frontier produced using the CS technique

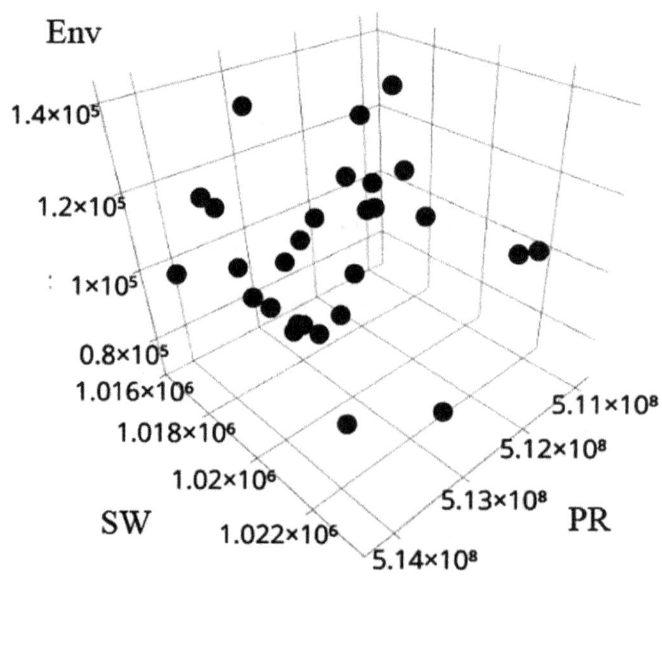

Table 5. Ranked individual solutions for the GSA technique

Description		Best	Median	Worst
weights	w_1	0.7	0.3	0.4
	w_2	0.2	0.4	0.3
	w_3	0.1	0.3	0.3
Objective functions	*PR*	208,859,000	155,942,000	124,922,000
	SW	1,004,340	1,004,780	1,005,040
	Env	334,224	378,267	351,498
Iterations	*t*	113	109	103
Metric	*HVI*	6887.16	5814.94	4321.9

Figure 3. Pareto frontier produced using the GSA technique

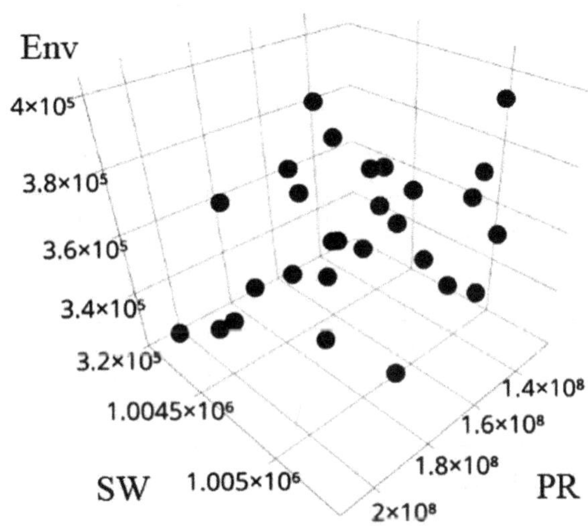

Table 6. Ranked individual solutions for the PSO technique

Description		Best	Median	Worst
weights	w1	0.1	0.5	0.3
	w2	0.2	0.2	0.6
	w3	0.7	0.3	0.1
Objective functions	PR	215,783,000	127,843,000	91,770,000
	SW	1,003,790	1,004,500	1,004,770
	Env	313,382	400,193	377,767
Iterations	t	137	116	113
Metric	HVI	6667.77	5035.63	3402.03

Figure 4. Pareto frontier produced using the PSO technique

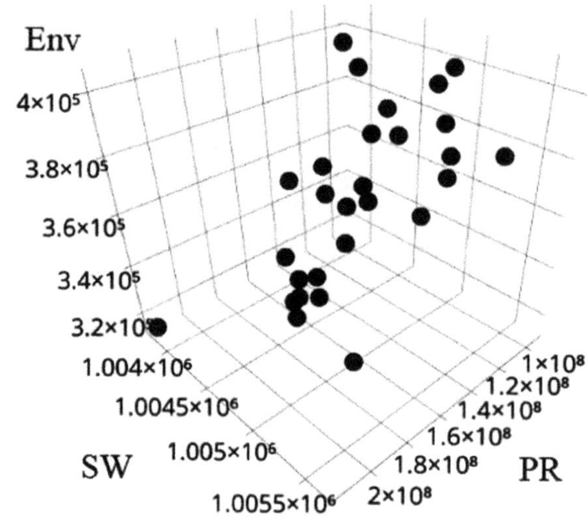

Table 7. Ranked individual solutions for the Lévy-GSA technique

Description		Best	Median	Worst
weights	w_1	0.2	0.2	0.4
	w_2	0.7	0.4	0.5
	w_3	0.1	0.4	0.1
Objective functions	*PR*	355,546,000	355,754,000	355,673,000
	SW	1,000,030	1,000,030	1,000,030
	Env	530,329	495,880	447,273
Iterations	*t*	300	300	300
Metric	*HVI*	18580.0395	17381.03197	15670.29223

Figure 5. Pareto frontier produced using the Lévy-GSA technique

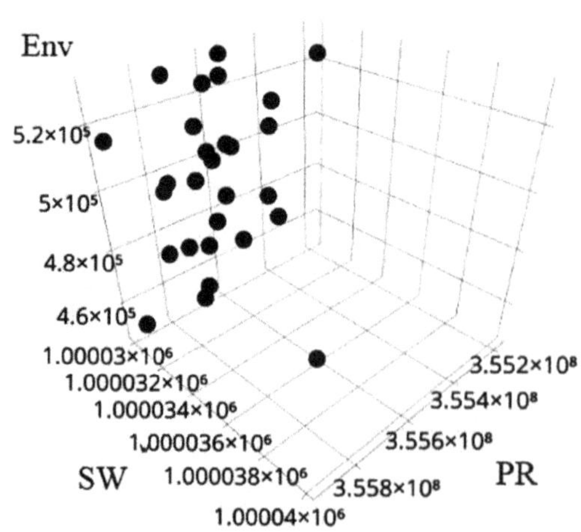

In this work, the PSO, CS and GSA techniques were employed to solve the MO biofuel supply chain problem. The level of dominance for the entire Pareto frontier produced by each of the techniques were measured. Observing that the GSA technique produced the highest level of dominance, the mentioned technique was further improved (Lévy-GSA). The dominance values for the entire frontier produced by all the techniques employed in this work are shown in Figure 6:

In Figure 6, it can be seen that the Pareto frontier generated using the Lévy-GSA technique is significantly more dominance as compared all the other techniques. The frontier produced by Lévy-GSA technique outperforms all frontiers produced by other techniques. The Lévy-GSA performed almost twice as efficient compared to all the other techniques employed in this work (in terms of frontier dominance). The Lévy-GSA technique boosted the GSA technique by 200.03%, generating the most dominant Pareto frontier. This is followed by the frontiers generated by the GSA, CS and PSO techniques respectively.

In Figures 2 - 4, it can be observed that the CS, GSA and PSO approaches produce frontiers which are very diverse covering the objective space sparsely. Thus it is possible that many solutions were not concentrated in the optimal regions as compared to the Lévy-GSA technique. In frontier generated by the Lévy-GSA approach (Figure 5) it is observed that the solutions seem more focused in certain regions of the objective space; which is the most optimal location.

The frontier dominance generated by the techniques also reflects the dominance levels of its best individual solution. This can be observed in Tables 4 - 6; where the best individual solutions produced by the CS, GSA and PSO techniques are 7307.18, 6887.16 and 6667.77. Similar to the dominance level of the entire frontier produced by the Lévy-GSA technique, the dominance of its individual solution far exceeds those of the other approaches by a very large margin, 18580.04 (see Table 6). The computational time (in seconds) taken by each technique employed in this work to construct the entire Pareto frontier is shown in Figure 7:

The Pareto frontier generated using the Lévy-GSA technique takes the longest computational time followed by the GSA, PSO and CS techniques respectively. This behavior is correlated to the dominance levels of the solutions generated by the techniques. Since the Lévy-GSA and GSA approaches produce the most dominant solutions, the region of the objective space explored is very wide. Thus, longer time taken to scrutinize the large number of solutions. On the contrary PSO and CS techniques take lesser time since the region of exploration is much smaller. Additionally it is also possible that these techniques often prematurely converge at sub-optimal regions in the objective space.

Figure 6. Dominance levels for Pareto frontiers generated by all four techniques

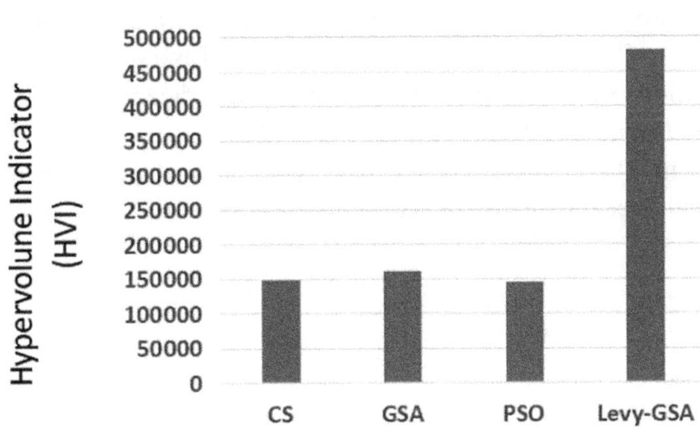

Figure 7. Computational time taken by each technique to construct the entire Pareto frontier

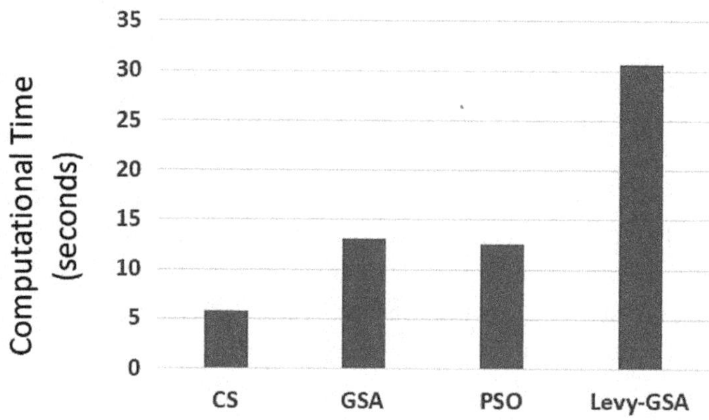

In terms of algorithmic complexity, the Lévy-GSA technique is not significantly more complex than other techniques utilized in this work. Besides the change in its stochastic generator, no additional computational steps have been added into the ones existing in the GSA technique. Thus the extended computational time may not have been caused by the increase in algorithmic complexity. In this work, all techniques produced converged solutions and no divergence issues were observed during program execution. All solutions produced were feasible where no constraints were violated and the decision variables (solutions) obtained were non-negative. The fitness requirements defined in all the techniques were respected during program execution. All techniques employed in this work performed stable computations during program execution.

CONCLUSION AND RECOMMENDATIONS

The Lévy-GSA approach outperformed all the other swarm intelligence tehcniques employed in this work in terms of Pareto frontier dominance. This was followed by the frontiers produced by the GSA, CS and PSO techniques. The individual solutions constituting the Pareto frontier produced by the Lévy-GSA technique was observed to concentrate on the optimal region in the objective space (see Figure 5). With respect to the best individual solution, the CS approach outperforms the GSA and PSO techniques while the Lévy-GSA approach proves to be far superior than all the other algorithms (Tables 4 -7).

In future works, the Lévy flight mechanism could be employed as a stochastic generator to other metaheuristics. This modification to the original metaheuristic may significantly contribute to the improvement of its performance. In addition, the hybrid approaches could be employed to facilitate the integration of the Lévy flight component into the metaheuristic. The NBI approach could also be employed (Ahmadi et al., 2015) to further enhance the quality of the solutions obtained.

REFERENCES

Abdelsalam, H. M., & Elassal, M. M. (2014). Joint economic lot sizing problem for a three—Layer supply chain with stochastic demand. *International Journal of Production Economics, 155*, 272–283. doi:10.1016/j.ijpe.2014.01.015

Ahi, P., & Searcy, C. (2013). A comparative literature analysis of definitions for green and sustainable supply chain management. *Journal of Cleaner Production, 52*, 329–341. doi:10.1016/j.jclepro.2013.02.018

Ahmadi, A., Kaymanesh, A., Siano, P., Janghorbani, M., Nezhad, A. E., & Sarno, D. (2015). Evaluating the effectiveness of normal boundary intersection method for short-term environmental/economic hydro-thermal self-scheduling. *Electric Power Systems Research, 123*, 192–204. doi:10.1016/j.epsr.2015.02.007

Awudu, I., & Zhang, J. (2013). Stochastic production planning for a biofuel supply chain under demand and price uncertainties. *Applied Energy, 103*, 189–196. doi:10.1016/j.apenergy.2012.09.025

Ba, B. H., Prins, C., & Prodhon, C. (2016). Models for optimization and performance evaluation of biomass supply chains: An Operations Research perspective. *Renewable Energy, 87*, 977–989. doi:10.1016/j.renene.2015.07.045

Brandenburg, M., Govindan, K., Sarkis, J., & Seuring, S. (2014). Quantitative models for sustainable supply chain management: Developments and directions. *European Journal of Operational Research, 233*(2), 299–312. doi:10.1016/j.ejor.2013.09.032

Bringmann, K., & Friedrich, T. (2013). Approximation quality of the hypervolume indicator. *Artificial Intelligence, 195*, 265–290. doi:10.1016/j.artint.2012.09.005

Castillo-Villar, K. K. (2014). Metaheuristic algorithms applied to bioenergy supply chain problems: Theory, review, challenges, and future. *Energies, 7*(11), 7640–7672. doi:10.3390/en7117640

Chen, Y. W., Wang, L. C., Wang, A., & Chen, T. L. (2017). A particle swarm approach for optimizing a multi stage closed loop supply chain for the solar cell industry. *Robotics and Computer-integrated Manufacturing, 43*, 111–123. doi:10.1016/j.rcim.2015.10.006

Corney, J. F., & Drummond, P. D. (2004). Gaussian Quantum Monte Carlo Methods for Fermions and Bosons'. *Physical Review Letters, 93*(26), 260401. doi:10.1103/PhysRevLett.93.260401

De Meyer, A., Cattrysse, D., Rasinmäki, J., & Van Orshoven, J. (2014). Methods to optimise the design and management of biomass-for-bioenergy supply chains: A review. *Renewable & Sustainable Energy Reviews, 31*, 657–670. doi:10.1016/j.rser.2013.12.036

Divyesh, V. R., Sahu, C., Kirubakaran, V., Radhakrishnan, T. K., & Guruprasath, M. (2017). Energy optimization using metaheuristic bat algorithm assisted controller tuning for industrial and residential applications. *Transactions of the Institute of Measurement and Control.*

Dong, N., Fang, X., & Wu, A. G. (2016). A Novel Chaotic Particle Swarm Optimization Algorithm for Parking Space Guidance. *Mathematical Problems in Engineering.*

Ganesan, T., Aris, M. S., & Elamvazuthi, I. (2018b). Multiobjective Strategy for an Industrial Gas Turbine: Absorption Chiller System. In Handbook of Research on Emergent Applications of Optimization Algorithms (pp. 531–556). Hershey, PA: IGI Global. doi:10.4018/978-1-5225-2990-3.ch023

Ganesan, T., Aris, M. S., & Vasant, P. (2018a). Extreme Value Metaheuristics for Optimizing a Many-Objective Gas Turbine System. *International Journal of Energy Optimization and Engineering*, *7*(2), 76–96. doi:10.4018/IJEOE.2018040104

Ganesan, T., Elamvazuthi, I., & Vasant, P. (2015). Multiobjective design optimization of a nano-CMOS voltage-controlled oscillator using game theoretic-differential evolution. *Applied Soft Computing*, *32*, 293–299. doi:10.1016/j.asoc.2015.03.016

Ganesan, T., Elamvazuthi, I., & Vasant, P. (2016b). Swarm Intelligence for Multiobjective Optimization of Extraction Process. In Handbook of Research on Modern Optimization Algorithms and Applications in Engineering and Economics (pp. 516-544). Hershey, PA: IGI Global. doi:10.4018/978-1-4666-9644-0.ch020

Ganesan, T., Vasant, P., & Elamvazuthi, I. (2016a). *Advances in Metaheuristics: Applications in Engineering Systems*. CRC Press. doi:10.1201/9781315297651

Gao, J., & You, F. (2015). Shale gas supply chain design and operations toward better economic and life cycle environmental performance: MINLP model and global optimization algorithm. *ACS Sustainable Chemistry & Engineering*, *3*(7), 1282–1291. doi:10.1021/acssuschemeng.5b00122

Gao, M., Ding, P., & Yang, Y. (2015, September). Time-Optimal Trajectory Planning of Industrial Robots Based on Particle Swarm Optimization. *Proceedings of the 2015 Fifth International Conference on Instrumentation and Measurement, Computer, Communication and Control (IMCCC)* (pp. 1934-1939). IEEE. 10.1109/IMCCC.2015.410

Ghaderi, H., Pishvaee, M. S., & Moini, A. (2016). Biomass supply chain network design: An optimization-oriented review and analysis. *Industrial Crops and Products*, *94*, 972–1000. doi:10.1016/j.indcrop.2016.09.027

Hong, Y. Y., Beltran, A. A., & Paglinawan, A. C. (2016). A Chaos-Enhanced Particle Swarm Optimization with Adaptive Parameters and Its Application in Maximum Power Point Tracking. *Mathematical Problems in Engineering*.

Jiang, S., Zhang, J., Ong, Y. S., Zhang, A. N., & Tan, P. S. (2015). A simple and fast hypervolume indicator-based multiobjective evolutionary algorithm. *IEEE Transactions on Cybernetics*, *45*(10), 2202–2213. doi:10.1109/TCYB.2014.2367526

Joshi, A. S., Kulkarni, O., Kakandikar, G. M., & Nandedkar, V. M. (2017). Cuckoo Search Optimization-A Review. *Materials Today: Proceedings*, *4*(8), 7262–7269.

Kumar, R. S., Kondapaneni, K., Dixit, V., Goswami, A., Thakur, L. S., & Tiwari, M. K. (2016). Multi-objective modeling of production and pollution routing problem with time window: A self-learning particle swarm optimization approach. *Computers & Industrial Engineering*, *99*, 29–40. doi:10.1016/j.cie.2015.07.003

Lin, T., Rodríguez, L. F., Shastri, Y. N., Hansen, A. C., & Ting, K. C. (2014). Integrated strategic and tactical biomass–biofuel supply chain optimization. *Bioresource Technology*, *156*, 256–266. doi:10.1016/j.biortech.2013.12.121

Ling, S. H., Chan, K. Y., Leung, F. H. F., Jiang, F., & Nguyen, H. (2016). Quality and robustness improvement for real world industrial systems using a fuzzy particle swarm optimization. *Engineering Applications of Artificial Intelligence*, *47*, 68–80. doi:10.1016/j.engappai.2015.03.003

Mareli, M., & Twala, B. (2017). *An adaptive Cuckoo search algorithm for optimisation.* Applied Computing and Informatics.

Marufuzzaman, M., Eksioglu, S. D., & Huang, Y. E. (2014). Two-stage stochastic programming supply chain model for biodiesel production via wastewater treatment. *Computers & Operations Research*, *49*, 1–17. doi:10.1016/j.cor.2014.03.010

Mastrocinque, E., Yuce, B., Lambiase, A., & Packianather, M. S. (2013). A multi-objective optimization for supply chain network using the bees algorithm. *International Journal of Engineering Business Management*, *5*, 38. doi:10.5772/56754

Mattos, C. L., Barreto, G. A., Horstkemper, D., & Hellingrath, B. (2017, June). Metaheuristic optimization for automatic clustering of customer-oriented supply chain data. *Proceedings of the 2017 12th International Workshop on Self-Organizing Maps and Learning Vector Quantization, Clustering and Data Visualization (WSOM)* (pp. 1-8). IEEE. 10.1109/WSOM.2017.8020025

Mousavi, S. M., Sadeghi, J., Niaki, S. T. A., & Tavana, M. (2016). A bi-objective inventory optimization model under inflation and discount using tuned Pareto-based algorithms: NSGA-II, NRGA, and MOPSO. *Applied Soft Computing*, *43*, 57–72. doi:10.1016/j.asoc.2016.02.014

Naidu, K., Mokhlis, H., & Bakar, A. H. A. (2014). Multiobjective optimization using weighted sum artificial bee colony algorithm for load frequency control. *International Journal of Electrical Power & Energy Systems*, *55*, 657–667. doi:10.1016/j.ijepes.2013.10.022

Ogunbanwo, A., Williamson, A., Veluscek, M., Izsak, R., Kalganova, T., & Broomhead, P. (2014). Transportation network optimization. In Encyclopedia of Business Analytics and Optimization (pp. 2570–2583). Hershey, PA: IGI Global. doi:10.4018/978-1-4666-5202-6.ch229

Osmani, A., & Zhang, J. (2013). Stochastic optimization of a multi-feedstock lignocellulosic-based bioethanol supply chain under multiple uncertainties. *Energy*, *59*, 157–172. doi:10.1016/j.energy.2013.07.043

Osmani, A., & Zhang, J. (2014). Economic and environmental optimization of a large scale sustainable dual feedstock lignocellulosic-based bioethanol supply chain in a stochastic environment. *Applied Energy*, *114*, 572–587. doi:10.1016/j.apenergy.2013.10.024

Pan, I., & Das, S. (2016). Fractional order fuzzy control of hybrid power system with renewable generation using chaotic PSO. *ISA Transactions*, *62*, 19–29. doi:10.1016/j.isatra.2015.03.003

Poudel, S. R., Marufuzzaman, M., & Bian, L. (2016). Designing a reliable bio-fuel supply chain network considering link failure probabilities. *Computers & Industrial Engineering*, *91*, 85–99. doi:10.1016/j.cie.2015.11.002

Seuring, S. (2013). A review of modeling approaches for sustainable supply chain management. *Decision Support Systems*, *54*(4), 1513–1520. doi:10.1016/j.dss.2012.05.053

Shi, J., Zhang, W., Zhang, Y., Xue, F., & Yang, T. (2015). MPPT for PV systems based on a dormant PSO algorithm. *Electric Power Systems Research*, *123*, 100–107. doi:10.1016/j.epsr.2015.02.001

Sottinen, T., & Tudor, C. A. (2006). On the Equivalence of Multiparameter Gaussian Processes. *Journal of Theoretical Probability*, *19*(2), 461–485. doi:10.100710959-006-0022-5

Srivastav, A., & Agrawal, S. (2017). Multi-objective optimization of slow moving inventory system using cuckoo search. In Intelligent Automation & Soft Computing. Academic Press.

Tan, Q., Wang, T., Zhang, Y., Miao, X., & Zhu, J. (2017). Nonlinear multi-objective optimization model for a biomass direct-fired power generation supply chain using a case study in China. *Energy*, *139*, 1066–1079. doi:10.1016/j.energy.2017.08.050

Yang, X. S. (2013). *Optimization and metaheuristic algorithms in engineering. In Metaheuristic in Water Geotechnical and Transport Engineering* (pp. 1–23). Waltham: Elsevier.

Yeh, W. C., Lin, W. T., Lai, C. M., Lee, Y. C., Chung, Y. Y., & Lin, J. S. (2016, July). Application of simplified swarm optimization algorithm in deteriorate supply chain network problem. *Proceedings of the 2016 IEEE Congress on Evolutionary Computation (CEC)* (pp. 2695-2700). IEEE. 10.1109/CEC.2016.7744127

Yue, D., You, F., & Snyder, S. W. (2014). Biomass-to-bioenergy and biofuel supply chain optimization: Overview, key issues and challenges. *Computers & Chemical Engineering*, *66*, 36–56. doi:10.1016/j.compchemeng.2013.11.016

Zhang, J., Osmani, A., Awudu, I., & Gonela, V. (2013). An integrated optimization model for switchgrass-based bioethanol supply chain. *Applied Energy*, *102*, 1205–1217. doi:10.1016/j.apenergy.2012.06.054

Zhao, F., Lei, W., Ma, W., Liu, Y., & Zhang, C. (2016). An Improved SPEA2 Algorithm with Adaptive Selection of Evolutionary Operators Scheme for Multiobjective Optimization Problems. *Mathematical Problems in Engineering*.

Zhou, C., Hou, C., Wei, X., & Zhang, Q. (2014). Improved hybrid optimization algorithm for 3D protein structure prediction. *Journal of Molecular Modeling*, *20*(7), 2289. doi:10.100700894-014-2289-2

APPENDIX

Nomenclature and Abbreviations

Biofuel Supply Chain Parameters

AC: abatement cost of carbon dioxide [yuan/kg]
$CEicb_{ik}$: increment of carbon dioxide emissions with loading each additional ton of biomass fuel per kilometer when broker k collects biomass fuel i [kg/t and km]
$AQmax_{i,t}$: maximum available quantity of local biomass fuel i in month t [t/month]
ACS: average electricity consumer surplus [yuan/kWh]
$CEitp_k$: increment of carbon dioxide emissions with loading each additional ton of fuel per kilometer when broker k transports biomass fuel to biomass power plant [kg/t and km]
$CEncb_{ik}$: carbon dioxide emissions per kilometer when broker k collects biomass fuel I with no-load conveyance [kg/km]
$CEntp_k$: carbon dioxide emissions per kilometer when broker k transports biomass fuel to biomass power plant with no load conveyance [kg/km]
CET: carbon dioxide emissions of thermal power plant for unit power generation [kg/kWh]
Dcb_{ik}: average transport distance when broker k collecting biomass fuel i [km]
Dtp_k: transport distance between broker k and biomass power plant [km]
E: efficiency of biomass power plant [decimal fraction]
EC: electricity consumption rate of biomass power plant [decimal fraction]
$extraY1$: first extra cost of excessive biomass power plant fuel inventory [yuan/month]
$extraY2$: second extra cost of excessive biomass power plant fuel inventory [yuan/month]
ER_k: expected return of broker k [decimal fraction mass/year]
FCb_k: fixed cost of broker k [yuan/year]
FCp: fixed cost of biomass power plant [yuan/year]
GC: unit generation cost of biomass power plant [yuan/kWh]
GS: government subsidies to biomass power generation [yuan/kWh]
GT: government tax revenues from biomass power plant [yuan/year]
HV_i : heat value of biomass fuel i [kJ/kg]
HVe: heat value of electricity [kJ/kWh]
HVmax: maximum heat value of mixed fuel [kJ/kg]
HV_{min}: minimum heat value of mixed fuel [kJ/kg]
$IQ_{i,0}$: inventory quantity of biomass fuel i at the beginning of month 1 [tonnes]
$CEcb$: carbon dioxide emissions during collecting biomass fuel [kg]
$CEtp$: carbon dioxide emissions during transporting biomass fuel to biomass power plant [kg]
EIC: extra inventory cost of biomass power plant [yuan]
$IQ_{i,t}$: inventory quantity of biomass fuel i at the end of month t [tonnes]
PP_i: purchase price of biomass fuel i from brokers [yuan/t]
$PQ_{ik,t}$: purchase quantity of biomass fuel i by broker k in month t [t]
q_t: electricity generation of biomass power plant in month t [kWh/month]
R_t: conversion rate from biomass fuel to electricity in month t [kg/kWh]

IQ_{max}: maximum inventory quantity of biomass power plant [t]
IL: rate of inventory loss [decimal fraction/month]
$LCcb_{ik}$: load capacity of conveyance when broker k collects biomass fuel i [t]
$LCtp_{ik}$: load capacity of conveyance when broker k transports biomass fuel i to biomass power plant [t]
$MCmax_i$: maximum moisture content of biomass fuel i required by biomass power plant [decimal fraction mass]
$MCorii_t$: original moisture content of biomass fuel i in month t [decimal fraction mass]
$MCaft_{ik}$: moisture content of biomass fuel i after processing by broker k [decimal fraction mass]
P: on-grid price of biomass power plant [yuan/kWh]
$PQ_{max, k}$: maximum purchasing quantity of biomass fuel by broker k [t/month]
q_{max} : maximum monthly electricity generation quantity of biomass power plant [kWh/month]
Q_{max}: maximum annual electricity generation quantity of biomass power plant [kWh/year]
Q_{min}: minimum annual electricity generation of biomass power plant [kWh/year]
$RIub_1$: first upper bound of reasonable fuel inventory [t]
$RIub_2$: second upper bound of reasonable fuel inventory [t]
$SIlb_i$: lower bound of safety inventory for biomass fuel i [t]
SC: unit storage cost of biomass power plant [yuan/month]
$SQmin_{ik}$: minimum supply quantity of biomass fuel i from broker
$TCcb_{ik}$: average unit transportation cost of broker k when collecting biomass fuel i [yuan/ km]
$TCtp_{ik}$: average unit transportation cost of broker k when transporting biomass fuel i to biomass power plant [yuan/km]
$WR_{ik,t}$: ratio of the weight of biomass fuel i after processing to the weight before processing by broker k in month t [decimal fraction mass]
$AP_{ik,t}$: average price of broker k buying biomass fuel i in month t [yuan/t]
$BC_{i,t}$: biomass fuel i consumption in month t [t]
$BR_{i,t}$: blending ratio of biomass fuel i in mixed fuel in month t [decimal fraction mass]
CER: carbon dioxide emissions reduction [kg]
CET_{eq}: carbon dioxide emissions of thermal power plant for power generation equal to biomass power plant [kg]
CEB: carbon dioxide emissions of biomass power plant [kg]
$SQ_{ik,t}$: supply quantity of biomass fuel i by broker k in month t [t]
VCp: total variable cost of biomass power plant [yuan/year]
$Y1t$: binary variable to determine whether the inventory is over $RIub1$ at the end of month t
$Y2t$: binary variable to determine whether the inventory is over $RIub2$ at the end of month t

GSA and Lévy-GSA

$G(t)$: Gravitational constant
n: Number of mass agents
α , ε : Constant parameter
$R_{ij}(t)$: Distance between agents
$F_i^d(t)$: Total force of each agent
$a_i^d(t)$: Acceleration of each agent

$v_i(t)$: Velocity of each mass agent
$x_i(t)$: Position of each mass agent
T: Function Evaluations

PSO

w, c_1, c_2, r_1, r_2: Constant Parameters
$v_i(t)$: Velocity of each particle
$x_i(t)$: Position of each particle
T_{max}: Maximum limit of function evaluations
s_i: Initial Social Influence
p_i: Initial Individual Influence

CS

β : Relaxation factor
t: Random Variable
y^t_i: Candidate solution
$Levy(\lambda)$: Lévy Distribution

Gaussian Distribution

$G_X(x)$: Gaussian Distribution Function
σ : Standard deviation
μ : Mean value

Chapter 13
Multi-Criteria Evaluation of Reconstruction Strategies for Distribution Power Networks Designed for Rural Power Supply

Tamara Leshchinskaya
Russian State Agrarian University – Moscow Timiryazev Agricultural Academy, Russia

Pavel Podobedov
Federal State Budgetary Scientific Institution, Federal Scientific Agroengineering Center VIM, Russia

Pavel Maslennikov
Federal State Budgetary Scientific Institution, Federal Scientific Agroengineering Center VIM, Russia

Anton Nekrasov
Federal State Budgetary Scientific Institution, Federal Scientific Agroengineering Center VIM, Russia

Alexey Nekrasov
Federal State Budgetary Scientific Institution, Federal Scientific Agroengineering Center VIM, Russia

ABSTRACT

The rural distribution network has deteriorated. This is due to the high failure rate of electrical equipment, high maintenance costs, reduced power quality, and the increased duration of power outages in agricultural production. This leads to a short supply of electricity, downtime of processing equipment, loss of production or production of low-grade products, as well as excessive energy losses during transmission. The important issue is the development of advanced methods for assessing the feasibility and effectiveness for component replacement of power transmission equipment with newer and more modern, reducing electrical energy loss in the distribution network. To solve these problems, various strategies have been developed and studied to improve the reliability of 10 kV overhead power lines by using modern insulators, wires, and supports.

DOI: 10.4018/978-1-7998-1216-6.ch013

Copyright © 2020, IGI Global. Copying or distributing in print or electronic forms without written permission of IGI Global is prohibited.

INTRODUCTION

Because of high wear level of power distribution networks, particularly that of network equipment (to 75%) and due to the worsening of maintenance quality, duration of power supply outages at sites is as high as 100 h to 120 h per year, while electric energy loss attains 20% to 30%. Voltage deviation is estimated to be in the range of –15% to +15%, at consumer's side. Asymmetry of voltages and currents, as well as harmonic distortion factors of voltage and current exceed essentially their highest permissible values which leads to undersupply of electricity and, therefore, to additional loss of electric energy and to low-quality products manufacture (Gerasimenko & Fedin 2008).

The major direction of reliability enhancement of 10 kV overhead lines is renovation of electric equipment fleet, improvement of operation performance, cost reduction of components, as well as the development of methods making it possible to evaluate the expediency of network components replacement. An important issue is reducing electric power loss, in 10 kV distribution lines, owing to the application of up-to-date equipment.

Recently, concepts and methods have been proposed enabling to optimize performance of power supply systems (PSS) based on the principles of decision analysis and multicriteriality (Knyazev 2005, Leschinskaya & Volkova 2018, Leschinskaya & Knyazev, 2016).

Method of multi-criteria efficiency evaluation of various 10 kV OHL configurations application has been developed, and the most viable re-equipment strategies were selected on its basis.

Purpose of research was to develop multi-criteria evaluation methods for the proposed reconstruction strategies for 10 kV OHLs to define optimal re-equipment options, based on the reliability indicators of electric components used in power distribution network sections.

DATA AND METHODS

Based on results of expert surveys followed by their analysis and on those of sampling statistical data related to 10 kV power distribution networks, reliability levels of currently operated lines have been studied. These studies made it possible to define the most perspective equipment for 10 kV OHL renovation. Multi-criteria evaluation was made for two options of partial criteria combinations followed by their convolution into an entire composed evaluation function. These results were obtained based on Bayes's criterion with the use of fuzzy-set theory. In the frames of multi-criteria model, parameters optimization was performed in accordance with the following algorithm:

- Development of strategies for operation performance improvement of 10 kV power distribution lines with the purpose of power supply reliability enhancement and electric energy quality perfection
- Selection and scientific description of partial criteria for practicability evaluation of electric equipment application designed to enhance power supply reliability and to reduce energy loss, in 10 kV power distribution networks
- Development of mathematical models for partial-criteria-based evaluation including analytical equations for partial criteria
- Selection of uncertainty factor for environment expressed in form of growth factor for perspective load extension (Leshchinskaya & Knyazev 2006)

- Partial criteria calculating for strategies under consideration, for various values of load growth factor
- Selecting a method for partial criteria convolution into an entire combined function
- Calculating values of matrix for evaluation functional
- Selecting optimal options in accordance with Bayes's criterion with the use of fuzzy-set theory

RESULTS AND DISCUSSION

Private Criteria Assessment

Evaluation criteria shall be based on the effectiveness and requirements for power supply system (Fishburn, 1964; Hurwicz, 1951). The following evaluation criteria have been specified for the expediency of network electric equipment replacement (Nekrasov & Podobedov, 2018):

- Reliability of power supply
- Electric power quality
- Economical effectiveness

The following partial evaluation criteria have been selected for the expediency of electric equipment replacement, in 10 kV OHLs:

- Electric energy undersupply (kWh) (see formula 1)
- Aggregate damage, in power networks (rubles) (see formula 4)
- Voltage drop (V) (see formula 5)
- Voltage asymmetry (%) (see formula 7)
- Cost of network re-equipment in accordance with the suggested configuration (rubles) (see formula 8)

Average annual electric energy undersupply due to possible failure of PSS components.

Electric energy undersupply (formula 1) is calculated based on the results of calculation of power supply outage duration (formula 2).

$$W_{under} = \sum S \cos\phi \left(T_{outage}\right) \tag{1}$$

$$T_{outage} = \sum_{i}^{n} \left(\frac{(\omega\tau + \gamma\mu t)}{100} \right) L \tag{2}$$

where

ω - is frequency of component failures (average annual number of failures per 100 km of network),
t - is duration of one outage (hours),

γ - is coefficient that takes the account of scheduled power shutdowns that are associated with lower damage,
μt - is annual duration of scheduled power shutdowns (hours),
S - is maximum aggregate power transmitted via 10 kV OHL (kW),
L - is length of 10 kV OHL (km).

On the basis of statistical data on failures of power network components and that of evaluations made by experts corrections were introduced into the equation for calculation of power supply outage time. With the use of database by Neklepaev (Neklepaev & Kryuchkov, 1989) relative failure rates for various components of 10 kV OHL have been defined that are presented in Table 1.

Expression for average annual failure rate (number of outages per 100 km) in 10 kV power distribution overhead lines has the following form (formula 3):

$$\omega = \omega_{pole} + \omega_{insulator} + \omega_{cable} + \omega_{fixture} \quad (3)$$

where

ω - is frequency of outages due to failure of components of 10 kV OHL,
ω_{pole} - is frequency of outages due to failure of pole,
$\omega_{insulator}$ - is frequency of outages due to failure of insulator,
ω_{cable} - is frequency of outages due to failure of cable,
$\omega_{fixture}$ - is frequency of outages due to failure of fixture.

Lack of input data for failure rate calculations is one of the major problems that were defined and solved, in the course of our studies. Corrections introduced into the calculation equation for failure rate of 10 kV OHL made it possible to not only separate available values corresponding to particular components but also to take account of the obtained initial data (see Tables 2 and 3), while evaluating suggested re-equipment strategies.

One more indicator of power supply reliability is damage due to electric energy undersupply (formula 4) (Levin & Leshchinskaya, 1989). It makes it possible to define damage in power networks, while engineering design of distribution lines and during their reconstruction, in case that enterprises do not provide corresponding input information.

$$D = D'W_{under} \quad (4)$$

Table 1. Failure rate for particular components

	Annual Failure Rate per 100 km	
Component Type	**Failure Rate Data by B. Neklepaev**	**%**
ω_{pole}	1.68	14
$\omega_{isolator}$	5.52	46
ω_{cable}	3.36	28
$\omega_{fixture}$	1.44	12

Table 2. Failure rate for various insulator grades per 100 km of 10 kV OHL

Insulator Grade	$\Omega_{insulator}$ (Average Annual Number of Failures per 100 km)
ShS-10	5.52
ShF -10	5.52
ODSK 16-20-B	2.76
OLK 12,5-10	3.56
ShS-20	5.02
ShTIZ-20 UO	4.41
ShPF-20A	4.41

Table 3. Failure rate for various cable grades per 100 km of 10 kV OHL

Cable Grade	ω_{cable} (Average Annual Number of Failures per 100 km)
AS	3.36
A	4.36
SIP-3	2.334
PZV	2.14
ASPT (k)	2.4

where

D′ - is specific average damage due to electric energy undersupply (rubles/kWh),
W_{under} - is annual electric energy undersupply (kWh/year),
D - is aggregate damage due to electric energy undersupply (rubles).

Voltage deviation at consumer side is one of the most important indicators of electric energy quality. This indicator depends on voltage drop in power distribution line. Voltage prop is calculated in accordance with the following formula (formula 5):

$$\Delta U_{max} = \frac{Pr_0 l + Qx_0 l}{U_{nom}} \tag{5}$$

where

r_0, x_0 - are active and reactive components of cable impedance (Ω),
P - is active capacity (W),
Q - is reactive capacity (VAr),
l - line length (km),
U_{nom} - is line rated voltage (kV).

Voltage deviation depends on voltage drop while the latter is defined by consumer loads. While designing 10 kV OHL such parameters as voltage asymmetry in phases, harmonic distortion, etc. are not normally taken into consideration.

For practical purposes, voltage drop is taken into account in combination with the method by French researchers A. Aiere and P. Gossen designed for estimating voltage quality, in point i of line, based on the second initial moment of distribution for voltage deviation in time (formula 6) (Budzko & Levin, 1985).

$$H_i = (1/T) \int_{t}^{t+T} \left[V(t)\right]^2 dt \tag{6}$$

where

$V(t)$ - is value of voltage deviation changing in time (%),
T - is observation period.

In accordance with correlation-regression analysis described in thesis by P. S. Pereverzev values of voltage dispersion for consumers is defined by the following coefficients (formula 7):

$$H=16.5 +0.078\Delta U^2_{\max OHL10} \tag{7}$$

Voltage dispersion value is considered to be an integral criterion of voltage quality.

Cost of complete re-equipment of 10 kV OHL is accepted as quality criterion that estimates economical effectiveness of electric equipment, рассчитываемая по formulae (formula 8).

$$C_{kl} =(C_{insul}\times K_{insul})+(C_{insul\ inst} \times K_{insul})+(C_{cab}\times L\times 3)+(C_{pol}\times K_{pole})+(C_{instpole}\times K_{pole})+(C_{fix}\times K_{pole}) \tag{8}$$

where

C - is cost of one 10 kV OHL component,
C_{instl} - is installation cost of one 10 kV OHL component,
K - is number of components of one type in 10 kV OHL,
L - is length of 10 kV OHL.

The aggregate cost of re-equipment shall be estimated with the account of component cost and that of their installation. Average market value was accepted as the cost of components and that of their installation.

Load growth ratio was involved as the indicator of uncertainty of environment where 10 kV OHL functions.

ANALYZABLE 10KV HIGH VOLTAGE LINES

Three configurations of power distribution overhead lines were selected as examples of multi-criteria optimization model application. One of them is that of Moscow high-voltage networks located in Podolsk Region that is a spur line of ‘Molchanovo’ distribution substation. Figure 1 shows its diagram of settlements connections. The second one is on the asset register of Orel Regional Power Distribution System. It has a large length and, apart from a number of settlements, supplies milk processing plant ‘Protasovo collective enterprise’. This network is fed from ‘Podlesnaya’ substation (see Figure 2). The third line is that of Lipetsk Regional Power Distribution System of Zadonsk Region that, apart from residential loads, provides power supply services to such consumers as 'Chmelenetsky' mine and large poultry farm (see Figure 3). Final selection of 10 kV OHL configuration depends on volume of damage in power distribution networks (see Table 4).

ANALYZABLE ELEMENTS OF COMPONOVATIONS OF HIGH-VOLTED LINES 10 KV

Components of 10 kV OHL used to build network configuration were selected based on the results of operation practice, availability of the required modern equipment, in Russian market, and recommendations by experts that proved perspectives of their application in networks (see Table 5 and Figures 5 to 7). For example, such enterprises as, LLC 'MZBA', LLC 'INSTA', LLC 'VOLTA' and JSC 'YU.M.E.K.', exposed their products at the exhibition 'Power Networks of Russia 2016', at their cooperative bench PO 'Forenergo' (see Figure 4).

In Table 4, values of damage due to electricity undersupply are given calculated for enterprises consuming electric energy from 10 kV OHLs under consideration.

Figure 1. Podolsk 10 kV OHL of Moscow High-Voltage Network

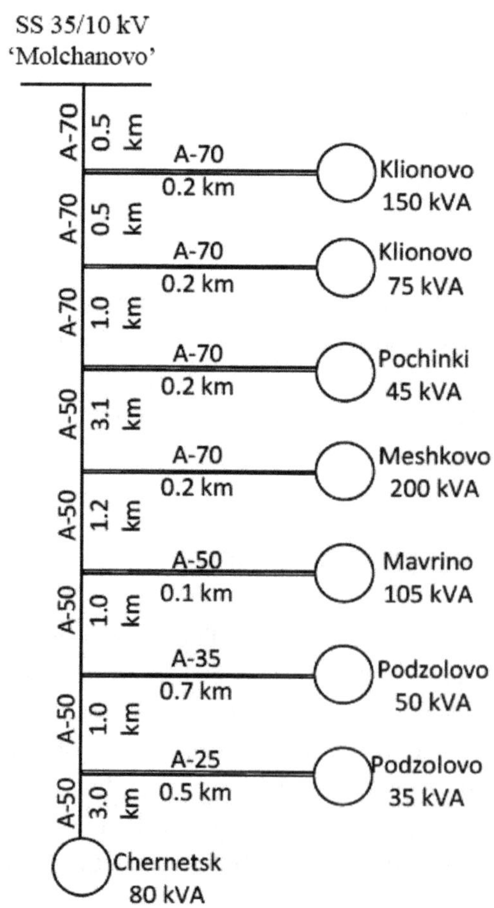

Table 4. Economic damage in power transmission lines

	Network	Section	Damage (thousand rubles)
1	Orel RES	'Podlesnaya' (Protasovo)	435
2	Podolsk MHVS	'Molchanovo' (Chernetskoe)	90
3	Lipetsk RES (mine)	'SS No.5' (Rudnik 3)	2559
4	Lipetsk RES (poultry farm)	'SS No.5' (Donskaya 3)	838

Figure 2. Diagram of 10 kV OHL of Orel RES

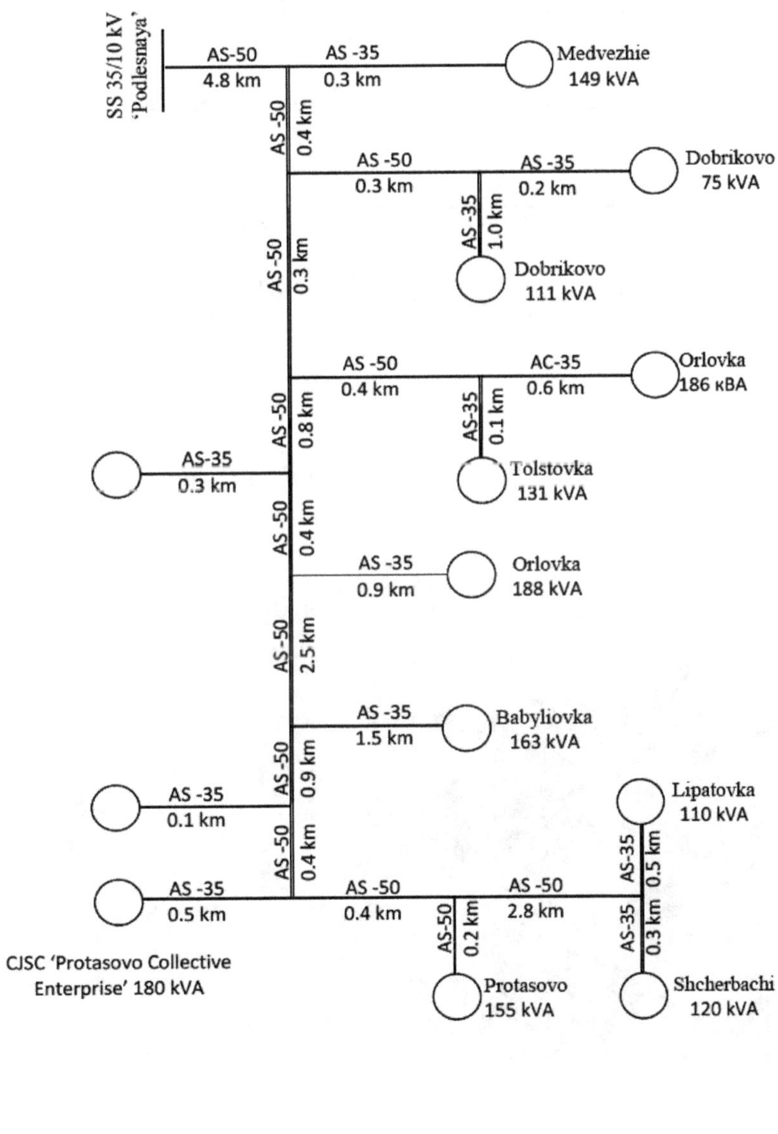

Figure 3. Diagram of 10 kV OHL of Lipetsk RES

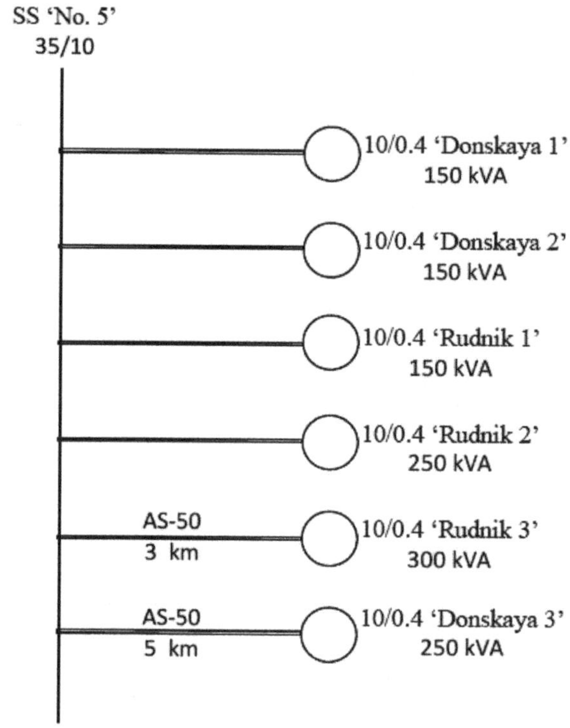

Figure 4. Cooperative exposition bench PO 'Forenergo' at the exhibition 'Power Networks of Russia 2016'

As it is clear from Table 4, damage in networks may substantially differ for supplied enterprises that differ in their production capacity which depends on requirements to reliability level of power supply, for different consumers. For each situation, its own specific configuration 10 kV OHL corresponds that has to be implemented by utilities, in the course of re-equipment, on the basis of available resources and with the account of possible aggregate damage.

Further growth of electric power consumption in Russia makes it necessary to perform re-equipment and of existing power distribution networks and construction of additional ones which is associated with the increase of load and, therefore, application of cables of larger ratings. Thus, a considerable volume of expenditures is required for cables replacement. This is not always technically possible since steel reinforced aluminium cables are rather heavy, while old-design poles are often not capable to bear cables of large cross-section area. That is why electric equipment market of Russia was investigated in order to specify optimal cable options for 10 kV OHLs.

Commonly used method of enhancement of cable attachment fixture reliability, in 6kV to 10 kV OHL, is application of pin type insulators of 20 kV rating, as well as new-type insulators that entered the market recently. These feature advanced reliability and improved technical characteristics such as mechanical strength, breakdown voltage, weight, cost.

Based on the analysis of their technical parameters insulator types were selected that are characterized by experts as the most perspective components for power distribution networks.

10kV OHL poles on the basis of SV115-7 stands have been designed with the aim to improve the reliability and durability of OHLs ensuring 5 to 10 times probability reduction of failure, in OHL, in case of ice and wind overloading and essential reduction of maintenance and operation cost, in stringent weather conditions.

Based on the expert evaluation and sampled statistical data, aggregate failure rate values have been derived for two existing and seven proposed 10 kV OHL configurations that can be used for further calculation of power supply outage duration. These data are given in Table 6.

Table 5. Configuration options of 10 kV OHL considered in multi-criteria evaluation model

Configuration No.	Pole Type	Insulator Type	Cable Type
1	Wood (impregnated)	ShS-10	AS
2	ZhB-SV 95-3s	ShF-10	AS
3	ZhB-SV 115-7	ShTIZ-20(UO)	SIP-3
4	ZhB-SV 115-7	ShPF -20A	PZV
5	ZhB-SV 115-7	OLK-12,5-10	PZV
6	ZhB-SV 115-7	OLSK 16-20-B	SIP-3
7	ZhB-SV 115-7	ShS-20D	ASPT(k)
8	ZhB-SV 115-7	OLSK 16-20-B	PZV

Table 6. Annual failure rate for OHL configurations under consideration

Configuration No.	Annual Failure Rate per 100 km
1	12
2	13
3	8.8
4	8.6
5	7.7
6	7.1
7	9.4
8	6.7
9	9.2

Figure 5. Perspective cables (from the left to the right): SIP-3, PZV, ASPT

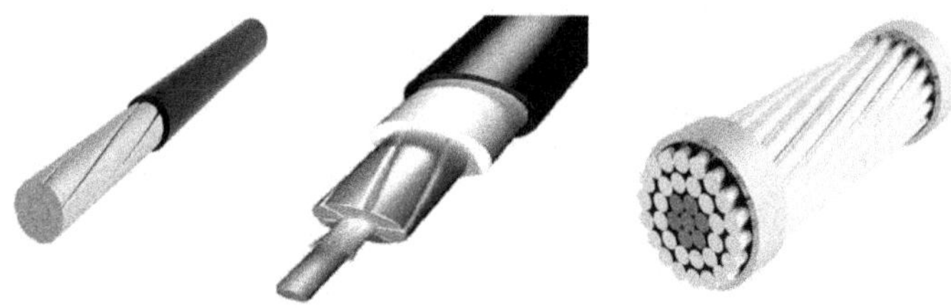

Figure 6. Perspective types of insulators (from the left to the right): ShTIZ-20; ShPF-20A, OLK-12.5-20, OLSK-16-20-4

RESULTS OF CONDUCTED MULTI-CRITERIA ESTIMATION OF OPTIMIZATION OPTIONS FOR 10 KV HIGH-VOLTAGE LINES

Apart from corrected initial data and equation reliability criterion calculation, uncertainty of environment has to be taken into account in which PSS functions. Load growth ratio was selected as the indicator of environment uncertainty (Zorin, Krushelnitsky & Fedosenko, 1977). In the course of analysis of information obtained as a result of expert survey, with the account of expected GDP, that in accordance with economic forecast amounts to 1.24%, values 1.03 and 1.06 of load growth ratio were assumed, for

Figure 7. Perspective pole of SV 115-7 type for 10 kV OHL

the next 5 years (Budzko, Levin & Leshchinskaya, 1987; Budzko, Levin, Terepko & Pereverzev, 1987; Zul & Lyzhko, 1986).

Formulas for two options of partial criteria combinations, for 10 kV OHL re-equipment evaluation strategies, are presented below (formulas 9 and 10):

$$F = \prod_{t=1}^{n} f_i = W_{under} \cdot U_{volt.\ drop} \cdot C_{comp.\ cost} \tag{9}$$

where

W_{under} - is electric energy undersupply (MWh),
$U_{volt.\ drop}$ - is voltage drop in networks (V),
$C_{comp.\ cost}$ - is aggregate cost of re-equipment with the account of installation works for selected configuration.

$$F = \prod_{t=1}^{n} f_I = D_{agr.\ dam} \cdot H_{volt.\ asym} \cdot C_{comp.cost} \tag{10}$$

where

$D_{agr.\ dam}$ - aggregate damage to consumer due to possible failure of electric equipment (thousand rubles),
$H_{volt.\ asym}$ - is voltage asymmetry in relation to its rated value (%),
$C_{comp.\ cost}$ - is aggregate cost of re-equipment with the account of components installation, for a particular configuration.

After calculating partial criteria of evaluation and after convolution of their matrixes into an integrated functional we can estimate Bayes's criterion (formula 11) designed to define most perspective options of 10 kV OHL re-equipment (Podobedov, 2014). The essence of Bayes's criterion is as follows: the optimal solution as that one to which minimal value of mathematical expectation for evaluation functional corresponds (Neimane, 2001) (see Table 7).

$$B_{\varphi 1} = \min \sum_{i=1}^{n} \left(P \cdot F \right) \tag{11}$$

where

p - is probability of load growth,
F - is value of evaluation functional.

It is seen from Table 7 that 10 kV OHL configurations 3 and 6 are the most perspective since they correspond to the case when the values of all partial criteria tend to their minimum. In certain conditions, configuration 8 provides rather high effectiveness. Selection of optimal options of re-equipment has to be carried out by enterprises that are responsible for particular network with the account of the balance between a required reliability level of power supply and available budget allocated to the activities on equipment replacement. To make the selection principle of network re-equipment strategy more descriptive, let us plot the graph for Bayes's criterion dependence on 10 kV OHL configuration selection, for various values of damage in network (see Figure 8).

It is clear from the graph that multi-criteria evaluation yields different results for networks of different types, i.e. the greater is possible damage the more effective become suggested renovation configurations. On all occasions, the most effective is configuration No. 6.

Table 7. Results of Bayes's criterion calculations, for 10 kV OHLs under consideration

Network	Orel RES		Moscow HVN		'Rudnik 3'		'Donskaya 3'	
Criteria Combination	No. 1	No. 2	No. 1	No. 2	No. 1	No. 2	No. 1	No. 2
φ1	14.2	0.97	7.4	1.1	67	32	43	86
φ2	14.4	1.03	7.5	1.1	66	31	42	85
φ3	15.1	1.11	8.2	1.25	7.06	9.1	2.14	10.7
φ4	24.2	1.70	13.52	1.93	12.2	15.6	3.74	18.9
φ5	22.1	1.54	10.96	1.76	11.1	14	3.45	18.4
φ6	13.1	0.92	7.10	1.0	6.5	8.5	1.94	9.6
φ7	19.1	1.69	8.3	1.7	17.0	36	6.90	62.4
φ8	19.9	1.42	7.6	1.6	14.3	12.9	3.12	16
φ9	17.69	1.26	6.38	1.3	11.1	16.9	4.75	43.1

Figure 8. Dependence of Bayes's criterion on the choice of re-equipment strategy, for 10 kV OHL under consideration, for various damage values

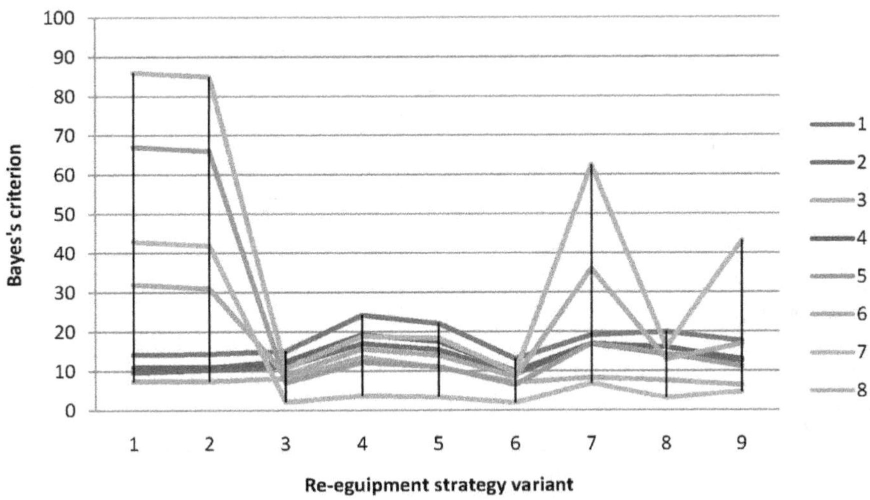

Fuzzy sets theory developed by (Zadeh, 1974; Zul & Lyzhko, 1986; De Luca & Termini, 1972; *Electrical world,* 1975; Kaufmann, 1975; Neimane, 2001) may appear a certain alternative to the methods based on the theory of chances. Indefinite factors of environment include perspective growth of load for a period of 5 to 10 years (Zorin, Krushelnitsky & Fedosenko, 1977). Apart for sampling and analyzing statistical data, it is advisable, in certain cases, to use expert evaluations treated in a relevant way to obtain additional information and to reduce the level of environment uncertainty (Levin & Leshchinskaya, 1989). Application of fuzzy sets theory provides a possibility to define a new approach to handling data of expert surveys developed by Beshelev and Gurevich. Since load values are not defined lets us perform evaluation process with the use of fuzzy sets theory. Effectiveness of expert procedure depends, to a substantial extent, on accuracy of questionnaire that have to not only make forecasted indicators more precise but also to contribute to evaluating expert's competence and objectiveness (Podobedov, Maslennikov & Nekrasov, 2017). Eleven experts representing various utilities took part in surveys (see Table 8).

In accordance with the method by Zadeh, membership function (MF) defines the extent to which element x within the load growth values interval (see Table 9) belongs to set of elements A. Values of this parameter are in the range of 0 to 1.

Membership function (MF), for fuzzy set, was defined for perspective load growth, for 5-year period (see Table 10 and Figure 9).

Final fuzzy set carriers have been determined depending on load growth ratio ($K_{l.g.}$) and PSS development strategies under consideration (see Tables 11 to 13).

Fuzzy sets for cost of 10 kV OHL re-equipment $\mu_{Cr.c.}$, as well as those of voltage drop $\mu_{Uv.d.}$ and electric energy undersupply μ_{Wunder} are functions of fuzzy sets for load growth comprising indefinite relationships described by their own membership functions $\mu \tilde{R}_1$, $\mu \tilde{R}_2$, $\mu \tilde{R}_3$ (see formula 12).

$$\mu_{\tilde{R}_{\bullet}}(K, S)=K(S)\,/K(S)_{max} \tag{12}$$

Table 8. Experts that took part in survey

Position	Experience (years)	Entire Length of Service (years)
Principal Metrology Engineer	5	10
Head of CL Maintenance Dpt.	8	17
Senior Maintenance Expert	4	15
Deputy Director, Substation Group '1'	14	28
Maintenance Engineer	12	24
Chief Foreman, Maintenance Service	10	18
Chief, Substation Group '2'	6	14
Deputy Chief, Substation Group '2'	2	6
Director, ZAO OVES	8	12
Chief Engineer, ZAO OVES	16	22
Deputy Chief, Substation Group '3'	5	8

Table 9. Integrated values of load growth

Interval No.	Interval Values
1	1.02 to 1.03
2	1.03 to 1.035
3	1.035 to 1.04
4	1.04 to 1.055
5	1.055 to 1.06
6	1.06 to 1.07
7	1.07 to 1.075
8	1.075 to 1.085

Figure 9. Membership function for fuzzy set, for 5-year perspective load growth period

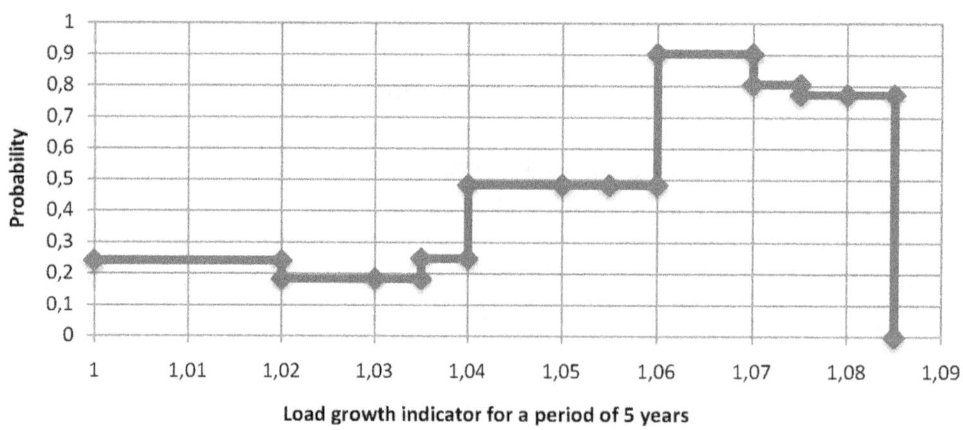

Table 10. Load growth factors and their probability values

NF $\mu\tilde{s}$	0.2	0.3	0.9
$K_{l.g.}$	1	1.03	1.06

Table 11. Membership function for uncertain target: electric energy undersupply (reliability of power supply).

Uncertain Target	$K_{l.g.}$	φ_1	φ_2	φ_3	φ_4	φ_5	φ_6	φ_7	φ_8	φ_9
μ_{Wunder} (MWh)	1	18.6	20.1	13.8	13.5	12.2	11.2	14.7	10.6	14.4
μ_{Wunder} (MWh)	1.03	19.2	20.7	14.2	13.9	12.5	11.6	15.1	10.9	14.8
μ_{Wunder} (MWh)	1.06	19.7	21.3	14.6	14.3	12.9	11.9	15.6	11.3	15.2

Table 12. Membership function for uncertain target: voltage drop (electric energy quality)

Uncertain Target	$K_{l.g.}$	φ_1	φ_2	φ_3	φ_4	φ_5	φ_6	φ_7	φ_8	φ_9
$\mu_{Uv.d.}$ (V)	1	685	685	643	643	643	643	685	643	685
$\mu_{Uv.d.}$ (V)	1.03	376	376	349	349	349	349	376	349	376
$\mu_{Uv.d.}$ (V)	1.06	395	395	361	361	361	361	395	361	395

Table 13. Membership function for uncertain target: cost of re-equipment (technical-economic indicator)

Uncertain Target	$K_{l.g.}$	φ_1	φ_2	φ_3	φ_4	φ_5	φ_6	φ_7	φ_8	φ_9
$\mu_{Cr.c.}$(mln. rub.)	1	7.7	6.8	12.5	19.7	20.1	13.6	16.7	20.7	12.8
$\mu_{Cr.c.}$ (mln. rub.)	1.03	9.1	8.8	13.9	23.2	23.6	15.0	18.7	24.2	14.8
$\mu_{Cr.c.}$ (mln. rub.)	1.06	9.1	8.8	13.9	23.2	23.6	15.0	18.7	24.2	14.8

Principal problems of setting up and solving multi-criteria optimization targets, in conditions of indefiniteness, are associated with uncertainty of both initial data and optimization criteria and limitations. In case of indefinitely specified problem, it is reasonable to talk not just about attaining a target but about the extent to which it can be attained. Besides, relevant limitations have to be taken into account (Glushkov, 1980). Membership functions for uncertain targets can be defined from the following equations (see formulas 13 to 15 and Table 14):

$$\mu_{W_u} = max[min\mu_{Wunder}, \mu\tilde{s}] \quad (13)$$

$$\mu_{\Delta u} = max[min\mu_{Uv.d.}, \mu\tilde{s}] \quad (14)$$

$$\mu_{\tilde{c}} = max[min\,\mu_{Cr.c.}, \mu\tilde{s}] \quad (15)$$

Table 14. Membership function for uncertain targets

Strategy	φ_1	φ_2	φ_3	φ_4	φ_5	φ_6	φ_7	φ_8	φ_9
μ_{Wunder}	0.9	0.9	0.68	0.68	0.6	0.55	0.73	0.53	0.71
$\mu_{Uv.d.}$	0.9	0.9	0.9	0.9	0.9	0.9	0.9	0.9	0.9
$\mu_{Cr.c.}$	0.37	0.36	0.57	0.9	0.9	0.61	0.77	0.9	0.61

Table 15. Membership function for common solution

Strategy	φ_1	φ_2	φ_3	φ_4	φ_5	φ_6	φ_7	φ_8	φ_9
Value	0.37	0.36	0.57	0.68	0.6	0.55	0.73	0.53	0.61

Membership functions for the common solution (see formula 16) are presented in Table 15.

$$\mu_D = \max\left[\min\left(\mu_{W_{under}}, \mu_{\Delta u}, \mu_{\tilde{c}}\right)\right] \tag{16}$$

Value 0.73 is the highest one and, therefore, configuration φ_7 has to be supposed to be the most effective.

Further studies will be focused on extending functional of the method, as well as that of the electric equipment inventory by including more new components. Electric equipment parameters of distribution substations will be studied with the aim of evaluation of power networks in their integrity.

CONCLUSION

1. Failure rate for various components of 10 kV OHL has been defined making it possible to adjust the expression for calculating power supply outage duration. The share of poles, insulators, cables and fixtures was found to be, respectively, 14%, 46%, 28% and 12% to the total time of outages.
2. Based on the data on failures by Neklepaev (12 outage per year), in 10 kV OHLs equipped with standard network components (ShS -10, AS, wood poles), and taking the account of share of particular components defined as the result of expert evaluation, quantitative indicators of failure rate have been determined, for components considered within configurations selected for analysis, as perspective ones.
3. Forecasted load growth rate was selected as environment uncertainty factor which made it possible to define effectiveness of certain 10 kV OHL configurations, for changing parameters of power consumption. Selection was performed by handling results of expert surveys and by constructing membership functions. Selected values of load growth rate are 3% and 6% during the next 5 years.
4. Partial criteria of evaluation were calculated with the account of environment uncertainty factor. Results of calculations proved power supply reliability enhancement and electric power quality improvement in accordance with the corresponding network configurations. For the case of load growth, the total annual power outage time is expected to decrease from 16.81 h to 8.9 h, in Orel

RES networks РЭС, for 10 kV OHL configuration using insulators of grade OLSK and enhanced-reliability poles SV 115-7.

5. Multi-criteria evaluation of the developed re-equipment strategies has been made with the use of fuzzy sets theory tools. Multi-criteria evaluation was carried out with the application of two methods using two combinations of partial evaluation criteria. In the first case, those were electric energy undersupply, re-equipment cost and voltage drop. In the second one, those were total damage due to power outage, voltage asymmetry and re-equipment cost.
6. Based on the data obtained as a result of convolution of evaluation functional matrix, Bayes's criterion was calculated which made it possible to select the most perspective options of 10 kV OHL re-equipment. i.e. configuration numbers from 3 to 9. Besides, configurations evaluation has been performed with the use of fuzzy sets theory to prove the practicability of application of the developed network re-equipment strategies.

With the use of fuzzy sets theory tools it was found out that strategies from φ_3 to φ_9 are practically equivalent. That is why decision makers are free to make the final selection of configuration. Responsible officers may base their choice on either power supply reliability and power quality criteria or on economic expedience depending on their preferences or consumer requirements.

REFERENCES

Aizerman, M. A., & Malishevsky, A. B. (1980). *Problems of logical substantiation of general decision theory. Part 2. Criteria of classical rationality and their application to the selection analysis*. IPU. (in Russian)

Budzko, I.A. & Levin, M.S. (1985). *Power supply of agricultural enterprises and settlements*. Moscow, Agropromizdat Publ. (in Russian)

Budzko, I. A., Levin, M. S., & Leshchinskaya, T. B. (1987). Evaluation of effectiveness of activities on electric energy loss reduction in operated 10 kV to 110 kV distribution networks [Abstract]. *Proceedings of the National Sci. Conf.*, Moscow (pp. 83-83). Academic Press. (in Russian)

Budzko, I. A., Levin, M. S., Terepko, O. A., & Pereverzev, P. S. (1987). *Complex evaluation of technical status of 10 kV and 0.38 kV rural networks. Power stations, 12*, 49. (in Russian)

De Luca, A., & Termini, S. (1972). Algebraic properties of fuzzy sets. *Journal of Mathematical Analysis and Applications*, *40*(2), 373–386. doi:10.1016/0022-247X(72)90057-1

Fishburn, P. C. (1964). *Decision and value theory. NY*: Wiley.

Gerasimenko, A. A., & Fedin, V. T. (2008). Electric energy transmission and distribution. Rostov-on-Don. Phoenix Publ. (in Russian).

Glushkov, V. M. (1980). *On system optimization*. Kiev: Kibernetika Publ. (in Russian)

Hurwicz, L. (1951). *Optimality criteria for decision making under ignorance*. Cowles Commission Discussion Papers.

Kaufmann, A. (1975). *Introduction to the theory of fuzzy subsets*. N.Y.: Academic Press.

Knyazev, P. V. (2005). *Selection and evaluation of power supply sources for remote rural areas [Extended abstract of dissertation]*. Moscow. V. P. Goryachkin. (in Russian)

Leschinskaya, T. B., & Knyazev, V. V. (2016). Ranking of the priority for the reconstruction of rural electrical networks 10 KV. Bulletin of the Federal State Educational Institution of Higher Professional Education, *4(74)*, 60-65. (in Russian)

Leschinskaya, T. B., & Volkova, E. A. (2018). The choice of the optimal variant of development of distribution air networks of 6-10 KV on a multi-criteria model Energy security and energy saving. Energy security and energy saving, 5, 23-30. (in Russian)

T. B. Leshchinskaya & V. V. Knyazev. (2006). *Multi-criteria evaluation of technical-economic status of power distribution networks*. FGOU VPO MGAU. (in Russian)

Levin, M.S. & Leshchinskaya, T.B. (1989). *Methods of decision theory in problems of power supply system's optimization*. Moscow, VIP Kenergo. (in Russian)

Neimane, V. (2001). On development of electricity distribution networks [Doctoral Dissertation].

Neklepaev, B. N., & Kryuchkov, I. P. (1989). The electrical part of power plants and substations: Reference data for course and diploma design. In Training manual for universities (4th ed.). Moscow: Energoatomizdat. (in Russian)

Nekrasov, A. I., & Podobedov, P. N. (2018). Multi-criteria evaluation of application of components of 10 kV overhead power distribution lines. *Bulletin of FGOU VPO Moscow State Agroengineering University, 1*(83), 63–68.

Podobedov, P. N. (2014). *Evaluation methods for selected designs of insulators and cables in multi-criteria model*. Innovations in agriculture, 4*(9)*, 246–247. (in Russian)

Podobedov, P. N., Maslennikov, P. A., & Nekrasov, A. I. (2017). *Expert evaluation of operation reliability of distribution substations electric equipment*. Electric equipment: Maintenance and repair, 11, 67–74. (in Russian)

Prokhorov, Yu. K., & Frolov, V. V. (2011). *Management decisions (2nd ed.). Saint-Petersburg: ITMO*. (in Russian)

Zadeh, L. A. (1971). Similarity relations and fuzzy orderings. *Information Sciences*, *3*(2), 166–200. doi:10.1016/S0020-0255(71)80005-1

Zadeh, L. A. (1974). *Basics of new approach to the analysis of complex systems and processes of decision selection*. Moscow: Znanie Publ. (in Russian)

Zorin, V. V., Krushelnitsky, A. Z., & Fedosenko, I. N. (1977). *Influence of uncertainty of initial information on 0.38 kV networks on evaluation of voltage quality. Problems of technical electrodynamics, 64*, 54–57. (in Russian)

Zul, N. M., & Lyzhko, V. M. (1986). Application of fuzzy sets theory to formalization of rational decision selection. *Proc. of MIISP*, Saint-Petersburg (pp. 62-67). Academic Press. (in Russian)

KEY TERMS AND DEFINITIONS

Criterion Bayesian: Criterion of maximum average win.

Fuzzy Set Theory: Is a branch of applied mathematics devoted to methods of analysis of uncertain data, in which the description of uncertainties of real phenomena and processes is carried out using the concept of sets that do not have clear boundaries.

Multi-Criteria Evaluation: Evaluation using several specific criteria.

Power Quality: The extent to which the parameters of electrical energy their established values.

Reliability of Power Supply: Is the ability of the electrical system to provide connected consumers with electrical energy of a given quality at any time interval.

The Fuzzy Set Membership Function: Is a generalization of the indicator (or characteristic) function of the classical set.

Chapter 14
Videocapsule Management of Agro-Technological Processes

Alexey Mikhailovich Bashilov
Moscow Aviation Institute (National Research University), Russia

Vladimir Alexandrovich Korolev
Moscow Aviation Institute (National Research University), Russia

ABSTRACT

The chapter reveals new opportunities for the use of digital video surveillance in the agro-industrial sector. It presents a methodological approach and practical guidelines for modeling and managing changes in agrotechnology, in the conditions of digital transformation of large amounts of data. The proposed technical solutions are aimed at creating better products and gaining competitive advantages, thanks to the improved relationship of specialists and the transition to predictive data analysis. Video surveillance as the most accessible way to obtain information involves the creation of large amounts of data, their long-term storage, fast retrieval, and diverse analysis. The possibilities of video surveillance systems widely used to ensure the safety of industrial buildings and the protection of areas equipped with automatic video analytics modules and integrated computer programs are considered. The organization of managing geographically-distributed agricultural production using mobile and remote video surveillance systems is shown. By using the proposed project of integrating digital video surveillance into a big data system, the organization can be transformed to improve strategy, make system decisions, improve marketing and sales results, improve efficiency and productivity of business processes, improve the performance of production personnel, minimize unprofitable risks, and maximize management efficiency, which will lead to the emergence of new opportunities.

DOI: 10.4018/978-1-7998-1216-6.ch014

Copyright © 2020, IGI Global. Copying or distributing in print or electronic forms without written permission of IGI Global is prohibited.

INTRODUCTION

Digital economy, as the basis of a new stage of technical and technological development of modern civilization, has a significant impact on all areas of human activity, including agricultural production. To successfully and effectively solve the problems of "digitalization", it is not enough to introduce any one information and control technology(Robert D. Fiete, 2012). In agricultural production, such tasks include the creation and implementation of accurate agricultural production technologies. Modern ideas concerning the implementation of agricultural technologies consider production systems a synthesis of natural and man-made systems –agrocenoses(Krausp,1980 & Kudrin, 2006). They include technical devices for providing and supporting agrotechnological processes (technocenoses) and natural structures (biocenoses), including living biological structures capable of self-development and self-regulation (plants, animals). The functioning of agrocenoses unites a large number of technological and biophysiological processes, distributed in time and space. Only an integrated system-oriented approach with a consistent and simultaneous use of several key information and communication technologies gives the expected effect (Norman S. Kopeika, 1998). This approach consistently and adaptively, in a cost-effective way, links the reality of agriculture (land, infrastructure, settlements and agricultural production) with their objectively measured digital representation. One of the most important components of digital control systems in agricultural production are vision systems and video surveillance for agrotechnological processes.

The purpose of the chapter is the development of a system-structured approach and the definition of practical guidelines for the design of modern information and control video surveillance systems in agricultural production, agrotechnical and agricultural enterprises based on video-digital integration, and transformation of large amounts of video data.

TECHNICAL AND TECHNOLOGICAL FEATURES OF DEVELOPMENT

Closed Circuit Video Surveillance in Agricultural Production

The approach developed in the article is aimed at creating better products and gaining competitive advantages by improving the relationship of agricultural production specialists and switching to operational analysis of video data that accurately and informatively reflects changes in agrotechnological processes.

In an effort to provide the most comfortable work for their users, manufacturers of various "smart" devices and developers of "digital" programs ensure their work in conjunction with each other. It can be traced even in everyday life: for example, through the Internet a modern mobile phone easily communicates with users on social networks and creates contacts and photos in an e-book. When building an agribusiness management system, combining different video surveillance tools into a single environment will help to multiply the efficiency and usability.

Technological processes in agriculture have significant differences from industry, they are associated with biological objects. These objects have the ability to self-organization and self-development. The most difficult problem is getting information about the behavior of biological objects and interpreting it through technical information and analytical tools for human understanding and decision making. The use of video surveillance, technical, machine and computer vision in the management of agrotechnological processes can be an effective way to improve agricultural production.

The key advantages are a significant increase in the information content, reliability, clarity, identity, efficiency of managing live, developing biotechnical, human-machine systems in a single language of video digital images with semantic-linguistic accompaniment. Video surveillance as the most accessible way to obtain valuable information involves the creation of large amounts of video data, their long-term storage, quick retrieval and diverse analysis.

Consider the capabilities of video surveillance systems, widely used for the safety of industrial buildings and the protection of territories (Damianovsky, 2013), which uses automatic video analytics modules and computer programs for integrating information and control flows (Soyfer, 2003 & Forsyth, 2012).

CLASSIFICATION OF VIDEO ANALYTICAL MEANS AND COMPUTER INTEGRATION PROGRAMS

The study of video surveillance technologies in foreign and domestic security and safety systems, as well as in applications for agricultural production(Bashilov, 2015), allowed us to classify the main modules of digital video analytics and software modules of computer integration that can be used to further improve the management systems of agricultural enterprises and agricultural production.

Fig. 1 shows a generalized classification scheme for typical video analytics modules that are integrated into a single digital management system for an agribusiness (agro-production).

Figure 1. A generalized classification scheme for typical video analytics modules

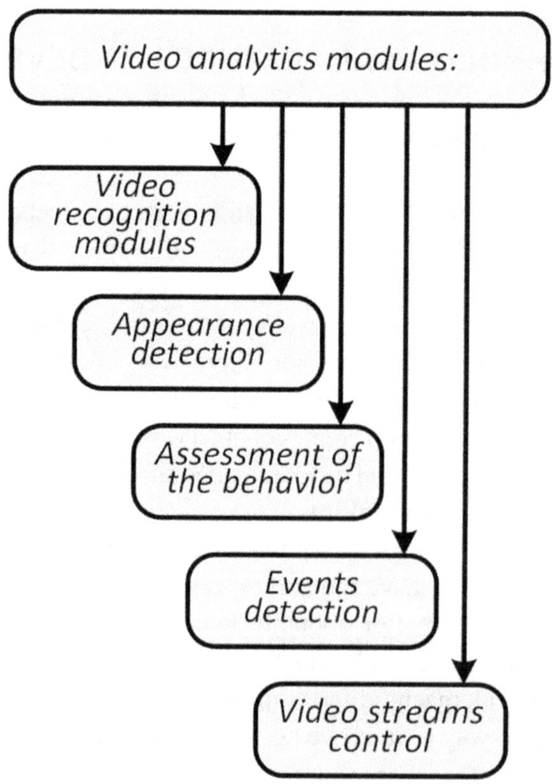

The text below shows an extended classification of typical video analytics modules with additional clarifications of their practical application, as well as a generalized classification of typical computer software modules integrating video analytics modules into existing objects of automated control of agrotechnological processes.

Expanded Classification of Typical Video Analytics Modules With Additional Clarifications for Use in Agricultural Production

1. Object recognition.
    a. Mobile vehicles (controlled automated access to the territory of the agricultural enterprise).
    b. The identification number of the animal attached to a specific location.
    c. Labels, symbols, barcodes on the process equipment, technical equipment, on the conveyor object of interest.
2. Appearance detection.
    a. Detects and enters into the database informative signs of a person.
    b. 3D monitoring (appraisal of animals, description of varietal and species characteristics of crops).
3. Assessment of the behavior and activity of mobile objects.
    a. Personnel monitoring (monitoring the presence and activity of personnel at the workplace).
    b. Animal control (monitoring the presence and routes of movement of animals in the herd).
    c. Plant control (monitoring of growth, development and fruiting).
4. Event detection
    a. Detection of animals, the queue length (counts the number of individuals and notifies when the threshold is exceeded).
    b. Smoke and fire detector (detects signs of smoke and fire in the field of view of the camera).
    c. Sabotage detector (detects camera malfunctions or deliberately disables them).
    d. Detector of departed objects (detects the departed object and the one who departed it).
    e. Loud sound detector (captures noise that exceeds a predetermined threshold).
5. Control of video cameras and video streams.
    a. Approaches, focuses, deletes, adjusts the mode of video surveillance.
    b. Processing audio streams (records and transmits sound in both directions from a surveillance camera to a specialist).
    c. Scanning an image (viewing a panorama divided into 2 or 4 frames).
    d. Tracking (monitors the perimeter of the selected area, border, line).
    e. Backup video channel with the image (creates a continuous video stream, records, regardless of force majeure factors).
    f. Interactive search and interchamber tracking (tracks the movement of an object between cameras, finds the object of interest in the archive).

Generalized Classification of Modules for Computer Software Integration of Video Analytics Modules

1. With messengers, thanks to this integration, it is possible to "communicate" with all video surveillance systems: receive images and alarm notifications from cameras, view the status of video servers. With the proliferation of fast and affordable mobile Internet, as well as high-end mobile phones, a new generation of instant messengers has emerged that focus primarily on mobile users.
2. With a single data storage and processing center for officials and authorities exercising their authority in accordance with their competencies.
3. With automatic mechanical devices (barriers, turnstiles, bolts) for their control.
4. With intercoms, with audio recorders, with outdoor day and night cameras.
5. With software and hardware system for recognizing numbers of auto tractor vehicles, with weighing equipment, with a cash control system, with a single dispatcher service system, with a cash flow system, with an access control system and fire alarm systems - to manage the safety of people and property.
6. With programs for the launch and development of automated process control systems and production.

Thus, on the basis of existing achievements of science and technology in the field of digital video surveillance, we can move on to a new innovative model of non-contact agricultural production (agriculture, animal husbandry). It is based on information integration, which results in the interpenetration of human knowledge and data on the collective self-organizing behavior of the objects of interest (plants, animals). Such a model will allow complementary, beneficial biotechnical cooperation based on mutually beneficial actions and results.

DESIGNING VIDEO SURVEENING TECHNOLOGIES AND VIDEO MANAGEMENT SYSTEMS FOR AGRICULTURAL OBJECTS OF RENEWABLE AGRICULTURAL PRODUCTIONS

Systematization and Modeling of Self-Organizing Agricultural Production

In agro-industrial systems, in order to maintain accurate (purposeful) agriculture, it is not enough just to observe the development of events; it is necessary to make wide and comprehensive use of knowledge about adaptive behavior. With the help of accumulated knowledge, the algorithm of continuous improvement is implemented, which can be easily fixed (saved) and replicated (inherited) with repeated similar applications. As a rule, automated control systems for agroprocesses and agroproductionare overgrown with biosensor attributes, becoming complex and expensive. However, they are worth using, because a systematic approach to solving a problem proves that the efficiency of a whole aspects integrated into a system is more than just the sum of the efficiencies of each particular element.

Fig. 2 shows a generalized diagram of the topological representation of a targeted, system-adaptive model of the technology of reproduction of agricultural products. It includes: consistently nested system-forming levels (biological, technological, environmental, etc.).Each system-level is represented in the form of a disc-shaped region with external and internal boundaries. Inside the topological area is filled with a functional environment and fixed objects. The functional environment in the active (adaptive)

Figure 2. Topological representation of a targeted, system-adaptive model of self-organizing agricultural production

T1, T2, T3 - components, structures, boundaries of different-purpose technologies; BIO - biological level of technology; TECHNO - the technical level of technology; IVF - ecological level of technology; A.Z. - adaptive zone; - - - - neutral interlevel border; 1 - external adaptive level boundary; 2 - internal adaptive level boundary; 3 - level object; 4 - level environment; 5 - border adaptive zone; 6 - post-page adaptive zone.

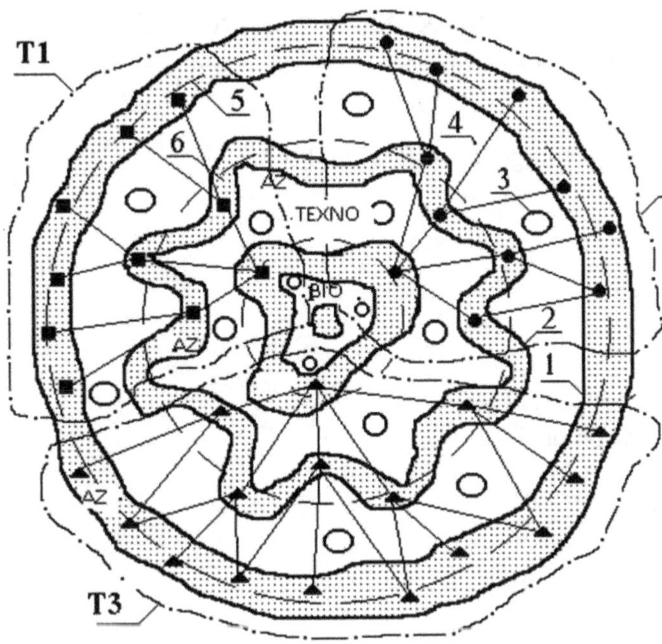

state can be compressed to the inner boundary of the disk or expand from the center of the disc to the outer boundary or perform these actions jointly. In the passive (non-adaptive) state, the functional environment does not change the outlines of the external and internal boundaries.

Objects (objects, signs, processes) of each level are immersed in their special environment, interacting with each other directly or through the nearest environment, adapt to a specific goal-oriented technology, for example: T1 - for processing, T2 - for long-term storage, T3 - for seeds.

Systemetry is necessary for system self-identification and determination of target adaptive functional capabilities in the context of the changing dynamics of the natural and social environment processes. Self-identification is an assessment of the nature-conformity, bio-adequacy and goal adaptability of the behavior of the objects of the agricultural system when fulfilling the designed goals(Bruce J. Schachter, 2017). Self-identification occurs by monitoring the objects of the natural and social environment and establishing their interconnection with the objects of the agricultural system.

Fig. 3 is a diagram of the process for improving the activities of agricultural production. The functioning of self-organizing agrarian production is informatively sensitive to the perception (measurement) of its own systemic complexity and is purposefully controlled based on the analysis of the results of system measurement.

The process of establishing the most important relationship between the objects of the agrobiotechnical system can be represented as a functional aiming at the ideal:

Figure 3. Scheme of the cycle of reusable creative acts of improving the activities of agricultural production.

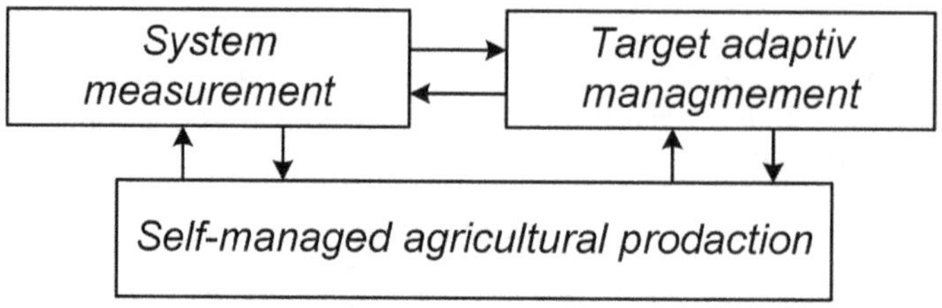

$$S = F\left(A, R, Z, C, T\right) \text{ ideal of “synergetics of EZN and IL”} \quad (1)$$

where

S is the structure of the target;
A is the set of objects of the system;
R– is the set of connections between objects;
Z is the set of connectedness compositions;
C – Wednesday;
T– period of system existence;
EZN– natural life;
IL – artificial life.

A person's ability to observe and recognize the functional, structural, dynamic complexity of an agrobiotechnical system is very limited due to: the lack of the ability to continuously and continuously monitor the behavior of the system's objects; inaccessibility to observe hidden, structurally branched objects of the system; inaccuracies in the perception of information about objects and the unreliability of its preservation; limited intellectual abilities for analysis, synthesis and systematization; lack of high sensitivity to the detection of synergistic effects and their evolution.

Most agro-industrial systems can be described as poorly organized (diffuse). In such a system, the task is not to determine all the components taken into account and their links with the system objectives. The diffuse system is controlled by a certain set of macroparameters, which are detected on the basis of selective analysis and subsequently distributed to the entire system. For a long time, the values of these macroparameters are not corrected and are not compared with the final results.

The ideal "replica" of a biotechnical system will be its full display and reproduction in a multidimensional structurally-organized information space (Sijiong Zhang, Changwei Li, &Shun Li, 2017). Each object of the system and each act of mutual relations should become carriers and transmitters of information "about themselves", receivers and transmitters of information "about the other", providing a multimedia and multiagent connection of "all with all." A biotechnical system, global in scale, total in its completeness, and volumetric in terms of accumulated information funds, may not be optimal for solving narrowly specialized tasks, but it will always be open for solving a varied "in life" spectrum of tasks, for active use in organization-reorganization, for training and management.

The Main Levels and Principles of Building Automated Video Surveillance Systems of the Objects of Interest in Agricultural Production Areas

CCTV Automation Levels

The agro-industrial zone, as a concentration of biological organisms, material flows and people, is the most difficult place in the technological chain of production of an agricultural enterprise. 70% of all agrotechnological operations are carried out according to an event scenario with human participation, 30% through self-organization and random influences. Automation of agrotechnological processes, minimization of losses of agricultural products and damage in the agricultural production zone is one of the priorities that the service personnel of the agricultural enterprise faces (Bashilov & Zaghinaylov, 2005).

In solving the problem of controlling agrotechnological operations and personnel actions, management may rely on video surveillance systems(Mironova & Kraiski, 2016). The construction of such systems is based on the synchronization of the video camera image with the agrotechnological schedule. The video stream is accompanied by information about the time of the implementation of the agrotechnological scenario. Registration of additional textual or other biometric information from the nodal agrotechnological operation, integrated with the video image, allows for the analysis of events. However, with long-term video monitoring, the video recording volume increases significantly and the viewing of the image archive can take a significant amount of time and, thus, the decision to correct or regulate the agrotechnological process is delayed or untimely. The first level of image analysis and decision making can be defined as poorly automated(Kopeika, 1998).

The complex behavior of biological objects, unprofessional actions of personnel and problems associated with ensuring the minimization of quality loss, maximizing productivity and energy saving, requires a more rapid response to negative situations. The factors mentioned above predetermined the second level of the development of video control systems for agrotechnological operations (Joseph Shamir,1999). They are based on the principle of describing the format of an event occurring at an agrotechnological video terminal (operation). Each of them corresponds to the moment of video recording only during "alarm events". That allows the dispatcher to get the opportunity in the shortest possible time to perform the necessary actions and his work becomes more operational. In the first case, the video surveillance system is integrated with the terminal; violations that are not visible by the camera fall out of its field of view. In the second case, the integration occurs directly with alarms that may come from external sensors controlling the agrotechnological process, for example: a sound sensor, a displacement sensor, a technological parameter control sensor, a moving object motion sensor. For events that have received the status of "alarming", the operator's attention is automatically drawn, and the video information is also automatically marked and structured by them. Analysis of the archived flow of information can now occur more purposefully, without a long look at all the information, but only from the moment the alarm signal appears.

The video monitoring system, automated at the third, higher level, significantly differs from the previously reviewed systems in terms of functionality. It provides for the detection and recognition of alarm situations based on internal information obtained from continuous analysis of the flow of images, registration of parameters of controlled objects and their morphological characteristics (Willers, 2013).

In an automated video monitoring system, you can define the desired set of actions with any object, with any set of morphological features of the object in any event in the agricultural production zone. Everything that happens in the frame of the agricultural production zone - the movement or movement of an object, the appearance of a new object or its disappearance, a change in the shape and size of an object, a change in texture and color, and the background of the environment - is an event. All that can be done in an automated video monitoring system is to turn on the camera to record or turn it off, issue a warning to the operator, turn on the alarm, put / remove sensors from surveillance, prohibit the passage, turn on / off the actuator, etc. - is a video administrative action.

Using a special scripting language, you can define an arbitrarily complex reaction of the video monitoring system to events, including the ability to determine the state and behavior of objects in space and time. In the automated control system there should be no restrictions on the description of the operation logic and the scale of the control system (the area of the object and the number of objects in the agro-industrial zone can be increased or changed).

The Principles of Building Video Surveillance Systems

The principle of motion detection. Video motion detection allows the operator to draw attention to movements in the agricultural production zone. This frees the operator from monotonous work and reduces the likelihood that an important event will go unnoticed.

The principle of automatic tracking of moving objects. This is the work of the system with PTZ cameras through the guidance program, as well as tracking of objects detected by a fixed camera with a mobile camera.

The principle of detection of morphological features of the object of interest. The main identification morphological features of objects are: shape, size, area, perimeter, orientation, color, surface uniformity, contrast spots of defects, texture, background.

The principle of video broadcast from adjacent or parallel objects. Video communication is carried out from two or more remote objects of the agricultural production zone via a local network or the Internet.

The principle of modularity - provides system flexibility. The system has a central core and a set of functional modules. Modules provide video cameras and sensors, service functions - sending messages, automatic redial, provide interfaces to the user.

The principle of a multilevel model - all functions of the system are divided into levels: the level of control of the system equipment; level of digital image processing; level of management and monitoring.

The principle of open architecture - a modular architecture, a multi-level functional model and open interfaces make it easy to increase the functions of the system, both by developing new modules and by integrating the system with already existing computerized systems of an agricultural enterprise.

The application of the aforementioned principles of formation allows the transition from a small autonomous system, while maintaining the potential for transformation into a large system as the needs of the consumer grow(Herbert Kaplan, 2007).

The basis for the use of the video network for the control of agro-production zones can be a system designed to address the safety and security of enterprises of certain domestic and foreign developers(Arnaud Darmont, 2013).The systems work with many types of cameras - analog, digital, stationary, PTZ, IP cameras.

Organization of the Process of Video Surveillance of Agricultural Objects Based on the Ethological Model

The main method in studying the behavior of animals (plant growth and development) is direct observation of each individual herd (population). Of which the best method is audio-visual surveillance. In addition, modern means of visual observation allows the direct observation of moving objects, to automatically recognize the behavioral acts of animals and plants. Long continuous observation is necessary in order to have an objective understanding of the dynamics of characteristic behavioral reactions of an object of interest .

An important point in the process of targeted audio-video surveillance at an agricultural farm is the early formation of a virtual model of the behavior of an object.

Let us consider, for example, an ethological model of animal behavior, which can be displayed as stereotypes and patterns of behavior known in zooengineering science (Bashilov&Legheza, 2011).

$$E = SBR + IR + AR^2 \tag{2}$$

where

E-ethological model of animal behavior;
SBR- specific behavioral reactions (cows, pigs, poultry);
IR -innate reflexes (instincts, individual reactions);
AR- acquired reflexes (learning, training).

The ethological model forms a system of reliable biological knowledge, patterns and mechanisms of behavioral responses of animals.

Knowledge of the species ethological features of each animal, the observation and fixation of instinctive behavioral acts and purposeful restructuring of animal behavior scenarios - this is the preparatory, programmable, anticipatory management function of zooengineering, which will provide a more advanced form (technology) of animals. The creation of such an innovative technology will: accelerate the growth and development of the animal, reduce energy losses from weakening emotional reactivity and stress, increase reproductive capacity, ensure continuous age-related behavior correction, carry out timely prevention of diseases, reduce labor costs and increase productivity.

Thus, the ethological model of animal behavior is the sum of complexes of functional actions:

$$E = \sum_{i=1}^{I} CFAi \, , \tag{3}$$

where

CFA_i - a complex of functional actions,
i– number of observed behavioral complexes.

I is the number of observed behavioral complexes.

$$CFAi = \sum_{j=1}^{J} GBAj \,, \tag{4}$$

where

GBA_j – goal-oriented behavioral act,
j–number of observables targeted behavioral acts.

Each focused behavioral act is characterized by postures (standing, lying, in motion), gestures (turning the head, position of the tail and ears), facial expressions (blinking eyes, opening the mouth, moving the lips).

$$GBAj = \sum_{m=1}^{M} PGFm \tag{5}$$

PGF_m – elements of the external manifestation of behavioral acts in form of poses (*P*), gestures (*G*), and facial expression (*F*).
m - the number of characteristic elements observed in a targeted behavioral act.

In order to have an objective understanding of the dynamics of characteristic functional actions of an animal, selective, long-term or continuous (with a given frame rate) observation of behavioral acts and their elements in time and space of habitat is necessary.

The use of video surveillance on livestock farms can significantly change the relationship between humans and animals, creating the effect of additional human contact with animals (the effect of the presence of a specialist on the farm when remotely monitoring animals), creating the possibility of closer attention, constant and complete control of animal behavior.

For any particular case on a livestock farm, you can create a vision system far exceeding the capabilities of the human eye, and sometimes a person as an image analyzer. At the same time, the use of special algorithms for processing the resulting image makes it possible to achieve solutions that are completely unexpected in terms of efficiency (with a large population of animals, accurately assess the condition of each).

To manage animals on farms requires a systematic look at each animal in the herd, placed in a stall or pasture. A systemic view can be formed by an "all-seeing eye," encompassing the external articles of the animal and its behavioral responses, reflecting the internal state and health of the animal.

Continuous, panoramic-volumetric tracking of each animal can provide a video network that can distinguish between an animal in the field of view of video cameras. Selective video recording can occur for animals going to the milking, in the aisles or passages, to pasture, on the scales, in the selection gate, during the distribution of food. The observation can be made from a stationary or mobile intelligent video camera that distributes the video stream to the database (video archive) of the personal data of the identified animal(Sinha, 2012).

As a result of systematically organized video for each animal, a video-time series of tracked behavioral acts can be compiled. From the analysis of the video stream, you can get information about the identification code (number, bar code, label, characteristic spot), geometric parameters (stats and animal postures), kinematic parameters (trajectory and speed of movement). Subsequently, this will allow for high-precision biometrics, testing, volumetric modeling, the formation of an ethological behavior model, and the diagnosis of the physiological state. Animal management teams can be carried out by sorting, redistributing, separating, carrying out preventive and agrotechnological measures aimed at ensuring the conditions of detention, health, productivity and quality of output.

The analysis of video images, video clips and video streams of animal behavioral acts can be performed visually by a specialist on the monitor screen by scrolling through the video archive. With a higher level of analysis of the personal video archive of the animal, some informative features can be automatically extracted from the plots viewed by special algorithms programmed by the video controller. With the highest level of video analysis, after automatic recognition and three-dimensional modeling, it is possible to build a tester (Kevin L. Priddy& Paul E. Keller, 2005)and ethological model of an animal with a matrix of biometric data. Such mutually integrated, highly accurate, virtual models will be available for quick study of the individual characteristics of each animal, especially when changing staff. The database will contain video sessions of each or interested animal sorted by the main technological processes: for milking, feeding, pasture, in a group, in a stall, etc. Looking at the personal data of the animal in manual, semi-automatic or automatic analysis mode, a complete picture of the behavior is formed animal daily, seasonal or perennial production cycles.

When designing an extensive, intelligent network of video surveillance on a livestock farm, several scientific and technical problems and problems arise. Where to install cameras? How to identify animals? How to distribute video information in a video archive? How to analyze the flow of video digital images? In what cases to make these or other decisions? How to manage the behavior of animals?

Let us consider possible solutions to these problems in the general structural diagram of an intelligent video surveillance network on a livestock farm (Fig. 4 and 5).

A multichannel encoder (matrix switcher) is capable of sending a signal to any output from any input and grouping or distributing video with audio and an identification control code.

Figure 4. General block diagram of a video surveillance system at a livestock farm

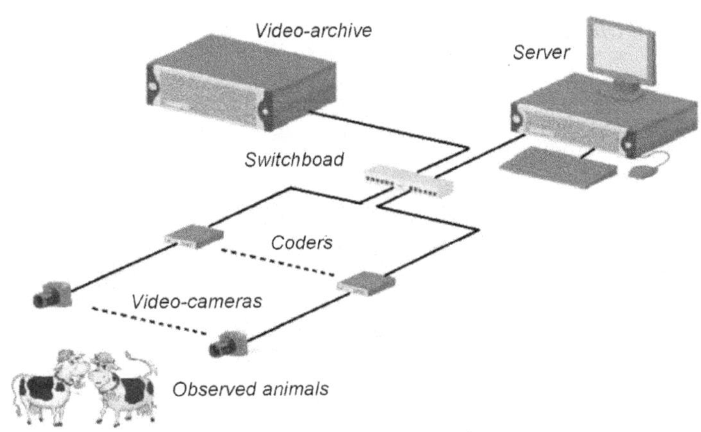

Figure 5. Structural and functional diagram of the video surveillance system on a livestock farm

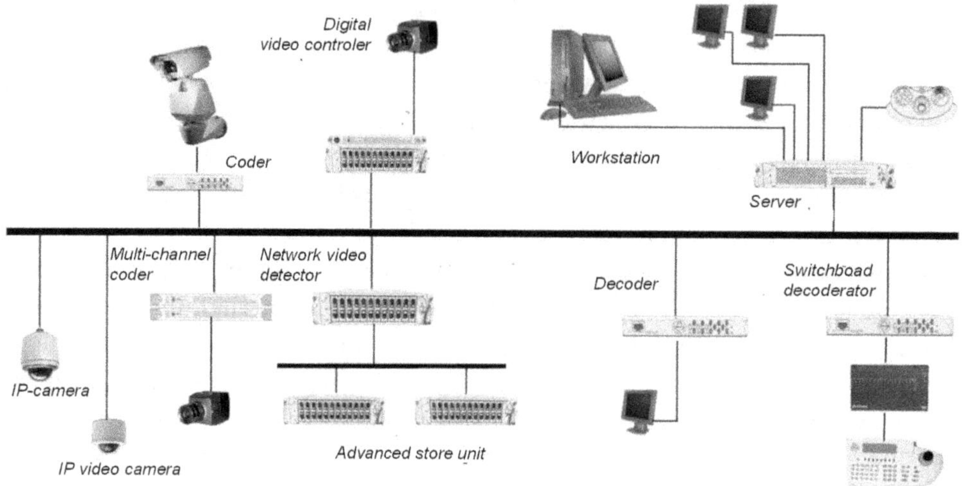

The task of finding behavioral patterns cannot be solved automatically. This process involves a close relationship with the expert. It is the person who must analyze the work of the algorithms, conduct different experiments with different parameters and analyze the results. Based on the work of some algorithms, the updated hypotheses can be confirmed or disproved(Lawrence A. Klein, 2012).

The control software can be configured to pool mobile objects in automatic mode - at certain intervals, or according to a predetermined schedule, or the control operator can perform manual polling - selecting an object or a group of objects from a list of moving objects prepared in advance in the program (Fig. 6).

As a result of the formation of a personalized video archive and analysis of the behavior of an individual animal, the efforts of specialists should be directed both to the satisfaction of biophysiological needs and to maximize milk production.

Using a video network to monitor animals will require the involvement of specialists trained to perform the functions of a video surveillance operator, the engineer and administrator will also need new competencies (Fig. 7).

Figure 6. The distribution scheme of the video information of the identified animal depending on the time and place of its content on the farm

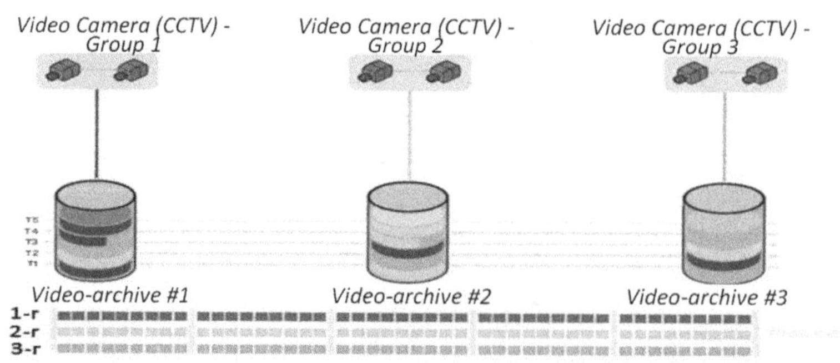

Figure 7. Basic video surveillance system management functions

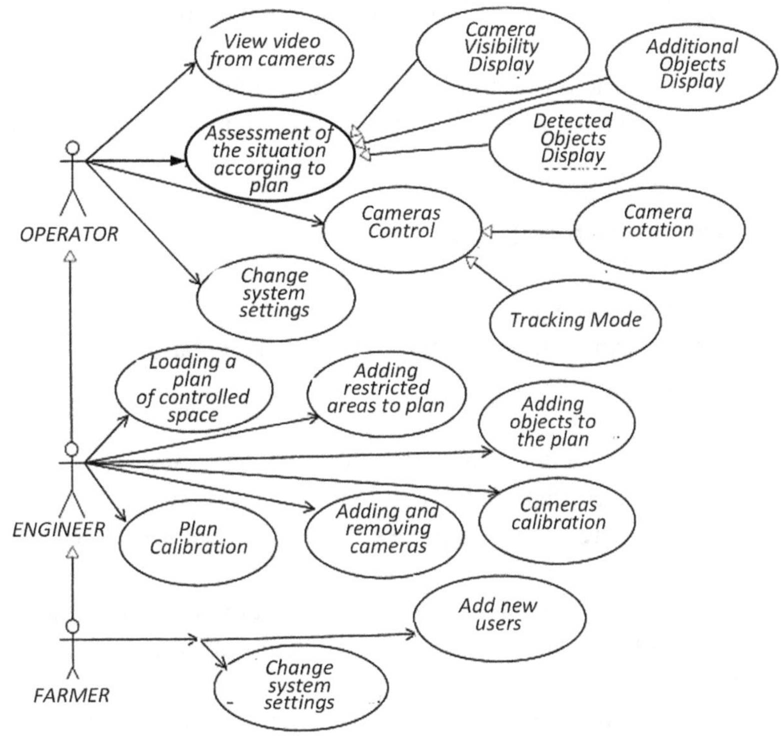

Problems solved using the video network:

- Visual observation of the real location, movement and behavior of animals and other moving objects.
- Storing the history of coordinates and videos of finding animals on the farm for a specified period of time.
- Reproduction of the route of movement on the digital map of the farm.
- Formation of statistical and reference data on the deviation of the animal from the route and violations of the schedule.
- Alarm in real time when deviating from the route or normal behavior.
- Determination of the general state of the herd of animals (the number of animals, the crowding and isolation of animals, the activity and anxiety of animals, the struggle for leadership).
- Inspection of the herd during grazing, milking, feeding, watering and rest (identification of weak and strong animals, healthy and sick).
- Monitoring the conditions of animals and the work of technological equipment.
- Determination of the general condition of the animal (body position in space, build, condition of coat and skin, the presence or absence of discharge from the nose, eyes, vagina).
- Observation of the indicated animal and determination of the structure of behavior (duration of lying, standing, feeding, watering, movement in the stall, movement to pasture).

- Detection of a sick animal by selective examination and observation of behavioral reactions.
- Detection of detailed signs of sexual activity and insemination of animals.
- Observation of prenatal signs and childbirth of an animal in a special room.
- Determining the nature of the disease by carefully examining the parts of the animal (with sufficient lighting, in the prescribed sequence: head, neck, chest, abdomen, udder, uterus, pelvis, limbs).

The Benefits of System Implementation

The introduction of modern technologies for determining the location, nature of the movement and behavioral acts of the animal during remote access to the video database significantly increases the efficiency of activities, including:

- Accelerates the process of communication and important decisions;
- Significantly reduces the response time to situations;
- Thanks to wireless communications employees receive remote access to databases with access rights;
- The real-time location of each animal is displayed on a digital map of the farm;
- The route of movement of animals and other mobile means is clearly recorded;
- Increases the efficiency of working time;
- Creates a reliable and sustainable mechanism for interaction of staff with animals.

The equipment and technology of the video surveillance system used ensure high work efficiency, increase the responsiveness, and productivity of the farm control center.

The introduction of the system effectively leads to a real reduction in the level of zootechnical disorders, as well as injuries among animals. The inevitability of sanctions for those who commit violations of standards will increase the level of discipline among farm operators.

A large number of cameras, a large number of users, horizontal and vertical integration of video surveillance equipment of various scales requires a careful approach when choosing a manufacturer of modern high-tech equipment for video surveillance systems for the farm(Levshin, Bashilov & Golovko, 2011). In fig. 8 shows an intelligent video surveillance system of animal behavior and vehicle movement.

The system of intellectual video surveillance is based on the integration of three components: the video subsystem; the subsystem for determining the local location; and the satellite system of global navigation. The intelligent video surveillance system based on location data from the positioning subsystem automatically determines the active video camera (the object is in view of which is located) and selects the video recording route. The video subsystem receives the video stream from the active camera and sends it to the video archive along the selected route or sends it to the operator. When using PTZ cameras, continuous monitoring of the object of observation is carried out throughout its movement. Through the use of a positioning system and video image analysis, an intelligent video surveillance system monitors a specific object without the involvement of an operator.

Figure 8. Structural and functional scheme of an intelligent video surveillance system for the behavior of animals and vehicles

1 - observed animals; 2 - livestock farm; 3 - a walking platform or pasture; 4 - vehicle; 5 - video cameras; 6 - access points, radio frequency identification base stations; 7 - tags, radio frequency identification tags; 8 - satellite global positioning system; 9 - server location subsystem; 10 - video subsystem server; 11 - video archive; 12 - automated workplace; 12 - the operator.

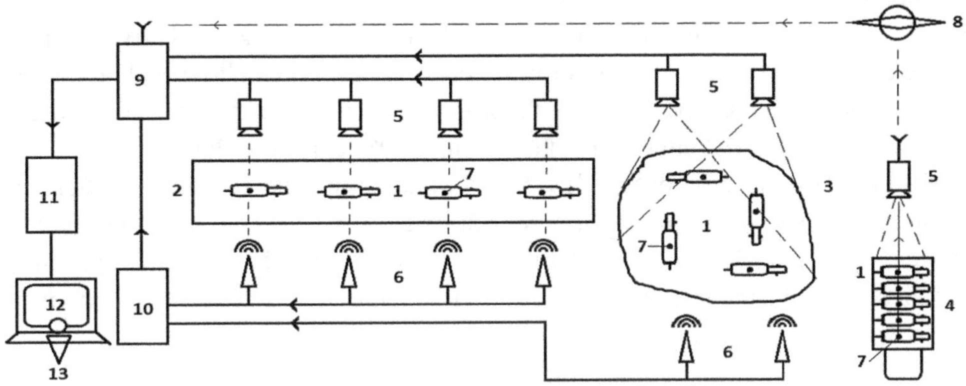

Integrated Video Monitoring and Management of Farms, Agricultural Areas and Renewable Agricultural Production

In recent decades, Russian agriculture has become, in some cases - highly industrialized and specialized, and in others - has been threatened with neglect, abandonment and afforestation of agricultural land. In the first case, there are environmental challenges to prevent damage from the intensification of agricultural production. In the second case, there are challenges related to the harmonized use of local resources that correspond to the climatic and geographical features of the area.

Obviously, sustainable agriculture is on a long-term basis between these two alternatives. Identifying and assessing the quality of agricultural areas and industries with a high environmental value in specific conditions is the actual, science-intensive task of designing a sustainable, high precision, energy-saving way of organizing enterprises of the agro-industrial complex (Strebkov,Kuznetsov,&Bashilov, 2010).

One of the most important problems in improving agricultural production is the logistical challenge of effectively visualizing and distributing information to professions concerning the search, discovery, allocation of material and energy resources (lost or unclaimed) so that they can make informed decisions concerning their use(Olsen, 2016).

The experience of using varietal, zoned, differentiated technologies for the production of agricultural products convincingly shows the existence of opportunities for progressive energy-saving optimization of agricultural technologies and agrotechnological processes. In accordance with this real-life pattern in agrotechnical (agro-production) patterns, the search for other optimization options has not been completed. There should be still unrevealed and unused opportunities, you need to look for multi-quality and different quality (spotting, rhythm, fusion, dispersion), its zones and boundaries (limits, shapes, orientations) of distribution in space and time.

Geodesic and geophysical data, terrain, hydrography, location of settlements and social infrastructure, production facilities and related elements, transport infrastructure, vegetation and soils, forest, arable land and crops, administrative units can provide valuable information in solving this problem.

Agroecological potential should be disclosed in a hierarchically differentiated form, taking into account the spatially distributed agro-industrial zones with renewable production cycles of agroproduction. It is advantageous to produce on the specifically selected territory what is most adapted to agroecological conditions or, if eco-resources are involved in production, or the same agrotechnological process can be realized with the lowest energy resources costs (Strebkov, Bashilov & Korolyov, 2011).

When building modern management systems for agricultural territories and production facilities, the key role is played by the use of geographic information systems, which allow evaluating the validity of the decisions made based on the modeling of technological processes, taking into account the conditions of the natural and agro-industrial environment (Kolesnikov,& Avanesov, 2007).

The concept of managing distributed GIS, based on adapting the structure of their multi-service communication networks to the structure of agro-industrial tasks, is to provide opportunities for quick information access to the accumulated and current data on the state (behavior) of agricultural objects, and their visualization at the request of specialists. Today, at the present level of development of science and technology, we have:

1. Distributed geo-information models and terrain models.
2. Satellite-based navigation stations and remote control of spatially distributed objects.
3. Systems of space, airborne, ground video monitoring of agricultural facilities and territories.
4. Video and multimedia methods and means of dispatching.
5. Innovative energy-saving projects of agricultural technologies and agricultural production.

In any technology of electron-optical monitoring and management of agricultural production, one of the main places is occupied by the video surveillance system, since it provides the most complete information about the identified object and its behavior (Bashilov & Korolyov, 2009). Fig. 9 shows a scheme for organizing the management of geographically distributed agricultural production using mobile and remote video surveillance systems.

Video cameras (4, 5, 8, 9, 10) directed to the objects of agricultural production give out the video signal (video frames, video series, video streams) at the output. Via coaxial cables and wireless radio channels, the video information arrives at the operator's workplace and through the switchboards(6) and (7) are displayed on the monitor screen. Additionally, various automated modes can be implemented: organizing and archiving video information with the possibility of repeated viewing; detection of the trajectory of movement of moving objects; object recognition by morphological features; focusing the operator's attention on video cameras that record deviations in the behavior of objects and agrotechnological parameters.

The local computer network or the Internet is usually used as a network, which allows you to deploy a video surveillance system based on an existing network. With the development of multiservice communication networks and their capabilities for processing video data in real time, the information and control capabilities of the agricultural production control system are significantly expanded, namely:

- Remote mobile video surveillance;
- Distributed geoinformation modeling;
- Prompt update of geospatial and navigation information;
- Geospatial video and video conferencing systems;
- Fixed, mobile and mobile complexes of technical equipment for video telephone communication;
- Means of aerial photography using an unmanned aerial vehicle.

Figure 9. Organization of management of geographically distributed agricultural production using mobile and remote video surveillance systems

1– multiservice communication network; 2– device of video-control with power unit and commutation unit; 3– on-board mobile video-complex; 4 – on-board video-camera; 5 – on-board video-camera on turning arm; 6 – commutator with packet-mode and high speed; 7 – controller of connection channels; 8 – portable complex of video-monitoring; 9 – mobile complex of video-monitoring; 10 – stationary stand of video-monitoring and management.

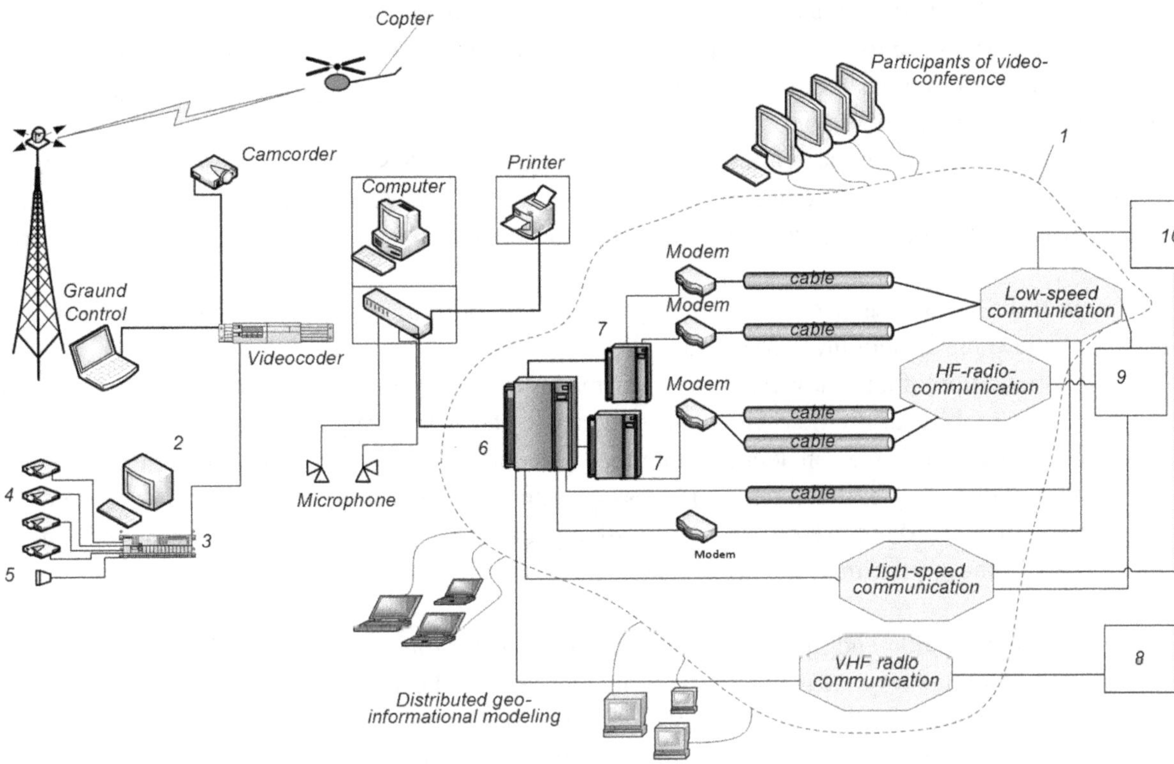

The structure of video surveillance complexes can include both their own means of creating a multiservice network and means that allow interfacing with territorial multiservice networks.

Video surveillance can be carried out both from ground cameras located on mobile vehicles and using video cameras located on an unmanned aerial vehicle(Barbara G. Grant, 2016). Ground-based video cameras make it possible to realize a circular observation mode for objects within a radius of up to 100 m, using (4) video cameras or objects 1.5 x 1.5 m in size up to 1.5 km away using a safety luminaire (5) mounted on a rotator.

The unmanned aerial vehicle included in the mobile complex provides video surveillance from a height of 100–400 m in a radius of up to 10 km from the mobile complex, thus providing a wider range of view compared with ground-based video cameras, as well as monitoring objects accessed by ground route is difficult.

Video information from a camera is transmitted via radio to a ground mobile complex. The UAV is controlled by the operator from the ground. The preparation of the camera for flight is not more than 5 minutes. Flight time - up to 1 hour. The launch and landing of the camera does not impose any special requirements.

Video information from video surveillance cameras through a video coder enters the system, is viewed by the operator and, if desired, any of the video cameras can be connected to the transmission system via communication channels to the stationary control station (10). The choice of a video camera can also be made by the operator of the stationary computer control tower.

At the stationary points of video surveillance and control, video information is collected from several mobile complexes. To display it, a specialized monitor, programs for analyzing and processing streams of video-digital images are used. The monitor allows you to display information from all directions of video surveillance, take control information from video registrars, accompanying sound and light signals or commands to automatic actuators.

The proposed technical solutions allow, in general, to build mobile systems for remote monitoring of agricultural production objects and territories of any length and size. At the same time, the constructed systems will meet the most modern requirements of the operational management of the agricultural enterprise.

The main tasks of the organization of energy-saving video-digital management of agricultural resources and renewable agricultural production.

The implementation of an innovative project of managing a geographically distributed agricultural production using mobile and remote video surveillance systems will require significant material and financial costs. However, if we consider the untapped possibilities, potentially existing, but unclaimed, then we have a deliberately collateral effectiveness of their use. Therefore, knowing that economic efficiency will be in demand in the long term, innovative goals and objectives should be predicted. Let us designate them in the approximate version of their final form:

1. Creation, organization and management of master plans for the development of the infrastructure of an agricultural enterprise using topographic, geodetic and cartographic information.
2. Improving the efficiency of functional planning infrastructures of an agricultural enterprise.
3. Monitoring the status of working areas and management of engineering infrastructure.
4. Operational observation of moving objects and vehicle identification.
5. Energy-saving inventory of territories, agricultural land, real estate.
6. Ensuring the safety of agricultural facilities.
7. Analysis of the terrain (search, detection, recognition, identification of lesions in crops) and identification of agro-ecological resources.
8. Video and photo observation of the operational situation (seasonal changes in terrain, distribution of green plantings by state categories, control of germination and survival rate of young plantings).
9. Monitoring and inspection of the state of the fields, the sharing of satellite images and digital aerial photographs obtained from RPV.
10. Providing reference and search functions and the provision of information from databases in a structural and descriptive form.
11. Designing two and three-dimensional models of the external environment with the help of long-distance and video digital vision systems.
12. Exclusion of the human factor, extension of the range of control of geographically dispersed objects.
13. Reduced administration costs, increased competitiveness.

SOLUTIONS AND RECOMMENDATIONS

The bibliography lists the most important domestic developments that reveal new opportunities for the use of digital video surveillance integrated into biotechnical and information management systems of agricultural production.

For the practical maintenance of accurate agrarian production, new robotized technical means are necessary, especially in the conditions of geographically dispersed location of agricultural facilities and in the absence of sufficient labor-related human resources.

In animal husbandry, for the individual care of animals and effective herd management, the creation of a "video shepherd" based on a multifunctional event-signaling mobile video surveillance unit, with self-learning and self-regulation operations, is promising.

In crop production, when implementing the technology of differentiated farming, which takes into account the state of each plant, it is necessary to create a "video technology field" based on the development of a multifunctional mechatronic-executive mobile video surveillance unit, with elements of homing and self-organization.

In nature management, especially in forestry, fish and hunting farms, it is advisable to create a "video inspectorate of territories" based on the development of a multifunctional reconnaissance-search mobile video surveillance unit, with self-learning and self-organizing algorithms.

Directions for Future Research

Other examples of the use of both local and integrated video surveillance systems in agricultural production are possible, providing prompt and clear presentation of information on the health and behavior of animals, on plant growth and development, on functional features and operation of technical and technological devices, on the interaction of geographically dispersed objects. The current level of development of video surveillance technology allows you to automate the process of managing video data based on typical video analytics modules and computer integration, carrying out operational processing, long-term storage and analysis of video data, the detection of negative events, proactive response to possible negative manifestations, visualization of predictive decisions.

Sharing video surveillance systems provides many options for obtaining synergistic effect: in assessing the individual state of the animal (identifying and determining the location of the animal in the herd, individual control and consideration of the parameters of the animal, keeping the calendar and history of the animal), in the process of milking (control of the operator's work and the behavior of the animal), during feeding feed, increase in live weight), in the process of insemination (identification of mate selection, observation of the calving of the animal), in assessing the mobility of the animal (control of the exercise, motor activity animal awnings, behavioral signs), during veterinary activities (appraisal, identification of diseases, formation of a calendar of veterinary activities).

CONCLUSION

So far agricultural production has unclaimed information and control resources available for the development and improvement of agrotechnological processes and agricultural enterprises, through the wider use and integration of video surveillance tools into a single self-organizing system of agricultural production.

To further improve agrotechnological processes and increase the speed and accuracy of management, the most promising potential tool is the development of control systems for robotic agrotechnological complexes using mobile remote automated video surveillance systems, video analytics, and video administration. New opportunities to improve the ergonomics and intelligence facilities of labor can be provided by designing management systems geographically distributed agricultural production using mobile and remote video surveillance, unmanned video service automation, video robots, remote web clients - included in a single integrated enterprise self-organization. Having been outlined in this chapter, the scientific approach to managing agricultural land resources and renewable agricultural production using mobile remote video surveillance systems is considered the most promising; its use in organizing agricultural production will be in demand.

REFERENCES

Bashilov, A.M. (2015). Innovational laser, optical and optic-electronic technologies in agrarian production. *Moscow, Technique and equipment for villages, 2*(212), 2-6.

Bashilov, A.M., &Korolyov, V.A. (2009). Video-monitoring and navigation in systems of precise arable farming. *Bulletin of MGAU Named After V. P. Goryachkin, 3*(34), 7-11.

Bashilov, A.M., & Legheza, V.N. (2011). Remote video-monitoring of animals' behavior in herd. *Series "Technique and equipment for villages", 12*, 24-26.

Bashilov, A.M., Sventitsky, I.I., & Korolyov, V.A. (2013). *Universal model of agro-technologies.* Moscow: Energy Engineering and Automatics.

Bashilov, A. M., & Zaghinaylov, V. I. (2005). Visualization and observation of systemic complexity of precise arable farming. *Proceedings of International scientific-technical conference "Intellectual systems".* Moscow: Publishing House "PhysMathLit".

Damianovsky, V. (2013). *From Light to Pixels.* Edition Butterworth-Heinemann.

Darmont, A. (2013). High Dynamic Range Imaging: Sensors and Architectures. Academic Press. doi:10.1117/3.903927

Fiete. (2012). Formation of a Digital Image: The Imaging Chain Simplified. Academic Press. doi:10.1117/3.919698

Forsyth, D. A., & Ponce, J. (2012). *Computer Vision: A Modern Approach* (2nd ed.). Moscow: Pearson Education, Prentice Hall.

Grant. (2016). Getting Started with UAV Imaging Systems: A Radiometric Guide. Academic Press. doi:10.1117/3.2239237

Kaplan, H. (2007). Practical Applications of Infrared Thermal Sensing and Imaging Equipment (3rd ed.). Academic Press. doi:10.1117/3.725072

Klein. (2012). Sensor and Data Fusion: A Tool for Information Assessment and Decision Making (2nd ed.). Academic Press. doi:10.1117/3.928035

Kolesnikov & Avanesov. (2007). Concept of creation of geo-spatial systems of video-connection based on new possibilities of multiservice networks of information change. *Scientific-Technical Magazine "Information and Space", 4,* 56-60.

Kopeika. (1998). A System Engineering Approach to Imaging. Academic Press. doi:10.1117/3.2265069

Krausp, V. R. (1980). Complex automatization in area of industrial animal breeding. Moscow: Publishing House "Machinostroyeniye".

Kudrin, B.I. (2006). Classics of technical coenoses. *Series "Investigations in cenology", 31.*

Levshin, A. G., Bashilov, A. M., & Golovko, V. A. (2011). Automatic piloting and introduction of centralized control of mobile aggregates. *Bulletin of Federal State Educational Institution of Higher Professional Education "Moscow State Agro-Engineering University named after V.P. Goryachkin, 2,* 18–22.

Mironova, T. V., & Kraiski, A. (2016). How to Use a Digital Camera as a Metering Device. Academic Press. doi:10.1117/3.2256725

Olsen, R. C. (2016). Remote Sensing from Air and Space (2nd ed.). Academic Press. doi:10.1117/3.2234477

Priddy & Keller. (2005). Artificial Neural Networks: An Introduction. Academic Press. doi:10.1117/3.633187

Schachter. (2017). Automatic Target Recognition (2nd ed.). Academic Press. doi:10.1117/3.2276891

Shamir, J. (1999). Optical Systems and Processes. Academic Press. doi:10.1117/3.336094

Sinha, P. K. (2012). Image Acquisition and Preprocessing for Machine Vision Systems. Academic Press. doi:10.1117/3.858360

Soyfer, A.V. (2003). *Methods of computer processing of pictures.* Moscow: Publishing House "PHYSMATHLIT".

Strebkov, D.S., Bashilov, A.M., & Korolyov, V.A. (2011). Principles of construction and versions of realization of power supply systems, navigation and movement management of promising agro-aggregates. *Polzunov's Bulletin, 2*(2), 280-284.

Strebkov, D.S., Kuznetsov, I.M., & Bashilov, A.M. (2010). Improvement of technical base of management of agricultural production on innovational-informational basis. *Series: Power supply and energy saving in agriculture. Moscow, State Scientific Institution VIESH, 5,* 49-65.

Willers. (2013). Electro-Optical System Analysis and Design: A Radiometry Perspective. Academic Press. doi:10.1117/3.1001964

Zhang, S., Li, C., & Li, S. (2017). Understanding Optical Systems through Theory and Case Studies. Academic Press. doi:10.1117/3.2267707

KEY TERMS AND DEFINITIONS

Agro-Industrial Self-Organizing System: A set of functionally, structurally and parametrically interconnected objects, with an evolutionary and purposefully changing form of adaptive, adaptive and executive-transformative actions, the implementation of which carried out on the basis of fundamental laws and regularities reflecting the effective interaction of the objects of the system in achieving a practical result, the most but satisfying the real social needs and the external conditions of the environment.

Agrocenoses: The functioning of agrocenoses unites a large number of technological and biophysiological processes distributed in time and space, they include technical devices for providing and supporting agrotechnological processes (technocenoses) and natural structures (biocenoses), including living biological structures capable of self-development, self-regulation (plants animals).

CCTV: Application technical, machine and computer vision in the management of agrotechnological processes to improve agricultural production.

Ethological Features: Characterize regularly or randomly occurring dynamic interactions of animals with each other or with objects of the environment, are recorded during continuous or event-based video surveillance in real time or at certain intervals of time.

Integrated Video Surveillance: Allows one person to simultaneously be in several areas of the object, personally observe from one point or in different places the development of events and, having an idea of the whole situation, make expedient decisions and remotely manage the situation on the object.

Presence Effect: It is provided by recording and displaying video images with high quality (accompanied by sound, color, text), which creates the effect of presence (more careful and prolonged attention) for an observer during actions that take place on a distant object.

Systemetry: An objective-subjective information-analytical mapping of the materially embodied agricultural production self-organizing system at different stages of its existence.

Video Administration: Management of agrotechnological processes, facilities, personnel of an agricultural enterprise on the basis of decision-making on the results of video surveillance obtained in manual, semi-automatic and automatic modes.

Video Analytics: Algorithmic processing of digital images (video-time series of tracked behavioral acts), from the analysis of which you can get information about the identification code of the object of interest, geometric data and kinematic parameters for the implementation of high-precision biometrics, 3D modeling, the formation of ethological behavior patterns, diagnostics of the physiological state of animals and plants.

Chapter 15
Modern Methods of Optimization in Models of Hydrothermal Coordination and Emission of Contaminating Particles in Power-Generating Plants

Gilberto Pérez Lechuga
Universidad Autónoma del Estado de Hidalgo, Mexico

Ugo Fiore
https://orcid.org/0000-0003-0509-5662
Parthenope University of Napoli, Italy

Francisco Martínez Sánchez
Universidad Autónoma del Estado de Hidalgo, Mexico

ABSTRACT

The problem of hydrothermal coordination is a classic example of nonlinear and large-scale dynamic mathematical optimization that involves the problem of the economic dispatch plus other constraints associated with the coordination of a pair of systems integrated into a network (i.e., the production of energy from thermoelectric and hydroelectric plants). A large number of methods and models that represent this problem, from classical methods of optimization to modern methods based on metaheuristics, exist. In this chapter, the classic structure of the problem is analyzed and new constraints are proposed, including those based on the level of customer service. Some recent modern formulations of the model are shown, and the existing literature is reviewed showing the modern methods of optimization based on heuristics. The authors illustrate it by optimizing an instance consisting of three thermo plants and three hydro plants through random search and genetic algorithm.

DOI: 10.4018/978-1-7998-1216-6.ch015

Copyright © 2020, IGI Global. Copying or distributing in print or electronic forms without written permission of IGI Global is prohibited.

INTRODUCTION

The problem of hydrothermal coordination is a classic problem in the literature of both nonlinear mathematical programming and electrical engineering. From an optimization point of view, the problem presents several areas of opportunity due to the complex nature of the same and the fact that new randomness inherent in models such as maintenance shutdowns, repair failures and others, are especially challenging for mathematicians and engineers building a master production plan under imperfect information conditions.

From an economic perspective, the problem has a fundamental importance due to the repercussions in terms of expenditures. A large number of industrially developed countries make great efforts to optimize their energy production and distribution systems through a highly competitive market due to the existence of a multiplicity of public and private companies that are dedicated in this area. It is well known that proper planning of the production and distribution of electricity can result in millions of dollars in benefits to companies that use modern optimization techniques for decision making in their respective companies.

For years, the mathematical optimization of processes has been a fundamental tool in the construction of adequate decisions for real and large-scale problems, immersed in an environment with little or no information on which to base oneself. The development of technology and the growing demand for products and services make it necessary to look for better methods to produce techniques and processes which are profitable and sustainable. In particular, the production and distribution of electricity is a field of opportunity for the development of models and modern optimization techniques capable of solving stochastic instances in an environment with little information, dynamic by nature and characteristically very large due to the great number of variables involved, restrictions and requirements for convergence and precision in reasonable times in their implementation into real systems. The use of exhaustive search algorithms does not give an exact solution and if the search space is very wide, the necessary calculation time can be unapproachable (combinatorial explosion) belying its usefulness.

Today, metaheuristics using random optimization, local search, iterated local search, grey wolves, voracious algorithms, acceptance by thresholds, ascent of hills with random reinitialization, simulated annealing, optimization based on ant colonies, swarm algorithms, taboo search, genetic algorithms, memetic algorithms, GRASP, search by stochastic broadcast, extreme optimization, and penguins search optimization are methods that support the exact methods developed in the past in order to provide practical solutions in time and form with a fairly reasonable accuracy and a speed of convergence useful for making decisions in real contexts.

Associated with the optimization of the operation of the model is also the minimization of the volume of particles emitted to the environment as a consequence of the operation of the thermoelectric plants. Notwithstanding the economic and mathematical importance of the problem addressed, in this analysis we are also interested in estimating and minimizing the volume of pollutant particles emitted into the atmosphere as a consequence of the operation of a group of thermoelectric plants associated with the system. To do this, we make an approximation to the state of the art in terms of mathematical models associated with the production of polluting particles emitted during the process of generating electrical energy in a thermoelectric system in order to incorporate such models into the general structure of the optimization model proposed here. For the above mentioned, we also focus our analysis on estimating and optimizing, through an indirect method, the quantity of polluting particles emitted into the atmosphere as a consequence of the operation of thermoelectric power plants.

In our proposal we intend to describe succinctly the structure of a mathematical model that defines the popular problem of planning the operation of a hydrothermal system in the short term. In particular, we are interested in addressing some of the different ways in which it is presented in real contexts, the authors and the modern methods used for its optimization. One area of special emphasis is the constraints associated with it as well as its meaning. Finally, we present an application case or instance where we compare two totally opposite methods for the solution the same: 1) a simple finite search algorithm and 2) the well-known search method using Genetic Algorithms. We also make an approximation to the particle emission model consistent with the system operation plan.

The motivation of this work is based on the interest that exists for the classical model of hydrothermal programming since, in addition to being a matter of practical importance in engineering, it constitutes a challenge for mathematicians and operations researchers. Every new technological constraint must be incorporated every day to reflect the complexity of the real model including the uncertainty of the incomplete information, and each new restriction implies new mathematical models that must be incorporated into the already difficult original problem. To deal with these large-scale problems, new methods for optimizing them, methods based on heuristics or metaheuristics, modifications or improvements to classical optimization methods and / or hybrid models that make use of both tendencies appear every day.

In this chapter we fully describe the original model of hydrothermal coordination, mentioning the most recent methods used to optimize the wide variety of existing models in the literature in an effort to superficially mark the evolution of the problem.

This proposal includes a review of the literature, the development of the mathematical model, some models to evaluate the contamination caused by the operation of the system and a restriction that guarantees the level of customer service. The proposal makes use of an instance to optimize a system formed by 9 generating plants. Two algorithms are implanted for its optimization and the comparative aspects of it are discus. We hope to contribute to the subject.

LITERATURE REVIEW

The efforts focused on minimizing the emission of polluting particles into the atmosphere have been reported in many different areas, especially in engineering and optimization. Normally a thermoelectric plant uses coal, diesel or gas, the most polluting being the mineral coal. Recent studies show the serious situation of pollution generated by the burning of fuels such as mineral coal (Shilin, Yufeng, Houzhang, Meng, Xuebin, Lituo, Chenping, Jianhong, Ting, Min, & Hongkian, 2016), (Wang, Hassan, Morgan, & Brown, 2011).

The following authors and methods are relevant. In (Tahir, Salman & Saaquib, 2016) the authors propose an improved hybrid approach based on the application of chaos theory in a differential evolution (DE) algorithm for the solution of a bio-objective constrained optimization problem. In their proposal, self-adjusted parameter setting in DE is obtained by using chaotic sequences. Then, a chaotic hybridized local search mechanism is embedded in DE to avoid trapping at local optima and to enhance its search space exploring ability. New heuristic strategies are developed to effectively handle the complex hydraulic and thermal constraints. An interesting example of mathematical programming applied to these types of models is found in (Zeynal & Eidiani, 2014). In their proposal, the authors develop a mixed integer linear programming model (MILP) to represent the problem. They propose an elaborate model based on pumped-storage and river system constraints and make use of the piecewise approximation approach and the branch and cut method.

Another interesting approach to the problem is found in (Cerisola, Latorre & Ramos, 2012), where the authors apply stochastic dual dynamic programming decomposition to a nonconvex multistage stochastic hydrothermal model in which the nonlinear water head effects on production and the nonlinear dependence between the reservoir head and the reservoir volume are modeled. They use nonconvex constraints to represent the production function of a hydro plant, as well as binary variables to represent the disjunctive programming.

Interesting approaches to solve nonconvex economic dispatch problems help explore the search space effectively. For example, in (Selvakumar & Thanushkodi, 2007) the authors use a particle swarm optimization (PSO) method to test three systems having nonconvex solution spaces. They report good results in large-scale models (with many variables). An interesting portrayal using the same technique can be seen in (Sun & Lu, 2010). In their proposal the authors present an improved quantum-behaved particle swarm optimization (IQPSO) for short-term combined economic emission hydrothermal scheduling. It is formulated as a bi-objective problem: (i) minimizing fuel cost and (ii) minimizing emission cost. They employ heuristic strategies in order to handle the equality constraints especially water dynamic balance constraints and active power balance constraints. It is important to note that this contribution contains a detailed analysis of the economic emission scheduling (EES) and combined economic emission scheduling (CEES) in hydrothermal scheduling.

In (Zoumas, Bakirtzis, Teocharis & Petridis, 2004) a genetic algorithm solution to the problem is presented. The generation scheduling of the hydro production system is formulated as a mixed-integer, nonlinear optimization problem and solved with an enhanced genetic algorithm featuring a set of problem-specific genetic operators, the solved problem comprising 13 hydro plants and 28 thermal units. In turn, (Provas, 2013) the paper presents a novel teaching-learning based optimization (TLBO) to solve the problem. It allows for nonlinearities like valve point loading effects of the thermal unit and the prohibited discharge zone of the water reservoirs of the hydro plants. The proposal is based on two basic concepts of education namely teaching phase and learning phase. In the first phase, learners improve their knowledge or ability through the teacher's methodology and in the second part learners increase their knowledge by interactions among themselves. The algorithm does not require any algorithm-specific parameters making the algorithm robust.

A perspective of the problem from the point of view of safety in the operation of the system is found in (Karami, Shaya, Aghaei &Ahmadi, 2013). This paper presents the application of the Mixed-Integer Programming (MIP) approach for solving the security-constrained daily hydrothermal generation scheduling which considers the intermittency and volatility of wind power generation. Their objective was to determine an optimum schedule of generating units for minimizing the cost of supplying energy and ancillary services while considering network security constraints. Here, the hydro plants multi performance curve with spillage and time delay between reservoirs were considered.

A version very close to the material developed in this chapter is found in (Fleten & Kristofferesen, 2008). They develop a multi-stage mixed-integer linear stochastic program for determining the short-term production plan for a price-taking hydropower plant operating under uncertainty. They use uncertainty when planning the operation day and the water that must be allocated among the reservoirs in order to strike a balance between current profits and expected future profits.

With regard to very short-term programming (one week), the sensitivity of the method is important. In (Dashti, Conejo, Jiang, & Wang, 2016), the proposal consists of short-term forecasting the precipitation in a medium-term horizon (e.g., 1 week). They study a two-stage robust scheduling approach for a hydrothermal power system considering water inflow uncertainty and employ a vector autoregressive (VAR) model to represent its seasonality and accordingly construct an uncertainty set in the robust optimization approach. They use a Benders' decomposition algorithm to solve this problem. Finally, a satisfactory review of the models used up to the year 2009 can be found in (Farhat & El-Hawary, 2009).

Recent applications of metaheuristic algorithms to the problem of hydrothermal coordination are as follows: Simulated annealing (Wong & Wong, 1994), (Christober, 2011), (Basu, 2005), (Pavlyukevich, 2007); Evolutionary programming (Yang, Yang & Wang, 1996), (Sinha, Chakrabarti & Chattopadhyay, 2003); Artificial Bee Colony algorithm (ABC) and Bath Algorithm (Suman, Nalin, & Sinha, 2018), (Gosh, Kaur, Bhullar, & Karar, 2019); Cuckoo search (Jubaer & Zainal, 2014), (Basu & Chowdhury, 2013), (Nguyen, Vo, Dinh, 2016), (Yang & Deb, 2010), (Walton, Hassan, Morgan & Brown, 2011), (Xiangtao, & Minghao, 2013); Neural networks (Basu, 2003), (Anastasiadis, Kondylis, Vokas, & Papageorgas, 2016); Swarm optimization methods (Nima, & Hassan, 2010), (Mandal & Chakraborty, 2011), (Chen, 2008), (Binghui, Xiaohui, & Jinwen, 2007), (Mandal, Basu & Chakraborty, 2008), (Hinojosa & Leyton, 2012), (Songfeng,Chengfu & Zhengding, 2010); Genetic algorithm (Sasikala & Ramaswamy, 2010), (Fogel, 1995), (Zoumas, Bakirtzis, Teocharis, Petridis, 2004); Benders decomposition (Lopez, Año & Ojeda, 2016), (Sifuentes & Vargas, 2007); Lagrangian duality (Ferrerira, 1992).

THE MATHEMATICAL MODEL AND ALTERNATE FORMULATIONS

In this proposal it is assumed that there exist m thermoelectric plants producing energy and n hydroelectric plants also connected to the system, Figure (1). Let the planning horizon be discretized into $t \in T$ uniform subintervals. Let S and H define the set of thermoelectric and hydroelectric plants that integrate the system respectively. Thus, the variable $P_i \in S, i = 1,2,\ldots,m$ represents the electrical power generated in the thermoelectric plant i and the variable $p_i \in H, j = 1,2,\ldots,n$ represents the electric power delivered by the hydroelectric plant p_j for all $t \in T$.

Figure 1. The three stages of construction of the model

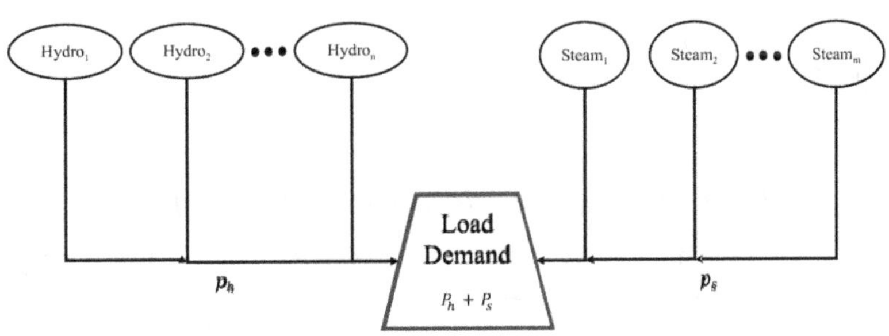

THE OBJETIVE FUNCTION

The objective function is normally written as the minimization of the sum of the partial cost functions associated with each thermoelectric production unit, i.e.,

$$\min_p g(p) = \min_{p_i} \left[f_1(p_1) + f_2(p_2) + \ldots + f_n(p_n)\right], f_i(p_i) \in C^2 \tag{1}$$

where $C(\mathbb{R}^2)$ is the set of all functions with second-order derivatives in $C(\mathbb{R}^2)$. Normally, $f_i(p_i)$ takes the form $f(p) = ap^2 + bp + c$ or similar functions; however, usually it is also customary to integrate into the objective function the operating costs of the hydroelectric plants as a function of time, i.e., by means of a function $h(\varphi)$.

Although the costs associated with the operation of a thermoelectric plant do not change abruptly, it is customary to write them as random variables. For example, if the costs associated with the operation of a thermoelectric plant can be expressed through the convex function in p, a typical stochastic cost function could be

$$g(p) = \min E_\theta\left[\tilde{g}(p)\right] = \min E_\theta\left[\sum_{i\in S} f_i(\theta_i, p_i)\right] + \sum_{j\in H} \gamma(\varphi) \tag{2}$$

where $E(\cdot)$ is the mathematical expectation operator, $\theta_i = (a_i, b_i, c_i), i \in S$, where a_i, b_i and c_i are random variables and the function $^3(\varphi)$ expresses the dependence of the operating costs of elements such as the size of the system, the head and the type of turbine (Dominique & Acker, 2018), (Renewable energy technologies: Costs analysis series, 2012). For example, some operational costs as a function of the output power are shown in Table 1.

Additionally, the objective function can also be enriched by placing fixed operating costs k during the system planning period T. It is important to mention that, some authors also include a complementary sinusoidal function to the objective function of the type $\left|e \sin[f(p^- - p]\right|$, [27]. Here, p^- is the lowest value of the range of operation of the thermoelectric plant.

An interesting form of the objective function is to write this in terms of binary variables as follows

$$\min_p g(p) = \min_{p_i} \left[\left(\sum_{i\in S} f_i(p_i) + k_i\right) z_i \middle| z_i \in \{0,1\}\right] + \sum_{j\in H} \gamma(\varphi) \tag{3}$$

Table 1. Costs of operating hydroelectric power plants based on the required power

Power Out (Kw)	5	25	50	100	250	500
Costs (USD)	2,869.89	5,217.98	8,218.32	14,349.44	32,612.38	63,007.11

Here, the variable z_i has the function of deciding if the thermoelectric plant $p_i \in S$ operates during the interval $t \in T$, $f_i(p_i)$ are the variable costs and k is the fixed cost associated with the model. The variable z_i is defined as follows

$$z_i = \begin{cases} 1, & \text{if the plant } i \text{ operates during } t, p_i \in S \\ 0, & \text{otherwise} \end{cases} \tag{4}$$

CONSTRAINTS ASSOCIATED WITH THE SYSTEM

The first group of constraints of the model are the obvious ones that must satisfy the demand of load requested online by the user, i.e., for all $t \in T$ we have

$$\sum_{i \in S} p_i + \sum_{j \in H} p_j + p_l \geq D \tag{5}$$

where p_l is the total loss of energy in the system (usually by Joul effect). The sum indicates that there must be a compensation in the production of energy to fully comply with the demand imposed D.

Another important group of equations have to do with the limiting values of power production by thermoelectric plant. Thus, for a given period $t \in T$

$$p_i^- \leq p_i \leq p_i^+, p_i \in S \tag{6}$$

Equation (6) limits the operation of thermo plants to the values of the interval in which they can operate. If the model must decide on the operation or not of the thermoelectric plant $p_i \in S$ during $t \in T$ then, Equations (4) and (6), must be linked in the following way

$$z_i p_i^- \leq p_i \leq p_i^+ z_i, p_i \in S, z_i \in \{0,1\} \tag{7}$$

A similar result is used in the case of hydroelectric plants in order to restrict the delivered power values. Here, p_i^+ is the maximum power value technically allowed in the thermoelectric plant i.

$$p_i^- \leq q_i(p_j) \leq p_j^+, p_j \in H \tag{8}$$

here, the limitation of the power delivered by the hydroelectric plants is given by the minimum q_j^- and the maximum water volumes q_j^+ that can be discharged from the dam j during a period t, i.e.,

$$q_j^- \leq q_j \leq q_j^+, \tag{9}$$

The water balance conditions in the reservoirs establish that there must be a quantity of water available at all times in order to satisfy fishing, recreational or simply technicalities of the system. This means that it is not possible to extract large volumes all the time but the quantities should be limited to certain intervals, then (Pérez-Lechuga, Medina-Marín, Tuoh-Mora, Mora-Cárdenas, 2010)

$$w_t = w_{t-1} - \left(q_t - s_t + r_t\right)t, w_t \geq \tilde{w}_t \quad (10)$$

This means that the water level w in the reservoir at the end of the period t (in m3/hr) must be equal to the residual volume of the previous period minus the volume of water q extracted during the current period plus the spillage rate over the reservoir during t minus the natural runoff due to rivers, rain, springs, etc. during the period t. Here, $\tilde{w}_t$ is the maximum level of water that can exist in the reservoir at the end of the period t, and $q\left(p\right)$ can be given by (Wood, 1996), (Pérez-Lechuga, Medina-Marín, Tuoh-Mora, Mora-Cárdenas, 2010).

$$q\left(p\right) = \beta_0 + \beta_1 p + \beta_2 p^2 \quad (11)$$

where the constants $\beta_0, \beta_1, \beta_2$ must be determined, because they are specific to each system.

In (Rajan, 2011) the power generated from a hydro plant $p\left(q\right)$, is related to the reservoir characteristics as well as the water discharge rate. The authors maintain that the hydro generator power output is a function of reservoir volume, v, and the rate of water discharge, q and propose that the model can also be written in terms of reservoir volume instead of the reservoir net head as follows

$$p = \left[c_1 v + c_3 q + c_4\right]v + \left[c_2 q + c_5\right]q + c_6, p \in S \quad (12)$$

where the electrical power $p \in S$ generated by any hydro plant can be obtained by knowing the power coefficients of this $c_i, i = 1, 2, \ldots, 6$ when there exists a volume v at the beginning of period t in the respective reservoir discharging a volume q of water.

An unusual constraint in this class of models is to ensure that the customer receives the quantity of product requested, that is, satisfies the demand in a timely manner. This concept is known as the level of customer service and it is defined as the probability that the required demand is satisfied through an adequate production plan. If we assume that the demand for electrical energy is a random variable D with known density given by $f_D\left(\xi\right) \forall t \in T$, then we must meet (Pérez-Lechuga, Medina-Marín, Tuoh-Mora, Mora-Cárdenas, 2010)

$$P\left(D \leq p_n + p_l\right) = \int_0^{\rho} dF_D\left(\xi\right) = \left(1 - \alpha\right), \alpha \in \left(0,1\right) \quad (13)$$

where $\rho = p_n + p_l$. In particular if $D = N\left(\mu_D, \sigma_D^2\right)$ then Equation (13) can be written as follows

$$P\left[D \leq \left(p_n + p_l\right)\right] = P\left[Z \leq \frac{\left(p_n + p_l\right) - \mu_D}{\sigma_D}\right] = \left(1-\alpha\right), \alpha \in \left(0,1\right)$$

where $Z = N\left(0,1\right)$. If K_α represents the standard value such that $F_D\left(k_\alpha\right) = \left(1-\alpha\right)$ therefore, Equation (13) is satisfied if

$$\frac{\left(p_n + p_l\right) - \mu_D}{\sigma_D} \geq k_\alpha$$

or equivalently

$$p_n + p_l \geq \sigma_D k_\alpha + \mu_D = \phi$$

which constitutes a deterministic constraint that can easily be included in the optimization code of the model. Note also that, the Equation (13) can be applied independent of the density function associated with demand D for theoretical or empirical probability density functions.

MODELS FOR THE EMISSION OF POLLUTING PARTICLES

Thermal power plants have been a major source of air pollution for a long time. Despite many new alternatives which produce significantly low emissions to generate electricity such as by oil, natural gas, hydro power and nuclear power etc., coal has remained the major source for electricity generation due to indigenous availability (Ayub, & Kumar, 2011).

In relation to the polluting particles emitted to the air by the thermoelectrical plants, the sulfur dioxide and nitrogen oxides are the gases that receive the most attention because they are the most harmful. Equation (14) devotes great attention to the measuring of emission of oxides of nitrogen ε in parts per million) (Rui, Jianzhong, Wang & Yongkiang, 2012), (Moorthy, Sangameswaraju, Ganesan & Subramanian, 2013).

$$\varepsilon = t\left(\lambda_1 + \lambda_2 p + \lambda_3 p^2 + \lambda_4 e^{\lambda_5 p}\right) \tag{14}$$

where $\lambda_1, \lambda_2, \lambda_3, \lambda_4$ and λ_5 are parameters to be determined in each plant emitting particles. Note that Equation (14) could be included in Equation (1) as part of the objective function.

Empirical statistical approaches for the modeling of particle emission based on chemical mass balance methods, multivariate models, deterministic models, Lagrangian models, Gaussian plume model, Eulerian models, modeling chemically reactive compounds and more, are reported in (Rusell, 1998).

Currently available control technologies in use for reduction of emissions of air toxics from coal-fired power plants are reported in (Emission of hazardous air pollutants from coal-fired power plants, 2019). A very comprehensive treatise on modeling air pollution can be found in (Varma, 2014).

SOME OTHER APPROACHES OF THE PROBLEM

Sometimes it is possible to express the costs of a thermoelectric plant based on the heat required for its operation. Let d_t be the heat required during period t, then if there were no limitations in the use of the resource, the following models were proposed in (Schaumburg, 2018).

$$\min \sum_{t\in T}\sum_{s\in S} \phi^s \left(\left(m_{1t}^u \delta_{1t}^s + m_{2t}^u \delta_{2t}^s \right) c^{kv} + m_{kt}^s c^k - \pi^s c_m \left(m_{1t}^u \delta_{1t}^s + m_{2t}^u \delta_{2t}^s \right) \right) \tag{15}$$

s.t.

$$m_{1t}^u \delta_{1t}^s + m_{2t}^u \le K \tag{16}$$

$$\sum_{t\in T} m_{1t}^u \delta_{1t}^s + m_{2t}^u \delta_{2t}^s + m_{kt}^s = \sum_{t\in T} d_t, \forall s \in S \tag{17}$$

$$m_{1t}^u, m_{2t}^u, m_{kt}^s, \ge 0, \forall s \in S, t \in T \tag{18}$$

as always, the objective function expresses the cost by heat production. Equation (16) says that the amount of heat produced does not exceed the amount generated, while the Equation (17) says that the total heat production is equal to the demand of this one. The rest are the non-negativity conditions of the model.

An alternate model described in (Shamburg, 2018) specifies conditions when it is possible to store the heat produced in the boilers of the system. In this, Equations (15), (16) and (18) remain the same but the following constraints are added to indicate that system cooling is not allowed.

$$V_{t+1}^s = V_t^s + m_{1t}^u \delta_{1t}^s + m_{2t}^u \delta_{2t}^s + m_{kt}^s - d_t, \forall t \in T \tag{19}$$

$$V_1^s = V_T^s + m_{1t}^u \delta_{1t}^s + m_{2t}^u \delta_{2t}^s + m_{kt}^s - d_t, \tag{20}$$

$$0 \le V_t^s \le V_{\max} \tag{21}$$

In this case the variable V_t^s represents the heat storage reserve at the beginning of period and t and $V_{\max}$ is the head storage limit.

Another formulation of this problem is found in the multi-phase programming models. The mathematical structure is as follows (Ermoliev & Wets, 1998)

$$\min_y Z = c_1 x_1 + E_{\xi_2} \left[\min_{x_2} c_2 x_2 + E_{\xi_3} \left[\min_{x_3} c_3 x_3 + \ldots \right] \right]$$

s.t.

$$A_{x_t} \leq b_t, t = 1, \ldots T$$

$$B(x_t) = d_t, t = 1, \ldots T \tag{22}$$

$$T_{x_{t-1}} + W_{x_t} = h(\xi_t), t = 2, \ldots T$$

where t are the stages and the set $A_{x_t} \leq b_t, t$ represents the constraints in each period. The constraints $B(x_t) = d_t$ are the McCormic approximations (McCormic, 1976) to a bilinear function $z = xy, x \in [\alpha_1 \alpha_2], y \in [\beta_1 \beta_2]$. An illustrative example of the use of linear functions in the construction of short-term hydrothermal planning models is found in (Cerisola & Ramos, 2012). In this proposal, authors express the P output power (in Mw) of a hydro unit as $P = gqh\eta$, where g stands for the product of the gravitational acceleration in mt / sec^2, and the water density in kg / mt^3, q is the flow released in (mt^3 / sec), h is the net water head in meters, and $\eta = \eta(h)$ is the efficiency of the unit formed by the turbine and the generator.

A CASE STUDY

This section is interested in solving an instance formed by 3 hydrothermal plants and 3 hydro plants. The model optimizes an operation plan in a 24-hour horizon. The minimum emissions of polluting particles found in the model are also reported. The model is as follows:

$$\min_{p_i} \left[\sum_{i \in S} a_{it} + b_{it} p_{it} + c_{it} p_{it}^2 + g_i(t) + \sum_{j \in H} \gamma_j(\varphi_j) + \varepsilon(p_{it}) \right]$$

s.t.

$$P\left(D_t \leq \sum_{i \in S} p_{it} + \sum_{j \in H} p_{jt} \right) = 1 - \alpha$$

$$w_t = w_{t-1} - (q_t - s_t + r_t) t, w_t \leq \tilde{w}_t$$

$$q_{jt} = \beta_0 + \beta_1 p_{jt} + \beta_2 p_{jt}^2$$

$$p_i^- \leq p_{ij} \leq p_i^+, p_i \in S, \forall t \in T$$

$$D_t \sim \mathcal{N}\left(\mu_{D_t}, \sigma_{D_t}\right), \forall t \in T$$

$$\varepsilon\left(p_{it}\right) = t\left(\lambda_1 + \lambda_2 p_{it} + \lambda_3 p_{it}^2\right), \forall t \in T$$

Table 2 shows the corresponding values used for a, b, c, p_i^-, p_i^+ and $g_i\left(t\right)$ the fixed cost of operation in each plant for all $t \in T$.

In the same way, Table 3 shows the values used in the hydroelectric plant.

The intervals and demand parameters used in this instance are given in Table 4.

The runoff of water into the reservoir will be simulated through a gamma distribution with probability density function, mean and variance given by

$$F_X\left(x, \alpha, \theta\right) = x^{\alpha-1} \frac{e^{-x/\theta}}{\theta^\alpha \Gamma\left(\alpha\right)}, x > 0,$$

$$E\left(X\right) = \alpha\theta, Var\left(X\right) = \alpha\theta^2$$

Table 2. Parametric values for the thermoelectric plant

$p_i \in S$	a	b	c	p_i^-	p_i^+	$g_i\left(t\right)$
1	561	8	0.0016	200	400	79,282
2	310	7.85	0.00194	300	400	105,665
3	780	7.97	0.00482	100	200	20,750

Table 3. Parametric values for the hydroelectric plant

$p_j \in H$	β_0	β_1	β_2	p_j^-	p_j^+
1	51216.863	1173.829	16.382	10	100
2	50834.983	1082.417	15.551	10	100
3	49816.928	1168.097	12.052	50	150

Table 4. Parametric values used per interval for $\alpha = 0.05$ and $k_{\alpha} = 1.96$

t	1	2	3	4	5	6	7	8	9	10	11	12
μ_D	500	520	550	560	610	670	730	790	820	850	900	1000
σ_D	20	21	30	20	22	21	15	15	18	24	25	20
ϕ	540	562	609	600	654	712	760	820	856	898	949	1040
t	13	14	15	16	17	18	19	20	21	22	23	24
μ_D	1100	1150	1200	1210	1215	1225	1200	1150	1090	980	800	750
σ_D	16	18	18	12	10	10	14	14	15	17	19	20
ϕ	1132	1186	1236	1234	1235	1245	1228	1178	1120	1014	838	790

and to project the simulated value, we use the product $F_X^{-1}(u)\psi$, with $\psi = 3600$. In this proposal $F_X^{-1}(u) \times u$ represents the inverse transformation of the cumulative distribution function of gamma density.

Table 5 shows the parametric values used in the gamma function in order to simulate the runoffs r of the process for all $t = 1, \ldots, 23$.

SOLUTION STRATEGIES

To solve the proposed instance, two different algorithms were used. The method of Random Search and Genetic Algorithm. The Non-adaptive random search algorithm (RSA) works as follows:

Table 5. Parametric values considered in the hydroelectric plant

j	$\tilde{w}_t$	q_j^-	q_j^+	w_{j0}	w_{jt}	$w_{j,24}$	α_j	θ_j	ψ
1	5.2×10^7	60000	340000	450000	40000	40000	1.41	47.92	3600
2	2.1×10^7	60000	320000	16000	10200	10200	1.62	47.92	3600
3	5.1×10^7	135000	500000	21000	14000	14000	1.28	42.80	3600

1. Generate a sequence of N random vectors $X_1, X_2, \ldots, X_N$ uniformly distributed over D.
2. Compute $Y_k = g\left(X_k\right), k = 1, \ldots, N$
3. Estimate g^* by

$$M_n = \min\left(Y_1, \ldots, Y_N\right)$$

In this algorithm it is clear that if a large number of vectors are generated at random over any set of finite measure D, then the probability that any particular subset about D with positive measures contains at least one of such vectors and tends to unify as the number of generated vectors tends to infinity. The corresponding convergence test can be found in (Rubienstein, 1992).

In order to consider the optimal trajectories achieved as an incremental process. Thus, for $\sigma = 1$ (white noise). The following improved version of the previous algorithm (Rubienstein, 1992) were used:

1. Find $x_1 \in D$
2. Compute $g\left(x_1\right)$
3. $i \leftarrow 1$
4. Generate Y_i from $\mathcal{N}\left(0, \sigma^2\right)$
5. If $\left(x_i + Y_i\right) \neq\in D$, set $x_{i+1} = x$ and go to step 4
6. Compute $g\left(x_i + Y_i\right)$.
7. $x_{i+1} = \begin{cases} x_i + y_i, & if\ g\left(x_1 + Y_i\right) < g\left(x_1\right) \\ x_i, & otherwise \end{cases}$
8. $i \leftarrow i + 1$
9. Go to step 4

Genetic Algorithm (GA) is a search-based optimization technique based on the principles of Genetics and Natural Selection. It is used to find optimal or near-optimal solutions to difficult problems which would otherwise take a lifetime to solve.

In this proposal, a value of μ_1 for crossover and μ_2, μ_3 for mutation and point mutation probabilities were used, respectively. To start the algorithm, a set of feasible problem solutions is generated randomly $p_i \in S$ and $p_j \in H$.

In this analysis it is necessary to find the optimal values of the vector

$$\xi = \left\{\xi_t : p_{it} \in S, p_{jt} \in H\right\}_{t=1}^{T}$$

such that they satisfy the fixed restrictive set. In this way, each solution represents a sequence of decisions for 24 hours given by the process ξ. If a non-feasible solution is produced, it is discarded and another random one is obtained and the process is repeated until a feasible solution is generated. Then, the capacities of reservoirs are updated for the next hour, and the same procedure is archived to yield other random solution for the period $t+1$.

Each hour every power unit is described by 11 bits; where the first 10 bits represents the power required for this unit and the last one is the binary flag indicating if the unit is on line in this period. Thus, the complete solution in each period has associated 66 bits; and every solution is composed by 1584 bits. In this way, 40 solutions are generated and evaluated in the objective function. Next a tournament is generated to generate a new set of refined solutions in which α crossover is performed with a probability μ_1 between two random solutions, using a single-point crossover.

For the above, the new chains of bits that are generated are also evaluated to determine their feasibility, otherwise, they are discarded and another solution is sought, following this process until a feasible solution is found. The new feasible solution replaces the previously found parents.

The mutation process was done in a similar way. Each solution is found with a probability μ_2, hence a bit of this solution is inverted with a probability $\mu_3 = 1/1584$, (Haupt & Haupt, 2004). When a solution is mutated, its feasibility is checked and if not, the mutation is reversed, shown in Table 6.

The results obtained and the comparative aspects of both algorithms are shown in Table (6) and Figure (2) shows two comparative diagrams of the algorithms used in this proposal. Note that, there is a clear superiority of the GA algorithm over the RSA algorithm in both precision and speed. Both procedures were programmed in MATLAB, and while GA required 135 seconds to achieve 40 iterations with 40 solutions, RSA needed 100,000 iterations in approximately 620 seconds with an average cost of 4.1386 × 105 during the entire planning period of the process.

Finally, Table (7) also shows the emission values of particulate pollutants to the atmosphere due to the operation of the three thermoelectric plants used with the characteristics already described. Here the following values were used $\lambda_1 = 4, \lambda_2 = 0.12$, and $\lambda_3 = 0.05$. Figure (4) shows the volumes of particles emitted in ppm/hr issued with both solutions of the proposed algorithms. In this one the clear difference of the GA in relation to the random search is observed. Given that the model is stochastic, GA solutions provide the highest power to cover the required demand, forcing more to the thermoelectric plants and emitting with it a larger quantity of pollutants.

CONCLUSION

Although the systematic coordination of the operation of a system of hydroelectric generation plants is usually complex, this has been a very interesting problem for engineers, mathematicians, physicists and experts in soft computing.

The reason is both simple and important. There are no two similar hydrogeneration systems in the world and each represents a challenge to optimize its operation at minimum cost while efficiently controlling the number of polluting particles that it emits into the atmosphere, regardless of its own economic importance.

In this document we have exposed some general features of the problem that are most frequently addressed in the literature. Given that the problem of hydrothermal coordination represents a clear example of mathematical optimization, we wanted to show the trends that today exist in this specialty mainly through the use of metaheuristics. The various forms that a mathematical model relate to this problem can take anywhere from a simple structure formed by an objective function (or utility function) that is

Table 6. Results obtained from the model

Algorithm	t	Hydro System			Thermal System			Total Power Delivered	Particles Emited (ppm/hr)	Minimal Cost $\times 10^5$	Better Option
		p_1	p_2	p_3	p_1	p_2	p_3	ρ	$\epsilon(p)$	$g(p)$	
RSM	1	88	97	132	209	303	0	829	3446.53	4.598	
GA		0	0	148	0	337	125	610	18682.20	2.2144	✳
RSM	2	119	791	141	368	0	0	707	25081.29	3.5341	
GA		99	0	0	264	0	350	713	25508.01	2.2209	✳
RSM	3	112	94	140	340	0	0	686	23616.12	3.5378	✳
GA		0	71	78	373	371	196	1089	59430.73	3.9545	
RSM	4	120	92	135	362	0	0	709	25223.13	3.5419	
GA		0	43	0	0	397	180	620	19298.4	3.2252	✳
RSM	5	96	97	134	240	0	184	751	28294.17	4.5937	
GA		0	0	125	349	0	194	668	22395.36	1.9606	✳
RSM	6	76	87	145	213	312	0	833	34798.41	4.5998	
GA		6	0	0	149	396	400	951	45338.17	3.0525	✳
RSM	7	91	88	139	219	310	0	847	35976.09	4.6005	
GA		0	100	0	340	341	0	781	30595.77	2.816	✳
RSM	8	103	94	137	201	383	0	918	42250.36	4.6046	
GA		0	0	150	400	391	0	941	44390.97	2.8260	✳
RSM	9	103	90	143	260	337	0	933	43640.41	4.6092	
GA		71	0	0	287	394	137	889	39626.73	3.0433	✳
RSM	10	96	98	145	238	378	0	955	45719.85	4.6096	
GA		0	0	0	350	369	185	904	40973.28	2.1511	✳
RSM	11	118	77	142	251	396	0	984	448534.88	4.6133	
GA		100	94	0	393	387	0	974	47554.68	3.7250	✳
RSM	12	120	75	138	248	385	0	966	46777.72	4.6110	
GA		99	0	147	247	400	196	1089	59430.73	3.9460	✳
RSM	13	115	69	142	321	397	0	1044	54626.08	4.6252	
GA		95	0	149	400	344	172	1160	67423.20	3.9521	✳
RSM	14	112	98	141	331	369	0	1051	55360.17	4.6244	
GA		100	0	100	400	400	200	1200	72148.96	3.9601	✳
RSM	15	119	98	94	222	361	0	894	40073.08	4.6056	
GA		0	96	150	400	400	197	1243	77405.61	3.9598	✳

continued on the following page

Table 6. Continued

Algorithm	t	Hydro System			Thermal System			Total Power Delivered	Particles Emited (ppm/hr)	Minimal Cost $\times 10^5$	Better Option
		p_1	p_2	p_3	p_1	p_2	p_3	ρ	$\epsilon(p)$	$g(p)$	
RSM	16	116	79	150	326	398	0	1069	57270.33	4.6262	✳
GA		100	99	145	400	387	198	1329	88475.53	4.8586	
RSM	17	90	93	140	360	384	195	1262	79787.64	5.7052	
GA		100	100	144	298	398	200	1240	77032.80	4.8507	✳
RSM	18	108	96	149	348	396	152	1249	78153.93	5.6999	
GA		77	100	150	375	398	150	1250	78279.85	4.8528	✳
RSM	19	103	91	146	347	382	183	1252	78529.44	5.7014	
GA		100	100	150	400	400	200	1350	91291.00	4.8601	✳
RSM	20	117	70	149	388	373	186	1283	82462.41	5.7088	
GA		99	97	125	325	385	175	1206	72870.52	4.8494	✳
RSM	21	115	82	107	396	400	198	1298	84399.96	5.7141	
GA		0	0	150	387	400	198	1135	64551.45	3.0587	✳
RSM	22	118	97	142	392	345	185	1279	81949.53	5.7069	
GA		0	0	148	398	392	172	1110	61742.20	3.0564	✳
RSM	23	107	86	136	228	392	189	1138	64892.76	5.6828	
GA		97	0	0	316	342	193	948	45052.96	3.0464	✳
RSM	24	119	100	116	375	369	0	1079	58345.53	4.6327	
GA		0	99	124	0	394	184	801	32180.17	3.1253	✳

Figure 2. Pseudo codes for the RSA and GA algorithms

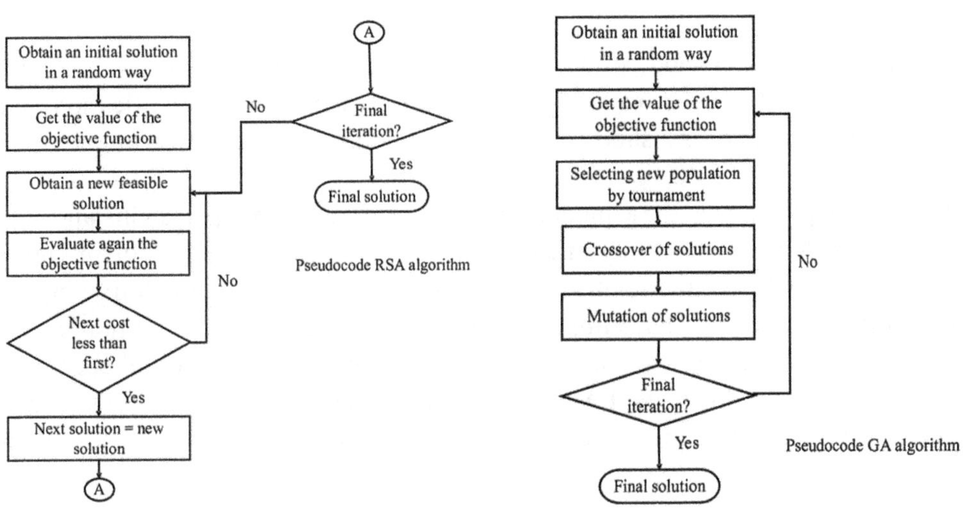

Table 7. Volumes (ppm/hr) emitted from particles in each algorithm used

RMS		GA	
Power deliveredE(p)Power deliveredE(p)	$e(p)$	**Power deliveredE(p)Power deliveredE(p)**	$e(p)$
829	34465.53	610	18682.2
707	25081.29	713	25508.01
686	23616.12	1089	59430.73
709	25223.13	620	19298.4
751	28294.17	668	22395.36
833	34798.41	951	45338.17
847	35976.09	781	30595.77
918	42250.36	941	44390.97
933	43640.41	889	39626.73
955	45719.85	904	40973.28
984	48534.88	974	47554.68
966	46777.72	1089	59430.73
1044	54626.08	1160	67423.2
1051	55360.17	1200	72148
894	40073.08	1243	77405.61
1069	57270.33	1329	88475.53
1262	79787.64	1240	77032.8
1249	78153.93	1250	78279
1252	78529.44	1350	91291
1283	82462.41	1206	72870.52
1298	84399.96	1135	64551.45
1279	81949.53	1110	61742.2
1138	64892.76	948	45052.96
1079	58345.53	801	32180.17

convex in some of its parameters and defined on a convex set, a problem that can be treated from the perspective of the traditional calculation, to highly complex, dynamic and stochastic models that require the optimization of the model over a short, medium or long term planning horizon requiring more sophisticated techniques to face them.

To illustrate the complexity that these models may have, we have developed an instance of short-term planning (24 hours) including uncertainty in energy demand which turns the problem into a stochastic structure. We have also included the runoff of water into the reservoirs as a random variable that gives more realism to the model and we have implemented two algorithms for its solution, a variant of the random search algorithm (RSA) and the use of a genetic algorithm. Our proposal confronts two techniques developed at different times; the RMS around the 60s and GA in the mid-70s being in force in its implementation. The objective of both proposals is to analyze the precision and rapid convergence of both as well as their practicality in the solution of complex large-scale models and include a simple mathematical model of contamination by emission of oxides of nitrogen. Although in this example there are only three hydro plants and three thermo plants, in practice these quantities multiply, creating instances with hundreds or thousands of variables and a large number of constraints without our analysis losing value for such examples.

The development of methods and models inspired in nature (bioinspired) has fortunately resulted in an exponential development as alternative techniques of analysis and optimization of highly complex instances and day by day create variants and new versions of these for engineering practitioners, applied and special mathematics in computer programming and numerical methods. Some of these are: Ant colony optimization, Artificial bee colony, Artificial immune systems, Artificial neural networks, Automatic computing, Bacterial foraging, Biological computing, Chaos optimization, Cloud Computing, Combinatorial Optimization, Computational intelligence, continuous optimization, Cultural algorithms, Differential evolution, Direct Search, Evolutionary computing, Fuzzy optimization, Genetic algorithms, Granular computing, Hybrid algorithms, Local and global search, Memetic algorithms, Meta-heuristic methods, Particle swarm optimization, Pattern search, Simulated annealing, Simulation and modelling, Soft Computing techniques, Support vector machines, Swarm intelligence, Taboo search, and Variable neighborhood search.

Important extensions of the problem can be achieved by including hydrothermal planning with reserve constraints, with the existence of cascading hydro plants, with models of water re-pumping, multi-objective systems, reconfiguration models, particle emission and more. The generation of electric power is an important industry in the economic development of countries.

The passage from industry 4.0 to energy 4.0 is an important factor of change that should be inserted in this new revolution since electric power has been a decisive factor in every industrial revolution. Energy 4.0 refers to the digitization of the electric sector in order to take advantage of alternative sources of energy in an optimal way. The elimination of nuclear energy and materials such as lignite, hard coal, oil and gas are processes that we will see in the short term. Therefore, the design of new transmission networks, distribution and storage systems are a priority in the new industrial era. The electrical industry is becoming increasingly complex and will necessarily be affected by industry 4.0. The key elements of this will be the cyberphysical systems composed of physical entities and controlled and monitored by computer-based algorithms.

In the near future the problems of generation, transmission and distribution, commerce, retail, and consumption will be strongly related to Cloud computing, Big data, IT and data security, Smart everything, Internet of things, Outsourcing, Standardization and Data protection; solving problems of the type: data and consumer protection, digitization of energy change, development prospects for distribution networks, critical infrastructure protection and smart energy, protection of intellectual property, outsourcing, E-commerce, and Liability for product and system failure.

REFERENCES

Anastasiadis, A., Kondylis, G. A., Vokas, G., & Papageorgas, P. (2016). Hydrothermal coordination in power systems with large-scale integration of renewable energy sources. *Management of Environmental Quality*, *27*(Issue: 3), 246–258. doi:10.1108/MEQ-04-2015-0054

Ayub, S., & Kumar, S. S. (2011). *Particulate matter emission from thermal power plant: parent material, formation mechanism, health concerns, control devices - a review*. Global Journal Engineering and Applied Sciences.

Basu, M. (2003). Hopfield neural networks for optimal scheduling of fixed head hydrothermal power systems. *Electric Power Systems Research*, *64*(1), 11–15. doi:10.1016/S0378-7796(02)00118-9

Basu, M. (2005). A simulated annealing-based goal-attainment method for economic emission load dispatch of fixed head hydrothermal power systems. *International Journal of Electrical Power & Energy Systems*, *27*(2), 147–153. doi:10.1016/j.ijepes.2004.09.004

Basu, M., & Chowdhury, A. (2013). Cuckoo search algorithm for economic dispatch. *Energy*, *60*, 99–108. doi:10.1016/j.energy.2013.07.011

Binghui, Y., Xiaohui, Y., & Jinwen, W. (2007). Short term hydrothermal scheduling using particle swarm optimization method. *Energy Conversion and Management*, *47*(8), 1902–1908.

Cerisola, S., Latorre, J. M., & Ramos, A. (2012). Stochastic dual dynamic programming applied to nonconvex hydrothermal models. *European Journal of Operational Research*, *218*(3), 687-697.

Chen, P. H. (2008). Pumped-storage scheduling using evolutionary particle swarm optimization. *IEEE Transactions on Energy Conversion*, *23*(1), 291–301. doi:10.1109/TEC.2007.914312

Dashti, H., Conejo, A.J., Jiang, R. & Wang, J. (2016). Weekly two-stage robust generation scheduling for hydrothermal power systems. *IEEE Transactions on Power Systems*, *31*(6).

Dominique, M., Bain, D. M., & Acker, T. L. (2018). Hydropower impacts on electrical system production costs in the Southwest United States. *Energies*, *11*(2), 368. doi:10.3390/en11020368

Emission of hazardous air pollutants from coal-fired power plants. Environmental Health and Engineering. (2019). Retrieved from https://www.lung.org/assets/documents/healthy-air/emissions-of-hazardous-air.pdf

Ermoliev, Y., & Wets, R. (1988). *Numerical Techniques for Stochastic Optimization*. Berlin: Springer Series in Computational Mathematics. doi:10.1007/978-3-642-61370-8

Farhat, I. A., & El-Hawary, M. E. (2009). Optimization methods applied for solving the short-term hydrothermal coordination problem. *Electric Power Systems Research*, *79*(9), 1308–1320. doi:10.1016/j.epsr.2009.04.001

Ferreira, L. A. F. M. (1992). A theoretical analysis of the classic hydro-thermal optimization algorithm in power system scheduling. *IEEE Proceedings International Symposium on Circuits and Systems*. 10.1109/ISCAS.1992.230627

Fleten, S.-E., & Kristoffersen, T. K. (2008). Short-term hydropower production planning by stochastic programming. *Computers & Operations Research*, *35*(8), 2656–2671. doi:10.1016/j.cor.2006.12.022

Fogel, D. B. (1995). A comparison of evolutionary programming and genetic algorithms on selected constrained optimization problems. *Simulation*, *64*(6), 397–404. doi:10.1177/003754979506400605

Ghosh, S., Kaur, M., Bhullar, S., & Karar, V. (2019). Hybrid ABC-BAT for Solving Short-Term Hydrothermal Scheduling Problems. *Energies*, *12*(3), 551. doi:10.3390/en12030551

Haupt, R. L., & Haupt, S. A. (2004). *Practical Genetic Algorithms*. John Wile & Sons, Inc.

Hinojosa, V. H., & Leyton, C. (2012). Short-term hydrothermal generation scheduling solved with a mixed binary evolutionary particle swarm optimizer. *Electric Power Systems Research*, *92*, 162–170. doi:10.1016/j.epsr.2012.06.006

Jubaer, A., & Zainal, S. A. (2014). A Maximum Power Point Tracking (MPPT) for PV system using Cuckoo Search with partial shading capability. *Applied Energy*, *119*, 118–130. doi:10.1016/j.apenergy.2013.12.062

Karami, M., Shayanfar, H. A., Aghaei, J., & Ahmadi, A. (2013). Scenario-based security-constrained hydrothermal coordination with volatile wind power generation. *Renewable & Sustainable Energy Reviews*, *28*, 726–737. doi:10.1016/j.rser.2013.07.052

Mandal, K. K., Basu, M., & Chakraborty, N. (2008). Particle swarm optimization technique based short term hydrothermal scheduling. *Applied Soft Computing*, *8*(4), 1392–3399. doi:10.1016/j.asoc.2007.10.006

Mandal, K. K., & Chakraborty, N. (2011). Short-term combined economic emission scheduling of hydrothermal systems with cascade reservoirs using particle swarm optimization technique. *Applied Soft Computing*, *11*(1), pp1295–pp1302. doi:10.1016/j.asoc.2010.03.006

McCormic, G. (1976). Computability of global solutions to factorable non-convex programs: Part I Convex underestimating problems. *Mathematical Programming*, *10*, 146–175.

Moorthy, V., Sangameswararaju, P., Ganesan, S., & Subramanian, S. (2013). Hydrothermal scheduling using ABC algorithm considering pollutant emission. *International Journal of Electrical Engineering*, *20*(5), 203–217.

Nguyen, T. T., Vo, D. N., & Dinh, B. H. (2016). Cuckoo search algorithm using different distributions for short-term hydrothermal scheduling with reservoir volume constraint. *Journal on Electrical Engineering and Informatics*, *8*(1), 76–92. doi:10.15676/ijeei.2016.8.1.6

Nima, A., & Hassan, R. S. (2010). Daily hydrothermal generation scheduling by a new modified adaptive particle swarm optimization technique. *Electric Power Systems Research*, *80*(6), 723–732. doi:10.1016/j.epsr.2009.11.004

Pavlyukevich, I. (2007). Lévy flights, non-local search and simulated annealing. *Journal of Computational Physics*, *226*(2), 1830–1844. doi:10.1016/j.jcp.2007.06.008

Pérez-Lechuga, G., Medina-Marín, J., Tuoh-Mora, J. C., & Mora-Cárdenas, G. E. (2010). The stochastic short-term hydrothermal scheduling problem. *Proceedings of the 15th annual International Conference on Industrial Engineering theory applications and practice*.

Provas, K. R. (2013). Teaching learning-based optimization for short-term hydrothermal scheduling problem considering valve point effect and prohibited discharge constraint. *International Journal of Electrical Power & Energy Systems*, *53*, 10–19. doi:10.1016/j.ijepes.2013.03.024

Rajan. (2011). Hydro-thermal unit commitment problem using simulated annealing embedded evolutionary programming approach. Electrical Power and Energy Systems, 939-946.

RE Technologies Cost Analysis Series. (2012). *Volume 1: Power Sector, Issue 3/5, Hydropower*. International Renewable Energy Agency. Retrieved from https://www.irena.org/documentdownloads/publications/re_technologies_cost_analysis-hydropower.pdf

Rubinstein, R. Y. (1992). *Monte Carlo Optimization, Simulation and Sensitivity of Queuing Networks*. John Wiley & Sons, Inc.

Rui, Z., & Jianzhong, Z., Wang, & Yongqiang, W. (2012). Multi-objective optimization of hydrothermal energy system considering economic and environmental aspects. *International Journal of Electrical Power & Energy Systems*, *42*(1), 384–395.

Russell, A. G. (1988). *Mathematical Modeling of the Effect of Emission Sources on Atmospheric Pollutant Concentrations, Air Pollution, the Automobile, and Public Health*. Academic Press.

Salgado, L., Josue, C., Año, O., Esteybar, O., & Mauricio, D. (2016). Hydrothermal scheduling with variable head hydroelectric plants: Proposed strategies using benders decomposition and outer approximation. Power and Energy Conference at Illinois (PECI) 2016 IEEE, 1-8.

Sasikala, J., & Ramaswamy, M. (2010). Optimal gamma based fixed head hydrothermal scheduling using genetic algorithm. *Expert Systems with Applications*, *37*(4), 3352–3357. doi:10.1016/j.eswa.2009.10.015

Schaumburg-Müller, C. (2018). *Mathematical models and methods for analysis of distributed power generation on market conditions*. IMM-PHD-2008-177. Recovered from http://orbit.dtu.dk/files/4963367/phd177_csm-web.pdf

Selvakumar, I. A. & Thanushkodi, K. (2007). A New Particle Swarm Optimization Solution to Nonconvex Economic Dispatch Problems. *IEEE Transactions on Power System*, *22*(1).

Shilin, Z., Yufeng, D., & Houzhang, T. (2016). Migration and Emission Characteristics of Trace Elements in a 660 MW Coal-Fired Power Plant of China. *Energy & Fuels*, *30*(7), 5937–5944. doi:10.1021/acs.energyfuels.6b00450

Sifuentes, W. S., & Vargas, A. (2007). Hydrothermal Scheduling Using Benders Decomposition: Accelerating Techniques. *IEEE Transactions on Power Systems*, *2*(3), 1351–1359. doi:10.1109/TPWRS.2007.901751

Sinha, N., Chakrabarti, R., & Chattopadhyay, P. K. (2003). Fast evolutionary programming techniques for short-term hydrothermal scheduling. *Electric Power Systems Research*, *66*(2), 97–103. doi:10.1016/S0378-7796(03)00016-6

Songfeng, L., Chengfu, S., & Zhengding, L. (2010). An improved quantum-behaved particle swarm optimization method for short-term combined economic emission hydrothermal scheduling. *Energy Conversion and Management*, *51*(3), 561–571. doi:10.1016/j.enconman.2009.10.024

Suman Sutradhar, S., Nalin, B., Dev Choudhury, N., B., & Sinha, N. (2018). Parallel ABC optimization algorithm for hydrothermal scheduling problems. AIP Conference Proceedings, 1. doi:10.1063/1.5049111

Sun, Ch., & Lu, S. (2010). Short-term combined economic emission hydrothermal scheduling using improved quantum-behaved particle swarm optimization. *Expert Systems with Applications*, *37*(6), 4232–4241. doi:10.1016/j.eswa.2009.11.079

Tahir, N. M., Salman, Z., & Saaqib, H. (2016). Short-term economic emission power scheduling of hydrothermal systems using improved chaotic hybrid differential evolution. *Turkish Journal of Electrical Engineering and Computer Sciences*, *24*, 2654–2670. doi:10.3906/elk-1403-298

Varma, S. A. K. (2014). *Air Pollution Modelling for Thermal Power Plants, Monitoring and Modelling*. LAP Lamber Academic Publishing.

Walton, S., Hassan, O., Morgan, K., & Brown, M. R. (2011). Modified cuckoo search: A new gradient free optimization algorithm. *Chaos, Solitons, and Fractals*, *44*(9), 710–718. doi:10.1016/j.chaos.2011.06.004

Wang, J., Zhang, Y., Liu, Z., Norris, P., Romero, C. E., Xu, H., & Pan, W.-P. (2017). Effect of coordinated air pollution control devices in coal-fired power plants on arsenic emissions. *Energy & Fuels*, *31*(7), 7309–7316. doi:10.1021/acs.energyfuels.7b00711

Wong, K. P., & Wong, Y. W. (1994). Short-term hydrothermal scheduling. II. Parallel simulated annealing approach. *IEE Proceedings. Generation, Transmission and Distribution*, *141*(5), 502. doi:10.1049/ip-gtd:19941351

Wood, A. J. (1996). *Power Generation, Operation, and Control*. John Wiley & Sons.

Xiangtao, L., & Minghao, Y. (2013). A hybrid cuckoo search via Lévy flights for the permutation flow shop scheduling problem. *International Journal of Production Research*, *15*, 4732–4754.

Yang, H. T., Yang, P. C., & Huang, C. L. (1996). Evolutionary programming based economic dispatch for units with non-smooth fuel cost functions. *IEEE Transactions on Power Systems*, *11*(1), 112–117. doi:10.1109/59.485992

Yang, X-S. & Deb, S. (2010). Engineering Optimization by Cuckoo Search. *International Journal of Mathematical Modeling and Numerical Optimization*, *1*(4).

Zeynal, H., & Eidiani, M. (2014). Hydrothermal scheduling flexibility enhancement with pumped-storage units. *Iranian Conference on Electrical Engineering (ICEE)*. 10.1109/IranianCEE.2014.6999649

Zoumas, C. E., Bakirtzis, A. G., Teocharis, J. B., & Petridis, V. (2004). A genetic algorithm solution approach to the hydrothermal coordination problem. *IEEE Transactions on Power Systems*, *19*(3), 1356–1364. doi:10.1109/TPWRS.2004.825896

ADDITIONAL READING

Abbas, G., Gu, J., Farooq, U., Asadi, U. M., & El-Hawary, M. (2017). Solution of an economic dispatch problem through particle swarm optimization: A detailed survey – Part I. *IEEE Access: Practical Innovations, Open Solutions*, *5*, 15105–15141. doi:10.1109/ACCESS.2017.2723862

Khosravi, A., & Yazdani-Asrami, M. (2015). Application of swarm-based optimization algorithms for solving dynamic economic load dispatch problem. *J. Adv. Comput. Res*, *6*(4), 13–26.

Lim, S., Montakhab, M., & Nouri, H. (2009). Economic dispatch of power system using particle swarm optimization with constriction factor. *Int. J. Innov. Energy Syst. Power*, *4*(2), 29–34.

Sharma, J., & Mahor, A., A. (2013). Particle swarm optimization approach for economic load dispatch: A review. *Int. J. Eng. Res. Appl*, *3*(1), 13–22.

Sheng, X., & Xu, W. (2015). Solving the economic dispatch problem with Q-learning quantum-behaved particle swarm optimization method. *Proc. 14th Int. Symp. Distrib. Comput. Appl. Bus. Eng. Sci. (DCABES)*, 98–101 10.1109/DCABES.2015.32

Sun, Y., & Wang, Z. (2016). Improved particle swarm optimization based dynamic economic dispatch of power system. *Procedia Manuf*, *7*, 297–302. doi:10.1016/j.promfg.2016.12.074

Zakaria, Z. N., Azmi, A., Laili, M. S., Jamalil, S. A. S., & Sulaiman, M. H. (2013). An extension of particle swarm optimization (E-PSO) algorithm for solving economic dispatch problem. *Proc. 1st Int. Conf. Artif. Intell., Modelling Simulation*, 157–161. 10.1109/AIMS.2013.32

KEY TERMS AND DEFINITIONS

Emission of Particles: Refers to particles solid or liquid emitted into the air by combustion of coal oil or diesel in a thermoelectric plant.

Genetic Algorithm: Is a metaheuristic inspired by the process of natural selection that belongs to the larger class of evolutionary algorithms (EA).

Hydrothermal Coordination: Is the optimal coordination of hydro and thermal plants to meet the system load demand at minimum possible operational cost of the thermal plants while satisfying the system constraints.

Integer Linear Programming: It is a branch of mathematical programming in which the objective function and the constraints (other than the integer constraints) are linear and some or all of the decision variables are discrete.

Level of Customer Service: It is the probability that a customer receives the quantity of product requested in the demand in time and with the required quality.

Random Search: Is a family of numerical optimization methods. These can hence be used on functions that are not continuous or differentiable. Such optimization methods are also known as direct-search, derivative-free, or black-box methods.

Stochastic Optimization: Refers to a collection of methods for minimizing or maximizing an objective function when randomness is present.

Chapter 16

A Model for Planning the Sowing of Agricultural Crops and Raising Animals Through Two-Stage Mathematical Programming

Gilberto Pérez Lechuga
Universidad Autónoma del Estado de Hidalgo, Mexico

Otilio A. Acevedo Sandoval
Universidad Autónoma del Estado de Hidalgo, Mexico

Karla N. Madrid Fernández
Universidad Autónoma del Estado de Hidalgo, Mexico

Raúl Román Aguilar
Universidad Autónoma del Estado de Hidalgo, Mexico

ABSTRACT

In this chapter, the authors present a mathematical model to calculate the exact quantities of animals that must compose a herd given the water, food, and land conditions available for their breeding. In the same model, the optimal calculation of vegetables that can be cultivated in the available land spaces is incorporated considering the same restrictions of water, nutrients, and area available for planting. Both models focus primarily from a deterministic perspective. Subsequently, the randomness of the same is uncovered through the uncertainty in the availability of water. Therefore, the first part of the proposal is made through a simple model of linear mathematical programming. The stochastic model is constructed from a two-phase mathematical programming model. The novelty of the proposal and its contribution consists of illustrating, step by step, the construction and solution of the scenarios of the stochastic model for a problem related to agriculture and animal husbandry.

DOI: 10.4018/978-1-7998-1216-6.ch016

Copyright © 2020, IGI Global. Copying or distributing in print or electronic forms without written permission of IGI Global is prohibited.

INTRODUCTION

Stochastic integer programming problems combine the difficulty of stochastic programming with integer programming. These problems involve the uncertainty that usually accompanies real problems. Sometimes, the parameters are known only within certain bounds, and an approach to tackling such problems is called robust optimization. On the other hand, in stochastic optimization one finds that there is not sufficient information about one or several parameters of the problem or that the information is scattered, diffuse or in the form of sample statistics. Under this scheme, the mathematical model is usually constructed using probability distributions to represent the uncertainty inherent in it. If this is the situation, the model is somewhat more accessible to solve using closed mathematical programming techniques or using metaheuristics based on simple rules that provide acceptable solutions in short times.

Mathematical modeling and the corresponding optimization of instances using models of two or more stages is one of the most challenging problems due to the discrete nature of the model and the non-convexity of the model. However, the richness of its applications makes it a necessity today where it is used in a wide number of fields. Traditionally, two totally different approaches are found for the process of decision making under the uncertainty environment. The first is called "Here and now" and consists of making a decision, waiting to see what nature offers and making the corresponding adjustments to the mistakes made. The second is called "Wait and observe," and as its name implies, the decision is made once the results that nature provides for the analyzed situation are presented.

In this document a proposal is developed which finds the optimal scenarios associated with a problem of planting vegetables and raising animals under an environment of uncertainty. The model is developed under different contexts explaining the models derived from each decision under the "Here and now" scheme. The results of the model are reported and the models illustrate the way they are obtained. The uncertain parameters of yield amount and price level are calculated using real data.

Usually a decomposition algorithm, which is enhanced with specialized Benders cuts for solving the stochastic problem must be used. However, we present a simple solution for the size of the instance managed through the different scenarios via mixed integer linear programming. The proposal begins with a deterministic model that seeks to maximize the utility of the investment measured as the difference between the expenses of the project and its income in a limited land area. Subsequently, the uncertainty is introduced through a simple probability distribution to illustrate the rain probabilities of the model. Variants of it is shown to illustrate the possibilities of the case. Following is our proposal. Our proposal is applicable in any known area of land or with any type of animal and/or vegetable. The model developed here is universal. The instance proposed here has already been tested in Mexican lands where the water conditions described in the model are presented71

Like any mathematical model, this proposal is based on a set of assumptions, for example proportionality in linear programming models, and the ease of constructing a probability distribution from sample data for the two-stage programming model. Naturally, any failure to comply with these assumptions leaves out the theory developed in this regard. This proposal is a mathematical model built to help in making decisions regarding the availability of inputs. It is not a proposal to build any engineering contraption to carry out the planting of vegetables or the raising of animals. These tools are applied posteriori, once the optimal decisions to make and the alternatives offered by the mathematical model are known.

The proposed instance and models with these characteristics can be in any of the following situations: a) There is a solution and it is unique, b) there is no solution, c) There are several solutions.

Generally, the inexistence of a solution is due to a poorly developed model, that is, with inconsistencies from a mathematical perspective. If the solution exists, the algorithm leads to a local optimum with a thoroughly tested convergence and precision with one or more solutions when the famous simplex method is applied. In a well-made model, any other optimization methodology (including the heuristic), will eventually converge to a very approximate solution in a reasonable time.

BACKGROUND

The original problem treated here was called the two-stage problem, although it is usually referred to as the second-stage problem and its formal conceptualization goes back to the year 1955, (Dantzig, 1955), (Radner, 1955), (Radner 1959), (Beale, 1955), (Babbar, 1955), (Tintner, 1955), (Bellman, 1957), (Charnes, 1959), (Bellman, 1970), (Horner, 1999).

Recently in the literature abundant references of application of the model to instances related to agriculture and water are found. For example, in (Galán-Martín, Pozo, Guillén-Goálbez, Antón Vallejo, & Laureano Jiménez Esteller, 2015) the authors present a decision-support tool based on a multi-stage linear programming model that identifies optimal crop plan decisions under the new Common Agricultural Policy. The model can also be used to calculate the minimum subsidy value that would make the implementation of greening rules economically appealing, thereby promoting the widespread adoption of more sustainable agricultural practices. In both cases, their analysis is an interesting challenge for engineers, mathematicians and practitioners of this specialty because of the inherent complexity of the developed models. The difficulty lies essentially in that for each result of a lottery of random variables, there may be a solution for it and the idea is to find the average representative value of such solutions.

In this document it is proposed to develop a model of optimal scenarios associated with a problem of planting vegetables and raising animals under an environment of uncertainty. The model is developed under different contexts and derived from each decision under the "here and now" scheme. The results of the model are reported and the models are widely used to illustrate the way they are obtained.

In (Li, Li, Li, & Huang, 2010) an inexact two-stage water management model is developed for planning agricultural irrigation in the Zhangweinan River Basin, China. The model is derived from the incorporation of interval-parameter programming (IPP) within a two-stage stochastic programming (TSP) framework and four decision scenarios associated with different water-resources management policies are examined. Targeted incomes, recourse costs, and net system benefits under different scenarios are optimized.

In turn, in (Fan, Huang, Guo, & Yang, 2012) an inexact two-stage stochastic partial programming method is developed for tackling uncertainties presented as intervals and partial probability distributions. A scenario-based interactive algorithm is proposed to solve the model. The developed method is applied to a case study for water-resources planning. The modeling results generate a series of decision alternatives under various system conditions and uncertainty. Another interesting contribution to the double-phase modeling is the water allocation in (Bekri, Disse, & Yannopoulos, 2015). In this reference, the authors propose a methodology that combines an ordinary multi-stage stochastic program with uncertainties expressed as fuzzy-boundary intervals and upper and lower-bound solution intervals for optimized water allocation targets. The probabilistic water allocations and shortages are estimated under a baseline scenario and four water and agricultural policy future scenarios accounting for optimistic and a pessimistic attitude of the decision makers.

An important reference on the application of these models is found in (Han, Huang, & Li, 2008). In this study, an interval-parameter multi-stage stochastic chance-constrained mixed integer programming method has been developed for inter-basin water resources management systems under uncertainty. By incorporating the chance-constrained programming techniques and mixed integer programming within an interval-parameter multi-stage stochastic programming framework, the model improves upon the interval-parameter multi-stage stochastic programming and incorporates uncertainties expressed as not only possibility and probability distributions but also as discrete intervals and pre-defined water policies as part of its optimization process.

In relation to the planning for the production of plants, a proposal very close to ours can be found in (Darby-Dowman, Barker, Audsley, & Parsons, 2000) where a two-stage stochastic programming with a recourse model for the problem of determining optimal planting plans for a vegetable crop is presented. Uncertainty caused by factors such as weather and its major influence on yields for many systems arising in horticulture are considered. The first stage of this model relates to finding a planting plan which is common to all scenarios and the second stage is concerned with deriving a harvesting schedule for each scenario. Solutions are obtained for a range of risk aversion factors that not only result in greater expected profit compared to the corresponding deterministic model but also are more robust.

A model with an excellent stochastic approach to irrigation in crops is found in (Tim, & Lall). There the decision problem of a farmer associated with allocating his land among different crops with varying water requirements is considered. The problem includes a decision to acquire a certain amount of irrigation water capacity prior to the season and to allocate this capacity as irrigation water applied during the season to each of the crops selected. Rainfall in the growing season and the market price of each crop at the end of the season are considered to be random variables. Two stochastic programming models are developed to consider facets of this problem and to understand how contracts that reduce market price uncertainty from the problem changes the farmer's decision.

Applications to modeling and optimization of investment portfolios can be found in (Xiaodong Ji, Shushang Zhu, Shouyang Wang & Shuzhong Zhang, 2005), for continuous casting planning see (Bellabdaoui & Teghem, 2006), for life-cycle asset allocation and consumption (Geyer, Hanke, & Weissensteiner, 2009), electric grid operations (Shi, & Oren, 2016), location under uncertainty (Xie, Li, & Ouyang, 2015), capacity expansion for manufacturing (Barahona, Berman, Gunluk, & Holmes, 2005), chain supply (Lokesh Nagar & Karuna Jain, 2008), (Paksoy, & Pehlivan, 2012), (Ahumada, & Villalobos, 2009), pump scheduling (Puleo, Morley, Freni, & Savi, 2014), approximate solution of multi parametric mixed-integer linear programming problems (Wittmann-Hohlbein, & Pistikopoulos, 2012), power production planning (Römisch & Vigerske, 2010), electricity contracts (Shi & Oren, 2016). Finally, for an adequate understanding of the difficulties and challenges presented by the two-stage stochastic integer programming it is recommended to see (Ahmed & Ahmed, 2011).

The multi-stage programming model under uncertainty

Consider the classical model of linear mathematical programming

$$\text{Min } c^t x : Ax = b, x \geq 0 \tag{1}$$

where $c, x \in \mathbb{R}^n, A_{m\times n}$, and c^t means c transposed. Suppose also that the information contained in the vector of availabilities b is given in terms of a distribution function with finite mean $E(b)$. That is, in principle, for each value of b there is a minimum for c^t provided that the set

$$S = \{x : Ax = b, x \geq 0\} \neq \varnothing .$$

In general, there will always be a discrepancy between Ax and b defined in terms of a random variable whose density depends on x. So, the problem can now be written as minimizing the penalty assigned to the difference ξ between both quantities, i.e.,

$$\text{Min [Min } c^t x + E[Min_\xi d^t]] : Ax + B\xi = b, x, y \geq 0 \quad (2)$$

where d and $\xi \in \mathbb{R}^l$ and $B_{m\times l}$ and $E(\cdot)$ is the mathematical expectation operator. This approach is called "here and now" because we must choose a decision variable x before knowing the values that the vector b will take. And that once known we should find a resource ξ from the following second-stage program

$$\text{Min [} c^t x + \mathcal{Q}(x)] : \mathcal{Q}(x)] = E(\mathcal{Q}(x, b)), x \geq 0 \quad (3)$$

where $\mathcal{Q}(x, b) = \infty$ if the constraints are inconsistent, and $\mathcal{Q}(x, b) = -\infty$ if the solutions of the constraints are unbounded below (Vajda, 1972).

The natural extension of the multistage model is as follows

$$Minc_1 x_1 + E\left(c_2 x_2^{w_2} + \ldots + E\left(c_{t-1} x_{T-1}^{w_{T-1},\ldots,w_2} + E\left(c_T x_T^{w_T,\ldots w_2}\right)\right)\ldots\right)$$

s.t.

$$A_1 x_1 = b_1$$

$$-B_1^{w_2} x_1 + A_2 x_2^{w_2} x_2 = b_2^{w_2}$$

$$\ldots \ldots \ldots \ldots \ldots \ldots \ldots \ldots \ldots \ldots$$

$$-B_{T-1}^{w_T} x_{T-1}^{w_{T-1},\ldots,w_2} + A_T x_{T-1}^{w_T,\ldots,w_2} = b_T^{w_T}$$

$$x_1, x_2^{w_2}, \ldots, x_{T-1}^{w_{T-1},\ldots w_2}, x^{w_T,\ldots,w_2} \geq 0$$

$$w_t \in \Omega_t \quad (4)$$

Note that, in this model the decision maker takes some action in the first stage. Then, a random event occurs affecting the outcome of the first-stage decision. An adjustment in the recourse decision is made in the second stage that compensates for any bad effects that might have been experienced as a result of the first-stage decision. The optimal policy from such a model is a single first-stage policy and a collection of recourse decisions (a decision rule) defining which second-stage action should be taken in response to each random outcome. The random nature of it usually occurs in the availability vector or in the matrix of technological coefficients. An excellent introductory document to the subject can be found in (Küçükyavuz & Suvrajeet, 2019).

NUMERICAL EXAMPLE

In this section we present an instance that illustrates the use of the two-phase program of allocation applied to the distribution of land and the allocation of livestock raising areas. The variants presented here correspond, firstly, to deterministic linear structures for the case of the management of animal diets and, secondly, to linear stochastic programming models in two phases for the problem of land distribution and allocation of areas for the raising of cattle. The uncertainty of the model is given in terms of rainfall and the availability of water and food for the crop and cattle respectively.

The section shows the analysis and optimization of a stochastic linear programming model that represents the problem of optimal allocation of crop surfaces and allocation of areas for livestock fattening (in several types thereof). For the development of this model we will proceed now to the case description. In this problem we want to obtain the optimum amount of land in square meters that must be assigned to the cultivation of the following products: a) walnut, b) lechuguilla, c) prickly pear. Likewise, it is desired to know the optimum land surface assigned to the fattening of the following species of animals: a) pigs, b) calves, c) sheep. The objective of the proposal is to maximize the utility (difference between sale prices and production costs) obtained by the commercial practice of these elements.

TECHNICAL INFORMATION OF THE MODEL

The walnut is the fruit of the walnut tree, edible, oval in shape and divided into two symmetrical halves that enclose the seed which is formed by four segments described as little united, white inside, oily and sweet taste.

The prickly pear is a cactus plant with very fleshy stems formed by a series of oval pallets with thorns representing the leaves, large flowers with many petals and fruit (prickly pear or nopal) that has yellow-green bark and sweet edible orange or greenish color pulp. The prickly pear is native from Mexico.

The lechuguilla is the smallest agave and usually grows in arid and semi-arid zones from Chihuahua to Tamaulipas in Mexico. The leaves of the lechuguilla, known as "pencas", are strong and stiff with hardened and very sharp tips which can easily penetrate clothes, skin and even the tires of an all-terrain vehicle. This plant contains substances such as saponins, steroids, and glycosides, among others. The water stored in this plant is rich in salts and minerals. The fibers of the lechuguilla are called ixtle and are used in the manufacture of brushes, sacks, covers for bales of cotton and other elements of the textile industry. The stem is used to reduce side pain, as a treatment for all kinds of strokes and to combat diabetes. Because of its high concentration of saponins you also get products that are used in the manu-

Table 1. Plant used quantities in this proposal

Parameter	Walnut	Lechuguilla	Prickly Pear
Yield (Ton/Ha)	2.5	19	5
Planting costs (MXN/Ha)	290	230	250
Sale price (MXN/Ton) X 1000	30	65	21
Area requirements (Plants/Ha)	120	15000	17000
Maintenance costs (MXN/Ha-year)	20	300	350
Nitrate (Kg/Tree-year)	1.8	0.1	0.14
Phosphorus Oxide (Kg/Tree-year)	0.495		
Potassium Oxide (Kg/Tree-year)	0.44		
Clay (Kg/Tree-year)			1
Manure (Kg/Ha)	50		5100
Water requirements (Mt^3/Ha-year)	65	21	57

facture of soaps. The root is used to combat dermatitis caused by some fungi. Alcove is made from the heart. Currently it is also used as a biological control agent against insects. The lechuguilla liquid is commercialized and it is taken, fermented with a little water and sugar and it becomes refreshing drink

With regard to the breeding of animals, in this proposal the following three are considered: sheep, pigs, calves. Tables (1) and (2) show the figures of animal and plant requirements used. Tables (3) and (4) show the costs and availability of inputs for the cultivation of plants and animals.

THE DETERMINISTIC VERSION OF THE MODEL

The variables used in the stochastic version of this proposal are as shown in Box 1.

The function of gross income for marketing the products is as follows (only one example is shown, the others are similar)

(MXN 30, 000/Ton) (2.5 Ton/Ha) (1 Ha/120 Plants) = MXN 65/Plant

Executing all the calculations for the rest of the crops, we obtain

$$625x_1 + 82.3x_2 6.18x_3$$

Revenue from female sheep is as follows

(MXN1300/Sheep) (x_{4h} Sheeps) = 1300 x_{4h}

The calculations for the other animals are similar and the equation of income from the sale of animals is given by

Table 2. Animal used quantities in this proposal

Parameter	Lamb	Pork	Calf
Sale price females (MXN/Animal)	1300	1500	5000
Sales price males (MXN / Animal)	1000	1300	4000
Sale price stallions (MXN/Animal)	3400	3200	7000
Leather sale price (MXN/Animal)	70	100	70
Meat yield (Kg/Animal)	40	100	400
Food costs (MXN/Day-Animal)	6.75	7.5	7.5
Area requirements (Animals /Mt^2)	1	1	1.5
Balanced food (Kg/Male animal - year)	30	45	60
Balanced food (Kg/Female animal - year)	80	50	70
Mineral salts (Kg./Animal-year)	20	15	45
Forage (Kg/Animal-year):	730	750	1000
Oats (Kg/year)	130	150	300
Straw (Kg/year)	300	300	350
Alfalfa (Kg/year)	300	300	
Corn (Kg/year)		200	
Sorghum (Kg/year)		50	
Water requirements (Mt^3/Animal-year)	0.5	2.3	1.6

Table 3. Availability of inputs for the cultivation of plants

Input	Cost (MXN/ Kg)	Availability (kg/ year)
Nitrate	2.60	600
Phosphorus Oxide	1.90	800
Potassium Oxide	5.20	700
Clay	0	∞
Manure	10	4000
Water (cost in mt3)	20	600 mt3

Table 4. Food for animals, costs and availability

Type of Food	Cost (MXN/Kg)	Availability (kg/year)
Balanced feed for sheep	10	1500
Forage (oats)	3	1000
Forage (straw)	2	1000
Forage (Lucerne)	2	1000
Mineral salts	2.50	5000
Balanced feed for pigs	11	1500
Corn	1.65	480
Sorghum	0.90	500
Balanced feed for calves	11	1500

Box 1.

x_1	$\equiv$	Number of plants of walnut to be planted
x_2	$\equiv$	Number of plants of lechuguilla to be planted
x_3	$\equiv$	Number of plants of prickly pear to be planted
x_4	$\equiv$	Total number of sheep in the herd
x_{4h}	$\equiv$	Number of sheep female ewes in the herd
x_{4m}	$\equiv$	Number of sheep male ewes in the herd
x_{4s}	$\equiv$	Amount of sheep stallions in the herd
x_5	$\equiv$	Total number of pigs in the herd
x_{5h}	$\equiv$	Amount of female pigs in the herd
x_{5m}	$\equiv$	Amount of male pigs in the herd
x_{5s}	$\equiv$	Amount of pigs stallions in the herd
x_6	$\equiv$	Amount of calves in the herd
x_{6h}	$\equiv$	Amount of female calves in the herd
x_{6m}	$\equiv$	Amount of male calves in the herd
x_{6s}	$\equiv$	Amount of calves stallions in the herd

$[1300\ x_{4h} + 1000x_{4m} + 3400\ x_{4s}] + [1500\ x_{5h} + 1300\ x_{5\mathrm{m}} + 3200\ x_{5\mathrm{s}}] + [5000\ x_{6h} + 4000\ x_{6\mathrm{m}} + 7000\ x_{6\mathrm{s}}]$

Therefore, the gross income equation $I(x), x \in \mathbb{R}^{12}$

$$I(x) = 1300\ x_{4h} + 1000x_{4m} + 3400\ x_{4s} + 1500\ x_{5h} + 1300\ x_{5\mathrm{m}}$$

$$+ 3200\ x_{5\mathrm{s}} + 5000x_{6h} + 4000x_{6\mathrm{m}} + 7000 + 625x_1 + 82.34x_2 + 6.18x_3 \tag{5}$$

The expense function for the 6 products is as follows. For reasons of space only the case of the walnut is explained. Thus, the input costs for walnut cultivation are as follows

1. Cost of nitrate

(1.80 Kg/Tree) (MXN 2.60/Kg) = MXN 4.68/Tree

2. Cost of phosphorus oxide

(0.44 Kg/Tree) (MXN 5.20/Kg) = MXN 2.2880/Tree

3. Cost of Potassium Oxide

(0.495 Kg/Tree) (MXN 1.90/Kg) = MXN 0.9405/Tree

4. Cost of manure

(50 Kg/Ha) (MXN 10/Kg) (1Ha/120 Trees) = MXN 4.166/Tree

5. Cost of water

(50 Kg/Ha) (MXN 10/Kg) (1Ha/120 Trees) = MXN 4.166/Tree

6. Total cost = MXN 22.9084/Tree

The costs for the inputs of the lechuguilla and prickly pear are obtained in a similar way. The equation of total costs for inputs is given by

$$22.9084\,x_1 + 0.2880\,x_2 + 0.4774\,x_3$$

Using the information shown in Figures (3) and (4) the expenses for the raising of animals are now obtained. The costs of maintenance of a stallion are similar to the maintenance cost of a female of that species, therefore the calculations are not repeated. Again, for reasons of space, only the breeding calculations for the sheep are replicated. The rest of the animals are similar.

The costs of the inputs for raising the sheep are (the maintenance period is 121 days)

1. Cost for balanced feed for sheep for males

(10 MXN/Kg) (30 Kg/Sheep-year) (1 Year/365 Days) (121 Days) = MXN 99.4520/Sheep

2. Cost for balanced feed for sheep for females

(MXN10/Kg) (80 Kg/Sheep-year) (1 Year/365 Days) (121 Days = MXN265.2054/Sheep

3. Cost of mineral salts for males and females

(2.50 MXN/Kg) (20 Kg/Sheep-year) (1 Year/365 Days) (121 Days) = MXN 16.5753/Sheep

4. Cost of oats for males and females

(3 MXN/Kg) (130 Kg/Sheep-year) (1 Year/365 Days) (121 Days) = MXN 129.2876/Sheep

5. Cost of straw for males and females

(2 MXN/Kg) (300 Kg/Sheep-year) (1 Year/365 Days) (121 Days) = MXN 198.9041/Sheep

6. Cost of lucerne for males and females

(2 MXN/Kg) (300 Kg/Sheep-year) (1 Year/365 Days) (121 Days) = MXN 198.9041/Sheep

7. Cost of water for males and females

(20 MXN/Mt3) (0.5 Mt3 /Sheep-year) (1 Year/365 Days) (121 Days) = MXN 3.3150/Sheep

8. Total cost = MXN 546.9808/Sheep.

Proceeding in a similar way you get the following results. For pigs with a maintenance period of 150 days: MXN 1069.92/Male pig and MXN 1092.5298/Female pig. For calves MXN 3402.1913/Male calf and MXN 3522.7393/Female calf.

Combining the above information, the function of production costs without including the costs of operation is given by

$$G(x) = 22.9000x_1 + 0.2880x_2 + 0.4774x_3 + 546.9086x_{4m} + 911.6389x_{4h} + 911.6389x_{4s}$$

$$+1069.9271x_{5m} + 1092.6298x_{5s} + 3402.1913x_{6m} + 3522.7393x_{6h} + 3522.7393x_{6s} \quad (6)$$

With the above, the utility function of the model is given by the difference between equations, that is to say $U(x) = I(x) - G(x)$.

The constraints associated with the deterministic model are as follows. Requirements and availability of water for the walnut

(65 Mt3/Ha-year) (1Ha/120Plants) (x_1 Plants) = 0.5416 x_1 (Mt3/Year)

For the rest of water requirements in vegetables and animals the calculations are similar. So the restrictive water equation is as follows

$$0.5416\ x_1 + 0.0014\ x_2 + 0.0034\ x_3 + 0.5\ x_4 + 2.3\ x_5 + 1.6\ x_6 \leq 600 \tag{7}$$

The nitrate requirements for plant products are as follows. The walnut needs 1.8 (Kg / Tree-year), therefore

$$(1.8\ \text{Kg/Tree-year})\ (x_1\ \text{Trees}) = 1.8\ x_1\ (\text{Kg/Year})$$

The calculation for the rest of the plant products is similar. Then, the restrictive equation of nitrate availability is

$$1.8\,x_1\ + 0.10\,x_2\ + 0.14\ x_3 \leq 600$$

Similarly, the requirements of Potassium Oxide are. The restrictive equation for this compound is then

$$1.8\,x_1\ + 0.10\,x_2\ + 0.14\ x_3 \leq 600$$

Analogously, the requirements of Phosphorus oxide and Potassium Oxide in walnut determine the following constraint

$$0.495\ x_1 \leq 800\ ,\ 0.440\ x_1 \leq 700$$

The manure requirements in walnuts is as follows

$$(50\ \text{Kg/Ha-year})\ (1\text{Ha/120 Tree})\ (x_1\ \text{Tree}) = 0.417\ x_1\ (\text{Kg/Year})$$

The calculation for prickly pear is similar, then

$$0.417\,x_1 + 0.3\,x_2 \leq 4000$$

In relation to the feeding requirements in animals and for the balanced feed for sheep we have

$$(30\ \text{Kg/Sheep-year})\ (x_4\ \text{Sheep}) = 30\ x_4\ (\text{Kg/Year})$$

Therefore

$$30x_4 \leq 1500$$

For pigs and calves the corresponding equations are

$$45x_5 \leq 1500, 60x_6 \leq 1500$$

To determine the constraint associated with the consumption of mineral salts for the sheep, proceed as follows

(20Kg/Sheep-year) (x_4 Sheep) = 20 x_4 (Kg/Year)

The calculation for the rest of the animals is similar, so we have

$$20x_4 + 15x_5 + 45\ x_6 \leq 1500$$

The area requirement for the walnut is shown below

(70mt2/Tree) (x_1 Tree) = 70 x_1 (mt2)

The calculation for the rest of vegetables and animals is similar. The equation that determines the requirements of the available land area is (the total available area is 20 hectares = 200,000 square meters)

$$70x_1 + 1.5x_2 + 2x_3 + x_4 + x_5 + 1.5x_6 = 200{,}000$$

The following constraints indicate that for every 10 females in the herd of animals there must be a stallion

$$x_{4h} = 10x_{4s}, x_{5h} = 10x_{5s}, x_{6h} = 10x_{6s}$$

In the same sense, the production of prickly pear is at least 20 percent of the production of all the vegetables, that is to say

$$x_3 \geq 0.20x_2 + 0.20x_3$$

Finally, the disaggregation equations of species are written

$$x_{4h} + x_{4m} + x_{4s} - x_4 = 0,$$

$$x_{5h} + x_{5m} + x_{5s} - x_5 = 0, x_{6h} + x_{6m} + x_{6s} - x_6 = 0,$$

As a result of the above, we have the following linear integer program

$$MaxU\left(x\right)=602.10\ x_1+82.05\ x_2+5.70\ x_3+388.36\ x_{4h}+456.01\ x_{4m}+2488.36\ x_{14s}+407.47\ x_{15h}$$
$$+230.07\ x_{5m}+2107.47\ x_{5s}+1477.26\ x_{16h}+597.80\ x_{16m}+3477.26\ x_{16s}$$

s.t.

$$0.54\ x_1+1.5\ x_2+2\ x_3+\ x_4+\ x_5+1.5\ x_{16}=200,000$$

$$1.8\ x_1+0.10\ x_2+0.14\ x_3\le 600$$

$$20\ x_4+1.5\ x_{25}+45\ x_{63}\le 5000$$

$$0.417\ x_4+0.3\ x_3\le 4000$$

$$x_{4\mathrm{h}}+\ x_{4\mathrm{m}}+\ x_{4\mathrm{s}}\ x_4=0$$

$$x_{5\mathrm{h}}+\ x_{5\mathrm{m}}+\ x_{5\mathrm{s}}\ x_5=0$$

$$x_{6\mathrm{h}}+\ x_{6\mathrm{m}}+\ x_{6\mathrm{s}}\ x_6=0$$

$$x_3\ge 0.20x_2+0.20x_1, 0.440x_1\le 700$$

$$30x_4\le 1500, 45x_5\le 1500, 60x_6\le 1500$$

$$x_{4\mathrm{h}}-10x_{4\mathrm{s}}=0, x_{5\mathrm{h}}-10x_{5\mathrm{s}}=0, x_{6\mathrm{h}}-10x_{6\mathrm{s}}=0,$$

$$x_1, x_2, x_3, x_4, x_5, x_{16}\ge 0,\ \text{Integer}$$

$$x_{1h}, x_{4m}, x_{4s}, x_{5h}, x_{5m}, x_{6h}, x_{6m}, x_{6s}, \ge 0,\ \text{Integer} \tag{9}$$

The optimum solution is as follows

$$x_1=254, x_2=1049, x_3=263, x_4=50, x_5=33, x_{4h}=40,$$

$$x_{4s}=4, x_{5h}=30,\ x_{5s}=33, x_{6h}=20, x_{6m}=3, x_{6s}=2, U\left(x\right)=369,883$$

THE DOUBLE STAGE MODEL WITH LIMITED RESOURCES

The two-stage stochastic programs of generalized recourse can be written in simplified form as follows

$$Min_x \left\{\phi_\xi \left\{f\left(x,\xi\right)\right\} / f\left(x,\xi\right) = f_1\left(f_{11}\left(x,\xi\right), Q_1\left(x,\xi\right)\right), x \in C_x\left(\xi\right) = \left\{x \in \mathbb{R}^{n_1} / g_1\left(x,\xi\right) o0\right\} a.s.\right\}$$

Were

$$Q_1\left(x,\xi\right) = \inf_{y(\xi)} \left\{q(x, y\left(\xi\right); \xi / y\left(\xi\right) \in C_y\left(x,\xi\right) = \left\{y\left(\xi\right) \in \mathbb{R}^{n_2} / g_2\left(x, y\left(\xi\right) : \xi\right) o0a.s\right\}\right\}$$

where in this formulation we add stage indicating indices every time it is necessary. In our case, ξ is a random element with a finite discrete distribution. We denote

$$\forall s \in \mathcal{S} : y_s = y\left(\xi^s\right), p_s = P\left(\xi = \xi^s\right) \geq 0, \sum_{s=1}^{S} p_s = 1.$$

and the mathematical expectation by

$$E_\xi = \left\{Q\right\}\left(x,\xi\right) = \sum_{s=1}^{S} p_s Q\left(x,\xi^s\right)$$

Thus, for nonlinear underlying program we

$$Min_X \left\{f\left(X\right) + \sum_{s=1}^{S} p_s Q\left(X,\xi^s\right) / x \in C_x\right\}$$

$$Q\left(X,\xi^s\right) = Min_{y_s} \left\{q\left(x,y_s\right) \mid y_s \in C_y\left(x,\xi^s\right)\right\}$$

This mathematical model can be seen as a scenario tree, Figure (1). The node 1 in the top level represents the first-stage program, arcs denote realizations ξ^s, and nodes in the bottom level correspond to the second-stage programs, have (Popela, Novotny, Roupec, Hrabec & Olstad, 2018)

Resuming the previous ideas our model will be constructed assuming that the available water is not a deterministic quantity and that the yield of crops and livestock directly depend on this compound. To simplify the analysis, only 3 scenarios are considered: a) A bad year (drought), b) An average year (sufficient amount of water for our process), c) A good year (water in abundance that does not generate uncertainty).

Figure 1. Schematic representation of the decision-making process

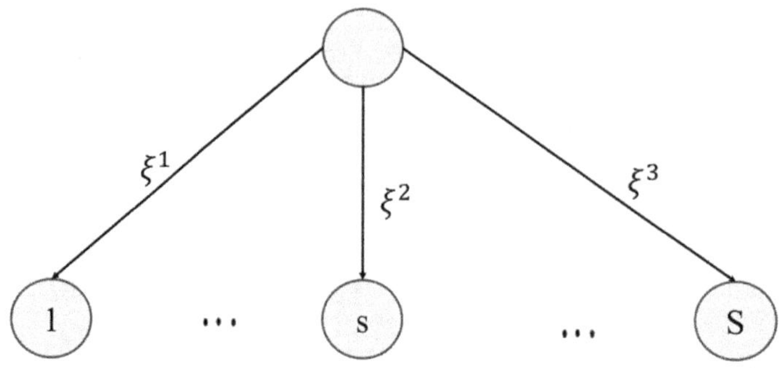

The following values are considered as measurement parameters. 20 percent above the average value will be associated with a good year, and 20 percent below the average is a bad year. Table (5) shows the variation of crop yields according to available water.

For scenario 1, the corresponding functions of gross income (by plant and global) of gross and net profit are respectively

$$500x_1 + 65.87x_2 + 4.95x_3$$

$$I(x) = 960x_{4h} + 960x_{4m} + 3400x_{4s} + 1400x_{5h} + 1400x_{5m} + 3200x_{5s} + 5000x_{6h}$$

$$+4000x_{6m} + 7000x_{6s} + 500x_1 + 65.87x_2 + 4.95x_3$$

$$U(x) = 476.73x_1 + 65.87x_2 + 4.473x_3 + 159.55x_{4h} + 316.26x_{4m} + 2600x_{4s} + 309x_{5h}$$

$$+330x_{5m} + 2108.55x_{5s} + 1477.20x_{6h} + 598x_{6m} + 3477.20x_{6s}$$

$$U_1^n(x) = 476.73x_{1v} + 65.87x_{2v} + 4.473x_3 + 159.55x_{4h} + 316.26x_{4m} + 2600x_{4s} + 309x_{5h}$$

$$+330x_{5m} + 2108.55x_{5s} + 1477.20x_{6h} + 598x_{6m} + 3477.20x_{6s} - 0.1660x_1 - 0.02x_2 - 0.0205x_3$$

Table 5. Crop yields (Ton/Ha) according to available water

Scenario	Walnut	Lechuguilla	Prickly Pear	Water Available (mt3)
Bad year	2	15.2	4	480
Normal year	2.5	19	5	600
Good year	3	22.8	6	720

For scenario 2 we have

$$625x_1 + 82.34x_2 + 6.18x_3$$

$$I(x) = 960x_{4h} + 960x_{4m} + 3400x_{4s} + 1400x_{5h} + 1400x_{5m} + 3200x_{5s} + 5000x_{6h} + 4000x_{6m} + 7000x_{6s} + 500x_1 + 65.87x_2 + 4.95x_3$$

$$U(x) = 601.73x_1 + 82.026x_{2v} + 5.703x_{3v} + 159.55x_{4h} + 316.26x_{4m} + 2600x_{4s} + 309x_{5h} + 330x_{5m} + 2108.55x_{5s} + 1477.20x_{6h} + 598x_{6m} + 3477.20x_{6s}$$

$$U_2^n(x) = 601.73x_{1v} + 82.06x_{2v} + 4.473x_{3v} + 159.55x_{4h} + 316.26x_{4m} + 2600x_{4s} + 309x_{5h} + 330x_{5m} + 2108.55x_{5s} + 1477.20x_{6h} + 598x_{6m} + 3477.20x_{6s} - 0.1660x_1 - 0.02x_2 - 0.0205x_3$$

And for scenario 3 we have

$$750x_1 + 98.80x_2 + 7.41x_3$$

$$I(x) = 960x_{4h} + 960x_{4m} + 3400x_{4s} + 1400x_{5h} + 1400x_{5m} + 3200x_{5s} + 5000x_{6h} + 4000x_{6m} + 7000x_{6s} + 750x_1 + 98.80x_2 + 7.41x_3$$

$$U(x) = 726.73x_1 + 98.50x_{2v} + 6.933x_3 + 159.55x_{4h} + 316.26x_{4m} + 2600x_{4s} + 309x_{5h} + 330x_{5m} + 2108.55x_{5s} + 1477.20x_{6h} + 598x_{6m} + 3477.20x_{6s}$$

$$U_3^n(x) = 726.73x_{1v} + 95.50x_{2v} + 5.703x_{3v} + 159.55x_{4h} + 316.26x_{4m} + 2600x_{4s} + 309x_{5h} + 330x_{5m} + 2108.55x_{5s} + 1477.20x_{6h} + 598x_{6m} + 3477.20x_{6s} - 0.1660x_1 - 0.02x_2 - 0.0205x_3$$

As can be seen the solutions are not altered in any of the cases of bad year and good year with respect to the average year[1]. It would then be expected that the revenues in the three scenarios range from 291, 297.7 to 379,071.6 units. Note that the quantities have a range of: 379,071.6 - 291, 297.7 = 87,773.90.

Thus, to reflect the fact that scenarios can be presented with a certain probability value, it is now appropriate to introduce a measure of uncertainty ρ_i for each of them. The analysis of this situation is made from the average scenario. It will be assumed that scenario 1 can be presented with a probability of 0.25, scenario 2 with a probability of 0.40 and the scenario 3 with 0.35. Now the new variables are: $x_{1i} \equiv$ number of walnut trees planted under scenario $i, i = 1,2,3$; $x_{2j} \equiv$ number of lechuguilla plants planted under scenario $j, j = 1,2,3$; $x_{3k} \equiv$ number of prickly pear plants planted under scenario $k, k = 1,2,3$.

Table 6. Reproduces the solutions to the proposed scenarios

Variable	Scenario 1	Scenario 2	Scenario 3
x_1	260	260	260
x_2	800	800	800
x_3	250	250	250
x_4	34	34	34
x_5	33	33	33
x_6	22	22	22
x_{1v}	247	247	247
x_{2v}	772	772	772
x_{3v}	240	240	240
x_{4m}	1	1	1
x_{4s}	3	3	3
x_{5h}	30	30	30

contines in next column

Table 6. Continued

Variable	Scenario 1	Scenario 2	Scenario 3
x_{5s}	3	3	3
x_{5m}	0	0	0
x_{6h}	20	20	20
x_{6m}	0	0	0
x_{6s}	2	2	2
x_{1r}	13	13	13
x_{2r}	28	28	28
x_{3r}	10	10	10
U_1^n	291,297.7		
U_2^n		335,183.5	
U_3^n			379,071.6

Then, using the Linear Integer Program as a general model and decomposing to the objective function of the problem $Z = Cx$ two partial functions in the following way:

$$\bar{Z} = Cx = f_0\left(x\right) + \sum_{i=1}^{3} \rho_i f_i\left(x\right) \tag{10}$$

where

$$f_0\left(x\right) = 159.55x_{4h} + 316.25x_{4m} + 2600x_{4s} + 309x_{5h} + 330x_{5m}$$
$$+2108.55x_{55} + 1477.20x_{6h} + 598x_{6m} + 3477.20x_{6s}$$

represents the elements of Cx that do not depend on any probabilistic value (in this case they are the variables associated with the animals x_4, x_5, x_6. The rest represent the variables whose dependence on a probability value is now the reason for analysis. The calculation of $\rho_i f_i$ fi is as follows.

1. For scenario 1, using $\rho = 0.25$, we have

$\rho_1 f_1$ (x) = 0.25 [601.73 x_{11} + 82.026 x_{21} + 5.703 x_{31}] – 0.25[0.1660 x_{11} + 0.02 x_{21} + 0.0205 x_{31}]

The constraints associated with this scenario are as follows2

$$0.4333x_{11} + 0.001112x_{21} + 0.00272x_{31} + 0.5x_4 + 2.3x_5 + 2.3x_5 + 1.6x_6 \leq 480$$

$$70x_{11} + 1.5x_{21} + 2x_{31} + x_4 + x_5 + 1.5x_6 \leq 200,000$$

$$x_{11} + 0.10x_{21} + 0.14x_{31} \leq 600$$

$$20x_4 + 15x_{15} + 45x_6 \leq 5000$$

$$0.417x_{11} + 0.3x_{31} \leq 4000$$

$$x_{4h} + x_{4m} + x_{4s} - x_4 = 0$$

$$x_{5h} + x_{5m} + x_{5s} - x_5 = 0$$

$$x_{6h} + x_{6m} + x_{6s} - x_6 = 0$$

$$x_{31} \geq 0.20x_{21} + 0.20x_{11}$$

$$0.440x_{11} \leq 700$$

$$30x_4 \leq 1500$$

$$45x_5 \leq 1500$$

$$60x_6 \leq 1500$$

$$x_{4h} = 10x_{5s}$$

$$x_{6h} = 10x_{6s}$$

$$x_{11}, x_{21}, x_{31}, x_4, x_5, x_6 \geq 0, \text{Integer}$$

$$x_{4h}, x_{4m}, x_{4s}, x_{5h}, x_{5m}, x_{5s}, x_{6h}, x_{6m}, x_{6s} \geq 0, \text{Integer}$$

2. For scenario 2, using $\rho = 0.40$, we have

$$\rho_2 f_2(\mathrm{x}) = 0.40\,[601.73\ x_{12} + 82.026\ x_{22} + 5.703\ x_{32}] - 0.40\,[0.1660\ x_{12} + 0.02\ x_{22} + 0.0205\ x_{32}]$$

The constraints associated with this scenario are as follows

$$0.5416x_{12} + 0.0014x_{22} + 0.0034x_{32} + 0.5x_4 + 2.3x_5 + 2.3x_5 + 1.6x_6 \leq 600$$

$$70x_{12} + 1.5x_{22} + 2x_{32} + x_4 + x_5 + 1.5x_6 \leq 200,000$$

$$1.8x_{12} + 0.10x_{22} + 0.14x_{32} \leq 600$$

$$20x_4 + 15x_{15} + 45x_6 \leq 5000$$

$$0.417x_{12} + 0.3x_{32} \leq 4000$$

$$x_{4h} + x_{4m} + x_{4s} - x_4 = 0$$

$$x_{5h} + x_{5m} + x_{5s} - x_5 = 0$$

$$x_{6h} + x_{6m} + x_{6s} - x_6 = 0$$

$$x_{32} \geq 0.20x_{22} + 0.20x_{12}$$

$$0.440x_{12} \leq 700$$

$$30x_4 \leq 1500$$

$45x_5 \leq 1500$

$60x_6 \leq 1500$

$x_{4h} = 10x_{4s}$

$x_{5h} = 10x_{5s}$

$x_{6h} = 10x_{6s}$

$x_{12}, x_{22}, x_{32}, x_4, x_5, x_6 \geq 0$, Integer

$x_{4h}, x_{4m}, x_{4s}, x_{5h}, x_{5m}, x_{5s}, x_{6h}, x_{6m}, x_{6s} \geq 0$, Integer

3. For scenario 3, using $\rho = 0.35$, we have

$\rho_2 f_2$ (x) = 0.35 [601.73 x_{12} + 82.026 x_{22} + 5.703 x_{32}] – 0.35 [0.1660 x_{12} + 0.02 x_{22} + 0.0205 x_{32}]

The constraints associated with this scenario are as follows

$0.649x_{13} + 0.00168x_{23} + 0.00408x_{33} + 0.5x_4 + 2.3x_5 + 2.3x_5 + 1.6x_6 \leq 720$

$0.5416x_{12} + 0.0014x_{22} + 0.0034x_{32} + 0.5x_4 + 2.3x_5 + 2.3x_5 + 1.6x_6 \leq 600$

$70x_{13} + 1.5x_{23} + 2x_{33} + x_4 + x_5 + 1.5x_6 \leq 200,000$

$1.8x_{13} + 0.10x_{23} + 0.14x_{33} \leq 600$

$20x_4 + 15x_{15} + 45x_6 \leq 5000$

$0.417x_{13} + 0.3x_{33} \leq 4000$

$x_{4h} + x_{4m} + x_{4s} - x_4 = 0$

$x_{5h} + x_{5m} + x_{5s} - x_5 = 0$

$$x_{6h} + x_{6m} + x_{6s} - x_6 = 0$$

$$x_{33} \geq 0.20x_{23} + 0.20x_{13}$$

$$0.440x_{13} \leq 700$$

$$30x_4 \leq 1500$$

$$45x_5 \leq 1500$$

$$60x_6 \leq 1500$$

$$x_{4h} = 10x_{4s}$$

$$x_{5h} = 10x_{5s}$$

$$x_{6h} = 10x_{6s}$$

$$x_{13}, x_{23}, x_{33}, x_4, x_5, x_6 \geq 0, \text{Integer}$$

$$x_{4h}, x_{4m}, x_{4s}, x_{5h}, x_{5m}, x_{5s}, x_{6h}, x_{6m}, x_{6s} \geq 0, \text{Integer} \tag{11}$$

Joining the three scenarios we have the following model of Linear Integer Programming. whose solution is given in Table (7)

$$\text{Max}\bar{Z} = 150.43\,x_{11} + 20.50\,x_{21} + 1.42\,x_{31} + 240.69\,x_{12} + 32.81\,x_{22} + 2.28\,x_{32} + 210.60\,x_{13} + 28.70\,x_{23} + 1.99\,x_{33} + 159.55\,x_{4h} + 316.25\,x_{4m} + 2600\,x_{4s} + 309\,x_{5h} + 330\,x_{5m} + 2108.55\,x_{5s} + 1477.20\,x_{6h} + 598\,x_{6m} + 3477.20\,x_{6s} - 0.1660\,x_{11} - 0.1660\,x_{12} - 0.1660\,x_{13} - 0.02\,x_{21} - 0.02\,x_{22} - 0.02\,x_{23} - 0.0205\,x_{31} - 0.0205\,x_{32} - 0.0205\,x_{33}$$

s.t.

$$0.4333x_{11} + 0.001112x_{21} + 0.00272x_{31} + 0.5x_4 + 2.3x_5 + 2.3x_5 + 1.6x_6 \leq 480$$

$$0.5416x_{12} + 0.0014x_{22} + 0.0034x_{32} + 0.5x_4 + 2.3x_5 + 2.3x_5 + 1.6x_6 \leq 600$$

$$0.649x_{13} + 0.00168x_{23} + 0.00408x_{33} + 0.5x_4 + 2.3x_5 + 2.3x_5 + 1.6x_6 \leq 720$$

$$70x_{11} + 1.5x_{21} + 2x_{31} + x_4 + x_5 + 1.5x_6 \leq 200,000$$

$$70x_{12} + 1.5x_{22} + 2x_{32} + x_4 + x_5 + 1.5x_6 \leq 200,000$$

$$70x_{13} + 1.5x_{23} + 2x_{33} + x_4 + x_5 + 1.5x_6 \leq 200,000$$

$$1.8x_{11} + 0.10x_{21} + 0.14x_{31} \leq 600$$

$$1.8x_{12} + 0.10x_{22} + 0.14x_{32} \leq 600$$

$$1.8x_{13} + 0.10x_{23} + 0.14x_{33} \leq 600$$

$$20x_4 + 15x_{15} + 45x_6 \leq 5000$$

$$0.417x_{11} + 0.3x_{31} \leq 4000$$

$$0.417x_{12} + 0.3x_{32} \leq 4000$$

$$0.417x_{13} + 0.3x_{33} \leq 4000$$

$$x_{4h} + x_{4m} + x_{4s} - x_4 = 0$$

$$x_{5h} + x_{5m} + x_{5s} - x_5 = 0$$

$$x_{6h} + x_{6m} + x_{6s} - x_6 = 0$$

$$x_{31} \geq 0.20x_{21} + 0.20x_{11}$$

$$x_{32} \geq 0.20x_{22} + 0.20x_{12}$$

$$x_{33} \geq 0.20x_{23} + 0.20x_{13}$$

$0.440x_{11} \leq 700$

$0.440x_{12} \leq 700$

$0.440x_{13} \leq 700$

$30x_{4} \leq 1500$

$45x_{5} \leq 1500$

$60x_{6} \leq 1500$

$x_{4h} = 10x_{4s}$

$x_{5h} = 10x_{5s}$

$x_{6h} = 10x_{6s}$

$x_{11}, x_{21}, x_{31}, x_{4}, x_{5}, x_{6} \geq 0$, Integer

$x_{12}, x_{22}, x_{32}, x_{13}, x_{23}, x_{33} \geq 0$, Integer

$x_{4h}, x_{4m}, x_{4s}, x_{5h}, x_{5m}, x_{5s}, x_{6h}, x_{6m}, x_{6s} \geq 0$, Integer

Note that the results are similar in all three scenarios. The reason is that there is enough water to keep the crops within the requested ranges. That is, no matter how much water is raised or lowered in each crop, there will always be enough water for its development. The solution of the model shows that if the requirements were lowered to a limit of 6 mt3 on average (and using 4 mt3 and 7 mt3) as lower and upper limits, then the recommended amounts of crop for vegetables in each scenario vary substantially.

The new result is achieved by modifying the restrictive water equations in the model already proposed as follows

$0.4333x_{11} + 0.001112x_{21} + 0.00272x_{31} + 0.5x_{4} + 2.3x_{5} + 2.3x_{5} + 1.6x_{6}4$

$0.5416x_{12} + 0.0014x_{22} + 0.0034x_{32} + 0.5x_{4} + 2.3x_{5} + 2.3x_{5} + 1.6x_{6} \leq 6$

$0.649x_{13} + 0.00168x_{23} + 0.00408x_{33} + 0.5x_{4} + 2.3x_{5} + 2.3x_{5} + 1.6x_{6} \leq 7$

Table 7. Results per scenario of the model

Variable	Scenario 1	Scenario 2	Scenario 3	Any Scenario 1
x_{11}	0			
x_{21}	4686			
x_{31}	938			
x_{12}		0		
x_{22}		4686		
x_{32}		938		
x_{13}			0	
x_{23}			4686	
x_{33}			938	
x_{4h}				40
x_{4m}				6
x_{4s}				4
x_{5h}				30
x_{5m}				0
x_{5s}				3
x_{6h}				20
x_{6m}				3
x_{6s}				2
$\overline{Z}$				461,864.80

Table 8. Results per the modified scenario of the model

Variable	Scenario 1	Scenario 2	Scenario 3	Any Scenario 1
x_{11}	0			
x_{21}	2403			
x_{31}	481			
x_{12}		0		
x_{22}		2884		
x_{32}		577		
x_{13}			0	
x_{23}			2804	
x_{33}			561	
x_{4h}				0
x_{4m}				0
x_{4s}				0
x_{5h}				0
x_{5m}				0
x_{5s}				0
x_{6h}				0
x_{6m}				0
x_{6s}				0
$\overline{Z}$				227,280.3

The new solution is shown in Table (8)

Note that although there are changes in the amount recommended per crop, the solution also advises not raising any type of animal. Note also that the gain in this case is much less than the model if you consider breeding animals and farming.

A DISAGGREGATED MODEL OF STOCHASTIC LINEAR PROGRAMMING

An interesting analysis of the model consists of separating the plants that will be dedicated to sale from those that will be dedicated to the reserve for future crops. To start this section, consider the following definition of variables. Let $x_{1vi}, i = 1,2,3$ be the number of walnut trees planted for sale under scenario i; $x_{2vj}, j = 1,2,3$ the number of lechuguilla plants planted for sale under scenario j; and $x_{3vk}, k = 1,2,3$ the number of prickly pear plants planted for sale under scenario k. Then, to link the probabilistic values of the problem, the following calculations of land requirements are obtained in relation to yield per hectare.

1. For the walnut: The information given shows that the yield per hectare of the fruits is 2.5 tons, which is equivalent to approximately 120 trees per hectare. Proceeding in linear form, we obtain a yield of 2 is (120/2.5) (2.5) = 96 Trees/Ha. For a yield of 3, there must be (120/2.5) (3) = 144 Trees/Ha. Then, using the following relationships, the equation of the area occupied by cultivation can be constructed:

(1 Ha/96 Trees) (x_{1v1} Trees) = 0.01042 x_{1v1} Ha (Scenario 1)

The rest is obtained similarly. So, we have 0.00833 x_{1v2} Ha, (Scenario 2) and 0.00694 x_{1v3} Ha, (Scenario 3).

2. For the lechuguilla we have that 19 tons of material obtained in each harvest is equivalent to approximately 15,000 plants meaning that in 15.2 tons there will be 12,000 plants. Equivalently, 22.8 tons will have 18,000 plants. So:

(1 Ha/12000 Plants) (x_{2v1} Plants) = 0.000083333 x_{2v1} Ha (Scenario 1)

Similarly, we have 0.00006666 x_{2v2} Ha, (Scenario 2) and 0.00005555 Ha, (Scenario 3).

3. For the prickly pear the calculation is as follows: 5 tons of material obtained in each harvest is equivalent to approximately 250 plants. In 4 tons there will be 200 plants., in 6 tons there will be 300 plants. So:

(1 Ha/200 Plants)(x_{3v1} Plants) = 0.005 x_{3v1} Ha (Scenario 1)

The rest is defined as 0.004 x_{3v2} Ha, (Scenario 2) and x_{3v3} 0.00333 Ha (Scenario 3).

Associating the equations obtained by scenario, we now have the following group of constraints

$$0.01043x_{1v1} + 0.000083333x_{2v1} + 0.005x_{3v1} \leq 20\left(\text{Scenario1}\right),$$

$$0.00833x_{1v2} + 0.00006666x_{2v2} + 0.004x_{3v2} \leq 20 \text{ (Sccenario 2)},$$

$$0.00694x_{1v3} + 0.00005555x_{2v3} + 0.00333x_{3v3} \leq 20 \text{ (Sccenario 3)},$$

Now, replacing[3] Equation (11) for each of the above constraints according to the desired scenario, and including the corresponding probabilities in the objective function and corresponding constraints we have the following stochastic linear programming models. Here, the restrictive set is such that $x_{1r} = 0.05x_1$ and $x_{1v} = 0.95x_1$, or equivalently $x_1 = 20x_{1r}$ and $x_1 = 1.052x_{1v}$.

Adding the last two expressions and dividing by two, we obtain that which can be put as

$$x_{1k} = 10x_{1rk} + 05263x_{1vk}$$

Analogously, the value of x_2 is given by $x_{2v} = 0.965x_2$ and $x_{2r} = 0.035x_2$ and conveniently clearing x_2 adding up the resulting terms and dividing by the k scenario, x_2 can be set as

$$x_{2k} = 14.285x_{2rk} + 0.5181x_{2vk}$$

Finally, for the case of x_3 this is expressed as $x_3 = 0.004x_3$ and $x_{3v} = 0.960x_3$. Proceeding as before, we have

$$x_{3k} - 12.5x_{3rk} + 05208x_{3vk}$$

These equations will be used in the scenarios that will be constructed for each case.

1. For scenario 1, using

$$x_{11} = 10x_{1r1} + 0.5263x_{1v1}, x_{21} = 14.285x_{2r1} + 0.5181x_{2v1}, x_{31} = 12.5x_{3r1} + 0.5208x_{3v1}, k = 1,$$

y $\rho = 0.25$.

$$\rho_1 f_1\left(x\right) = 0.25\left[601.73x_{11} + 82.026x_{21} + 5.703x_{31}\right] - 0.25\left[0.1660x_{11} + 0.02x_{21} + 0.0205x_{31}\right] =$$

$$0.25[601.73\left(10x_{1r1}+0.5263x_{1v1}\right)+82.026\left(14.285x_{2r1}+0.5181x_{2v1}\right)+$$

$$5.703\left(12.5x_{3r1}+0.5208x_{3v1}\right)]-$$

$$0.25[0.1660\left(10x_{1r1}+0.5263x_{1v1}\right)+0.02\left(14.285x_{2r1}+0.5181x_{2v1}\right)+$$

$$0.0205\left(12.5x_{3r1}+0.5208x_{3v1}\right)]=$$

$$1503.98x_{1r1}+79.15x_{1v1}+292.86x_{2r1}+10.62x_{2v1}+17.75x_{3r1}+0.74x_{3v1}$$

Substituting conveniently the values of x_1, x_2 and x_3 for their new values of x_{11}, x_{21} and x_{31}. and applying the corresponding decrease applied to water availability we arrived to

$$4.333(10x_{1r1}+0.5263x_{1v1})+0.00112\left(14.285x_{2r1}+0.5181x_{2v1}\right)+$$

$$0.00272(12.5x_{3r1}+0.5208x_{3v1})+0.5x_4+2.3x_5+1.6x_6\leq 480,$$

or equivalently

$$0.00141x_{3v1}+0.5x_4+2.3x_5+1.6x_6\leq 480,$$

also, the value of x_1, x_2 and x_3 is substituted in the rest of the equations with its corresponding value obtaining

$$700x_{1r1}+36.84x_{1v1}+21.42x_{2r1}+0.777x_{2v1}+25x_{3r1}$$

$$+1.041x_{3v1}+x_4+x_5+1.5x_6\leq 200000,$$

$$18x_{1r1}+0.95x_{1v1}+1.428x_{2r1}+0.0518x_{2v1}+1.75x_{3r1}+0.0729x_{3v1}\leq 600$$

$$20x_4+15x_5+45x_6\leq 5000$$

$$x_{4h}+x_{4m}+x_{4s}-x_4=0$$

$$x_{5h}+x_{4m}+x_{5s}-x_5=0,\ x_{6h}+x_{6m}+x_{6s}-x_6=0,$$

$$12.5x_{3r1} + 0.5208x_{3v1} \geq 2.857x_{2r1} + 0.1036x_{2v1} + 2x_{1r1} + 0.1052x_{1v1},$$

$$4.40x_{1r1} + 0.2315x_{1v1} \leq 700,$$

$$30\,x_4 \leq 1500, \quad 45x_5 \leq 1500, \quad 60x_6 \leq 1500,$$

$$x_{4h} = 10x_{4s}, \; x_{5h} = 10x_{5s}, x_{6h} = 10x_{6s},$$

$$x_{1v1}, \; x_{1r1}, \; x_{2v1}, x_{2r1}, \; x_{3v1}, \; x_{3r1}, \; x_4, \; x_5, \; x_6, \geq 0, \text{ Integer}$$

$$x_{4h}, \; x_{4m}, \; x_{4s}, \; x_{5h,} \; x_{5m}, \; x_{5s}, \; x_{6h}, \; x_{6m}, x_{6s}, \geq 0, \quad \text{Integer}$$

2. In this, the following relationships are given by

$$x_{12} = 10x_{1r2} + 0.5263x_{1v2}, x_{22} = 14.285x_{2r2} + 0.5181x_{2v2}, x_{32} = 12.5x_{3r2} + 0.5208x_{3v2}, \; k = 2,$$

y $\rho = 0.40$.

The function $\rho_2 f_2(x)$ is given by

$$\rho_2 f_2(x) = 0.40\left[601.73x_{12} + 82.026x_{22} + 5.703x_{32}\right] - 0.40\left[0.1660x_{12} + 0.02x_{22} + 0.0205x_{32}\right] =$$

$$0.40[601.73\left(10x_{1r2} + 0.5263x_{1v2}\right) + 82.026\left(14.285x_{2r2} + 0.5181x_{2v2}\right) +$$

$$5.703\left(12.5x_{3r2} + 0.5208x_{3v2}\right)] - 0.40[0.1660\left(10x_{1r2} + 0.5263x_{1v2}\right) +$$

$$0.02\left(14.285x_{2r2} + 0.5181x_{2v2}\right) + 0.0205\left(12.5x_{3r2} + 0.5208x_{3v2}\right)] =$$

$$2105.47x_{1r2} + 110.81x_{1v2} + 410x_{2r2} + 14.87x_{2v2} + 24.86x_{3r2} + 1.035x_{3v2}$$

In this case, an available water volume of 600 m3 is considered. Proceeding as before we have

$$5.416x_{1r2} + 0.2850x_{1v2} + 0.2x_{2r2} + 0.00072x_{2v2} + 0.0425x_{3r2} +$$

$$0.00177x_{3v2} + 0.5x_4 + 2.3x_5 + 1.6x_6 \leq 600,$$

$$700x_{1r2} + 36.84x_{1v2} + 21.42x_{2r2} + 0.777x_{2v2} + 25x_{3r2} +$$

$$1.041x_{3v1} + x_4 + x_5 + 1.5x_6 \leq 200000$$

$$18x_{1r2} + 0.95x_{1v2} + 1.428x_{2r2} + 0.0518x_{2v2} + 1.75x_{3r2} + 0.0729x_{3v2} \leq 600$$

$$20x_4 + 15x_5 + 45x_6 \leq 5000$$

$$4.17x_{1r2} + 0.22x_{1v2} + 3.75x_{3r2} + 0.1562x_{3v2} \leq 4000,$$

$$x_{4h} + x_{4m} + x_{4s} - x_4 = 0,$$

$$x_{5h} + x_{4m} + x_{5s} - x_5 = 0, \ x_{6h} + x_{6m} + x_{6s} - x_6 = 0$$

$$x_{5h} + x_{4m} + x_{5s} - x_5 = 0, \ x_{6h} + x_{6m} + x_{6s} - x_6 = 0$$

$$12.5x_{3r2} + 0.5208x_{3v2} \geq 2.857x_{2r2} + 0.1036x_{2v2} + 2x_{1r2} + 0.1052x_{1v2}$$

$$4.40x_{1r2} + 0.2315x_{1v2} \leq 700$$

$$30\,x_4 \leq 1500, \quad 45x_5 \leq 1500, \quad 60x_6 \leq 1500$$

$$x_{4h} = 10x_{4s}, \ x_{5h} = 10x_{5s}, \ x_{6h} = 10x_{6s},$$

$$x_{1v2}, \ x_{1r2}, \ x_{2v2}, \ x_{2r2}, \ x_{3v2}, \ x_{3r2}, \ x_{4,} \ x_5, \ x_6 \ \geq 0 \ \text{Integer}$$

$$x_{4h}, x_{4m}, \ x_{4s}, \ x_{5h,} x_{5m}, x_{5s}, \ x_{6h}, \ x_{6m}, x_{6s} \geq 0, \ \text{Integer}$$

Finally, to obtain scenario 3, the following relationships are used

$$x_{13} = 10 \ x_{1r3} + 0.5263 \ x_{1v3},$$

$$x_{23} = 14.285 \ x_{2r3} + 0.5181, \ x_{2v3}\text{x33} = 12.5\text{x3r3} + 0.5208\text{x3v3},$$

$$\text{k} = 3, \ \text{and Á} = 0.35.$$

The function $\rho_3 f_3$ (x) is given by

$$\rho_3 f_3\left(x\right) = 0.35\left[601.73x_{13} + 82.026x_{23} + 5.703x_{33}\right] - 0.35\left[0.1660x_{13} + 0.02x_{23} + 0.0205x_{33}\right] =$$

$$0.35[601.73\left(10x_{1r3} + 0.5263x_{1v3}\right) + 82.026\left(14.285x_{2r3} + 0.5181x_{2v3}\right) +$$

$$5.703\left(12.5x_{3r3} + 0.5208x_{3v3}\right)] -$$

$$0.35[0.1660\left(10x_{1r3} + 0.5263x_{1v3}\right) + 0.02\left(14.285x_{2r3} + 0.5181x_{2v3}\right) +$$

$$0.0205\left(12.5x_{3r3} + 0.5208x_{3v3}\right)] =$$

$$95.61x_{1r3} + 126.64x_{1v3} + 468.58x_{2r3} + 17x_{2v3} + 28.41x_{3r3} + 1.183x_{3v3}$$

Scenario 3 considers an availability of 720 m3. Proceeding as before

$$6.490x_{1r3} + 0.3415x_{1v3} + 0.240x_{2r3} + 0.00087x_{2v3} + 0.06x_{3r3} +$$

$$0.00212x_{3v3} + 0.5x_4 + 2.3x_5 + 1.6x_6 \leq 720,$$

$$700x_{1r3} + 36.84x_{1v3} + 21.42x_{2r3} + 0.777x_{2v3} + 25x_{3r3}$$

$$+1.041x_{3v3} + x_4 + x_5 + 1.5x_6 \leq 200000,$$

$$18x_{1r3} + 0.95x_{1v3} + 1.428x_{2r3} + 0.0518x_{2v3} + 1.75x_{3r3} + 0.0729x_{3v3} \leq 600,$$

$$20x_4 + 15x_5 + 45x_6 \leq 5000,$$

$$4.17x_{1r3} + 0.22x_{1v3} + 3.75x_{3r3} + 0.1562x_{3v3} \leq 4000,$$

$$x_{4h} + x_{4m} + x_{4s} - x_4 = 0,$$

$$x_{5h} + x_{4m} + x_{5s} - x_5 = 0, \; x_{6h} + x_{6m} + x_{6s} - x_6 = 0,$$

$$12.5x_{3r3} + 0.5208x_{3v3} \geq 2.857x_{2r3} + 0.1036x_{2v3} + 2x_{1r3} + 0.1052x_{1v3},$$

$$4.40x_{1r3} + 0.2315x_{1v3} \leq 700,$$

$$30\,x_4 \leq 1500,\ 45x_5 \leq 1500,\ 60x_6 \leq 1500,$$

$$x_{4h} = 10x_{4s},\ \ x_{5h} = 10x_{5s}, x_{6h} = 10x_{6s},$$

$$x_{1v3}, x_{1r3}, x_{2v3}, x_{2r3}, x_{3v3}, x_4 x_5, x_6, \geq 0,\ \text{Integer}$$

$$x_{4h}, x_{4m}, x_{4s}, x_{5h,} x_{5m}, x_{5s}, x_{6h}, x_{6m}, x_{6s}, \geq 0,\ \text{Integer}$$

The stochastic Linear Integer Programming model that represents the average scenario is obtained as follows

$$\text{Max}\ \ \bar{Z} = [159.55x_{4h} + 316.25x_{4m} + 2600x_{4s} + 309x_{5h} + 330x_{5m} + 2108.55x_{5s}$$

$$+1477.20x_{6h} + 598x_{6m} + 3477.20x_{6s}] + [1503.98x_{1r1} + 79.15x_{1v1} +$$

$$292.86\,x_{2r1} + 10.62x_{2v1} + 17.75x_{3r1} + 0.74x_{3v1}] + 2105.47x_{1r2} +$$

$$[110.81x_{1v2} + 410x_{2r2} + 14.87x_{2v2} + 24.86x_{3r2} + 1.035x_{3r2}] + [95.61x_{1r3} +$$

$$126.64x_{1v3} + 468.58x_{2r3} + 17x_{2v3} + 28.41x_{3r3} + 1.183x_{3v3}]$$

s.t.

$$4.333x_{1r1} + 0.22280x_{1v1} + 0.016x_{2r1} + 0.00058x_{2v1} + 0.034x_{3r1} +$$

$$0.00141x_{3v1} + 0.5x_4 + 2.3x_5 + 1.6x_6 \leq 480,$$

$$5.416x_{1r2} + 0.2850x_{1v2} + 0.2x_{2r2} + 0.00072x_{2v2} + 0.0425x_{3r2} +$$

$$0.00177x_{3v2} + 0.5x_4 + 2.3x_5 + 1.6x_6 \leq 600,$$

$6.490x_{1r3} + 0.3415x_{1v3} + 0.240x_{2r3} + 0.00087x_{2v3} + 0.06x_{3r3} +$

$0.00212x_{3v3} + 0.5x_4 + 2.3x_5 + 1.6x_6 \leq 720,$

$700x_{1r1} + 36.84x_{1v1} + 21.42x_{2r1} + 0.777x_{2v1} + 25x_{3r1}$

$+1.041x_{3v1} + x_4 + x_5 + 1.5x_6 =\leq 200000,$

$700x_{1r3} + 36.84x_{1v3} + 21.42x_{2r3} + 0.777x_{2v3} + 25x_{3r3}$

$+1.041x_{3v2} + x_4 + x_5 + 1.5x_6 \leq 200000,$

$18x_{1r1} + 0.95x_{1v1} + 1.428x_{2r1} + 0.0518x_{2v1} + 1.75x_{3r1} + 0.0729x_{3v1} \leq 600,$

$18x_{1r2} + 0.95x_{1v2} + 1.428x_{2r2} + 0.0518x_{2v2} + 1.75x_{3r2} + 0.0729x_{3v2} \leq 600,$

$18x_{1r3} + 0.95x_{1v3} + 1.428x_{2r3} + 0.0518x_{2v3} + 1.75x_{3r3} + 0.0729x_{3v3} \leq 600,$

$4.17x_{1r1} + 0.22x_{1v1} + 3.75x_{3r1} + 0.1562x_{3v1} \leq 4000,$

$4.17x_{1r2} + 0.22x_{1v2} + 3.75x_{3r2} + 0.1562x_{3v2} \leq 4000,$

$4.17x_{1r3} + 0.22x_{1v3} + 3.75x_{3r3} + 0.1562x_{3v3} \leq 4000,$

$12.5x_{3r1} + 0.5208x_{3v1} \geq 2.857x_{2r1} + 0.1036x_{2v1} + 2x_{1r1} + 0.1052x_{1v1},$

$12.5x_{3r2} + 0.5208x_{3v2} \geq 2.857x_{2r2} + 0.1036x_{2v2} + 2x_{1r2} + 0.1052x_{1v2},$

$12.5x_{3r3} + 0.5208x_{3v3} \geq 2.857x_{2r3} + 0.1036x_{2v3} + 2x_{1r3} + 0.1052x_{1v3},$

$4.40x_{1r1} + 0.2315x_{1v1} \leq 700,$

$4.40x_{1r2} + 0.2315x_{1v2} \leq 700,\ 4.40x_{1r3} + 0.2315x_{1v3} \leq 700,$

$$x_{4h} + x_{4m} + x_{4s} - x_{4} = 0,$$

$$x_{5h} + x_{4m} + x_{5s} - x_{5} = 0,\ x_{6h} + x_{6m} + x_{6s} - x_{6} = 0,$$

$$30x_{4} \leq 1500, 45x_{5} \leq 1500, 60x_{6} \leq 1500,$$

$$x_{4h} = 10x_{4s},\ x_{5h} = 10x_{5s}, x_{6h} = 0x_{6s},$$

$$x_{1v1}, x_{1r1}, x_{2v1}, x_{2r1}, x_{3v1}, x_{3r1} \geq 0,\ \text{Integer}$$

$$x_{1v2}, x_{1r2}, x_{2v2}, x_{2r2}, x_{3v2}, x_{3r2}, \geq 0,\ \text{Integer}$$

$$x_{1v3}, x_{1r3}, x_{2v3}, x_{2r3}, x_{3v3}, x_{4} x_{5}, x_{6}, \geq 0,\ \text{Integer}$$

$$x_{4h}, x_{4m}, x_{4s}, x_{5h,} x_{5m}, x_{5s}, x_{6h}, x_{6m}, x_{6s}, \geq 0,\ \text{Integer}$$

The solution to this instance is given by (Table 9). Note that in this model different solutions appear for crops in each proposed scenario.

THE DOUBLE STAGE MODEL WITH LIMITED RESOURCES

From the above results a general formulation of the stochastic model will now be obtained. As can be seen, the model consists of taking a set of decisions in which there is not complete information about some random events. These decisions are called first stage and, in this case, they are represented by the vector $\left(x_{4}, x_{5}, x_{6},\right)$. Here, it is desired to determine the decision of which animals would be devoted to breeding (divided into males, females and stallions). Subsequently, complete information is received about the realizations (results of an experimentation) of some vector ξ. Then, once this has happened, a series of corrective actions y must be taken in a second stage. In the case of interest in this proposal, the random vector is the set of yields of the land as a function of water availability. Corrective actions are the decisions of which crop should be sown.

Rewriting the equation (3), the problem is

$$\min C^{t} x + E_{\xi}\left[Q\left(x, \xi\right)\right] : Ax = b,\ x \geq 0,$$

where

Table 9. Results by scenario for the disaggregated model

Variable	Scenario 1	Scenario 2	Scenario 3	Any Scenario
x_{1v1}	0			
x_{2v1}	3			
x_{3v1}	1800			
x_{1v2}		0		
x_{2v2}		0		
x_{3v2}		0		
x_{1v3}			0	
x_{2v3}			9048	
x_{3v3}			1800	
x_{4h}				40
x_{4m}				6
x_{4s}				4
x_{5h}				30
x_{5m}				0
x_{5s}				3

contines in next column

Table 9 Continued

Variable	Scenario 1	Scenario 2	Scenario 3	Any Scenario
x_{6h}				20
x_{6m}				3
x_{6s}				2
x_{4}				50
x_{5}				33
x_{6}				25
x_{1r1}				0
x_{1r2}				0
x_{2r1}				328
x_{1r3}				0
x_{2r3}				0
x_{2r2}				0
x_{3r3}				0
x_{3r2}				0
$\overline{Z}$				325, 934.90

$$Q\left(x,\xi\right) = \text{Min}\left\{q^t y \middle| \, Wy = h - Tx, y > 0\right\}$$

Here, ξ the vector formed by the components of q^t, h and T. E_ξ denotes the mathematical expectation with respect to ξ. In the proposed model it is assumed that W it is fixed (fixed resources). In this proposal, the random vector is a discrete variable with only three different values. Here the matrix T is random. In this formulation, a second stage problem for a particular scenario $k = 3$ can be given as

$$Q(x_3,\xi) = \text{Min}[95.61x_{1r} + 126.64x_{1v} + 468.58x_{2r} + 17x_{2v} + 28.41x_{3r} + 1.183x_{3v}],$$

s.t.

$$t_1(3)[x_{1r} + x_{1v}] + t_2(3)[x_{2r} + x_{2v}] + t_3(3)[x_{3r} + x_{3v}] + 0.5x_4 + 2.3x_5 + 1.6x_6 \le 720,$$

$$700x_{1r} + 36.84x_{1v} + 21.42x_{2r} + 0.777x_{2v} + 25x_{3r} + 1.041x_{3v} + x_4 + x_5 + 1.5x_6 \le 200000,$$

$$18x_{1r} + 0.95x_{1v} + 1.428x_{2r} + 0.0518x_{2v} + 1.75x_{3r} + 0.0729x_{3v} \le 600,$$

$$20x_4 + 15x_5 + 45x_6 \le 5000,$$

$$4.17x_{1r} + 0.22x_{1v} + 3.75x_{3r} + 0.1562x_{3v} \le 4000,$$

$$x_{4h} + x_{4m} + x_{4s} - x_4 = 0,$$

$$x_{5h} + x_{4m} + x_{5s} - x_5 = 0,\ x_{6h} + x_{6m} + x_{6s} - x_6 = 0,$$

$$12.5x_{3r} + 0.5208x_{3v} \ge 2.857x_{2r} + 0.1036x_{2v} + 2x_{1r} + 0.1052x_{1v},$$

$$4.40x_{1r} + 0.2315x_{1v} \le 700,$$

$$30x_4 \le 1500,\ 45x_5 \le 1500,\ 60x_6 \le 1500,$$

$$x_{4h} = 10x_{4_s},\ x_{5h} = 10x_{5_s},\ x_{6h} = 10x_{6_s},$$

$$x_{1v}, x_{1r}, x_{2v}, x_{2r}, x_{3v}, x_{3r}, x_4, x_5, x_6 \ge 0,\ \text{Integer}$$

$$x_{4h}, x_{4m}, x_{4s}, x_{5h}, x_{5m}, x_{5s}, x_{6h}, x_{6m}, x_{6s}, \ge 0,\ \text{Integer}$$

where $t_i(k)$ represents crop yield i under the k scenario (or k state of nature). In this formulation the random vector $\xi = (t_1, t_2, t_3)$ is formed of three returns and that ξ can take one of three different values, that is, ξ_1, ξ_2, or ξ_3 which can be represented as

$$\left[t_1(1), t_2(1),\ t_3(1),\ t_1(2), t_2(2),\ t_2(3), t_1(3), t_1(2), t_1(3)\right]$$

An alternate representation will be specifying that the random vector $\xi(k)$ depends on the k scenario, which takes three values. In the way in which we represent the land assignment in this formulation it is known as implicit representation of the stochastic program. An implicit condensed representation is obtained by defining the function $\psi(x) = E_\xi\left[Q(x,\xi)\right]$ as the function value or function resource which can be written as

$$\min C^t x + \psi(x) : Ax = b,\ x \geq 0,$$

CONTINUOUS RANDOM VARIABLES

A special case of this type of mathematical models is the case in which the probability distribution of crop yields is considered continuous. In the simplest case, consider the case where the yield of each crop i can be adequately described through a random variable uniformly distributed in the range $\left[\alpha, \beta\right]$. For example, α can be considered as 80% of average yield and β as 120% of that performance, so the expectation falls on the value used for the deterministic case. Again, we are interested in obtaining the quantities of animals to be raised in a first stage. In the second stage, we must decide what quantities of vegetables to plant (for sale and replacement) once land yields are known as a function of available water. We will use the notation $\psi(x) = E_\xi\left[Q(x,\xi)\right]$ where $Q(x,\xi)$ is the value of the second stage for a random vector realization.

Then

$$E_\xi\left[Q(x,\xi)\right] = \sum_{i=1}^{3} E_\xi\left[Q_i(x_i,\xi)\right] = \sum_{i-1}^{3} \psi_i(x_i),$$

where $Q_i(x_i,\xi)$ is the optimal value of the second stage for vegetable crops.

Obtaining an algorithm to evaluate the solution of this model is somewhat complex, however it will be explained in the simplest possible way. For a given value $t_3(\xi)$ of the yield of the vegetables, the following representation of the second stage is obtained:

$$Q(x_3,\xi) = \text{Min}[95.61 x_{1r}(\xi) + 126.64 x_{1v}(\xi) + 468.58 x_{2r}(\xi) +$$
$$17 x_{2v}(\xi) + 28.41 x_{3r}(\xi) + 1.183 x_{3v}(\xi)],$$

s.t.

$$t_1(3)\left[x_{1r}(\xi) + x_{1v}(\xi)\right] + t_2(3)\left[x_{2r}(\xi) + x_{2v}(\xi)\right] +$$

$$t_3(3)\left[x_{3r}(\xi)+x_{3v}(\xi)\right]+0.5x_4(\xi)+2.3x_5(\xi)+1.6x_6(\xi)\leq 720,$$

$$700x_{1r}(\xi)+36.84x_{1v}(\xi)+21.42x_{2r}(\xi)+0.777x_{2v}(\xi)+25x_{3r}(\xi)$$

$$+1.041x_{3v}(\xi)+x_4+x_5+1.5x_6\leq 200000,$$

$$18x_{1r}(\xi)+0.95x_{1v}(\xi)+1.428x_{2r}(\xi)+0.0518x_{2v}(\xi)+1.75x_{3r}(\xi)+0.0729x_{3v}(\xi)\leq 600,$$

$$20x_4+15x_5+45x_6\leq 5000,$$

$$4.17x_{1r}(\xi)+0.22x_{1v}(\xi)+3.75x_{3r}(\xi)+0.1562x_{3v}(\xi)\leq 4000,$$

$$x_{4h}+x_{4m}+x_{4s}-x_4=0,$$

$$x_{5h}+x_{4m}+x_{5s}-x_5=0,\;\; x_{6h}+x_{6m}+x_{6s}-x_6=0,$$

$$12.5x_{3r}(\xi)+0.5208x_{3v}(\xi)\geq 2.857x_{2r}(\xi)+0.1036x_{2v}(\xi)+2x_{1r}(\xi)+0.1052x_{1v}(\xi),$$

$$4.40x_{1r}(\xi)+0.2315x_{1v}(\xi)\leq 700,$$

$$30x_4\leq 1500,\; 45x_5\leq 1500,\; 60x_6\leq 1500,$$

$$x_{4h}=10x_{4_s},\;\; x_{5h}=10x_{5_s},\;\; x_{6h}=10x_{6_s},$$

$$x_{1v}(\xi),x_{1r}(\xi),x_{2v}(\xi),x_{2r}(\xi),x_{3v}(\xi),x_{3r}(\xi),x_4,x_5,x_6\geq 0,\;\text{Integer}$$

$$x_{4h},x_{4m},x_{4s},x_{5h,}x_{5m},x_{5s},x_{6h,}x_{6m},x_{6s},\geq 0,\;\text{Integer}$$

Let $x_3^* \in \mathbb{R}^{18}$ be the optimal solution to the problem defined as

$$x_3^*=[x_{1v}^*(\xi),x_{1r}^*(\xi),x_{2v}^*(\xi),x_{2r}^*(\xi),x_{3v}^*(\xi),x_{3r}^*(\xi),x_4^*,$$

$$x_5^*,x_6^*,x_{4h}^*,x_{4m}^*,x_{4s}^*,x_{5h}^*,x_{5m}^*,x_{5s}^*,x_{6h}^*,x_{6m}^*,x_{6s}^*]$$

then the objective function is expressed as

$$Q\left(x_3^*,\xi\right)=\text{Min}\begin{bmatrix}95.61x_{1r}^*\left(\xi\right)+126.64x_{1v}^*\left(\xi\right)+468.58x_{2r}^*\left(\xi\right)\\+17x_{2v}^*\left(\xi\right)+28.41x_{3r}^*\left(\xi\right)+1.183x_{3v}^*\left(\xi\right)\end{bmatrix}, \quad \text{(III.46)}$$

Now, as each crop i is supposed to be uniformly distributed between α and β, then it is true that

$$\alpha x_1 \leq 720 \leq \beta x_1$$

$$\alpha x_2 \leq 720 \leq \beta x_2$$

$$\alpha x_3 \leq 720 \leq \beta x_3$$

Therefore, the mathematical expectation $\mathbf{E}\left[Q\left(x_3\right)\right]$ associated with the function is obtained as

$$\int_\alpha^\beta\int_\alpha^\beta\int_\alpha^\beta\left[95.61x_{1r}^*\left(\xi\right)+126.64x_{1v}^*\left(\xi\right)+468.58x_{2r}^*\left(\xi\right)+17x_{2v}^*\left(\xi\right)+28.41x_{3r}^*\left(\xi\right)+1.183x_{3v}^*\left(\xi\right)\right]$$
$$f_1\left(\hat{t}_1\right)f_2\left(\hat{t}_2\right)f_3\left(\hat{t}_3\right)\ d\hat{t}_1\,d\hat{t}_2\,d\hat{t}_3=\psi_3\left(x_3\right)$$

where the $f_i\left(\tilde{i}\right)$ denote the corresponding density functions associated with each variable In this case it is assumed that all three are equal.

The calculation for $\psi_1\left(x_1\right)$ and $\psi_2\left(x_2\right)$ is done in a similar way. In each case, the most difficult problem is to find the optimal solution of these, x_1^* and x_2^* .The nature of these problems lies beyond the scope of this document, and here it is mentioned that they should be the steps to solve this problem when using continuous probability distributions.

SOME NUMERICAL RESULTS

Below are the results of some runs of the model in order to study the changes in the optimal solution under different scenarios with different probabilities using Equation (10). We propose the following changes in the probability distribution with the following results, Table (10).

The results obtained for each ρ are shown in the Table[4] (11).

From the table (11) it is observed that the best utilities are obtained when the probabilities associated with scenarios two and three are higher. The above is the result of expecting good rainfall that generally gives better yields to the farmland. Figure (2) reproduces the results of the nine runs shown in the Table (11).

Table 10. Reevaluated model coefficients

Run Number	ρ_1	x_{11}	x_{21}	x_{31}	x_{11}	x_{21}	x_{31}
1	0.20	30.086	4.1	0.284	0.0332	0.004	0.0041
2	0.30	45.129	6.15	0.426	0.0498	0.006	0.00615
3	0.40	60.172	8.2	0.568	0.0664	0.008	0.0082
4	0.50	75.215	10.25	0.71	0.083	0.01	0.01025
5	0.60	90.258	2.3	0.852	0.0996	0.012	0.0123
Run Number	ρ_2	x_{12}	x_{22}	x_{32}	x_{12}	x_{22}	x_{32}
1	0.50	120.345	16.405	1.14	0.083	0.01	0.01
2	0.40	96.276	13.124	0.912	0.0664	0.008	0.0082
3	0.40	96.276	13.124	0.912	0.0664	0.008	0.0082
4	0.40	96.276	13.124	0.912	0.0664	0.008	0.0082
5	0.30	72.207	9.843	0.684	0.4098	0.006	0.00615
Run Number	ρ_3	x_{13}	x_{23}	x_{33}	x_{13}	x_{23}	x_{33}
1	030	63.18	8.61	0.597	0.0498	0.006	0.0061°5
2	0.30	63.18	8.61	0.597	0.0498	0.006	0.00615
3	0.20	42.12	5.74	0.398	0.0332	0.004	0.0041
4	0.10	21.06	2.87	0.199	0.0166	0.002	0.00205
5	0.10	21.06	2.87	0.199	0.0166	0.002	0.00205

Table 11. Optimal values associated with the change of the probability vector

Run Number	ρ_1	ρ_2	ρ_3	α	β	γ	Z^*
1	0.20	0.50	0.30	480	600	720	461,864.80
2	0.30	0.40	0.30	480	600	720	204,934.10
3	0.40	0.40	0.20	480	600	720	201,038.10
4	0.50	0.40	0.10	480	600	720	197,142.10
5	0.60	0.30	0.10	480	600	720	191,292.90
6	0.10	0.45	0.45	480	600	720	462,177.90
7	0.20	0.40	0.40	480	600	720	462,177.90
8	0.30	0.35	0.35	480	600	720	462,177.90
9	0.40	0.30	0.30	480	600	720	462,177.90

In the previous values, it is observed that there is a great waste of land for sowing. Trying to improve the use of this good, the same scenarios are now analyzed, varying the right side α, β, γ . Table (12) summarizes these results. The Figure (3) also presents them schematically

Again, the congruence in the results with respect to the optimal values is observed. Note that these are obtained for the lowest probability value (scenario 1) and distributing the remainder of the probability in two and three.

It is to be expected that if the highest probability value is assigned to scenario three, the gain will increase even more. Although there are notable differences (for example, there are not several minis, in fact there is only one) in this type of experiment with respect to the previous one, the same tendency observed with respect to the probabilistic assignment is manifested.

It is suggested to the reader to continue making a sensitivity study of the model for the disaggregated case of the same

CONCLUSION

In this document a two-phase mathematical programming model has been presented to establish the optimum amounts of vegetable cultivation and animal breeding assuming variability in water availability.

First, is a comparative basis with the stochastic model. A solution of both approaches is promoted using simple methods of analysis, purposely avoiding the technicalities that would involve the use of advanced techniques such as the decomposition of Benders for the analysis of these.

The model presented shows the wealth of applications that can be made with this technique in the area of agriculture, assuming randomness in some parameters such as the availability vector or the matrix of technological constraints. Although there are several alternate ways to abbreviate the problem, in this

Figure 2. Optimal values from 9 runs

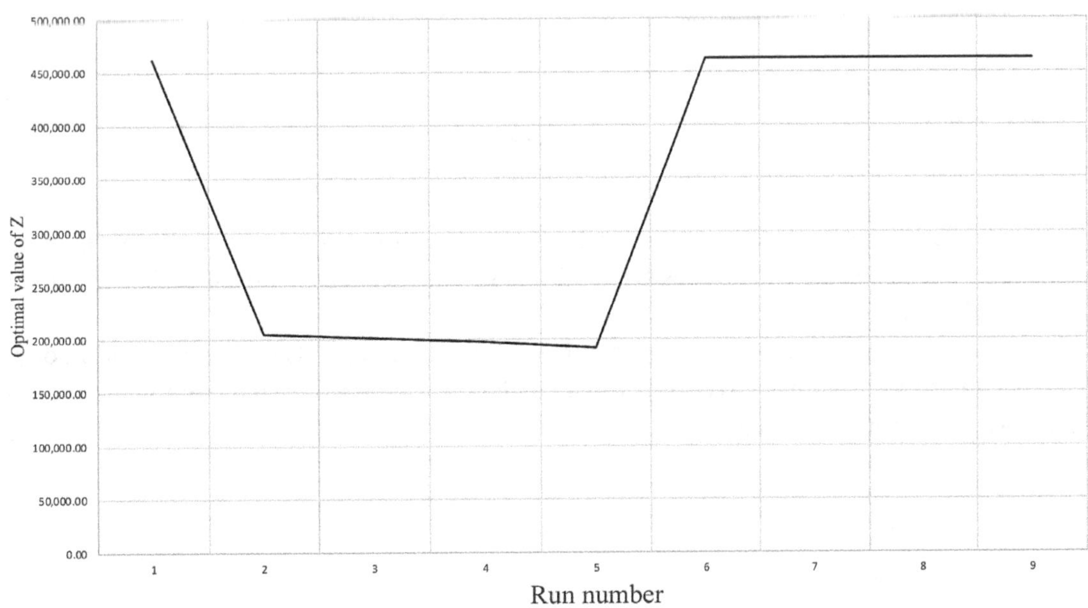

Table 12. Optimum values associated with the change on the right side

Run Number	ρ_1	ρ_2	ρ_3	α	β	γ	Z^*
1	0.20	0.50	0.30	4	6	7	85,265.42
2	0.30	0.40	0.30	4	6	7	79,668.55
3	0.40	0.40	0.40	4	6	7	73875.32
4	0.50	0.40	0.10	4	6	7	68082.09
5	0.60	0.30	0.10	4	6	7	62485.22
6	0.10	0.45	0.45	4	6	7	232891.50
7	0.20	0.40	0.40	4	6	7	229324.60
8	0.30	0.35	0.35	4	6	7	225757.60
9	0.30	0.35	0.35	4	6	7	222190.70

Figure 3. Optimal values resulting from the 9 runs, Table (11)

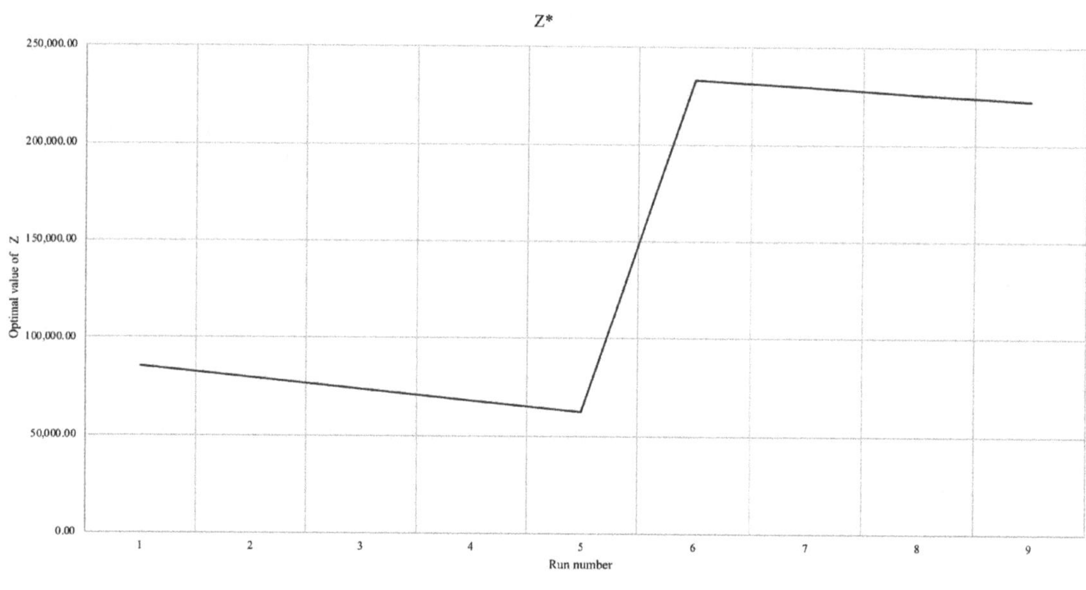

proposal we have preferred to show the applicability of the technique in cases of vegetable planting and animal husbandry where limited resources can be found or with scattered or existing information in the form of probability distributions.

As it is to notice the precision in the application of the model depends on the calculation of probabilities associated to the vector ξ. By itself this calculation represents a challenge when there is not enough historical information about it.

REFERENCES

Ahmed, S., & Ahmed, S. (2011). *Two-Stage Stochastic Integer Programming: A brief introduction*. Wiley on Line Library. DOI.org/10.1002/9780470400531.eorms0092

Ahumada, O., & Villalobos, R. J. (2009). Application of planning models in the agri-food supply chain: A review. *European Journal of Operational Research, 196*(1), 1–20. doi:10.1016/j.ejor.2008.02.014

Babbar, M. M. (1955). Distribution of solutions of a set of linear equations (with an application to linear programming). *Journal of the American Statistical Association, 50*(271), 854–869. doi:10.1080/01621459.1955.10501971

Barahona, F., Berman, S., Gunluk, S., & Holmes, S. (2005). Robust capacity planning in semiconductor manufacturing. *Naval Research Logistics, 52*(5), 459–468. doi:10.1002/nav.20086

Beale, E. M. L. (1955). On minimizing a convex function subject to linear inequalities. *Journal of the Royal Statistical Society. Series B. Methodological, 17*(2), 173–184. doi:10.1111/j.2517-6161.1955.tb00191.x

Bekri, E., Disse, M., & Yannopoulos, P. (2015). Optimizing Water Allocation under Uncertain System Conditions for Water and Agriculture Future Scenarios in Alfeios River Basin (Greece)-Part B: Fuzzy-Boundary Intervals Combined with Multi-Stage Stochastic Programming Model. *Water (Basel), 7*(11), 6427–6466. doi:10.3390/w7116427

Bellabdaoui, A., & Teghem, J. (2006). A mixed-integer linear programming model for the continuous casting planning. *International Journal of Production Economics, 104*(2), 260–270. doi:10.1016/j.ijpe.2004.10.016

Bellman, R., & Zadeh, L. A. (1970). Decision-making in a fuzzy environment. *Management Science, 17*(4), 141–161. doi:10.1287/mnsc.17.4.B141

Bellman, R. E. (1957). *Dynamic Programming*. Princeton University Press.

Charnes, A., & Cooper, W. W. (1959). Chance-constrained programming. *Management Science, 6*(1), 73–79. doi:10.1287/mnsc.6.1.73

Dantzig, G.B. (1955). Linear programming under uncertainty. *Management Science*, 197-206.

Darby-Dowman, K., Barker, S., Audsley, E., & Parsons, D. (2000). A two-stage stochastic programming with recourse model for determining robust planting plans in horticulture. *The Journal of the Operational Research Society, 51*(1), 83–89. doi:10.1057/palgrave.jors.2600858

Fan, Y. R., Huang, G. H., Guo, P., & Yang, L. (2012). Inexact two-stage stochastic partial programming: Application to water resources management under uncertainty. *Stochastic Environmental Research and Risk Assessment, 26*(2), 281–293. doi:10.100700477-011-0504-6

Galán-Martín, A., Pozo, C., Guillén-Goálbez, G., Antón Vallejo, A., & Laureano Jiménez Esteller, L. (2015). Multi-stage linear programming model for optimizing cropping plan decisions under the new Common Agricultural Policy. *Land Use Policy, 48*, 515–524. doi:10.1016/j.landusepol.2015.06.022

Geyer, A., Hanke, M., & Weissensteiner, A. (2009). Life-Cycle Asset Allocation and Consumption Using Stochastic Linear Programming. *Journal of Computational Finance*, *12*(4), 29–50. doi:10.21314/JCF.2009.203

Han, Y. C., Huang, G. H., & Li, C. H. (2008). An interval-parameter multi-stage stochastic chance-constrained mixed integer programming model for inter-basin water resources management systems under uncertainty. *2008 Fifth International Conference on Fuzzy Systems and Knowledge Discovery Shandong*. 10.1109/FSKD.2008.465

Horner, P. (1999). Planning under uncertainty. Questions & answers with George Dantzig. *OR/MS Today*, *26*, 26-30.

Ji, X., Zhu, S., Wang, S., & Zhang, S. (2005). A stochastic linear goal programming approach to multistage portfolio management based on scenario generation via linear programming. *IIE Transactions*, *37*(10), 957–969. doi:10.1080/07408170591008082

Küçükyavuz, S., & Suvrajeet, S. (2019). *An Introduction to Two-Stage Stochastic Mixed-Integer Programming*. Retrieved from http://faculty.washington.edu/simge/Preprints/SMIP_Tutorial.pdf

Lechuguilla - Una pequeña planta con larga historia. (n.d.). Accessed on May 2019, recovered from: http://suculentasycactus.com/lechuguilla-pequena-planta-larga-historia/

Li, W., Li, Y. P., Li, C. H., & Huang, G. H. (2010). An inexact two-stage water management model for planning agricultural irrigation under uncertainty. *Agricultural Water Management*, *97*(11), 1905–1914. doi:10.1016/j.agwat.2010.07.005

Nagar, L., & Jain, K. (2008). Supply chain planning using multi-stage stochastic programming. *Supply Chain Management*, *13*(Issue: 3), 251–256. doi:10.1108/13598540810871299

Paksoy, T., & Pehlivan, Y. N. (2012). A fuzzy linear programming model for the optimization of multi-stage supply chain networks with triangular and trapezoidal membership functions. *Journal of the Franklin Institute*, *349*(1), 93–109. doi:10.1016/j.jfranklin.2011.10.006

Popela, P., Novotny, J., Roupec, J., Hrabec, D. H., & Olstad, A. (2018). *Two-stage stochastic programming for engineering problems*. Accessed in November 2018, recovered from: http://www.engineering-mechanics.cz/pdf/21_5_335.pdf

Puleo, V., Morley, M., Freni, G., & Savi, D. (2014). Multi-stage Linear Programming Optimization for Pump Scheduling. *Procedia Engineering*, *70*, 1378–1385. doi:10.1016/j.proeng.2014.02.152

Radner, R. (1955). The linear team: An example of linear programming under uncertainty. *2nd Symp. Lin. Prog.*, 381-396.

Radner, R. (1959). The application of linear programming to team decisions problem. *Management Science*, *5*(2), 143–150. doi:10.1287/mnsc.5.2.143

Römisch, W., & Vigerske, S. (2010). Recent Progress in Two-stage Mixed-integer Stochastic Programming with Applications to Power Production Planning. In P. Pardalos, S. Rebennack, M. Pereira, & N. Iliadis (Eds.), *Handbook of Power Systems I. Energy Systems*. Berlin: Springer. doi:10.1007/978-3-642-02493-1_8

Sen, S. L. S., Yu, L., & Genc, T. (2006). A stochastic programming approach to power portfolio optimization. *Operations Research*, *54*(1), 55–72. doi:10.1287/opre.1050.0264

Shi, J., & Oren, S. S. (2016). Wind power integration through stochastic unit commitment with topology control recourse. *19th Power Systems Computation Conference, (PSCC 2016)*, Genova, Italy.

Tim, H. W., & Lall, U. (2013). Optimal crop choice, irrigation allocation, and the impact of contract farming. *Production and Operations Management*, *22*(5), 1126–1143.

Tintner, G. (1955). Stochastic linear programming with applications to agricultural economics. In H. A. Antosiewicz (Ed.), *Proceedings of the Second Symposium in Linear Programming*, (pp. 197-228). National Bureau of Standards.

Vajda, S. (1972). *Probabilistic Programming*. Academic Press.

Wittmann-Hohlbein, M., & Pistikopoulos, E. N. (2012). Two-Stage Method for the Approximate Solution of General Multiparametric Mixed-Integer Linear Programming Problems. *Industrial & Engineering Chemistry Research*, *51*(23), 8095–8107. doi:10.1021/ie201408p

Xie, S., Li, X., & Ouyang, Y. (2015). Decomposition of general facility disruption correlations via augmentation of virtual supporting stations. *Transportation Research Part B: Methodological*, *80*, 64–81. doi:10.1016/j.trb.2015.06.006

KEY TERMS AND DEFINITIONS

Lechuguilla: Is an agavacea plant from the arid zones and is one of the species that for generations has been the main source of income for a large number of peasant and low-income families from the desert areas of northern Mexico.

Lucerne: It is a plant that is widely used as grass and for this purpose it is cultivated intensively in the whole world. It has a life cycle of between five and twelve years, depending on the variety used, as well as the climate; in benign conditions it can reach twenty years.2 It reaches a height of 1 meter, developing dense clusters of small purple flowers. Its roots are usually very deep and can measure up to 4.5 meters.

Matrix Random: Is a matrix in which some or all elements are random variables.

Mixed-Integer Programming: Is the set of linear programming problems for which all or part of its variables belong to the integers.

MXN: International symbol to define the Mexican peso.

Stochastic Programming: Is a framework for modeling optimization problems that involve uncertainty.

Walnut: Tree high and strong trunk tree, smooth and gray bark, large round cup, thick branches, deciduous, large, green and bright, unisexual flowers and fruit (nut) in edible drupa. It can reach up to 30 m in height.

ENDNOTES

1 This is surely due to the fact that in that range the optimal solution does not change, and which could also have been obtained using sensitivity analysis.

2 The first constraint is related to water requirements. Using the results of the corresponding we have, for example, the water consumptions per walnut plant are respectively for scenarios 1, 2 and 3: 0.433, 0.5416 and 0.647 m^3. Similarly, for lechuguilla and the prickly pear, the estimated consumption per scenario are: 0.00112, 0.0014, 0.00168; 0.00272, 0.0034, 0.00408. These calculations were obtained assuming linearity in the consumptions.

3 This is done to avoid inconsistency with respect to the area used by cultivation.

4 Here, the first three constraints of the model will be substituted in their right value by α, β and γ, respectively.

Chapter 17
Theoretical Study of the Effect of an Elastic-Damping Mechanism in the Tractor Transmission on a Machine-Tractor Unit Performance While Sowing

Sergey Senkevich
https://orcid.org/0000-0001-6354-7220
Federal State Budgetary Scientific Institution, Federal Scientific Agroengineering Center VIM, Russia

Vladimir Kravchenko
FSBEI HE, Don State Agrarian University, Russia

Pavel Lavrukhin
FSBEI HE, Don State Agrarian University, Russia

Pavel Ivanov
FSBEI HE, Don State Agrarian University, Russia

Anna Senkevich
FSBEI HE, Don State Agrarian University, Russia

ABSTRACT

The chapter deals with a numerical simulation of a machine-tractor unit with an elastic-damping mechanism installed between an engine and a tractor transmission while fulfilling an agricultural performance (sowing). The mechanism is a complex system. The description of an elastic-damping mechanism shows all its advantages. There has been given a theoretical substantiation of a transparency level effect on a machine-tractor unit performance while sowing. The purpose of the work is to conduct theoretical study to identify the effect of an elastic-damping mechanism in a tractor transmission on working characteristics of a sowing machine. There has been obtained a regression equation and identified optimum parameters of an elastic-damping mechanism installed in a tractor transmission. There has been also given a substantiation of an expediency to install an elastic-damping mechanism in a tractor transmission for sowing grain crops.

DOI: 10.4018/978-1-7998-1216-6.ch017

Copyright © 2020, IGI Global. Copying or distributing in print or electronic forms without written permission of IGI Global is prohibited.

INTRODUCTION

For promising and modernized tractors there are being developed various transmissions. The results of experimental studies of the existing types indicate a high dynamic load during such processes as starting from a standstill, shifting gears, as well as during steady modes of tractor motion. Taking into account a high cost of modern transmissions and additional costs of fixing them, the problem to reduce dynamic load in tractor transmissions is of great relevance.

It is believed that the introduction of additional elastic damping mechanisms (EDM) as active vibration filters at the entrance and exit of a transmission can provide a required level of protection of an engine and transmission elements from fluctuations of an external traction load. This makes a thorough study aimed at reducing dynamic loading especially important.

It is time consuming to carry out experiments with real objects or with physical models, so it's reasonable to develop numerical simulations (mathematical models) using the results of simpler experiments. To simulate all the processes occurring in the transmission when the tractor is working is a relevant problem.

The known numerical simulations do not allow sufficiently taking into account the real design features, conditions and modes of a controlled tractor motion, their interaction with an environment, with an intensity of changes and their features.

We have proposed a tractor model with an elastic-damping mechanism (EDM) installed in a tractor transmission. An EDM is needed to reduce vibrations in the engine-transmission system. It is closely installed to the engine and well protects the engine and transmission from fluctuations of external loads. An EDM installation reduces a number of external effects, thereby protects an engine and transmission from high dynamic loads, and allows decreasing a possible drive wheel travel reduction (slip).

The proposed simulation (model) allows to study the effect of external traction load oscillations on the working characteristics of an engine, transmission, a whole unit. This model has made it possible to conduct a multi-factor experiment to find optimum parameters of the elastic-damping mechanism. The conducted study has proved the effectiveness of an EDM to reduce dynamic load of transmission units.

BACKGROUND OF THE PROBLEM

The Effect of Transmission Elasticity on an MTU Performance

There are many constructive solutions aimed to reduce dynamic loads in a tractor transmission and to improve engine working conditions (Filin, et al 2016; Gabdrafikov, &Safina, 2017). The source of oscillations frequency of an engine crankshaft rotation can be not only fluctuations of traction resistance, as the studies have shown (Kostyuchenkov, & Plaksin, 2010; Babanin, & Polivaev, 2015; Zech, et al 2019), but also vibrations of transmission shafts; therefore, to reduce dynamic loads on an engine it is necessary to increase transmission flexibility (Shekhovtsov, et al 2016; Zech, et al 2019). However, among the researchers there is no consensus concerning the installation location of elastic elements in a transmission.

The fluctuation compensation of a resistance moment is the most reasonable way to optimize the loading modes of a machine-tractor engine. A lot of researchers have proven the effectiveness of introducing various elastic mechanisms into the elements and links of a MTU (Feuerhuber, 2013; Polivaev, 2017).

The theoretical and experimental study (Polifke, 2016; Polifke, 2017; Nekhoroshev, 2014; Nekhoroshev, & Nekhoroshev 2015) found that the improvement of dynamic characteristics of a tractor can be achieved by installing elastic elements in a tractor transmission.

To reduce torsional vibration levels in power transmissions of high power tractors, these are installed elastic and friction dampers (springs, etc.) in the clutch driven discs (Polifke, 2016; Polifke, 2017). The effect of main parameters of such elastic-friction dampers on torsional vibrations of a tractor power transmission has been studied by many scientists (Kalmykov, 2014; Polivaev, 2018).

The positive effect of elastic elements on the dynamic characteristics of tractors is noted in all listed papers. To reduce torsional vibrations and reduce dynamic loads in a transmission, special elastic damping elements installed in the clutch driven discs or flexible drives in the driving wheels are of the widest application (Polifke, 2016; Polifke, 2017; Polivaev, 2018). However, an elastic drive in the wheels and a damper in the clutch do not adequately dampen the transmission and clutch vibrations that occur when performing a technological performance.

V.P. Goryachkin, when studying a spring installed between an engine and a tool, noted that if there was an elastic connection, an impact duration on a tool would stretch over time, as a result of which an average traction effort would be less than that in an inelastic impact (Goryachkin, 1965).

The other researchers (Polivaev, 2010; Polivaev, 2011; Polivaev, & Vedrinsky, 2013; Polivaev, & Vedrinsky, 2014; Polivaev, 2016) considered the effect of an elastic drive of driving wheels on some MTU working characteristics. The installation of elastic elements on semi-axes of driving wheels reduces the vibrations in a tractor transmission by 30 - 40%, and the installation of elastic damping elements decreases it by 40 - 70%, while the peak values of the moment on the semi-axis with a sudden load change decrease by 40 - 50%, and the peak values of the moment on the engine shaft reduce by 25 - 36%.

The installation of an elastic driving gear of driving wheels also results in decrease of a MTU energy loss per unit path length by 10 - 15% and an increase in the tractor traction efficiency, which contributes to an improvement of a MTU efficiency.

Energy loss decrease with an elastic driving gear is due to a more stable loading of a tractor, an increase in smoothness, a decrease in travel reduction (slip) (up to 13 - 16%) and a decrease in speed fluctuations. The increase in tractor traction efficiency is due to the decrease of load unevenness and rotation speed of power transmission elements.

Along with the advantages, it should be noted that elastic-damping elements mounted on a wheel produce significant torques, therefore such devices are bulky. In addition, they cannot protect an engine from torsional vibrations in a tractor transmission.

To improve the acceleration characteristics of 14 kN class tractors V.V. Kotlyarov and Yu.I. Deyanov (Kotlyarov & Deyanov, 1973) suggested installing a special device between an engine and an input shaft of a gearbox (Figure 1), based on a tractor torque increaser, which has a flexible joint instead of a freewheel clutch.

The authors found a significant reduction in dynamic loads on an engine and an increase in a transmission efficiency factor. Unfortunately, the authors studied only the effectiveness of the proposed flexible device at MTU acceleration. In addition, rotation of an elastic element occurs at a sufficiently large angle, which results in a significant increase in device dimensions.

The search of the methods to improve quality of vibration dampers resulted in the appearance of non-metallic elements for damping vibrations: rubber, air, liquid.

Figure 1. Diagram of the damping device installed between the engine and the input shaft of the gearbox
J is an engine inertia moment; J_2 is a transmission inertia moment; J_3 is a wheel inertia moment; S is a satellite; P is a planet carrier; S1 and S2 are central gears of a planetary reduction gear

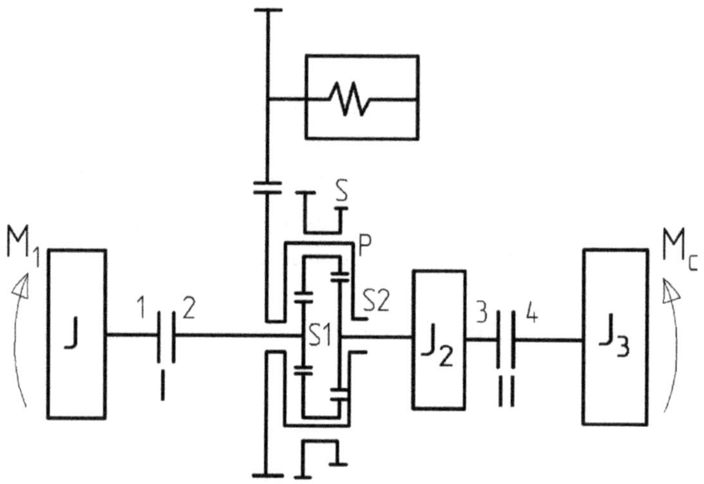

However, when using a pneumatic damper, it is necessary to use powerful dampers. Friction absence results in a decrease in oscillation damping degree, therefore such elastic elements increase oscillation frequency of the system.

It is necessary to use comprehensive dampers combining sufficient vibration damping with low mass and volume, i.e. it's necessary to introduce viscous and elastic elements (Bespamyatnova, 2002). If periodic oscillations of resistance to motion are overcome by the kinetic energy of an MTU and are compensated using various devices, then when an average value of resistance is changed to ensure efficient working characteristics, it is necessary to use gears maneuvering. The possibilities to improve MTU working characteristics on the basis of wheeled tractors are limited. The research work supervised by the professor V.A. Kravchenko showed that a tractor had to have a device accumulating potential energy and giving it away when a MTU accelerates.

At the same time, this device should smooth load fluctuations and reduce its increase when a MTU starts moving (Kravchenko, 2010). One of the most important issues in tractor dynamics is to determine the parameters of this device and the place where it should be installed to ensure the use of potential energy reserves (Senkevich, et al, 2019a).

The analysis of the above mentioned works showed that the existing vibration dampers and damping devices have a number of significant drawbacks: they change transmission rigidity within small limits, have a linear characteristic, do not provide infinitely variable transmission range, etc.

The elastic-damping mechanism (Figure 2), the design of which is protected by the patent (Patent 2222440 C1 7 V 60 K 17/10) does not have the above disadvantages.

A detailed description of the mechanism and its performance is given in the references (Senkevich, et al, 2019b; Patent 2398147; Patent 2222440, 2004).

Let's identify the main performance modes of the device itself. When a machine-tractor unit accelerates, an EDM can operate in three basic modes which ensure its smooth starting and acceleration.

Figure 2. Basic diagram of the EDM in a tractor transmission (Patent RU 2398147)
1. a planetary reduction gear; 2. a hollow gear of a planetary reduction gear; 3. a solar gear of a planetary reduction gear; 4. a drive of an oil pump; 5. an oil pump; 6. movable loads to change inertia moments; 7. satellites of a planetary reduction gear; 8. a regulator of cargo position; 9. a suction port (passage); 10. a delivery port (passage); 11. a primary shaft of a gear change box; 12. a gear change box; 13. a clutch; 14. an internal combustion engine; 15. an adjustable throttle; 16. a safety valve; 17. an adjustable valve; 18. an inlet to a valve and a throttle; 19. a hydraulic flow distributor; 20. a safety valve, a throttle valve; 21. a hydraulic tank; 22. a pneumatic hydro accumulator (a hydro pneumatic accumulator); 23. a cylinder; 24. a movable piston; 25. a hydraulic drive (a regulator of cargo position); 26. Brakes of a solar gear of a planetary reduction gear; 27, 28, 29. an oil tube

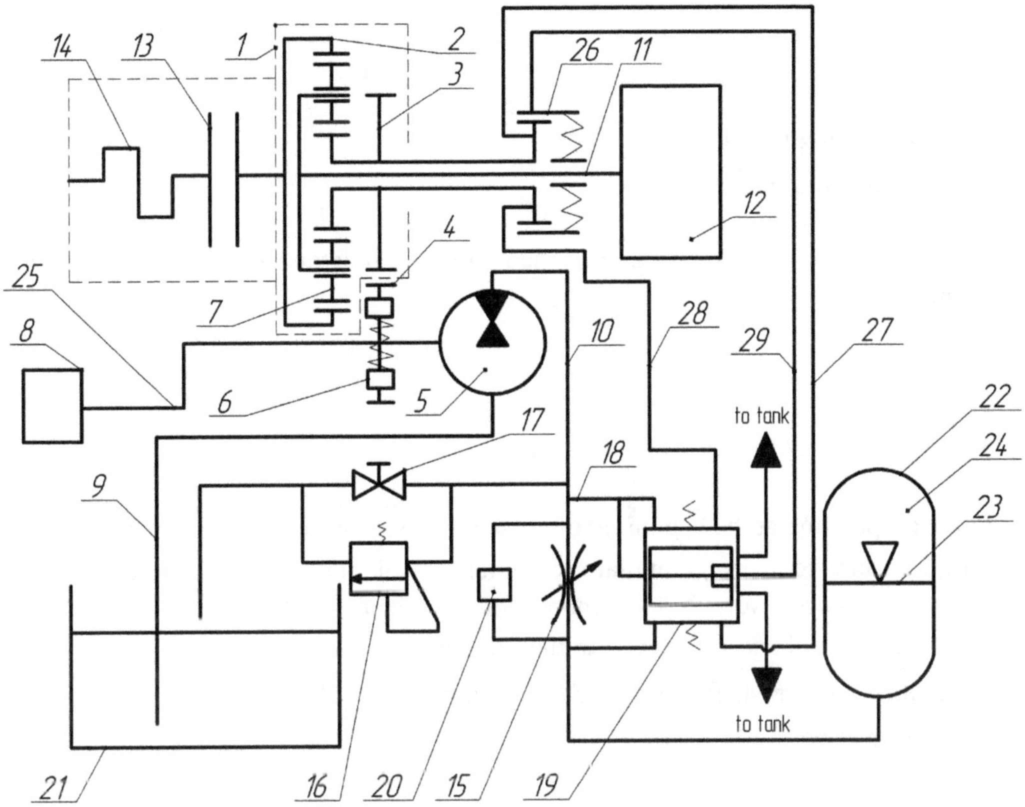

The first mode is when an elastic-damping mechanism is activated simultaneously with a friction clutch, significantly reducing transmission rigidity.

In **the second mode** an elastic-damping mechanism is used both as a clutch to start transmission and as a flexible joint when a friction clutch of a corresponding tractor gearbox is closed.

In **the third mode** an elastic-damping mechanism ensures a smooth change in an engine torque and MTU's speed when moving from one gear to another without stopping a tractor. The third mode of a planetary reduction gear performance is similar to the first mode, though it characterizes the process of stage-by-stage acceleration of a MTU with a minimum travel reduction (slip) of a wheeled running gear.

Effect of a Transmission Transparency Degree on a MTU Performance

During the performance, a certain amount of oil is supplied to the accumulator, which depends on the total number of pump cycles. A torque required to turn pumps is determined by the well-known formula (Senkevich, 2006; Kravchenko, 2010):

$$M_P = P_1 \cdot V_p, \tag{1}$$

where

P_1 is pressure of a working liquid;
V_p is a working volume of pumps;

The pressure of a working liquid in a pneumatic hydroaccumulator raises in accordance with the air pressure principle:

$$P_1 = P_0 \left(\frac{V_0}{V_1} \right)^n, \tag{2}$$

where P_0, P_1 are air pressure at the beginning of pressure and at the actual moment respectively; V_0, V_1 are volumes of a pneumatic hydroaccumulator compression chamber at the beginning of pressure and at the current moment respectively; n is a compression exponent.

The moment M_p raises in accordance with the same principle (1).

The ratio between a pump shaft moment M_p and a moment M_s, applied to a central gear can be determined by the equation:

$$M_p = M_s \cdot j_d \cdot \eta_d, \tag{3}$$

where j_d, η_d are a gear ratio and an efficiency factor of a pump drive of an EDM.

On the other hand, it can be written as:

$$M_s = M_1 \left(j_p - 1 \right) \eta_p \text{ and} \tag{4}$$

$$M_v = M_1 \cdot j_p \cdot \eta_p, \tag{5}$$

where

$j_p, \eta_p,$ are a gear ratio (final) and an efficiency factor of a planetary reduction gear;

M_v is a moment on a carrier of a planetary reduction gear.

After these transformations we get:

$$\mathrm{M}_v = \frac{\mathrm{K} \cdot V_p \cdot \mathrm{P}_1 \cdot j_p}{j_d \cdot \eta_d \cdot \left(j_p - 1\right)} \tag{6}$$

Since

$$V_1 = V_0 - V_p \cdot h_{sum} \cdot \eta_p, \tag{7}$$

Taking into account the equation 2, we get:

$$M_v = K \cdot V_p \cdot \mathrm{P}_0 \left(\frac{1}{1 - \frac{V_p}{V_0} \cdot h_{sym} \cdot \eta_p} \right)^h \frac{j_p}{j_d \cdot \eta_d \left(j_d - 1\right)}. \tag{8}$$

In the equation (8) the ratio V_p / V_0 is a maximum number of pump cycles, when an accumulator is completely filled with oil. Then the ratio of a current number of pump cycles to a maximum number is referred to as *a relative number of pump cycles*.

Then finally:

$$M_v = K \cdot V_p P_0 \left(\frac{1}{1 - j_r \cdot \eta_p} \right)^h \frac{j_p}{j_d \eta_d \left(j_d - 1\right)} \tag{9}$$

where

j_r is a relative number of pump cycles;

η_p is a coefficient of pump flow.

Figure 3 (Senkevich, 2006; Kravchenko, 2010) shows dependence of the torque M_v on a relative number of pump cycles. The graph shows that a driving moment applied to a carrier initially increases smoothly. Therefore, in the initial period of MTU starting, a solar gear of a planetary reduction gear and a pump begin to rotate, pumping oil into a pneumatic hydroaccumulator. Since the moment of resistance on a drive shaft of a gear change box is important, a tractor stands still until the resistance on a solar gear is less than on a carrier.

With further pressure increase in a pneumatic hydroaccumulator, a tractor starts accelerating smoothly, and pumps gradually stop. Practically, a tractor motion begins at point 1, where a driving moment on a shaft of a gear change box is equal to a resistance moment. The rate of driving moment increase can be regulated by a throttle.

At the end of the first phase of acceleration there is a sharp increase in a driving moment and, to prevent dynamic loads in a tractor transmission, a safety valve is provided. In Figure 5, at point 2, the safety valve is triggered and the driving moment is limited to this.

In the second phase of acceleration, a driving moment gradually decreases to the value of a resistance moment. At the steady speed mode of a MTU, a pump works in the mode "stop" and a transmission doesn't practically differ from a serial one.

When external load oscillations appear, force F_0 on the satellites axis (Figure 6 (a)) changes by ΔF_0. Since force F_0 is equal to the sum of satellite's circumferential force, then $\Delta F_0 = \Delta F_1 + \Delta F_s$ and $\Delta F_1 = \Delta F_s$. Perturbation ΔF_1 are transferred to an engine, perturbation ΔF_s are transferred to a pump.

Figure 4 *(b, c, d)* shows velocities (Senkevich, 2006; Kravchenko, 2010), in which the vectors depict the velocities during external load oscillations. With a fixed gear 3 (position b), the lines drawn through the ends of the vector V_0, $\pm \Delta V_0$ and the point C determine a speed increase $\pm\Delta V_{1m}$. These increases will be maximum, making the engine shaft speed vary with a maximum amplitude.

When gear 3 (position *c*) is rotated under the effect of force ΔF_s increase, the velocity design changes, i.e. as velocity amplitude $\pm\Delta V_3$ increases, amplitude $\pm\Delta V_1$ decreases.

At position *d*, when the velocity amplitude of gear 3 $\pm\Delta V_3$ is maximum, there are no any velocity oscillations. This means that with oscillations of load and a carrier shaft, the changes in motor shaft rotation speed will be of a zero, i.e. an engine will be isolated from external oscillations.

Figure 3. Diagram of the dependence of a driving torque on a relative number of pump cycles

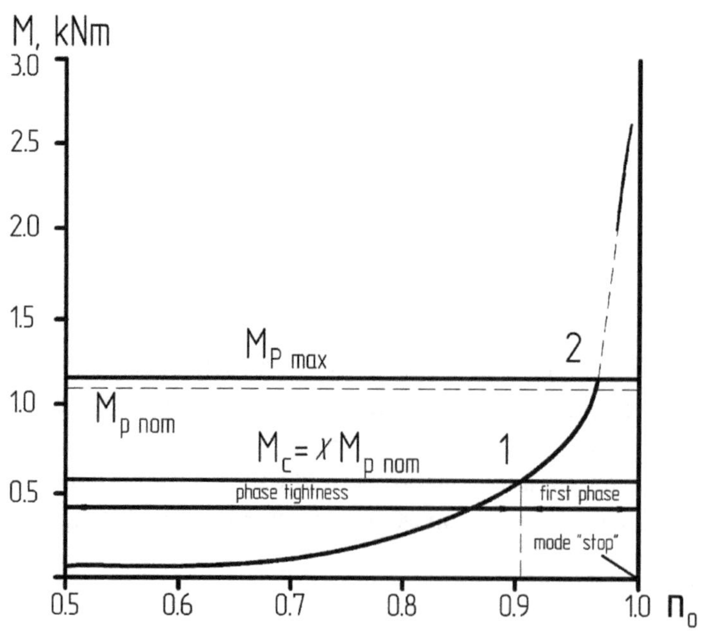

Figure 4. Designs of forces and velocities of elements of a planetary reduction gear at external load oscillation: a is a force design; Designs of velocities: b is a velocity at an absolute transparency, c is a velocity at a partial transparency, d is a velocity at an absolute lack of transparency

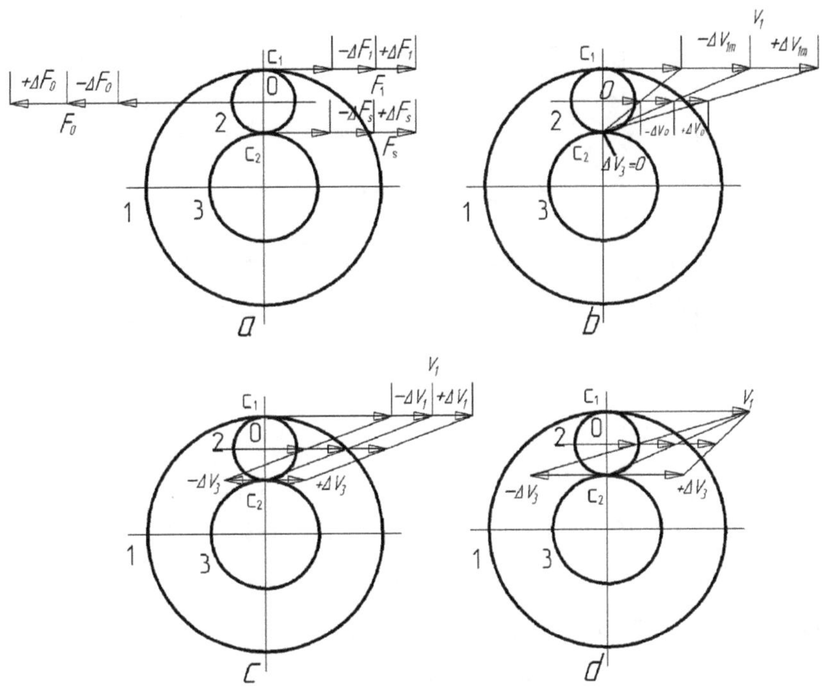

According to Figure 4 (*b,c,d*), $\Delta\omega_{1m}$ will have a maximum when

$$\Delta\omega_s = 0\left(b\right), \pm\Delta\omega_{1m} = \pm\Delta\omega_v \cdot j_p$$

The current value of $\Delta\omega_1$ when

$$\Delta\omega_s = 0\left(c\right), \pm\Delta\omega_1 = \pm\Delta\omega_v \cdot j_p = \pm\Delta\omega_s\left(j_p - 1\right)$$

The maximum value of $\Delta\omega_s$ when

$$\Delta\omega_1 = 0\left(d\right), \pm\Delta\omega_{sm} = \Delta\omega_v \frac{j_p}{j_p - 1}$$

Lack of a reduction gear transparency will be $0 < \Delta\omega_s < \Delta\omega_{sm}$, i.e. a gearbox partially transfer vibrations to the engine. With an absolute non-transparency a reduction gear will quench any oscillations, when $\Delta\omega_s = \Delta\omega_{sm}$.

Figure 5. Graph of a dependence of a transparency degree on a traction load oscillation frequency: values λ_s 1. is 0.03; 2. is 0.35; 3. is 0.04; 4. is 0.045; 5. is 0.05; 6. is 0.055

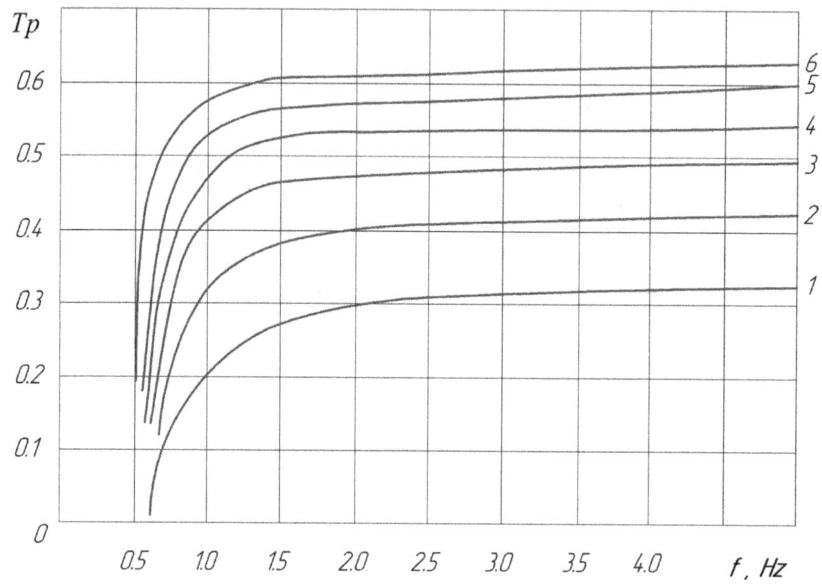

Figure 6. Scheme of a serial MTU performance

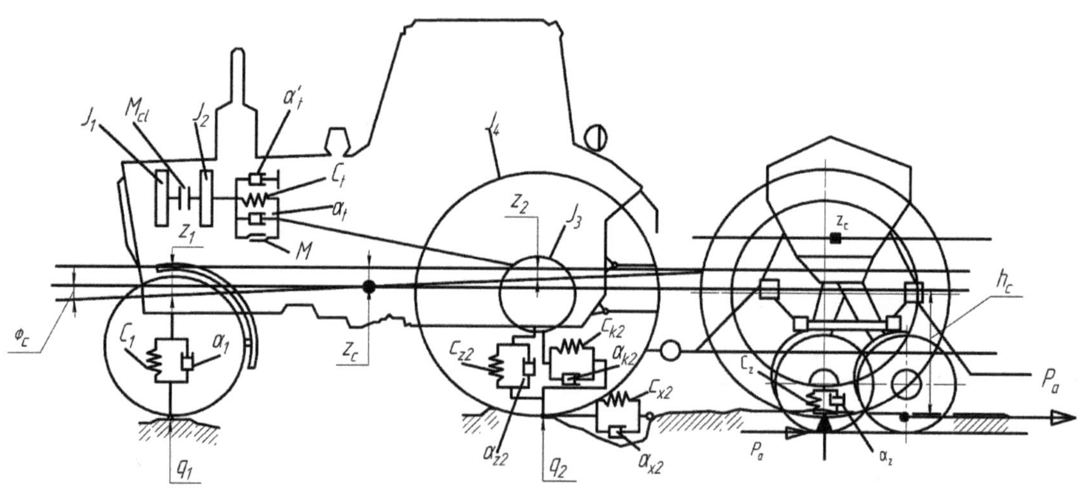

To assess protective qualities of a mechanism, there is proposed a characteristic "a transparency degree" in the ratio of a current oscillation amplitude of an engine shaft rotation frequency to its maximum value, i.e.:

$$Tp = \frac{\Delta\omega_1}{\Delta\omega_{1m}}, \tag{10}$$

or

$$Tp = 1 - \frac{\Delta\omega_s}{\Delta\omega_{1s}}\left(j_p - 1\right), \tag{11}$$

where

Tp is a transparency degree;
$\Delta\omega_1, \Delta\omega_{1s}$ are a current and a maximum values of an engine rotation frequency;
$\Delta\omega_s$ is a current value of a gear 3 rotation frequency amplitude;
j_p is a reduction gear ratio when a gear 3 is fixed.

When *Tp=0* a reduction gear transparency is of a zero, i.e. it's absolutely non-transparent and will quench oscillations on an engine. When *Tp=1* a reduction gear is absolutely transparent and an engine is not protected from load oscillations.

To determine an amplitude $\Delta\omega_{1m}$, it is necessary to solve a differential equation of motion of equivalent inertia moments when a tractor traction load oscillates. Having accepted the change in an equivalent inertia moment of a traction load according to the theorem of the sines, we obtain the following differential equation:

$$\left(J_1 - J\right) \cdot \Delta\omega_1 = \Delta M_r = \frac{1}{2} M_u^r \delta_u \sin t \tag{12}$$

and an amplitude value

$$\Delta\omega_{1m} = \frac{M_a^r \delta_a}{P\left(J_1 - J\right)} \cdot j_{p,} \tag{13}$$

where

M_a^r is an average value of a traction load torque driven to a carrier shaft;
δ_a is a degree of traction load oscillations;
P is a traction load oscillation frequency;
J_1, J are inertia moments of an engine and a MTU.

An amplitude $\Delta\omega_s$ of the gear 3 can be identified when solving the differential equation of rotating mass increase of the gear 3, an oil pump and its drive J_s. The weight J_s is applied the force $F_{s0} \pm \Delta F_s$ and a braking moment of an oil pump connected with a pneumatic hydroaccumulator pipeline, i.e.:

$$J_s \cdot \Delta\omega_s = -c \cdot \Delta\phi_3 + \Delta F_{3} \cdot r_3 \tag{14}$$

$$\Delta\phi_s + K^2 \Delta\phi_s = h \cdot \sin p \cdot t, \tag{15}$$

where

$K = \sqrt{\frac{c}{j_s}}$ is a cyclic frequency of free oscillations;

$h = \frac{F_{s0} \cdot r_3 \cdot \delta_a}{2J_s}$ is a module of a perturbation amplitude;

$P = \frac{2\pi}{T}$ is a cyclic frequency of forced oscillations;

T is a period of forced oscillations;

C is a rigidity coefficient of a braking system (a pneumatic hydroaccumulator).

The obtained equation is an equation of forced oscillations ΔF_s

The forced oscillation amplitude is

$$\Delta\omega_s = \frac{F_{s0} \cdot r_3 \cdot \delta_a}{2J_a} \cdot \frac{P}{P^2 - K^2} \text{ when } P > K \tag{16}$$

$$\Delta\omega_s = \frac{1}{2} \cdot \frac{M_r^2 \delta_a}{P \cdot J_s} \cdot \frac{1}{1 - \frac{K^2}{P^2}} \cdot \frac{j_p - 1}{j_p} \tag{17}$$

Then we get:

$$Tp = 1 - \frac{\left(j_p - 1\right)}{2\left(1 - \frac{K^2}{P^2}\right) \cdot \lambda_s \cdot j_p^2} \tag{18}$$

where $\lambda_s = \frac{J_s}{J_1 - J_u}$ is a relative weight of a damping mechanism.

Transparency is primarily determined by the ratio of frequencies of a damper mechanism *K* and an external load *P* .

As it is known (Lurie, 1969; Bespamyatnova, 2002; Senkevich, 2006; Kravchenko, 2010), in the basic agricultural performance a traction force has a range of oscillation frequencies and periods *P* = *3.14 – 42* $\sec^{-1}$. *T=2.0 – 0.15 c.*

To reduce a transparency degree of a damping mechanism, it is necessary the function $1 - K^2 / P^2$ be positive, i.e. forced oscillations would have a nature of a large frequency, when $P < K$.

Then a damping mechanism should have a minimum natural frequency of oscillations when plowing. A transparency degree is affected by inertia moments of all mass of a mechanism, including a mass of a damping mechanism.

When a coefficient λ_s decreases, a transparency degree *Tp* decreases and protective forces of a mechanism increase. It is possible to achieve such a positive effect with an inertia decrease J_3 and with an increase in a mass of a mechanism $J_1 + J$, i.e. when a tractor unit operates at high velocities.

Figure 5 (Senkevich, 2006; Kravchenko, 2010) shows a graph of a dependence of a transparency degree in various agricultural operations on a traction load oscillation frequency *f*. When p < 10 c^{-1}, a mechanism transparency decreases sharply for all values λ_s. At other frequencies, a transparency degree slightly varies with each value.

Purpose of Research

The purpose of the work is to conduct theoretical study to identify the effect of an elastic-damping mechanism in a tractor transmission on working characteristics of a sowing machine. Since it is time consuming to carry out experiments with real objects or with physical models (with an EDU in transmission), we have designed numerical simulations (mathematical models). This simulation replaces a real object and allows studying all the processes occurring in the transmission when the tractor operates.

Based on the analysis of the study, the purpose of this work is to:

1. To conduct theoretical study of the effect of an elastic-damping mechanism installed in a tractor transmission on dynamic parameters of a sowing unit;
2. To determine the effect of a traction load oscillation frequency on tractor working characteristics;
3. To determine optimum parameters of an elastic-damping mechanism installed in a tractor transmission for sowing, taking into account traction load oscillations;
4. To substantiate the expediency to install an elastic-damping mechanism in a tractor transmission when sowing grain crops.

A THEORETICAL STUDY OF A MTU PERFORMANCE

Accepted Assumptions

The program of theoretical research provides solution of the following issues:

1. To develop a dynamic simulation scheme of a sowing MTU, which operates in actual conditions.
2. To generate differential equations of motion of the elements of a dynamic simulation of a sowing MTU, taking into account the vertical and longitudinal oscillations of a sowing machine.
3. To determine a traction load oscillation effect on MTU working characteristics; to conduct a multifactor study and to identify optimum parameters of an elastic-damping mechanism for that purpose.

When designing a numerical simulation of a MTU there have been accepted the following assumptions:

- Tractor frame and a driver's cabin are given as a solid body
- A MTU motion is progressive and steady;
- Oscillations of structural elements are considered from a static equilibrium with the origin of coordinates in the center of a tractor mass;
- Elastic elements have a linear characteristic;
- Inertia forces of a fuel pump regulator are insignificant;
- A power impact on the right and left wheel is the same and simultaneous;
- Vertical oscillations of a tractor frame are believed not to affect longitudinal oscillations of a unit;
- A tractor with a sowing machine is connected through a joint;
- A sowing machine oscillates only vertically and horizontally without any angular oscillations;
- Viscous friction forces in tire and transmission elements are proportional to relative velocities, and inelastic resistance forces are proportional to acting forces.

A Dynamic Simulation of a Serial Machine-Tractor Unit

A MTU must be considered as a complex dynamic system, taking into account an interaction of all its elements, the motion of which occurs under the effect of external perturbing and controlling impacts described by random functions (Senkevich, et al, 2019b; Grebnev, Poliaev, & Vorokhobin, 2011; Karabanitsky, & Levshukova, 2014).

A tractor transmission is an oscillatory system consisting of flywheel masses with elastic and friction connections. In the study of a MTU dynamics, a dynamic simulation (model) which replaces a real MTU is usually designed.

Progressively moving and rotating masses of a MTU are replaced by dynamically equivalent flywheels, inertia moments of which are chosen because of equality of flywheel kinetic energy and a mass of a machine-tractor unit replaced by it.

The scheme of a real MTU performance attached to a wheel tractor of 14 kN class can be represented as in Figure 6.

The scheme of performance is a schematic depiction of a MTU and a dynamic simulation. It allows visualizing the location of the replaced elements.

A numerical simulation (mathematical model) of a real machine-tractor unit with a wheel 14 kN class tractor is easier to generate based on a dynamic simulation (model). The developed dynamic model of the serial machine-tractor unit is shown in Figure 7.

The starting process of a MTU is divided into the following phases: the first acceleration phase (when clutch discs reduce travel (slip) and a crankshaft angular velocity ω_1 is not equal to an angular velocity of a transmission primary shaft ω_2), the second acceleration phase and a steady motion. For each phase there has been generated its own system of equations.

Figure 7. Scheme of a dynamic simulation equivalent to a machine-tractor unit with a serial transmission: a) is a dynamic simulation of a serial MTU at the beginning of motion (starting) $\omega_1 > \omega_2$; b) is a dynamic simulation of a serial MTU at a closed clutch wheel gear $\omega_1 = \omega_2$

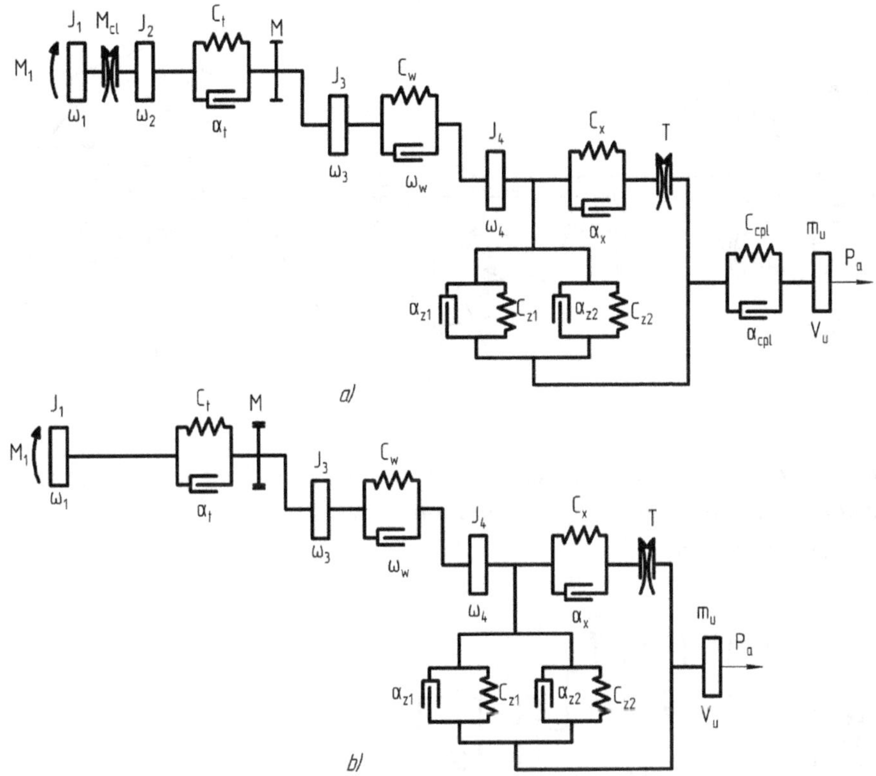

A Dynamic Simulation of a MTU With an EDM Installed in a Tractor Transmission

The scheme of a MTU performance with an EDM in a tractor transmission can be represented as in Figure 8. The general scheme of a MTU (which is in Figure 6) is added with a planetary reduction gear which establishes two stages of freedom.

A dynamic simulation of an agricultural MTU with an EDM in a tractor power train is shown in its general view in Figure 9.

The inertia moments of an engine (Senkevich, 2006; Kravchenko, 2010) have been replace with a flywheel with an inertia moment J_1 ; the inertia moment of a planetary reduction hollow gear has been replaced with an inertia moment J_K ; the inertia moment of a planetary reduction gear carrier has been replaced with an inertia moment J_v; the inertia moments of a central gear of a planetary mechanism and an oil pump drive have been replaced with an inertia moment J_s; the inertia moment of a tractor power train has been replaced with an inertia moment J_3 ; the inertia moments of a running gear and wheels have been replaced with an inertia moment J_4 ; progressively moving masses of a tractor and a machine, not connected with a tractor power train have been replaced with m_u . In the presented dynamic simula-

Figure 8. A functional scheme of a MTU with an EDM in a tractor power train

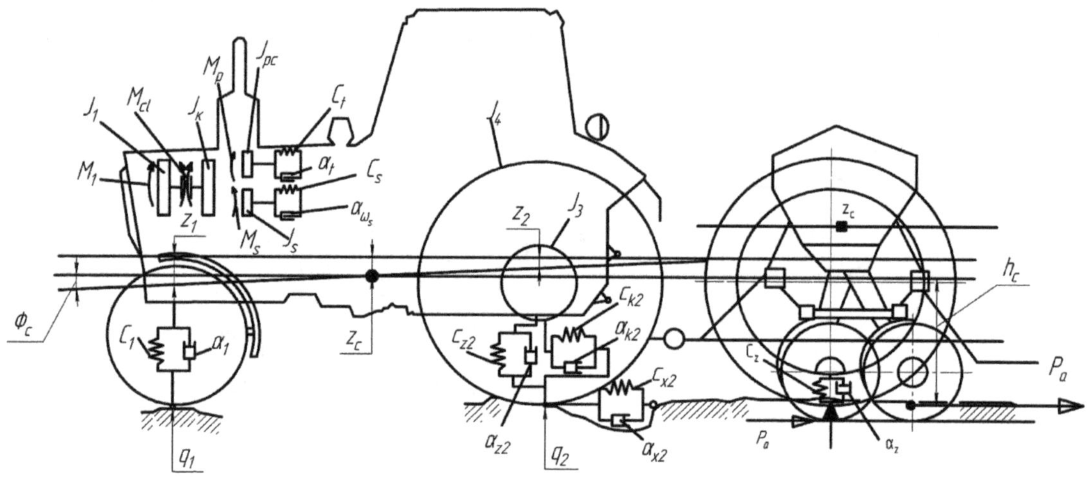

Figure 9. Scheme of a dynamic simulation, equivalent to a MTU with an EDM in a tractor power train: a) is a dynamic simulation at the beginning of motion (starting) $\omega_1 > \omega_2$; b) is a dynamic simulation at a closed clutch wheel gear $\omega_1 = \omega_2$

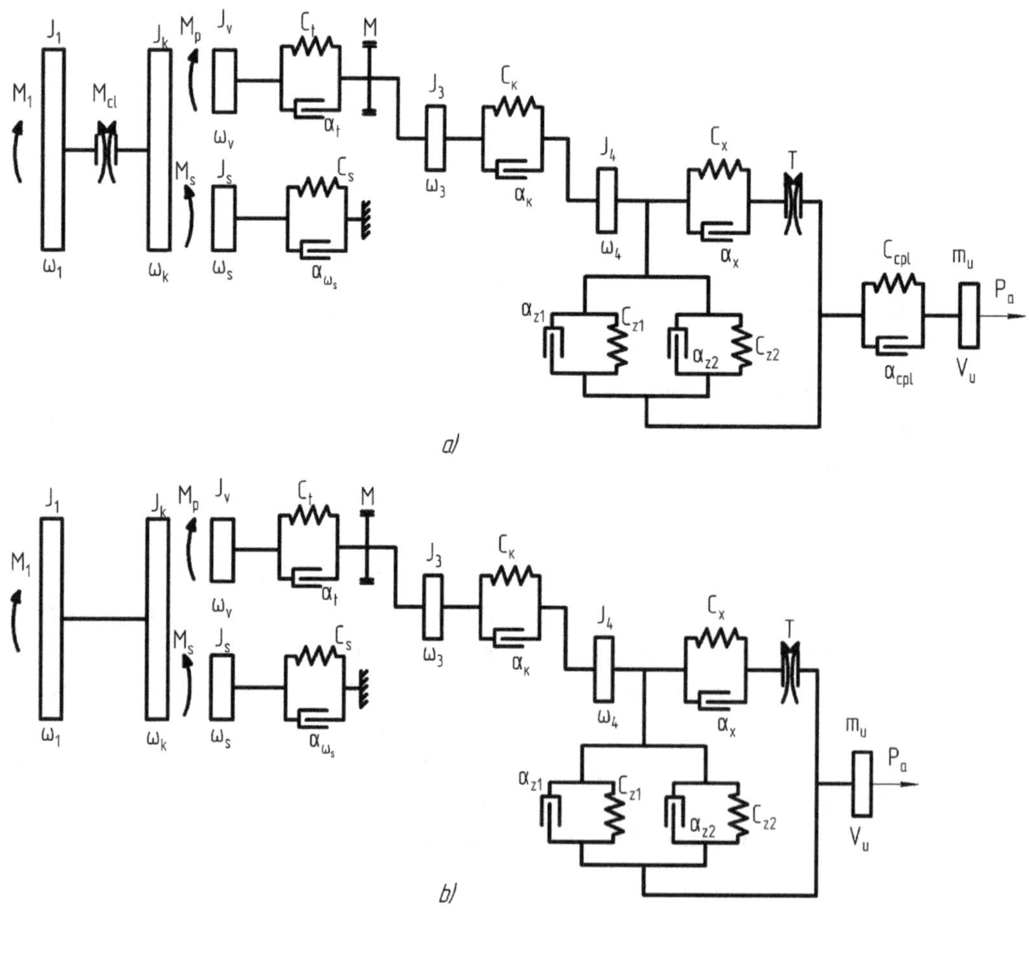

tion of a MTU, angles of engine crankshaft rotation φ_1, a clutch shaft φ_{cl}, a primary transmission shaft φ_2, drive wheel axis φ_3, a tire tread φ_4, and a position coordinate of a MTU x are taken as generalized coordinates.

Since diesel engines of agricultural tractors are equipped with all-mode crankshaft velocity regulators that have mass m_r, and some have turbochargers with an inertia moment J_r, all above listed generalized coordinates are necessary to be added a position of a rail or a fuel pump meter z_p and a rotation angle of a turbocharger rotor φ_r. Then, angular velocities of an engine crankshaft ω_1, a clutch shaft ω_{cl}, a primary transmission shaft ω_2, tractor drive wheel axle ω_3, a tire tread ω_4, a turbocharger rotor ω_r, as well as a MTU progressive motion velocity V_u and a displacement of a controller clutch are the generalized velocities of a simulation, as derivatives of generalized time coordinates. An MTU motion occurs as a result of a joint action of the following moments and forces: an engine torque M_1, a clutch friction moment M_{cl}, a moment of elastic forces in a tractor transmission M_{fr}, a tire torque M_w, a moment resulted from tire elasticity and damping M_{bw}, a force resulted from a tire tread contact $T(\lambda)$ and resistance of machine working parts P_a. Travel reduction (slip) of a running gear, which has a significant impact on a MTU motion, is identified by a travel reduction (slip) coefficient δ. A longitudinal deformation of a pneumatic tire, which is a generalized coordinate, is denoted by λ. To ensure physical similarity of a dynamic simulation (model) to a real agricultural MTU, the calculation of forces and moments takes into account dry and viscous friction in a transmission α_t; circumferential α_k, radial α_z and horizontal α_x damping in tires; transmission rigidity c_t; circumferential c_k, radial c_z and longitudinal c_x tire rigidity; travel reduction (slip) of a running gear δ.

In the presented simulations, there can be identified the following individual parts: an engine with a regulator and a turbocharger (if any), a clutch, a planetary reduction gear, an accumulator, power train friction clutches, a power train, driving wheels, an external load.

Differential equations taking into account all factors, describe the behavior of the parts of a dynamic simulation (an engine, a power train, driving wheel and a load) are the equations of a MTU motion in real conditions.

Equations of Motion of MTU Dynamic Simulation Parts

Equations of an Internal Combustion Engine Motion

A steady mode of an internal combustion engine (ICE) performance is characterized by the constancy of an angular velocity of a crankshaft at a given level of load. Maintenance of such a regime is possible under the condition when an engine torque M_1 is equal to a resistance moment of external forces M_f (Grebnev, Poliaev, & Vorokhobin, 2011; Karabanitsky, & Levshukova, 2014).

When a perturbation is introduced into the system as an increase of a resistance moment, a system equilibrium is disturbed. A resulting difference in the moments causes acceleration or deceleration of a crankshaft rotation.

For an engine in the first acceleration phase (starting) of a MTU, an external load is the moment developed by a clutch M_{cl}, and in the second phase it is the moment of elasticity in a power train M_t. (Senkevich, 2006; Kravchenko, 2010), i.e.:

$$J_1 \cdot \dot{\omega}_1 = M_1 - M_{cl}\left(M_t\right). \tag{19}$$

In a diesel engine mounted with a regulator and a turbocharger, a torque M_1 is presented as a function depending on three variables: an engine crankshaft rotation velocity ω_1, a turbocharger rotor rotation velocity ω_r and a position of a rail or a high-pressure fuel pump meter:

$$M_1 = f\left(\omega_1, \omega_r, z\right). \tag{20}$$

For convenient analytical calculations, the equation (20) is presented as a dependence of an engine torque deviation on an initial equilibrium mode upon an increase of crankshaft and turbocharger angular velocity, as well as a position of a rail or a high-pressure fuel pump meter:

$$\Delta M_1 = f\left(\Delta\omega_1, \Delta\omega_r, \Delta z\right)e^{i\theta}. \tag{21}$$

The deviation of an engine torque from an initial equilibrium mode can be determined according to the dependence:

$$\Delta M_1 = \frac{\partial M_1}{\partial \omega_1}\Delta\omega_1 + \frac{\partial M_1}{\partial \omega_r}\Delta\omega_r + \frac{\partial M_1}{\partial z}\Delta z \tag{22}$$

or

$$J_1 \cdot \Delta\dot{\omega}_1 = a_1 \cdot \Delta\omega_1 + a_2 \cdot \Delta\omega_r + a_3 \cdot \Delta z - M_{cl}, \tag{23}$$

where $a_1 = \dfrac{\partial M_1}{\partial \omega_1}$, $a_2 = \dfrac{\partial M_1}{\partial \omega_r}$, $a_3 = \dfrac{\partial M_1}{\partial z}$ are the experimental coefficients.

The equation of a turbocharger rotor rotation can be written similarly:

$$J_r \cdot \Delta\dot{\omega}_r = b_1 \cdot \Delta\omega_1 + b_2 \cdot \Delta\omega_r + b_3 \cdot \Delta z, \tag{24}$$

where b_1, b_2, b_3 are the experimental coefficients.

The change in a position of a rail or a high-pressure fuel pump meter can be described as follows:

$$m_r \cdot \Delta\ddot{z} + \beta \cdot \Delta\dot{z} + c_1 \cdot \Delta z = c_2 \cdot \Delta\omega_1, \tag{25}$$

where

β is a coefficient of viscous friction;

c_1, c_2 are the experimental coefficients.

Equations of a Clutch Motion

In the first acceleration phase when a clutch is switched on, an angular velocity of its driving wheels ω_1 and driven wheels ω_2 are different, and besides $\omega_1 > \omega_2$.

During this period, a resistance moment for an engine is a friction moment of a clutch M_{cl}, which depends on a number of pairs of friction surfaces, a force developed by compression springs, radii of discs, a value of a friction coefficient between discs, and a rate of a clutch starting.

$$M_{cl} = n \cdot P_{\max} \cdot r_{mid} \cdot \left(1 - e^{-k_1 \cdot t}\right) \cdot \left\{\mu_{\max} - \Delta\mu \cdot \left[1 - e^{-k_2 \cdot (\omega_1 - \omega_2)}\right]\right\}, \tag{26}$$

where

n and r_{mid} are a number of pairs of friction surfaces and a middle radius of discs;
P_{max} is a maximum compression force of discs by clutch disc springs;
t is time of a clutch starting;
k_1, k_2 are experimental indicators of an exponent;
$\mu_{\max}$ is a maximum value of a friction coefficient;
$\Delta\mu$ is a relative change in a friction coefficient.

Most agricultural tractors are provided with a clutch only. Some tractors have friction clutches installed instead of a missing clutch, performing its functions.

In general, a clutch and friction clutches can serve together to transfer energy of a transmission engine of domestic and foreign tractors.

In a first approximation, the equation of a clutch shaft rotation can be represented as:

$$J_2 \cdot \dot{\omega}_2 = M_{cl} - M_2. \tag{27}$$

The moment created by a friction clutch M_{cl} depends on a number of friction pairs (discs), their geometrical dimensions, a friction coefficient μ and a pressure of a working P fluid in a booster.

A friction coefficient of friction clutch discs is determined with sufficient practical accuracy according to:

$$\mu = 1 - \upsilon \cdot \left[1 - e^{-k_2 \cdot (\omega_{cl} - \omega_2)}\right], \tag{28}$$

where υ is a relative change of a friction coefficient:

$$\nu = \left(\mu_{\max} - \mu_{\min}\right) / \mu_{\max}.$$

Equations of a Power Train Motion

In accordance with a dynamic simulation (model) of a MTU, a tractor transmission can be represented by a torsional-oscillating system with an inertia moment J_2. A tractor transmission has certain values of rigidity and damping, which are characterized by coefficients c_t and a_t (viscous friction) of a driven part of a drive shaft of a gear speed box. Based on a dynamic simulation, a tractor transmission motion, taking into account the initial conditions (ω_2=ω_3=0), can be described by the following relationship (Senkevich, 2006; Kravchenko, 2010):

$$J_2 \cdot \dot{\omega}_2 = M_{cl} - \left(M_t + \alpha' \cdot \omega_2\right), \tag{29}$$

where M_t is a moment of elasticity in a tractor power train, which takes into account its elastic properties, dry friction and damping $\alpha_t^{/} \cdot \omega_2$ is a parameter that characterizes energy costs to overcome a viscous friction at a constant rotation frequency of power train parts.

For the considered dynamic simulation, the moment value M_t will be determined by the dependence:

$$M_t = c_t \cdot \left(\varphi_2 - i_t \varphi_3\right) + \alpha_t \left(\dot{\varphi}_2 - i_t \cdot \dot{\varphi}_3\right) + M \cdot sign\left(\dot{\varphi}_2 - i_t \cdot \dot{\varphi}_3\right), \tag{30}$$

where M is a moment of dry friction; i_t is a transmission ratio.

Equations of a Planetary Reduction Gear Motion

A planetary mechanism in a transition mode has two degrees of freedom and is a differential mechanism. Its motion will be described if we analyze forces and velocities. On the basis of Figure 10, neglecting inertia moments of satellites, we can write down (Senkevich, 2006; Kravchenko, 2010):

$$\begin{aligned} M_k &= T \cdot (2r_c + r_s) \\ M_v &= 2T \cdot (r_c + r_s) \\ M_s &= T \cdot r_s \end{aligned} \tag{31}$$

On the other hand, a moment on a central gear wheel M_s and oil pressure P_p, created by a pump during a steady motion, are related by the dependence:

$$M_s = P_p \cdot V_p \cdot k_p \tag{32}$$

where V_p is a working volume of a pump; k_p is a coefficient of oil pump capacity.

Figure 10. Designs of forces and velocities of a planetary mechanism during acceleration

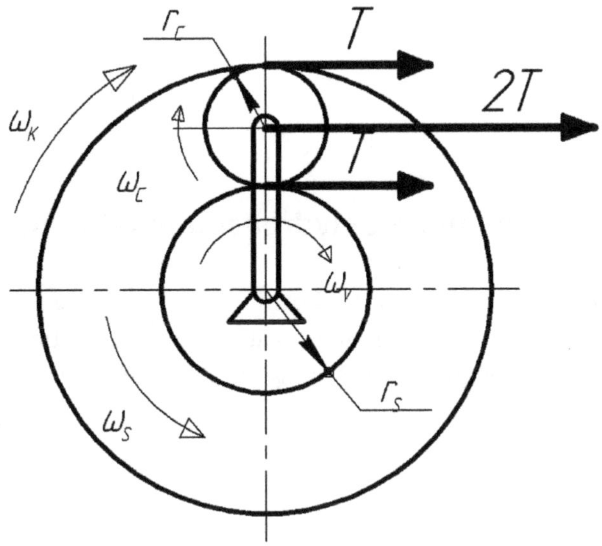

Thus, from the analysis of the expressions (26), (29), (30), (31), (32) taking into account the inertia moments of rotation J_{npu}, reduced to a central gear, it follows:

$$\begin{aligned} J_k \cdot \frac{d\omega_k}{dt} &= M_{cl} - T \cdot (2 \cdot r_c + r_s) \\ J_s \cdot \frac{d\omega_s}{dt} &= T \cdot r_s - P_p \cdot V_p \cdot k_p \\ J_v \cdot \frac{d\omega_v}{dt} &= 2T \cdot (r_c + r_s) - M_t \end{aligned} \tag{33}$$

To describe the motion of rotating parts of a planetary mechanism the equations (33) must be supplemented by the equation of kinematic dependence in a mechanism during a transition process. Из условия баланса мощностей имеем: According to a power balance we have:

$$N_k = N_s + N_v \tag{34}$$

or

$$M_k \cdot \omega_k = M_s \cdot \omega_s + M_v \cdot \omega_v \tag{35}$$

Substituting the data from the expression (32) we get:

$$T \cdot (2r_c + r_s) \cdot \omega_k = T \cdot r_s \cdot \omega_s + 2T \cdot (r_c + r_s) \cdot \omega_v \tag{36}$$

Dividing right and left sides of the expression (36) by T and expressing ω_k we obtain the expression for a kinematic dependence:

$$\omega_k = \frac{r_s \cdot \omega_s + 2(r_c + r_s) \cdot \omega_v}{2r_c + r_s} \tag{37}$$

Equations of Motion of a Pneumatic Hydroaccumulator and a Throttle

The pneumatic hydroaccumulator, schematically shown in Figure 11, is a cylinder with air and oil cavities separated by a piston with a rod. If we denote pressure before a throttle P_p and after a throttle P, the equation of oil consumption Q_{th} through a throttle cross section A_{th} can be expressed, using the Bernoulli formula, as follows:

$$\begin{aligned} Q_p &= Q_{th} = \nu \cdot A_{th} \cdot \sqrt{\frac{2}{\gamma}(P_p - P)}, \\ Q_p &= V_p \cdot \frac{\omega_s}{2\pi} \end{aligned} \tag{38}$$

where ν is a coefficient of an oil discharge and compression ratio; γ is oil density.

When a safety valve is closed, oil consumption through a throttle is determined by a displacement of an accumulator piston h and an area A:

$$Q_{th} = A \cdot \frac{dh}{dt} = \frac{V_p}{2\pi} \cdot \frac{d\omega_s}{dt} \tag{39}$$

If pressure in a discharge line reaches its maximum value, then an oil consumption will be carried out through a safety valve, and at $P_p = P_{max}$ it can be expressed as follows:

$$\begin{aligned} Q_{th} &= 0, \quad \frac{dh}{dt} = 0, \\ Q_p &= \nu \cdot A_v \cdot \sqrt{\frac{2}{\gamma}(P_{\max} - P_{sl})} \end{aligned} \tag{40}$$

where A_v is a safety valve cross section; P_{sl} is an oil pressure at a discharge.

The basis for this assumption is the following calculation of a dependence of a piston displacement upon a pressure in an accumulator, due to an isothermal process. The process taking place in a pneumatic hydroaccumulator can be classified as isothermal because an accumulator volume is large enough and a piston velocity is relatively small.

Taking into account air pre-compression in an accumulator to a pressure P_0 and an initial volume of an air cavity $V_o = A_o \cdot h_o$ (Figure 11), we obtain the following isotherm equation:

Figure 11. Scheme of the pneumatic hydroaccumulator at two positions of a piston

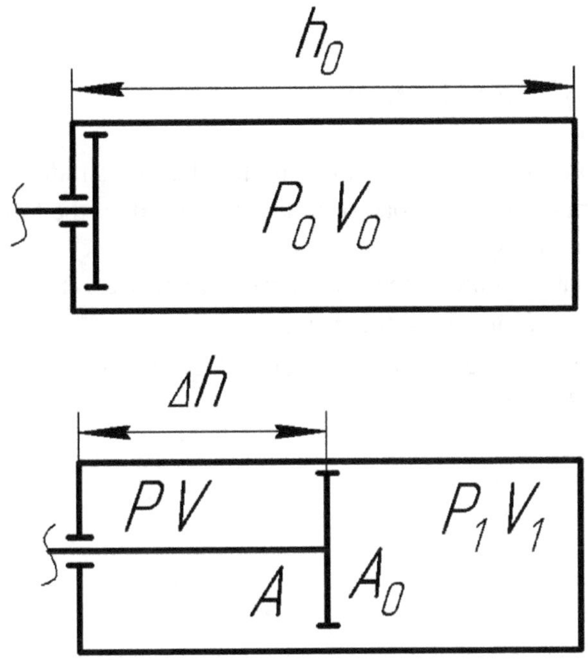

$$(P_0 + P_{atm}) \cdot V_0 = (P_1 + P_{atm}) \cdot V_1 . \tag{41}$$

from where:

$$P_1 = \frac{P_0 \cdot h_0 + P_{atm} \cdot h}{h_0 - h}, \tag{42}$$

and thus an excess oil pressure is:

$$P = P_1 \cdot \frac{A_0}{A} . \tag{43}$$

When the inertia forces of progressive accumulator masses and the forces of viscous friction in it are neglected due to their small values, the accumulator equation is written in the following form:

$$P \cdot A = \frac{P_0 \cdot h_0 + P_{atm} \cdot h}{h_0 - h} \cdot A_0 , \tag{44}$$

where we get (on the basis of 39):

$$h = \frac{V_p}{2\pi A} \cdot \int \omega_s dt . \tag{45}$$

Equations of a Driving Wheel Motion

A driving wheel tire when moving under a tractor weight undergoes a radial, longitudinal and tangential deformation. It slightly dampens traction load oscillations caused by the forces of working mechanism resistance and by a road roughness.

A tire in a dynamic simulation is presented in a form of two masses, characterized by inertia moments of a wheel rim with a tire part J_3 and a tire tread J_4.

The motion of a tractor running gear can be described by the following system of equations (Senkevich, 2006; Kravchenko, 2010):

$$\left.\begin{array}{l} J_3 \cdot \dot{\omega}_3 = M_t \cdot i_t - M_w \\ J_4 \cdot \dot{\omega}_4 = M_w - M_{ew} \end{array}\right\}, \tag{46}$$

where

M_w is a tire torque of a running gear;
M_{ew} is a moment from a longitudinal deformation of a running gear tire.

A tire torque of a running gear M_w can be calculated as follows:

$$M_w = c_{k2} \cdot \left(\varphi_3 - \varphi_4\right) + \alpha_{k2} \cdot \left(\omega_3 - \omega_4\right), \tag{47}$$

where c_{k2}, α_{k2} is a torsional rigidity and damping of a drive wheel.

A moment from a longitudinal deformation of a running gear tire M_{ew} is stipulated by tire rigidity and damping, therefore it can be expressed by dependence:

$$M_{ew} = (c_x \cdot x + \alpha_x \cdot \dot{x}) \cdot r_d + R \cdot a_2, \tag{48}$$

where c_x, α_x are coefficients of longitudinal rigidity and viscous friction in a tire; a is a displacement of a vertical reaction of a road contact R relative to a wheel axis; r_d is a dynamic wheel radius.

At a wheel rolling, tire elements are deformed in tangential and longitudinal directions at a road contact.

Some elements come out of a road contact and new tire elements obtain this deformation when coming into a road contact. Therefore, when simulating a dynamic MTU, instead of a tangential rigidity of a pneumatic tire we used a longitudinal tire rigidity c_x which is usually included in a general rigidity of a power train.

During a road contact, a tractor drive wheel covers an area equal to the difference in a tire imprint length l and an amount of a tire longitudinal deformation x. This explains the fact that a rolling radius of a running gear, which is determined on a given area according to rotations of a driving or a driven wheel, is always larger than that defined statically.

It should also be noted that due to a tire longitudinal deformation, a tire center on the road displaces forward relative to a drive wheel axis. Moreover, a displacement value raises with an increase of an impelling power of a drive wheel, which is determined by the formula:

$$T = c_x \cdot x + \alpha_x \cdot \dot{x}. \tag{49}$$

A normal axle load and a road response create additional reason that counteracts a drive wheel rolling. As it was established by the tractor traction tests carried out by the North Caucasus Machine Testing Station, the resistance to a running gear rolling raises with increasing a torque to a wheel axis, despite of a constant normal load. With a small error, a rolling arm $a_ш$ due to a tire deformation x is determined by the formula:

$$a_w = f \cdot r_c + \frac{c_x \cdot x + \alpha_x \cdot \dot{x}}{c_z}, \tag{50}$$

where

f is a coefficient of resistance to a free wheel rolling;
r_c is a free rolling radius.

A normal drive wheel reaction can be given as:

$$R_2 = c_z \cdot z_w + \alpha_z \cdot \dot{z}_w + c_z \cdot z_0, \tag{51}$$

where

c_z, α_z are radial rigidity and a coefficient of a radial viscous friction of running gear tires respectively;
z_w is a radial deformation of a running gear tire caused by vertical oscillations of a tractor frame;
z_0 is a static deflection of a running gear tire.

Taking into account a dynamic radius of a drive wheel being equal to a distance from a wheel axis to a road, we can write:

$$r_d = r_0 - z_0 - z_w. \tag{52}$$

Equations of a Tractor Frame Motion

When performing an agricultural performance of a MTU, there are oscillations in a longitudinal and vertical planes due to road roughness and oscillations of MTU traction resistance. Taking into account these oscillations, a tractor frame motion is described by a system of differential equations (Senkevich, 2006; Kravchenko, 2010):

$$\left.\begin{aligned}&m_c \cdot \ddot{z}_c = c_{z1} \cdot (z_1 - q_1) + \alpha_{z1} \cdot (\dot{z}_1 - \dot{q}_1) + c_{z2} \cdot (z_2 - q_2) + \alpha_{z2} \cdot (\dot{z}_2 - \dot{q}_2);\\ &J_c \cdot \ddot{\varphi}_c = \left[c_{z1} \cdot (z_1 - q_1) + \alpha_{z1} \cdot (\dot{z}_1 - \dot{q}_1)\right] \cdot a - \left[c_{z2} \cdot (z_2 - q_2) + \alpha_{z2} \cdot (\dot{z}_2 - \dot{q}_2)\right] \cdot b + \\ &+\left[c_{x2} \cdot x_2 + \alpha_{x2} \cdot \dot{x}_2\right]\end{aligned}\right\} \tag{53}$$

where

J_c is an inertia moment of a tractor relative to a mass center;
z_c is a vertical deviation of a mass center from an equilibrium;
φ_c is an angle of a tractor frame rotation relative to a transverse-horizontal axis passing through a mass center;
z_1, z_2 are vertical deviations from an equilibrium of tractor front and rear axles;
q_1, q_2 are coordinates of a micro profile under front and rear wheels;
a, b is a projection along a horizontal axis of a distance from a gravity center to front and rear axles;
h_c is a distance along a vertical axis from a traction power action line to a tractor mass center;
c_{z1}, c_{z2} is a reduced rigidity of elastic elements of front and rear suspensions and tires;
α_{z1}, α_{z2} are damping coefficients of front and rear suspensions and tires.

Representing a radial deformation of a tire of a wheeled tractor by the dependence

$$z_w = z - q, \tag{54}$$

and vertical deviations from an equilibrium of mass centers of front and rear axles by the equations:

$$z_1 = z_c + a \cdot \varphi_c \tag{55}$$

and

$$z_2 = z_c - b \cdot \varphi_c, \tag{56}$$

it is possible to determine the dynamic rolling radius of a wheel tractor running gear:

$$r_d = r_0 - b \cdot \varphi_c - q + z_0. \tag{57}$$

Many researchers have noted that there is a traction resistance increase when a tractor speed raises. In addition, many studies have shown that a traction resistance due to a soil density heterogeneity is random in nature. To study tractor traction dynamics a random function of a traction resistance is considered as stationary and ergodic at a steady motion of a MTU. On the other hand, a certain interest is caused by a periodic effect on a dynamic system with a given frequency.

Equations of a MTU Motion and an External Load

Denoting a MTU mass as m_u, and a resistance power as P_a, we obtain (Senkevich, 2006; Kravchenko, 2010):

$$m_u \cdot \frac{dV}{dt} = T(\lambda) - P_a = C_x \cdot \lambda + \alpha_x \cdot \lambda - P_a(t, V) \tag{58}$$

The mass m_a includes translational and reduced to them rotational masses that are not involved in a transference of an engine torque.

The speed of MTU agricultural performance is determined by values of forces and moments applied to a wheel tractor running gear. On the basis of the Keldysh theory of rolling wheels it can be written

$$V_u = \left[\omega_4 \cdot r_0 \cdot \left(1 - \varepsilon_1 \cdot R_2 - \varepsilon_2 \cdot \frac{x}{r_0} \cdot sign\omega_4\right) - \dot{x}\right] \cdot \left(1 - \delta\right), \tag{59}$$

where

r_0 is a free wheel radius;

ε_1, ε_2 are coefficients which characterize the deformation of drive wheel tires from vertical and longitudinal powers, respectively;

δ is a coefficient of a drive wheel travel reduction (slip).

According to the methods described in GOST 7057-2001 "Agricultural tractors. Methods of testing.", a travel reduction (slip) value is defined as a result of joint sliding of a contact imprint and a so-called false travel reduction (slip), therefore a coefficient value δ_g obtained during the tests must be corrected to distinguish net travel reduction (slip):

$$\delta = \delta_g - x / l, \tag{60}$$

where l is a length of a tire contact with the soil.

Travel reduction (slip) of energy running gears can be represented as a functional dependence on a pushing force T obtained in a wheel contact with a surface:

$$\delta = \begin{cases} 0, if\, T \le \varphi_0 R \\ -\dfrac{\ln\left[\dfrac{(\varphi \cdot R - T)}{(\varphi - \varphi_0) \cdot R}\right]}{K_6}, if\, \varphi_0 R \angle T \le \varphi R, \\ 0.9, if\, T \ge \varphi R \end{cases} \tag{61}$$

where φ_0, φ are coefficients of static friction and a drive wheel adhesion with the surface.

The resistance power of a MTU motion is a function of time and an actual translational speed of a MTU:

$$P_a(t) = f(t, V)$$

Then the force of resistance to a MTU motion can be represented as a functional dependence:

$$P_a(t) = \left[P_{a0} + \Delta P_a \cdot (V_u - V_0)\right] \cdot \left(1 - e^{-k_u V_u}\right) + F(t)\; P_c(t) = P^0_{:@}\left(1 - e^{-k_{BA} V_{Æ}}\right) + P_{:@}(t), \tag{62}$$

where $P^0_{:@}$ is a numerical simulation of a traction power generated by an agricultural machine at the MTU velocity V_0; ΔP_u is a regularity of a traction power increase generated by an agricultural machine with an actual velocity increase from V_0 to V_u; $F(t) = A_P \sin(2\pi f t)$ is a random stationary function, which characterizes changes in the machine technological resistance; k_u k_{BA} is an experimental coefficient characterizing a traction power increase generated by an agricultural machine.

Thus, a MTU motion is described by the equations and functional dependencies (19 - (62). Their joint solution will allow determining the optimum parameters and MTU performance modes which ensure maximum performance with an economical fuel consumption.

The coefficients values of these equations are precisely described in the work of Senkevich (2006) and Kravchenko (2010) and are determined on the basis of previous work and our experimental studies.

System of Differential Equations of a MTU Motion With a Serial Tractor Transmission

According to the dynamic simulation (model) (Figure 7a), there is no planetary reduction gear and pneumatic hydroaccumulator in a serial transmission. The system of differential equations of a MTU motion will include only formulas (23-30; 46-59; 61; 62). The numerical simulation (mathematical model) of a serial MTU in the first acceleration phase at $\omega_1 \neq \omega_2$ has the form:

$$J_1 \cdot \Delta\dot{\omega}_1 = a_1 \cdot \Delta\omega_1 + a_2 \cdot \Delta\omega_r + a_3 \cdot \Delta z - M_{cl}$$

$$M_{cl} = n \cdot P_{\max} \cdot r_{mid} \cdot \left(1 - e^{-k_1 \cdot t}\right) \cdot \left\{\mu_{\max} - \Delta\mu \cdot \left[1 - e^{-k_2 \cdot (\omega_1 - \omega_2)}\right]\right\}$$

$$m_r \cdot \Delta\ddot{z} + \beta \cdot \Delta\dot{z} + c_1 \cdot \Delta z = c_2 \cdot \Delta\omega_1$$

$$J_1 \cdot \dot{\omega}_1 = M_1 - M_{cl}\left(M_t\right)$$

$$J_2 \cdot \dot{\omega}_2 = M_{cl} - \left(M_t + \alpha' \cdot \omega_2\right)$$

$$M_t = c_t \cdot \left(\phi_2 - i_t \phi_3\right) + \alpha_t \left(\dot{\phi}_2 - i_t \cdot \dot{\phi}_3\right) + M \cdot sign\left(\dot{\phi}_2 - i_t \cdot \dot{\phi}_3\right)$$

$$\dot{\varphi}_3 = \omega_2 - i_t \cdot \omega_3$$

$$J_3 \cdot \dot{\omega}_3 = M_t \cdot i_t - M_w$$

$$J_4 \cdot \dot{\omega}_4 = M_w - M_{ew}$$

$$M_w = c_{k2} \cdot \left(\varphi_3 - \varphi_4\right) + \alpha_{k2} \cdot \left(\omega_3 - \omega_4\right)$$

$$M_{ew} = (c_E \cdot x \mid \alpha_E \cdot \dot{x}) \cdot r_d + R \cdot a_2$$

$$m_c \cdot \ddot{z}_c = c_{z1} \cdot \left(z_1 - q_1\right) + \alpha_{z1} \cdot \left(\dot{z}_1 - \dot{q}_1\right) + c_{z2} \cdot \left(z_2 - q_2\right) + \alpha_{z2} \cdot \left(\dot{z}_2 - \dot{q}_2\right)$$

$$\begin{aligned} J_c \cdot \ddot{\varphi}_c = &\left[c_{z1} \cdot \left(z_1 - q_1\right) + \alpha_{z1} \cdot \left(\dot{z}_1 - \dot{q}_1\right)\right] \cdot a - \left[c_{z2} \cdot \left(z_2 - q_2\right) + \alpha_{z2} \cdot \left(\dot{z}_2 - \dot{q}\right)\right] \cdot b + \\ &+ \left[c_{x2} \cdot x_2 + \alpha_{x2} \cdot \dot{x}_2\right] \cdot \left(r_d + h_c\right) - P_c \cdot h_c \end{aligned}$$

$$z_1 = z_c + a \cdot \varphi_c$$

$$z_2 = z_c - b \cdot \varphi_c$$

$$r_d = r_0 - b \cdot \varphi_c - q + z_0$$

$$m_u \cdot \dot{V}_u = c_x \cdot x + \alpha_x \cdot \dot{x} - P_a - m_c \cdot g \cdot \varphi_c$$

$$_a = \left[P_{a0} + \Delta P_a \cdot \left(V_u - V_0\right)\right] \cdot \left(1 - e^{k_u \cdot V_u}\right) + F\left(t\right)$$

$$V_u = \left[\omega_4 \cdot r_0 \cdot \left(1 - \varepsilon_1 \cdot R_2 - \varepsilon_2 \cdot \frac{x}{r_0} \cdot sign\omega_4\right) - \dot{x}\right] \cdot \left(1 - \delta\right)$$

$$\delta = \begin{cases} 0,\ if\ '' \leq \varphi_0 R \\ -\dfrac{\ln\left[\dfrac{\left(\varphi \cdot R - T\right)}{\left(\varphi - \varphi_0\right) \cdot R}\right]}{K_6},\ if\ \varphi_0 R \angle T \leq \varphi R \\ 0.9\ ,\ if\ '' \geq \varphi R \end{cases}$$

These limits of wheel travel reduction (slip) imposed on the system of equations are preserved.

When locking the driving and driven parts of a friction clutch (Figure 7*b*), the numerical simulation (mathematical model) has the following form (the 2-nd acceleration phase $\omega_1 = \omega_2$):

$$J_1 \cdot \dot{\omega}_1 = a_1 \cdot \omega_1 + a_2 \cdot \omega_r + a_3 \cdot z - M_t$$

$$J_r \cdot \dot{\omega}_r = b_1\omega_1 + b_2\omega_4 + b_3 z$$

$$m_r \cdot \ddot{z} + \beta \cdot \dot{z} + c_1 \cdot z = c_2 \cdot \omega_1$$

$$M_t = c_t\left(\varphi_2 - i_t\varphi_3\right) + \alpha_t\left(\dot{\varphi}_2 - i_t\dot{\varphi}_3\right) + M \times sign\left(\dot{\varphi}_2 - i_t\dot{\varphi}_3\right)$$

Further, the system of equations will be identical to the condition of the first acceleration phase at $\omega_1 \neq \omega_2$.

System of Differential Equations of a Sowing Machine with an EDM in a Tractor Transmission

The numerical simulation (mathematical model) with an EDM has the following form (Figure 9 *a*) in the first acceleration phase at $\omega_1 \neq \omega_2$:

$$J_1 \cdot \dot{\omega}_1 = a_1 \cdot \omega_1 + a_2 \cdot \omega_r + a_3 \cdot z - M_{cl}$$

$$M_{cl} = n \cdot P_{\max} \cdot r_{mid} \cdot \left(1 - e^{-k_1 \cdot t}\right) \cdot \left\{\mu_{\max} - \Delta\mu \cdot \left[1 - e^{-k_2 \cdot (\omega_1 - \omega_k)}\right]\right\}$$

$$J_r \cdot \dot{\omega}_r = b_1\omega_1 + b_2\omega_4 + b_3 z$$

$$m_r \cdot \ddot{z} + \beta \cdot \dot{z} + c_1 \cdot z = c_2 \cdot \omega_1$$

$$\omega_k = \frac{r_s}{(2r_{sat} + r_A)} \cdot \omega_s + \frac{2 \cdot (r_{sat} + r_s)}{(2r_{sat} + r_s)} \cdot \omega_v$$

$$J_{eq} \cdot \dot{\omega}_s = \frac{r_s}{(2r_{sat} + r_s)} \cdot M_{cl} - k \cdot V_p \cdot P_p$$

$$P_p = \frac{V_p^2 \cdot \gamma \cdot \omega_s^2}{4\pi^2 \cdot 2v^2 \cdot F_{th}^2} + \frac{P_0 \cdot h_0 + \frac{V_p \cdot D}{2\pi \cdot A}}{h_0 + \frac{V_p \cdot D}{2\pi \cdot A}} \cdot \frac{A_0}{A}$$

$$J_{eq} = (J_s \cdot \frac{1}{i_{eq}^2} + J_{pump}^{\text{rot.}part} + J_{gdop} + J_{ex.\text{w}})$$

$$\dot{D} = \omega_s$$

$$J_v \cdot \dot{\omega}_v = \frac{2(r_{sat} + r_s)}{(2r_{sat} + r_s)} \cdot M_{cl} - M_t$$

$$J_2 \cdot \dot{\omega}_2 = M_{cl} - \left(M_t + \alpha' \cdot \omega_2\right)$$

$$\dot{\varphi}_2 = \omega_v - i_t \cdot \omega_3$$

$$M_t = A_t \cdot \left(\varphi_v - i_t \varphi_3\right) + \alpha_t \left(\dot{\varphi}_v - i_t \cdot \dot{\varphi}_3\right) + M \cdot sign\left(\dot{\varphi}_v - i_t \cdot \dot{\varphi}_3\right)$$

$$\dot{\varphi}_3 = \omega_2 - i_t \cdot \omega_3$$

$$J_3 \cdot \dot{\omega}_3 = M_t \cdot i_t - M_w$$

$$J_4 \cdot \dot{\omega}_4 = M_w - M_{ew}$$

$$M_w = c_{:2} \cdot (\varphi_3 - \varphi_4) + \alpha_{k2} \cdot (\omega_3 - \omega_4)$$

$$M_{ew} = (c_x \cdot x + a_x \cdot \dot{x}) \cdot r_d + R \cdot a_2$$

$$m_c \cdot \ddot{z}_c = c_{z1} \cdot (z_1 - q_1) + \alpha_{z1} \cdot (\dot{z}_1 - \dot{q}_1) + c_{z2} \cdot (z_2 - q_2) + \alpha_{z2} \cdot (\dot{z}_2 - \dot{q}_2)$$

$$\begin{aligned} J_c \cdot \ddot{\varphi}_c = & \left[c_{z1} \cdot \left(z_1 - q_1\right) + \alpha_{z1} \cdot \left(\dot{z}_1 - \dot{q}_1\right)\right] \cdot a - \\ & - \left[c_{z2} \cdot \left(z_2 - q_2\right) + \alpha_{z2} \cdot \left(\dot{z}_2 - \dot{q}\right)\right] \cdot b + \\ & + \left[c_{x2} \cdot x_2 + \alpha_{x2} \cdot \dot{x}_2\right] \cdot \left(r_d + h_c\right) - P_c \cdot h_c \end{aligned}$$

$$z_1 = z_c + a \cdot \varphi_c;$$

$$z_2 = z_c - b \cdot \varphi_c;$$

$$r_d = r_0 - b \cdot \varphi_c - q + z_0$$

$$m_a \cdot \dot{V}_u = c_x \cdot x + \alpha_x \cdot \dot{x} - P_a - m_c \cdot g \cdot \varphi_c$$

$$P_a = \left[P_{a0} + \Delta P_a \cdot \left(V_u - V_0\right)\right] \cdot \left(1 - e^{k_u \cdot V_u}\right) + F\left(t\right)$$

$$V_u = \left[\omega_4 \cdot r_0 \cdot \left(1 - \varepsilon_1 \cdot R_2 - \varepsilon_2 \cdot \frac{x}{r_0} \cdot sign\omega_4\right) - \dot{x}\right] \cdot \left(1 - \delta\right)$$

$$\delta = \begin{cases} 0, \ if \ ''(\lambda) \leq \varphi_0 R \\ -\dfrac{\ln\left[\dfrac{\varphi R - T(\lambda)}{(\varphi - \varphi_0)R}\right]}{K_6}, \ if \ \varphi_0 R \angle T(\lambda) \leq \varphi R \\ 0.9, \ if \ ''(\lambda) \geq \varphi R \end{cases}$$

At $\omega_1 = \omega_K$ which means the end of a travel reduction (slip) of clutch discs we go to a modified system. The initial conditions of this system will be the final values of unknown values of the previous system, satisfying the condition $\omega_1 = \omega_K$.

The numerical simulation with an EDM has the following form (Figure 9 *b*) in the second acceleration phase at $\omega_k = \omega_2 \cdot i$:

$$J_1 \cdot \dot{\omega}_1 = a_1 \cdot \omega_1 + a_2 \cdot \omega_r + a_3 \cdot z - \frac{2r_{sat} + r_s}{2 \cdot \left(r_{sat} + r_s\right)} \cdot M_t$$

$$J_r \cdot \dot{\omega}_r = b_1\omega_1 + b_2\omega_4 + b_3 z$$

$$m_r \cdot \ddot{z} + \beta \cdot \dot{z} + c_1 \cdot z = c_2 \cdot \omega_1$$

$$J_{eq} \cdot \dot{\omega}_s = \frac{r_s}{2 \cdot (r_{sat} + r_s)} \cdot M_t - k \cdot V_p \cdot P_p$$

$$\omega_v = \frac{(2r_{sat} + r_s)}{2(r_{sat} + r_s)} \cdot \omega_1 - \frac{r_s}{2(r_{sat} + r_s)} \cdot \omega_s$$

$$P_p = \frac{V_p^2 \cdot \gamma \cdot \omega_s^2}{4\pi^2 \cdot 2\nu^2 \cdot F_{th}^2} + \frac{P_0 \cdot h_0 + \frac{V_p \cdot D}{2\pi \cdot A}}{h_0 + \frac{V_p \cdot D}{2\pi \cdot A}} \cdot \frac{A_0}{A}$$

$$J_{eq} = (J_s \cdot \frac{1}{i_{eq}^2} + J_{pump}^{\text{rot.}part} + J_{gdop} + J_{ex.\text{w}})$$

$$\dot{D} = \omega_s$$

$$\dot{\varphi}_3 = \omega_2 - i_t \cdot \omega_3$$

$$M_t = A_t \cdot (\varphi_2 - i_t\varphi_3) + \alpha_t(\dot{\varphi}_v - i_t \cdot \dot{\varphi}_3) + M \cdot sign(\dot{\varphi}_v - i_t \cdot \dot{\varphi}_3)$$

$$J_3 \cdot \dot{\omega}_3 = M_t \cdot i_t - M_w$$

$$J_4 \cdot \dot{\omega}_4 = M_w - M_{ew}$$

$$M_w = c_{k2} \cdot (\varphi_3 - \varphi_4) + \alpha_{k2} \cdot (\omega_3 - \omega_4)$$

$$M_{ew} = (c_x \cdot x + a_x \cdot \dot{x}) \cdot r_d + R \cdot a_2$$

Further, the system of equations will be identical to the condition of the first acceleration phase at $\omega_1 \neq \omega_K$.

Methodology of Solution of MTU Motion Equations

To solve the system of equations, we apply the Runge–Kutta method, which allows obtaining the necessary accuracy of numerical calculations. The equations of motion are reduced to a canonical form, where coefficient values are determined according to the results of our own tests and the study of a number of the mentioned above researchers. The solution of a MTU numerical simulation (model) is based on the method of numerical integration of a system of ordinary differential equations. The solution of the system of differential equations has been carried out in the MAPLE system, which allows obtaining numerical solutions of given quantities after a certain period of time.

Methodology of a Simulated Experiment of a Sowing Machine Performance: Regression Analysis

On the basis of the work of the professors A. B. Lurie and Bespamyatnova N.M. (Lurie, 1969; Bespamyatnova, 2002) we consider the frequency-traction characteristics of a MTU to be an additional affecting factor. Experimental tests have determined an oscillation range in a traction load during sowing, which ranges from 0.3 to 2.5 Hz (Lurie, 1969; Bespamyatnova, 2002). To identify how a load oscillation frequency affects an EDM performance, we conducted simulated experiments on the numerical simulation given above. The simulation (model) has been supplied with MTU parameters in the 14 kN class tractor and a sowing machine with a width of 5.4 meters.

According to the literature and previous study (Senkevich, 2006; Kravchenko, 2010; Senkevich, et al, 2019a), there have been selected the factors that significantly affect a traction load damping, including a load oscillation frequency. When there have been chosen a design of the experiment, main levels and intervals of variation of factors, we proceed to the experiment itself. Each matrix line is an experiment condition (Spiridonov, 1981).

For the study there has been chosen a rotatable design of a second order with a core in the form of a semi-replica with a defining contrast of $1=X_1 \cdot X_2 \cdot X_3 \cdot X_4 \cdot X_5$. Matrix design includes 32 experiments. This design allows expanding the studied range of factors due to experiments in magnitudes. This greatly increases the opportunity to explore a large area of optimization function presence. Matrix X of the experiment, the procedure of tests and the method of experimental data processing (regression analysis) have been presented in detail in Spiridonov (1981).

As an optimization criterion (a response function), the indicator "a transmission transparency degree" has been defined by Formula 10, which characterizes transmission damping properties. The purpose of the experiment is to determine optimal parameters of an elastic damping mechanism in a tractor transmission.

In the study, the following factors have been taken as variables: $S_{др}$ is a throttle cross-sectional area, $V_{пга}$ is a volume of pneumatic hydroaccumulator (PGA), $P_в$ is an air pressure in PGA, $J_{дг}$ is an inertia moment of additional load of an oil pump driving gear, f is an oscillation frequency of a traction load. The values of the factors in the encoded and natural form are presented in Table 1.

All calculations of the regression analysis, including the graph plotting have been performed in the MathCAD system.

Table 1. Levels and intervals of variation of factors

N	Factors	Factor Symbol, meas.	Code Mark	Intervals of Variation	Natural Values Corresponding to Coded Levels of the Factors		
					Upper (+1)	Base (0)	Lower (-1)
1	A throttle cross-sectional area	S_{th}, m^2	X_1	1.1×10^{-4}	3.524×10^{-4}	2.424×10^{-4}	1.324×10^{-4}
2	Volume of a pneumatic hydroaccumulator (HPA)	V_{hpa}, m^3	X_2	1.23×10^{-3}	5.079×10^{-3}	3.848×10^{-3}	2.617×10^{-3}
3	Air pressure in HPA	P_a, Pa	X_3	1×10^5	5×10^5	4×10^5	3×10^5
4	Inertia moment of an additional load	J_{th}, $kg{\cdot}m^2$	X_4	2.32×10^{-3}	6.96×10^{-3}	4.64×10^{-3}	2.32×10^{-3}
5	Oscillation frequency of a traction load	f, Hz	X_5	0.5	1.8	1.3	0.8

Results of the Simulated Experiment of a Sowing MTU Performance

The Matrix of experiment and the matrix of results (responses) is presented in Figure 12.

A regression equation in general will be:

$$y = b_0 + b_1x_1 + b_2x_2 + b_3x_3 + b_4x_4 + b_{12}x_1x_2 + b_{13}x_1x_3 + b_{14}x_1x_4 + \\ +b_{23}x_2x_3 + b_{24}x_2x_4 + b_{34}x_3x_4 + b_{11}x_1^2 + b_{22}x_2^2 + b_{33}x_3^2 + b_{44}x_4^2 \tag{63}$$

Equation coefficients of the regression (of the simulation) have been identified by the formula:

$$B = (X'' \times X)^{-1} \times X'' \times Y \tag{64}$$

The obtained equation of regression has the following form:

$$y = 0.722 - 6,325 \cdot 10^{-3} \cdot x_1 - 6.657 \cdot 10^{-3} \cdot x_2 + 0.033 \cdot x_3 - 5.526 \cdot 10^{-3} \cdot x_4 - \\ -3.072 \cdot 10^{-3} \cdot x_5 + 6.981 \cdot 10^{-3} \cdot x_1 \cdot x_2 + 6.851 \cdot 10^{-3} \cdot x_1 \cdot x_3 + \\ +5.709 \cdot 10^{-3} \cdot x_1 \cdot x_4 - 7.045 \cdot 10^{-3} \cdot x_1 \cdot x_5 + 8.462 \cdot 10^{-3} \cdot x_2 \cdot x_3 + \\ +7.245 \cdot 10^{-3} \cdot x_2 \cdot x_4 - 5.554 \cdot 10^{-3} \cdot x_2 \cdot x_5 + 7.115 \cdot 10^{-3} \cdot x_3 \cdot x_4 + \\ +2.725 \cdot 10^{-3} \cdot x_3 \cdot x_5 - 6.826 \cdot 10^{-3} \cdot x_4 \cdot x_5 + 2.905 \cdot 10^{-3} \cdot x_1^2 + \\ +2.466 \cdot 10^{-3} \cdot x_2^2 - 7.031 \cdot 10^{-3} \cdot x_3^2 + 3.445 \cdot 10^{-3} \cdot x_4^2 + 0.038 \cdot x_5^2 \tag{65}$$

Figure 12. Rotatable matrix design of second order X and matrix of responses Y

X =

	0	1	2	3	4	5	6	7	8	9	10	11	12	13	14	15	16	17	18	19	20
0	1	-1	-1	-1	-1	1	1	1	1	-1	1	1	-1	1	-1	-1	1	1	1	1	1
1	1	1	-1	-1	-1	-1	-1	-1	-1	-1	1	1	1	1	1	1	1	1	1	1	1
2	1	-1	1	-1	-1	-1	-1	1	1	1	-1	-1	-1	1	1	1	1	1	1	1	1
3	1	1	1	-1	-1	1	1	-1	-1	1	-1	-1	1	1	-1	-1	1	1	1	1	1
4	1	-1	-1	1	-1	-1	1	-1	1	1	-1	1	1	-1	-1	1	1	1	1	1	1
5	1	1	-1	1	-1	1	-1	1	-1	1	-1	1	-1	-1	1	-1	1	1	1	1	1
6	1	-1	1	1	-1	1	-1	-1	1	-1	1	-1	1	-1	1	-1	1	1	1	1	1
7	1	1	1	1	-1	-1	1	1	-1	-1	1	-1	-1	-1	-1	1	1	1	1	1	1
8	1	-1	-1	-1	1	-1	1	1	-1	1	1	-1	1	-1	1	-1	1	1	1	1	1
9	1	1	-1	-1	1	1	-1	-1	1	1	1	-1	-1	-1	-1	1	1	1	1	1	1
10	1	-1	1	-1	1	1	-1	1	-1	-1	-1	1	1	-1	-1	1	1	1	1	1	1
11	1	1	1	-1	1	-1	1	-1	1	-1	-1	1	-1	-1	1	-1	1	1	1	1	1
12	1	-1	-1	1	1	1	1	-1	-1	-1	-1	-1	-1	1	1	1	1	1	1	1	1
13	1	1	-1	1	1	-1	-1	1	1	-1	-1	-1	1	1	-1	-1	1	1	1	1	1
14	1	-1	1	1	1	-1	-1	-1	-1	1	1	1	-1	1	-1	-1	1	1	1	1	1
15	1	1	1	1	1	1	1	1	1	1	1	1	1	1	1	1	1	1	1	1	1
16	1	2	0	0	0	0	0	0	0	0	0	0	0	0	0	0	4	0	0	0	0
17	1	-2	0	0	0	0	0	0	0	0	0	0	0	0	0	0	4	0	0	0	0
18	1	0	2	0	0	0	0	0	0	0	0	0	0	0	0	0	0	4	0	0	0
19	1	0	-2	0	0	0	0	0	0	0	0	0	0	0	0	0	0	4	0	0	0
20	1	0	0	2	0	0	0	0	0	0	0	0	0	0	0	0	0	0	4	0	0
21	1	0	0	-2	0	0	0	0	0	0	0	0	0	0	0	0	0	0	4	0	0
22	1	0	0	0	2	0	0	0	0	0	0	0	0	0	0	0	0	0	0	4	0
23	1	0	0	0	-2	0	0	0	0	0	0	0	0	0	0	0	0	0	0	4	0
24	1	0	0	0	0	2	0	0	0	0	0	0	0	0	0	0	0	0	0	0	4
25	1	0	0	0	0	-2	0	0	0	0	0	0	0	0	0	0	0	0	0	0	4
26	1	0	0	0	0	0	0	0	0	0	0	0	0	0	0	0	0	0	0	0	0
27	1	0	0	0	0	0	0	0	0	0	0	0	0	0	0	0	0	0	0	0	0
28	1	0	0	0	0	0	0	0	0	0	0	0	0	0	0	0	0	0	0	0	0
29	1	0	0	0	0	0	0	0	0	0	0	0	0	0	0	0	0	0	0	0	0
30	1	0	0	0	0	0	0	0	0	0	0	0	0	0	0	0	0	0	0	0	0
31	1	0	0	0	0	0	0	0	0	0	0	0	0	0	0	0	0	0	0	0	0

Y =

	0
0	0.812
1	0.755
2	0.623
3	0.892
4	0.936
5	0.894
6	0.945
7	0.821
8	0.724
9	0.857
10	0.915
11	0.623
12	0.907
13	0.934
14	0.826
15	0.965
16	0.609
17	0.609
18	0.636
19	0.576
20	0.67
21	0.55
22	0.616
23	0.599
24	0.758
25	0.979
26	0.609
27	0.609
28	0.609
29	0.609
30	0.609
31	0.609

Figure 13. Graphs of dependence of a transmission transparency degree on a traction load oscillation and an inertia moment of an EDM driving gear

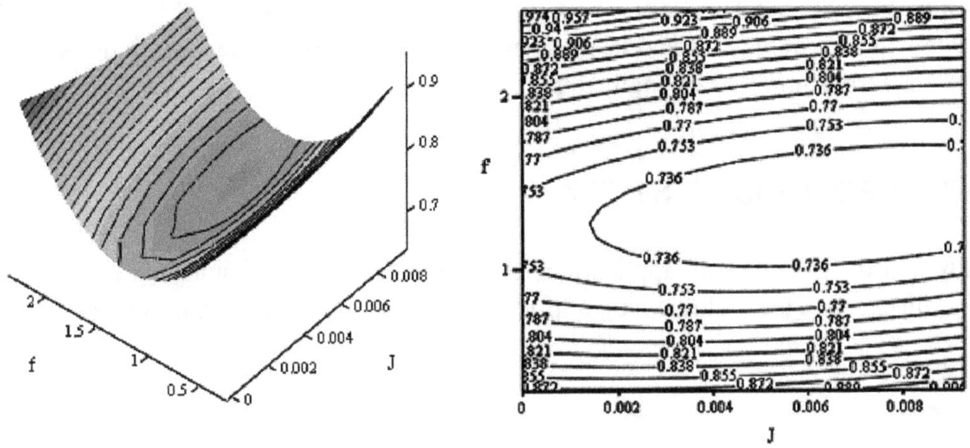

where

$$x_1 = \frac{S_{th} - 2.424 \times 10^{-4}}{1.1 \times 10^{-4}};$$
$$x_2 = \frac{V_{pga} - 3.848 \times 10^{-3}}{1.231 \times 10^{-3}};$$
$$x_3 = \frac{P_u - 4 \times 10^5}{1 \times 10^5}; \qquad (66)$$
$$x_4 = \frac{J_{th} - 4.64 \times 10^{-3}}{2.32 \times 10^{-3}};$$
$$x_5 = \frac{f - 1.3}{0.5}.$$

Substituting (65) into (66) and analyzing the obtained equation, we have received response surfaces (a transmission transparency degree) in physical terms (Figure 13).

CONCLUSION

During the study we can make the following conclusions:

1. The regression equation analysis obtained in a simulated experiment has shown, that in the range of 0.3-2.5 Hz of load oscillation frequency, the inertia moment of an additional driving gear loads should be of 0-0.009 kg $\times$ m^2.
2. The optimum parameters of an elastic-damping mechanism, obtained in a simulated experiment with the most possible oscillation frequency of a traction load *f=1.2-1.3 Hz* have the following

values: a throttle cross-sectional area $S_{th} = 2.16$ cm^2; a volume of a pneumatic hydroaccumulator $V_{hpa} = 4.04$ dm^3; an air pressure in a pneumatic hydroaccumulator $P_a = 0.575$ MPa; an inertia moment of additional load $J_{th} = 2.093 \times 10^{-3}$ kg $\times$ m^2; a response value in this case is $Tp = 0.754$.

3. The obtained data show that an elastic-damping mechanism reduces oscillations of an external traction load transmitted to an engine on 15% on average.

FURTHER DIRECTIONS OF STUDY

Further study and further practical application of an EDM are aimed at studying its effect on a MTU performance during such technological operations as plowing and transportation works.

REFERENCES

Babanin, N. V. (2015). Uluchshenie plavnosti hoda mashinno-traktoronogo agregata na baze traktora klassa 1.4 s uprugodempfiruyushchim privodom na vedushchih kolesah [Improvement of the smoothness of the machine and tractor unit on the basis of a class 1.4 tractor with an elastic damping mechanism on driving wheels]. *Current Problems of Science and Education*, *22*, 46.

Babanin, N. V., & Polivaev, O. I. (2015). Eksperimentalnie issledovaniya na plavnost hoda, proizvoditelnost i toplivnuyu ekonomichnost mashinno-traktornogo agregata na baze traktora klassa 1.4, oborudovannogo gazogidravlicheskim uprugodempfiruyushchim privodom [Experimental researches on smoothness of a course, productivity and fuel economy of the machine-tractor unit on the basis of a class 1.4 tractor, equipped with a gas-hydraulic elastic-damping drive]. *Bulletin of the Voronezh State Agrarian University*, *3*(46), 112–118.

Bespamyatnova, N.M. (2002). *Nauchno-metodicheskie osnovi adaptatsii pochvoobrabativayushchih mashin* [Scientific and methodological fundamentals of adaptation of soil cultivating and sowing machines]. Rostov-on-Don, Russia: LLC “Terra”.

Deboli, R., Calvo, A., & Preti, C. (2017). Whole-body vibration: Measurement of horizontal and vertical transmissibility of an agricultural tractor seat. *International Journal of Industrial Ergonomics*, *58*, 69–78. doi:10.1016/j.ergon.2017.02.002

Feuerhuber, K., Lindert, S. O., & Schlacher, K. (2013). *Vibration attenuation by semi-active dampers.* Paper presented at 14th International Conference on Computer Aided Systems Theory (EUROCAST), Las Palmas, Spain. DOI: 10.1007/978-3-642-53862-9_35

Filin, V. A., Pronin, V. Yu., Kryuchkov, S. V., & Filin, A. V. (2016). Obosnovanie stepeni zagruzki traktornogo dvigatelya pri vypolnenii selskokhozyastvennykh rabot [Substantiation of the degree of loading of the tractor engine in the performance of agricultural works]. In Proceedings of Energy-Efficient and Resource-Saving Technologies and Systems. Saransk, Russia: FSBEU of Higher Professional Education of the Moscow State University named after N.P. Ogaryova.

Gabdrafikov, F. Z., & Safina, R. R. (2017). Povyshenie tehniko-ekonomicheskih pokazateley traktornyh dizeley na chastichnyh i neustanovivshihsya rezhimah raboty [Increase of technical and economic indicators of tractor diesel engines on partial and unsteady operating modes]. *Russian Electronic Scientific Journal*, *2*(24), 95–109.

Goryachkin, V. P. (1965). Zemledel'cheskaya mekhanika [Agricultural mechanics]. In Sobraniye sochineniy. Moscow, Russia: Kolos.

Grebnev, V. P., Poliaev, O. I., & Vorokhobin, A. V. (2011). *Traktory i avtomobili: teoriya i ekspluatatsionniye svoystva: uchebnoye posobie dlya studentov vuzov, obuchaushchihsya po napravleniyu «Agroingeneriya»* [Tractors and automobiles: theory and operational properties: a teaching aid for university students studying in the direction of "Agroengineering"]. Voronezh, Russia: KnoRus.

Kalmykov, A. V. (2014). *Snizheniye dinamicheskoy nagruzhennosty silovoy peredachi traktora za schet izmeneniya krutilnoy zhestkosti reaktivnogo zvena* [Reduction of the dynamic loading of the power transmission of the tractor due to the change in the torsional stiffness of the reactive link] (PhD dissertation). Volgograd State Agrarian University, Volgograd, Russia.

Karabanitsky, A.P., &Levshukova, O.A. (2014). *Teoreticheskoe obosnovanie parametrov energosberegaushchikh mashinno-traktornykh agregatov: uchebnoe posobie* [Theoretical justification of the parameters of energy-saving machine-tractor units: manual]. Krasnodar, Russia: KubSAU.

Kostyuchenkov, N. V., & Plaksin, A. M. (2010). Ekspluatatsionniye svoystva mobilnyh agregatov [Operational properties of mobile units]. AAstana, Kazakhstan: KATU named after S. Seifullina.

Kotlyarov, V.V., & Deyanov, Y.U.I. (1973). Issledovaniyc protsessa razgona MTA s podatlivoy svyaz'yu v transmissii traktora [Investigation of the process of acceleration of the machine-tractor unit with a malleable connection in the tractor transmission]. *Mekhanizatsiya i elektrifikatsiya s.-kh. proizvodstva*, *16*, 47-57.

Kravchenko, V. A. (2010). *Povishenie dinamicheskih i ekspluatatsionnyh pokazateley selskohozyaystvennyh mashinno traktornyh agregatov* [Increase of dynamic and operational parameters of agricultural machine and tractor units] (PhD dissertation). FSEE HE ABSSAEA, Zernograd, Russia.

Kravchenko, V. A., Senkevich, S. E., Senkevich, A. A., Goncharov, D. A., & Duryagina, V. V. (2008). *Ustroystvo dlya snizheniya zhestkosti transmissii mashinno-traktornogo agregata* [Device for reducing the stiffness of the transmission of the machine-tractor unit]. RU Patent 2398147 F16H47/04.

Kravchenko, V. A., Senkevich, S. E., Senkevich, A. A., Yarovoy, V. G., & Tolstoukhov, Yu. S. (2004). *Ustroystvo dlya snizheniya zhestkosti transmissii transportnogo sredstva* [Device to reduce the rigidity of the vehicle's transmission]. RU Patent 2222440, B60K 17/10.

Lurie, A. B. (1969). *Dinamika regulirovaniya navesnykh selskokhozyaystvennykh agregatov* [The dynamics of regulation of mounted agricultural units]. Leningrad, Russia: Mechanical Engineering.

Nekhoroshev, D. A. (2014). *Stabilizatsiya rezhimov MTA izpolzovaniem pnevmogidravlicheskoy mufty stsepleniya* [Stabilization of MTU modes using pneumatic hydraulic clutch] (PhD dissertation). Volgograd State Agrarian University, Volgograd, Russia.

Nekhoroshev, D. D., & Nekhoroshev, D. A. (2015). Energeticheskoe sredstvo v transmissii kolesnogo traktora [Power means in the transmission of the wheeled tractor]. *Unique Research of the XXI Century*, *12*(12), 40–41.

Pobedin, A. V., Dolotov, A. A., & Shekhovtsov, V. V. (2016). Decrease of the Vibration Load Level on the Tractor Operator Working Place by Means of Using of Vibrations Dynamic Dampers in the Cabin Suspension. *Procedia Engineering*, *150*, 1252–1257. doi:10.1016/j.proeng.2016.07.136

Polifke, G. (2016). *Hydrodamp - hydraulic torsional vibration damper for tractors*. Paper presented at Conference on Agricultural Engineering, Koln, Germany.

Polifke, G. (2017). *Hydrodamp - hydraulic torsional vibration damper for tractors and construction machines design for a race-sensitive drive train*. Paper presented at VDI Conference on Couplings and Clutch Systems in Drives 2017 / 2nd VDI Conference on Vibration Reduction in Mobile Systems, Ettlingen, Germany.

Polivaev, O.I., & Kutkov, A. Y. (2010). Effektivnost ispolzovaniya transportnykh sredstv s uprugodempfiruyushchim privodom vedushchikh koles pri tormozhenii [Efficiency of the use of transport units with an elastic-damping drive of driving wheels during braking]. *Machinery in Agriculture*, *5*, 30-32.

Polivaev, O. I., Gorban, L. K., Vorohobin, A. V., & Vedrinsky, O. S. (2018). Decrease of dynamic loads in mobile energy means. *Conference Series: Materials Science and Engineering (IOP)*, 327(4). 10.1088/1757-899X/327/4/042083

Polivaev, O. I., Pilayev, S. N., & Kutkov, A. Yu. (2011). Obosnovanie parametrov i effektivnosti ispolzovaniya uprugodempfiruyushchego privoda vedushchikh koles traktorno-transportnogo agregata v rezhime tormozheniya [Substantiation of parameters and efficiency of using the elastic-damping drive of the driving wheels of the tractor-transport unit in braking mode]. *Bulletin of the Voronezh State Agrarian University*, *3*, 48–52.

Polivaev, O. I., & Vedrinsky, O. S. (2013). *Vliyanie uprugodempfiruyushchego privoda vedushchikh koles traktora na dinamicheskie nagruzki v transmissii pri razgone MTA* [Influence of the elastic-damping drive of the tractor's driving wheels on the dynamic loads in the transmission when accelerating the MTU]. Innovative Technologies and Technical Means for Agroindustrial Complex.

Polivaev, O. I., & Vedrinsky, O. S. (2014). Analiz vliyaniya uprugodempfiruyushchego privoda koles na dinamicheskie nagruzki v transmissii traktora pri razgone [Analysis of the effect of the elastic-damping drive of the wheels on the dynamic loads in the tractor's transmission during acceleration]. *The Bulletin of the Don Agrarian Science*, *28*, 5–9.

Polivaev, O. I., Vedrinsky, O. S., & Derkanosova, N. M. (2016). Povishenie dolgovechnosti stsepleniya traktorov za schet uprugofriktsionnogo dempfera [Increasing the durability of tractor traction due to elastic-friction damper]. *Science and Education in Modern Conditions*, *12*, 226–230.

Polivaev, O. I., Zhidenko, K. S., Maksimov, I. I., & Levkin, I. G. (2017). *Uprugo-dempfiruyushchiy privod vedushchih koles traktora* [Elastic-damping drive of the tractor's driving wheels]. Paper presented at Modern Trends in the Development of Technology and Technical Means in Agriculture, Voronezh, Russia.

Senkevich, S., Kravchenko, V., Duriagina, V., Senkevich, A., & Vasilev, E. (2019a). Optimization of the Parameters of the Elastic Damping Mechanism in Class 1.4 Tractor Transmission for Work in the Main Agricultural Operations. In P. Vasant, I. Zelinka, & G. W. Weber (Eds.), *Intelligent Computing & Optimization. ICO 2018. Advances in Intelligent Systems and Computing* (Vol. 866, pp. 168–177). Cham: Springer. doi:10.1007/978-3-030-00979-3_17

Senkevich, S. E. (2006). *Povysheniye effektivnosti funktsionirovaniya kul'tivatornogo mashinno-traktornogo agregata na baze traktora klassa 1.4* [Improving the efficiency of the cultivating machine-tractor unit on the basis of a class 1.4 tractor] (PhD dissertation). FSEE HE ABSSAEA, Zernograd, Russia.

Senkevich, S. E., Sergeev, N. V., Vasilev, E. K., Godzhaev, Z. A., & Babayev, V. (2019b). Use of an Elastic-Damping Mechanism in the Tractor Transmission of a Small Class of Traction (14 kN): Theoretical and Experimental Substantiation. In *Advanced Agro-Engineering Technologies for Rural Business Development* (pp. 149–179). IGI Global. doi:10.4018/978-1-5225-7573-3.ch006

Shekhovtsov, V. V., Sokolov-Dobrev, N. S., & Potapov, P. V. (2016). Decreasing of the Dynamic Loading of Tractor Transmission by means of Change of the Reactive Element Torsional Stiffness. *Procedia Engineering, 150*, 1239–1244. doi:10.1016/j.proeng.2016.07.129

Spiridonov, A. A. (1981). *Planirovanie eksperimenta pri issledovanii tehnologicheskikh protsessov* [Planning an experiment in the study of technological processes]. Moscow, Russia: Mechanical Engineering.

Zech, P., Plöger, D. F., & Rinderknecht, S. (2019). Active control of planetary gearbox vibration using phase-exact and narrowband simultaneous equations adaptation without explicitly identified secondary path models. *Mechanical Systems and Signal Processing, 120*, 234–251. doi:10.1016/j.ymssp.2018.10.030

KEY TERMS AND DEFINITIONS

Degree of Transmission Transparency: Is a filtration degree of oscillations from an external load to a power unit (engine). A very rigid transmission is called transparent, i.e. one that passes a reactive external load through itself without any changes.

Elastic-Damping Mechanism (EDM): Is a mechanism that dampens vibrations and dissipates energy inside the system under the effect of cyclic loads. In this case, the mechanical energy of the oscillations is converted into the thermal energy.

Gear Change Box (GB): Is used to change a torque and to transfer it from engine to wheels.

Machine-Tractor Unit (MTU): Is a tractor connected to a working machine or machines. A tractor in the unit is a source of mechanical energy which helps the machine to move around the field, for example when a tractor pulls a cultivator.

Multi-Factor Experiment: Is an experiment that involves varying several factors, possibly affecting on the studied phenomenon.

Planetary Reduction Gear: Is a mechanism comprising several planetary gears, rotating around a central (solar) gear. Planetary gears (satellites) are fixed by a carrier. A planetary gear includes another outside hollow gear geared inside with satellites (planetary gears).

Regression Analysis: Is a statistical study of the effect of one or several independent variables on a dependent variable.

Chapter 18
Energy-Saving Systems Using Photovoltaic Modules

Pavel Valentinovich Tikhonov
Federal Scientific Agroengineering Center VIM, Russia

Vladimir Aleksandrovich Mayorov
Federal Scientific Agroengineering Center VIM, Russia

Konstantin Sergeevich Morenko
Federal Scientific Agroengineering Center VIM, Russia

ABSTRACT

The chapter presents the results of the development of two systems. The first is a photovoltaic system parallel to the power supply network of LED lamps. The algorithms of the system operation for both working and emergency lighting are shown. The basic operating modes of the system are considered taking into account the criterion of the minimum cost of electricity generated. These modes provide the most complete use of solar energy in the working day with minimal additional costs, allowing the consumer to save on electricity and increase the reliability of the emergency lighting system. The second system is a solar photovoltaic module built into a standard double-glazed window sash size 730x700 (mm), which is designed to charge a block of lithium-ion batteries with a capacity of 6.8 Ah with an output voltage of 5.25 V, the energy of which can be used to power any device having a USB 2.0 connector. The results of calculation of the required peak photovoltaic power of the module are presented; the technology of its sealing is described.

DOI: 10.4018/978-1-7998-1216-6.ch018

Copyright © 2020, IGI Global. Copying or distributing in print or electronic forms without written permission of IGI Global is prohibited.

INTRODUCTION

The use of renewable energy sources began with ground-based systems of low power. They were located in regions where there were no centralized power supply networks to provide the minimum necessary energy for small social and industrial facilities. However, as a result of successful work in a reducing the cost of produced energy, as well as in connection with the growing environmental problems and to ensure the energy security of countries over the past 15-20 years, the world is almost explosive growth in the use of renewable energy sources. Access to reliable and sustainable energy stimulates investment, including public investment, to developing a wide range of renewable energy converters into consumer energy formats in order to fully realize the potential of renewable energy of a country. Renewable energy systems included in the centralized energy networks of countries provide up to 30% of energy consumption (Murdock, Adib et al., 2018). Currently, renewable sources occupy a leading place in terms of growth of their share in world energy production. Among the various types of renewable energy sources, solar energy occupies a leading position in terms of growth rates, reaching a total (worldwide) capacity of more than 500 GW (Beauvais, Chevillard, Paredes, Heisz, Rossi & Schmela, 2019).

Today, the energy produced from photovoltaic modules is competitive with fossil fuel generating plants (Beauvais, Chevillard, Paredes, Heisz, Rossi & Schmela, 2019). However, there are still problems hindering their mass distribution. Among the restraining reasons, first of all, we note the significant initial capital investments, the need for the use of batteries for guaranteed power supply during periods of energy shortage and the complexity of effective systems based on smart grid networks. If the first reason is mainly solved by the state providing stimulating investments, the rest require the search for new solutions to simplify the system without significantly reducing its functionality. To solve such problems, an energy-saving photovoltaic system is designed, which allows led lamps to work in parallel with photovoltaic modules and General-purpose power supply.

BACKGROUND

The development of solar energy naturally occurred in the direction of low-power power systems to power plants with a capacity of tens and hundreds of megawatts. However, currently photovoltaic modules with a capacity of hundreds of watts have not lost their relevance. Especially widely autonomous systems are used in agriculture both in animal husbandry (Tursunov, Sabirov, Hwoan, Turdiev, Shirmatov & Yuldoshev, 2013) and in crop production (Ismanzhanov & Tashiev, 2016). The use of low-power systems in urban environments is due not least to the advanced development of energy-efficient low-current electronics and lighting equipment at a constant current. Solar photovoltaic modules provide electricity of this kind without losses for conversion. All this directly contributes to the development of systems where a significant part of the network operates on direct current (Dragičević, Vasquez & Guerrero, 2016). Switching to DC networks, reduces conversion losses and avoids additional devices, increasing reliability and reducing cost. Therefore, the power supply of the load with direct current from photovoltaic panels, it is expedient to carry out directly. Accordingly, in the presented photovoltaic system, the work of led lamps is carried out directly on a direct current, but in parallel to the led modules, the energy from the grid is also supplied. The paper presents in detail the approaches to such parallel operation of led lamps from photovoltaic modules and power supply. The approach of the proposed solution differs significantly from the developed similar solutions, which are based on supercapacitors, batteries as buffer tanks, controllers on a high-frequency transformer (Vieira & Mota, 2010; Ali, Lange, Elrayyah, Sozer, De Abreu-Garcia & Mpanda, 2018). The described solution allows to create cheap and reliable in comparison with analogs lighting system based on photovoltaic modules.

A team of scientists from the University of California, Los Angeles (UCLA) has developed a new transparent solar cell that is able to generate electricity and transmit light, which opens up the prospect of creating energy-generating windows. The new generation of polymer solar panels produces energy by absorbing predominantly infrared radiation rather than the visible part of the spectrum. At the same time, the photoactive solar cell has almost 70% transparency, which will integrate it into ordinary window glass.

Sharp introduced translucent solar panels in the domestic market of Japan. The design is made in the form of laminated glass without the use of metal. Solar cells partially transmit sunlight, and between the layers of glass can leave an air gap to improve the insulating properties. One panel of standard size 138x100x1.0 cm provides up to 95 W under favorable insolation conditions. This is about three times worse than traditional solar panels, but the relatively low power is offset by the fact that new panels can be used instead of window panes and even instead of walls in office buildings.

In the present work describes the results of the same PV module, but with the standard photoelectric converters. The module is a variant of the design of the energy-generating window with the shading of the solar energy flow, which can be used in tropical and subtropical climate. The study of the characteristics of the solar photovoltaic module mounted in the sash of the window frame and equipped with a block of lithium-ion batteries with a scheme of its precharge from this module.

ENERGY-SAVING SYSTEM OF WORKING AND EMERGENCY LIGHTING BASED ON PHOTOVOLTAIC MODULES

The use of photoelectric converters is growing from year to year. The widespread introduction of photovoltaic systems is mainly due to the reduction of their cost and the growth of electricity tariffs from traditional energy sources. The development of energy-efficient electronics and lighting devices with direct current also has a positive effect. When powered by photovoltaic devices, DC consumers do not lose energy on conversion, in contrast to the power from grid through the inverter. Therefore, if the consumers need direct current supply, it can be efficiently realized from semiconductor photoelectric converters. There is no doubt that this concept will gain in popularity as the cost of photovoltaic energy decreases.

A significant part of household devices, office equipment and other electrical equipment is powered by AC power, but it operates using direct current from built-in power converters. Technologically, these devices can operate using direct current, but in reality this possibility is not laid by manufacturers. Currently, the devices that can get supply from both the AC network and the DC network are rare, but the demand for such devices will grow. It is important that from a technological point of view, this does not require significant upgrading of devices. At the same time, these devices should work with parallel power supply from photovoltaic cells and the power grid. The work of such systems can occurs using different concepts, one of which is discussed in more detail below. The lighting system based on LED lamps acts as electrical load. This load is selected taking into account several factors: gaining popularity of this device as a light source, DC operation, the scheme of industrial lamps can be converted to DC power without significant effort, an additional energy source will increase the reliability of emergency lighting.

To assess the possibility of sharing photovoltaic and lighting devices, consider the structure of a lighting system. According to regulatory documents, artificial lighting is divided into the following types: working, emergency and standby lighting (Figure 1), which are functionally different from each other. Emergency lighting will be considered in detail because of it seems to be the most promising direction of use the developed system. This is due to the need for low-power light sources, since the requirements for unevenness and the illumination of it are significantly lower than the working light, as well as with the requirement for an additional power source. According to regulatory documents emergency lighting is provided in case of a power failure of the main (working) lighting and connects to the power supply, regardless of the working light. What is especially important for facilities whose power supply is from one source, i.e. of third category of power supply reliability, in connection with which there is a need for an additional power source. From the above, it follows that the developed system is optimally suited to the requirements.

Emergency lighting is divided into two types: evacuation and reserve lighting. We will not describe types of emergency lighting, they described in detail in the relevant regulatory documents. Let us consider in more detail the norms that should apply to the developed lighting system with regard to its use for emergency lighting. These standards include: normalized illumination on a horizontal surface; extreme unevenness of illumination; the order of inclusion and the duration of the emergency lighting.

The analysis of these data made it possible to note the fact that, reserve lighting places high demands on the battery capacity from types of emergency lighting, since the duration of its operation in the emergency is depend on the restoring of main lighting. This time in the worst case is determined by the arriving of the operational team to eliminate the consequences of the accident. As a result, this leads to the need of increasing the capacity of batteries for the developed system. In addition, it is necessary to allocate and excessive requirements for normalized illumination compared with other types of emergency lighting. The use of the developed lighting system as a backup lighting is not recommended, it can be effectively used for other types of emergency lighting. Accordingly, it is recommended to use the system for emergency evacuation lighting (Figure 1).

For evacuation emergency lighting networks, such a system will power emergency lighting with electricity from photovoltaic modules (PVM) and batteries. Nevertheless, evacuation emergency mode is a rare situation; therefore, it is necessary to ensure a more complete use of solar energy. For this, the system will have an energy-saving function. This function is the power supply of emergency lighting fixtures, working in conjunction with working lighting and some of the working lighting fixtures as on figure 2.

Figure 1. Classification of artificial lighting by function

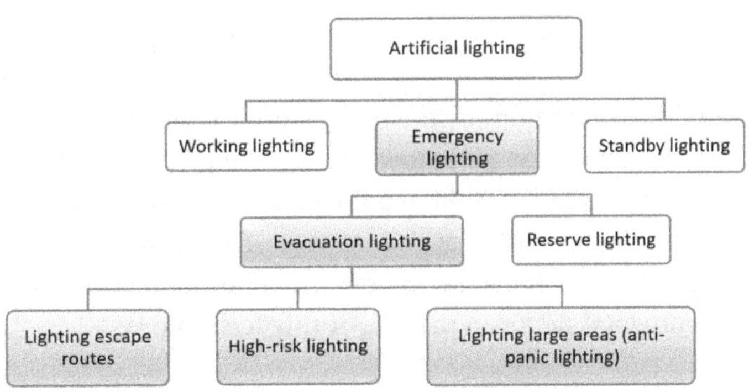

Figure 2. Work and emergency lighting system

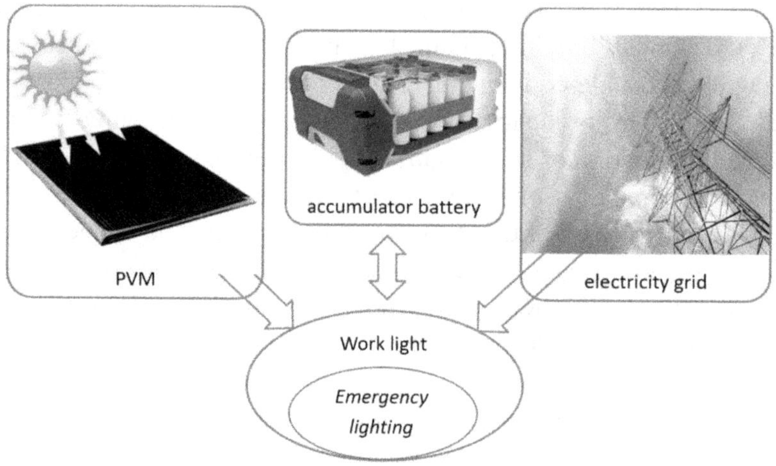

These lamps will work in parallel from the mains and from the PVM. With such a connection, it is important that the voltage from the PVM is higher than the operating voltage from the grid power supply unit. In this case, the power from the PVM will be spent on the lighting, when the output voltage decreases and in case of lack of power is compensated from the mains supply unit.

The work of lamps in the system can be organized in several ways. The first option is to directly parallel the connection of the PVM with the network power supply, in this case, a certain coordination of loadand generator powers, the output voltage of the PVM, and some other parameters is necessary. The scheme can be effective only in certain options. Another option is possible with the use of traditional AB charge controllers from PVM for powering LED light sources, but there may be problems with adjusting the output voltage for consistent work with the grid. The most effective option is to use a controller with MPPT technology (Maximum Power Point Tracking), specially designed for parallel operation of the PVM and the grid to the load with a constant current, but there are no such devices in the free market. An option that has many opportunities for the implementation of this task is the use of step-up and step-down converters with voltage and current stabilizers. There may be different variations of this scheme depending on the types of converters used, LED strips and other parameters.

Knowing the requirements for the developed system and the functionality of it work, we set its characteristics. In this type of system, the characteristics depend on many factors, the main of which are the operating modes, the criterion for choosing is the load power and the climatic factor characterized by the region of use. Consider these factors in more detail. Initially, we consider the choice of electrical power based on the criterion of the minimum cost of electrical energy. In this connection, we estimate the potential of energy sources, taking into account their cost. Then the system will be dominated by those sources that will provide a minimum of the energy cost. Currently, the cost of energy produced by solar panels in some regions competes with energy from the grid, providing a more attractive price. Taking into account the long-term trend of reducing the cost of photovoltaic modules and the growth of tariffs, the advantage of such systems will grow every year. Therefore, in such systems, the maximum should be provided with lighting from the solar energy. In addition to cost trends, it is important not to take into account the environmental factor in the use of this technology.

Based on the foregoing, the sun energy will be used effectively in the developed lighting systems. In the peaks of solar energy supply, all the lamps will work on solar energy. But peak production and peak consumption occur in different periods of the year, i.e. during production peaks (the period of the summer maximum) only part of the lamps work, therefore it is necessary to take into account this feature when calculating a certain load factor for the device during the summer period. The capacity of the batteries in the system will be minimized due to the inability to compete with the cost of electricity from the PVM and the power grid. The cost of electricity will be several times more expensive even when using of lithium iron-phosphate batteries, which are the most efficient in terms of the specific cost of the charge-discharge cycle.

Next, we consider the modes of operation and the influence of the climatic factor on the described system. The modes of operation of the system will be considered taking into account the energy balance equations (1), which will reflect the climatic resource of the system in a specific area. In order to avoid cluttering the equations, they do not reflect the energy losses for various conversions and currents of self-discharge of AB.

$$\begin{cases} E_{user(t)} \leq E_{PV(t)}; when: t_{s.w} < t < t_{en.w}; E_{PV(t)} \geq E_{user(t)} \\ E_{user(t)} = E_{PV(t)} + E_{net(t)}; when: t_{s.w} < t < t_{en.w}; E_{PV(t)} < E_{user(t)} \\ E_{user(t)} = E_{PV(t)} + E_{net(t)}; when: t_{s.w} < t < t_{en.w}; E_{PV(t)} < E_{user(t)} \\ E_{bat(t)} = E_{PV(t)} + E_{net(t)}; when: t_{en.em.w} < t < t_{en.ch.w}; c \leq c_{min} \\ E_{bat(t)} \leq E_{PV(t)}; when: t_{en.w} < t < t_{s.w}; c < c_{max} \end{cases}$$
$$in\ general: \quad (1)$$
$$C_{EE} \rightarrow min; c \geq c_{min.es}$$

where

$E_{user(t)}$ – electrical load for each hour t;
$E_{PV(t)}$ – electricity generated by photovoltaic devices for hour t;
$E_{net(t)}$ – electric power consumed from the power supply network of general purpose for hour t;
$E_{bat(t)}$ – battery charge (discharge) for hours t;
$E_{e.user(t)}$ – electrical load of emergency light sources for hour t;
$t_{s.w}$ – start time of the working day;
$t_{en.w}$ – end time of the working day;
$t_{s.em.w}$ – start time of the emergency mode;
$t_{en.em.w}$ – end time of the emergency mode;
$t_{en.ch.w}$ – end time of batteries charging after emergency mode;
C_{EE} – cost of electrical energy;
c – battery capacity;

$c_{\min}$ – minimum battery capacity that is sufficient for lights operation in emergency mode;

$c_{\min.es}$ – estimated minimum battery capacity that is sufficient for lights operation in emergency mode, taking into account the reduction of its characteristics at the end of life;

$c_{\max}$ – maximum battery capacity.

We will describe the operating modes of the system in the day period, we will start from the beginning of the work day, when the working room lighting starts to function. Together with the light, the developed system starts working. At the same time, during periods of high insolation, the controller directs the electricity generated from the PVM to the powered lamps. When the amount of solar energy is significantly reduced, the energy deficit is replenished from the power grid, due to parallel operation. The energy balance of these modes is described by the first two expressions of the equations system(1). In case of emergency in the power grid, the power supply of working lights is terminated, and the PVM and AB starts to supply to the operation of emergency lights. This mode corresponds to the third expression of the equation system (1). After the returning of grid power supply, the process of charging AB from the PVM begins. If its power is not enough, then the deficit is filled by the power grid. In case of unforeseen situations, the same charging occurs when their capacity is below the permissible level. The energy balance of such a mode is described by the fourth expressions of the equation system (1). The next mode of operation begins after the end of the work day and lasts till the start of the work day. There is no energy consumption, and the energy generated by the solar cell is sent to the battery for compensating the self-discharge. The energy balance of this mode is described by the fifth expression of the equation system (1).

Thus, above we have considered fundamental modes of system operation, taking into account the criterion of the minimum energy cost. These modes provide the most complete use of solar energy during a working day with minimal additional cost, allowing the consumer to save on electricity and increase the reliability of the emergency lighting system.

In the discussed above operation modes, the potential of solar energy is not fully realized. By little increasing the capacity of the batteries, the potential can be brought closer to the maximum. The equations describing this operation mode are represented by the equations system (2).

$$\begin{cases} E_{user(t)} = E_{PV(t)}; when: t_{s.w} < t < t_{en.w}; \\ E_{user(t)} = E_{PV(t)} + E_{bat(t)}; when: t_{s.w} < t < t_{en.w}; E_{PV(t)} \left\langle E_{user(t)}; c \right\rangle c_{min} \\ E_{user(t)} = E_{PV(t)} + E_{net(t)}; when: t_{s.w} < t < t_{en.w}; E_{PV(t)} < E_{user(t)}; c \leq c_{min} \\ E_{user(t)} + E_{bat(t)} \leq E_{PV(t)} when: t_{s.w} < t \left\langle t_{en.w}; E_{PV(t)} \right\rangle E_{user(t)} \\ E_{e.user(t)} \leq E_{bat(t)} + E_{PV(t)} when: t_{s.em.w} < t < t_{en.em.w}; \\ E_{bat(t)} = E_{PV(t)} + E_{net(t)}; when: t_{en.em.w} < t < t_{en.ch.w}; c \leq c_{min} \\ E_{bat(t)} \leq E_{PV(t)}; when: t_{en.w} < t < t_{s.w}; c < c_{max} \end{cases}$$

$$in\ general: \qquad (2)$$

$$P_{PV} \rightarrow max; C_{PV} \rightarrow min.$$

where

P_{PV} – potential of usage the generated photovoltaic energy;
! $_{PV}$ –cost of photovoltaic energy.

The difference with previous version is in the operation mode of the system, covering the period of time before and after the work day. In the period from sunrise to the beginning of the working day, electricity from the solar cell is supplied to the battery, as well as in the period after the end of the working day and before the sun goes down. The energy balance of this mode is described by the seventh expression of the equation system (2). For saving energy in this mode, the battery capacity must be increased. The stored energy in AB energy during periods of lack of energy from the PVM will be supplied partially or completely from the AB. This mode is described by the second expression of the equations system (2). When the capacity of the battery decreases to a certain level, the lack of energy will be replaced by the power grid, which is described by the third expression of the equations system (2). The minimal battery capacity below which there will be no discharge is determined by the energy that is needed to operate the emergency lighting system. During periods when the power generated by the photovoltaic modules will exceed the load, the excess energy will be supplied to the batteries. This mode is described by the fourth equation in the equations system (2). It is worth noting that the increased battery capacity compared with the first option allows you to compensate for fluctuations in the generated and consumed power.

The remaining modes that is not described by the system of equations (2) are similar to the modes of the equations system (1).

Thus, there will be a system that realizes the full potential of solar energy generation by photovoltaic panels, subject to the minimum cost of photovoltaic energy and guaranteed power supply of emergency lights. But it is worth noting that this option will be more expensive than the first one, taking into account both the initial capital costs and the cost of electricity, due to the high cost of AB.

To solve optimization problems that reflect the operation modes for (1) and (2)equations systems, it is necessary to determine the average daily parameters of the system for a particular object, then calculate the monthly average. Further, knowing the average data of the solar energy potential for a certain area, it is necessary to solve the optimization problem for the object under consideration, taking into account the selected criteria and based on the selected mode of an energy-saving system of working and emergency lighting based on photovoltaic modules. Solving such a task requires a lot of data and a large amount of labor resources.

However, the analysis of current data allows summarizing some characteristics for the noted use cases without solving an optimization problem. Let's start with the minimum power of photovoltaic conversion plants. It will depend on the cost of the photovoltaic module and controller, which will implement the algorithm of a certain variant of operation. The cost of the installed system based on such a controller and PVM must be competitively capable with the cost of the power grid energy. It is also recommended that the minimum power in standard insolation conditions provide the rated charging current of the used type of battery. In terms of the minimum battery capacity of, the capacity should be sufficient for the minimum duration of the emergency lamp operation time, taking into account the decrease in the battery capacity at the end of its life. For most of the options recommended above, the duration of emergency lighting will be one hour, and in the case of lighting of high-risk areas, this time is usually tens of minutes.

Next, we will take a closer look at the work algorithms and approaches for building a parallel operation of a photovoltaic module together with a electrical grid. The work of such systems can be implemented in various ways, but the most effective way is to use a special controller that will take the maximum power from the PVM (due to MPPT technology) and synchronize the power of two sources, one of which is the power grid. It is not possible to find such controllers in the free market; therefore, to evaluate the effective operation of such lighting systems, to work out algorithms and modes of operation with a grid, we developed a scheme of operation without the controller described above (see Figure 3).

In the developed scheme (Figure 3) the power is supplied to the lamp in parallel from the PVM and from the network power supply. It is important with this connection that the voltage on the PVMis higher than the voltage of the power supply unit. In this case, the PVM will supply the lamp and when the output fails, the voltage drops and the grid power supply unit compensates the lack of power.

In such an operation algorithm, we select the operation mode in which during periods of low load of the lighting system and high PVM electric power generation there are moments when it is necessary to reduce the voltage on the lamps in order to protect them from burnout. To this end, the circuit uses a voltage regulator at the output of the PVM (see Figure 3). It is desirable to select a pulse voltage stabilizer, because of the efficiency of it is significantly higher. At the LEDs supply voltage of 12 V, it is better to select the stabilizer with input parameters up to 24 V in order to increase the energy generation by the PVM on a bright sunny day, when the output voltage on them reaches 17-18 V.

Photovoltaic modules, as a backup power source, should be used primarily for the emergency lighting, therefore the original circuit was transformed for such a network. Figure 4 shows the schematic diagrams of the system based on the approach proposed in Figure 3.

Figure 3. Schematic diagram of the combined working lighting system

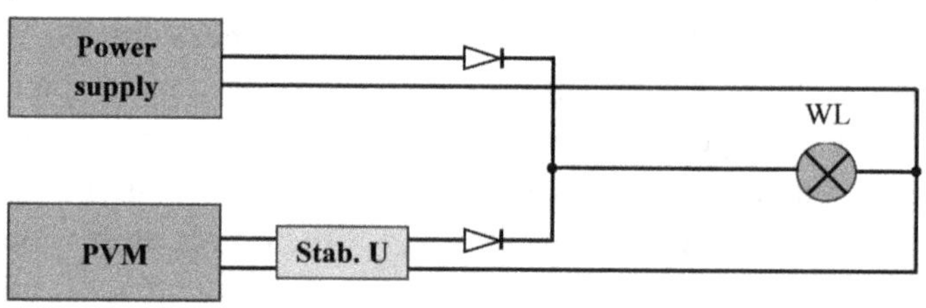

Figure 4. Schematic diagram of the combined emergency lighting system

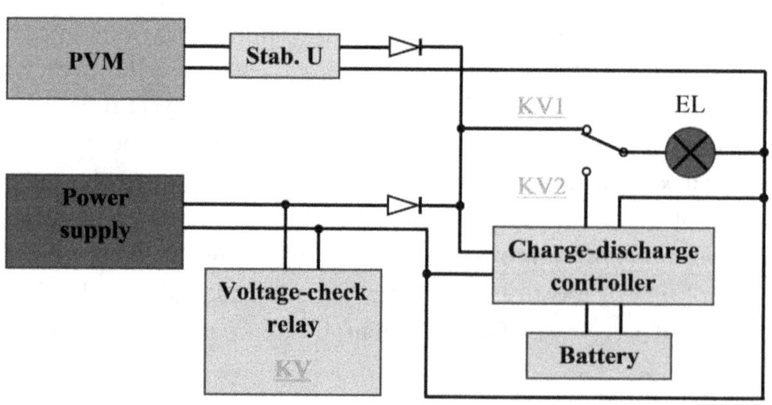

Due to the presence in the circuit of the voltage control relay, the lamp operates in emergency mode when there is no voltage from the main source (Figure 4). If the grid voltage fails, the relay disconnects the direct power supply from the PVM and the power supply grid and connects to the power supply from the battery. At such times, in the case of PVM power generation, the emergency lighting is supplied from the charge-discharge controller of the battery, thereby increasing the reliability of the power supply to the emergency lighting system. During periods when the power supply is restored, the relay switches and restores the working power supply circuit of the lamps from the PVM and the power grid with a parallel charging of the battery.

In the case of the substantial capacity of a battery of photovoltaic modules, it is possible to energize together the system of working and emergency lighting (Figure 5).

Power of emergency and working lighting in the developed system is implemented as described below. The operation of the circuit is constructed in such a way that the energy produced by the photovoltaic cells is directed primarily to the work of emergency lightings. During the period of high insolation, the electric power is supplied to the emergency and working lamps. This is achieved by slightly lowering the output voltage on the power supply unit No. 1 for emergency lighting (see Figure 5) in comparison with the power supply unit No. 2. A separate power supply unit should be used for emergency lighting also in terms of safety requirements. Since the power supply of work lamps and emergency lighting is implemented from different inputs inside the building, in the case of the object power supply in the second category of reliability.

The use of rechargeable batteries in such systems (Figure 3, 4, 5) is limited, since the cost of energy stored in them cannot yet compete with energy either from PVM or from the power grid. In this regard, AB in the system are used for uninterrupted power supply of emergency lighting when operating in emergency mode. They provides the minimum working time of the emergency lamps depending on the type of emergency situation.

It is worth noting that the diagram figure 3, 4, 5 uses a typical 12 V power supply to power a low-power LED strip that does not require a current-stabilized power source. For higher power systems, other power supplies are used with output current stabilizers, in which the voltage depends on the connected load. Circuit with power supply with stabilized current requires a slightly different approach, its principle of operation will be reflected in the following works.

Figure 5. Schematic diagram of the combined system of working and emergency lighting

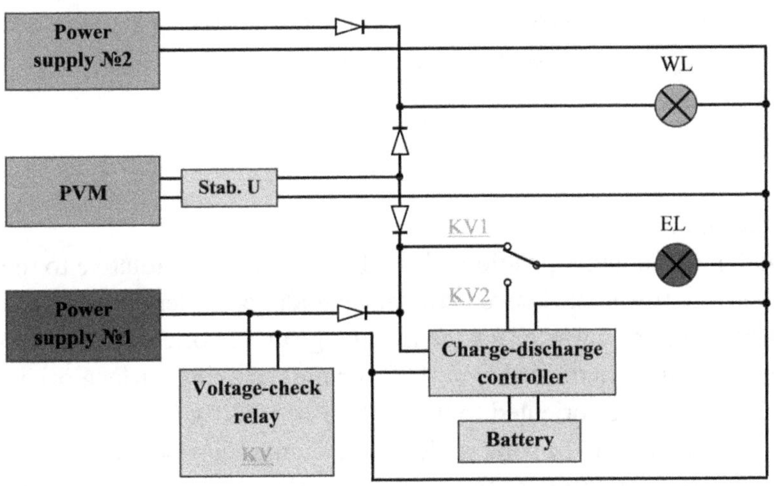

Based on the described algorithm and the principle of operation of the system, it is necessary to single out the simplicity of the organization of such systems and the possibility of efficient operation of photoconverters with a power grid without batteries, while using minimal additional devices. Such a system ensures the removal of solar energy at the level of power generation in the classical system with a PWM controller and AB. If we take into account that during periods of high battery charge the energy consumption from photovoltaic cells is greatly reduced, then in the developed system the load is relatively constant, in this connection, the efficiency of using of solar energy in the proposed solution will be higher. It is necessary to properly evaluate the power of the lighting load and photovoltaic modules. It is desirable to select the load relatively constant during the day, as a rule, these are lamps installed in rooms where natural light is limited for various reasons (shopping centers, basements, etc.).

Despite the advantages of the proposed solution, the issue of complete power removal from photovoltaic cells remains relevant. Solving this issue requires the use of specialized controllers based on the technology of searching for the maximum power point, the so-called MMPT (Maximum Power Point Tracking) controllers, but their advantages are significant only on high power systems, because of their high price. For the same cost, it will be more profitable for small systems to increase the power of the photovoltaic part due to additional photoconverters than will increase it with the use of the MPT controller. In this regard, we note that if there is a low-power system, then the work according to the developed algorithm will be optimal. If the system is over a dozen kW, then it is worth considering the possibility of using such a controller. We note that for such specific load thr controllers is not available in free market and its development is necessary.

Thus, the advantages of using the proposed algorithms of work are in the efficient use of solar energy at the minimum cost of its implementation.

THE DESIGN OF THE WINDOW PHOTOVOLTAIC MODULE

As photoelectric solar radiation converters in the module, one-sided flexible solar cells (ESS) of 125x125 (mm) with a peak power of 3.4 W were used.

The arrival of solar radiation on variously oriented inclined surfaces is calculated using methods for determining the angle of sunlight incidence on the plane at a given point in time - j (Duffie & Beckman, 1991; Erbs Klein & Duffie, 1982; Bortsov & Potovskaya, 2013):

$$\begin{aligned}\cos j = \sin\delta\cdot\sin\varphi\cdot\cos S - \sin\delta\cdot\cos\varphi\cdot\sin S\cdot\cos y + \cos\delta\cdot\cos\varphi\cdot\cos S\cdot\cos\omega \\ + \cos\delta\cdot\sin\varphi\cdot\sin S\cdot\cos y\cdot\cos\omega + \cos\delta\cdot\sin\gamma\cdot\sin S\cdot\sin\omega\end{aligned} \quad (3)$$

where

φ is the latitude of the area;

δ is the solar declination (the angular position of the Sun at solar noon relative to the equatorial plane);

S is the slope of the surface (the angle between the plane under consideration and the horizontal plane);

γ is the azimuth of the receiving surface (for the receiving surface oriented to the south $\gamma = 0°$, for the receiving surface oriented to the east $\gamma = -90°$, for the receiving surface oriented to the west $\gamma = 90°$, for the receiving surface oriented to the north $\gamma = \pm 180°$);

ω is the hour angle (measured from the solar midday, the angular displacement of the Sun relative to the local meridian, due to the rotation of the Earth and corresponding to 15 ° per hour).

Substituting the values of optical efficiency-η_{opt} and receiver efficiency -η_{pr}, generated energy of solar cells takes the following expression:

$$W_{PV} = E_0 \cdot \tau \cdot \cos j_{av} \cdot \eta_{opt} \cdot \eta_{pr} \quad (4)$$

where

E_0 is the standard solar illumination (AM-1.5),
τ is the working time of the solar cell,
j_{av} - the average angle of incidence of sunlight on the plane under consideration.

To calculate the annual rate of monthly performance of vertical photovoltaic panels (table 1, in kWh per kW of installed peak power of the solar module, designation kWh, kW), the developed computer model was used to estimate the insolation of various static or tracking surfaces. The desire for a detailed account of the dynamics of changes in insolation and the possibility of applied actinometry determined the choice of the hourly interval for modeling the illumination conditions of the working surface of the solar battery.

A computer program allows you to simulate the daily variation of the direct, diffuse, and reflected components of solar radiation for the average day of the month based on monthly sums of direct and diffuse (or even just total) insolation of a horizontal surface. In this case, the average long-term sums of solar radiation for the Moscow region were used. An assessment was made for various orientation schemes for the magnitude and time of solar energy entering the panel surface. For comparison, in table. 2 there are shown the calculated data on the performance of horizontal and inclined photovoltaic modules. Similar calculations can be carried out for other regions (Rakhimov, Sadullaeva, Kolomiets, Tashmatov & Usmonov, 2017; Natsheh, 2016).

To calculate the annual rate of monthly performance of vertical photovoltaic panels (table 1, in kWh per kW of installed peak power of the solar module, designation as kWh, kW), the developed computer model was used to estimate the insolation of various static or tracking surfaces. The desire for a detailed account of the dynamics of changes in insolation and the possibility of applied actinometry determined the choice of the hourly interval for modeling the illumination conditions of the working surface of the solar battery.

When developing a computer program, the calculation method was used in accordance with equation (1), which determines the arrival of solar radiation on variously oriented surfaces during the hour, day, week, month, year, depending on the angle of inclination, latitude of the terrain with the introduction of the long-term solar average radiation for the Moscow region. The tables show the average long-term values of the energy illumination, hourly, daily, monthly and annual amounts of radiation and on a horizontal surface the statistical characteristics of the temporal variability of the amounts of radiation. Hourly and daily amounts received on the schedules of the daily rate. Monthly sums were calculated as the product of the daily value by the number of calendar days in the month, the annual is as by summing the monthly sums. The difference between the sum of total and direct radiation can be calculated as sum of the scattered radiation. The error in calculating the correlation coefficient for the used length of the observation series was 15–25%.

A computer program allows you to simulate the daily variation of the direct, diffuse, and reflected components of solar radiation for the average day of the month based on monthly sums of direct and diffuse (or even just total) insolation of a horizontal surface. In this case, the average long-term sums of solar radiation for the Moscow region were used. The program features allow you to evaluate various orientation schemes by the magnitude and time of solar energy entering the panel surface.

For comparison, Table 2 shows the calculated data on the performance of horizontal and inclined modules.

Estimates of monthly power generation allow approximately determine the size of the photovoltaic module for charging the battery during daylight hours depending on the orientation of the module and the features of the battery charging scheme. For practical application and implementation of specified technical requirements, the data in Table 1 were recalculated (Figure 6) into the required area of the photovoltaic module (left ordinate axis). The number of the solar cells corresponding to this area is shown on the right ordinate axis.

These dependences show that in the period from March to September, six solar cells provide the required power generation with the module in a vertical position almost independently of the geographical orientation in the sector 180 degrees from east to west across the south. In the remaining months, the dependencies diverge quite significantly and in December for the module of southern orientation it is necessary to use of 22 solar cells, while for the east (or west) directions the vertical photovoltaic module should contain at least 47 solar cells. A comparison of the data in Tables 1 and 2 shows that for an attic window, the flaps of which may be at a more favorable angle to the front of the solar radiation, the situation is noticeably improving. In this case, you can proportionally reduce the area of the photo module or get more photovoltaics.

Table 1. Estimated monthly power generation (kWh, kW) by verticalsolar panels for the Moscow region

Panel Orientation	1	2	3	4	5	6	7	8	9	10	11	12	Year
South orientation	13.5	36.1	66.0	60.1	71.2	66.5	70.7	66.4	54.3	36.7	24.9	15.9	582.3
Southeast (Southwest)	12.3	30.9	58.7	58.0	74.9	71.9	74.7	66.3	48.8	30.6	19.7	12.9	559.8
East (West)	9.8	22.1	45.8	49.1	71.8	73.2	73.3	58.8	37.8	20.3	10.8	7.4	480.3

Table 2. Estimated monthly power generation (kWh, kW) by horizontal solar panels and installed at an angle to the horizon for the Moscow region

Panel Orientation	1	2	3	4	5	6	7	8	9	10	11	12	Year
Horizontal surface	11.5	24.2	55.5	77.7	112.9	116.7	116.4	91.0	58.0	29.0	13.0	8.2	714.2
South orientation, incline 35°–40°	13.5	33.6	70.2	86.4	115.3	114.3	116.2	98.4	69.1	39.5	22.2	13.6	792.4

Figure 6. Monthly dependence of the size of the photovoltaic module for charging a block of lithium-ion batteries to 1/3 of its capacity during one light day; on the right shows the corresponding number of solar cells

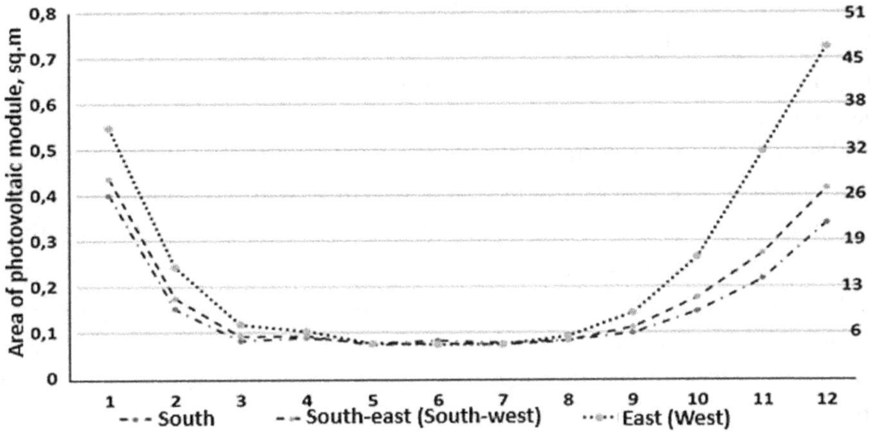

The decision on the choice of the number of used solar cells obviously meets conflicting requirements - on the one hand, the need to maintain the charge of the battery during the winter months requires an increase in their number, but on the other, it reduces the light opening through the sash. In addition, to operate the battery charging circuit with a double lithium-ion battery voltage, it is necessary to have a photo module output voltage not lower than 9.2 V, i.e. 16 elements must be connected in series. Based on the analysis carried out for the developed design of the window photovoltaic module, 16 solar cells with a size of 125x62 (mm) were used, the total area of which is 8 standard solar cells. This allowed, as can be seen from Figure 1, to meet the requirement of charging the battery from February to October for most orientations, with the exception of those close to the east and west.

SEALING MODULE ELEMENTS

The choice of technology for sealing a photovoltaic module mounted in a sash is one of the most important structural decisions. The main factor here is the chemical composition of the sealing material. This technology must meet the following requirements:

- Have reliable adhesion to the materials of the surface of the ESS to ensure high-quality optical and thermal contact in order to minimize reflection from the front surface of the ESS and a stable heat sink;
- Be light and heat resistant when exposed to solar radiation and heat reheat;
- Have no structural elements and background impurities that have a destructive effect on the materials of the photoconverter surface and switching elements, reducing its service life;
- Protect of the photo transducer from mechanical effects and moisture;
- Be environmentally safe.

The most widely used currently received thermoplastic filler based on a copolymer of ethylene vinyl acetate (EVA). However, this material has drawbacks in almost all the above requirements, either immediately or after a certain period of operation. For these reasons, in the developed design, we propose a two-component liquid polysiloxane compound for sealing, which is structured at room temperature by hydrosylation in a low-modulus gel. The basics of this technology were developed by a group of I.S. Persits under the direction of D.S. Strebkov (Persits, Potapov, Strebkov& Chekhunina 2006). Studies showed the advantage of this technology over EVA. As a basis for the developed technological process, the technology of manufacturing a vacuum double-glazed window with a thermoplastic spacer along the perimeter was adopted. The pre-vacuumed cavity is filled with an optical medium in the form of a two-component liquid polysiloxane compound, knocked at room temperature by the mechanism of hydrosilication into a low-temperature gel.

The essence of the proposed sealing method in the developed design is that the polysiloxane compound fills the space with solar elements placed in it between the back side of the outer window glass and the polycarbonate sheet attached to it with thermoplastic spacer. Polycarbonate sheet (optional technologic operation) is attached to the rear side of the photoconverters to increase the gap between the glasses and does not affect the photoelectric characteristics. The scheme of the described construction is shown in Figure 7.

Figure 7. Scheme of a glass pane window sash with photovoltaic module: 1 - external glass pack; 2 - polysiloxane gel; 3 - polycarbonate; 4 - internal glass; 5 - hermetic double glazing; 6 - line of photoconverters (end); 7 - thermoplastic spacer; A - the direction of solar radiation falling on the solar cell; B - the direction of solar radiation entering the light aperture of window sash.

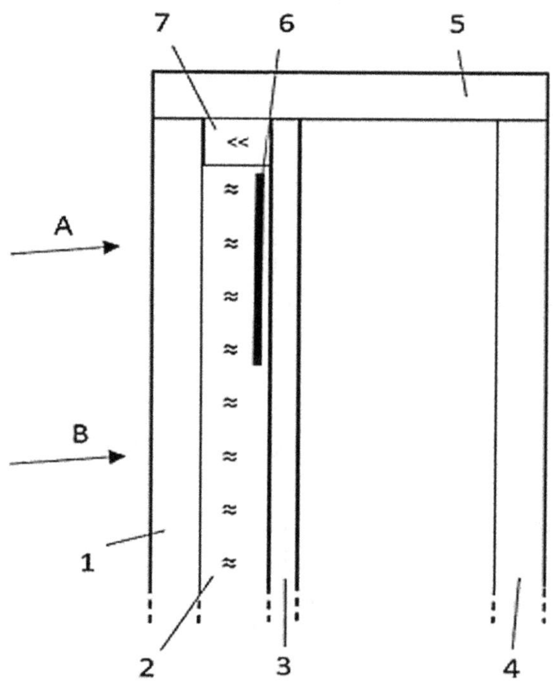

From the above diagram, it is clear that the gel envelops the ESS and is located throughout the entire area of the light opening. This allows for efficient heat sink (thermal conductivity coefficient of a polysiloxane gel is of 0.18 W / (m K) against of 0.13 W / (m K) of EVA. Gel significantly better damps mechanical stresses with thermal cycles — linear thermal expansion coefficient is of 2.5 • 10-4 K-1 and 4 • 10-4 K-1, respectively, and elastic modulus is of 0.006 N / mm^2 and 10 N / mm^2.

OPTICAL CHARACTERISTICS OF THE GLASS IN THE SPECTRAL RANGE OF SOLAR RADIATION

The optical characteristics of this multilayer structure are of obvious interest. For this, the reflection coefficient in the direction A and the transmittance in the direction B were measured. The results of measurements on the SPECORD 205 spectrophotometer are shown in Figures 8 and 9.

It can be seen from the spectral dependences that the reflection coefficient of the "external glass – gel – photoconverter" structure is 3%; therefore, more than 90% of the solar radiation energy is absorbed by the photo transducer (the long-wavelength self-absorption edge of silicon at room temperature reaches a wavelength of approximately λ = 1050 nm). Figure 9 shows the transmission spectrum in direction B (curve 1) and the transmission spectrum of a standard single-chamber glass unit (two glasses, curve 2).

The main difference between these curves is the shift of the short-wave edge of the spectrum to the long-wave side in the region of λ = 400 nm for a structure with a polysiloxane gel and polycarbonate. However, this does not affect the illumination inside the room, since this shift occurs in the ultraviolet region of the spectrum, where the photosensitivity of the human eye is almost zero. The eye sensitivity is less than one percent at λ = 400 nm, only 3-4% at λ = 450 nm and increases to 100% in the green region of the spectrum at λ = 550 nm.

Figure 8. The reflection spectrum in the direction A of the window sash

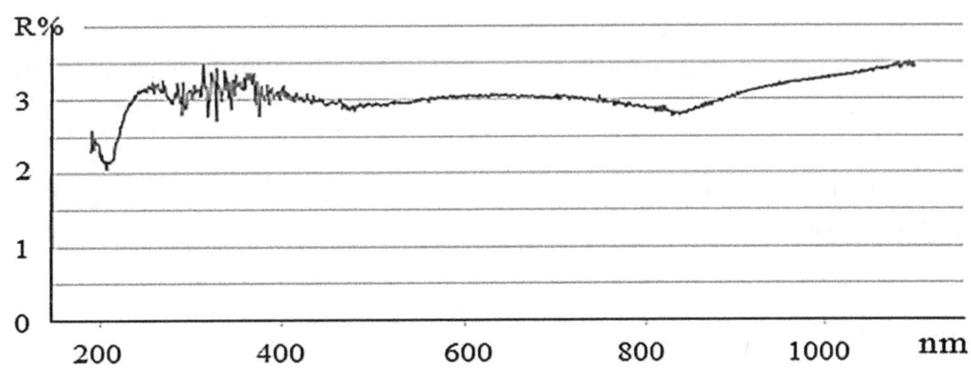

Figure 9. Transmission spectrum: 1 - in the direction of the B sash with a photovoltaic module; 2 - standard single-chamber glass unit

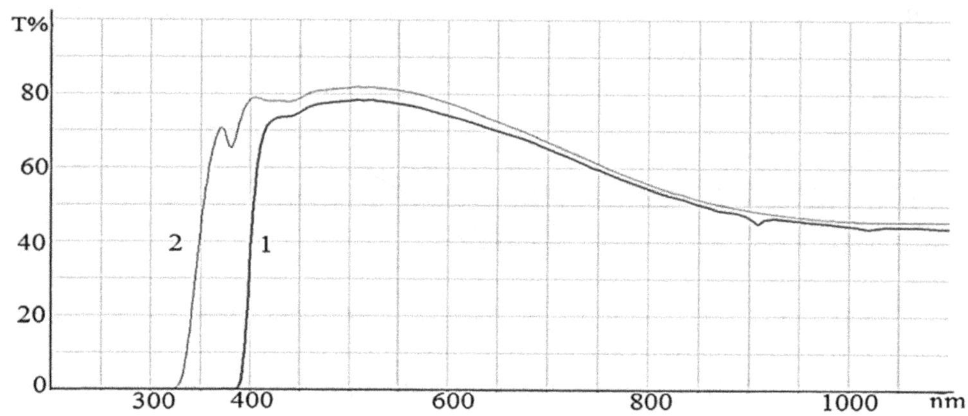

THE DESIGN OF ELECTRONIC CIRCUITS FOR RECHARGING A LITHIUM-ION BATTERY AND POWER SUPPLY

In order to fulfill the technical requirements for ABP in terms of output voltage (a pair of batteries) and capacity (two such pairs connected in parallel), NCR 1850B batteries were used.

To convert the current and voltage of a solar module, an electronic circuit has been designed and manufactured (Figure 10) consisting of two electronic modules (Ali, Lange, Elrayyah, Sozer, De Abreu-Garcia & Mpanda, 2018) of a pulse conversion and a battery. The voltage of the photoelectric converter 1 is 9.0 - 9.3 V, it supplies the electronic module 2, the main function of which is to charge the battery 3 operating in a buffer mode with a voltage of 7.4 - 8.4 V. The electronic module 4 converts constant battery voltage to provide a load 5 with a stabilized output voltage of 5.25 V. The maximum load current is 3 A.

Electronic modules 2 and 4 are made according to an identical electrical circuit (Figure 11) and are adjustable switching regulators, the level of output voltage is regulated by a variable resistor R1

A resistor of 47 kΩ sets the standard voltage level of the lithium-ion battery of 4.2 V per cell.

Figure 10. Block diagram: 1 - photovoltaic module; 2 - charging scheme; 3 - battery; 4 - load power circuit; 5 - USB connector

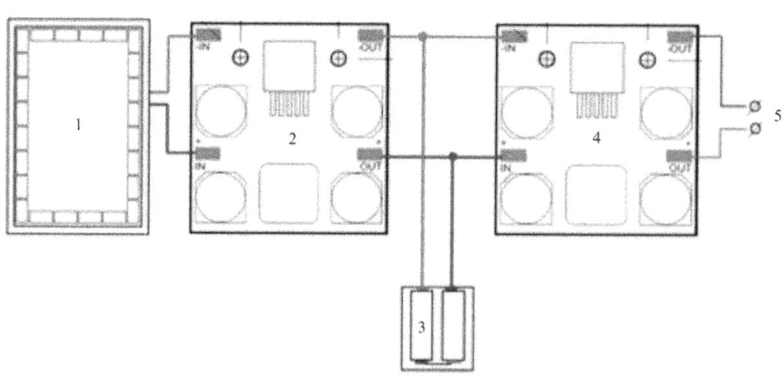

Figure 11. Electrical circuit of the electronic module 2 with variable resistor of 47kΩ

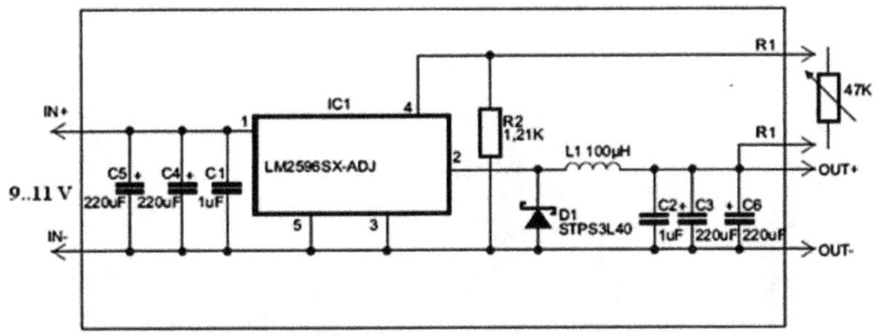

To charge the batteries of mobile phones, smartphones, tablets, etc., a USB 2.0 connector with an output voltage level of 5.2 - 5.25 V is used. For this, in the circuit (Figure 11) of the electronic module 4 a resistor R1 with a nominal resistance of 3.6 - 3.9 kΩ is used.

The modules are made on a PCB with an aluminum substrate, which allows removing the output current up to 2 A without additional heat sink, unlike linear stabilizers. The device has thermal protection and output current limiting from 3 to 4 A depending on temperature. Output voltage cannot exceed the input voltage. Table 3 shows the nominal technical characteristics of the electronic module, which can be applied to a wide range of photovoltaic power capacities and battery voltages.

After assembling the entire system, studies were made of the dependence of the charging current of the ABP on the illumination P (W / m2) at a different level of the initial voltage at ABPU (V). The flux density of solar radiation was measured with a BenningSUN2 tester. The results of these studies are shown in Fig. 8. As expected, the charge current of ABP decreases with increasing of initial charge of the battery. However, with a low level of charge ABP in low light conditions, the charge current is sufficient to maintain the charge at the level of 1/3 of the capacity. Thus, technical requirements are met. The design of a photovoltaic module embedded in a standard sash of a window frame is shown in figure 12, where a) is the view from the outside, b) is the view from the inside.

Table 3. Technical characteristics of the electronic module

Parameter	Value
Input voltage	Less than 40 V
Output voltage	1,2 – 37 V
Output current over the entire voltage range	Less than 3 A
Output current limits	3 – 4 A
Frequency conversion	150 kHz
Efficiency when Uin = 25V, Uout = 12 V, Iout = 3 A	90%
Operating temperature	-40 – 105°C
Module size	43x40x12 mm
Module weight	15 g.

Figure 12. Window shutter with built-in photovoltaic module: a - view from the outside; b - view from the inside; 1 - solar cells - view from the front and the rear side; 2 - LED strip; 3 - electronic unit with built-in battery and USB connector

The module is an ESS tape mounted around the perimeter of a window sash and works to charge a lithium-ion rechargeable battery located in the electronics unit along with an electronic charging circuit. The LED strip is fixed on the frame, which is switching on by the pushbutton on the electronic unit. A sensor for charging the battery is mounted here. A significant advantage of this development is that the photovoltaic module being placed in a glass unit, does not require additional constructions for sealing and housing design, as well as devices for placement and fastening. During the tests, the system showed the following characteristics.

1. Peak photovoltaic power 24 W
2. Maximum battery capacity 6.8 Ah (3 smartphones)
3. Output voltage 5,2 - 5,25 V
4. Light shading by solar cells 37%
5. The dependence of battery charging time from 1/3 of its capacity during different weather and time of year is shown in Figure 13.

Figure 13. Experimental dependence of battery charging time from 1/3 of its capacity during different weather and time of year

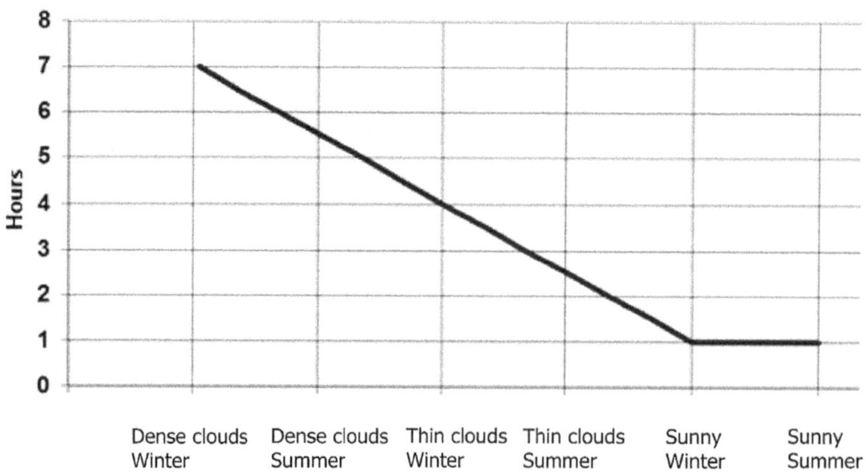

The tests of the developed system performed in natural conditions (built-in double-glazed windows in the window unit) confirm the calculated data for the time of battery charging during different weather and time of year, as it shown in Figure 13.

Taking into account the deviations of total and direct radiation in the tables of long-term values of energy illumination, temperature characteristics affecting the generation of electricity for supplying the battery charging scheme, the number of solar cells in real systems with different geographic orientation is increased by 1.5-2 times.

FUTURE RESEARCHES

The use of photovoltaic energy as an additional source for led lamps requires a thorough analysis of the effectiveness of the developing system. In this analysis, it is necessary to consider the placement of the building, the shading of windows and the rooms, taking into account external factors, the load factor of lighting and the lamps power. These factors will be used during the calculation of power of the entire photovoltaic system parallel to the power supply network of led lamps. Such an analysis will make it possible to prepare a methodology by which projects on such systems will be developed.

In the future, a pilot study of the parallel power supply system of lamps will be carried out to confirm the correctness of the developed technical solutions and identify weaknesses. These studies will help to work out the optimal modes of operation and the need for further adjustments in the system.

An important component of future work relates to the territorial potential of such a system. It is necessary to assess in which areas the use of the system is appropriate and will bring economic benefits.

For the photovoltaic thermal module embedded in the standard double-glazed window sash it should performed experimental studies in tropical and subtropical climate to assess the prospects of its application in this climatic conditions.

CONCLUSION

The concept of parallel operation is also applicable to other consumers operating using direct current. For example, various chargers can be connected to this system. The ability of the system to work together with the inverter is interesting. In such a bundle, some of the energy will be spent directly on DC consumers, and a part of it will be transformed through an inverter into a grid.

The developed system, taking into account the development trends of photovoltaic cells and rising energy prices, already today has good prospects for application at various enterprises. It is advisable to use it as an emergency and working lighting system, increasing the potential of using solar energy and the reliability of power supply for lighting installations, as well as reducing energy costs. It is necessary to competently approach the choice of load, which should be the building with the rooms where artificial lighting is necessary during the working day. As an example, we note shopping centers and buildings with all sorts of basement rooms. In addition, the use of additional energy sources increases the reliability of both emergency and working lighting.

The climatic factor of using systems based on renewable energy needs to determine the characteristics of the developed lighting system, taking into account this factor and the characteristics of consumers. In addition, the characteristics of the system affect the modes of operation. For the system, two variants of modes have been developed that will ensure the effective use of the produced photoelectric energy under the selected criteria.

During periods when the arriving of solar energy is significantly reduced, the energy deficit is replenished from the power grid, due to parallel operation. For some regions, in the winter period, it is necessary to provide for the compensation of the battery charge at night from the power grid during the period of the discount rate.

During production peaks (the period of summer maximum loads), only a part of the lamps work on some objects; therefore, it is necessary to take this feature into account when applying a certain load factor for an enterprise in the summer period. The capacity of the batteries in the system will be minimized due to the inability to compete with the cost of electricity from the PVM and the power grid.

The designed solar photovoltaic module built into a standard double-glazed window sash size 730x700 (mm) is designed for charging a block of lithium-ion batteries with a capacity of 6.8 Ah with an output voltage of 5.25 V, whose energy can be used to power any device that has a USB 2.0 connector. The technology of its sealing is described.

The required peak photovoltaic power of the module are of 24 W, the maximum capacity of the battery pack is 6.8 Ah, the output voltage is 5.2 - 5.25 V (3 smartphones) when the light is shaded by solar cells 37%. Dependence of battery charging time on 1/3 capacity from weather and time of year.

The algorithm of development, creation and research of built-in solar photovoltaic modules used in this work can be applied to other low-power systems.

Thus, the possibility of creating a design of window photovoltaic modules for low power supply systems located in virtually any geographic orientation in regions with year-round sunrise and sunset, i.e. standard window acquires new useful quality.

REFERENCES

Ali, A., Lange, J., Elrayyah, A., Sozer, Y., De Abreu-Garcia, J. A., & Mpanda, A. (2018). A Hybrid Flyback LED Driver with Utility Grid and Renewable Energy Interface. In *33nd Annual IEEE Applied Power Electronics Conference and Exposition*. (pp. 3377-3384). San Antonio, TX: IEEE. 10.1109/APEC.2018.8341588

Beauvais, Chevillard, Guillén Paredes, Heisz, Rossi, & Schmela. (2019). *Global Market Outlook For Solar Power*. Brussels, Belgium: Solar Power Europe.

Bortsov A.V. & Potovskaya E.I. (2013). Opredeleniye intensivnosti solnechnogo izlucheniya dlya naklonnoy prinimayushchey poverkhnosti [Determination of solar radiation intensity for an inclined receiving surface]. *Visnik NTU "KHIII"*, *59*(1032).

Dragičević, T., Lu, X., Vasquez, J. C., & Guerrero, J. M. (2016). DC Microgrids-Part I: A Review of Control Strategies and Stabilization Techniques. *IEEE Transactions on Power Electronics*, *31*(7), 4876–4891.

Dragičević, T., Lu, X., Vasquez, J. C., & Guerrero, J. M. (2016). DC Microgrids–Part II: A Review of Power Architectures, Applications and Standardization Issues. *IEEE Transactions on Power Electronics*, *31*(5), 3528–3549. doi:10.1109/TPEL.2015.2464277

Duffie, J. A., & Beckman, W. A. (1991). *Solar Engineering of Thermal Processes*. New York: Wiley Interscience.

Erbs, D. G., Klein, S. A., & Duffie, J. A. (1982). Estimation of the diffuse radiation fraction for hourly, daily and monthly-average global radiation. *Solar Energy*, *28*(4), 4. doi:10.1016/0038-092X(82)90302-4

Ismanzhanov, A. I., & Tashiev, N. M. (2016). Development And Research Of The Technology For Powdering Agricultural Products Using Solar Energy. *Applied Solar Energy*, *52*(4), 256–258. doi:10.3103/S0003701X16040101

Murdock, H. E., & Adib, R. (2018). *Renewables 2018 Global Status Report*. Paris: Academic Press.

Natsheh, E. M. (2016). Power Generation of Solar PV Systems in Palestine. *Applied Solar Energy*, *52*(3), 193–196. doi:10.3103/S0003701X16030105

Persits, I.S., Potapov, V.N., Strebkov, D.S., & Chekhunina, G.S. (2006). Solnechnyy fotoelektricheskiy modul' i sposob ego izgotovleniya [Solar photovoltaic module and method of its manufacture]. *FIIP Bull, 26*.

Rakhimov, E. Y., Sadullaeva, S. E., Kolomiets, Y. G., Tashmatov, K. K., & Usmonov, N. O. (2017). Analysis of the solar energy potential of the republic of Uzbekistan. *Applied Solar Energy*, *53*(4), 344–346. doi:10.3103/S0003701X17040120

Tursunov, M. N., Sabirov, K., Hwoan, P. J., Turdiev, B. M., Shirmatov, A. R., & Yuldoshev, I. A. (2013). Self-contained photovoltaic unit as a source of power supply for a milking parlor on a cattle farm. *Applied Solar Energy*, *49*(4), 232–234. doi:10.3103/S0003701X13040166

Vieira, J. A. B., & Mota, A. M. (2010). Implementation of a Stand-Alone Photovoltaic Lighting System with MPPT Battery Charging and LED Current Control. In *IEEE International Conference on Control Applications Part of 2010 IEEE Multi-Conference on Systems and Control*. (pp. 185-190). Yokohama, Japan: IEEE. 10.1109/CCA.2010.5611257

Compilation of References

2. Prakhovnik, A. V., Popov, V. A., & Nakhodov, V. F. (2010). Risk and uncertainty assessment in the development of strategy for forming the potential of dispatchable generation in the national power system of Ukraine. International Journal Global Energy Issues, 34(1-4), 132-138. doi:10.1504/IJGEI.2010.037400

Abdelsalam, H. M., & Elassal, M. M. (2014). Joint economic lot sizing problem for a three—Layer supply chain with stochastic demand. *International Journal of Production Economics*, *155*, 272–283. doi:10.1016/j.ijpe.2014.01.015

Abramov, Yu. M. (2016). *Elektricheskie stantsii i podstantsii [Electrical stations and substations]*. Russia: Don State Technical University. Retrieved from: https://studfiles.net/preview/6023617/

Adadurov, E. A. (2005). Povyshenie effektivnosti ispol'zovaniya akkumulyatorov teploty s vozobnovlyaemymi istochnikami energii [Increasing the efficiency of using heat accumulators with renewable energy sources] [Dissertation]. (in Russian)

Ahi, P., & Searcy, C. (2013). A comparative literature analysis of definitions for green and sustainable supply chain management. *Journal of Cleaner Production*, *52*, 329–341. doi:10.1016/j.jclepro.2013.02.018

Ahmad, A. S., Hassan, M. Y., Abdullah, M. P., Rahman, H. A., Hussin, F., Abdullah, H., & Saidur, R. (2014). A review on applications of ANN and SVM for building electrical energy consumption forecasting. *Renewable & Sustainable Energy Reviews*, *33*(1), 102–109. doi:10.1016/j.rser.2014.01.069

Ahmadi, A., Kaymanesh, A., Siano, P., Janghorbani, M., Nezhad, A. E., & Sarno, D. (2015). Evaluating the effectiveness of normal boundary intersection method for short-term environmental/economic hydrothermal self-scheduling. *Electric Power Systems Research*, *123*, 192–204. doi:10.1016/j.epsr.2015.02.007

Ahmed, S., & Ahmed, S. (2011). *Two-Stage Stochastic Integer Programming: A brief introduction*. Wiley on Line Library. DOI.org/10.1002/9780470400531.eorms0092

Ahumada, O., & Villalobos, R. J. (2009). Application of planning models in the agri-food supply chain: A review. *European Journal of Operational Research*, *196*(1), 1–20. doi:10.1016/j.ejor.2008.02.014

Aizerman, M. A., & Malishevsky, A. B. (1980). *Problems of logical substantiation of general decision theory. Part 2. Criteria of classical rationality and their application to the selection analysis.* IPU. (in Russian)

Aksenov, A. G., Izmaylov, A. I., Dorokhov, A. S., & Sibirev, A. V. (2018). Onion bulbs orientation during aligned planting of seed-onion using vibration-pneumatic planting device. *INMATEH - Agricultural Engineering*, *55*(2), 63–70.

Aksjonov, S.I. (2004). *Voda i ejo rol' v reguljacii biologicheskih processov [Water and its role in the regulation of biological processes]*. Moskva: Institut komp'juternyh issledovanij. (in Russian)

Al Harbi, Y., Eugenio, N. N., & Al Zahrani, S. (1998). Photovoltaic-thermal solar energy experiment in Saudi Arabia. *Renewable Energy*, *15*(1-4), 483–486. doi:10.1016/S0960-1481(98)00209-2

Aleksandrov, V. Y. A., & Kislyuk, I. M. (1994). Reakciya kletok na teplovoj shok: Fiziologicheskij aspekt. [Cell reaction to heat shock: physiological aspect]. *Citologiya. T., 36*(1), 5.

Ali, A., Lange, J., Elrayyah, A., Sozer, Y., De Abreu-Garcia, J. A., & Mpanda, A. (2018). A Hybrid Flyback LED Driver with Utility Grid and Renewable Energy Interface. In *33nd Annual IEEE Applied Power Electronics Conference and Exposition.* (pp. 3377-3384). San Antonio, TX: IEEE. 10.1109/APEC.2018.8341588

Anastasiadis, A., Kondylis, G. A., Vokas, G., & Papageorgas, P. (2016). Hydrothermal coordination in power systems with large-scale integration of renewable energy sources. *Management of Environmental Quality, 27*(Issue: 3), 246–258. doi:10.1108/MEQ-04-2015-0054

Aniskin, V. I., & Rybaruk, V. A. (1972). *Teoriya i tekhnologiya sushki i vremennoj konservacii zerna aktivnym ventilirovaniem [Theory and technology of drying and temporary preservation of grain by active gates]*. Moskva: Kolos. (in Russian)

ANSYS - Engineering Simulation & 3D Design Software. (n.d.). Retrieved from http://www.ansys.com/

Asanov, M.M., Bekirov, E.A. & Voskresenskaya, S.N. (2014). Snizhenie vliyaniya nagreva poverhnosti fotoehlementa na ehffektivnost' ego raboty (Reducing the influence of the heating of the photocell surface on the efficiency of its work). *Stroitel'stvo i tekhnogennaya bezopasnost' [Construction and Technogenic Safety]*, 51, 92-97. (in Russian)

Avdeeva V.N., Starodubtseva G.P., Lyubaya S.I. (2008). Predposevnaya obrabotka semyan pshenitsy ozonom [Preseeding treatment of wheat seeds with ozone]. *Agrarian Science, 5*, 19-20. (in Russian)

Awudu, I., & Zhang, J. (2013). Stochastic production planning for a biofuel supply chain under demand and price uncertainties. *Applied Energy, 103*, 189–196. doi:10.1016/j.apenergy.2012.09.025

Ayub, S., & Kumar, S. S. (2011). *Particulate matter emission from thermal power plant: parent material, formation mechanism, health concerns, control devices - a review.* Global Journal Engineering and Applied Sciences.

Ba, B. H., Prins, C., & Prodhon, C. (2016). Models for optimization and performance evaluation of biomass supply chains: An Operations Research perspective. *Renewable Energy, 87*, 977–989. doi:10.1016/j.renene.2015.07.045

Babaev, B. D. (1996). Himicheskie prevrashcheniya i fazovye ravnovesiya sistemy [Chemical transformations and phase equilibria of the system Li,Na,Ca,Ba//F,MoO4] [Dissertation]. (in Russian)

Babaev, B. D. (2016b). Razrabotka i issledovanie energosistem na osnove vozobnovlyaemyh istochnikov s fazoperekhodnym akkumulirovaniem tepla [Development and investigation of power systems based on renewable sources with phase-transition heat accumulation] [Dissertation]. (in Russian)

Babaev, B.D. & Volshanik, V.V. (2012). Prognozirovanie ekologicheskoj obstanovki razrabotannoj programmoj EVM vyyavleniya himicheskih vzaimodejstvij [Predicting the environmental situation developed by the computer program to detect chemical interactions]. *Vestnik RUDN, seriya: Inzhenernye issledovaniya [Bulletin of RUDN, a series: Engineering researches]*, 4, 135-140.

Babaev, B.D. (2014). Principy teplovogo akkumulirovaniya i ispol'zuemye teploakkumuliruyushchie materialy [Principles of thermal accumulation and used heat storage materials]. *Teplofizika vysokih temperatur [Thermophysics of High Temperatures]*, 52(5), 760-776. (in Russian).

Babaev, B.D. (2015). *Patent No. 2547680. Russian Federation.* (in Russian)

Babaev, B.D., & Halilullaev, G.M. (2005). *Certificate of state registration of computer program No. 2005610201 "Description of thermochemical reactions in multicomponent reciprocal systems "Thermal effect depending on temperature."*

Babaev, B.D., Kharchenko, V., Panchenko, V., & Vasant, P. (2019). Materials and Methods of Thermal Energy Storage in Power Supply Systems. In *Renewable Energy and Power Supply Challenges for Rural Regions* (pp. 115-135). Academic Press. doi:10.4018/978-1-5225-9179-5.ch005

Babaev, B. D. (2016a). Akkumulyatory tepla pri ispol'zovanii vozobnovlyaemyh istochnikov energii [Heat accumulators using renewable energy sources]. *Energetik [Power engineer]*, *3*, 19–22.

Babaev, B. D., & Danilin, V. N. (2002). *Energoakkumuliruyushchie ustanovki [Energy storage facilities]*. Makhachkala. (in Russian)

Babanin, N. V. (2015). Uluchshenie plavnosti hoda mashinno-traktoronogo agregata na baze traktora klassa 1.4 s uprugodempfiruyushchim privodom na vedushchih kolesah [Improvement of the smoothness of the machine and tractor unit on the basis of a class 1.4 tractor with an elastic damping mechanism on driving wheels]. *Current Problems of Science and Education*, *22*, 46.

Babanin, N. V., & Polivaev, O. I. (2015). Eksperimentalnie issledovaniya na plavnost hoda, proizvoditelnost i toplivnuyu ekonomichnost mashinno-traktornogo agregata na baze traktora klassa 1.4, oborudovannogo gazogidravlicheskim uprugodempfiruyushchim privodom [Experimental researches on smoothness of a course, productivity and fuel economy of the machine-tractor unit on the basis of a class 1.4 tractor, equipped with a gas-hydraulic elastic-damping drive]. *Bulletin of the Voronezh State Agrarian University*, *3*(46), 112–118.

Babbar, M. M. (1955). Distribution of solutions of a set of linear equations (with an application to linear programming). *Journal of the American Statistical Association*, *50*(271), 854–869. doi:10.1080/01621459.1955.10501971

Baishev, A. V., & Toropov, A. S. (2018). Assessment of the possibility of using waste heat of power transformers for heating and hot water supply of electrical substations [Otsenka vozmozhnosti ispolzovaniia sbrosnogo tepla silovykh transformatorov dlia otopleniia i goriachego vodosnabzheniia elektricheskikh podstantsii]. *Bulletin of the Khakass State University*, *24*, 5–7.

Bajdukin, J.A., Pershin, A.F., & Fedorov, A.V. (1988). Jelektroozonirujushhaja ustanovka dlja dezinfekcionnyh kamer pticefabrik [Electrozoniruyuschaya installation for disinfection chambers of poultry farms] *Nauka – APK [Science agribusiness]*, *2*, 42-46. (in Russian)

Bakhmutov, V. A. (1980). Razmeshcheniye semyan po ploshchadi pri ryadovykh posevakh [seeds placement on the area at ordinary crops]. Mechanization and electrification of agriculture, 5, 24-27. (in Russian)

Bakker, M., Zondag, H. A., & van Helden, W. G. J. (2002). Demonstration on a dual flow photovoltaic/thermal combined panel. *Proceedings of PV in Europe-From PV Technology to Energy Solutions Conference and Exhibition*, Rome, Italy, October 7-11. Academic Press.

Banaj, A., Šumanovac, L., Hefer, G., Tadić, V., Banaj, Đ., & Kovačev, I. (2017). Yield of corn grain by sowing in twin rows with MaterMacc-2 planter. *Proceedings of the 45th International Symposium on Agricultural Engineering, Actual Tasks on Agricultural Engineering*, Opatija, Croatia, February 21-24 (pp. 141-152).

Bannykh, O.P. (2012). *Osnovnye konstruktsii i teplovoi raschet teploobmennikov* [The basic design and thermal calculation of heat exchangers]. Russia: SPbNUU ITMO.

Bansal, N., Dhaliwal, A. S., & Mann, K. S. (2016). Dielectric characterization of rapeseed (Brassica napus L.) from 10 to 3000 MHz. *Biosystems Engineering*, *143*, 1–8.

Barahona, F., Berman, S., Gunluk, S., & Holmes, S. (2005). Robust capacity planning in semiconductor manufacturing. *Naval Research Logistics*, *52*(5), 459–468. doi:10.1002/nav.20086

Barbosa-Canovas, G. V., Vega-Mercado, H., & Gongora-Nieto, M. M. (2001). Dehydration of foods. *Past, present and future Conference: 2nd Inter-American Drying Conference (IADC),* Boca del Rio, Mexico (pp. 43-55). Academic Press.

Barulin, E.P., Isaev, V.N., & Sokolsky, A.I. (2009). *Raschety teploizoliruiushchego oborudovaniia* [Calculations of insulating equipment]. Ivanovo, Russia: Ivan state chemical-tehnol. un-t.

Bashilov, A. M., & Zaghinaylov, V. I. (2005). Visualization and observation of systemic complexity of precise arable farming. *Proceedings of International scientific-technical conference "Intellectual systems"*. Moscow: Publishing House "PhysMathLit".

Bashilov, A.M. (2015). Innovational laser, optical and optic-electronic technologies in agrarian production. *Moscow, Technique and equipment for villages, 2*(212), 2-6.

Bashilov, A.M., & Legheza, V.N. (2011). Remote video-monitoring of animals' behavior in herd. *Series "Technique and equipment for villages", 12*, 24-26.

Bashilov, A.M., &Korolyov, V.A. (2009). Video-monitoring and navigation in systems of precise arable farming. *Bulletin of MGAU Named After V. P. Goryachkin, 3*(34), 7-11.

Bashilov, A.M., Sventitsky, I.I., & Korolyov, V.A. (2013). *Universal model of agro-technologies.* Moscow: Energy Engineering and Automatics.

Basu, M. (2003). Hopfield neural networks for optimal scheduling of fixed head hydrothermal power systems. *Electric Power Systems Research, 64*(1), 11–15. doi:10.1016/S0378-7796(02)00118-9

Basu, M. (2005). A simulated annealing-based goal-attainment method for economic emission load dispatch of fixed head hydrothermal power systems. *International Journal of Electrical Power & Energy Systems, 27*(2), 147–153. doi:10.1016/j.ijepes.2004.09.004

Basu, M., & Chowdhury, A. (2013). Cuckoo search algorithm for economic dispatch. *Energy, 60*, 99–108. doi:10.1016/j.energy.2013.07.011

Baum, A. E., & Rezchikov, V. A. (1983). *Sushka zerna* [Drying grain]. Moskva: Kolos. (in Russian)

Beale, E. M. L. (1955). On minimizing a convex function subject to linear inequalities. *Journal of the Royal Statistical Society. Series B. Methodological, 17*(2), 173–184. doi:10.1111/j.2517-6161.1955.tb00191.x

Beauvais, Chevillard, Guillén Paredes, Heisz, Rossi, & Schmela. (2019). *Global Market Outlook For Solar Power.* Brussels, Belgium: Solar Power Europe.

Bekirov, E.A. & Asanov, M.M. (2017). Analiz sistem vodyanogo ohlazhdeniya fotoehlektricheskih panelej (Analysis of water cooling systems for photovoltaic panels). *Stroitel'stvo i tekhnogennaya bezopasnost' [Construction and Technogenic Safety], 6*(58), 55-59. (in Russian)

Bekirov, E.A. & Karkach, D.V. (2018). Dvumernaya model' teplovyh processov v solnechnom kollektore i ee ehksperimental'naya proverka (Two-dimensional model of thermal processes in a solar collector and its experimental verification). *Stroitel'stvo i tekhnogennaya bezopasnost' [Construction and Technogenic Safety]*, 10 (62), 191-201. (in Russian)

Bekri, E., Disse, M., & Yannopoulos, P. (2015). Optimizing Water Allocation under Uncertain System Conditions for Water and Agriculture Future Scenarios in Alfeios River Basin (Greece)-Part B: Fuzzy-Boundary Intervals Combined with Multi-Stage Stochastic Programming Model. *Water (Basel), 7*(11), 6427–6466. doi:10.3390/w7116427

Bellabdaoui, A., & Teghem, J. (2006). A mixed-integer linear programming model for the continuous casting planning. *International Journal of Production Economics*, *104*(2), 260–270. doi:10.1016/j.ijpe.2004.10.016

Bellman, R. E. (1957). *Dynamic Programming*. Princeton University Press.

Bellman, R., & Zadeh, L. A. (1970). Decision-making in a fuzzy environment. *Management Science*, *17*(4), 141–161. doi:10.1287/mnsc.17.4.B141

Belov V.V., Matkin A.Ju., & Semenova M.A. (2018). Problemy hranenija ovoshhej v hozjajstvah [Problems of storage of vegetables in farms]. (in Russian)

Belov, A. A., Toporkov, V. N., & Vasilyev, A. N. (2018). Planning and conducting screening experiment to study the production of fertilizers in the electrohydraulic treatment of solutions. *International Technical and Economic Journal*, *5*, 22–28.

Bergene, T., & Lovvik, O. M. (1995). Model calculations on a flat-plate solar heat collector with integrated solar cells. *Solar Energy*, *55*(6), 453–462. doi:10.1016/0038-092X(95)00072-Y

Bernardon, D. P., Sperandio, M., Garcia, V. J., Russi, L. N., Canha, J., Abaide, A. R., & Daza, E. F. B. (2011). Methodology for allocation of remotely controlled switches in distribution networks based on a fuzzy multi-criteria decision making algorithm. Electric Power Systems Research, 81(2), 414-420. doi:10.1016/j.epsr.2010.10.010

Bertling, L. (2002). Reliability Centred Maintenance for Electric Power Distribution Systems. Thesis submitted in accordance with academic requirements for the Degree of Doctor of Philosophy, Stockholm: Royal Institute of Technology (KTH).

Bertrand, Y., & Laurichesse, D. (2007, November). Comparison of the oxidation stabilities of vegetable based and mineral insulating oils. *Proceedings of MATPOST conference*. Academic Press.

Bespamyatnova N.M. (2010). *Vibratsii v tekhnologicheskikh protsessakh [Vibrations in technological processes]*. Rostov-on-Don, Russia: LLC "Terra".

Bespamyatnova, N.M. (2002). *Nauchno-metodicheskie osnovi adaptatsii pochvoobrabativayushchih mashin [Scientific and methodological fundamentals of adaptation of soil cultivating and sowing machines]*. Rostov-on-Don, Russia: LLC "Terra".

Bian, D. C., Yan, D., Zhao, J. C., & Niu, S. Q. (2018). Experimental Study of Pulsed Discharge Underwater Shock-Related Properties in Pressurized Liquid Water. *Advances in Materials Science and Engineering*.

Binghui, Y., Xiaohui, Y., & Jinwen, W. (2007). Short term hydrothermal scheduling using particle swarm optimization method. *Energy Conversion and Management*, *47*(8), 1902–1908.

Bishop, D. J. (1994). Application of new techniques to CA storage. *Proceedings of the International Institute of Refrigeration International Symposium*, Istanbul, Turkey, June 8–10 (pp. 323-329). Academic Press.

Bishop, D. (1990). Controlled atmosphere storage. In C. J. V. Dellino (Ed.), *Cold and Chilled Storage Technology* (pp. 66–98). London, UK: Blackie.

Blonbou, R. (2011). Very short-term wind power forecasting with neural networks and adaptive Bayesian learning. *Renewable Energy*, *36*(3), 1118–1124. doi:10.1016/j.renene.2010.08.026

Bobovich, B. B. (2014). Polimernye konstruktsionnye materialy [Polymeric construction materials]. Moscow, Russia: FORUM: INFRA-M.

Borisov, G. S. (2010). *Osnovnye processy i apparaty himicheskoj tehnologii* [*Basic processes and apparatuses of chemical technology*]. Moscow: Al'jans. (in Russian)

Borodin I.F., Pershin A.F., Evdoseeva A.Yu., Fedorov A.V. (1989). Perspektivy ispol'zovaniya koronnogo razryada v sel'skokhozyaystvennykh elektroozoniruyushchikh ustanovkakh *[Prospects for the use of corona discharge in agricultural electro-zoning installations]*. Collection of scientific papers MIISP. P.3-9. (in Russian)

Bortsov A.V. & Potovskaya E.I. (2013). Opredeleniye intensivnosti solnechnogo izlucheniya dlya naklonnoy prinimayushchey poverkhnosti [Determination of solar radiation intensity for an inclined receiving surface]. *Visnik NTU "KHIII"*, *59*(1032).

Borzov, V. P., & Shabalina, L. N. (2002). Sbornik zadach po teplotekhnike [Collection of tasks for heat engineering]. Kostroma, Russia: KSAA.

Bosanac, M., Sørensen, B., Katic, I., Sørensen, H., Nielsen, B., & Badran, J. (n.d.). Photovoltaic/Thermal Solar Collectors and Their Potential in Denmark. Retrieved from https://www.ecn.nl/fileadmin/ecn/units/egon/pvt/pdf/EFP1713_00-0014.pdf

Brandenburg, M., Govindan, K., Sarkis, J., & Seuring, S. (2014). Quantitative models for sustainable supply chain management: Developments and directions. *European Journal of Operational Research*, *233*(2), 299–312. doi:10.1016/j.ejor.2013.09.032

Bringmann, K., & Friedrich, T. (2013). Approximation quality of the hypervolume indicator. *Artificial Intelligence*, *195*, 265–290. doi:10.1016/j.artint.2012.09.005

Brown, R. E. (2009). *Electric Power Distribution Reliability*. CRC Press.

Budnikov, D. A., & Vasiliev, A. N. (2018). *The Use of Microwave Energy at Thermal Treatment of Grain Crops*. In V. Kharchenko & P. Vasant (Eds.), *Handbook of Research on Renewable Energy and Electric Resources for Sustainable Rural Development* (pp. 475–499). Hershey, PA: IGI Global. doi:10.4018/978-1-5225-3867-7.ch020

Budzko, I. A., Levin, M. S., & Leshchinskaya, T. B. (1987). Evaluation of effectiveness of activities on electric energy loss reduction in operated 10 kV to 110 kV distribution networks [Abstract]. *Proceedings of the National Sci. Conf.*, Moscow (pp. 83-83). Academic Press. (in Russian)

Budzko, I.A. & Levin, M.S. (1985). *Power supply of agricultural enterprises and settlements*. Moscow, Agropromizdat Publ. (in Russian)

Budzko, I. A., Levin, M. S., Terepko, O. A., & Pereverzev, P. S. (1987). *Complex evaluation of technical status of 10 kV and 0.38 kV rural networks. Power stations, 12*, 49. (in Russian)

Buhan, S., & Çadırcı, I. (2015). Multistage wind-electric power forecast by using a combination of advanced statistical methods. *IEEE Transactions on Industrial Informatics*, *11*(5), 1231–1242. doi:10.1109/TII.2015.2431642

Buhan, S., Özkazanç, Y., & Çadırcı, I. (2016). Wind pattern recognition and reference wind mast data correlations with NWP for improved wind-electric power forecasts. *IEEE Transactions on Industrial Informatics*, *12*(3), 991–1004. doi:10.1109/TII.2016.2543004

Buhan, S., Terciyanli, E., Özkan, M. B., Küçük, D., Haliloğlu, A. B., Demirci, T., ... Ünsal, H. (2013, May). Verification of a very short term wind power forecasting algorithm for Turkish transmission grid. *Proceedings of the 4th International Conference on Power Engineering, Energy and Electrical Drives* (pp. 1217-1221). IEEE. 10.1109/PowerEng.2013.6635786

Buonomano, A., Calise, F., & Vicidimini, M. (2016). Design, Simulation and Experimental Investigation of a Solar System Based on PV Panels and PVT Collectors. *Energies*, *9*(7), 497. doi:10.3390/en9070497

Burdon, J., Lallu, N., Haynes, G., McDermott, K., & Billing, D. (2008). The effect of delays in establishment of a static or dynamic controlled atmosphere on the quality of 'Hass' avocado fruit. *Postharvest Biology and Technology*, *49*(1), 61–68. doi:10.1016/j.postharvbio.2008.01.002

Butko M.P., Volohov B.V., Pershin A.F., & Frolov V.S. (2010). Primenenie ozona na predprijatijah pishhevoj promyshlennosti [The use of ozone in the food industry] *Pererabotka i hranenie sel'hozsyr'ja [Processing and storage of agricultural products]*, 3, 8-11. (in Russian)

Butko, M.P. & Borisov, S.G. (1995). Patent 2608.2.03. Russia.

Butko, M. P., & Frolov, V. S. (2001). Ozon: sintez i yego primeneniye *[Ozone: synthesis and its application]*. Moscow. [in Russian]. *GNU VNIIVSGE, 2001*, 357s.

Buynyy, R. A., Podolnyi, S. V., & Zorin, V. V. (2004). Modyfikovanyi metod optymizatsii nadiinosti rozpodilchykh merezh 10 kV [Modified method reliability optimization of distribution 4. Begovic, M., Lambert, F.C. (2002). Case Studies: Recloser Positioning and Reliability Indices.

Castillo-Villar, K. K. (2014). Metaheuristic algorithms applied to bioenergy supply chain problems: Theory, review, challenges, and future. *Energies*, *7*(11), 7640–7672. doi:10.3390/en7117640

Cerisola, S., Latorre, J. M., & Ramos, A. (2012). Stochastic dual dynamic programming applied to nonconvex hydrothermal models. *European Journal of Operational Research*, *218*(3), 687-697.

Charnes, A., & Cooper, W. W. (1959). Chance-constrained programming. *Management Science*, *6*(1), 73–79. doi:10.1287/mnsc.6.1.73

Chelnokov, A. A., & Trukhacheva, T. V. (1984). Opyt i perspektivy kompleksnogo ispol'zovaniya ozona v mikrobiologicheskoy promyshlennosti [Experience and prospects for the integrated use of ozone in the microbiological industry]. Moscow: ONTIDI microbioprom. (in Russian)

Chen, P. H. (2008). Pumped-storage scheduling using evolutionary particle swarm optimization. *IEEE Transactions on Energy Conversion*, *23*(1), 291–301. doi:10.1109/TEC.2007.914312

Chen, Y. W., Wang, L. C., Wang, A., & Chen, T. L. (2017). A particle swarm approach for optimizing a multi-stage closed loop supply chain for the solar cell industry. *Robotics and Computer-integrated Manufacturing*, *43*, 111–123. doi:10.1016/j.rcim.2015.10.006

Chetvergov, N. A., Bezrukikh, N. S., & Bezrukikh, E. G. (2005). *Raschot dinamicheskikh kharakteristik kontsentratsii ozona v vozdukhe pomeshcheniy pri rabote v nikh ozone generatorov razlichnoy proizvoditel'nosti* [Calculation of the dynamic characteristics of the concentration of ozone in the indoor air when ozone generators of various capacities are working in them]. (pp. 111–115). Moscow: Technical Physics. (in Russian)

Cheung, K., Wang, X., Chiu, B. C., Xiao, Y., & Rios-Zalapa, R. (2010, January). Generation dispatch in a smart grid environment. In *2010 Innovative Smart Grid Technologies (ISGT)* (pp. 1-6). IEEE.

Chowdhury, A., & Koval, D. (2009). *Power distribution system reliability practical methods and applications*. IEEE.

Corney, J. F., & Drummond, P. D. (2004). Gaussian Quantum Monte Carlo Methods for Fermions and Bosons'. *Physical Review Letters*, *93*(26), 260401. doi:10.1103/PhysRevLett.93.260401

Cortes, E., & Rivera, W. (2010). Exergetic and exergoeconomic optimization of a cogeneration pulp and paper mill plant including the use of a heat transformer. *Energy*, *35*(3), 1289–1299. doi:10.1016/j.energy.2009.11.011

D'yakonov, V. P. (2008a). *MATLAB 7.*/R2006/2007 (tutorial)*. Moskva: DMK-Press. (in Russian)

D'yakonov, V. P. (2008b). *SIMULINK 5/6/7 (tutorial)*. Moskva: DMK-Press. (in Russian)

Dahlquist, E. (Ed.). (2013). *Technologies for Converting Biomass to Useful Energy: Combustion, Gasification, Pyrolysis, Torrefaction and Fermentation*. Sweden: Taylor & Francis. doi:10.1201/b14561

Damianovsky, V. (2013). *From Light to Pixels*. Edition Butterworth-Heinemann.

Dantzig, G.B. (1955). Linear programming under uncertainty. *Management Science*, 197-206.

Darby-Dowman, K., Barker, S., Audsley, E., & Parsons, D. (2000). A two-stage stochastic programming with recourse model for determining robust planting plans in horticulture. *The Journal of the Operational Research Society*, *51*(1), 83–89. doi:10.1057/palgrave.jors.2600858

Darmont, A. (2013). High Dynamic Range Imaging: Sensors and Architectures. Academic Press. doi:10.1117/3.903927

Dashti, H., Conejo, A.J., Jiang, R. & Wang, J. (2016). Weekly two-stage robust generation scheduling for hydrothermal power systems. *IEEE Transactions on Power Systems*, *31*(6).

Daus, A., Horev, B., Dvir, O., Ish-shalom, S., & Lichter, A. (2011). The efficacy of ultrasonic fumigation for disinfestation of storage facilities against postharvest pathogens. *Postharvest Biology and Technology, Amsterdam*, *62*(1), 310–313. doi:10.1016/j.postharvbio.2011.06.012

Davò, F., Alessandrini, S., Sperati, S., Delle Monache, L., Airoldi, D., & Vespucci, M. T. (2016). Post-processing techniques and principal component analysis for regional wind power and solar irradiance forecasting. *Solar Energy*, *134*, 327–338. doi:10.1016/j.solener.2016.04.049

de la Torre, M., Juberias, G., Dominguez, T., & Rivas, R. (2012, July). The CECRE: Supervision and control of wind and solar photovoltaic generation in Spain. *Proceedings of the 2012 IEEE Power and Energy Society General Meeting* (pp. 1-6). IEEE.

De Luca, A., & Termini, S. (1972). Algebraic properties of fuzzy sets. *Journal of Mathematical Analysis and Applications*, *40*(2), 373–386. doi:10.1016/0022-247X(72)90057-1

De Meyer, A., Cattrysse, D., Rasinmäki, J., & Van Orshoven, J. (2014). Methods to optimise the design and management of biomass-for-bioenergy supply chains: A review. *Renewable & Sustainable Energy Reviews*, *31*, 657–670. doi:10.1016/j.rser.2013.12.036

De Vita, P., Colecchia, S. A., Pecorella, I., & Saia, S. (2017). Reduced inter-row distance improves yield and competition against weeds in a semi-dwarf durum wheat variety. *European Journal of Agronomy*, *85*, 69–77. doi:10.1016/j.eja.2017.02.003

De Vrie, D. W. (1998). *Design of a photovoltaic/thermal combi-panel* [PhD thesis]. Eindhoven Technical University.

Deboli, R., Calvo, A., & Preti, C. (2017). Whole-body vibration: Measurement of horizontal and vertical transmissibility of an agricultural tractor seat. *International Journal of Industrial Ergonomics*, *58*, 69–78. doi:10.1016/j.ergon.2017.02.002

Dehghani, N., & Dashti, R. (2011). Optimization of Recloser Placement to Improve Reliability by Genetic Algorithm. Energy and Power Engineering, 3, 508-512. doi:10.4236/epe.2011.34061

Demirci, T., Kalaycıoglu, A., Küçük, D., Salor, Ö., Güder, M., Pakhuylu, S., ... Bilgen, S. (2011). Nationwide real-time monitoring system for electrical quantities and power quality of the electricity transmission system. *IET Generation, Transmission & Distribution*, *5*(5), 540–550. doi:10.1049/iet-gtd.2010.0483

Devyanin, S.N., Chumakov, V.L., & Markov, V.A. (2012). Biogas - alternative fuel for diesel engines. *Alternative Fuel Transport*, *2*, 68-73.

Dibirov, Y.A. (2012). Phase equilibria and thermophysical compositions in the system Ca//F,Cl,SO4,MoO4. Actual problems of development of renewable energy resources. *Proceedings of the V-th School of Young Scientists*. Academic Press.

Divyesh, V. R., Sahu, C., Kirubakaran, V., Radhakrishnan, T. K., & Guruprasath, M. (2017). Energy optimization using metaheuristic bat algorithm assisted controller tuning for industrial and residential applications. *Transactions of the Institute of Measurement and Control*.

Dominique, M., Bain, D. M., & Acker, T. L. (2018). Hydropower impacts on electrical system production costs in the Southwest United States. *Energies*, *11*(2), 368. doi:10.3390/en11020368

Dong, N., Fang, X., & Wu, A. G. (2016). A Novel Chaotic Particle Swarm Optimization Algorithm for Parking Space Guidance. *Mathematical Problems in Engineering*.

Dragičević, T., Lu, X., Vasquez, J. C., & Guerrero, J. M. (2016). DC Microgrids-Part I: A Review of Control Strategies and Stabilization Techniques. *IEEE Transactions on Power Electronics*, *31*(7), 4876–4891.

Dragičević, T., Lu, X., Vasquez, J. C., & Guerrero, J. M. (2016). DC Microgrids–Part II: A Review of Power Architectures, Applications and Standardization Issues. *IEEE Transactions on Power Electronics*, *31*(5), 3528–3549. doi:10.1109/TPEL.2015.2464277

Dubey, S., & Tay, A. A. O. (2012). Experimental Study of the Performance of Two Different Types of Photovoltaic Thermal (PVT) Modules under Singapore Climatic Conditions. *Journal of Fundamentals of Renewable Energy and Applications*, *2*, 1–6. doi:10.4303/jfrea/R120313

Dubey, S., & Tiwari, G. (2008). Thermal modeling of a combined system of photovoltaic thermal (PV/T) solar water heater. *Solar Energy*, *82*(7), 602–612. doi:10.1016/j.solener.2008.02.005

Duffie, J. A., & Beckman, W. A. (1991). *Solar Engineering of Thermal Processes*. New York: Wiley Interscience.

Duffy, J., & Beckmann, U. A. (2013). Fundamentals of solar thermal power. Educational - reference manual. 888 p.

Eberhart, S. A., & Russel, W. A. (1966). Stability parameters for comparing varieties. *Crop Science*, *6*(1), 36–40. doi:10.2135/cropsci1966.0011183X000600010011x

Egorov, G.A. (1985). *Tehnologicheskie svojstva zerna [Technological properties of grain]*. Moscow: Agropromizdat. (in Russian)

El Wakil, N., Chereches, N. C., & Padet, J. (2006). Numerical study of heat transfer and fluid flow in a power transformer. *International Journal of Thermal Sciences*, *45*(6), 615–626. doi:10.1016/j.ijthermalsci.2005.09.002

Emission of hazardous air pollutants from coal-fired power plants. Environmental Health and Engineering. (2019). Retrieved from https://www.lung.org/assets/documents/healthy-air/emissions-of-hazardous-air.pdf

Engelsht, V.S. & Murataliev, V.Z. (2013). Exothermic effect in the interaction of calcium and silicon oxides. *Thermal physics of high temperatures*, *51*(5), 717.

Erbs, D. G., Klein, S. A., & Duffie, J. A. (1982). Estimation of the diffuse radiation fraction for hourly, daily and monthly-average global radiation. *Solar Energy*, *28*(4), 4. doi:10.1016/0038-092X(82)90302-4

Erinmez, I. A., Kappenman, J. G., & Radasky, W. A. (2002). Management of the geomagnetically induced current risks on the national grid company's electric power transmission system. *Journal of Atmospheric and Solar-Terrestrial Physics*, *64*(5-6), 743–756. doi:10.1016/S1364-6826(02)00036-6

Ermoliev, Y., & Wets, R. (1988). *Numerical Techniques for Stochastic Optimization*. Berlin: Springer Series in Computational Mathematics. doi:10.1007/978-3-642-61370-8

Erokhov, V. I., & Karunin, A. L. (2005). Gas-diesel cars (design, calculation, operation). Moscow: Graf-Press.

Evdoseeva A.J., Pershin A.F., & Fedorov A.V. (1989). Issledovanie rezhimov raboty ozonirujushhej ustanovki s dezinfekcionnoj kameroj [Study of operating modes of the ozonating unit with a disinfection chamber]. *Jelektrotehnologii v sel'skohozjajstvennom proizvodstve [Electrotechnology in agricultural production]*, 73-78. (in Russian)

Ewbank, N., & Buckby, L. (2002). Lynx engineering enters seed drill market with the seedline system. *Landwards*, *82*(1), 153–155.

Fan, Y. R., Huang, G. H., Guo, P., & Yang, L. (2012). Inexact two-stage stochastic partial programming: Application to water resources management under uncertainty. *Stochastic Environmental Research and Risk Assessment*, *26*(2), 281–293. doi:10.100700477-011-0504-6

Farhat, I. A., & El-Hawary, M. E. (2009). Optimization methods applied for solving the short-term hydrothermal coordination problem. *Electric Power Systems Research*, *79*(9), 1308–1320. doi:10.1016/j.epsr.2009.04.001

Feng, J. W., Jiang, L., Zhu, D., Su, K. Z., Zhao, D. Y., Zhang, J. B., & Zheng, Z. (2016). Dielectric barrier discharge plasma induced degradation of aqueous atrazine. *Environmental Science and Pollution Research International*, *23*(9), 9204–9214. doi:10.100711356-016-6148-9 PMID:26832879

Feofanov, J.A. (2005). *Funkcional'nyj analiz tehnicheskih obektov [Functional analysis of technical objects].* Moscow: MGTU MAMI. (in Russian)

Ferreira, L. A. F. M. (1992). A theoretical analysis of the classic hydro-thermal optimization algorithm in power system scheduling. *IEEE Proceedings International Symposium on Circuits and Systems*. 10.1109/ISCAS.1992.230627

Feuerhuber, K., Lindert, S. O., & Schlacher, K. (2013). *Vibration attenuation by semi-active dampers.* Paper presented at 14th International Conference on Computer Aided Systems Theory (EUROCAST), Las Palmas, Spain. DOI: 10.1007/978 3-642-53862-9_35

Fiete. (2012). Formation of a Digital Image: The Imaging Chain Simplified. Academic Press. doi:10.1117/3.919698

Filin, V. A., Pronin, V. Yu., Kryuchkov, S. V., & Filin, A. V. (2016). Obosnovanie stepeni zagruzki traktornogo dvigatelya pri vypolnenii selskokhozyastvennykh rabot [Substantiation of the degree of loading of the tractor engine in the performance of agricultural works]. In Proceedings of Energy-Efficient and Resource-Saving Technologies and Systems. Saransk, Russia: FSBEU of Higher Professional Education of the Moscow State University named after N.P. Ogaryova.

Filippov, Yu. V., Voblikova, V. A., & Panteleev, V. I. (2008). Elektrosintez ozona [Electrosynthesis of Ozone] Moscow State University. (in Russian)

Fishburn, P. C. (1964). *Decision and value theory. NY*: Wiley.

Fleten, S. E., & Kristoffersen, T. K. (2008). Short-term hydropower production planning by stochastic programming. *Computers & Operations Research*, *35*(8), 2656–2671. doi:10.1016/j.cor.2006.12.022

Florschuetz, L. W. (1979). Extension of the Hottel–Whiller model to the analysis of combined photovoltaic/thermal flat plate collectors. *Solar Energy*, *22*(4), 361–366. doi:10.1016/0038-092X(79)90190-7

Fogel, D. B. (1995). A comparison of evolutionary programming and genetic algorithms on selected constrained optimization problems. *Simulation*, *64*(6), 397–404. doi:10.1177/003754979506400605

Fontenault, B. J., & Gutierrez-Miravete, E. (n.d.). Modeling a Combined Photovoltaic-Thermal Solar Panel. Retrieved from http://www.comsol.com/paper/download/150905/gutierrezmiravete_paper.pdf

Forsyth, D. A., & Ponce, J. (2012). *Computer Vision: A Modern Approach* (2nd ed.). Moscow: Pearson Education, Prentice Hall.

Freeman, J. E., & Mudd, J. B. (1981). Reaction of ozone with sulfhydryls of human erythrocytes. *Archives of Biochemistry and Biophysics*, *208*(1), 212–220. doi:10.1016/0003-9861(81)90142-9 PMID:7259179

Fujisawa, T., & Tani, T. (1997). Annual exergy evaluation on photovoltaic-thermal hybrid collector. *Solar Energy Materials and Solar Cells*, *47*(1-4), 135–148. doi:10.1016/S0927-0248(97)00034-2

Gabdrafikov, F. Z., & Safina, R. R. (2017). Povyshenie tehniko-ekonomicheskih pokazateley traktornyh dizeley na chastichnyh i neustanovivshihsya rezhimah raboty [Increase of technical and economic indicators of tractor diesel engines on partial and unsteady operating modes]. *Russian Electronic Scientific Journal*, *2*(24), 95–109.

Galán-Martín, A., Pozo, C., Guillén-Goálbez, G., Antón Vallejo, A., & Laureano Jiménez Esteller, L. (2015). Multi-stage linear programming model for optimizing cropping plan decisions under the new Common Agricultural Policy. *Land Use Policy*, *48*, 515–524. doi:10.1016/j.landusepol.2015.06.022

Gamburg, K. Z., Borovskii, G. B., & Voinikov, V. K. (2017). The study of the participation of heat shock proteins in the resistance to high and low temperatures with the use of thellungiella (thellungiellasalsuguinea) and transgenic lines of arabidopsis (arabidopsis thaliana). *Journal of Stress Physiology & Biochemistry*, *13*(1), 23–34.

Ganesan, T., Aris, M. S., & Elamvazuthi, I. (2018b). Multiobjective Strategy for an Industrial Gas Turbine: Absorption Chiller System. In Handbook of Research on Emergent Applications of Optimization Algorithms (pp. 531–556). Hershey, PA: IGI Global. doi:10.4018/978-1-5225-2990-3.ch023

Ganesan, T., Elamvazuthi, I., & Vasant, P. (2016b). Swarm Intelligence for Multiobjective Optimization of Extraction Process. In Handbook of Research on Modern Optimization Algorithms and Applications in Engineering and Economics (pp. 516-544). Hershey, PA: IGI Global. doi:10.4018/978-1-4666-9644-0.ch020

Ganesan, T., Aris, M. S., & Vasant, P. (2018a). Extreme Value Metaheuristics for Optimizing a Many-Objective Gas Turbine System. *International Journal of Energy Optimization and Engineering*, *7*(2), 76–96. doi:10.4018/IJEOE.2018040104

Ganesan, T., Elamvazuthi, I., & Vasant, P. (2015). Multiobjective design optimization of a nano-CMOS voltage-controlled oscillator using game theoretic-differential evolution. *Applied Soft Computing*, *32*, 293–299. doi:10.1016/j.asoc.2015.03.016

Ganesan, T., Vasant, P., & Elamvazuthi, I. (2016a). *Advances in Metaheuristics: Applications in Engineering Systems*. CRC Press. doi:10.1201/9781315297651

Gao, M., Ding, P., & Yang, Y. (2015, September). Time-Optimal Trajectory Planning of Industrial Robots Based on Particle Swarm Optimization. *Proceedings of the 2015 Fifth International Conference on Instrumentation and Measurement, Computer, Communication and Control (IMCCC)* (pp. 1934-1939). IEEE. 10.1109/IMCCC.2015.410

Gao, J., & You, F. (2015). Shale gas supply chain design and operations toward better economic and life cycle environmental performance: MINLP model and global optimization algorithm. *ACS Sustainable Chemistry & Engineering*, *3*(7), 1282–1291. doi:10.1021/acssuschemeng.5b00122

Gapeeva, N.A. & Zhilenko, O.B. (2018). Avtonomnoe teplosnabzhenie vysotnyh zdanij (Autonomous heat supply of high-rise buildings). *Stroitel'stvo i tekhnogennaya bezopasnost' [Construction and Technogenic Safety]*, *10*(62), 77-89. (in Russian)

Garg, H. P., & Agarwal, R. K. (1994). Study of a photovoltaic-thermal system-thermosyphonic solar water heater combined with solar cells. *Energy Conversion and Management*, *35*(7), 605–620. doi:10.1016/0196-8904(94)90044-2

Garg, H. P., & Agarwal, R. K. (1995). Some aspects of a PV/T collector/forced circulation flat plate solar water heater with solar cells. *Energy Conversion and Management*, *36*(2), 87–99. doi:10.1016/0196-8904(94)00046-3

Garkavi L.H., Kvakina E.B., Kuz'menko T.S., & Shihlyarova A.I. (2003). Antistressornye reakcii i aktivacionnaya terapiya. Reakciya aktivacii kak put' k zdorov'yu cherez process samoorganizacii [Antistress reactions and activation therapy activation reaction as a way to health through self-organization processes]. Ekaterinburg: Tom Chast' II.

Garkavi, L. H., Kvakina, E. B., & Ukolova, M. A. (1997). Adaptacionnye reakcii i rezistentnost' organizma [Adaptive reactions and body resistance.]. Izvestiya Rostovskoj-na-Donu gosudarstvennoj akademii stroitel'stva [News of the Rostov-on-Don State Academy of Civil Engineering].

Garkavi, L. H., Tolmachev, G. N., Mihajlov, N. Y. U., Esipov, Y. U. V., Benya, F. M., Zverinceva, M.M., … Plyaka, P. S. (2007). Adaptacionnye reakcii i urovn ireaktivnost i kak ehffektivnye diagnosticheskie pokazateli donozologicheskih sostoyanij [Adaptation reactions and levels of reactivity as effective diagnostic indicators of prenosological conditions.]. Vestnik Yuzhnogo nauchnogo centra RAN [Bulletin of the South Scientific Center of RAS], 3(1), 61-66.

Garkavi, L. H., Zhukova, G. V., Kapkina, N. N., Mashchenko, N. M., & Mihajlov, N. Y. U. (2008). Aktivacionnaya terapiya i kachestvo zhizni v processe reabilitacii [Activation therapy and quality of life in the process of rehabilitation]. Sibirskijonkologicheskijzhurnal [Siberian Oncology Journal], S2, 34-35.

Garkavi, L. Kh. (2015). Periodic system of unspecific adaptational reactions and its significance for theory and practice. Cardiometry, 7.

Gazizov, T. T. (2017). *Metody global'noy optimizatsii: uchebnoye posobiye* [*Global optimization methods: a tutorial*]. Tomsk: In-Spectrum. (in Russian)

Gellert, A., Florea, A., Fiore, U., Palmieri, F., & Zanetti, P. (2019). A study on forecasting electricity production and consumption in smart cities and factories. *International Journal of Information Management*.

Gerasimenko, A. A., & Fedin, V. T. (2008). Electric energy transmission and distribution. Rostov-on-Don. Phoenix Publ. (in Russian).

Geyer, A., Hanke, M., & Weissensteiner, A. (2009). Life-Cycle Asset Allocation and Consumption Using Stochastic Linear Programming. *Journal of Computational Finance*, *12*(4), 29–50. doi:10.21314/JCF.2009.203

Ghaderi, H., Pishvaee, M. S., & Moini, A. (2016). Biomass supply chain network design: An optimization-oriented review and analysis. *Industrial Crops and Products*, *94*, 972–1000. doi:10.1016/j.indcrop.2016.09.027

Ghahramani, A., Castro, G., Karvigh, S. A., & Becerik-Gerber, B. (2018). Towards unsupervised learning of thermal comfort using infrared thermography. *Applied Energy*, *211*, 41–49.

Ghosh, S., Kaur, M., Bhullar, S., & Karar, V. (2019). Hybrid ABC-BAT for Solving Short-Term Hydrothermal Scheduling Problems. *Energies*, *12*(3), 551. doi:10.3390/en12030551

Giebel, G., Brownsword, R., Kariniotakis, G., Denhard, M., & Draxl, C. (2011). The state-of-the-art in short-term prediction of wind power: A literature overview. ANEMOS.

Glushkov, V. M. (1980). *On system optimization*. Kiev: Kibernetika Publ. (in Russian)

Goldstein, B. D., & Balchum, O. J. (1967). Effect of ozone on lipid peroxidation in the red blood cell. *Proceedings of the Society for Experimental Biology and Medicine*, *126*(2), 356–358.

Golestaneh, F., Pinson, P., & Gooi, H. B. (2016). Very short-term nonparametric probabilistic forecasting of renewable energy generation—With application to solar energy. *IEEE Transactions on Power Systems*, *31*(5), 3850–3863. doi:10.1109/TPWRS.2015.2502423

Golicin, G. A., & Petrov, V. M. (2005). Informaciya i biologicheskie principy optimal'nosti: Garmoniya i algebra zhivogo. [Information and biological principles of optimality: Harmony and living algebra.] Izd. 2-e, stereotipnoe. – M.: KomKniga, - 128 p.

Goryachkin, V. P. (1965). Zemledel'cheskaya mekhanika [Agricultural mechanics]. In Sobraniye sochineniy. Moscow, Russia: Kolos.

Gosstroi Rossii. (2003). SNiP 41-01-2003. Otoplenie, ventiliatsiia i konditsionirovanie [SNiP 41-01-2003 Heating, ventilation and air conditioning].

GOST 12.1.007-76 Occupational safety standards system. Harmful substances. Classification and general safety requirements. (n.d.). (in Russian)

Grant. (2016). Getting Started with UAV Imaging Systems: A Radiometric Guide. Academic Press. doi:10.1117/3.2239237

Grebnev, V. P., Poliaev, O. I., & Vorokhobin, A. V. (2011). *Traktory i avtomobili: teoriya i ekspluatatsionniye svoystva: uchebnoye posobie dlya studentov vuzov, obuchaushchihsya po napravleniyu «Agroingeneriya»* [Tractors and automobiles: theory and operational properties: a teaching aid for university students studying in the direction of "Agroengineering"]. Voronezh, Russia: KnoRus.

Gridin, S. V., & Petrenko, A. F. (2013). Energoeffektivnost sposobov utilizatsii otrabotannogo tepla sistem okhlazhdeniia silovykh transformatorov [Energy Efficiency of Methods for Utilization of Waste Heat from Cooling Systems of Power Transformers]. *Energy Saving*, *7*, 9–13.

Grigorev, V. A. (2003). Razrabotka akkumulyatorov teploty s zernistym teplonositelem i metoda ih rascheta na osnove matematicheskogo modelirovaniya (Development of heat accumulators with a granular coolant and methods for calculating them on the basis of mathematical modeling) [Dissertation]. (in Russian)

Grishchenko, F. V. (1963). Osnovy teorii dvizheniya semyan i raspredeleniye ikh v pochve pri bezryadkovom poseve [Fundamentals of the theory of seeds movement and their distribution in the soil at sowing without rows]. *Ryazanskiy SKhI*, *10*, 31–33.

Grits I.V., Fomichev A.V. (1985). Zavisimost' stepeni antibakterial'nogo i antifagovogo deystviya ozona ot kontsentratsii kletok i fagovykh chastits v pitatel'noy srede [Dependence of the degree of antibacterial and antiphage action of ozone on the concentration of cells and phage particles in the nutrient medium]. *Microbiology*, *3*, 410-413. (in Russian)

Gruntovye teploobmenniki dlia geotermalnogo teplovogo nasosa [Ground heat exchangers for a geothermal heat pump] . (n.d.). Retrieved from http://solarsoul.net/gruntovye-teploobmenniki-dlya-geotermalnogo-teplovogo-nasosa

Gubernsky, Yu. D., & Litskevich, V. K. (1991). Zhilishche dlya cheloveka [Residence for a man Stroyizdd]. Academic Press. (in Russian)

Gunes, G., Liu, R. H., & Watkins, C. B. (2002). Controlled-atmosphere effects on postharvest quality and antioxidant activity of cranberry fruits. *Journal of Agricultural and Food Chemistry*, *50*(21), 5932–5938.

Gungor, V. C., Sahin, D., Kocak, T., Ergut, S., Buccella, C., Cecati, C., & Hancke, G. P. (2011). Smart grid technologies: Communication technologies and standards. *IEEE Transactions on Industrial Informatics*, *7*(4), 529–539. doi:10.1109/TII.2011.2166794

Han, Y. C., Huang, G. H., & Li, C. H. (2008). An interval-parameter multi-stage stochastic chance-constrained mixed integer programming model for inter-basin water resources management systems under uncertainty. *2008 Fifth International Conference on Fuzzy Systems and Knowledge Discovery Shandong*. 10.1109/FSKD.2008.465

Haupt, R. L., & Haupt, S. A. (2004). *Practical Genetic Algorithms*. John Wile & Sons, Inc.

Hendrie, S. D. (1979). Evaluation of combined photovoltaic/thermal collectors. *Proceedings of the ISES international congress* (Vol. 3, pp. 1865-1869). Academic Press.

Hinojosa, V. H., & Leyton, C. (2012). Short-term hydrothermal generation scheduling solved with a mixed binary evolutionary particle swarm optimizer. *Electric Power Systems Research*, *92*, 162–170. doi:10.1016/j.epsr.2012.06.006

Hong, Y. Y., Beltran, A. A., & Paglinawan, A. C. (2016). A Chaos-Enhanced Particle Swarm Optimization with Adaptive Parameters and Its Application in Maximum Power Point Tracking. *Mathematical Problems in Engineering*.

Hörbe, T. A. N., Amado, T. J. C., Reimche, G. B., Schwalbert, R. A., Santi, A. L., & Nienow, C. (2016). Optimization of Within-Row Plant Spacing Increases Nutritional Status and Corn Yield: A Comparative Study. *Agronomy Journal*, *108*(5), 1962–1971. doi:10.2134/agronj2016.03.0156

Horner, P. (1999). Planning under uncertainty. Questions & answers with George Dantzig. *OR/MS Today*, *26*, 26-30.

Horynski, M. (1999). The influence of electric field and mechanical stresses on the electric properties of the grain. *Proceedings of the International Conference on Trends in Agricultural Engineering Location*, Prague, Czech Republic (pp. 35-39). Academic Press.

Hosseini, R., Hosseini, N., & Khorasanizadeh, H. (2011). An Experimental study of combining a Photovoltaic System with a heating System. *Proceedings of the World Renewable Energy Congress*, Linkoping, Sweden (pp. 2993-3000). Academic Press. 10.3384/ecp110572993

Ho, T. B., Roberts, T. K., & Lucas, S. (2015). Small-scale household biogas digesters as a viable option for energy recovery and global warming mitigation – Vietnam case study. *Journal of Agricultural Science and Technology A*, *5*, 387–395.

Hu, Q., Zhang, R., & Zhou, Y. (2016). Transfer learning for short-term wind speed prediction with deep neural networks. *Renewable Energy*, *85*, 83–95. doi:10.1016/j.renene.2015.06.034

Hurwicz, L. (1951). *Optimality criteria for decision making under ignorance*. Cowles Commission Discussion Papers.

Ibrahim R.A., Al-Ahmadi S.S. (2014) Utilization of Ozone to Control Potato Tuber Moth, Phthorimaea operculella (Lepidoptera: Gelechiidae), in Storage. *African Entomology*, *22*(2), 330-336. doi:10.4001/003.022.0209

Ibrahim, A., & Jin, G.L. (2009). Hybrid photovoltaic thermal (PV/T) air and water based solar collectors suitable for building integrated applications. *American Journal of Environmental Sciences*, 618–624.

Ibrahim, A., Othman, M. Y., Ruslan, M. H., Mat, S., & Sopian, K. (2011). Recent advances in flat plate photovoltaic/thermal (PV/T) solar collectors. *Renewable & Sustainable Energy Reviews*, *15*(1), 352–365. doi:10.1016/j.rser.2010.09.024

Imai, S., Sakaguchi, Y., & Shirafuji, T. (2018). Simultaneous generation of acidic and alkaline water using atmospheric air plasma formed in water. *Japanese Journal of Applied Physics*, *57*(1).

Ishizaka, A., & Nemery, Ph. (2013). *Multi-criteria decision analysis: methods and software*. New Delhi: Times by Aptara Inc. doi:10.1002/9781118644898

Ismanzhanov, A. I., & Tashiev, N. M. (2016). Development And Research Of The Technology For Powdering Agricultural Products Using Solar Energy. *Applied Solar Energy*, *52*(4), 256–258. doi:10.3103/S0003701X16040101

Jafari, H., Kalantari, D., & Azadbakht, M. (2017). Semiindustrial continuous band microwave dryer for energy and exergy analyses, mathematical modeling of paddy drying and it's qualitative study. *ENERGY Vol*, *138*, 1016–1029. doi:10.1016/j.energy.2017.07.111

Jankowski, K. J., Dubis, B., Budzyński, W. S., Bórawski, P., & Bułkowska, K. (2016). Energy efficiency of crops grown for biogas production in a large-scale farm in Poland. *Energy*, *109*, 277–286. doi:10.1016/j.energy.2016.04.087

Javadian, S. A. M., & Haghifam, M.-R. (2010). Optimal placement of protective devices in distribution networks based on risk analysis. Proceedings of the IEEE/PES Transmission and Distribution Conference and Exposition: Latin America (T&D-LA), Sao Paulo, November 8-10 (pp. 147-154). IEEE. doi:10.1109/TDC-LA.2010.5762875

Javadian, S. A. M., & Massaeli, M. (2011). Impact of distributed generation on distribution system's reliability considering recloser-fuse miscoordination - A practical case study. *Indian Journal of Science and Technology*, *4*(10), 1279–1284.

Jiang, S., Zhang, J., Ong, Y. S., Zhang, A. N., & Tan, P. S. (2015). A simple and fast hypervolume indicator-based multiobjective evolutionary algorithm. *IEEE Transactions on Cybernetics*, *45*(10), 2202–2213. doi:10.1109/TCYB.2014.2367526

Ji, J., Lu, J., Chow, T., He, W., & Pei, G. (2007). A sensitivity study of a hybrid photovoltaic/thermal water-heating system with natural circulation. *Applied Energy*, *84*(2), 222–237. doi:10.1016/j.apenergy.2006.04.009

Ji, X., Zhu, S., Wang, S., & Zhang, S. (2005). A stochastic linear goal programming approach to multistage portfolio management based on scenario generation via linear programming. *IIE Transactions*, *37*(10), 957–969. doi:10.1080/07408170591008082

Joshi, A. S., Kulkarni, O., Kakandikar, G. M., & Nandedkar, V. M. (2017). Cuckoo Search Optimization-A Review. *Materials Today: Proceedings*, *4*(8), 7262–7269.

Jubaer, A., & Zainal, S. A. (2014). A Maximum Power Point Tracking (MPPT) for PV system using Cuckoo Search with partial shading capability. *Applied Energy*, *119*, 118–130. doi:10.1016/j.apenergy.2013.12.062

Jumah, RY, Raghavan, GSV. (2001). Analysis of heat and mass transfer during combined microwave-convective spouted-beddrying. *Drying technology, 19*(3-4), 485-506.

Kak vybrat teploobmennik dlia basseina? [How to choose a heat exchanger for the pool ?]. (n.d.). Retrieved from http://udobnovdome.ru/kak-vybrat-teploobmennik-dlya-bassejna/

Kalmykov, A. V. (2014). *Snizheniye dinamicheskoy nagruzhennosty silovoy peredachi traktora za schet izmeneniya krutilnoy zhestkosti reaktivnogo zvena* [Reduction of the dynamic loading of the power transmission of the tractor due to the change in the torsional stiffness of the reactive link] (PhD dissertation). Volgograd State Agrarian University, Volgograd, Russia.

Kalogirou, S. A. (2001). Use of TRYNSYS for modeling and simulation of a hybrid PV-thermal solar system for Cyprus. *Renewable Energy*, *23*(2), 247–260. doi:10.1016/S0960-1481(00)00176-2

Kamilov, A. G., Muminov, R. A., & Tursunov, M. N. (2008). Ocenka jeffektivnosti sistemy solnechnogo jelementa i kollektora v uslovih zharkogo klimata [Evaluation of system effectiveness solar cell and collector in a hot climate]. *Geliotehnika*, *2*, 32–35. (in Russian)

Kaplan, H. (2007). Practical Applications of Infrared Thermal Sensing and Imaging Equipment (3rd ed.). Academic Press. doi:10.1117/3.725072

Karabanitsky, A.P., &Levshukova, O.A. (2014). *Teoreticheskoe obosnovanie parametrov energosberegaushchikh mashinno-traktornykh agregatov: uchebnoe posobie* [Theoretical justification of the parameters of energy-saving machine-tractor units: manual]. Krasnodar, Russia: KubSAU.

Karami, M., Shayanfar, H. A., Aghaei, J., & Ahmadi, A. (2013). Scenario-based security-constrained hydrothermal coordination with volatile wind power generation. *Renewable & Sustainable Energy Reviews*, *28*, 726–737. doi:10.1016/j.rser.2013.07.052

Kariniotakis, G., Martí, I., Casas, D., Pinson, P., Nielsen, T. S., Madsen, H., . . . Brownsword, R. (2004, November). What performance can be expected by short-term wind power prediction models depending on site characteristics. *Proceedings of the European Wind Energy Conference EWEC* (pp. 22-25). Academic Press.

Karminskij, V. D. (2005). Tekhnicheskaya termodinamika i teploperedacha [Technical thermodynamics and heat transfer]. Moscow: Marshrut. (in Russian)

Karthick, S. P., & Helen Catherine, R. L. (n.d.). Modeling and Analysis of Solar Photovoltaic Thermal Collector. Retrieved from http://www.rroij.com/open-access/modelling-and-analysis-of-solar-photovoltaicthermal-collector.pdf

Kaufmann, A. (1975). *Introduction to the theory of fuzzy subsets*. N.Y.: Academic Press.

Kern, J. E. C., & Russell, M. C. (1978). Combined photovoltaic and thermal hybrid collector. *Proceedings of the 13th IEEE photovoltaic specialists* (pp. 1153-1157). IEEE.

Khanbari, O. S., & Thompson, A. K. (1996). Effect of controlled atmosphere, temperature and cultivar on sprouting and processing quality of stored potatoes. *Potato Research*, *39*(4), 523–531. doi:10.1007/BF02358471

Kharchenko, V. V. (2014). PVThermal modules. *Proceedings of XI International Annual Conference "Renewable and small energy"* (pp. 248-257). Academic Press.

Kharchenko, V., Panchenko, V., Tikhonov, P. V., & Vasant, P. (2018). Cogenerative PV Thermal Modules of Different Design for Autonomous Heat and Electricity Supply. In Handbook of Research on Renewable Energy and Electric Resources for Sustainable Rural Development (pp. 86-119). Academic Press. doi:10.4018/978-1-5225-3867-7.ch004

Kharchenko, V., Gusarov, V., & Bolshev, V. (2019). Reliable Electricity Generation in RES-Based Microgrids. In *Handbook of Research on Smart Power System Operation and Control* (pp. 162–187). IGI Global; doi:10.4018/978-1-5225-8030-0.ch006

Kharchenko, V., Nikitin, B., Tikhonov, P., & Gusarov, V. (2013). Investigation of experimental flat PV thermal module parametres in natural conditions. *Conference Proceeding - 5th International Conference, TAE 2013: Trends in Agricultural Engineering*, Prague, Czech Republic, September 2-3 (pp. 51-55).

Kharchenko, V., & Vasant, P. (Eds.). (2018). *Handbook of Research on Renewable Energy and Electric Resources for Sustainable Rural Development*. Hershey, PA: IGI Global.

Kiflemariam, R., Zevallos, F., Almas, M., & Lin, C. (n.d.). Enhanced Transient Modeling of Hybrid Photovoltaic Air (PVT) Module. Retrieved from https://www.comsol.com/paper/download/194257/kiflemariam_paper.pdf

Kjell, A. (2013). R.F. Patent No. 2234755. Patent Service of the Russian Federation.

Klein. (2012). Sensor and Data Fusion: A Tool for Information Assessment and Decision Making (2nd ed.). Academic Press. doi:10.1117/3.928035

Kljuev, S., & Zvjagincev, V. (2016). Kompleksnaja sistema upravlenija dlja ovoshhehranilishh [Integrated management system for vegetable stores]. *Avtomatizacija i Proizvodstvo [Automation and production]*, *1*(46), 16–17.

Knyazev, P. V. (2005). *Selection and evaluation of power supply sources for remote rural areas [Extended abstract of dissertation]*. Moscow. V. P. Goryachkin. (in Russian)

Kolesnikov & Avanesov. (2007). Concept of creation of geo-spatial systems of video-connection based on new possibilities of multiservice networks of information change. *Scientific-Technical Magazine "Information and Space", 4*, 56-60.

KOMPAS-3D v17. (n.d.). Retrieved from http://kompas.ru/

Konev, S. V., Matus, V. K., Melnikova, A. M., & Rudenok, A. N. (1982). Deystviye ozona na membranozavisimyye funktsii drozhzhevykh kletok kandidoza [The effect of ozone on the membrane-dependent functions of yeast cells of candidiasis]. *Microbiology Today, 51*(2), 220–228. (in Russian)

Kopeika. (1998). A System Engineering Approach to Imaging. Academic Press. doi:10.1117/3.2265069

Kostyuchenkov, N. V., & Plaksin, A. M. (2010). Ekspluatatsionniye svoystva mobilnyh agregatov [Operational properties of mobile units]. AAstana, Kazakhstan: KATU named after S. Seifullina.

Kotlyarov, V.V., & Deyanov, Y.U.I. (1973). Issledovaniye protsessa razgona MTA s podatlivoy svyaz'yu v transmissii traktora [Investigation of the process of acceleration of the machine-tractor unit with a malleable connection in the tractor transmission]. *Mekhanizatsiya i elektrifikatsiya s.-kh. proizvodstva, 16*, 47-57.

Kovrikov, I. T., Nogtikov, A. A., & Sazonov, S. N. (1995). Povysheniye Effektivnosti Ispolzovaniya Kombinirovannykh Pochvoobrabatyvayushche-posevnykh mashin [Increasing the efficiency of combined tillage and sowing machines using]. Sovershenstvovaniye tekhnologiy i tekhn. sredstv dlya mekhanizatsii s.-kh. protsessov, 82, 80-81.

Kozyrskiy, V. V., Kuz'menko, B. V., & Guy A. V. (2005). Matematicheskaya model' razmeshcheniya kommutatsionnykh apparatov v usloviyakh rynochnykh otnosheniy mezhdu elektrosnabzhayushchey kompaniyey i potrebitelem [Mathematical model of switching devices in the conditions of market relations between the power supply company and the consumer]. *Elektrifikatsiya ta avtomatizatsiya sil's'kogo gospodarstva, 3*, 40-52.

Kozyrskiy, V. V., Kuz'menko, B. V., & Guy, A. V. (2005). Podkhod k opredeleniyu mest razmeshcheniya sektsionnykh kommutatsionnykh apparatov [Approach to determining of the location of sectional switching devices]. *Elektryfikatsiia ta avtomatyzatsiia silskoho hospodarstva, 1*, 32-44.

Kozyrskyi, V. V. & Guy, O. V. (2004). Vybir optymalnoi kilkosti sektsionuiuchy prystroiv dlia rozpodilnykh merezh napruhoiu 10 kV [Choosing of the optimal number of partitioning devices for 10 kV distribution networks]. *Elektryfikatsiia ta avtomatyzatsiia silskoho hospodarstva, 2*, 12-20.

Krasnik, V. V. (2011). *Ekspluatatsiia elektricheskikh podstantsii i raspredelitelnykh ustroistv* [Operation of electrical substations and switchgear]. Russia: NTS ENAS.

Krasnoshchekov, N.V. (2002). O proyektirovanii tekhnologiy proizvodstva selsko-khozyaystvennoy produktsii [About design of production of agricultural products technologies*]. Prioritety mekhanizatsii rasteniyevodstva i zhivotnovodstva, 138*, 11–20.

Krausp, V. R. (1980). Complex automatization in area of industrial animal breeding. Moscow: Publishing House "Machinostroyeniye".

Kravchenko, V. A. (2010). *Povishenie dinamicheskih i ekspluatatsionnyh pokazateley selskohozyastvennyh mashinno-traktornyh agregatov* [Increase of dynamic and operational parameters of agricultural machine and tractor units] (PhD dissertation). FSEE HE ABSSAEA, Zernograd, Russia.

Kravchenko, V. A., Senkevich, S. E., Senkevich, A. A., Goncharov, D. A., & Duryagina, V. V. (2008). *Ustroystvo dlya snizheniya zhestkosti transmissii mashinno-traktornogo agregata* [Device for reducing the stiffness of the transmission of the machine-tractor unit]. RU Patent 2398147 F16H47/04.

Kravchenko, V. A., Senkevich, S. E., Senkevich, A. A., Yarovoy, V. G., & Tolstoukhov, Yu. S. (2004). *Ustroystvo dlya snizheniya zhestkosti transmissii transportnogo sredstva* [Device to reduce the rigidity of the vehicle's transmission]. RU Patent 2222440, B60K 17/10.

Krivopishin, I. P. (1988). *Ozon v promyshlennom ptitsevodstve* [Ozone in industrial poultry farming]. Moscow: Rosagropromizdat. (in Russian)

Küçükyavuz, S., & Suvrajeet, S. (2019). *An Introduction to Two-Stage Stochastic Mixed-Integer Programming*. Retrieved from http://faculty.washington.edu/simge/Preprints/SMIP_Tutorial.pdf

Kudrin, B.I. (2006). Classics of technical coenoses. *Series "Investigations in cenology", 31.*

Kumar, R. S., Kondapaneni, K., Dixit, V., Goswami, A., Thakur, L. S., & Tiwari, M. K. (2016). Multi-objective modeling of production and pollution routing problem with time window: A self-learning particle swarm optimization approach. *Computers & Industrial Engineering*, *99*, 29–40. doi:10.1016/j.cie.2015.07.003

Kupfer, K. (1999). Drying oven method, infrared- and microwave drying as reference methods for determination of material moisture. *Technisches messen, 66(6)*, 227–237.

Lagland, H., & Kauhaniemi, K. (2007). Outage Cost Comparison of Different Medium Voltage Networks. *Proceedings of the CIRED 19th International Conference on Electricity Distribution*, Vienna, May 21-24. Academic Press.

Lavrentiev, S. V. (2018). Primenenie teplovykh nasosov dlia utilizatsii teplovykh poter v transformatornykh podstantsiiakh [Application of heat pumps for utilization of heat losses in transformer substations]. *Proceedings of the X International Student Scientific Conference "Student Scientific Forum*". Retrieved from https://scienceforum.ru/2018/article/2018003793

Lavrukhin, P.V. (2003). *Issledovaniye sovremennykh sposobov i skhem razmeshcheniya semyan po ploshchadi polya. [Study of modern methods and schemes of seed placement on the area of the field].* Otchet o NIR. FGOU VPO AChGAA, Zernograd, Russia.

Lechuguilla - Una pequeña planta con larga historia. (n.d.). Accessed on May 2019, recovered from: http://suculentasycactus.com/lechuguilla-pequena-planta-larga-historia/

Lee, S.J., Ma, S.H., Hong, Y.C., & Choi, M.C. (2018). Effects of pulsed and continuous wave discharges of underwater plasma on Escherichia coli. *Separation and Purification Technology*, *193*, 351–357. doi:10.1016/j.seppur.2017.10.040

Lei, Y., Jia, F., Lin, J., Xing, S., & Ding, S. X. (2016). An intelligent fault diagnosis method using unsupervised feature learning towards mechanical big data. *IEEE Transactions on Industrial Electronics*, *63*(5), 3137–3147. doi:10.1109/TIE.2016.2519325

Leschinskaya, T. B., & Volkova, E. A. (2018). The choice of the optimal variant of development of distribution air networks of 6-10 KV on a multi-criteria model Energy security and energy saving. Energy security and energy saving, 5, 23-30. (in Russian)

Leschinskaya, T. B., & Knyazev, V. V. (2016). Ranking of the priority for the reconstruction of rural electrical networks 10 KV. Bulletin of the Federal State Educational Institution of Higher Professional Education, *4(74),* 60-65. (in Russian)

Leshchinskaya, T. B. (2003). Primeneniye metodov mnogokriterial'nogo vybora pri optimizatsii sistem elektrosnabzheniya sel'skikh rayonov [Application of multicriteria choice methods for optimizing of rural power supply system]. *Elektrichestvo*, *1*, 14-22.

Levin, B.D. (2004). *Protsessy i apparaty pishchevykh proizvodstv. Teplovye protsessy [Processes and equipment for food production. Thermal processes].* Krasnoyarsk, Russia: Krasnoyars. state Agrarian Un.

Levin, M.S. & Leshchinskaya, T.B. (1989). *Methods of decision theory in problems of power supply system's optimization.* Moscow, VIP Kenergo. (in Russian)

Levshin, A. G., Bashilov, A. M., & Golovko, V. A. (2011). Automatic piloting and introduction of centralized control of mobile aggregates. *Bulletin of Federal State Educational Institution of Higher Professional Education "Moscow State Agro-Engineering University named after V.P. Goryachkin, 2*, 18–22.

Ling, S. H., Chan, K. Y., Leung, F. H. F., Jiang, F., & Nguyen, H. (2016). Quality and robustness improvement for real world industrial systems using a fuzzy particle swarm optimization. *Engineering Applications of Artificial Intelligence, 47*, 68–80. doi:10.1016/j.engappai.2015.03.003

Lin, T., Rodríguez, L. F., Shastri, Y. N., Hansen, A. C., & Ting, K. C. (2014). Integrated strategic and tactical biomass–biofuel supply chain optimization. *Bioresource Technology, 156*, 256–266. doi:10.1016/j.biortech.2013.12.121

Liu, Y., Li, Z.Y., Li, X.D., Zhou, G.Y., Li, H., Zhang, Q., & Lin, F.C. (2017). Energy Transfer Efficiency Improvement of Liquid Pulsed Current Discharge by Plasma Channel Length Regulation Method. *IEEE Transactions on Plasma Science, 45*(12 part 2), 3231-3239.

Li, W., Li, Y. P., Li, C. H., & Huang, G. H. (2010). An inexact two-stage water management model for planning agricultural irrigation under uncertainty. *Agricultural Water Management, 97*(11), 1905–1914. doi:10.1016/j.agwat.2010.07.005

Lobo, M. G., & Sanchez, I. (2012). Regional wind power forecasting based on smoothing techniques, with application to the Spanish peninsular system. *IEEE Transactions on Power Systems, 27*(4), 1990–1997. doi:10.1109/TPWRS.2012.2189418

Long, L. C. P., & Cheng, Y. (2011). Patent CN102204588, China, IPC A23B 7/154.

Lukashevich, V.K. (2001). *Osnovy metodologii nauchnyh issledovanij [Fundamentals of research methodology].* Jelajda. (in Russian)

Lurie, A. B. (1969). *Dinamika regulirovaniya navesnykh selskokhozyaystvennykh agregatov* [The dynamics of regulation of mounted agricultural units]. Leningrad, Russia: Mechanical Engineering.

Lykov, A. V. (1963). Teoriya teplo- i massoperenosa [Theory of heat and mass transfer]. Moscow: Gosehnergnoizdat. (in Russian)

Lykov, A. V., & Maksimov, G. A. (1957). Issledovanie processa sushki v pole vysokoj chastoty: Teplo-i massoobmen v kapillyarno-poristyh telah [Heat and mass transfer in capillary-porous bodies]. Moscow: Gosehnergoizdat. (in Russian)

Lyubushko, N.I., Zvolinskiy, V.N., (2001). Zernovyye seyalki na rubezhe XXI veka [Grain drills at the turn of the XXI century]. *Traktory i selskokhozyaystvennyye mashiny, 2*, 4-7.

Machnev, A.B. (2012). Soshnik s napravitelem-raspredelitelem semyan dlya poseva zernovykh kultur [Coulter with guide-distributor of seeds for sowing of crops] *Traktory i selkhozmashiny, 7*, 42-43.

Magomedov, M. M. (2000). Kompleksnaya metodologiya razrabotki fazoperekhodnyh teploakkumuliruyushchih materialov na osnove mnogokomponentnyh system [Complex methodology of development of phase-transition heat-accumulating materials on the basis of multicomponent systems] [Dissertation]. (in Russian)

Mahoney, W. P., Parks, K., Wiener, G., Liu, Y., Myers, W. L., Sun, J., ... Haupt, S. E. (2012). A wind power forecasting system to optimize grid integration. *IEEE Transactions on Sustainable Energy, 3*(4), 670–682. doi:10.1109/TSTE.2012.2201758

Makoklyuyev, B. I., Pavlikov, V. S., Vladimirov, A. I., & Fefelova, G. I. (2003). Vliyaniye metereologicheskikh faktorov na elektropotrebleniye energoob"yedineniy [The influence of meteorological factors on the power consumption of energy interconnections]. *Energetik, 6*, 45-47.

Maksimov, G. A. (1956). Issledovanie processov teplo- i massoobmena pri vnutrennem istochnike: avtoref.

Mandal, K. K., Basu, M., & Chakraborty, N. (2008). Particle swarm optimization technique based short term hydrothermal scheduling. *Applied Soft Computing*, *8*(4), 1392–3399. doi:10.1016/j.asoc.2007.10.006

Mandal, K. K., & Chakraborty, N. (2011). Short-term combined economic emission scheduling of hydrothermal systems with cascade reservoirs using particle swarm optimization technique. *Applied Soft Computing*, *11*(1), pp1295–pp1302. doi:10.1016/j.asoc.2010.03.006

Mareli, M., & Twala, B. (2017). *An adaptive Cuckoo search algorithm for optimisation*. Applied Computing and Informatics.

Markov, V. A., Devyanin, S. N., Zykov, S. A., & Gaidar, S. M. (2016). Biofuels for internal combustion engines. Moscow: SIC Engineer.

Markov, V.A., Devyanin, S.N., & Shimchenko, S.P. (2014). Experimental uses of a diesel generator set operating on biogas. *Truck*, *4*, 41-47.

Markov, V. A., Bashirov, R. M., & Gabitov, I. I. (2002). *Toxicity of exhaust gases of diesel engines*. Moscow: Moscow State Technical University.

Martinson, L. K., & Malov, Y. U. I. (2011). Differencial'nye uravneniya matematicheskoj fiziki [Differential equations of mathematical physics]. Moscow: Izd-vo MGTU im. N.Eh. Baumana. (in Russian)

Marufuzzaman, M., Eksioglu, S. D., & Huang, Y. E. (2014). Two-stage stochastic programming supply chain model for biodiesel production via wastewater treatment. *Computers & Operations Research*, *49*, 1–17. doi:10.1016/j.cor.2014.03.010

Mastrocinque, E., Yuce, B., Lambiase, A., & Packianather, M. S. (2013). A multi-objective optimization for supply chain network using the bees algorithm. *International Journal of Engineering Business Management*, *5*, 38. doi:10.5772/56754

Mattos, C. L., Barreto, G. A., Horstkemper, D., & Hellingrath, B. (2017, June). Metaheuristic optimization for automatic clustering of customer-oriented supply chain data. *Proceedings of the 2017 12th International Workshop on Self-Organizing Maps and Learning Vector Quantization, Clustering and Data Visualization (WSOM)* (pp. 1-8). IEEE. 10.1109/WSOM.2017.8020025

McCormic, G. (1976). Computability of global solutions to factorable non-convex programs: Part I Convex underestimating problems. *Mathematical Programming*, *10*, 146–175.

McEniry, J., Allen, E., Murphy, J. D., & O'Kiely, P. (2014). Grass for biogas production: The impact of silage fermentation characteristics on methane yield in two contrasting biomethane potential test systems. *Renewable Energy*, *63*, 524–530. doi:10.1016/j.renene.2013.09.052

Metodicheskie rekomendacii po matematicheskomu modelirovaniyu processa sushki i ohla-zhdeniya zerna v ustanovkah plotnogo sloya [Guidelines for mathematical modeling of the drying and cooling of grain in dense layer installations]. (1977). Moscow: VIEHSKH. (in Russian)

Mironova, T. V., & Kraiski, A. (2016). How to Use a Digital Camera as a Metering Device. Academic Press. doi:10.1117/3.2256725

Mishra, P. & Mishra, B. (2017). Dielectric Study of pure and insecticide treated rice seeds (ORYZA SATIVA L.). *Journal of science and arts, 4,* 823-828.

Mitrofanova, S.A. (2016). Vliyanie vida otrazhayushchih ehlementov ploskih solnechnyh kollektorov na obluchennost' priemnika [The influence of the type of reflecting elements of flat solar collectors on the irradiation of the receiver]. *Stroitel'stvo i tekhnogennaya bezopasnost' [Construction and Technologic Safety]*, *4*(56), 25-28. (in Russian).

Miyato, T., Maeda, S. I., Ishii, S., & Koyama, M. (2018). Virtual adversarial training: A regularization method for supervised and semi-supervised learning. *IEEE Transactions on Pattern Analysis and Machine Intelligence*. PMID:30040630

Mocanu, E., Nguyen, P. H., Gibescu, M., & Kling, W. L. (2016). Deep learning for estimating building energy consumption. Sustainable Energy. *Grids and Networks*, *6*, 91–99.

Mohandes, M., Rehman, S., & Rahman, S. M. (2011). Estimation of wind speed profile using adaptive neuro-fuzzy inference system (ANFIS). *Applied Energy*, *88*(11), 4024–4032. doi:10.1016/j.apenergy.2011.04.015

Monteiro, C., Bessa, R., Miranda, V., Botterud, A., Wang, J., & Conzelmann, G. (2009). *Wind power forecasting: State-of-the-art 2009*. Argonne National Lab.

Moorthy, V., Sangameswararaju, P., Ganesan, S., & Subramanian, S. (2013). Hydrothermal scheduling using ABC algorithm considering pollutant emission. *International Journal of Electrical Engineering*, *20*(5), 203–217.

Mostofizadeh, C., & Kulick, C. (1998). Use of a new type of heat transformer in process industry. *Applied Thermal Engineering*, *18*(9-10), 857–874. doi:10.1016/S1359-4311(97)00115-4

Mousavi, S. M., Sadeghi, J., Niaki, S. T. A., & Tavana, M. (2016). A bi-objective inventory optimization model under inflation and discount using tuned Pareto-based algorithms: NSGA-II, NRGA, and MOPSO. *Applied Soft Computing*, *43*, 57–72. doi:10.1016/j.asoc.2016.02.014

Mufuta, J. M., & Van den Bulck, E. (2000). Modelling of the mixed convection in the windings of a disc-type power transformer. *Applied Thermal Engineering*, *20*(5), 417–437. doi:10.1016/S1359-4311(99)00034-4

Murdock, H. E., & Adib, R. (2018). *Renewables 2018 Global Status Report*. Paris: Academic Press.

Nagar, L., & Jain, K. (2008). Supply chain planning using multi-stage stochastic programming. *Supply Chain Management*, *13*(Issue: 3), 251–256. doi:10.1108/13598540810871299

Nagornaya, N. V. (2007). *Ekonomika energetiki* [*Energy Economics*]. Vladivostok, Russia: FESTU Publishing House.

Naidu, K., Mokhlis, H., & Bakar, A. H. A. (2014). Multiobjective optimization using weighted sum artificial bee colony algorithm for load frequency control. *International Journal of Electrical Power & Energy Systems*, *55*, 657–667. doi:10.1016/j.ijepes.2013.10.022

Natsheh, E. M. (2016). Power Generation of Solar PV Systems in Palestine. *Applied Solar Energy*, *52*(3), 193–196. doi:10.3103/S0003701X16030105

Neimane, V. (2001). On development of electricity distribution networks [Doctoral Dissertation].

Nekhoroshev, D. A. (2014). *Stabilizatsiya rezhimov MTA izpolzovaniem pnevmogidravlicheskoy mufty stsepleniya* [Stabilization of MTU modes using pneumatic hydraulic clutch] (PhD dissertation). Volgograd State Agrarian University, Volgograd, Russia.

Nekhoroshev, D. D., & Nekhoroshev, D. A. (2015). Energeticheskoe sredstvo v transmissii kolesnogo traktora [Power means in the transmission of the wheeled tractor]. *Unique Research of the XXI Century*, *12*(12), 40–41.

Neklepaev, B. N., & Kryuchkov, I. P. (1989). The electrical part of power plants and substations: Reference data for course and diploma design. In Training manual for universities (4th ed.). Moscow: Energoatomizdat. (in Russian)

Nekrasov, A. I., & Podobedov, P. N. (2018). Multi-criteria evaluation of application of components of 10 kV overhead power distribution lines. *Bulletin of FGOU VPO Moscow State Agroengineering University*, *1*(83), 63–68.

Nguyen, T. T., Vo, D. N., & Dinh, B. H. (2016). Cuckoo search algorithm using different distributions for short-term hydrothermal scheduling with reservoir volume constraint. *Journal on Electrical Engineering and Informatics*, *8*(1), 76–92. doi:10.15676/ijeei.2016.8.1.6

Nielsen, H. A., Madsen, H., & Nielsen, T. S. (2006). Using quantile regression to extend an existing wind power forecasting system with probabilistic forecasts. Wind Energy. *An International Journal for Progress and Applications in Wind Power Conversion Technology*, *9*(12), 95–108.

Nima, A., & Hassan, R. S. (2010). Daily hydrothermal generation scheduling by a new modified adaptive particle swarm optimization technique. *Electric Power Systems Research*, *80*(6), 723–732. doi:10.1016/j.epsr.2009.11.004

Norena, C.P.Z. & Lescano-Anadon, C.E. (2016). Dielectric Properties of Importance in Operations of Post-harvest of Sorghum. *International journal of food engineering, 13*(4).

Norton, B., & Edmonds, J. E. J. (1991). Aqueous propylene–glycol concentrations for the freeze protection of thermosyphon solar energy water heaters. *Solar Energy*, *47*(5), 375–382. doi:10.1016/0038-092X(91)90031-Q

Novikov, A.M., & Novikov D.A. (2013). *Metodologija [Methodology]*. Moscow: SIN-TEG. (in Russian)

Novikov, D. A., & Taybakhtin, E. S. (2015). *Ispolzovanie tepla silovykh transformatorov dlia otopleniia proizvodstvennykh pomeshchenii podstantsii s tseliu snizheniia raskhoda elektroenergii [The use of heat power transformers for heating the industrial premises of substations in order to reduce power consumption]*. Retrieved from: http://gridology.ru/projects/453/documents

NTTS FSK EES [STC FGC UES]. (n.d.). Retrieved from http://www.ntc-power.ru/media/.pdf

Nussupbekov, B. R., Khassenov, A. K., Karabekova, D. Z., Stoev, M., Beysenbek, A. Z., & Kazankap, B. I. (2017). New technology for the recovery of heat exchangers. *Bulletin of the University of Karaganda-Physics*, *1*(85), 67 72. doi:10.31489/2017Ph1/67-72

O'Leary, G., Aggarwal, P., Calderini, D., Connor, D., Craufurd, P., Eigenbrode, S., ... Hatfield, J. (2018). Challenges and Responses to Ongoing and Projected Climate Change for Dryland Cereal Production Systems throughout the World. *Agronomy (Basel)*, *8*(4), 34. doi:10.3390/agronomy8040034

Odrin, V.M. (1989). *Metod morfologicheskogo analiza tehnicheskih system [Method of morphological analysis of technical systems]*. Moscow: VNIIPI. (in Russian)

Ogunbanwo, A., Williamson, A., Veluscek, M., Izsak, R., Kalganova, T., & Broomhead, P. (2014). Transportation network optimization. In Encyclopedia of Business Analytics and Optimization (pp. 2570–2583). Hershey, PA: IGI Global. doi:10.4018/978-1-4666-5202-6.ch229

Olsen, R. C. (2016). Remote Sensing from Air and Space (2nd ed.). Academic Press. doi:10.1117/3.2234477

Oquab, M., Bottou, L., Laptev, I., & Sivic, J. (2015). Is object localization for free-weakly-supervised learning with convolutional neural networks. *Proceedings of the IEEE Conference on Computer Vision and Pattern Recognition* (pp. 685-694). IEEE. 10.1109/CVPR.2015.7298668

Ortiz-Canavate, J., Hills, D. J., & Chancellor, W. J. (1981). Diesel Engine Modification to Operate on Biogas. *Transactions of the ASAE. American Society of Agricultural Engineers*, *24*(4), 808–813. doi:10.13031/2013.34342

Osmani, A., & Zhang, J. (2013). Stochastic optimization of a multi-feedstock lignocellulosic-based bioethanol supply chain under multiple uncertainties. *Energy*, *59*, 157–172. doi:10.1016/j.energy.2013.07.043

Osmani, A., & Zhang, J. (2014). Economic and environmental optimization of a large scale sustainable dual feedstock lignocellulosic-based bioethanol supply chain in a stochastic environment. *Applied Energy*, *114*, 572–587. doi:10.1016/j.apenergy.2013.10.024

Osório, G. J., Matias, J. C. O., & Catalão, J. P. S. (2015). Short-term wind power forecasting using adaptive neuro-fuzzy inference system combined with evolutionary particle swarm optimization, wavelet transform and mutual information. *Renewable Energy*, *75*, 301–307. doi:10.1016/j.renene.2014.09.058

Ospanov A.B., Vasiliev A.N., Budnikov D.A., Karmanov D.K., & Dautkanova D.R. (2016). Changing Parameters of the Microwave Field in the Grain. *Layer Journal of Engineering and Applied Sciences ll,* (Special Issue 1), 2915-2919.

Ospanov, A. B., Vasilyev, A. N., Budnikov, D. A., & Karmanov, D. K. (2017). Improvement of grain drying and disinfection process in the microwave field. Almaty: Nur-Print.

Othman, M. Y. H., Ruslan, H., Sopian, K., & Jin, G. L. (2009). Performance study of photovoltaic thermal (PV/T) solar collector with V-grooved absorber plate. *Sains Malaysiana*, 537–541.

Owens, M. J., Riley, P., & Horbury, T. S. (2017). Probabilistic solar wind and geomagnetic forecasting using an analogue ensemble or "similar day" approach. *Solar Physics*, *292*(5), 69. doi:10.100711207-017-1090-7

Özkan, M. B., Terciyanli, E., Küçük, D., Buhan, S., Demirci, T., Yıldız, C., & Günindi, M. (2013). Verification of a real-time wind power monitoring and forecast system for Turkey. *Proceedings of the IET Renewable Power Generation Conference*, Beijing, China. Academic Press.

Özkan, M. B., Guvengir, U., Küçük, D., Secen, A. U., Buhan, S., Demirci, T., ... Karagoz, P. (2017, September). Probabilistic Wind Power Forecasting by Using Quantile Regression Analysis. *Proceedings of the International Workshop on Data Analytics for Renewable Energy Integration* (pp. 72-82). Springer. 10.1007/978-3-319-71643-5_7

Özkan, M. B., & Karagoz, P. (2015). A novel wind power forecast model: Statistical hybrid wind power forecast technique (SHWIP). *IEEE Transactions on Industrial Informatics*, *11*(2), 375–387.

Ozkan, R., Smilanick, J. L., & Karabulut, O. A. (2011). Toxicity of ozone gas to conidia of Penicillium digitatum, Penicillium italicum, and Botrytis cinerea and control of gray mold on table grapes. *Postharvest Biology and Technology*, *60*(1), 47–51. doi:10.1016/j.postharvbio.2010.12.004

Paksoy, T., & Pehlivan, Y. N. (2012). A fuzzy linear programming model for the optimization of multi-stage supply chain networks with triangular and trapezoidal membership functions. *Journal of the Franklin Institute*, *349*(1), 93–109. doi:10.1016/j.jfranklin.2011.10.006

Panchenko, V.A. (2019). Patent No. 188073. Russia.

Panchenko, V.A., Strebkov, D.S., Polyakov, V.I. & Arbuzov, Y.D. (2015). Vysokovol'tnye solnechnye moduli s naprjazheniem 1000 V (High-voltage solar modules with a voltage of 1000 V). *Al'ternativnaya energetika i ekologiya [Alternative energy and ecology]*, *19*(183), 76-81. (in Russian)

Panchenko, V. A. (2018). Solar Roof Panels for Electric and Thermal Generation. *Applied Solar Energy*, *54*(5), 350–353. doi:10.3103/S0003701X18050146

Panchenko, V., Kharchenko, V., & Vasant, P. (2019). *Modeling of Solar Photovoltaic Thermal Modules. In Intelligent Computing & Optimization* (pp. 108–116). Springer Nature Switzerland. doi:10.1007/978-3-030-00979-3_11

Pan, I., & Das, S. (2016). Fractional order fuzzy control of hybrid power system with renewable generation using chaotic PSO. *ISA Transactions*, *62*, 19–29. doi:10.1016/j.isatra.2015.03.003

Pavlyukevich, I. (2007). Lévy flights, non-local search and simulated annealing. *Journal of Computational Physics, 226*(2), 1830–1844. doi:10.1016/j.jcp.2007.06.008

Peretyagin, S. P. (1995). Ozon v biologii i meditsine [Ozone in biology and medicine]. In Abstracts of the 2nd All-Russian Scientific and Practical Conference. Academic Press. (in Russian)

Pérez-Lechuga, G., Medina-Marín, J., Tuoh-Mora, J. C., & Mora-Cárdenas, G. E. (2010). The stochastic short-term hydrothermal scheduling problem. *Proceedings of the 15th annual International Conference on Industrial Engineering theory applications and practice.*

Perova, M.B. (1998). Regulirovaniye otnosheniy mezhdu sel'skimi potrebitelyami i energosnabzhayushchimi organizatsiyami [Regulation of relations between rural consumers and energy supplying organizations]. *Mekhanizatsiya i elektrifikatsiya sel'skogo khozyaystva, 4*, 18-19.

Perrier, C., & Beroual, A. (2009). Experimental investigations on insulating liquids for power transformers: Mineral, ester, and silicone oils. *IEEE Electrical Insulation Magazine, 25*(6), 6–13. doi:10.1109/MEI.2009.5313705

Pershin, A. F., Bogdanov, K. V., & Smirnov, A. A. (2012). Povysheniye mer bezopasnosti pri ekspluatatsii dezinfektsionnykh kamer s primeneniyem ozona [Improving security measures in the operation of ozone disinfection chambers]. *Proceedings of the 8th International Scientific and Technical Conference "Energy Saving and Energy Supply in Agriculture"* (pp. 317-321). Academic Press. (in Russian)

Pershin, A. F., Kazev, Yu. R., & Fedorov, A. V. (1993). RF patent N°2002413 IPC A 01 M 19/00 B.I. No. 39-49.

Pershin, A. F., Smirnov, A. A., & Bogdanov, K. V. (2012). K voprosu bezopasnoy ekspluatatsii dezinfektsionnykh kamer s primeneniyem ozona. [On the issue of safe operation of disinfection chambers with the use of ozone]. Innovation in agriculture, 1. 50-54. (in Russian)

Persits, I.S., Potapov, V.N., Strebkov, D.S., & Chekhunina, G.S. (2006). Solnechnyy fotoelektricheskiy modul' i sposob ego izgotovleniya [Solar photovoltaic module and method of its manufacture]. *FIIP Bull, 26.*

Pikulin, V. P., & Pohozhaev, S. I. (2004). Prakticheskij kurs po uravneniyam matematicheskoj fiziki [Practical course on the equations of mathematical physics 2nd ed.]. Moscow: MCNMO. (in Russian)

Pobedin, A. V., Dolotov, A. A., & Shekhovtsov, V. V. (2016). Decrease of the Vibration Load Level on the Tractor Operator Working Place by Means of Using of Vibrations Dynamic Dampers in the Cabin Suspension. *Procedia Engineering, 150*, 1252–1257. doi:10.1016/j.proeng.2016.07.136

Podobedov, P. N. (2014). *Evaluation methods for selected designs of insulators and cables in multi-criteria model.* Innovations in agriculture, 4*(9),* 246–247. (in Russian)

Podobedov, P. N., Maslennikov, P. A., & Nekrasov, A. I. (2017). *Expert evaluation of operation reliability of distribution substations electric equipment.* Electric equipment: Maintenance and repair, 11, 67–74. (in Russian)

Polifke, G. (2016). *Hydrodamp - hydraulic torsional vibration damper for tractors.* Paper presented at Conference on Agricultural Engineering, Koln, Germany.

Polifke, G. (2017). *Hydrodamp - hydraulic torsional vibration damper for tractors and construction machines design for a race-sensitive drive train.* Paper presented at VDI Conference on Couplings and Clutch Systems in Drives 2017 / 2nd VDI Conference on Vibration Reduction in Mobile Systems, Ettlingen, Germany.

Polivaev, O. I., Zhidenko, K. S., Maksimov, I. I., & Levkin, I. G. (2017). *Uprugo-dempfiruyushchiy privod vedushchih koles traktora* [Elastic-damping drive of the tractor's driving wheels]. Paper presented at Modern Trends in the Development of Technology and Technical Means in Agriculture, Voronezh, Russia.

Polivaev, O.I., & Kutkov, A. Y. (2010). Effektivnost ispolzovaniya transportnykh sredstv s uprugodempfiruyushchim privodom vedushchikh koles pri tormozhenii [Efficiency of the use of transport units with an elastic-damping drive of driving wheels during braking]. *Machinery in Agriculture*, *5*, 30-32.

Polivaev, O. I., Gorban, L. K., Vorohobin, A. V., & Vedrinsky, O. S. (2018). Decrease of dynamic loads in mobile energy means. *Conference Series: Materials Science and Engineering (IOP)*, 327(4). 10.1088/1757-899X/327/4/042083

Polivaev, O. I., Pilayev, S. N., & Kutkov, A. Yu. (2011). Obosnovanie parametrov i effektivnosti ispolzovaniya uprugodempfiruyushchego privoda vedushchikh koles traktorno-transportnogo agregata v rezhime tormozheniya [Substantiation of parameters and efficiency of using the elastic-damping drive of the driving wheels of the tractor-transport unit in braking mode]. *Bulletin of the Voronezh State Agrarian University*, *3*, 48–52.

Polivaev, O. I., & Vedrinsky, O. S. (2013). *Vliyanie uprugodempfiruyushchego privoda vedushchikh koles traktora na dinamicheskie nagruzki v transmissii pri razgone MTA* [Influence of the elastic-damping drive of the tractor's driving wheels on the dynamic loads in the transmission when accelerating the MTU]. Innovative Technologies and Technical Means for Agroindustrial Complex.

Polivaev, O. I., & Vedrinsky, O. S. (2014). Analiz vliyaniya uprugodempfiruyushchego privoda koles na dinamicheskie nagruzki v transmissii traktora pri razgone [Analysis of the effect of the elastic-damping drive of the wheels on the dynamic loads in the tractor's transmission during acceleration]. *The Bulletin of the Don Agrarian Science*, *28*, 5–9.

Polivaev, O. I., Vedrinsky, O. S., & Derkanosova, N. M. (2016). Povishenie dolgovechnosti stsepleniya traktorov za schet uprugofriktsionnogo dempfera [Increasing the durability of tractor traction due to elastic-friction damper]. *Science and Education in Modern Conditions*, *12*, 226–230.

Polovinkin, A.I. (1988). *Osnovy inzhenernogo tvorchestva [Fundamentals of engineering creativity]*. Moscow: Mashinostroenie.

Polshina, V. G. (2018). Sovremennaya tekhnologiya khraneniya plodoovoshchnoy produktsii [Modern technology of storage of fruits and vegetables]. (in Russian). Retrieved from http://web.snauka.ru/issues/2018/01/85522

Poncet, A. M., Fulton, J. P., McDonald, T. P., Knappenberger, T., Shaw, J. N., & Bridges, R. W. (2018). Effect of Heterogeneous Field Conditions on Corn Seeding Depth Accuracy and Uniformity. *Applied Engineering in Agriculture*, *34*(5), 819–830. doi:10.13031/aea.12238

Popel, O. S., & Fortov, V. E. (2015). Renewable energy in the modern world. Publishing house of Moscow Power Engineering Institute.

Popela, P., Novotny, J., Roupec, J., Hrabec, D. H., & Olstad, A. (2018). *Two-stage stochastic programming for engineering problems*. Accessed in November 2018, recovered from: http://www.engineeringmechanics.cz/pdf/21_5_335.pdf

Poudel, S. R., Marufuzzaman, M., & Bian, L. (2016). Designing a reliable bio-fuel supply chain network considering link failure probabilities. *Computers & Industrial Engineering*, *91*, 85–99. doi:10.1016/j.cie.2015.11.002

Poulek, V., Strebkov, D. S., Persic, I. S., & Libra, M. (2012). Towards 50 years lifetime of PV panels laminated with silicone gel technology. *Solar Energy*, *86*(10), 3103–3108. doi:10.1016/j.solener.2012.07.013

Prakash, J. (1994). Transient analysis of a photovoltaic thermal solar collector for cogeneration of electricity & hot air/water. *Energy Conversion and Management*, *35*(11), 967–972. doi:10.1016/0196-8904(94)90027-2

Priddy & Keller. (2005). Artificial Neural Networks: An Introduction. Academic Press. doi:10.1117/3.633187

Prokhorov, Yu. K., & Frolov, V. V. (2011). *Management decisions (2nd ed.). Saint-Petersburg: ITMO*. (in Russian)

Provas, K. R. (2013). Teaching learning-based optimization for short-term hydrothermal scheduling problem considering valve point effect and prohibited discharge constraint. *International Journal of Electrical Power & Energy Systems*, *53*, 10–19. doi:10.1016/j.ijepes.2013.03.024

Pruss, V. L., & Tislenko, V. V. (1989). *Povysheniye nadezhnosti sel'skikh elektricheskikh setey* [The reliability improving of rural electrical networks]. Leningrad: Energoatomizdat.

Pticyn, S. Yu. (1964). Znachenie temperatury sredy pri sushke i uvlazhnenie zerna. *[Proceedings of VIM]* (Vol. 34, pp. 5-28). VIM.

Puleo, V., Morley, M., Freni, G., & Savi, D. (2014). Multi-stage Linear Programming Optimization for Pump Scheduling. *Procedia Engineering*, *70*, 1378–1385. doi:10.1016/j.proeng.2014.02.152

Radner, R. (1955). The linear team: An example of linear programming under uncertainty. *2nd Symp. Lin. Prog.*, 381-396.

Radner, R. (1959). The application of linear programming to team decisions problem. *Management Science*, *5*(2), 143–150. doi:10.1287/mnsc.5.2.143

Rajan. (2011). Hydro-thermal unit commitment problem using simulated annealing embedded evolutionary programming approach. Electrical Power and Energy Systems, 939-946.

Rakhimov, E. Y., Sadullaeva, S. E., Kolomiets, Y. G., Tashmatov, K. K., & Usmonov, N. O. (2017). Analysis of the solar energy potential of the republic of Uzbekistan. *Applied Solar Energy*, *53*(4), 344–346. doi:10.3103/S0003701X17040120

Rakhym, U., Auyelbek, I., Tazhikhan, T., Taraba-yev, B., Maemerov, M., Iztayev, B., Dauletkeldi, Y., & Moldabekova, N. (2016). Influence of presowing ion-ozone cavitational processing and air ionization in the cultivation process on high-yielding and seed characteristics of grain crops. *IJPT*, *8*(2), 14317-14327

Rawat, P., Debbarma, M., Mehrotra, S., & Sudhakar, K. (2014). Design, development and experimental investigation of solar photovoltaic/thermal (PV/T) water collector system. *International Journal of Science. Environmental Technology*, *3*(3), 1173–1183.

RE Technologies Cost Analysis Series. (2012). *Volume 1: Power Sector, Issue 3/5, Hydropower*. International Renewable Energy Agency. Retrieved from https://www.irena.org/documentdownloads/publications/re_technologies_cost_analysis-hydropower.pdf

Revyakin, E. L., Tabashnikov, A. T., Samoylenko, E. M., & Dragaytsev, V. I. (2011). *Resource-saving technologies: state, prospects, efficiency*. Moscow: RosInformagrotekh.

Rezgo, G.J. (2011). Ozonirovanie kak innovacionnyj metod hranenija polukopchjonnyh kolbas [Ozonation as an innovative method of storage of smoked sausages]. *Tovaroved prodovol'stvennyh tovarov [Merchandise production goods]*, *2*, 35 - 39. (in Russian)

Rice, R. G. (1999). Ozone in the Unites States of America – States of the art. *Ozone Science and Engineering*, *21*(2), 99–118. doi:10.1080/01919519908547244

Rodriguez, J. M., Alonso, O., Duvison, M., & Domingez, T. (2008, July). The integration of renewable energy and the system operation: The Special Regime Control Centre (CECRE) in Spain. *Proceedings of the 2008 IEEE Power and Energy Society General Meeting-Conversion and Delivery of Electrical Energy in the 21st Century* (pp. 1-6). IEEE.

Römisch, W., & Vigerske, S. (2010). Recent Progress in Two-stage Mixed-integer Stochastic Programming with Applications to Power Production Planning. In P. Pardalos, S. Rebennack, M. Pereira, & N. Iliadis (Eds.), *Handbook of Power Systems I. Energy Systems*. Berlin: Springer. doi:10.1007/978-3-642-02493-1_8

Rubin, A. B. (1999). Biofizika: V 2-h t. T. 2.: Biofizikakletochnyh processov (2-e izd.). Moscow: Knizhnyjdom Univer-sitet.

Rubin, A., & Riznichenko, G. (2014). *Mathematical biophysics*. Springer Series in Chemical Physics. T. XV. doi:10.1007/978-1-4614-8702-9

Rubinstein, R. Y. (1992). *Monte Carlo Optimization, Simulation and Sensitivity of Queuing Networks*. John Wiley & Sons, Inc.

Rui, Z., & Jianzhong, Z., Wang, & Yongqiang, W. (2012). Multi-objective optimization of hydrothermal energy system considering economic and environmental aspects. *International Journal of Electrical Power & Energy Systems*, *42*(1), 384–395.

Ruma, H., Sakugawa, T., & Akiyama, H. (2018). The Role of Pulse Voltage Amplitude on Chemical Processes Induced by Streamer Discharge at Water Surface. *Catalysts*, *8*(5).

Russell, A. G. (1988). *Mathematical Modeling of the Effect of Emission Sources on Atmospheric Pollutant Concentrations, Air Pollution, the Automobile, and Public Health*. Academic Press.

Ruzavin, G.I. (1975). *Metody nauchnogo issledovanija [Research Methods]*. Moscow: Mysl'. (in Russian)

Salgado, L., Josue, C., Año, O., Esteybar, O., & Mauricio, D. (2016). Hydrothermal scheduling with variable head hydroelectric plants: Proposed strategies using benders decomposition and outer approximation. Power and Energy Conference at Illinois (PECI) 2016 IEEE, 1-8.

Samojlovich V.G., Gibalov V.I., & Kozlov K.V. (n.d.). Fizicheskaja himija bar'ernogo razrjada [*Physical chemistry of the barrier discharge*] Moscow State University. (in Russian)

Sandnes, B., & Rekstad, J. (2002). A photovoltaic/thermal (PV/T) collector with a polymer absorber plate: Experimental study and analytic model. *Solar Energy*, *72*(1), 63–73. doi:10.1016/S0038-092X(01)00091-3

Sasikala, J., & Ramaswamy, M. (2010). Optimal gamma based fixed head hydrothermal scheduling using genetic algorithm. *Expert Systems with Applications*, *37*(4), 3352–3357. doi:10.1016/j.eswa.2009.10.015

Schachter. (2017). Automatic Target Recognition (2nd ed.). Academic Press. doi:10.1117/3.2276891

Schaumburg-Müller, C. (2018). *Mathematical models and methods for analysis of distributed power generation on market conditions*. IMM-PHD-2008-177. Recovered from http://orbit.dtu.dk/files/4963367/phd177_csm-web.pdf

Schinabeck, T. M., Weigler, F., & Floeter, E. (2018). Variability in determination of the single kernel moisture content of grain by means of TD-NMR spectroscopy. *Quality assurance and safety of crops & foods, 10*(1), 75-82.

Sel'eGans. (1960) Ocherki ob adaptacionnom syndrome [Essays on Adaptation Syndrome]. Moscow: Gosudarstvennoe izdatel'stvo medicinskoj literatury.

Sel'eGans. (1972). Na urovne celogo organizma [At the level of the whole organism]. Moscow: Nauka.

Sel'eGans. (1979). *Stress bez distressa* [Stress without distress]. Moskva: Progress.

Selvakumar, I. A. & Thanushkodi, K. (2007). A New Particle Swarm Optimization Solution to Nonconvex Economic Dispatch Problems. *IEEE Transactions on Power System*, *22*(1).

Semenov, A. N. (1959). *Zernovyye seyalki* [*Grain drills*]. Moscow, Kiev: Mashgiz. Russia.

Semenov, Yu. P. (2015). *Teplotekhnika* [*Thermal Engineering*]. Russia: Infra-M.

Senkevich, S. E. (2006). *Povysheniye effektivnosti funktsionirovaniya kul'tivatornogo mashinno-traktornogo agregata na baze traktora klassa 1.4* [Improving the efficiency of the cultivating machine-tractor unit on the basis of a class 1.4 tractor] (PhD dissertation). FSEE HE ABSSAEA, Zernograd, Russia.

Senkevich, S. E., Sergeev, N. V., Vasilev, E. K., Godzhaev, Z. A., & Babayev, V. (2019b). Use of an Elastic-Damping Mechanism in the Tractor Transmission of a Small Class of Traction (14 kN): Theoretical and Experimental Substantiation. In *Advanced Agro-Engineering Technologies for Rural Business Development* (pp. 149–179). IGI Global. doi:10.4018/978-1-5225-7573-3.ch006

Senkevich, S., Kravchenko, V., Duriagina, V., Senkevich, A., & Vasilev, E. (2019a). Optimization of the Parameters of the Elastic Damping Mechanism in Class 1.4 Tractor Transmission for Work in the Main Agricultural Operations. In P. Vasant, I. Zelinka, & G. W. Weber (Eds.), *Intelligent Computing & Optimization. ICO 2018. Advances in Intelligent Systems and Computing* (Vol. 866, pp. 168–177). Cham: Springer. doi:10.1007/978-3-030-00979-3_17

Sen, S. L. S., Yu, L., & Genc, T. (2006). A stochastic programming approach to power portfolio optimization. *Operations Research, 54*(1), 55–72. doi:10.1287/opre.1050.0264

Seuring, S. (2013). A review of modeling approaches for sustainable supply chain management. *Decision Support Systems, 54*(4), 1513–1520. doi:10.1016/j.dss.2012.05.053

Sevela, P. & Olesen, B.W. (2013). Development and Benefits of Using PVT Compared to PV. *Sustainable Building Technologies,* 90-97.

Shaik, M. H., Sudheer, P. K., & Vijaya, K. R. (2013). Biogas–Diesel Dual Fuel Engine Exhaust Gas Emissions. *International Journal of Advanced Research in Engineering and Technology, 4*(3), 211–216.

Shakhov, A.B. (2010). Ecological and economic basis of energy conversion of agricultural waste. *Bulletin of the Federal State Educational Establishment of Higher Professional Education, 5*(44), 82-86.

Shakhov, A. V. (2011). *Organizational and economic basis for the implementation of bioenergy potential of agricultural production: monograph.* Moscow: MSAU.

Shamir, J. (1999). Optical Systems and Processes. Academic Press. doi:10.1117/3.336094

Shapovalenko, N. S., Dorovskih, V. A., Korshunova, N. V., Shtarberg, M. A., Slastin, S. S., & Nevmyvako, E. E. (2011). Vliyanie holodovogo stressa na intensivnost' perekisnogo okisleniya lipidov i antioksidantnuyu sistemu tkanej ehksperimental'nyh zhivotnyh [The effect of cold stress on the intensity of lipid peroxidation and the antioxidant system of experimental animal tissues]. Byulleten' fiziologii i patologii dyhaniya [Bulletin of physiology and pathology of respiration], 39, 22-25.

Sharma, D.N. & Sinh K. (2002). Uravneniya v chastnyh proizvodnyh dlya inzhenerov [Partial differential equations for engineers]. Moscow: Tekhnosfera. (in Russian)

Shekhovtsov, V. V., Sokolov-Dobrev, N. S., & Potapov, P. V. (2016). Decreasing of the Dynamic Loading of Tractor Transmission by means of Change of the Reactive Element Torsional Stiffness. *Procedia Engineering, 150*, 1239–1244. doi:10.1016/j.proeng.2016.07.129

Shennon, K. (1963). Raboty po teorii informacii i kibernetike [Works on information theory and cybernetics]. Moscow: Inostrannaya literatura.

Shesterin, I. V. (2004). Vliyaniye ozona i protraviteley na posevnyye kachestva i ozdorovleniye yarovoy pshenitsy [Influence of ozone and disinfectants on sowing qualities and improvement of spring wheat]. [Dissertation abstract]. (in Russian)

Shi, J., & Oren, S. S. (2016). Wind power integration through stochastic unit commitment with topology control recourse. *19th Power Systems Computation Conference, (PSCC 2016)*, Genova, Italy.

Shi, J., Zhang, W., Zhang, Y., Xue, F., & Yang, T. (2015). MPPT for PV systems based on a dormant PSO algorithm. *Electric Power Systems Research*, *123*, 100–107. doi:10.1016/j.epsr.2015.02.001

Shilin, Z., Yufeng, D., & Houzhang, T. (2016). Migration and Emission Characteristics of Trace Elements in a 660 MW Coal-Fired Power Plant of China. *Energy & Fuels*, *30*(7), 5937–5944. doi:10.1021/acs.energyfuels.6b00450

Shishkin, N. D. (2005). Kombinirovanie i effektivnoe ispol'zovanie istochnikov teplovoj energii v avtonomnyh teploenergeticheskih kompleksah: vklyuchaya vozobnovlyaemye istochniki [Combination and effective use of thermal energy sources in autonomous heat and power complexes, including renewable sources] [Dissertation]. (in Russian)

Shmal'gauzen, I. I. (1968). Kiberneticheskie voprosy biologii / Pod obshch. red. i s predisl. R. L. Berg i A. A. Lyapunova [Cybernetic questions of biology / Under total. ed. and with foreword. R. L. Berg and A. A. Lyapunov]. Novosibirsk, Nauka, sib. otdelenie.

Sifuentes, W. S., & Vargas, A. (2007). Hydrothermal Scheduling Using Benders Decomposition: Accelerating Techniques. *IEEE Transactions on Power Systems*, *2*(3), 1351–1359. doi:10.1109/TPWRS.2007.901751

Silikonovaia transformatornaia zhidkost Sofeksil-TSZH Rekomendatsii po primeneniiu [Sofaxil-TSZh Silicone Transformer Liquid. Recommendations for use]. (2016). Moscow, Russia.

Silovye transformatory. Katalog [Power transformers. Catalog]. (2018). Retrieved from: http://silovoytransformator.ru/110kv/tdtn-40000-110-u1.htm

Simon, J., & Johan, W. (2011). Cryogenic biogas upgrading using plate heat exchangers.

Sinha, P. K. (2012). Image Acquisition and Preprocessing for Machine Vision Systems. Academic Press. doi:10.1117/3.858360

Sinha, N., Chakrabarti, R., & Chattopadhyay, P. K. (2003). Fast evolutionary programming techniques for short-term hydrothermal scheduling. *Electric Power Systems Research*, *66*(2), 97–103. doi:10.1016/S0378-7796(03)00016-6

Smilanick, J. L. (2003). Use of ozone in storage and packing facilities. *Proc. 99th Meeting* (pp. 131-139). Washington State Horticultural Association.

Smith, D. L., Atungulu, G. G., Sadaka, S., & Rogers, S. (2018). Implications of micro-wave drying using 915MHz frequency on rice physico-chemical properties. *Cereal Chemistry*, *95*(2), 211–225. doi:10.1002/cche.10012

Sokut, L.D & Murovskaya, A.S. (2017). Perspektivy razvitiya sistem ehlektrosnabzheniya za schet podklyucheniya vetrovyh i solnechnyh ehlektrostancij s nakopitelyami ehnergii v obshchuyu ehnergosistemu [Prospects for the development of power supply systems due to the connection of wind and solar power plants with energy storage in a common power system]. *Stroitel'stvo i tekhnogennaya bezopasnost' [Construction and Technogenic Safety]*, 7 (59), 113-123. (in Russian)

Solimpeks. Volther Hybrid PV-T Panels. (n.d.). Retrieved from http://www.solimpeks.com

Solovyov, A.K. (2018). EHkonomiya ehnergii pri ehkspluatacii zdanij i passivnye sistemy ispol'zovaniya solnechnoj ehnergii [Energy savings in the operation of buildings and passive systems of solar energy use]. *Stroitel'stvo i tekhnogennaya bezopasnost' [Construction and Technogenic Safety]*, *10*(62), 179-191. (in Russian)

Son, E.E. (2013). Modern studies of the thermophysical properties of substances. *Thermophysics of High Temperatures*, *51*(3), 392.

Sondergaard, M. M., Ioannis, A., Fotidis, I. A., Kovalovszki, A., & Angelidaki, I. (2015). Anaerobic co-digestion of agricultural byproducts with manure for enhanced biogas production. *Energy & Fuels*, *29*(12), 8088–8094. doi:10.1021/acs.energyfuels.5b02373

Songfeng, L., Chengfu, S., & Zhengding, L. (2010). An improved quantum-behaved particle swarm optimization method for short-term combined economic emission hydrothermal scheduling. *Energy Conversion and Management*, *51*(3), 561–571. doi:10.1016/j.enconman.2009.10.024

Sorensen, B. (2011). Transformation, transmission and accumulation of energy. Academic Press.

Sottinen, T., & Tudor, C. A. (2006). On the Equivalence of Multiparameter Gaussian Processes. *Journal of Theoretical Probability*, *19*(2), 461–485. doi:10.100710959-006-0022-5

Soyfer, A.V. (2003). *Methods of computer processing of pictures.* Moscow: Publishing House "PHYSMATHLIT".

Spiridonov, A. A. (1981). *Planirovanie eksperimenta pri issledovanii tehnologicheskikh protsessov* [Planning an experiment in the study of technological processes]. Moscow, Russia: Mechanical Engineering.

Srivastav, A., & Agrawal, S. (2017). Multi-objective optimization of slow moving inventory system using cuckoo search. In Intelligent Automation & Soft Computing. Academic Press.

Stankus, S.V., Khairulin, R.A., Martynets, V.G., & Bezverkhiy, P.P. (2013). Studies of the thermophysical properties of substances and materials at the Novosibirsk Scientific Center of the SB RAS in 2002-2012. *Thermophysics of High Temperatures*, *51*(5), 769.

Steffen, H. (2007). Fruit and vegetable disinfection at SAMRO by means of ozone and UV Radiation. *Proceedings of the 18th World Ozone Congress*. Academic Press.

Strebkov, D. S., Kirsanov, A. I., & Panchenko, V. A. (n.d.). Patent for invention 17773387.0-1230PCT/IB2017051470. RU/28.03.16/RUA2016111201. Russia.

Strebkov, D.S., Bashilov, A.M., & Korolyov, V.A. (2011). Principles of construction and versions of realization of power supply systems, navigation and movement management of promising agro-aggregates. *Polzunov's Bulletin, 2*(2), 280-284.

Strebkov, D.S., Kuznetsov, I.M., & Bashilov, A.M. (2010). Improvement of technical base of management of agricultural production on innovational-informational basis. *Series: Power supply and energy saving in agriculture. Moscow, State Scientific Institution VIESH, 5,* 49-65.

Strebkov, D.S., Persits, I.S. & Panchenko, V.A. (2014). Solnechnye moduli s uvelichennym srokom sluzhby [Solar modules with extended service life]. *Innovacii v sel'skom hozyajstve [Innovations in agriculture],* 3(8), 154-158. (in Russian)

Strebkov, D.S., Polyakov, V.I. & Panchenko, V.A. (2013). Issledovanie vysokovol'tnyh solnechnyh kremnievyh modulej [Investigation of high-voltage solar silicon modules]. *Al'ternativnaya energetika i ekologiya [Alternative Energy and Ecology], 6*(2), 36-42. (in Russian)

Suman Sutradhar, S., Nalin, B., Dev Choudhury, N., B., & Sinha, N. (2018). Parallel ABC optimization algorithm for hydrothermal scheduling problems. AIP Conference Proceedings, 1. doi:10.1063/1.5049111

Sun, Ch., & Lu, S. (2010). Short-term combined economic emission hydrothermal scheduling using improved quantum-behaved particle swarm optimization. *Expert Systems with Applications*, *37*(6), 4232–4241. doi:10.1016/j.eswa.2009.11.079

Sunsystems. (n.d.). Retrieved from http://www.sunsystem.bg/en/fotovoltaika/PV-T/

Swift, G., Molinski, T. S., & Lehn, W. (2001). A fundamental approach to transformer thermal modeling. I. Theory and equivalent circuit. *IEEE Transactions on Power Delivery*, *16*(2), 171–175. doi:10.1109/61.915478

T. B. Leshchinskaya & V. V. Knyazev. (2006). *Multi-criteria evaluation of technical-economic status of power distribution networks*. FGOU VPO MGAU. (in Russian)

Taha, H. A. (2017). *Operations Research: An Introduction*. Pearson.

Tahir, N. M., Salman, Z., & Saaqib, H. (2016). Short-term economic emission power scheduling of hydrothermal systems using improved chaotic hybrid differential evolution. *Turkish Journal of Electrical Engineering and Computer Sciences*, *24*, 2654–2670. doi:10.3906/elk-1403-298

Tan, Q., Wang, T., Zhang, Y., Miao, X., & Zhu, J. (2017). Nonlinear multi-objective optimization model for a biomass direct-fired power generation supply chain using a case study in China. *Energy*, *139*, 1066–1079. doi:10.1016/j.energy.2017.08.050

Teplovye nasosy [Heat pumps]. (n.d.). Retrieved from http://www.nibe-evan.ru/nibedocuments/17337/M11540-1.pdf

Terciyanli, E., Demirci, T., Küçük, D., Buhan, S., Özkan, M. B., Köksoy, C. E., ... Günindi, M. (2013, May). The architecture of a large-scale wind power monitoring and forecast system. *Proceedings of the 4th International Conference on Power Engineering, Energy and Electrical Drives* (pp. 1162-1167). IEEE. 10.1109/PowerEng.2013.6635775

Terciyanli, E., Demirci, T., Küçük, D., Sarac, M., Cadirci, I., & Ermis, M. (2014). Enhanced nationwide wind-electric power monitoring and forecast system. *IEEE Transactions on Industrial Informatics*, *10*(2), 1171–1184. doi:10.1109/TII.2013.2294157

Thompson, A. K. (2010). *Controlled atmosphere storage of fruits and vegetables* (2nd ed.). Oxford, UK: CAB International. doi:10.1079/9781845936464.0000

Thompson, A. K. (2016). Controlled Atmosphere Storage. In *Fruit and Vegetable Storage*. Cham: Springer. doi:10.1007/978-3-319-23591-2_2

Tikhomirov, P. M. (2015). *Raschet transformatorov* [*Calculation of transformers*]. Russia: Tyumen State Oil and Gas University. Retrieved from https://studfiles.net/preview/5332633/page:76/

Tim, H. W., & Lall, U. (2013). Optimal crop choice, irrigation allocation, and the impact of contract farming. *Production and Operations Management*, *22*(5), 1126–1143.

Timonin, Yu. N., Gorelov, Yu. I., & Stepanov, V. M. (2013). *R.F. Patent No. 128776*. Patent Service of the Russian Federation.

Tintner, G. (1955). Stochastic linear programming with applications to agricultural economics. In H. A. Antosiewicz (Ed.), *Proceedings of the Second Symposium in Linear Programming*, (pp. 197-228). National Bureau of Standards.

Tislenko, V. V. (1991). Modeli i metody upravleniya nadezhnost'yu sistem elektrosnabzheniya sel'skokhozyaystvennogo naznacheniya [Models and methods for managing of the reliability of agricultural power supply system]. Kyiv: Instytut elektrodinamiki AN Ukrainy.

Tislenko, V. V., Kholmskiy, D. V., & Gurbich, R. F. (1975). Issledovaniye zavisimosti pokazateley nadezhnosti elektrosnabzheniya potrebiteley ot dliny vozdushnykh liniy elektroperedachi raspredelitel'noy seti [Study of the reliability indicators dependence of electricity supply on the length of the air-distribution network of power lines]. *Energetika i elektrifikatsiya*, *2*, 15-17.

Tkachenko, S. N., Golosman, E. Z., & Lunin, V. V. (2001). Tsementosoderzhashchiye katalizatory ochistki gazov ot ozona [Cement-containing catalysts for ozone gas purification]. *Catalysis in Industry*, (2), 52–55. (in Russian)

Transmission and Distribution Committee. (2003). IEEE Guide for Electric Power Distribution Reliability Indices. IEEE Std 1366.

Trockaja T.P. (2005) Osnovnye napravlenija isp. ozona v mjasomolochnoj promyshlennosti [The main directions isp. ozone in meat and dairy industry]. In *Conference materials: Modern technologies and complexes of technical means in agricultural production*. Academic Press. (in Russian)

Trukhacheva, T. V., Gavrilov, V. B., Malama, G. A., & Astakhov, V. A. (1992). Kineticheskiye zakonomernosti gibeli mikroorganizmov pod deystviyem ozona [Kinetic regularities of the death of microorganisms under the action of ozone]. *Microbiology Today*, *61*(4), 660–665. (in Russian)

Trunin, A. S. (1997). *Complex research methodology for multicomponent systems.*

Tsoi, A. D., Kopyrin, V. S., & Sosnovsky, O. G. (2013). *U. Patent No.* 2505. Patent Service of Ukraine.

Tursunov, M. N., Sabirov, K., Hwoan, P. J., Turdiev, B. M., Shirmatov, A. R., & Yuldoshev, I. A. (2013). Self-contained photovoltaic unit as a source of power supply for a milking parlor on a cattle farm. *Applied Solar Energy*, *49*(4), 232–234. doi:10.3103/S0003701X13040166

Twidell, D., & Weir, A. (1990). *Renewable energy sources.*

Tzortzakis, N. (2015). Gaseous ozone-enrichment for the preservation of fresh produce. *Acta Horticulturae*, (1079), 273–278. doi:10.17660/ActaHortic.2015.1079.32

Vaccaro, A., Popov, M., Villacci, D., & Terzija, V. (2011). An integrated framework for smart microgrids modeling, monitoring, control, communication, and verification. *Proceedings of the IEEE*, *99*(1), 119–132. doi:10.1109/JPROC.2010.2081651

Vajda, S. (1972). *Probabilistic Programming*. Academic Press.

Varma, S. A. K. (2014). *Air Pollution Modelling for Thermal Power Plants, Monitoring and Modelling*. LAP Lamber Academic Publishing.

Vasil'ev, A. N. (2009). Ehnergosberegayushchie ehlektrotekhnologii sushki i predposevnoj obrabotki zerna aktivnym ventilirovaniem [Energy-saving electrical technologies of drying and pre-sowing treatment of grain by active ventilation].

Vasil'ev, A. N., & Dzhanibekov, A. K. (2014a). Kak obespechit' predposevnuyu obrabotku semyan v bunkerah aktivnogo ventilirovaniya s trebuemym kachestvom [How to ensure presowing treatment of seeds in bunkers of active ventilation with the required quality]. Moscow: FGBNU VIM.

Vasil'ev, A. N., & Dzhanibekov, A. K. (2014b). O vozmozhnosti ispol'zovaniya vtorichnyh ionov dlya predposevnoj obrabotki semyan. Teoreticheskij i nauchno-prakticheskij zhurnal [Innovacionnoe razvitie APK Rossii na baze intellektu-al'nyh mashinnyh tekhnologij. *Innovacii v sel'skomhozyajstve [Innovation in agriculture]*, *1*(6). Retrieved from http://ej.viesh.ru/wp-content/uploads/2014/04/insel6.pdf

Vasil'ev, A. N., & Kononenko, A. F. (2008). Ehlektrotekhnologiyai upravlenie v realizaciiadaptivnyhrezhimovpredposevnojobrabotkizernaaktivnymventilirovaniem [Electrotechnology and control in the implementation of adaptive modes of presowing treatment of grain by active ventilation]. Rostov-na-Donu: Terra Print.

Vasil'ev, A. N., & Udincova, N. M. (2000). Analizparametrovvozdushno-teplovojobrabotkidlyauluchsheniyaposevnyhkachestvsemyan [Analysis of the parameters of air-heat treatment to improve the sowing qualities of seeds]. Mekhanizaciya i ehlektrifikaciya sel'skogo hozyajstva [Mechanization and electrification of agriculture], 1, 9-10.

Vasil'ev, A. N., Dzhanibekov, A. K., & Udincova, N. M. (2015). Principy predposevnoj obrabotki semyan [Principles of presowing seed treatment]. *Innovacii v sel'skomhozyajstve [Innovations in Agriculture]*, *4*(14), 29-33. Retrieved from http://ej.viesh.ru/wp-content/uploads/2016/02/insel14.pdf

Vasil'ev, A. N., Kononenko, A. F., & Ereshko, A. S. (2002). Predposevnaya obrabotka semyan yachmenya ehlektroaktivirovannym vozduhom [Presowing treatment of barley seeds with electroactivated air]. Materialy 41 Nauchno-tekhnicheskoj konferencii. Chelyabinsk [Materials 41 Scientific and Technical Conference. Chelyabinsk] (pp. 217-220). Academic Press.

Vasil'yev, A.P. & Guk, Yu.B. (2000). Obespecheniye nadezhnosti, bezopasnosti i energosberezheniya v usloviyakh energeticheskogo rynka [Ensuring reliability, safety and energy saving in the conditions of the energy market]. *Energetika i elektrifikatsiya*, *3*, 45-53.

Vasilchuk N.S. & Epstein V.A. (2007) Vliyaniye predposevnoy obrabotki semyan sistemnymi protravitelyami i ozonom na nachal'nyye rostovyye protsessy i produktivnost' ozimoy pshenitsy [Effect of presowing treatment of seeds with systemic disinfectants and ozone on the initial growth processes and productivity of winter wheat]. *Vestnik SSAU*, *2*, 70 -71. (in Russian)

Vasil'ev A.N. (2014). Ispol'zovanie zapasyonnoj ehnergii biologicheskih ob"ektov v tekhnologicheskih processah APK s pomoshch'yu ehlektrotekhnologij [The use of stored energy of biological objects in the technological processes of the agro-industrial complex with the help of electrical technologies]. *Vestnik VIEHCH [Herald VIESH]*, *1*(14), 28-35.

Vasiliev, A. N., Budnikov, D. A., Gracheva, N. N., & Smirnov, A. A. (2018). Increasing Efficiency of Grain Drying With the Use of Electroactivated Air and Heater Control. In *Handbook of Research on Renewable Energy and Electric Resources for Sustainable Rural Development*. Hershey, PA: IGI Global. doi:5225-3867-7.ch011 doi:10.4018/978-1-

Vasiliev, A. N., Vasiliev, A. A., Budnikov, D., & Samarin, G. (2019). Analysis of experimental research results obtained for grain drying with electrically activated air. doi:10.4018/978-1-5225-7573-3.ch009

Vasiliev, A. N., Vasiliev, A. A., Budnikov, D., Tikhomirov, D., & Ospanov, A. B. (2019). Control and Optimized Management of Grain Drying in Forced-Aerated Bins Advanced Agro-Engineering Technologies for Rural Business Development. doi:10.4018 / 978-1-5225-7573-3.ch008

Vasiliev, A. N., Budnikov, D. A., Gracheva, N. N., & Smirnov, A. A. (2018). Increasing efficiency of grain drying with the use of electroactivated air and heater control. In V. Kharchenko & P. Vasant (Eds.), *Handbook of Research on Renewable Energy and Electric Resources for Sustainable Rural Development* (pp. 255–282). Hershey, PA: IGI Global; doi:10.4018/978-1-5225-3867-7.ch011

Vasiliev, A. N., Ershova, I. G., Belov, A. A., Timofeev, V. N., Uhanova, V. Y., Sokolov, A. V., & Smirnov, A. A. (2018). Energy-saving system development based on heat pump. *Amazonia Investiga*, *7*(17), 219–227.

Vasiliev, A. N., Ospanov, A. B., Budnikov, D. K., Karmanov, D. K., Salginbayev, D. B., & Vasilyev, A. A. (2016). Controlling reactions of biological objects of agricultural production with the use of electrotechnology. *International Journal of Pharmacy & Technology*, *8*(4), 26855–26869.

Vasiliev, Yu. N., Zolotarevsky, L. S., & Ksenofontov, S. I. (1992). *Gas and Gas-Diesel Engines*. Moscow: RAO Gazprom.

Vasiljak, L. M., Kostjuchenko, S. V., & Kol'cov, G. V. (2008). Primenenie impul'snogo i nepreryvnogo uf-izluchenija dlja obezzarazhivanija vody i vozduha [The use of pulsed and continuous UV radiation for disinfecting water and air] [Sanitary engineering]. *Santehnika*, *3*, 75–78.

Vasilyev A.N. (2013). Elektrotechnologies as controlling influence on bio-logical objects in agricultural production. *Research in Agricultural Electric Engineering*, *3*(3), 96-102.

Vasilyev, A. N. (2016). Research Methodology for Microwave Convective Grain Processing. *Innovation in agriculture, 3*(18), 143-154. Retrieved from http://ej.viesh.ru/wp-content/uploads/2016/03/insel16.pdf

Vasilyev, A. N., & Severinov, O. V. (2014). Matematicheskaya model' teplo- i vlagoobmena v ehlementarnom sloe pri sushke zerna aktivnym ventilirovaniem. *AgroEhkoInfo,* (2). (in Russian) Retrieved from http://agroecoinfo.narod.ru/journal/STATYI/2014/2/st_13.doc

Vasilyev, A. N., & Severinov, O. V. (2015). K raschyotu teplo-i vlagoobmena v plotnom sloe zerna. *Teoreticheskij i nauchno-prakticheskij zhurnal Innovacii v sel'skom hozyajstve [Theoretical, scientific and practical journal "Innovations in Agriculture"]. 1*(11). (in Russian) Retrieved from http://ej.viesh.ru/journal/2015-01/

Vasilyev, A. N., & Vasilyev, A. A. (2018). Control of parameters of grain drying process in dense layer. *Proceedings of the International Russian Automation Conference RusAutoCon 2018.* IEEE. doi:10.1109/RUSAUTOCON.2018.8501729

Vasilyev, A. N., Ospanov, A. B., Budnikov, D. A., Vasilyev, A. A., & Karmanov, D. K. (2015). Kriterii optimal'nogo upravleniya sushkoj zerna. Mezhdunarodnaya agroinzheneriya [International Agroengineering], 2, 12-17. (in Russian)

Vasilyev, A. N., Severinov, O. V., & Makarova, Yu. M. (2016). Razrabotka komp'yuternoj modeli teplo- i vlagoobmena v plotnom sloe zerna/ *Vestnik NGEHI (tekhnicheskie nauki), 4*(59), 63-71. (in Russian) Retrieved from http://elibrary.ru/item.asp?id=26083913

Vasilyev, A. N., Vasilyev, A. A., & Budnikov, D. A. (2018). The Functional Dependencies of the Drying Coefficient for the Use in Modeling of Heat and Moisture-Exchange Processes. In P. Vasant & I. Zelinka & G.W. Weber (Eds.), *Intelligent Computing & Optimization ICO 2018.* Springer. doi:10.1007/978-3-030-00979-3_24

Vasilyev, A.N., Budnikov, D.A., & Vasilyev A.A. (2014). Vlijanie osobennostej uborki zerna na rezhimy ego obezzarazhivanija v pole SVCh [Influence of features of grain harvesting on modes of its disinfection in the microwave field]. In *Vestnik VNIIMZh /Serija: Mehanizacija, avtomatizacija i mashinnye tehnologii v zhivotnovodstve* (pp. 82-85). Moscow: GNU VNIIMZh. (in Russian)

Vasilyev, A.N., Ospanov A.B., Dautkanova D.R., Budnikov D.A., Karmanov D.K., Shalginbaev D.B., & Vasilyev A.A. (2016). Primenenie diehlektricheskogo nagreva dlya konvektivnoj sushki zerna/*Vestnik ETU, 2*, 16-20. (in Russian)

Vasilyev, A.N., Ospanov A.B., Dautkanova D.R., Budnikov D.A., Karmanov D.K., Shalginbaev D.B., Vasilyev A.A., & Moshanov K.A. (2016). Modelirovanie mikrovolnovogo oblucheniya v processah teplovoj obrabotki zernovyh kul'tur. *Issledovanie i rezul'taty [Research and Results], 4.* (in Russian)

Vasilyev, A. N., & Rudenko, N. B. (2009). *Model' teplovlagoobmena pri recirkulyacii zerna. In Vestnik agrarnoj nauki Dona* (pp. 37–41). ACHGAA. (in Russian)

Vedenev, A. G., & Vedeneva, T. A. (2011). *Guide to biogas technology*. Bishkek, Kyrgyzstan: DEMI.

Vieira, J. A. B., & Mota, A. M. (2010). Implementation of a Stand-Alone Photovoltaic Lighting System with MPPT Battery Charging and LED Current Control. In *IEEE International Conference on Control Applications Part of 2010 IEEE Multi-Conference on Systems and Control.* (pp. 185-190). Yokohama, Japan: IEEE. 10.1109/CCA.2010.5611257

Vinogradov, A., Vasiliev, A., Bolshev, V., Semenov, A., & Borodin, M. (2018). Time factor for determination of power supply system efficiency of rural consumers. In Handbook of Research on Renewable Energy and Electric Resources for Sustainable Rural Development (pp. 394–420). Hershey, PA: IGI Global. doi:10.4018/978-1-5225-3867-7.ch017

Vinogradov, A., Vasiliev, A., Bolshev, V., Vinogradova, A., Kudinova, T., Sorokin, N., & Hruntovich, N. (2019). Methods of Reducing the Power Supply Outage Time of Rural Consumers. In Renewable Energy and Power Supply Challenges for Rural Regions (pp. 370–392). Hershey, PA: IGI Global. doi:10.4018/978-1-5225-9179-5.ch015

Vozmilov, A.G., Astaf'ev, D.V., & Matveev, S.D. (2007). Primenenie ozona v tehnologicheskih processah pticevodstva i kriterii sravnitel'noj ocenki ozone generatorov [The use of ozone in poultry production processes and criteria for comparative evaluation of ozone generators]. *Mehanizacija i jelektrifikacija sel'skogo hozjajstva [Mechanization and electrification of agriculture]*, *3*, 13-16. (in Russian)

Walton, S., Hassan, O., Morgan, K., & Brown, M. R. (2011). Modified cuckoo search: A new gradient free optimization algorithm. *Chaos, Solitons, and Fractals*, *44*(9), 710–718. doi:10.1016/j.chaos.2011.06.004

Wang, J., Zhang, Y., Liu, Z., Norris, P., Romero, C. E., Xu, H., & Pan, W.-P. (2017). Effect of coordinated air pollution control devices in coal-fired power plants on arsenic emissions. *Energy & Fuels*, *31*(7), 7309–7316. doi:10.1021/acs.energyfuels.7b00711

Wang, S., Minku, L. L., & Yao, X. (2015). Resampling-based ensemble methods for online class imbalance learning. *IEEE Transactions on Knowledge and Data Engineering*, *27*(5), 1356–1368. doi:10.1109/TKDE.2014.2345380

Wang, X., Ma, P., Huang, Y., Guo, J., Zhu, D., Porter, A. L., & Wang, Z. (2017). Combining SAO semantic analysis and morphology analysis to identify technology opportunities. *Scientometrics*, *111*(1), 3–24. doi:10.100711192-017-2260-y

Wei L.S.; Liang X.; Zhang Y.F. (2018). Numerical investigation on the effect of gas parameters on ozone generation in pulsed dielectric barrier discharge. Plasma science & technology, 20(12). doi:10.1088/2058-6272/aadca6

Weiland, P. (2010). Biogas production: Current state and perspectives. *Applied Microbiology and Biotechnology*, *85*(4), 849–860. doi:10.100700253-009-2246-7 PMID:19777226

Willers. (2013). Electro-Optical System Analysis and Design: A Radiometry Perspective. Academic Press. doi:10.1117/3.1001964

Wills, R. B. H., Pitakserikul, S., & Scott, K. J. (1982). Effects of pre-storage in low O2 or high CO2 concentrations on delaying the ripening of bananas. *Australian Journal of Agricultural Research*, *33*, 1029–1036. doi:10.1071/AR9821029

Wittmann-Hohlbein, M., & Pistikopoulos, E. N. (2012). Two-Stage Method for the Approximate Solution of General Multiparametric Mixed-Integer Linear Programming Problems. *Industrial & Engineering Chemistry Research*, *51*(23), 8095–8107. doi:10.1021/ie201408p

Wojciech, J. (2006). US Patent 6991930, USA, IPC C12N 1/14 A01N 63/04.

Wolf, M. (1976). Performance analysis of combined heating and photovoltaic power systems for residences. *Energy Conversion*, *16*(1-2), 79–90. doi:10.1016/0013-7480(76)90018-8

Wong, K. P., & Wong, Y. W. (1994). Short-term hydrothermal scheduling. II. Parallel simulated annealing approach. *IEE Proceedings. Generation, Transmission and Distribution*, *141*(5), 502. doi:10.1049/ip-gtd:19941351

Wood, A. J. (1996). *Power Generation, Operation, and Control*. John Wiley & Sons.

Wozniak, A., Nowakowicz-Debek, B., & Stepniowska, A., with co-authors. (2016). Effect of ozonation on microbiological and chemical traits of wheat grain. *Plant, Soil and Environment*, *62*(12), 552–557.

Wu, J., Doan, H., & Cuenca, M. A. (2006). Investigation of gaseous ozone as an anti-fungal fumigant for stored wheat. *Journal of Chemical Technology and Biotechnology (Oxford, Oxfordshire)*, *81*(7), 1288–1293.

Xiangtao, L., & Minghao, Y. (2013). A hybrid cuckoo search via Lévy flights for the permutation flow shop scheduling problem. *International Journal of Production Research*, *15*, 4732–4754.

Xie, S., Li, X., & Ouyang, Y. (2015). Decomposition of general facility disruption correlations via augmentation of virtual supporting stations. *Transportation Research Part B: Methodological*, *80*, 64–81. doi:10.1016/j.trb.2015.06.006

Yang, H. T., Yang, P. C., & Huang, C. L. (1996). Evolutionary programming based economic dispatch for units with non-smooth fuel cost functions. *IEEE Transactions on Power Systems*, *11*(1), 112–117. doi:10.1109/59.485992

Yang, X. S. (2013). *Optimization and metaheuristic algorithms in engineering. In Metaheuristic in Water Geotechnical and Transport Engineering* (pp. 1–23). Waltham: Elsevier.

Yang, X-S. & Deb, S. (2010). Engineering Optimization by Cuckoo Search. *International Journal of Mathematical Modeling and Numerical Optimization*, *1*(4).

Yeh, W. C., Lin, W. T., Lai, C. M., Lee, Y. C., Chung, Y. Y., & Lin, J. S. (2016, July). Application of simplified swarm optimization algorithm in deteriorate supply chain network problem. *Proceedings of the 2016 IEEE Congress on Evolutionary Computation (CEC)* (pp. 2695-2700). IEEE. 10.1109/CEC.2016.7744127

Yue, D., You, F., & Snyder, S. W. (2014). Biomass-to-bioenergy and biofuel supply chain optimization: Overview, key issues and challenges. *Computers & Chemical Engineering*, *66*, 36–56. doi:10.1016/j.compchemeng.2013.11.016

Yu, R., Gao, J., Yu, M., Lu, W., Xu, T., Zhao, M., ... Zhang, Z. (2019). LSTM-EFG for wind power forecasting based on sequential correlation features. *Future Generation Computer Systems*, *93*, 33–42. doi:10.1016/j.future.2018.09.054

Yutkin, L. A. (1986). Electrohydraulic effect and its application in industry/Leningrad. *Engineering.*

Zadeh, L. A. (1971). Similarity relations and fuzzy orderings. *Information Sciences*, *3*(2), 166–200. doi:10.1016/S0020-0255(71)80005-1

Zadeh, L. A. (1974). *Basics of new approach to the analysis of complex systems and processes of decision selection.* Moscow: Znanie Publ. (in Russian)

Zamaldinova, G.I., Garkushin, I.K., Garkushin, A.I., Parfenova, S.N. (2012). Analytical description of the temperatures and compositions of low-melting eutectics in the series LiF–MF (M=Na,K,Rb,Cs) and prediction of the eutectic characteristics in the system LiF–FrF. *Journal of Inorganic Chemistry*, *57*(6), 961.

Zech, P., Plöger, D. F., & Rinderknecht, S. (2019). Active control of planetary gearbox vibration using phase-exact and narrowband simultaneous equations adaptation without explicitly identified secondary path models. *Mechanical Systems and Signal Processing*, *120*, 234–251. doi:10.1016/j.ymssp.2018.10.030

Zeynal, H., & Eidiani, M. (2014). Hydrothermal scheduling flexibility enhancement with pumped-storage units. *Iranian Conference on Electrical Engineering (ICEE).* 10.1109/IranianCEE.2014.6999649

Zhang, S., Li, C., & Li, S. (2017). Understanding Optical Systems through Theory and Case Studies. Academic Press. doi:10.1117/3.2267707

Zhang, J., Osmani, A., Awudu, I., & Gonela, V. (2013). An integrated optimization model for switchgrass-based bioethanol supply chain. *Applied Energy*, *102*, 1205–1217. doi:10.1016/j.apenergy.2012.06.054

Zhang, Y. F., Wei, L., Liang, S. X., Deng, H. Z., & Šimek, M. (2018). Characteristics of the Discharge and Ozone Generation in Oxygen-Fed Coaxial DBD Using an Amplitude-Modulated AC Power Supply. *Plasma Chemistry and Plasma Processing*, *38*(6), 1199–1208. doi:10.100711090-018-9922-2

Zhao, F., Lei, W., Ma, W., Liu, Y., & Zhang, C. (2016). An Improved SPEA2 Algorithm with Adaptive Selection of Evolutionary Operators Scheme for Multiobjective Optimization Problems. *Mathematical Problems in Engineering.*

Zharkov, S. V. (2014). Assessment and enhancement of the energy supply system efficiency with emphasis on the cogeneration and renewable as main direction for fuel saving. *International Journal of Energy Optimization and Engineering*, *3*(4), 1–20. doi:10.4018/ijeoe.2014100101

Zheng, JS. (2017). Inactivation of Staphylococcus aureus in water by pulsed spark discharge. *Scientific Reports*, *7*.

Zhou, C., Hou, C., Wei, X., & Zhang, Q. (2014). Improved hybrid optimization algorithm for 3D protein structure prediction. *Journal of Molecular Modeling*, *20*(7), 2289. doi:10.100700894-014-2289-2

Zhu, L., He, Z. H., Gao, Z. W., Tan, F. L., Yue, X. G., & Chang, J. S. (2014). Research on the influence of conductivity to pulsed arc electrohydraulic discharge in water. *Journal of Electrostatics*, *72*(1), 53–58. doi:10.1016/j.elstat.2013.11.004

Zorin, V. V., Popov, V. A., & Ekel', P. E. (1988). Modeli optimizatsii nadezhnosti raspredelitel'nykh elektricheskikh setey [Optimization models of reliability of electrical distribution networks]. *Energetika i elektrifikatsiya*, *3*, 46-49.

Zorin, V. V., Tislenko, V. V., Kleppel', F., & Adler, G. (1984). Nadezhnost' sistem elektrosnabzheniya [Reliability of power supply system]. Kyiv: Vishcha shkola.

Zorin, V. V., Krushelnitsky, A. Z., & Fedosenko, I. N. (1977). *Influence of uncertainty of initial information on 0.38 kV networks on evaluation of voltage quality. Problems of technical electrodynamics, 64*, 54–57. (in Russian)

Zoumas, C. E., Bakirtzis, A. G., Teocharis, J. B., & Petridis, V. (2004). A genetic algorithm solution approach to the hydrothermal coordination problem. *IEEE Transactions on Power Systems*, *19*(3), 1356–1364. doi:10.1109/TPWRS.2004.825896

Zul, N. M., & Lyzhko, V. M. (1986). Application of fuzzy sets theory to formalization of rational decision selection. *Proc. of MIISP*, Saint-Petersburg (pp. 62-67). Academic Press. (in Russian)

Zverev, S.V., & Zvereva N.S (2007). Fizicheskie svojstva zerna i produktov ego pererabotki [Physical properties of grain and its products]. Moscow: DeLi print. (in Russian)

About the Contributors

Valeriy Kharchenko is a researcher and consultant in a wide range of energy saving fields, including energy policy at national and international level, energy RTD strategy, RUE and RES. Education 1994 - 1995 Tacis Training Programme (Manager in energy) 1994 Tashkent State Technical University, Uzbekistan (professor) 1987 State Higher Training Courses on Patenting and Invention (Certificate with distinction) 1987 Institute of Electronics, Moscow (Doctor of Engineering Sciences Degree) 1969 Institute of Inorganic Chemistry, Novosibirsk (Candidate of Chemistry Degree) 1956 - 1961 Tashkent Polytechnic Institute (Technical University) (technology engineer) International activity experience: 1994-1996 –counterpart expert in the Project of EU "Monitoring and evaluation of the energy (non-nuclear) projects implementation in Uzbekistan" (carried out by European companies); 1998-2000 - the expert for evaluation of submitted proposals INTAS (ID 716); 1996-2000 –senior expert of the Central Asia Energy Advisory Group (a project under the EC DG XVII's SYNERGY Programme) with obligations included duties for formation of expert groups, peering and estimation of their reports and preparing general Interim and Final annual reports; 2000-2001 years -UNDP Program, Global Environmental Facility, Government of Uzbekistan (Joint Project UZB/98/G42/A/1G/99), expert at the premises. Among of obligations - evaluation and reception of reports of local expert groups on various aspects of the organization of heating and hot water supply system of Tashkent city Area of Scientific activity. Development of new technological approaches for superpure silicon fabrication, improvement of silicon epitaxial layers technology, application of silicon layers in various devices, including PV cells, application of installations on the basis of solar cells in practice, development of new approaches for production of silicon for solar cells, application of a solar energy for electricity and heat producrion; Participation in a number of international projects of the Eurocommission and UNDP on an energy policy, energy saving (RUE) and renewable energy sources (RES), works on increase of efficiency of use of energy in systems of heating and hot water supply.

Pandian Vasant is a senior lecturer at Department of Fundamental and Applied Sciences, Faculty of Science and Information Technology, Universiti Teknologi PETRONAS in Malaysia. He holds PhD (UNEM, Costa Rica) in Computational Intelligence, MSc (UMS, Malaysia, Engineering Mathematics) and BSc (2nd Class Upper-Hons in Mathematics, UM, Malaysia) in Mathematics. He has co-authored research papers and articles in national journals, international journals, conference proceedings, conference paper presentation, and special issues lead guest editor, lead guest editor for book chapters, edited books, and keynote lecture. (218 publications indexed in SCOPUS). In the year 2009, Dr. Pandian Vasant was awarded top reviewer for the journal Applied Soft Computing (Elsevier), awarded outstanding reviewer in the year 2015 for ASOC (Elsevier) journal and Top reviewer for Sentinels of Science: Computer Science (Oct. 2015 - Sept. 2016). He has 26 years of working experience at the various universities from 1989-2017. Currently he is Editor-in-Chief of IJCO and IJEOE, Member of American Mathematical Society, MERLIN, and NAVY Research Groups.

* * *

Pavel Maslennikov in 2014 graduated from the Non-government Educational Institution of Higher Professional Education "Economics and Energy Institute" with a degree in electric power systems in the network. Works as an engineer in the laboratory of "electricity and power supply and electrical safety." He is the author of 8 scientific papers.

Raul Roman Aguilar, PhD, Energy Engineering, is a research professor at the Autonomous University of the State of Hidalgo (UAEH), Mexico, on the Apan campus, was coordinator of the engineering degree in cold technology and is the leader of a research group where he conducts research in the area of energy systems applied to cold, contributing to the generation and application of knowledge of thermal cooling systems and the relationship of energy to biological systems, developing theoretical-experimental studies that allow the interpretation and compression of fluid flow phenomena, heat transfer for applications in the field of refrigeration and conservation of biological products.

Baba Babaev studied at the faculty of heat and power engineering of the Moscow State Engineering and Construction Institute, which graduated in 1982. After graduation worked at the Kola NPP. From 1990 to 1993, working in the Scientific Research Energy Institute G. M. Krzhizhanovskogo, Dagestan branch, increased scientific qualifications to a senior researcher. In 1996 defended the dissertation of the candidate of chemical sciences at the Kuban State Technological University. And in 2016 defended the dissertation of the doctor of technical sciences in "United Institute of High Temperatures". Specialist in the field of energy, development of energy systems based on renewable sources and energy storage, actively introduces new information technologies in various fields of energy in the educational process. More than 160 scientific works have been published, 17 patents and 8 author's certificates for fundamentally new developments in the field of energy, some of which are implemented in practice. At a high scientific and methodological level reads the special courses "Energy accumulators", "Physical principles of renewable energy sources", "Energy facilities of non-traditional renewable energy sources", holds open lectures, participates in the formation of the subject of course and diploma works, training of young specialists. Since 1996 has been working at the Federal State Budget Educational Educational Institution of Higher Education "Dagestan State University", Professor of the Department "Renewable Energy Sources".

Alexey Bashilov was born in 1951 in Smolensk region. He graduated from the Moscow Institute of agricultural engineers (MGAU. Goryachkin V. P.), faculty of electrification and automation of agriculture in 1973, Doctor of technical Sciences (2001), Professor (2006).

Vadim Bolshev is a researcher at Federal Scientific Agroengineering Centre VIM and specializes in research in the field of power supply for rural consumers. Vadim studied at the Orel Technical College as an electrical technician from 2004 to 2008, at the Orel State Agrarian University as an electrical engineer from 2008 to 2012. In 2016 he graduated from the postgraduate course at the Orel State Agrarian University. Vadim has a great professional experience in the field of working with electrical equipment. So he worked in the water and sewer utility "Orelvodokanal" for 3.5 years as an electrician (2010 - 2013), in the scientific and clinical multidisciplinary center named after Z.I. Krugloi for 3 years as an engineer (2013-2016) and 1.5 years as chief engineer (2016-2018). To date, he works in the Federal Scientific Agro-Engineering Center of VIM as a research fellow. The field of scientific activity is to develop methods and tools aimed at improving power supply efficiency. This includes, first of all, the development of methods and devices for monitoring power quality and the technical state of power supply system elements as well as the automation of these systems.

Dmitry Budnikov is Head of the Laboratory of Electrophysical Impacts FSAC VIM. Research interests: "Intensification of technological processes in agriculture due to the electrical effects and improved control algorithms." The bulk of research is made for drying grain in the dense layer using microwave fields.

Serkan Buhan received the B.Sc., M.Sc., and Ph.D. degrees in electrical and electronics engineering from Hacettepe University, Ankara, Turkey, in 2005, 2009, and 2016, respectively. From 2006 to 2012, he was in the Department of Power Electronics, Space Technologies Research Institute, Scientific and Technological Research Council of Turkey (TUBITAK), Ankara. He is now with the Energy Institute, Marmara Research Center (MRC), TUBITAK. His research interests include renewable energy systems, wind power forecasting, electric power quality, hydro power forecasting and basin optimization.

Sergei Chirskii graduated from Bauman Moscow State Technical University in Power Engineering department, piston engines chair. PhD thesis was dedicated to applying similarity theory on frictional losses modeling. Thesis defended in 2012. Main workplace is piston engines chair of Bauman Moscow State Technical University, associate professor. Specialized in: frictional losses measuring and modelling; computer modelling of parts strength and thermal conditions.

Turan Demirci received the B.Sc., M.Sc., and Ph.D. degrees in electrical engineering from the Middle East Technical University (METU), Ankara, Turkey, in 2000, 2002, and 2010, respectively. He was a Chief Senior Researcher in the Power Electronics Department, the Scientific and Technological Research Council of Turkey (TUBITAK) Space Research Institute, Ankara. He is currently with the Energy Institute, Marmara Research Center, Ankara. His research interests are software and hardware system development in power applications, wide area power quality monitoring, and wind power forecasting.

Sergei Devyanin was born in 1954, graduated from the Moscow Automobile and Road Con-struction State Technical University (MADI) in 1976. Doctor of Technical Sciences, Professor of the Automobile Transport Department of the Russian State Agrarian University (Moscow Timiryazev Agricultural Academy). Author of more than 200 scientific papers in the field of engine construction, operation and maintenance of automotive engineering, the use of alter-native fuels in piston engines, fuel supply in diesel engines.

Irina Ershova, PhD of Technical Sciences, is a Senior Researcher FSAC VIM. She has authored 60 articles in the best Russian newspapers. Her scientific works are devoted to the study of the effects of electrophysical effects on agricultural objects.

Elena Eventeva, higher education (mechanical engineer, English), teacher-researcher, graduated from the faculty of engineering of Agricultural Academy of Velikie Luki in 2001 and is currently working as a teacher in this organization. Research experience – more than 4 years in the direction of scientific work: "Optimization of design parameters of the ultrasonic radiator for peasant farms". Research is aimed at reducing the energy intensity of agricultural production, energy saving and improving the reliability of energy supply to the agro-industrial complex. The author has more than 20 scientific papers, 6 patents, a, regular participant of interregional and international conferences.

Karla N. Fernández is an Engineer in Cold Technology with interest in the applications of cryogenics to agriculture and livestock breeding and the conservation of biological systems and perishable products as well as high impact sports.

Ugo Fiore, Ph.D., is with the Department of Management Studies and Quantitative Methods at Parthenope University of Naples. He is also an adjunct professor at Federico II University of Naples. Earlier, he had been working for ten years in the telco industry. His research interests include nonlinear analysis, deep learning, anomaly detection, optimisation, energy-aware systems, covert communications, and security. He has authored or co-authored more than 45 papers on international peer-reviewed journals. He is serving as Associate Editor with two journals and is a member of the editorial board in three other journals. He has been invited as a keynote speaker in two international conferences, has participated to the organizing committees of numerous conferences, and serves as a reviewer for several journals. He regularly sits in PhD examination boards in foreign universities.

Timothy Ganesan is an active researcher in the fields of computational intelligence, machine learning and optimization. He holds degrees in Chemical Engineering with a doctorate in Process Optimization.

Sergey Gribkov is a candidate of Tech.Sci., Academician. Russian Academy of Engineering, Leading Researcher of Federal State Unitary Enterprise TsAGI The expert on a wind to power installations and complexes of the guaranteed power supply of low powers & small wind power.

Pavel Ivanov from 1998-2003 was a student of Azov-Black Sea state Agroengineering Academies. 2003-2006 post-graduate student of the Department "tractors and cars". 2006-2010 high school teacher at the Department "theoretical mechanics". 2010-2012 high school teacher at the Department "quality management in of Agroengineering". Since 2012 teacher of higher school at the Department "technical ser-vice in the agro-industrial complex". In 2012 defended thesis on competition for the degree of candidate of technical Sciences. Dissertation topic -Substantiation of parameters of the working body of the sub- soil sowing of cereal grains in the conditions of pneumatic transportation of seeds in to coulters. 2013 awarded the title of Associate Professor. Currently, Associate Professor of the "Technical Service Department in the Agro-Industrial Complex".

Pinar Karagoz is currently Professor in Computer Engineering Department, Middle East Technical University (METU). She received her Ph.D. from the same department in 2003. She worked as a visiting researcher in State University of New York (SUNY) at Stony Brook. Her research interests include data mining, web usage mining, social network analysis, information extraction from the web, semantic web services, web service discovery and composition. Dr. Karagoz has authored several publications in international journals and leading conferences. Some of her papers were published in journals such as IEEE TKDE, IEEE Industrial Informatics, ACM TWEB, Information Systems Journal, SIGMOD Record, Knowledge and Information Systems, Knowledge based Systems. Some of her research were presented and published in conferences including VLDB, CIKM, ASONAM, DAWAK, ICWS. She has also taken part in the organization committee of several conferences including ICDM and VLDB.

Bibigul Koshoeva, PhD of Technical Sciences, Associate Professor of the Kyrgyz State Technical University. Author of 35 articles, two monographs in leading Russian, Kyrgyz and foreign scientific journals. Her research papers are devoted to the automation of process control systems at agricultural facilities.

Volodymyr Kozyrskyi graduated department of the electrification of Ukrainian Agricultural Academy (now The National University of Life and Environmental Sciences of Ukraine) in 1978. He was appointed director of the research institute of engineering and technology in September 2005. He appointed director of the Education and Research Institute of Energetics, Automation and Energy Efficiency in 2009.

Vladimir Kravchenko is Doctor of Technical Sciences, Professor. Honored Worker of Higher Professional Education of the Russian Federation. In 1970 he graduated from the Azov-Black Sea Institute of Agricultural Mechanization, specialty "Agricultural Mechanization". In 1983 he defended his thesis for the degree of Candidate of Technical Sciences, in 2012 - the dissertation for the degree of Doctor of Technical Sciences in the specialty 05.20.01 "Technologies and means of agricultural mechanization in technical sciences". The academic title of professor in the department "Tractors and cars" awarded the Higher Attestation Commission in 2000. "Inventor of the USSR", a representative of the Institute in the International Association of Road Education.

Dilek Küçük, Ph.D, is an associate professor and chief researcher at the Energy Institute of TÜBİTAK Marmara Research Center (MRC). She is also the leader of Power Systems Information Technologies Group at the institute. Her group is the recipient of best research group award of TÜBİTAK MRC for the year 2017. She has obtained her B.S., M.S. and Ph.D. degrees in Computer Engineering all from Middle East Technical University in Ankara (Turkey), in 2003, 2005, and 2011, respectively. Between May 2013 and May 2014, she has studied as a post-doctoral researcher at European Commission's Joint Research Centre in Italy. Her research interests include energy informatics, data mining, social media analysis, natural language processing, and database applications in engineering domains. She is the author or co-author of 14 papers published at SCI-indexed journals, in addition to more than 35 papers presented at international conferences.

Pavel Lavrukhin from 1976 -1981 was astudent of Azov-Black sea Institute of agricultural mechanization. 1981 -1985 engineer of the North-Caucasus machinery testing station. 1985 – 1994 research associate, Department of mechanization field crops 1994 - 2001 senior researcher of the scientific-production company Agrotehnic in 2000 defended thesis for the degree of candidate of technical Sciences. 2001 - 2005 associate Professor of the Department Theoretical Mechanics. 2005-2018 associate professor of tractors and cars.

Tamara Leshchinskaya, Department: power supply and electrical engineering named after academician Budzko Academic degree: Doctor of technical Sciences Academic title: Professor List of subjects taught Power supply of agriculture, Electric power systems and networks. Education 1964, MYISP named after V. P. Goryachkin, qualification: electrical engineer: production, distribution and use of electricity in agriculture. 1972-candidate of technical Sciences, specialty 05.20.02-electrical Engineering and electrical equipment in agriculture 1990-doctor of technical Sciences, specialty 05.20.02-electrical Engineering and electrical equipment in agriculture 1992-Professor at the Department Of power supply of agriculture Labour activity Total work experience: 55 years Scientific and pedagogical experience: 50 years Publication activity A total of 252 published works, including 16 monographs and 15 textbooks and teaching.

Vladimir Markov was born in 1958, graduated in 1981 from the Moscow State Technical University named after N.E. Bauman. Doctor of Technical Sciences, Head of the "Piston en-gines" department of the Bauman Moscow State Technical University. Full member (acade-mician) of the Academy of Military Sciences of Russia. Author of more than 400 scientific papers in the field of engine construction, the use of alternative fuels in piston engines, diesel fuel equipment, control and regulation of internal combustion engines.

Vladimir Mayorov is a candidate of Technical Sciences, VIM Federal Scientific Agroengineering Center, Leading Researcher, Head of Solar Energy Laboratories. Graduated from the All-Union Correspondence Polytechnic Institute, majoring in Electronic Devices; Moscow Institute of Physics and Technology with a degree in Laser Engineering.

Andrey Musenko is a postgraduate student, Federal State Budgetary Scientific Institution Federal Scientific Agroengineering Center VIM, Moscow, Russia.

Alexey Nekrasov was born October 30, 1945, in 1973 graduated from the faculty of electrification of agriculture of the Kazakh GSHI on a specialty the electrical engineer of agricultural production. In 2007 he defended his doctoral thesis on the specialty 05.20.02 – Electro technology and electric equipment in agriculture. Currently conducts research on electrification of agriculture, for the effective use and operation of electrical equipment of agricultural enterprises. Directly involved in the development of a new resonance methods of transmission and use of electric energy, and also the implementation of other fundamental and most important applied scientific and technical programs. Nekrasov A.I. has over 300 scientific publications, including 20 books, monographs, guidelines and brochures, 55 patents and copyright certificates on inventions. Completed developments in agricultural production and recommended for use by scientists, design organizations and experts-energy, are used in the educational process.

Anton Nekrasov in 2004 graduated from the energy faculty of the Moscow state Agroengineering University. V. P. Goryachkina majoring in electrical engineer majoring in electrification and automation of agriculture. In 2015 he defended his thesis. He works as a senior researcher in the laboratory of power supply and electrical safety. He is the author of 100 published scientific works, including 3 guidelines, 17 patents.

Gai Oleksandr was born 07.04.1979 in Chernigov. Since 1986 studied in high school N°32 and graduated in 1996 with a silver medal. In 1996 entered the Chernihiv Technological Institute on specialty "Industrial electronics", graduated in 2001 and received a diploma in direction "Electronic systems". Since October 2001 entered graduate school on specialty 05.14.02 "Electric systems, systems and networks". Because of the death of V.V. Tislenko, since 2003 was transferred under the leadership of Doctor of Science, Professor V.V. Kozyrskiy. In April 2008 defended dissertation on "Optimizing the number and placement of sectionalizing devices at rural distribution networks". Have over 15 years of scientific and pedagogical experience. Along with studies in graduate school, since September 2001 taught at the Department "Electrical Engineering and Automation", which was later renamed the Department of "Electric systems and networks", as an assistant. Since September 2007, worked as an assistant of the Department "Power supply of agriculture" at the National Agrarian University. Since October 2009 working on the post of associate Professor of the Department of electricity supply named after Professor V.M. Sinkov National University of Life and Environmental Sciences. In January 2012, received the academic rank of assistant professor of electrical supply Department. From February 2012 to August 2015, was the acting head of the Department. Take an active part in University and international scientific conferences and carry out research, development and implementation work under contractual and budget subjects of the University. Was responsible performer on two themes N°110/449-pr "Smart Grid technologies in energy systems and in industries with biotech facilities" and N° 110/505-pr "Managing of energy consumption and generation at the agricultural enterprises based on the concept of intelligent technologies (Smart Grid)". In 2016, under my leadership Bodunov V.M. defended his thesis on "Management of regimes of lightly loaded electric networks with distributed generation sources"; repeatedly was official opponent on PhD theses. The results of scientific and research activity enshrined in more than 60 scientific works, including 2 monographs, tutorial, patent, recommendations to the Ministry of Agriculture and Food of Ukraine regarding the research systems based on SMART GRID technologies (for example, energy supply systems and systems with biotechnological facilities) and reports on contractual and budgetary topics. In 2009 passed the advanced training at the Educational-Scientific Institute of Postgraduate Education under the «Teaching and learning using information and communication technologies in the educational process", in 2010 - "Installation and commissioning of automation and remote control systems of power lines" and "Rational use of energy by Separate Subdivision of National University of Life and Environmental Sciences of Ukraine ", in 2011 - "Mentor student groups", in 2012 – advanced training at the educational-thematic plan for training of engineering-technical workers on the topic "Problems of resource and energy saving in poultry farming and ways of their implementation", in 2013 - "Energy saving and energy management according to the program "Energy Audit" and "Pedagogical skills teaching in English for teachers", in 2014 - "Advisers and experts-advisers". Have gratitude from the Minister of Agrarian Policy and Food of Ukraine and repeatedly awarded diplomas of international agricultural exhibitions. Fluent in English and Ukrainian. Fluent in computer (MS Office, email, etc.). Actively participate in the learning experience of the implementation of SMART GRID technologies in the European Union based on the "IEC Smart Grid Standardization Roadmap" and planning the harmonization of this document with the existing regulatory framework in Ukraine.

Mehmet Baris Ozkan received the B.Sc. and M.Sc. degrees in computer engineering from the Middle East Technical University (METU), Ankara, Turkey, in 2011 and 2014, respectively, where he is currently pursuing the Ph.D. degree. He has been working as a senior researcher with the Scientific and Technological Research Council of Turkey (TUBITAK), Marmara Research Center (MRC) Energy Institute, Kocaeli, Turkey, since 2011. His research interests include power quality systems, renewable energy systems, and wind power forecasting.

Vladimir Panchenko from 2003 to 2009 studied at "Bauman Moscow State Technical University" at the faculty "Power Engineering " with qualification of engineer. From 2009 to 2012 studied at the graduate school of the ''All-Russian Scientific Research Institute of Agriculture Electrification'', specializing in "Renewable Energy Installations". Passed an internship program on the International Energy Program in the Solar Energy Center (SEC), the Certificate under the Indian Technical and Economic Cooperation (ITEC), the Program of the Ministry of External Affairs (MEA), the Government of India - India, Delhi, November-December 2011. Participated in the International Conference-Exhibition Intersolar Europe 2012, Germany, Munich, June 11 - 15, 2012 as part of the delegation of the Ministry of Energy of the Russian Federation. Ph.D. thesis "Development and research of solar photovoltaic thermal module with a paraboloid-type concentrator" was protected on July 2, 2013. From 2013 to 2017 head of the laboratory ''Solar photovoltaic modules with an experimental - technological site''. The Council of Young Scientists of the VIESH was elected the Chairman of the Council of Young Scientists and was appointed a member of the Academic Council of the VIESH in 2014 - 2017. In 2014 was appointed as a contact person and a member of the International Solar Energy Society of the Russian section of the community. Since September 2015 associate professor at the ''Russian University of Transport''. Since 2017, a senior researcher at the ''Solar Energy Laboratory'' of "Federal Scientific Agroengineering Center VIM". Passed the advanced training in 2017 in ''Russian University of Transport'' the program "Using e-learning and distance educational technologies in the implementation of educational programs on distance learning". In 2016 awarded the grant ''Young lecturer of Russian University of Transport''. In 2018 awarded the Scholarship of the President of the Russian Federation for 2018 - 2020 for young scientists and graduate students engaged in advanced research and development in priority areas of the modernization of the Russian economy. Has 3 medals of exhibitions, 3 diplomas of the winner of competitions, participated in more than 90 conferences and exhibitions of various levels, has more than 130 publications, including patents and educational publications. Area of interest includes three-dimensional modeling, use of solar energy and electrical transport.

Gilberto Pérez-Lechuga is an Industrial Engineering, Master in Science in Operations Research, Engineering Doctor in Operations Research and Applied Mathematics. Other studies: Reliability Engineering in Nuclear Plants, Mathematical Models for Supply Chain and Transportation Network Design. École des Hautes Études Commerciales de Montréal (Hec. Montreal), Montreal Canadá. Post-Doc Stay: 1. Math's Depth. Lerida University, 2. Math's Depth. Lerida University: Invited Professor of the course: Introduction to the Stochastic Optimization, 3. Visitor of the Germany ULM University, Math's. Depth.

Some Distinctions and Awards: a) Editorial reviewer of the journal Engineering Applications of Artificial Intelligence, The International Journal of Intelligent Real-Time Automation. A journal of IFAC, The International Federation of Automatic Control. ELSEVIER. b) Guest Editor of the Journal of the Operations Research Society of China, c) Guest Editor of IEEE Transactions on Fuzzy Systems Special Issue Smart Fuzzy Optimization in Operational Research and Renewable Energy: Modelling, Simulation, and Applications, d) Member of the International Advisory Council of the International Journal of Energy Optimization and Engineering (IJEOE). Ed. Igi-Global. National Researcher Level 1, issued by the National Systems of Researchers of Science and Technology Council, Mexico. Secretary of the science and technology commission of the southern center of the National Association of Universities and Institutions of Higher Education, Mexico ANUIES.

Pavel Podobedov in 2014 graduated from the Russian State Agrarian University - Moscow Agricultural Academy Timiryazev honors in the direction of electric power and electrical engineering. In 2017 he graduated from the graduate school of the Federal State Budgetary Scientific Institution "Federal Scientific Agroengineerring Center VIM" (FSAC VIM). Works as an engineer in the laboratory of "electricity and power supply and electrical safety." He is the author of 15 scientific papers and 1 patent.

Anatoly Rusakov is Cand.Tech.Sci., leading researcher of National Research University "Moscow Power Engineering Institute", Russia.

Gennady Samarin (1974) is doctor of technical Sciences (2009), associate Professor (2008), chief researcher of the laboratory of mechatronics and intelligent control systems of the Federal state budgetary scientific institution "Federal scientific Agroengineering center VIM" (2018), Laureate of the state prize of the Russian Federation for young scientists for outstanding work in the field of science and technology (2003).

Otilio Arturo Sandoval in 1998 received a Ph.D. degree in Biology with specialization in Edaphology by the Science Faculty of the Autonomous National University of Mexico (UNAM) and the Ph.D. degree in Chemistry by the Autonomous University of the State of Hidalgo (UAEH) in 2007. He has being member of the National Research System (CONACyT) for 20 years. During his academic career, he has been professor at the UNAM and at the Autonomous Metropolitan University - Iztapalapa (UAM). Since 1998, he is Research Professor at the UAEH. At these universities, he has direct different thesis of bachelor, masters and Ph.D. students. His work ranges the study of forest soils and the watershed management to establish programs focused on soil conservation and rehabilitation His research lines are the taxonomy, soil genesis and mineralogy of the argillic fraction of acid soils. In 2015, the Autonomous University of Queretaro (UAQ) decided to name its Special Lecture of Chemistry, Environment and Sustainable Development as Otilio Arturo Acevedo Sandoval.

Andrei Savastenko was born in 1960, graduated in 1983 from the Peoples Friendship University of the P. Lumumba. Candidate of Technical Sciences, Associate Professor of the Department "Heat engineering and automotive engines" of Moscow State Technical University (MADI). Author of more than 80 scientific works in the field of engine, diesel fuel equipment, the use of alternative fuels in piston engines and diesel fuel additives.

Vitaliy Savchenko graduated with honors in 2008 Department of Energy and Automation, National University of Life and Environmental Sciences of Ukraine in specialty "Automated process control." In 2012 he defended his thesis for the degree of candidate of technical sciences, specialty 05.09.03 - Electrotechnical complexes and systems on "electro-technological complex for potato pre-treatment in a magnetic field." Has over 150 scientific and educational works.

Sergey Senkevich. 1997-2002 student of the Agroengineering University. 2002-2005 postgraduate-course student of the Department of Tractors and Cars. 2006 defended thesis for the degree of candidate of technical sciences. 2009 awarded the title of associate Professor of the Department of Tractors and Cars. 2005-2013 associate Professor of the Department of Tractors and Cars. 2013-2019 senior researcher, Laboratory of Mobile Energy Facilities. From 2019 to the present, head of the laboratory for the automated drive of agricultural machinery.

Oleksandr Sinyavsky graduated in 1985 Faculty of Agricultural automation of Ukrainian Agricultural Academy. He worked at the department of electric drive and electro technologies from 1992. In 1992 he defended his thesis on "Automated electrical preparation of nutrient solution in hydroponic greenhouses». Has more than 230 scientific and educational works.

Alexander Smirnov, CSc. (Engineering), is a Senior Researcher FSAC VIM (2009-present). He finished the postgraduate school of All-Russian Research Institute of Electrification of Agriculture in 2009. He has authored 26 articles in the best Russian newspapers. His scientific works are devoted to the study of the effects of electrophysical effects (microwave, ozone, optical radiation) on agricultural objects.

Anatoly Sopov is currently a graduate student of the 1 course of Orlovsky State Agrarian University. N.V. Parahin. The field of scientific activity are the development of new ways and means of beneficial use of excess heat of power transformers, creating a power transformer with a combination of functions of voltage transformation and heating for the objects that are located in close proximity to a transformer substation. Anatoly used to be the engineer of the production and technical group of Orel District Electric Networks from 2016 to 2018. Since 2018 he has been a specialist of the capital construction management of the branch of IDGC of Center, PJSC - Orelenergo.

Anatoly Sugrobov is a Professor of National Research University "Moscow Power Engineering Institute", Russia.

Makarevich Svitlana graduated with honors in 2001 Department of Electrification and Automation of agricultural industry, National Agrarian University in speciality "Electrification and automation in agricultural industry." In 2013 she defended her thesis for the degree of candidate of technical sciences, specialty 05.09.03 - Electrotechnical complexes and systems on "Autonomous electromechanical complexes with compenced asynchronous machines " Has over 50 scientific and educational works.

Pavel Tikhonov is a researcher in the field of solar energy installations. From 2003 to 2008, he studied at Marit State University "MarSU" at the Faculty of Electric Power. In 2009 he entered postgraduate education the All-Russian Research Institute of Agriculture. He was engaged in researching the field of photovoltaic thermal modules and installations based on them. In 2014, he defended his Ph.D thesis on the topic: "The justification of the parameters of the photoelectric thermal module." Currently working in the laboratory of solar energy Federal Scientific Agro-engineering Center VIM. The main activity is connected with the development of cogeneration photovoltaic thermal modules and energy-saving systems based on photovoltaic module.

Victoria Ukhanova, CSc. (Engineering), is a senior research fellow at FSKN VIM (2009-present). Postgraduate course of all-Russian scientific research Institute of electrification of agriculture in 2009. Author of 26 articles in the best Russian and foreign scientific journals, author of 3 patents of the Russian Federation. Her scientific works are devoted to the study and forecasting of electric energy consumption at agricultural facilities.

Aleksey Vasiliev graduated from the graduate school in 1997 at the Moscow Institute of Agricultural Production Engineers. He defended his thesis in 1998. For many years he worked at the Azov-Black Sea Institute of Mechanization and Electrification of Agriculture. In 2009 he defended his doctoral thesis. Professor. He currently works at the VIM Federal Research Agroengineering Center as a head of research.

Alexander Vinogradov is a leading researcher at the Federal Scientific Agroengineering Center VIM and specializes in research into improving power supply system efficiency for rural consumers. Alexander studied at the Kostroma State Agricultural Academy from 1994 to 1999, majoring in Electrification and Automation of Agriculture. From 1999 to 2018 he worked as an assistant, senior lecturer, head of the department, associate professor of the Power Supply Department of the Orel State Agrarian University. Since 2018 he has been a leading researcher at the Federal Scientific Agroengineering Center VIM. Alexander has been the General Director of the LLC "Information and Energy Center AVPS-Innovation" since 2007 Alexander has experience in the preparation, organization and implementation of educational services, the development of teaching technologies, the development and creation of technical training facilitiesas well as experience in conducting research, developing theoretical and practical aspects of various kinds of interactions including social ones. He also has practical experience in conducting energy surveys, designing power supply systems, installing electrical equipment, managing the design organization and designing. He has 4 dan karate-do and 3 dan kobudo. Research interests: research is being carried out to develop new approaches to the creation of power supply systems, automation of electric networks, the development of energy accounting systems, the provision of quality of electricity supplied to consumers, energy saving, electrical modeling of human and society, educational technologies, use of renewable energy sources. He is an author and a co-author of more than 200 scientific and methodical works, including 33 patents, 9 monographs, 6 essays. He is also an author and a co-author of 4 collections of literary works. He is the author and the editor-in-chief of the newspaper "High Voltage Truth" and the editor-in-chief of the journal "Agrotechnics and energy supply".

Alina Vinogradova is a researcher at the Federal Scientific Agroengineering Center VIM and specializes in research in the field of power supply for rural consumers. Alina used to study at the Oryol State Agrarian University from 2003 to 2008. In 2015 she defended her thesis. Alina has great professional experience in the field of designing electrical networks. So she worked as a design engineer (electrician) in JSC Orelenergoremont in the period from 2007 to 2008; as an engineer for project work at JSC Orelenergoremont from 2008 to 2009; as an engineer of the design department of VL in JSC "Voronezhenergoproekt" from 2009 to 2010; as an assistant of the Department of "Power Supply" in OrelSAU from 2010 to 2011; as a senior lecturer of the department "Power supply" in OrelSAU from 2011 to 2016; as an associate professor of the Department of "Power Supply" in OrelSAU from 2016 to 2018. At the moment she works in the federal scientific agroengineering center VIM as a research assistant. The field of scientific activity is to develop methods and tools aimed at improving the efficiency of power supply. This includes, first of all, the development of methods and devices for monitoring the quality of electrical energy and the technical state of the elements of power supply systems as well as its automation.

Index

S

T

U

V

W

Purchase Print, E-Book, or Print + E-Book

IGI Global's reference books are available in three unique pricing formats:
Print Only, E-Book Only, or Print + E-Book.
Shipping fees may apply.

www.igi-global.com

Recommended Reference Books

ISBN: 978-1-5225-7727-0
© 2019; 335 pp.
List Price: $195

ISBN: 978-1-5225-5577-3
© 2019; 291 pp.
List Price: $195

ISBN: 978-1-5225-5291-8
© 2019; 252 pp.
List Price: $175

ISBN: 978-1-5225-5766-1
© 2019; 501 pp.
List Price: $195

ISBN: 978-1-5225-5909-2
© 2019; 308 pp.
List Price: $195

ISBN: 978-1-5225-8027-0
© 2019; 337 pp.
List Price: $205

Do you want to stay current on the latest research trends, product announcements, news and special offers?
Join IGI Global's mailing list today and start enjoying exclusive perks sent only to IGI Global members.
Add your name to the list at **www.igi-global.com/newsletters.**

Publisher of Peer-Reviewed, Timely, and Innovative Academic Research

IGI Global
DISSEMINATOR OF KNOWLEDGE

www.igi-global.com Sign up at www.igi-global.com/newsletters facebook.com/igiglobal twitter.com/igiglobal linkedin.com/igiglobal

Ensure Quality Research is Introduced to the Academic Community

Become an IGI Global Reviewer for Authored Book Projects

The overall success of an authored book project is dependent on quality and timely reviews.

In this competitive age of scholarly publishing, constructive and timely feedback significantly expedites the turnaround time of manuscripts from submission to acceptance, allowing the publication and discovery of forward-thinking research at a much more expeditious rate. Several IGI Global authored book projects are currently seeking highly-qualified experts in the field to fill vacancies on their respective editorial review boards:

Applications and Inquiries may be sent to:
development@igi-global.com

Applicants must have a doctorate (or an equivalent degree) as well as publishing and reviewing experience. Reviewers are asked to complete the open-ended evaluation questions with as much detail as possible in a timely, collegial, and constructive manner. All reviewers' tenures run for one-year terms on the editorial review boards and are expected to complete at least three reviews per term. Upon successful completion of this term, reviewers can be considered for an additional term.

If you have a colleague that may be interested in this opportunity,
we encourage you to share this information with them.

IGI Global Proudly Partners With eContent Pro International

Receive a 25% Discount on all Editorial Services

Editorial Services

IGI Global expects all final manuscripts submitted for publication to be in their final form. This means they must be reviewed, revised, and professionally copy edited prior to their final submission. Not only does this support with accelerating the publication process, but it also ensures that the highest quality scholarly work can be disseminated.

English Language Copy Editing

Let eContent Pro International's expert copy editors perform edits on your manuscript to resolve spelling, punctuaion, grammar, syntax, flow, formatting issues and more.

Scientific and Scholarly Editing

Allow colleagues in your research area to examine the content of your manuscript and provide you with valuable feedback and suggestions before submission.

Figure, Table, Chart & Equation Conversions

Do you have poor quality figures? Do you need visual elements in your manuscript created or converted? A design expert can help!

Translation

Need your documjent translated into English? eContent Pro International's expert translators are fluent in English and more than 40 different languages.

Hear What Your Colleagues are Saying About Editorial Services Supported by IGI Global

"The service was very fast, very thorough, and very helpful in ensuring our chapter meets the criteria and requirements of the book's editors. I was quite impressed and happy with your service."

– Prof. Tom Brinthaupt,
Middle Tennessee State University, USA

"I found the work actually spectacular. The editing, formatting, and other checks were very thorough. The turnaround time was great as well. I will definitely use eContent Pro in the future."

– Nickanor Amwata, Lecturer,
University of Kurdistan Hawler, Iraq

"I was impressed that it was done timely, and wherever the content was not clear for the reader, the paper was improved with better readability for the audience."

– Prof. James Chilembwe,
Mzuzu University, Malawi

Email: customerservice@econtentpro.com **www.igi-global.com/editorial-service-partners**

IGI Global
DISSEMINATOR OF KNOWLEDGE
www.igi-global.com

Celebrating Over 30 Years of Scholarly Knowledge Creation & Dissemination

InfoSci®-Books

A Database of Over 5,300+ Reference Books Containing Over 100,000+ Chapters Focusing on Emerging Research

InfoSci®-Books Database

The **InfoSci®-Books** database is a collection of over 5,300+ IGI Global single and multi-volume reference books, handbooks of research, and encyclopedias, encompassing groundbreaking research from prominent experts worldwide that span over 350+ topics in 11 core subject areas including business, computer science, education, science and engineering, social sciences and more.

Open Access Fee Waiver (Offset Model) Initiative

For any library that invests in IGI Global's InfoSci-Journals and/or InfoSci-Books databases, IGI Global will match the library's investment with a fund of equal value to go toward **subsidizing the OA article processing charges (APCs) for their students, faculty, and staff** at that institution when their work is submitted and accepted under OA into an IGI Global journal.*

INFOSCI® PLATFORM FEATURES

- No DRM
- No Set-Up or Maintenance Fees
- A Guarantee of No More Than a 5% Annual Increase
- Full-Text HTML and PDF Viewing Options
- Downloadable MARC Records
- Unlimited Simultaneous Access
- COUNTER 5 Compliant Reports
- Formatted Citations With Ability to Export to RefWorks and EasyBib
- No Embargo of Content (Research is Available Months in Advance of the Print Release)

**The fund will be offered on an annual basis and expire at the end of the subscription period. The fund would renew as the subscription is renewed for each year thereafter. The open access fees will be waived after the student, faculty, or staff's paper has been vetted and accepted into an IGI Global journal and the fund can only be used toward publishing OA in an IGI Global journal. Libraries in developing countries will have the match on their investment doubled.*

To Learn More or To Purchase This Database:
www.igi-global.com/infosci-books

eresources@igi-global.com • Toll Free: 1-866-342-6657 ext. 100 • Phone: 717-533-8845 x100

IGI Global
DISSEMINATOR OF KNOWLEDGE
www.igi-global.com

IGI Global
DISSEMINATOR OF KNOWLEDGE
www.igi-global.com

Publisher of Peer-Reviewed, Timely, and Innovative Academic Research Since 1988

IGI Global's Transformative Open Access (OA) Model: **How to Turn Your University Library's Database Acquisitions Into a Source of OA Funding**

In response to the OA movement and well in advance of Plan S, IGI Global, early last year, unveiled their OA Fee Waiver (Offset Model) Initiative.

Under this initiative, librarians who invest in IGI Global's InfoSci-Books (5,300+ reference books) and/or InfoSci-Journals (185+ scholarly journals) databases will be able to subsidize their patron's OA article processing charges (APC) when their work is submitted and accepted (after the peer review process) into an IGI Global journal.*

How Does it Work?

1. When a library subscribes or perpetually purchases IGI Global's InfoSci-Databases including InfoSci-Books (5,300+ e-books), InfoSci-Journals (185+ e-journals), and/or their discipline/subject-focused subsets, IGI Global will match the library's investment with a fund of equal value to go toward subsidizing the OA article processing charges (APCs) for their patrons.
   Researchers: Be sure to recommend the InfoSci-Books and InfoSci-Journals to take advantage of this initiative.
2. When a student, faculty, or staff member submits a paper and it is accepted (following the peer review) into one of IGI Global's 185+ scholarly journals, the author will have the option to have their paper published under a traditional publishing model or as OA.
3. When the author chooses to have their paper published under OA, IGI Global will notify them of the OA Fee Waiver (Offset Model) Initiative. If the author decides they would like to take advantage of this initiative, IGI Global will deduct the US$ 1,500 APC from the created fund.
4. This fund will be offered on an annual basis and will renew as the subscription is renewed for each year thereafter. IGI Global will manage the fund and award the APC waivers unless the librarian has a preference as to how the funds should be managed.

Hear From the Experts on This Initiative:

"I'm very happy to have been able to make one of my recent research contributions, 'Visualizing the Social Media Conversations of a National Information Technology Professional Association' featured in the *International Journal of Human Capital and Information Technology Professionals*, freely available along with having access to the valuable resources found within IGI Global's InfoSci-Journals database."

– **Prof. Stuart Palmer,**
Deakin University, Australia

For More Information, Visit: www.igi-global.com/publish/contributor-resources/open-access or contact IGI Global's Database Team at eresources@igi-global.com.

Are You Ready to **Publish Your Research?**

IGI Global
DISSEMINATOR OF KNOWLEDGE

IGI Global offers book authorship and editorship opportunities across 11 subject areas, including business, computer science, education, science and engineering, social sciences, and more!

Benefits of Publishing with IGI Global:

- Free one-on-one editorial and promotional support.
- Expedited publishing timelines that can take your book from start to finish in less than one (1) year.
- Choose from a variety of formats including: Edited and Authored References, Handbooks of Research, Encyclopedias, and Research Insights.
- Utilize IGI Global's eEditorial Discovery® submission system in support of conducting the submission and blind review process.
- IGI Global maintains a strict adherence to ethical practices due in part to our full membership with the Committee on Publication Ethics (COPE).
- Indexing potential in prestigious indices such as Scopus®, Web of Science™, PsycINFO®, and ERIC – Education Resources Information Center.
- Ability to connect your ORCID iD to your IGI Global publications.
- Earn royalties on your publication as well as receive complimentary copies and exclusive discounts.

Get Started Today by Contacting the Acquisitions Department at:

acquisition@igi-global.com

www.ingramcontent.com/pod-product-compliance
Lightning Source LLC
LaVergne TN
LVHW080405011026
849161LV00009B/2353
* 9 7 8 1 7 9 9 8 1 2 1 6 6 *